LUMINESCENCE OF CRYSTALS, MOLECULES, AND SOLUTIONS

LUMINESCENCE OF CRYSTALS, MOLECULES, AND SOLUTIONS

Proceedings of the International Conference on Luminescence held in Leningrad, USSR, August 1972

Edited by
FERD WILLIAMS

Assistant Editors
B. BARON
M. MARTENS
S. P. VARMA
Physics Department
University of Delaware
Newark, Delaware

PLENUM PRESS · NEW YORK–LONDON · 1973

Library of Congress Catalog Card Number 73-77339

ISBN 0-306-30736-7

© 1973 Plenum Press, New York
A Division of Plenum Publishing Corporation
227 West 17th Street, New York, N.Y. 10011

United Kingdom edition published by Plenum Press, London
A Division of Plenum Publishing Company, Ltd.
Davis House (4th Floor), 8 Scrubs Lane, Harlesden, London,
NW10 6SE, England

Printed in the United States of America

CONFERENCE COMMITTEES

INTERNATIONAL COMMITTEE

F. Williams (USA), Chm.
H. A. Klasens (Netherlands), Sec.
V. V. Antonov-Romanovskii (USSR)
D. Curie (France)
Th. Förster (BRD)
G. F. J. Garlick (UK)
R. M. Hochstrasser (USA)
A. Jablonski (Poland)
B. S. Neporent (USSR)
N. Riehl (BRD)
F. Seitz (USA)
Sh. Shionoya (Japan)
G. Szigeti (Hungary)

PUBLICATION COMMITTEE

F. J. Bryant (UK)
N. E. Geacintov (USA)
I. P. Ipatova (USSR)
H. A. Klasens (Netherlands)
M. Martens (USA)
D. S. McClure (USA)
R. B. Murray (USA)
J. Schanda (Hungary)
M. Sharnoff (USA)
Sh. Shionoya (Japan)
G. Szigeti (Hungary)
F. Vogel (USA)
R. Vogel (USA)
F. Williams (USA)

EDITORIAL COMMITTEE

B. Baron
P. DiBona
R. Ewing
S. Hillenius
C. Jarman
H. Luchinsky
M. Martens
A. Ray
A. Sinha
S. P. Varma
F. Williams

SOVIET ORGANIZING COMMITTEE

M. D. Galanin (Moscow), Chm.
P. P. Feofilov (Leningrad), V.Chm.
B. S. Neporent (Leningrad), V.Chm.
V. L. Ermolaev (Leningrad), Sec.
Z. L. Morgenshtern (Moscow), Sec.
V. V. Antonov-Romanovskii (Moscow)
N. G. Basov (Moscow)
N. A. Borisevich (Minsk)
A. A. Bundel (Moscow)
S. A. Fridman (Moscow)
A. N. Georgobiani (Moscow)
A. M. Gurvich (Moscow)
A. A. Kaplyanskii (Leningrad)
Ch. B. Lushchik (Tartu)
L. Ya. Markovskii (Leningrad)
M. M. Miroshnikov (Leningrad)
Yu. V. Morozov (Moscow)
N. P. Novoselov (Leningrad)
A. A. Rovnyakov (Leningrad)
M. T. Shpak (Kiev)
K. K. Schvarts (Riga)
V. I. Sorokin (Leningrad)
I. A. Vanin (Leningrad)

PROGRAM COMMITTEE

P. P. Feofilov, Chm.
A. M. Tkachuk, Sec.
N. G. Bakhshiev
V. L. Ermolaev
I. P. Ipatova
A. A. Kaplyanskii
V. P. Klochkov
Ch. B. Lushchik
L. Ya. Markovskii
B. S. Neporent
A. I. Ryskin
N. V. Starostin
B. P. Zakharchenya
N. D. Zhevandrov

SOCIAL COMMITTEE

I. P. Ipatova, Chw.
I. E. Obyknovennaya, V. Chw.

PREFACE

These Proceedings report the scholarly work presented in Leningrad at the largest conference ever held on luminescence. In addition to the large number of delegates, the Conference was distinguished by strong and balanced representation on the program of papers from capitalist and socialist nations. The Conference was sponsored by the International Union of Pure and Applied Physics and by the Academy of Science of the USSR. As noted in the Opening Ceremony, this Conference is in the series held approximately every three years since 1938. All branches of luminescence are included.

It was recognized, during the early stages of organization of the Conference, that there would be difficulties associated with the preparation of an English version of the Proceedings. Until just before the Conference, it was not evident whether translations of the Russian papers would be available and whether facilities would exist at the Conference for a working publication committee to edit the manuscripts. It was not possible, therefore, to make contractual arrangements before the Conference for publication. Translations and the facilities were indeed generously made available by the Soviet Organizers; and the Publications Committee, whose membership is listed with the Conference Committees, made a magnificent effort in Leningrad to edit the manuscripts. Several months following the Conference, it became clear that the problems associated with the large number of papers, the translations and the indexing necessitated further sustained editorial work by those knowledgeable both in luminescence research and in English idiom. At that time, arrangements were completed with the Plenum Publishing Corporation for photo-offset publication; and the final editorial work, preparation of camera-ready copy, proofing and indexing were then done at the University of Delaware. This work was in part assisted by discretionary funds remaining from the 1969 Delaware Luminescence Conference.

These Proceedings report most, but not all, of the material presented. Of the 135 papers presented, 94 appear herein. Almost a score were not submitted as complete manuscripts in English; a comparable number were not included for technical reasons, e.g. inadequate figures, length substantially in excess of that prescribed by the Conference Organizers, or inability of the Editors to penetrate and clarify the author's meaning through the available translation. The delay in completing arrangements for publication placed a time schedule on publication which obviated authors' proofs so that manuscripts which needed further attention by the authors were returned to them for revisions and subsequent publication elsewhere.

The Editorial Committee, consisting of my research group, devoted almost a month of their time to editing, supervising preparation of the camera-ready text and figures, proofing and indexing. The papers follow the approximate order of the presentation at the Conference, with some regrouping of subsections for logical coherence. The Committee considered its primary responsibility was to accuracy of scientific content and therefore only those improvements in style and grammar were made that we were reasonably certain did not risk alteration of technical substance. Some compromise was made between getting the job done fast and getting it done completely and without error.

These Proceedings are believed to represent the latest and much of the finest research on luminescence throughout the world, and therefore early availability to English-reading scientists was considered imperative. The Russian-reading scientists will have access to these works in a parallel publication prepared by the Soviet Organizers, which is scheduled to appear in Izvestiya Akademii Nauk SSSR, ser. fiz.

Ferd Williams, Editor and
Chairman of the International
Committee

February 12, 1973
Newark, Delaware, USA

CONTENTS

II-VI COMPOUNDS

IMPURITY CENTERS AND DEFECTS

LIST OF CONTRIBUTORS (AUTHOR INDEX)

OPENING CEREMONY

The 1972 International Conference on Luminescence was opened
by V.M. Tuchkevich, representative of the Presidium of the Academy
of Sciences of the USSR in Leningrad. F.A. Kokourov, vice-
chairman of the Council of the City of Leningrad welcomed the delegates and guests on behalf of the City.

P.P. Feofilov, corresponding member of the Academy of Sciences
of USSR, quoted from a letter from E.F. Bertaut, secretary of the
Solid State Commission of the International Union of Pure and Applied Physics, wishing the organizers "Good Luck for a successful
convention", and then read the following for L.A. Artsimovich,
Academician and Chairman of the National Committee of Soviet
Physicists:

"The National Committee of Soviet Physicists greets the participants and guests of the International Conference on Luminescence which are gathered in our country to discuss the state and
the perspectives of the development of this science.

"Luminescence is one of the most important branches of the
science of light and its interaction with matter. Luminescent
phenomena are of great scientific importance. Also, there is an
unlimited number of applications of luminescence to various branches of technology and industry. Therefore, luminescence is very
important for the progress of different branches of science and
industry.

"The Soviet physicists are certain that scientific contacts
and creative discussions of scientific problems not only promote
the development of science but also improve the mutual understanding between the peoples of different countries in their struggle
for peace.

"I wish for all the participants and guests of the International Conference on Luminescence fruitful work and new
creative achievements for the common welfare."

The chairman of the International Committee of the Conference,

1

F. Williams, then made the following remarks:

"I am happy to represent the International Committee at the opening ceremonies of this important conference. It is a pleasure for all of us to be in the beautiful city of Leningrad. However, we are here primarily to discuss scientific matters.

"This conference is in the sequence that goes back before World War II: to Oxford in 1938 and most recently to Newark, Delaware in 1969. It is most fitting and proper that this conference be held in the Soviet Union, particularly in Leningrad. The Soviet Union has a long history of excellent research on luminescence, for example, the pioneer work of the elder Vavilov. Leningrad has been and is distinguished by strong schools in both luminescence of organic materials and luminescence of inorganic materials, for examples, the schools of Terenin and Gross, respectively.

"The Soviet Organizing Committee, under Chairman Galanin, has worked hard to arrange facilities for the conference and to assemble a fine program. Now it is up to the speakers and delegates. Luminescence is a broad subject covering many branches: organics, inorganics; molecules, crystals; theory, experiment. For the International Committee I urge each speaker to make his, or her, presentation as clear as possible to those who are not specialists in that particular branch of luminescence. Also, I note that it is in the tradition of these conferences to have vigorous discussions, without rancor, following each presentation. All delegates are encouraged to participate. Good Luck!"

M.D. Galanin, chairman of the Soviet Organizing Committee, completed the Opening Ceremony with the keynote speech which is summarized as follows:

"In the sequence of the Luminescence Conferences the conviction has been created that the aim of these conferences is to consider all luminescence phenomena in whatever substances they occur.

"Indeed, only such a general approach which is also accepted by the Journal of Luminescence makes luminescence a specific branch of physics.

"Luminescence is a complex field and has connections with many different branches of physics. Two main goals are involved when luminescence is investigated. The first is the possibility to use luminescence to create new light sources or to transform one radiation into another, and the second, to obtain information about the structure of substances and about the relevant physical phenomena.

"I would like to mention that such a general approach to
luminescence phenomena has been developed in this country by S.I.
Vavilov. His works on luminescence yield, polarized luminescence,
energy transfer, luminescence quenching and many other subjects
are well known.

"I remind you that the discovery of Vavilov - Cherenkov radiation was a result of a deep understanding of luminescence phenomena.

"The classical papers by A.N. Terenin, especially on the triplet state, were another source of the development of luminescence
investigations in this country.

"It is better to speak about the scientific program of the
conference after it is finished. However, one can find some ten-
dencies, which are evident from our program. One of them is an
increased interest in phenomena at high excitation intensities.
This is quite natural, due to the development of quantum electronics and the possibility to utilize excitation with lasers. The
progress in quantum electronics has resulted in the fact that sometimes luminescence is considered only as a spontaneous noise preceding laser generation. But I believe that such an approach is
too narrow. In this connection I would like to mention one more
interesting direction, which occupies a considerable place in the
conference program. This is cooperative processes arising from
the summation of excitations in crystals. The phenomenon of cooperative sensitization is of interest not only as a new unusual
physical phenomenon but also is important for the technical problem of the transformation of infrared into visible radiation.
Many other current and interesting topics are included in the program of the conference.

"In preparing the program, the theoretical works having a
general importance, and the most new and interesting experimental
works have been given preference.

"The Program Committee has carried out a large amount of work
and has had great difficulties in selecting papers due to the large
number submitted.

"I would like to express the hope that the personal contacts
at our conference will contribute to a better understanding between
the scientists from different countries, and to the consolidation
of peace and progress.

"On behalf of the Soviet Organizing Committee I would like to
greet the foreign scientists participating in the conference. We
hope that their stay in Leningrad will be both useful and pleasant."

SPONTANEOUS AND STIMULATED EMISSION FROM EXCITONS AT HIGH

CONCENTRATION

S. Nikitine and H. Haken

Laboratoire de Spectroscopie et d'Optique du Corps

Solide, Université Louis Pasteur, Strasbourg, France

and Institut für theor. physik, Universität Stuttgart,

Stuttgart, BRD (Germany)

ABSTRACT

In this paper a short review of exciton spectra at high concentrations is given. Arguments in favor of the formation of
biexcitons are discussed. Different effects for blue and red
shifts of absorption and emission lines are discussed. It is
further shown that excitons do not emit stimulated emission unless
the emission and absorption are displaced from each other. Biexcitons can emit stimulated radiation. The competition and the
stimulated emission are discussed for the exciton-exciton and the
exciton-LO phonon effects.

INTRODUCTION

The spontaneous emission from excitons at low concentrations
is well-known and a list of processes involved in this kind of
luminescence can be given, keeping in mind that we consider
optically excited luminescence only, as follows (1-11):

$h\nu_o = E_g - E_{ex}$,   resonance emission from excitons;

$h\nu_1 = E_g - E_{ex} - \Delta E_{jj}$,   emission from forbidden exciton
state resulting from jj coupling
(para excitons for example);

$h\nu(D^{+}ex)=E_g-E_{ex}-E_{(ex3)}$,

emission from 3 particle bound excitons complexes (a similar process involves acceptors);

$h\nu(D^{o}ex)=E_g-E_{ex}-E_{(ex4)}$,

emission from 4 particle bound exciton complexes (a similar process involves acceptors);

$h\nu(ex-nLO)=E_g-E_{ex}-nE_{(LO)}$,

simultaneous emission of a photon and nLO phonons from an exciton; and

$h\nu(D^{i}-nLO)=E_{ex}-E_{exj}-nE_{(LO)}$,

simultaneous emission of a photon and nLO phonons from an exciton complex.

Here D^i means the ith donor state bound to the exciton, E_{exj} the binding energy of this state, and E_{LO} the energy of the LO phonon. At low temperatures, the excitation is collected mainly in the state of lowest energy from which the most important emission takes place.

SPONTANEOUS EMISSION AT HIGH CONCENTRATIONS OF EXCITONS

At high concentrations of excitons, some of the above processes will also take place. However, some of them are influenced by high concentrations of excitons. New emission processes may take place and their importance may become so great that processes observed at low concentration of excitons may become comparatively unimportant.

When the concentration becomes high, excitons may form molecules or biexcitons (12). We are going to concentrate our interest on CuCl and CuBr. The results on the biexcitons are well-known now; therefore, we are going to summarize the experiments and the interpretation.

A new line ν_B appears at high concentrations of excitons in some substances, in particular in CuCl and CuBr (Fig. 1). The intensity of the line I_B is related to the intensity i of the exciting line (Fig. 2),

$$\log I_B = m_B \log i + c^{te}, \tag{1}$$

but m_B varies with i and $m_B \approx 2$ for $i \ll i_0$, where i_0 is a characteristic constant given later, but $m \approx 1$ when $i \gg i_0$. The intensity I_0 of ν_0 depends linearly on i. The difference (where ν' means wave numbers) $\nu'_0 - \nu'_B$ is about 350 cm^{-1} for CuCl and 236 cm^{-1} for CuBr. This emission is interpreted as resulting from the process

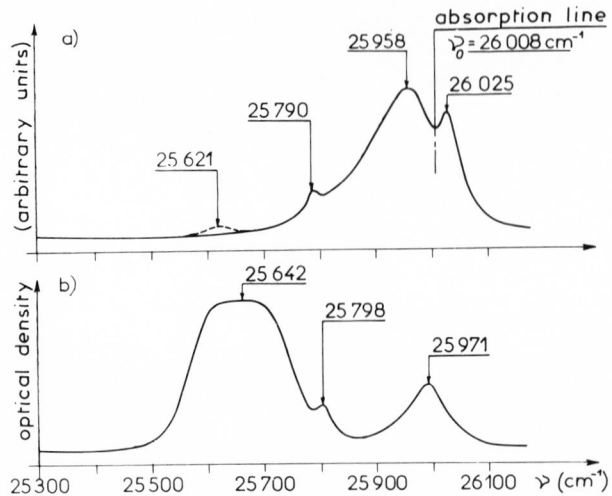

Fig. 1   Emission of CuCl at low temperatures: (a) spectrum at
         low excitation intensities; (b) spectrum at very high
         excitation intensities. The band at 25642 cm^{-1} is the
         ν_B band assigned to the recombination of biexcitons. In
         some crystals and more recent experiments a structure
         appears in this band at ν_0=25971 cm^{-1}.

involving the radiative recombinations of biexcitons:

$$(biex) \to h\nu_B + (ex) \tag{2}$$

this gives the following balance:

$$h\nu_B = E_g - E_{ex} - E_B + E_B^k - E_{ex}^k \tag{3}$$

Where E_B is the binding energy of the biexciton and the last two
terms are the kinetic energies resulting from the process.
Clearly:

$$E_B = h(\nu_0 - \nu_B) \tag{4}$$

The binding energy of biexcitons has been calculated by
different authors (13, 14, 15). All these theories have the defect
that they do not take into account the polarisability of the crystal. An approximate calculation has been made by the author and
Myzyrowicz (16). In an improved form, it leads to the formula
which is applicable only for $\sigma=m_e/m_h \ll 1$

$$E_B = (E_{ex}/E_H)[D - \sigma^{1/2}(M_p/m_o)^{1/2}E_o] - E^{ad} \tag{5}$$

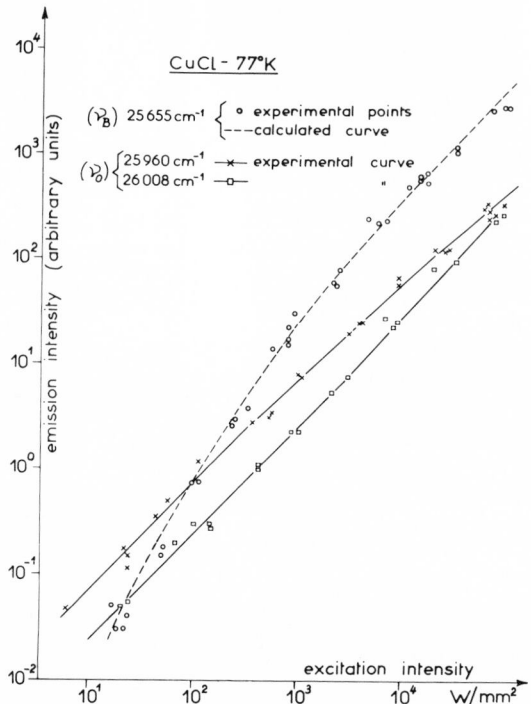

Fig. 2 The variation of the intensity I_B of the band ν_B (dotted
line) and I_O of the band ν_O taken for two different frequencies (solid line) as a function of i, the intensity
of excitation light on a log log scale .

This formula has the advantage of introducing E_{ex}/E_H which
is an empirical quantity known from exciton spectra and, to some
extent, includes the polarisability of the crystal. In Eq. 5,
E_H is the binding energy of the hydrogen atom, D the depth of Morse
potential well for H_2, E_O is the zero point vibration energy of H_2,
and E^{ad} is an adiabatic correction which can be negligible if
$\sigma << 1$ (see also (17)). The formula is applicable only for small
values of σ when the Born-Oppenheimer approximation is applicable.
The values of E_B calculated for several compounds are given in
Table 1.

It is to be noted that the calculations from (5) are in rather
good agreement with Hanamura's values which can, however, be obtained from the graph only with rather poor accuracy. It should
also be remembered that in recent measurements, a structure appears
in the biexciton bands which can come from the forbidden exciton
levels or rotational structures (18).

Table 1

Numerical values entering in the calculation
of E_B for different substances

	E_{ex} cm^{-1}	m_e/m_h	D_B cm^{-1}	Eo_B cm^{-1}	ΔE^{ad} cm^{-1}	$E_{Bcalc.}$ cm^{-1}	$E_{Bobs.}$ cm^{-1}	$E_{Bcalc.}$ b cm^{-1}
CuCl	1523	0.02	560	183	28	359	350	∿304
CuBr	873	0.01	320	74	8	238	236	∿210
ZnO	338	0.125	124	104	74 a			30.4
CdS	226	0.17	85	81.5	36 a			17
Ge	34	0.665	12.4	23.4	32 a			

(a) Born-Oppenheimer approximation is presumably not applicable.
(b) Evaluated from Hanamura's curve.

The kinetics of formation of biexcitons (12) has led to the
division into optical excitons $\underline{K}\approx0$ with concentration n_0 and
thermal excitons $\underline{K}>0$ with concentration n_1. If the concentration
of biexcitons is $\overline{n}_B$ the calculation gives the equations:

$$n_0 = A_0\tau i/(1 + \tau K); \quad n_1 = (K''/B)[(i/i_0 + 1)^{1/2} - 1]$$

$$n_B = [(K'')^2/2BC] \cdot [(i/i_0 + 1)^{1/2} - 1]^2$$

$$i_0 = [(K'')^2/2B] \cdot \{(1 + K\tau)/[A_1 + K\tau(A_0 + A_1)]\}$$

where K is the conversion rate of optical to thermal excitons; K"
the rate of radiationless decay of thermal excitons; B is a bimolecular collision coefficient; and C is the recombination coefficient of biexcitons. A_0 and A_1 are absorption constants for formation of optical and thermal excitons from initial carriers formed optically by absorption in the continuum. It can be seen from
Fig. 2 that for CuCl these formulae are in good agreement with
experiment. The agreement is also good for CuBr. I_0 is proportional to n_0, and therefore to i for all values of i. I_B is proportional to n_B. The variation has a slope of 2 for $i<<i_0$ and 1
for $i>>i_0$.

The question arises as to whether or not biexcitons can be
formed in any material and if not, for what reason. A final answer
to this question is not possible as the theory of other processes
which should be considered is not yet known. However, it can be
seen that for CuCl and CuBr the experiment is in good agreement
with the calculation concerning the biexciton. Biexcitons have
been suggested to exist in some other substances though a careful
confirmation is desirable. But in many substances biexcitons have
not been identified. Table 1 can give a reason for this. It can

be seen that E_B deduced from Hanamura's graph is very small for
these cases. In the case of CuCl and CuBr, the line breadth on
both sides of ν_B is considerable in our crystals. In better
crystals, this breadth will probably be smaller as shown by Goto
and Ueta (19). Nevertheless, this breadth (probably due to
fluctuation of the kinetic energy of the exciton) is of the order
of, or even larger than E_B, say for ZnO or CdS. This could mean
that biexcitons are not stable in such compounds and other processes may take place. The situation is to some extent comparable
to resonant decaying states in elementary particles physics.

As an alternative possibility of biexciton formation the
scattering of an exciton by another exciton has been considered
(20-24). It is possible that this effect is observable only when
biexcitons are not stable. This effect, as well as some other
effects, is treated in the paper by Levy, Bivas and Grun in this
Conference and will be only treated briefly here. The effect is
as follows:

$$(ex)_1 + (ex)_2 \rightarrow e + h + h\nu \quad (ex\text{-}ex) \qquad (6)$$

the energy balance gives:

$$h\nu \quad (ex\text{-}ex) = Eg - 2 E_{ex} - E_{eh} \qquad (7)$$

Here E_{eh} is the kinetic energy of electron and hole. The observed
line $\nu(ex\text{-}ex)$ is shifted by a band filling effect to lower energies
(red shift) with increasing concentration. This can be calculated
in a simple way (24):

$$\Delta\nu = [m_e^{-1} + m_h^{-1}](h/8)(3/8\pi)^{2/3}n^{2/3}. \qquad (8)$$

The masses, assuming ellipsoidal bands at $|K| = 0$, are $m_e = (m_{xe} \, m_{ye} \, m_{ze})^{1/3}$ and a similar definition for m_h. The agreement
with experiment is quite good as will be seen in the paper of Levy
et al.

The photon and LO phonon emission effect has been mentioned
in the section on emission at low concentration of excitons. The
effect of simultaneous emission of a phonon and a photon becomes
very important at high concentrations of excitons. We consider it
in the section on stimulated emission.

When the concentration of excitons becomes large, the exciton
absorption line $\nu_0(a)$ gradually shifts to higher energies (blue
shift) but the emission line $\nu_0(e)$ does not shift in an appreciable
way. This effect is analysed in the paper of Levy, Bivas and Grun
in this Conference (25). Different theories have been advanced.

One theory states that the dielectric constant changes on account of the high concentration of excitons (26a), which leads to a shift of the exciton levels. The theories of Hanamura (27) and of Kaldysh and Kozloff (28) either assume or are equivalent to Bose condensation of excitons. This again leads to a shift of the exciton levels. These two theories predict, however, the shift of the absorption line but not the absence of appreciable shift of the emission line, which is in contradiction with the experimental situation.

On this aspect, so far, the only satisfactory theory is based on the fact that at high concentrations of carriers in the formation of wave functions of new excitons, the summation over all the $|k|$ states is not possible but has to start from a lowest $|k_0|$ value, the lower states being occupied already. This leads to a shift of the exciton absorption line to higher energies when the concentration increases. It has to be emphasized that this is not the usual band filling effect of band to band transitions. The alternative case in which it is supposed that all carriers combine rapidly to excitons has also been treated (26b). A similar shift of the absorption line is obtained. Both cases are probably simultaneously realized and their relative importance depends only on the lifetime of the carriers with respect to the exciton formation process. It has to be emphasized that both effects concern non-equilibrium situations. As regards the emission line all the $\underline{k}$ states are available for the formation of the optical exciton states, which are the lowest states in the exciton band. On account of the selection rules, these excitons mostly will be involved in the emission. A small shift of the maximum of the emission band and broadening of the line are probable in this model. These theories are in qualitative agreement with experiment. Thus, it seems that the above experiments and their interpretation are opposed to the assumption equivalent to Bose condensation. The shift predicted in this theory, first alternative, is

$$\Delta E_{ex}/E_{ex} = 32\pi a^3 n_e, \tag{9}$$

where n_e is the concentration of carriers and a the radius of excitons. The n_e is proportional to $i^{1/2}$ and so is $\Delta \nu_0(a)$. In the second alternative a similar expression is obtained but the numerical factor is changed.

The bleaching of the absorption line n = 1 can have different origins (Fig. 3). They seem to be covered by a simple phenomenological theory which assumes that excitons cannot be formed with a higher concentration than n_0. For the steady state, the kinetic relation is:

$$\alpha i(n_0 - n_{ex}) - n_{ex}/\tau = 0 ;$$

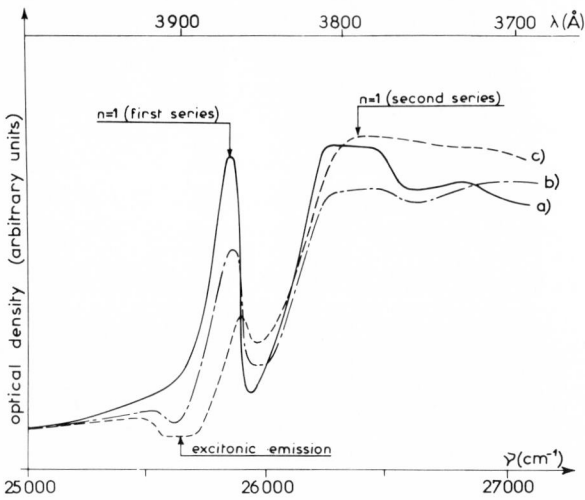

Fig. 3   Absorption spectrum of CuCl measured with a first source
         and at different irradiation intensities of a second
         source, showing the "blue" shift and the "bleaching" of
         the n = 1 line of the spectrum.

where n_{ex} is the concentration of excitons, α is an absorption
cross-section and τ is the lifetime. This gives for the absorption coefficient:

$$K = K_0/(1 + \alpha\tau i);\qquad\qquad\qquad (10)$$

K_0 is an absorption coefficient for t = 0. This theory has been
carried out supposing that the irradiation takes place in the
absorption line. If an irradiation takes place in the continuum
a correction should be introduced. Only qualitative data are
known at present (29). A reasonable value is $n_0 \propto 1/a^3$.

STIMULATED EMISSION FROM EXCITONS

It has been pointed out that stimulated emission involving
excitons can take place at high concentrations of excitons. A
complete theory is being worked out for different processes, but
it can be shown now that some of them do and some do not lead
to stimulated emission.

It can be shown that the kinetic equation for excitons is:

$$dn_p/dt = 2(g^2/\gamma)(n_{ex} - n_p) - 2Kn_p \qquad\qquad (11)$$

Here n_p is the concentration of photons, n_{ex} is the concentration
of excitons, g is a transition matrix element, γ is a damping constant of the exciton, and K is a loss coefficient of the photons.
It can be seen that the terms containing n_p are negative, which
means that no stimulated emission takes place. Therefore, no
stimulated emission is expected in the exciton resonance line ν_0.

It has been shown that the absorption line $\nu_0(a)$ is shifted
with respect to the emission line $\nu_0(e)$ and bleaches at high concentrations of excitons in CuCl. In this case, the above statement
is no longer rigorous and a stimulated emission seems not to be
ruled out. However, it has not yet been observed.

Fine structure levels of excitons (say para-exciton levels)
are shown to exist in different compounds. Some of them are lower
than the allowed ortho-exciton level. So at low temperatures, the
excitation may accumulate in such an exciton level. The absorption
transitions under usual conditions are forbidden however. This
could possibly lead to a stimulated emission if the line is not in
the wing of the allowed absorption line $\nu_0(a)$. The theory of these
effects has not been developed and no experiment in favor of such
an emission has been performed to the authors' knowledge.

The conditions for stimulated emission are favorable from
biexcitons which have been identified in CuCl and CuBr and also
recently in Cu_2O (30, 31). The rate equation for this emission
is:

$$dn_p/dt = n_p Gn_B + Gn_B - 2 Kn_p, \qquad (12)$$

where n_p is again the photon number per cm^3 in a given mode, G is
a gain coefficient, n_B is the concentration of biexcitons, and the
last term describes the losses. This equation shows that up from
a certain concentration, the emission is stimulated $n_B > 2K/G$. The
observation is not simple, however using another form of (12), it
can be seen that:

$$I_B = (I_{sp}/\alpha)[\exp(\alpha x)-1], \qquad (13)$$

where α is a net gain coefficient, I_{sp} is the intensity of spontaneous emission, and x is the length of the emitting material.
Recently such a dependence of x has been found experimentally by
Shaklee, Lehery and Nahovy (32). The gain was shown to be very
large. It is to be remembered that no process of optical formation
of biexcitons is known, which is in contrast to the case for excitons.

It can be shown that the (ex-LO) process very often observed
at low concentration gives rise to stimulated emission up from a

threshold concentration. This has been shown first by Haug (33).
The rate equation is:

$$dn_{LO}/dt = n_{LO}\ \beta n_{ex} + \beta n_{ex} - 2\ K\ n_{LO}. \tag{14}$$

Here n_{LO} is the number of photons emitted in this process, n_{ex} is
the concentration of excitons, β is a rate coefficient, and K is
a loss coefficient. It can be seen that stimulated emission is
again predicted.

It can be shown experimentally that the above process (ex-LO)
is in competition with the (ex-ex) process in some crystals. This
is seen from the next rate equation, in which no process other
than (ex-ex) are considered.

$$dn_p/dt\ (\text{ex-ex}) = \alpha n_{ex}^2 - 2\ K\ n_{p(\text{ex-ex})}. \tag{15}$$

Here $n_{p(\text{ex-ex})}$ is the number of photons emitted in the ex-ex process. It is seen that the coefficient of $n_{p(\text{ex-ex})}$ is negative
and; therefore, that no stimulated emission is predicted. A third
order effect due to stimulated emission could become important at
very high concentration of excitons and photons. (Such effects
may be at the origin of the observations of Prof. Pilkuhn who
communicated to the authors that stimulated emission has been observed in the ex-ex process.) It is clear that both equation (14)
and (15), being simultaneously valid, a pronounced competition must
take place when stimulated emission sets in for the process (ex-LO).
This is seen in Fig. 4 from Grun and Levy's experiments for CdS
(unpublished, after Levy and Grun). In these crystals some intermediate processes are observed which account for the shift of the
threshold of the stimulated emission with respect to the saturation
of the (ex-ex) effect. It is also seen that the intensity of
luminescence in this process increases with i^2. For very high
values of i, a stimulated emission sets in with a variety of
processes involving nLO emissions. It has been suggested that this
should be named an avalanche emission (34, 12). This effect is
believed to be induced by the great density of LO phonons (Fig. 5).

We have seen that exciton complexes participate in luminescence
in a pronounced manner. No stimulated emission is expected from
these complexes on account of the giant oscillator strength of the
corresponding absorption lines (35). However, if it happens that
the absorption is shifted with respect to emission, stimulated
emission could be possible. It can be seen from Fig. 5 that such
a stimulated emission takes place with cooperation of LO phonons.
A general theory of such effects does not seem to have been worked
out.

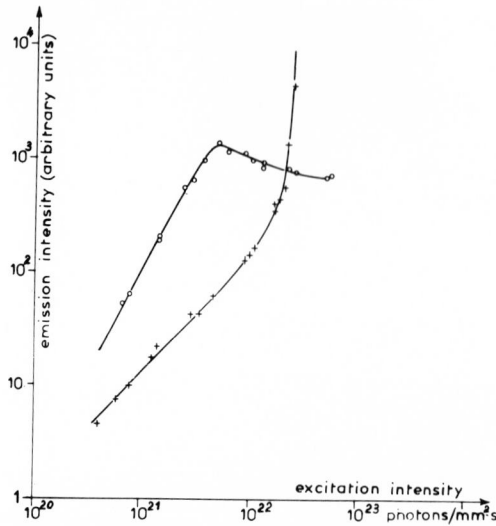

Fig. 4 Variation of the intensity of the ν(ex-ex) and of the
ν(ex-LO) lines for CdS as a function of i showing the
competition of both effects. Some intermediate effects
are seen in some crystals explaining the shift of the
stimulated emission threshold of ν(ex-LO) to the saturation effect of ν(ex-ex) after Levy and Grun (unpublished).

SUMMARY

    Different processes of exciton luminescence were considered,
in particular for high concentrations of excitons. The emission
from biexcitons was first discussed and then consideration given
to the emission from excitons when they interact and when simultaneous emission of phonons and photons takes place. The "blue
shift" and bleaching of the exciton absorption line and the "red
shift" of luminescence from collision of excitons with excitons
were discussed.

    It was shown that no stimulated emission is expected in the
exciton emission line, unless the absorption line is strongly
shifted out of the emission line, which has however been observed.
Biexcitons are shown to exhibit stimulated emission. When phonons
are simultaneously emitted with photons, stimulated emission is
expected above a threshold exciton concentration. Simultaneous
emission of photons and phonons was shown to be important in the
stimulated emission of exciton complexes.

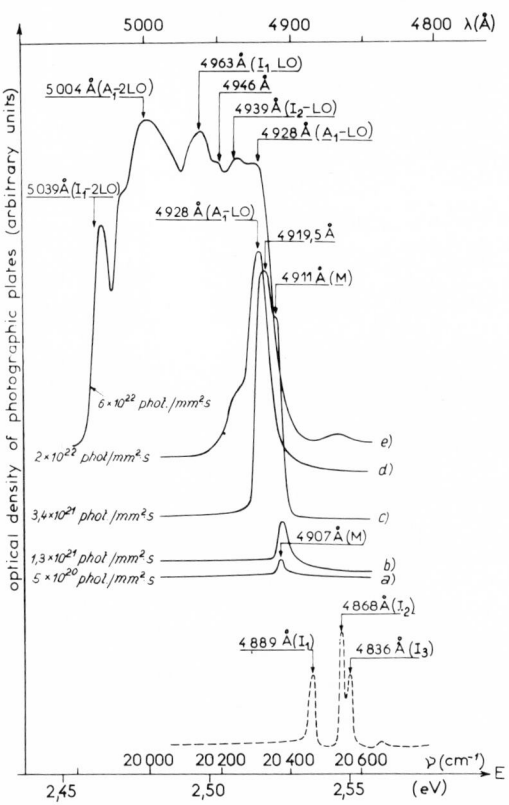

Fig. 5   The variation of spectra of CdS crystals as a function of
         i.   This figure shows some intermediate spectra between
         the saturation of ν(ex-ex) and the stimulated emission of
         ν(ex-LO) as well as more complicated stimulated emission
         processes.

REFERENCES

(1) E. F. Grillot, M. Bancie-Grillot, P. Pesteil, and A. Zmerli,
     C.R. Acad. Sc., Paris, 242, 1794 (1956).

     E. F. Grillot, J. Phys. Rad., 17, 822 (1956).

     S. Nikitine, G. Perny, M. Sieskind, R. Reiss, J. Phys. Rad.,
     17, 817 (1956).

     S. Nikitine and G. Perny, J. Phys. Rad., 17, 1017 (1956).

S. Nikitine and R. Reiss, C.R. Acad. Sc., Paris $\underline{244}$, 2788 (1956); $\underline{245}$, 52 (1957).

V. A. Arkhangelskaia and P. P. Feofilov, J. Phys. Rad. $\underline{17}$, 824 (1956).

(2) Y. Toyozawa, Suppl. Progr. Theor. Phys. $\underline{12}$, 93 (1959).

(3) S. Nikitine, J. Ringeissen and M. Certier, Acta Phys. Pol. $\underline{26}$, 745 (1964).

S. Nikitine, J. Ringeissen and J. Sennet, Conference on Phys. Semiconductors Radiative Recombination, 279 (1964).

(4) a) S. Nikitine and R. Reiss, J. Phys. Rad. Letters $\underline{20}$, 718 (1959).

b) A. Myzyrowicz, J. B. Grun, A. Bivas, R. Levy and S. Nikitine, Phys. Letters $\underline{25A}$, 286 (1967).

c) J. B. Grun, A. Myzyrowicz, F. Raga, A. Bivas, R. Levy and S. Nikitine, Phys. Stat. Sol. $\underline{22}$, K155 (1967).

d) R. Levy, A. Bivas and J. B. Grun, J. Phys. $\underline{31}$, 507 (1970).

e) See also J. J. Hopfield and D. G. Thomas.

(5) M. Lampert, Phys. Rev. Letters $\underline{1}$, 450 (1958).

(6) E. I. Rashba, Sov. Phys. JETP $\underline{4}$, 759 (1962).

(7) S. Nikitine, Phil. Mag., $\underline{4}$, 1 (1959); J. Chimie Phys. $\underline{55}$, 43 (1958).

(8) D. G. Thomas and J. J. Hopfield, Phys. Rev. $\underline{128}$, 2135 (1962).

(9) S. Nikitine and R. Reiss, J. Phys. Rad. Letters $\underline{20}$, 718 (1959).

R. Reiss, Thesis (Strasbourg), Cahiers de Phys. $\underline{13}$, 129 (1959).

(10) S. G. El Komoss and B. Stebe, in press, Nuovo Cimento, J. Phys.

S. G. El Komoss, J. Phys. Chem. Solid., $\underline{33}$, 750 (1972); Phys. Rev. $\underline{10}$, 3411 (1971).

(11)   S. Nikitine, J. Ringeissen and M. Certier, Acta Phys.
        Polon. 26, 745 (1964).

        R. Reiss, Thesis (Strasbourg), Cahiers de Phys. 13, 129
        (1959).

        S. Lewonczuk, J. Ringeissen and S. Nikitine, J. Phys.
        32, 941 (1971).

(12)   a) A. Myzyrowicz, J. B. Grun, R. Levy, A. Bivas and
           S. Nikitine, Phys. Letters 26A, 615 (1968).

        b) S. Nikitine, A. Myzyrowicz and J. B. Grun, Helv. Phys.
           Acta 41, 1058 (1968).

        c) A. Myzyrowicz, Thesis (Strasbourg) (1968).

        d) A. Bivas, R. Levy, S. Nikitine, and J. B. Grun,
           J. Phys. 31, 227 (1970).

        e) A. Bivas, Thesis (3ème Cycle) (Strasbourg) (1969).

        f) R. S. Knox, S. Nikitine and A. Myzyrowicz, Opt.
           Comm. 1, 19 (1969).

        g) J. B. Grun, S. Nikitine, A. Bivas and R. Levy, J. of
           Luminescence 1,2, 241 (1970).

(13)   R. R. Sharma, Phys. Rev. 170, 770 (1968).

(14)   R. K. Wehner, Solid State Comm. 9, 457 (1969).

(15)   O. Akimoto and Hanamura, Solid State Comm. 10, 253 (1972).

        See also, Adamovski, Bednarek, Suffczynki, Solid State
        Comm. 9, 2037 (1971).

(16)   S. Nikitine, J. B. Grun and A. Myzyrowicz, Helv. Phys.
        Acta 41,1058 (1968).

        A. Myzyrowicz, Thesis, Strasbourg (1969).

        S. Nikitine, Non-Linear Optics Colloquium, Titisee (1971).

(17)   Z. A. Kazamanyan, Sov. Phys. 1, 341 (1967).

(18)   J. B. Grun, S. Nikitine, A. Bivas and R. Levy, J. of
        Luminescence 1,2, 241 (1970).

A. Bivas, R. Levy, S. Nikitine and J. B. Grun, J. de Phys. 31, 227 (1970), see also (19).

(19) H. Souma, H. Koike, K. Kaoru, M. Ueta and Suzuki, J. Phys. Soc. Japan, in press.

See also, T. Goto and M. Ueta, J. Phys. Soc. Japan 24 656 (1968).

(20) D. Magde, H. Mahr, Phys. Rev. Letters 24, 890 (1970).

(21) M. Pilkuhn, Non-Linear Optics Conference, Titisee, (Sept. 1971).

(22) C. Benoit A LA Guillaume, J. Debever and F. Salvan, Phys. Rev. 177, 567 (1969).

See also C. Benoit A La Guillaume, F. Salvan and Voos, Intern. Confer. University of Delaware (1969), N. Holland, Amsterdam.

(23) R. Levy and J. B. Grun, in press.

(24) a) R. Levy, J. B. Grun, H. Haken and S. Nikitine, Solid State Comm. 10, 915 (1972).

b) R. Levy, A. Bivas and J. B. Grun, this conference.

(25) a) A. Myzyrowicz, J. B. Grun, A. Bivas, R. Levy and S. Nikitine, Phys. Lett. 25A, 286 (1967).

b) R. Levy, A. Bivas, J. B. Grun, J. Phys. 31, 507 (1970), see also (23).

(26) H. Haug, J. Appl. Phys. 39, 4681 (1968).

(27) a) A. Bivas, R. Levy, J. B. Grun, C. Comte, H. Haken and S. Nikitine, Optics Comm. 2, 227 (1970).

b) L. V. Keldysh and A. N. Kozlov, JETP 27, 521 (1968). E. Hanamura, J. Phys. Soc. Japan 29, 50 (1970).

c) H. Haken and Goll, to be published.

(28) R. Levy, A. Bivas and J. B. Grun, Phys. Letters 36A 159 (1971).

(29) S. Nikitine, R. Levy and J. B. Grun, to be published.

(30) J. R. Haynes, Phys. Rev. Letters 17, 860 (1966).

(31) E. F. Gross and F. I. Kreingold, JETP LETTERS 12, 68 (1970).

O. I. Lvov and P. P. Pavinskii, ZhETF Letters 14 253 (1971).

(32) K. L. Shaklee, R. F. Leheny and R. E. Nahovy, Phys. Rev.
Letters 26, 888 (1971).

(33) K. Era, and D. Langer, J. Appl. Phys. 42, 1021 (1971).

(34) T. Goto, D. W. Langer, Phys. Rev. Letters 27, 1004 (1971).

(35) A. Myzyrowicz, J. B. Grun, F. Raga and S. Nikitine, Phys.
Letters 24A, 335 (1967).

J. B. Grun, A. Myzyrowicz, F. Raga, A. Bivas, R. Levy and
S. Nikitine, Phys. Stat. Solidi 22, K155 (1967).

BIEXCITONS AND CONDENSATION OF EXCITONS IN SEMICONDUCTORS

A. A. Rogachev

Physico-Technical Institute of Academy of Sciences

Leningrad, USSR

ABSTRACT

A brief review is given of experimental data supporting the existence of biexcitons in germanium and silicon. Condensation of biexcitons is also discussed and it is shown that the density of the condensed phase is 2×10^{16} cm-3 in germanium.

Electrons and holes in semiconductors can be considered as particles with an effective mass m* interacting in accordance with the Coulomb law at distances greater than a few lattice constants. An inevitable consequence of the interaction is the possibility of forming electron-hole atoms (excitons), excitonic molecules (biexcitons) and a condensed electron-hole phase (1-21).

A theoretical estimate shows that the binding energy of excitons in excitonic molecules and the condensed phase is at least a few tenths of electron volts making it possible to observe such states at liquid helium temperatures. Experimental observation of biexcitons and electron-hole condensates have been reported by various authors (3-21). There are many contradictions between the experimental data obtained so far, but it is possible to distinguish between two major approaches to the experimental results on the collective properties of excitons in these semiconductors.

According to one approach, the broad, long-wave lines in the luminescence of pure germanium (0.709 ev) and silicon (1.08 ev) are due to radiative recombination of excitonic molecules (3-9), while the other approach suggests them to be a result of recombination of electron-hole condensates (13-21). From the first approach it would follow that the binding energy of excitonic molecules in germanium is equal to 3.5×10^{-3} ev and the density of electron-hole pairs in condensates is about 2×10^{16} cm^{-3}. In the second case the binding energy of the molecules is negligibly small and the density of the condensed phase is equal to 2.6×10^{17} cm^{-3}.

This paper is a brief review of experimental data supporting the conclusion that biexcitons in germanium and silicon do exist and that the luminescence band with maxima at 0.709 ev in germanium and 1.08 ev in silicon is due to their radiative recombination. The biexcitonic nature of the lines has been demonstrated by studying the dependence on temperature and concentration of the intensities of exciton and biexciton radiation (3-6).

There is a general relation between exciton and biexciton densities:

$$n_b \sim n_{ex}^2 \tag{1}$$

Its validity does not depend on whether the exciton-biexciton system is in thermal equilibrium. The intensities of recombination radiation are proportional to the concentrations of excitons and biexcitons, and can be measured experimentally. The experimental dependence of n_{ex} on n_b is shown in Fig. 1. To obtain these data two photon absorption of the radiation from a pulsed dye laser ($h\nu = 0.53$ ev) was used. As a result of the very small absorption constant, a practically uniform distribution of carriers was achieved (6). The intensity of the biexciton line is proportional to the square of the exciton line intensity in a wide range of concentrations: $(10^{12}-10^{15})$cm^{-3}.

If stationary excitation is used, the temperature dependence of n_b and n_{ex} is determined from reaction kinetics such that:

$$n_b = \frac{\sigma v \, n_{ex}^2}{\frac{1}{\tau} + \sigma v \, \frac{N_{ex}^2}{N_b} \exp(-E_M/kT)} \tag{2}$$

where E_M is the binding energy of the biexcitons, n_{ex} and n_b the effective densities of states in exciton and biexciton bands, τ and σ the lifetime and formation cross section of biexcitons, and v the thermal velocity.

If the temperature is high enough the time of thermal dissociation of biexcitons is much shorter than their lifetime with respect to interband recombination. In this temperature range the dependence of $n_{ex}^2 \, n_b^{-1}$ on temperature is exponential. Experimental results for germanium and silicon are shown in Fig. 2 and Fig. 3.

The binding energy of biexcitons found from these data is 3.5×10^{-3} ev for germanium and 5.5×10^{-3} ev for silicon.

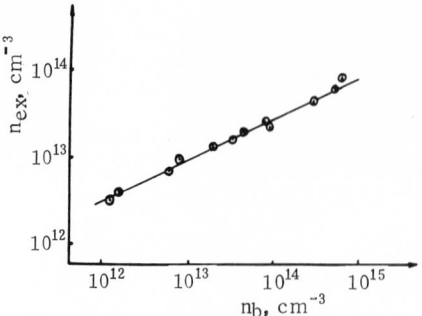

Fig. 1   Free exciton concentration as a function of biexciton
         concentration.

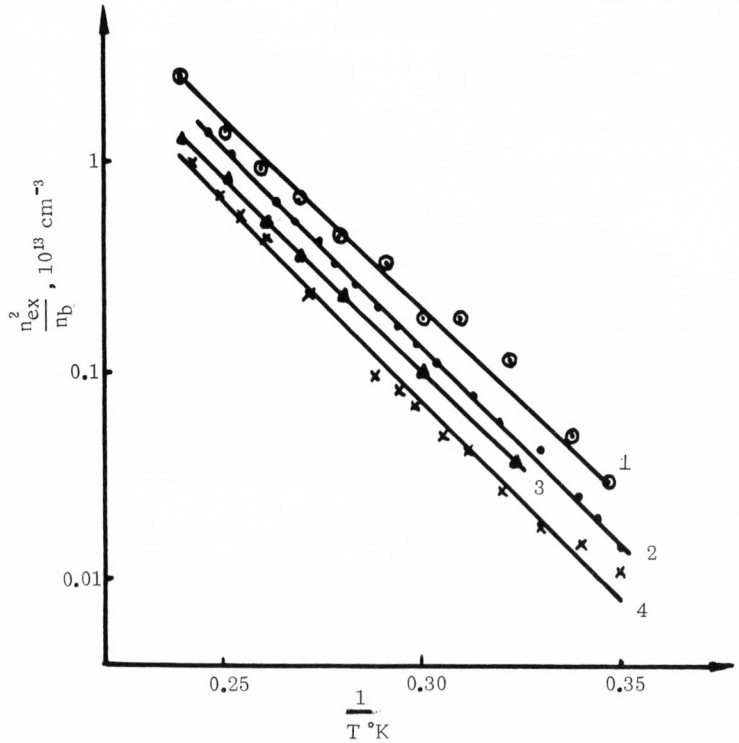

Fig. 2   Temperature dependence of $n_{ex}^{2} n_{b}^{-1}$ for germanium.

For very high excitation levels a linear dependence of n_{b} on
n_{ex} has been observed (4,5). To explain this it has been suggested
that the accumulation of long-wave phonons generated in the course
of thermalization of electrons and holes, which are created by

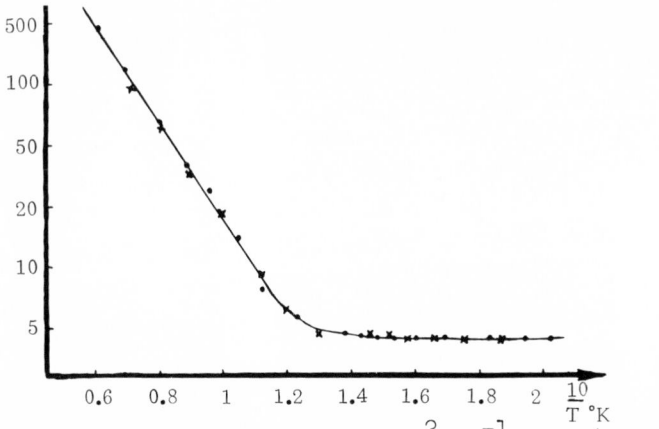

Fig. 3 Temperature dependence of $n_{ex}^2 \, n_b^{-1}$ for silicon.

light, takes place (5). The lifetime of the phonons is limited by surface scattering and can be some hundreds of microseconds. At low temperatures the intensity of biexciton radiation is much greater than that of all other luminescence lines, but no trace of metallic electron-hole condensates was found up to concentrations as high as 5×10^{14} cm^{-3}. Such condensates were found at the higher concentrations but their appearance was not accompanied by any essential change in the character of the recombination radiation. The first data of this kind were obtained in studying the modulation of the direct absorption edge in germanium due to the screening of the Coulomb interaction by free carriers (8,9). It was found that at liquid helium temperatures modulation is absent if the concentrations of electron-hole pairs is smaller than 5×10^{14} cm^{-3}. The modulation became measurable and its magnitude rose very sharply when the concentration exceeded 10^{15} cm^{-3}. The shape of the modulation and emission spectra remained unchanged at concentrations corresponding to the sharp rise in the modulation.

Variation of the number of electron-hole pairs in the sample can be observed either by varying the intensity of excitation (solid line in Fig. 4) or by making measurements during some interval of time after the end of excitation (dotted line in Fig. 4). Electron-hole pairs in the sample decay in accordance with a law which is nearly exponential: $n \sim n_o \exp(-t/\tau)$. Making the measurements at different times t it is possible to obtain data corresponding to different values of n. It is evident that both of the methods must give equal results for a one phase system. The analysis of modulation experiments shows that at excitation levels for which a sharp rise in modulation takes place there are two regions, one with a low and one with a high concentration of free carriers (8,9).

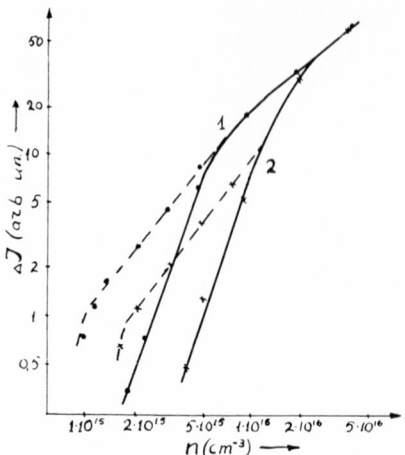

Fig. 4 Maximum modulation as a function of the average concentration of electron-hole pairs: 1: T=4.2K; 2: T=2.0K. The
solid line represents the data obtained by changing the
light pulse intensity. The dashed line indicates the
changing of the delay between the light pulse end and the
moment of measurement.

The concentration of electron-hole pairs in conducting regions
was estimated to be 2×10^{16} cm^{-3}. Another result of the experiments
is that no connection was found between the appearance of the long
wave radiation and the existence of metallic type condensates.
The condensates appear in the sample at excitation levels at least
three orders of magnitude greater than that needed for observation
of the long wave radiation.

The same results have been obtained in microwave absorption
measurements in silicon and germanium (22,23). It has been shown
that microwave absorption arises at concentrations greater than
10^{15} cm^{-3}, while the intensity of the long wave-length radiation
depends linearly on excitation for a very wide intensity range (23).

Investigation of the d-c conductivity shows that electron-hole
condensates fill up the whole sample at 2K with a concentration of
2×10^{16} cm^{-3} and a further rise of excitation leads to the growth of
regions of high concentrations (27). The absence of electron-hole
condensates at small concentrations and their appearance at concentrations greater than 5×10^{14} cm^{-3} was demonstrated by the observation of large fluctuations of photocurrent in n-p junctions (10-
12). The main idea of these experiments is that capturing of the
exciton condensates in the high field region of a n-p junction

leads to current fluctuations with total charge equal to the number of excitons in the condensate.

From the biexciton point of view the shape of the luminescence spectra is to be considered as follows: When radiative annihilation of one of the excitons in a molecule occurs, some of the energy is transmitted to another.[1] As a result of this process a broad emission band is formed. For the long-wave part of the band the mechanism originally proposed by Haynes(3) is most probable. In this case annihilation of one exciton is accompanied by ionisation of the other. This process is probably responsible for the low energy tail of the radiation (6).

Since the binding energy of biexcitons in germanium and silicon is of the same order of magnitude as that of the exciton, the effective radius of biexcitons is about equal to the exciton Bohr radius r_h. Then the average quasi-momentum of .electrons and holes in biexciton is r_h^{-1}, and the average recoil energy (which is equal to the half width of the biexciton band) is about equal to the binding energy of the exciton. Experimental data presented on Fig. 5 and Fig. 6 are in agreement with this conclusion.

The binding energy of a biexciton may be determined by measuring the distance between the center of the exciton line and the short wave-length side of the biexciton band. A comparison of binding energies, determined by this method, with "thermal" values discussed above is shown in Fig. 5 and Fig. 6. The agreement in germanium is satisfactory if the broadening of the biexciton band due to the finite resolution of the spectrometer is taken into account. In silicon the "optical" value is 30% greater than the thermal. Assuming the long-wave length radiation to be connected with electron-hole condensates, the binding energy of an exciton in the condensed phase is equal to the distance between the exciton maximum and the free exciton energy, which is greater by that of $3/5\ E_F$ than the low energy side of the band (E_F is the sum of Fermi energies of electrons and holes). Since the density of electron-hole pairs in condensates appears to be greater than the effective density of excitons, the correlation part of the kinetic energy is small and E_F can be calculated as the Fermi energy of an ideal gas. The binding energy obtained by such a procedure is equal to 15×10^{-3} ev for silicon and 5.5×10^{-3} ev for germanium. The binding energy in the condensed phase, E_{e-h}, may also be determined by the use of the virial theorem. Then,

$$(3/5)E_F - E_{ex} = E_{e-h}$$

and we obtain 2.5×10^{-3} ev for silicon and 1.5×10^{-3} ev for germanium.

---

[1]This process is possible in germanium and silicon where phonons take off main part of quasimomentum.

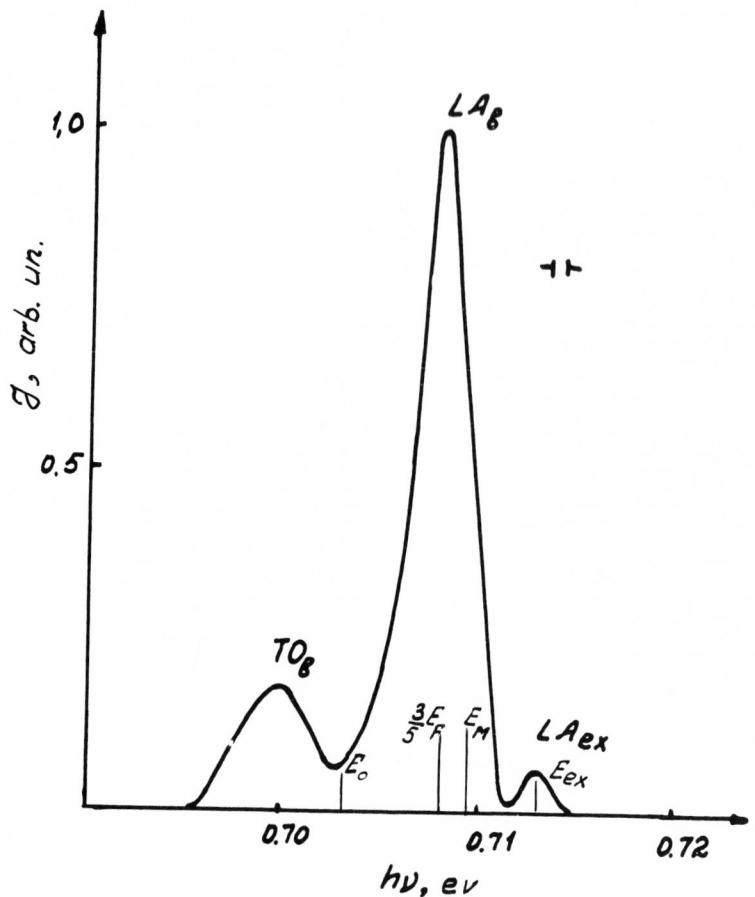

Fig. 5 Emission spectrum of pure germanium at T=4.2K.

It is possible to improve this model (14) of the condensate by taking into account the dependence of the exchange energy on the quasimomentum. In that case the E_{e-h} determined by the two methods are not identical.

The dependence of the long-wave length band shape on the number of electron-hole pairs in the system should be different for biexcitons than for electron-hole pairs. If the electron-hole condensates hypothesis is valid, the shape of the band must remain unchanged until the average concentration is smaller than the concentration in the condensates (i.e. $n=2.6 \times 10^{17}$ cm-3 (16)). Below this concentration, addition of electron-hole pairs leads only to an increase of the condensate volume and cannot produce a noticeable change in the shape of the spectra. If, however, the long-

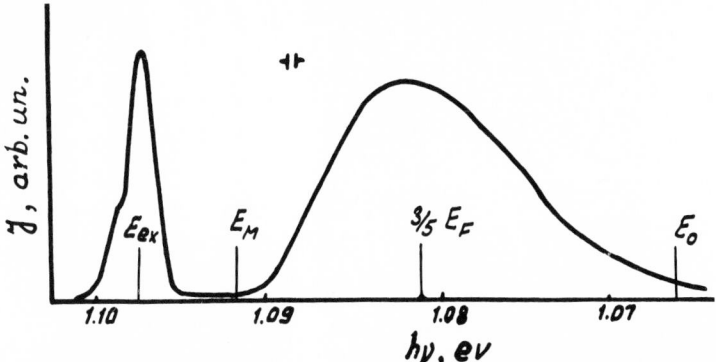

Fig. 6 Emission spectrum of pure silicon (16).

wave length bands arise from biexcitons, their broadening should
be observed at much lower concentrations. In any case, broadening
must be observable for n>2x10^{16} cm^{-3}(i.e. the concentration at the
metal-insulator transition (11)). The spectra of radiative recombination in germanium are shown in Fig. 7 for different levels of
excitation. To prevent overheating the 10^{-3} cm thick specimens
were immersed in liquid helium. Non-equilibrium carriers were
created by light pulses of 10^{-6} sec. The density of electron-hole
pairs was calculated from the known rate of generation estimated
by means of measurements of the photoconductivity of the same
sample at room temperature. Since the exact value of the concentration is critical for the experiments, another independent method
for measuring the concentration was used. To obtain the value of
electron-hole pair density in the sample, measurements of modulation
spectra were made under the same conditions as the luminescence.
At very high excitation levels the modulation of the absorption
edge is due mainly to the filling of the valence band states. The
width of the modulation spectra is a direct measure of the hole
concentration. Modulation spectra are shown in Fig. 8. High
energy sides of the spectra are smeared out by electron-hole interactions. Concentrations obtained from the photoconductivity and
the modulation experiments agree within 30%.

It follows from Fig. 7 and Fig. 8 that the shape of both
luminescence and modulation of absorption edge spectra depend on
the concentration for n>4x10^{16} cm^{-3}. This means that the concentration in the condensed phase is certainly smaller than this value.
Investigations of both reaction kinetics and spectral measurements
show that the long-wave length radiation is a result of biexciton
recombination. The condensed phase of the biexcitons has an
equilibrium density about equal to 2x10^{16} cm^{-3}.

A. A. ROGACHEV

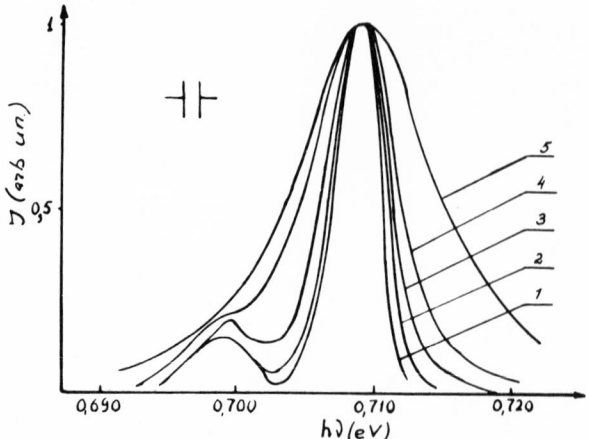

Fig. 7   Luminescent spectra of pure germanium at various excitation
levels (T=3K). 1, $n=2\times10^{16}$ cm^{-3}; 2, $n=5\times10^{16}$ cm^{-3};
3, $n=10^{17}$ cm^{-3}; 4, $n=2\times10^{17}$ cm^{-3}; 5, $n=4\times10^{17}$ cm^{-3}.

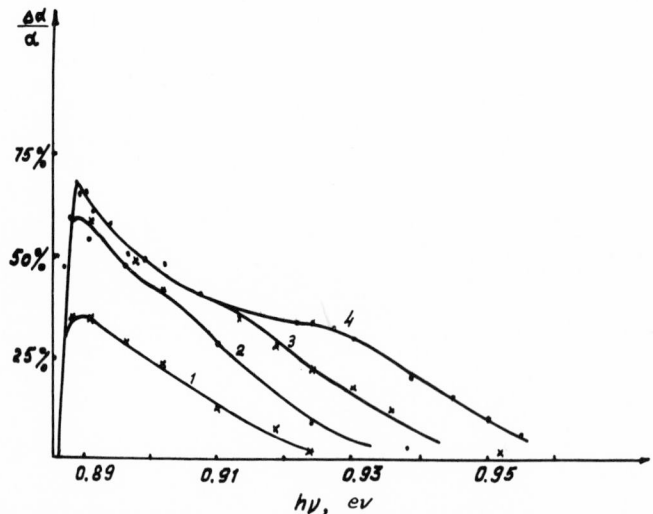

Fig. 8   Modulation spectra of the same sample as in Fig. 7; T=3K.
1, $n=5\times10^{16}$ cm^{-3}; 2, $n=10^{17}$ cm^{-3}; 3, $n=2\times10^{17}$ cm^{-3};
4, $n=4\times10^{17}$ cm^{-3}.

In principle two situations are possible: 1. The sublimation
energy of the condensed phase is much greater than kT. Normally a
gas-liquid type phase transition takes place. The transition is
possible at very low concentrations if the temperature is low
enough. 2. The sublimation energy is very small or is even of
opposite sign. The transition is still possible if the binding

energy of biexcitons become smaller when the concentration rises. At some density of biexcitons the difference between the binding energy of excitons in biexcitons and those in the condensate may be sufficient for the phase transition (26). The small broadening of the biexciton band near the metal-insulator transition point (11) suggests that, if the first situation occurs, the sublimation heat is smaller than 1.6×10^{-3} ev per molecule.

Present experimental data are not sufficient to make a conclusion about the nature of the condensate and further investigations are necessary.

The author would like to thank V. M. Asnin and N. I. Sablina for helpful discussions.

REFERENCES

1. E. F. Gross, Usp. Fiz. Nauk, 76, 433 (1962).

2. M. A. Lampert, Phys. Rev. Lett., 1, 451, (1958).

3. J. R. Haynes, Phys. Rev. Lett., 17, 860, (1966).

4. C. B. Guillaume, F. Salvan, and M. Voos, Journal of Luminescence, 1/2, 315 (1970).

5. V. M. Asnin, A. A. Rogachev, and N. I. Sablina, Fiz. Techn. Polupr., 4, 808, (1970).

6. V. M. Asnin, B. V. Zubov, T. M. Murina, A. M. Prokchorov, A. A. Rogachev, and N. I. Sablina, JETP, 62, 737, (1972).

7. L. V. Keldysh, Proc. IX Int. Conf. Phys. Semicond., 1384, Moscow (1968).

8. V. M. Asnin, A. A. Rogachev, Letters to JETP, 9, 415, (1969).

9. V. M. Asnin, A. A. Rogachev, Proc. III Int. Conf. on Photoconductivity, 13, Stanford University (1969).

10. V. M. Asnin, A. A. Rogachev, and N. I. Sablina, Letters to JETP, 11, 162, (1970).

11. V. M. Asnin, A. A. Rogachev, and N. I. Sablina, Fiz. Tverd. Tela, 14, 399, (1972).

12. C. B. Guillaume, M. Voos, F. Salvan, J. M. Laurant, and A. Bonnot, C. R. Acad. Sc., Paris, 272, 236, (1971).

13. Ya. E. Pokrovsky and K. I. Svistunova, Letters to JETP, $\underline{9}$, 435, (1969).

14. A. S. Kaminsky and Ya. E. Pokrovsky, Letters to JETP, $\underline{11}$, 381, (1970).

15. Ya. E. Pokrovsky and K. I. Svistunova, Fiz. Techn. Polupr., $\underline{4}$, 491, (1970).

16. Ya. E. Pokrovsky, A. S. Kaminsky, and K. I. Svistunova, Proc. X Int. Conf. Phys. Semicond., 504, Cambridge, Massachusetts, (1970), published by U.S. Atomic Energy Commission.

17. V. S. Bagaev, T. I. Galkina, O. V. Gogolin, and L. V. Keldysh, Letters to JETP, $\underline{10}$, 309 (1969).

18. V. S. Bagaev, T. I. Galkina, and O. V. Gogolin, Proc. X Int. Conf. Phys. Semicond., 500, Cambridge, Massachusetts, (1970).

19. V. S. Vavilov, V. A. Zayatz, and V. I. Murzin, Letters to JETP, $\underline{10}$, 304, (1969).

20. V. S. Vavilov, V. A. Zayatz, and V. I. Murzin, Proc. X Int. Conf. Phys. Semicond., 509, Cambridge, Massachusetts, (1970).

21. B. M. Ashkinadze, I. P. Kretzu, S. M. Ryvkin and I. D. Yaroschetzky, JETP, $\underline{56}$, 507, (1970).

22. B. M. Ashkinadze and V. V. Rojdestvensky, Letters to JETP, $\underline{15}$, 371, (1972).

23. P. Gladkov, B. P. Jurkin, and N. A. Penin, Fiz. Techn. Polupr., (to be published 1972).

24. O. Akimoto and E. Hanamura, Solid St. Commun., $\underline{10}$, 253, (1972).

25. R. S. Rice, Solid St. Commun., $\underline{10}$, (to be published 1972).

26. E. Hanamura, Report at XI Int. Conf. Phys. Semicond., Warsaw, (1972).

27. V. M. Asnin and A. A. Rogachev, Letters to JETP, $\underline{14}$, 494, (1972).

INTRAMOLECULAR RELAXATIONS AND LUMINESCENCE OF ORGANIC COMPLEX MOLECULES

B. S. Neporent

The State Optical Institute, Leningrad, USSR

ABSTRACT

The principal aim of this paper is the consideration of the relations between intramolecular relaxations and transformations of absorbed light energy in polyatomic molecules. An attempt will be made to connect the molecular characteristic governing these transformations with the experimental data on the vibronic spectra of molecules.

It is in common use to determine vibrational and electronic (or more precisely--vibronic) relaxations of polyatomic molecules as changes in their discrete, vibrational or, respectively, vibronic states. These relaxations, however, shorten the initial state lifetime and therefore increase their level width in wave numbers, which is:

$$\delta\nu = 5\times10^{-12}W = 5\times10^{-12}\theta^{-1} , \qquad (1)$$

where W is the relaxation probability, and θ is the characteristic relaxation lifetime of the state in question. These and related effects are discussed in our works (1-5). Byrne and Ross (6) suggested that relaxations were manifested in spectra providing $W \approx 10^{12}\text{sec}^{-1}$.

The broadened levels do not represent discrete states; therefore, those states are not Born-Oppenheimer ones especially if θ is sufficiently short. The theoretical treatment of a polyatomic molecule as a system of well defined vibrational and "pure" electronic states with transitions between them is the result of mathematical description difficulties. Instead of working out the problem of real mixed molecular states with a more or less

31

arbitrary form of vibrationally mixed electronic states and widened levels, contemporary theories deal with orthogonal vibrational coordinates, harmonic vibrations of molecules, and the Born-Oppenheimer approximation. These theories lead to idealized pure states with narrow levels. Accordingly, the theories of electronic relaxations, e.g., that which was suggested by Henry and Kasha (7), are also based on Born-Oppenheimer approximation and their results concern only relatively slow relaxations and slightly diffused vibronic spectra.

Being interested in both diffuse and continuous vibronic spectra, we cannot use the above theories. In the following we will treat the level width, $\delta\nu$, or relaxation probability, W, as empirical molecular parameters.

According to our model (3,4) a polyatomic molecule in any state of excitation (ground state, first excited state, etc.) has several partial electronic states, among them one equilibrium state represented by the deepest pit on a potential hypersurface, some quasi-stationary states with the pits of moderate depth, and some nonstationary states which are the canyons on the hypersurface. All these partial states are bounded by means of the mixed states, which are represented by the potential ridges (barriers) separating all the minima and canyons. Some illustrations of this are represented in Fig. 1. The stable partial states are well-known (cis-trans isomerism, photo- and thermochromy, etc.); the unstable ones, being by far much more common manifest themselves only indirectly in the vibronic spectra.

Pure vibronic states with sharp levels exist only at the bottoms of sufficiently deep pits. The higher parts of the pits are characterized by the uncertainty, $\delta\nu$, of electronic energy, ν_e. The only way to conserve the useful, pictorial model of such a molecule is to ascribe a thickness, $\delta\nu(R)$, to each point of a potential hypersurface, $\nu_e(R)$, R representing all normal coordinate values, i.e., the instantaneous configuration of a molecule.

None of the states shown in Fig. 1 are interacting. The surface III[*] is thickened due to interaction of state III[*] with the dissociation continuum. Fig.2 represents the sections of potential surfaces along the coordinates r_d^* and r_r^* for the excited states of Fig. 1. The degree of interaction between states under consideration increase from a to b and c. The potential surfaces are thickened near the crossing regions and the opposite sides. Dotted lines represent the section of cones which are the ways of electronic relaxations from the higher excited states. The greater the amount of molecular vibrational energy, Q, especially energies q_d and q_r of vibrations ν_d and ν_r, active in relaxations considered, the higher are the probabilities W_d^e and W_r^e of these relaxations.

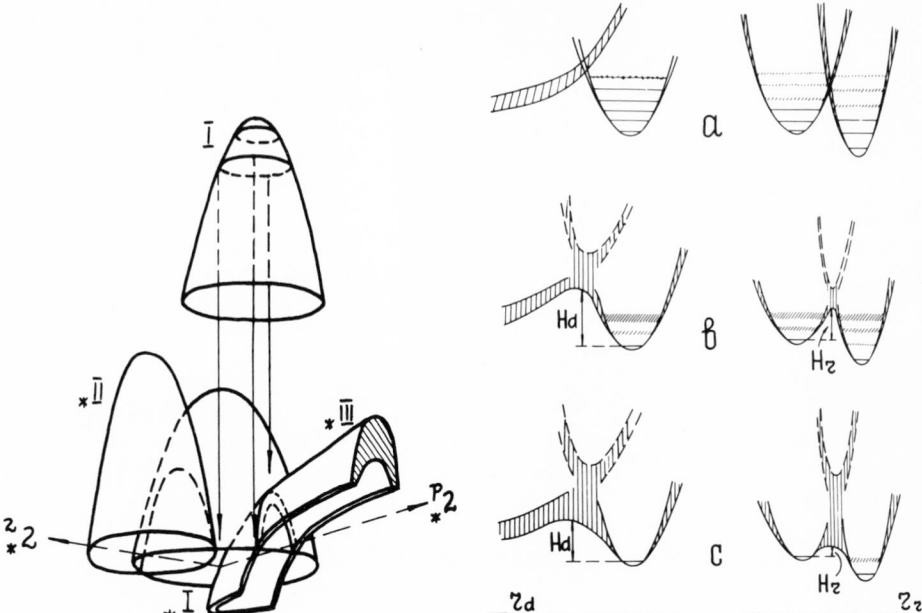

Fig. 1   The scheme of two normal coordinates (r_d and r_r) potential
surfaces for the ground and first excited states (see the
text).

Fig. 2   The excited state surfaces of Fig. 1 sections along the
coordinates r_d and r_r.

The cause of electronic relaxations is the discrepancy between
the nuclear configuration and the electronic structure of a molecule. This relaxation results either in the transition into the
self-consistent pure state or in the destruction of a molecule
through the unstable state.

Vibrational relaxations are possible owing to the nonlinearity
of the molecular vibration. This is because valence forces are only
quasi-elastic. The density of the states and the anharmonicity of
vibrations and, therefore, the probabilities of vibrational relaxations grow very fast while Q or energies q_e of active vibrations
ν_e are increasing. The region of high Q and q_e (see Fig. 3) is the
zone of "mixed vibrations" ($\sum$-type state) formed by superposition
of widened levels of strongly interacting vibrations. The region
of low Q and q_e is the zone of "independent vibration" (Δ-type state)
formed by the pure vibronic state with sharp vibrational levels.

A new type of intramolecular "parametric" relaxations (in addition to vibrational and electronic ones) was considered by us (2).

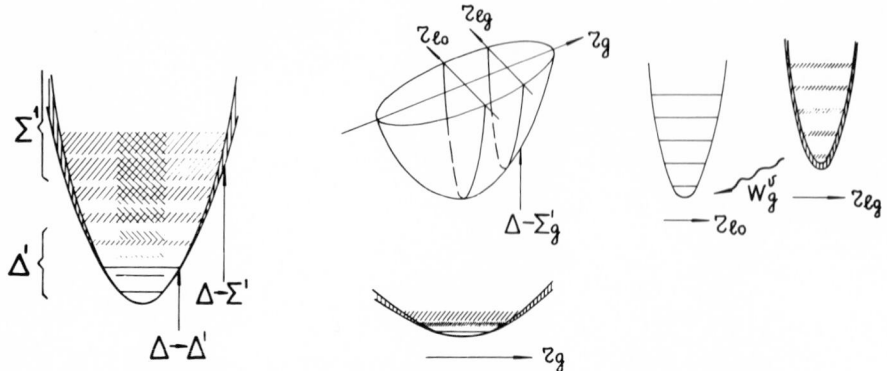

Fig. 3   The scheme of vibrational states of Δ- and $\sum$-types.   The
         potential surface section along the r_i-coordinate.   In the
         middle the levels of ν_k vibration, interacting with ν_i are
         seen.

Fig. 4   The scheme of parametric relaxations origin according to
         (2).

Evidently, some of the bending vibrations ν_g are known to disturb
easily the π-electronic states of an aromatic molecule, as it was
suggested by Terenin (8).   Such a vibration, ν_g, being more or less
excited, forms one partial $\sum$-state with some other low-frequency
vibrations and is therefore subjected to some distortion.   As a
result the distortion is translated to a π-electron cloud and all
vibronic levels become disturbed and widened.   The parametric relaxations spread to all vibronic levels and even to those belonging
to well isolated, nearly harmonic high-frequency vibrations ν_e (see
Fig. 4).   For instance, it is the exclusion of ν_g type vibrations in
molecules clamped in suitable rigid matrix that leads in our opinion
(2) to appearance of quasi-line Shpolsky spectra.   It is impossible
to classify the parametric relaxations even nominally as vibrational
or electronic ones.

The character of vibronic spectrum, i.e., its shape, width and
degree of diffusiveness, are determined by the mutual positions,
forms, and thicknesses of combining state hypersurfaces.   A great
number of polyatomic molecules are "complex" in our classification
(9,10).   That is, their vibrational relaxation probabilities are
much greater, than the lifetimes τ and, therefore, the initial states
of all the optical vibronic transitions are the equilibrium ones.
The final states may also be of the Δ'-type, but in most cases they
are of the $\sum'$-type or of the intermediate type.

Fig. 5 represents the various vibronic spectra diagrams (4).

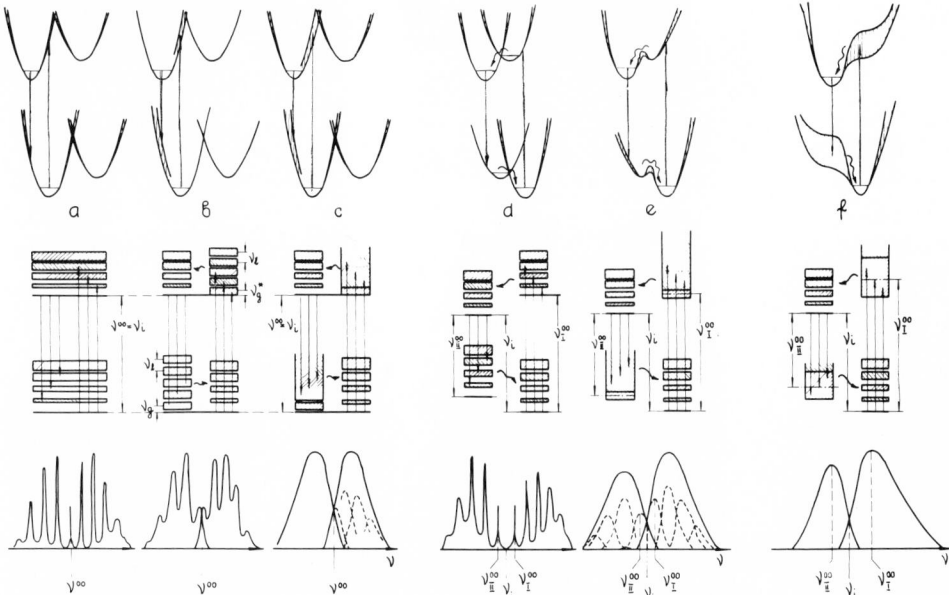

Fig. 5 The formation of vibronic spectra of different types
(schematically) according to (4).

Figures 5a-5c are the cases of moderate electronic structure
changes and therefore relatively small equilibrium configuration
alterations $\Delta R_O = R'_O - R_O$, accompanying an optical transition in a
molecule. Such a transition takes place between the equilibrium
ground and excited states of a molecule. Thus, the corresponding
spectra are described by two-level scheme. Absorption and diffuseness of the fluorescence spectra arise from the pure electronic frequency ν^{oo} (fig. 5a). This is due to the growth of
vibrational relaxation probabilities W^v of the Franck-Condon state
vibration levels induced by their vibrational energy increase. On
the other hand, diffuseness is constant in the spectra in Fig. 5b,
where only parametric relaxations are considered (the ν_g vibration
lies beyond the drawing plane). The last two-level spectra, 5c,
are continuous, inhomogeneously broadened, and possess a hidden
vibrational structure. Here, the Franck-Condon states of the
$\sum$-type are attained owing to either relatively large ΔR_O or large
intramolecular interactions.

The next spectral group 5d-5e belongs to the molecules under-
going the violent valence electron cloud changes accompanying the
optical transition. These transitions take place between the
equilibrium initial state and quasi-stationary Franck-Condon state,
which is relaxed into the equilibrium final state. As it was suggested by us (10), such spectra are to be described within the

four-level electronic state scheme. They contain two zero elec-
tronic frequencies ν_I^{oo}, ν_{II}^{oo} and two electronic transition integral
probabilities B_{eI}, B_{eII}.

The spectra 5d,5e as well as the 5a-5c are the configurational
ones, but their diffuseness depends not only on vibrational and
parametric relaxations but also on electronic relaxations. These
relaxations take place in Franck-Condon states. Furthermore, in
the cases 5d-5e the zero frequencies, ν_I^{oo} and ν_{II}^{oo}, do not coincide,
hence the fluorescence and absorption spectra are additionally
separated. These spectra may be characterized by the different
integral intensities B_{eI} and B_{eII} and different sets of vibrational
frequencies.

The spectra shown in Fig. 5f are of special interest. The re-
laxational width of the Franck-Condon levels exceeds the configurational distribution in the system under consideration. Therefore,
each vibronic relaxation spectrum consists of a homogeneously
broadened band (10). A relatively narrow configurational distribution appears only at these band edges. Similarly, the narrow relaxa-
tional broadening appears at the edges of the inhomogeneously
broadened configurational band.

This scheme is a means for qualitative analysis of vibronic
spectra; in some cases, quantitative. The quantitative description
of the vibronic spectra is based upon the combination of spectral
and integral relations, describing both absorption and fluorescence
spectra.

The general relation between the absorption and fluorescence
spectra within the two-level scheme was given by Stepanov (11).
The uncertain constants were excluded from this relation by the
author and Mazurenko (12). The relations of the integrated inte-
sities of these spectra are obtained in the most correct and useful
form by Strickler and Berg (13). It was assumed (11-13) that there
was a thermal equilibrium distribution of molecules among vibrational states. Suppose this distribution holds for the vibronic
levels of both partial systems M_I and M_{II} (see Fig. 6). Assume
also that the additive contributions of these systems are in the
formation of summary states and summary spectra (Fig. 6), as shown
in (14,5). Then we obtain the spectral and integral relations for
the summary spectra in the case of the four-level scheme:

$$\frac{\phi_\nu/\nu^3}{\eta_{a\nu}\alpha_\nu/\nu} = 2.89\times10^{-9}GSn_\nu^2(\tau/\eta_\phi)\int\phi_\nu d\nu 10^{-0.626[(\nu-\nu_i)/T]} \qquad (2)$$

$$\eta_\phi/\tau = 1/\tau_\phi = 1/\beta\tau_\alpha = 2.89\times10^{-9}\frac{G}{\beta}\frac{n_\phi^3}{n_\alpha}\frac{\int\phi_\nu d\nu}{\int(\phi_\nu/\nu^3)d\nu} \times \int\frac{\eta_{a\nu}\alpha_\nu}{\nu} d\nu, \qquad (3)$$

$$\frac{[1/\int(\phi_\nu/\nu^3)d\nu]\phi_\nu/\nu^3}{[1/\int(\eta_{\alpha\nu}\alpha_\nu/\nu)d\nu]\eta_{\alpha\nu}\alpha_\nu/\nu} = S\beta\,(\underline{n}^2_\nu\underline{n}_\alpha/\underline{n}^3_\phi)10^{-0.626[(\nu-\nu_i)/T]}. \quad (4)$$

$$\frac{[1/\tau_\phi\int(\phi_\nu/\nu^3)d\nu]\phi_\nu/\nu^3}{1/[\tau_\alpha\int(\eta_{\alpha\nu}\alpha_\nu/\nu)d\nu(\eta_{\alpha\nu}\alpha_\nu/\nu)]} = S\,(n^2_\nu n_\alpha/n^3_\phi)10^{-0.626[(\nu-\nu_i)/T]}. \quad (5)$$

Eq. (2) may be applied to all the vibronic spectra, but Eq. (3) and
(4) apply only to spectra allowed by symmetry (the latter are
characterized by the independence of either temperature or the
high intensities).

It is convenient to adopt the inverse lifetimes of the
excited state as a measure of integral intensities of vibronic
spectra. For the absorption spectra this lifetime τ_α is calculated
from the Strickler and Berg absorption integral (Eq. (3) with
$\beta=1$). For the fluorescence spectra $\tau_\phi = \tau/\eta_\phi$, where τ is the
experimental lifetime and η_ϕ--fluorescence quantum yield. This
procedure simplifies the necessary normalization of fluorescence
and absorption spectra by their areas F_ϕ and F_α (12,5),
$F_\phi/F_\alpha = \tau_\alpha/\tau_\phi$. In fact, there is no need to make such a normalization procedure, since the frequency ν_i is connected with the point
ν_c of the intersection of the arbitrary normalized spectra (the
areas F'_ϕ, F'_α) by the relation

$$\nu_i = \nu_c - 1.6\log_{10}\frac{F'_\phi}{F'_\alpha} \times \frac{\tau_\phi}{\tau_\alpha} \qquad (6)$$

The "absorption yield" $\eta_{\alpha\nu}$ is introduced into Eqs. (2)-(5) to
exclude the non-exciting absorption; n is the refractive index of
the system in consideration, n^{-1}_α and n^{-3}_ϕ are the values of n^{-1} and
n^{-3} averaged over absorption and fluorescence spectra. The parameters G,S, and β are respectively the ratios of electronic
statistical weights (multiplicities) g_e/g^*_e, vibrational partition
functions s/s^*, and optical transition probabilities B_e/B^*_e between
ground and excited states, including both systems M_I and M_{II} in
each of these states. These values for the two-level and four-
level schemes are listed in Table 1. The possibility is also
proposed for cross transitions $B_{eI\leftrightarrow II}$ between levels of the
partial states M_I and M_{II}. Some other cases and some comments
governing the relations (2)-(5) are given in Ref. (5).

It should be pointed out that the multiplicities g_{eI} and g_{eII}
of the partial states M_I and M_{II} may be different, i.e., the
G-value may be equal to 1/3, 1, 3. The S-value must be specifically
analyzed. We only mention here that a large departure from S=1

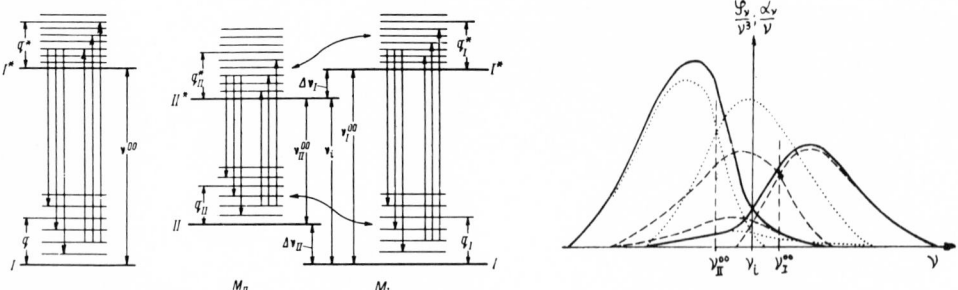

Fig. 6   The two- and four-level schemes of electronic states (left
         part). The origin of total vibronic spectra from the ones
         of M_I (broken line) and M_{II} (dotted line) partial states.
         (right part).

occurs if one of the initial state vibrations is converted into
one vibration of the final state (or vice versa) as a result of an
optical transition followed by relaxation. We will adopt below
G=1, S=1. When β=1, Eqs. (2)-(4) are transformed into the known
ones for the two-level scheme (11-13). Therefore, the breakdown
of relation (3) with β=1,i.e., the difference in the values of τ_α
and τ_ϕ implies the necessity of the application of a four (or
three-) level scheme.

Let us consider some examples. A very simple case is represented in Fig. 7 by the spectra of acenaphtene in heptane, reconstructed from the data of Berlman (15) according to which τ_ϕ=77 nsec.

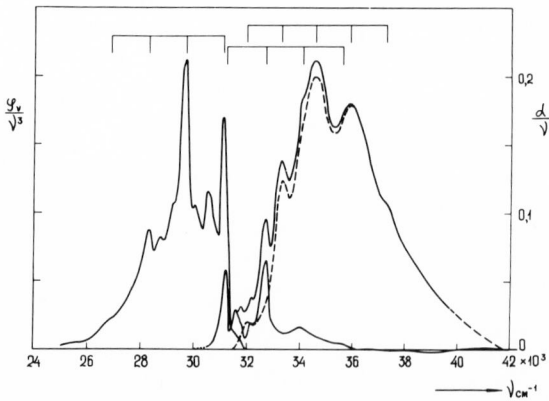

Fig. 7   The acenaphtene in heptane spectra, reconstructed from the
         data of (15).

Table 1

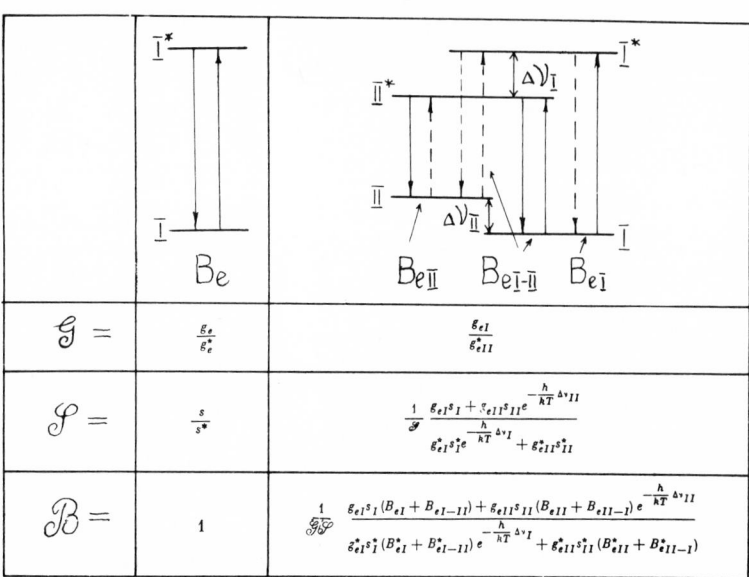

The τ_α value calculated from the whole absorption band is $\tau_\alpha = 7.5$ nsec, however, this band is easily divided into two components, the first of which is characterized by $\tau_\phi = 80$ nsec. This gives the value $\nu_I^{00} = 31100$ cm^{-1}, which is determined by the point of intersection of ϕ_ν/ν^3 and α_ν/ν in the normalized spectra. Thus, this system is a trivial two-level scheme with the second absorption band displaced by 800 cm^{-1} and originating, apparently, from the same ground state as the first one.

The same case is represented by the continuous spectra of triphenylbenzene (15). The dividing of the absorption band into two components on the basis of the integral relation (3) is no less precise than in the preceeding case of a structured spectra. The spectra in Fig. 8 are characterized by $\tau_\alpha = 1.1$ nsec (whole absorption band), $\tau_{\alpha_I} = 160$ nsec (low freq. first absorption band), and $\tau_f = 158$ nsec (fluorescence band (15)), and by $\nu_I^{00} = 31800$ cm^{-1}.

The possible determination of all four electronic levels may be illustrated by the spectra of some diphenylpolyene solutions. These spectra, which possess a diffuse vibrational structure, have been reconstructed from the data of Nikitina, Ter-Sarkisyan, et al (16), Birks and Dayson (17), and Berlman (15) and are given in Fig. 9. The broken lines represent the normalized fluorescence spectra and the solid lines are the above mentioned correct normalization by integral intensities of the absorption and fluorescence spectra, i.e., by their areas F_ϕ, F_α.

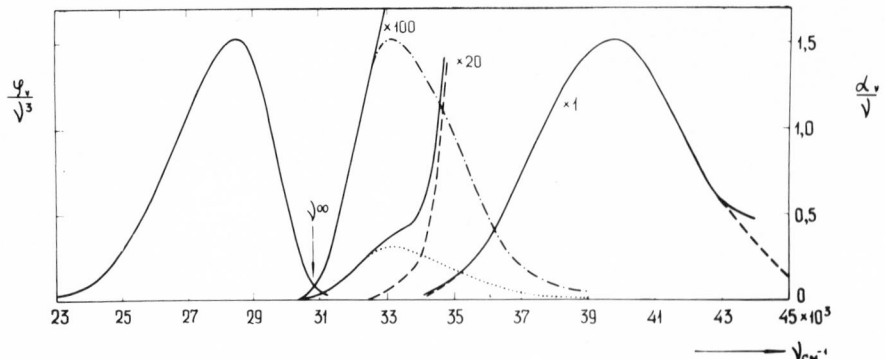

Fig. 8   The triphenyl benzene in heptane spectra, reconstructed
         from the data of (15).

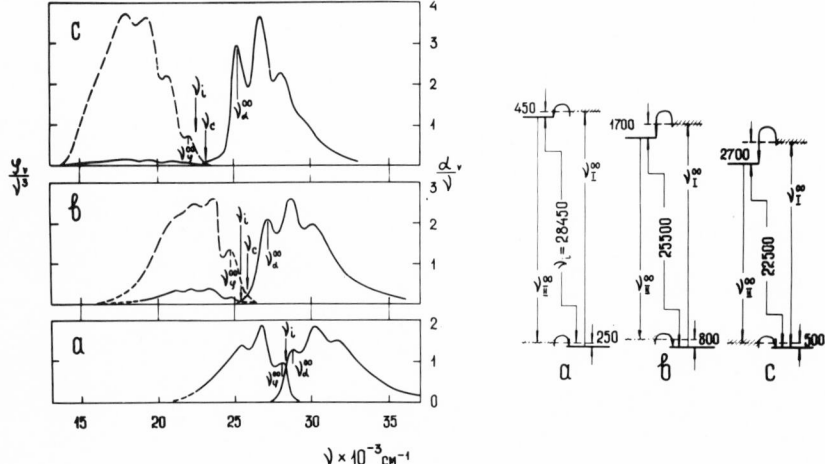

Fig. 9   The diphenylpolynes in heptane and cyclohexane spectra re-
         constructed from the data of (15-17).   Right side--the four-
         level schemes of electronic states.   (a) diphenylbutadiene,
         (b) diphenylhexatriene, (c) diphenyloctatetraene.

Table 2

Substance	ν_i cm^{-1}	ν_I^{00} cm^{-1}	ν_{II}^{00} cm^{-1}	$\tau_\alpha \times 10^9$ sec.	$\tau_\phi \times 10^9$ sec	β
Diphenylbutadiene (a)	28450	28900	28200	1.5	1.8	1.2
Diphenylhexatriene (b)	25500	27200	24700	2.3	16.3	7.1
Diphenyloctatetraene (c)	22500	25200	22000	3.7	89	18.7

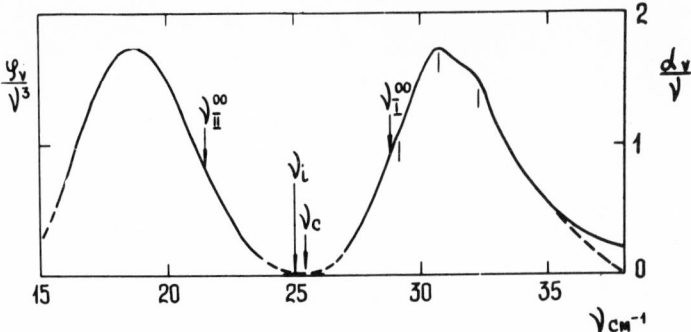

Fig. 10   The all-trans-retinol in isopentane spectra, reconstructed
from the data of (18). The ν_I^{OO} and ν_{II}^{OO} values are calculated according to (19).

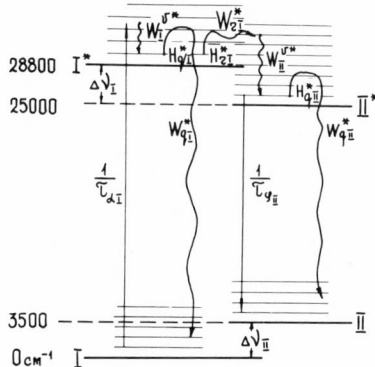

Fig. 11   The four-level scheme for the trans-retinol molecule in
the solution.

The τ_ϕ values are taken from (15-17) and the τ_α values are
calculated from absorption spectra according to Eq. (3). Thus the
ν_i values are derived which form the four-level scheme represented
by the right side of Fig. 9. The ν_I^{OO} and ν_{II}^{OO} values are expected
to coincide with the first subbands maximum of vibrational struc-
ture in the absorption and fluorescence spectra.

Here the cross-transitions between levels of the partial
states M_I and M_{II} are probably the Franck-Condon forbidden ones.
All the ν, τ, and β values for these spectra are given in Table 2.
It is evident that the spectra considered are described very well
by the four level scheme, eliminating all the contradictions caused
by the discrepancies between the ν_I^{OO} and ν_{II}^{OO} with regard to the
B_{eI} and B_{eII} values.

An interesting complicated case is represented by the trans-
retinol molecule, in which the electronic relaxations cause not
only integral intensity differences in the absorption and fluorescence spectra, but also lead to some complicated temperature dependencies of the fluorescence quantum yield. The spectra of trans-
retinol solutions (Fig. 10), according to Dalle and Rosenberg (18),
are characterized by a severe difference of τ_α and τ_ϕ values:
τ_α = 2.7 nsec being practically independent of temperature, τ_ϕ
increasing from 40 nsec to 200 nsec with a linear dependence on
temperature from 70 K to 300 K. This τ_ϕ growth is associated (18)
with the temperature redistribution of excited molecules, the
smaller probabilities of vibronic transitions being ascribed to
higher vibrational levels.

All the experimental data (18) can be explained in terms of the
four-level scheme given in Fig. 11 for the system under consideration. According to this scheme the fluorescence quantum yield η_ϕ
measured in (18) is determined by two competing processes, the first
being realized in the Franck-Condon state I^*; the second, in stationary state II^*, that is:

$$\eta_\phi = \eta_{rI}\eta_{\phi II} = [W_{rI}^*/(W_{rI}^*+W_{qI}^*)][A_{II}^*/(A_{II}^*+W_{qII}^*)]. \qquad (7)$$

Here, the probabilities W_{qI}^* and W_{qII}^* represent the relaxations
causing the deactivation of the states I^* and II^*. The electronic
rearrangement populates the level II^* with the probability W_{rI}^*.
The values of ν_I^{oo} and ν_{II}^{oo}, determined by the Klochkov and Korotkov
method (19), show both the direct and inverse vibronic transitions
to occur completely in different M_I and M_{II} systems, since
$\Delta\nu_I, \Delta\nu_{II} \gg kT$. In all calculations the assumption was made that the
trans-retinol molecule is complex in our classification and the
probabilities of all electronic relaxations depend on temperature
exponentially, governed by their barrier heights. In all cases,
molecular characteristics having complex exponential temperature
dependencies as derived from expression (7) are fulfilled satisfactorily. On the basis of these considerations, the values of
$\tau_{\phi II}$ as well as the values of probabilities of electronic relaxations and the heights of corresponding barriers have been calculated
and summarized in Table 3.

Thus the interpretation suggested describes all complex
experimental data completely and with no contradictions. Moreover,
a number of constants are obtained for the trans-retinol molecules.

All the results given above refer to solvated molecules, i.e.,
the four-level scheme may be due not only to intramolecular, but
also partially to intermolecular interactions. The τ_ϕ/τ_α value may
cause the intermolecular effects to diminish the mutual separation
of the fluorescence and absorption spectra caused by intramolecular

Table 3

ν_i cm^{-1}	ν_I^{oo} cm^{-1}	ν_{II}^{oo} cm^{-1}	$A_I^* = 1/\tau_\alpha$ sec^{-1}	$(W_{qI}^*/W_{rI}^*)_\infty$	
25000	28800	21500	3.7×10^8	20	
$H_{qI}^* - H_{rI}^*$ cm^{-1}	$A_{II}^* = 1/\tau_{\phi II}$ sec^{-1}		$(W_{qII}^*)_\infty$ sec^{-1}	H_{qII}^* cm^{-1}	β
150	4.9×10^7		7.5×10^9	480	7.5

interactions. Expressions (2)-(5) are certainly applicable to all four-level systems. It is suggested, however, that our considerations may prove more fruitful in the case of vapor phase studies than in the investigation of liquid phases, because the former offer more possibility of varying the excitation energy of an isolated molecule and thus controlling the influences upon it (20-22).

The considerations and data given above demonstrate that the process of the interaction of light with complex molecules includes their relaxing states and, hence, is complicated. In the general case none of these processes can be described adequately, or at least completely, in terms of the usual two-level scheme and the Born-Oppenheimer approximation, which are applicable only to the pure stationary states of polyatomic molecules or to processes in the simplest molecules. The approach suggested here of taking into account all relaxations following every stage of absorption, transformation, and emission of light allows a more complete description of the complex system of the polyatomic molecules and also provides much more information regarding the properties of these molecules.

REFERENCES

1-5. B. S. Neporent, Opt. i Spectr. 32, 38, 252, 458, 670, 880 (1972).

6. I. P. Byrne, I. G. Ross, Austr. J. Chem. 24, 1107 (1971).

7. B. Henry and M. Kasha, Ann. Rev. Phys. Chem. 19, 161 (1968).

8. A. N. Terenin, Zhurn. Phys. Khimii 18, 1 (1944).

9. B. S. Neporent, Zhurn. Phys. Khimii 21, 1111 (1947).

10. B. S. Neporent, Zhurn. Experim. Theor. Phys. 21, 172 (1951).

11. B. I. Stepanov, Dokl. Acad. Nauk. SSSR 112, 839 (1957); Izv. Acad. Nauk SSSR, ser. phys. 22, 1372 (1958).

12. B. S. Neporent, Izv. Acad. Nauk SSSR, ser. phys. 22, 1372
    (1958). Yu. T. Mazurenko and B. S. Neporent, Opt. i Spectr.
    12, 571 (1962).

13. S. J. Strickler, R. A. Berg, J. Chem. Phys. 37, 814 (1962).

14. B. S. Neporent, Molecular Photonics, L., Publ. "Nauka", p. 18
    (1970).

15. I. Berlman, Handbook of Fluorescence Spectra of Aromatic
    Molecules, Academic Press, N.Y.-London, (1965).

16. A. N. Nikitina, G. S. Ter-Sarkisyan, B. M. Mikhailov and L.
    E. Minchenkova, Opt. i Spectr. 14, 655 (1963).

17. J. B. Birks and B. J. Dayson, Proc. Roy. Soc. 275A, 136 (1963).

18. J. P. Dalle and B. Rosenberg, Photochemistry and Photobiology
    12, 151 (1970).

19. V. P. Klochkov and V. A. Korotkov, Opt. i Spectr. 20, 582
    (1966).

20. B. A. Neporent, Izv. Acad. Nauk SSSR, ser. phys. 15, 533 (1951).

21. B. S. Neporent, N. G. Bakhshiev and Yu. T. Mazurenko,
    Elementary Photoprocesses in Molecules, Publ. "Nauka", L.
    p. 80 (1966).

22. N. A. Borisevich, and B. S. Neporent, Jurn. Prikl. Spectr. 8,
    377 (1968).

TRIPLET STATE PROPERTIES FROM PHOSPHORESCENCE MICROWAVE DOUBLE RESONANCE STUDIES

M. A. El-Sayed

Department of Chemistry

University of California

Los Angeles, California 90024, USA

ABSTRACT

The production of the state of spin alignment of molecules in the lowest triplet state at low temperatures is discussed. The conditions under which microwave radiation of resonance frequencies with the zerofield (zf) transitions of these molecules would result in an observed change in the phosphorescence intensity are specified. The basic equations relating these intensity changes to unimolecular rate constants are derived. The application of phosphorescence-microwave double resonance (PMDR) spectroscopy to determine the different properties of the triplet state is summarized. A new and convenient magnetic axis system for discussing the zf levels of Π,Π^* triplet states is introduced. The more recent applications of PMDR in determining triplet state geometry as well as the mechanisms of inter- and intramolecular energy transfer processes are discussed in detail.

I.   INTRODUCTION

In 1925, the effect of microwave or radio frequency radiation on the intensity and polarization of atomic emission was determined (1). Since then, the field of optical detection of magnetic transitions of atoms, ions and simple molecules has been greatly developed (2). Extending these methods to determine the magnetic Zeeman levels of the triplet state was first demonstrated in 1967 (3-5). The optical determination of zerofield (zf) transitions of the triplet state in zero field was first accomplished in 1968 (6).

In the literature, many abbreviations are used by different authors, e.g., MODR (microwave optical double resonance), OMDR (optical microwave double resonance), ODESR (optical detection of ESR) as well as PMDR (phosphorescence microwave double resonance). The word "optical" could refer to fluorescence, phosphorescence, or even absorption. In general, MODR also implies the presence of a magnetic field in the experiment. MODR (or OMDR) and ODESR are thus used when the magnetic properties, measured optically, of the Zeeman levels are of interest in the studies. As will be shown in this review, zerofield studies on the triplet state of large phosphorescent molecules could yield other important properties, e.g., radiative, nonradiative, and structural, in addition to the magnetic properties. PMDR seems to be more general and yet more descriptive of the studies in this field. PMDR would thus be reserved for zf double resonance studies involving the triplet state of phosphorescent molecules.

In this article, the principles of the PMDR methods are first discussed. The applications of these methods are then summarized in Table 1 and the principles of new applications are given.

II. PRINCIPLES OF PMDR METHODS

Due to the anisotropy in the spin orbit perturbation involved in the intersystem crossing and radiative process of the triplet state, the lowest triplet state of molecules could be formed with its zf levels having large differences in their population at low temperatures, when the spin lattice relaxation (SLR) processes are absent. This results in a population distribution which is different from the Boltzman distributions. Pumping of the triplet state could take place by direct singlet-singlet absorption followed by the $S_1 \rightarrow T_1$ intersystem crossing (ISC) process (7) by triplet-triplet energy transfer from a donor or by direct singlet-triplet absorption. Since molecules in different spin levels have their electrons in different planes in the molecule, unequal population would mean a preferred spin direction in the molecular framework. In this case, the triplet state is said to be in a state of spin polarization or spin alignment (7). If the decay constants of the different zf levels are different from one another, the state of spin alignment can be detected by a change of the decay curve from exponential with one component to a nonexponential decay of more than one component (up to three, if the three zf levels are all radiative). Another method of detecting the state of spin alignment is from the large effects of magnetic fields or of microwaves of resonance frequencies on the emission spectrum, on the shape of the decay curve and sometimes on the polarization characteristics of bands originating from different zf levels of symmetric molecules.

The actual degree of spin alignment is related to the population ratios of the zf levels. In the absence of the SLR processes, the value of this ratio would depend on the mode of preparation of the triplet state. Three extreme cases could be realized. In the first case, the system is being continuously pumped and decays continuously, i.e., the system is in a steady state. By equating the rates of pumping of any zf level with its rate of decay, one obtains the following ratio:

$$n_j/n_i = K_j \, k_i/K_i k_j \qquad (1)$$

where n, k and K are the population, decay constants and pumping rate constants, respectively, of the zf levels τ_i and τ_j K_i could be the $S_1 \rightarrow T_1$ intersystem crossing rate if pumping is carried out by singlet-singlet absorption followed by ISC to τ_i, it could be the radiative rate constant of the $S_o \rightarrow T_i$ absorption process if singlet-triplet absorption is used for pumping; or K_i could be a sum of the ISC rate constants to the zf levels of the donor each multiplied by magnetic mapping parameters if pumping is carried out by triplet-triplet energy transfer from a donor to the molecule under examination.

On the other hand, if after reaching the steady state, the excitation is cut off, then the ratio given by (1) decays according to the following equation:

$$(n_j/n_i)_t = (K_j k_i/K_i k_j) \, \exp[-(k_j-k_i)t] \qquad (2)$$

Thus, if the value of Eq. (1) is close to unity, i.e., the degree of spin alignment is small, larger values for the spin alignment could be created by letting the system, whose decay constants k_i and k_j are different, decay (8).

If the K values for the different zf levels are different, a large degree of spin alignment could be obtained before the system reaches its steady state. This is accomplished experimentally by using pulsed excitation (9). The population ratio is simply equal to the ratio of the pumping rates, if the rate of pumping of the zf levels is much faster than their decay rates (a situation which is true at short times after short pulsed excitation) (9). In this case the following equation holds:

$$n_j/n_i = K_j/K_i \qquad (3)$$

The most important requirement for observing PMDR signals

is that the application of microwave radiation of resonance
frequencies changes the net number of radiating molecules in the
system, thus causing a net change in the intensity of the emission
monitored. The ratio of the intensity in the presence of resonant
microwaves (I^ν) to that in the absence of the microwaves (I)
depends on the degree of spin alignment (which depends on the
excitation mode), the manner in which the microwave is applied
(pulsed, rapidly or slowly swept) and the microwave power used.
If the rate of sweep is very slow so as to satisfy the steady
state requirements, saturation of the zf levels is the maximum that
can be accomplished when using microwaves of sufficient power.
Then I^ν/I is given by (11):

$$I^\nu/I = n_i^\nu/n_i = [(K_i+K_j)/K_i] \times [k_i/(k_i+k_j)] \qquad (5)$$

The sweeping of the microwaves across resonance can be used
in all three extreme cases whose population ratios are given by
Eq. (1) through (3). The sweep rate across resonance should be
fast compared with the decay rate of the zf levels. If the
microwaves saturate the $T_i \leftrightarrow T_j$ transition, then the intensity
ratio of the emission from zf level τ_i is given by:

$$I^\nu/I = (1/2)(1 + n_j/n_i) \qquad (6)$$

Depending on the mode of excitation, n_j/n_i would be related to the
rate constants involving the τ_i and τ_j levels as given by
Eq. (1) through (3). Equations have been derived for systems
in which the emission originates from two zf levels (12).

If the microwave radiation is swept in a very short time with
enough power to completely invert the population of the τ_i and τ_j
levels, then:

$$I^\nu/I = n_j/n_i \qquad (7)$$

Substituting Eq. (1) through (3) in Eq. (7), one obtains the
corresponding intensity ratio expressed in terms of pumping and
decay constants. It is thus obvious that the signal to noise
ratio obtained in an optical detection method for a given system
greatly depends on the mode of excitation, the mode of exposure
to the microwaves and the mode of detection.

III. SUMMARY OF PREVIOUS APPLICATIONS OF PMDR

Table 1 gives a summary of the applications and historical
development of PMDR techniques together with references to the
original work. Detailed review articles have been written in
this field and the reader is referred to these reviews. For
general PMDR techniques, the reader is referred to reference (13)

Table 1

Dates and References of the
Important Applications of PMDR
and MODR Techniques

TOPIC	DATE	REFERENCES
MAGNETIC PROPERTIES		
ESR Detection of Spin Alignment	'66	24
MODR of ESR Signals of Triplet State	'67	3,4,5
PMDR of Zerofield Signals of Triplet State	'68	6
ENDOR	'69	25,26
EEDOR	'70	27
Polarization of Zerofield Magnetic Transitions	'70	28
PMDR Detection of Transfer Hyperfine	'70	29
MODR of Level Anticrossing	'70	30
Coherent Coupling	'71	31
PMDR Determination of Relative Signs of Zerofield Parameters	'71	32
PMDR Zerofield Signal at $77^{O}K$	'72	33
Spin Echo	'72	34
Hole Burning in PMDR Signals	'72	35
OPTICAL AND STRUCTURAL PROPERTIES		
Optical Detection of Spin Alignment	'63, '67	36,37
PMDR Effects on Phosphorescence Spectrum	'69	38
PMDR Effects on Polarization	'70	39
PMDR Assignment of Triplet States	'71	18,32
Spin Alignment and Phosphorescence Spectroscopy	'71	40
PMDR and Geometry of Triplet State	'72	19,here
Sensitivity of Pseudo-Jahn Teller Distortion to Solvent Using PMDR	'72	Table 3
NONRADIATIVE PROPERTIES		
Intersystem Crossing Relative Rates:		
- From Decay Curves	'69	41
- From PMDR Signals in Steady State	'70	42
- From PMDR Signals for Decaying Systems	'70	8b
- From PMDR Signals Using Pulsed Excitation	'71	43
PMDR and Spin Conservation in Triplet-Triplet Energy Transfer	'69	22
MODR and Spin Conservation in Triplet-Triplet Energy Transfer	'71	44
Triplet Exciton Coherent Length	'71	45
Distinction Between S-S and T-T Transfer in Mixed Crystals	'72	23

and (14). The experimental techniques of PMDR are carefully
examined in the detailed articles given in Reference (15).
The experimental methods of the MODR techniques are discussed in
reference (16). The application of PMDR to the understanding
of the nonradiative processes, in particular the intersystem
crossing in aromatics and N-heterocyclics, is discussed in detail
in reference (12). These, together with the references in Table 1,
cover the important contribution of the PMDR and MODR techniques.
In the section below, more applications of the PMDR methods are
discussed. In these methods, symmetry properties are combined with
the known important spin orbit interactions in aromatic molecules
to predict further properties of the triplet state.

IV. NEW AND CONVENIENT MAGNETIC AXIS SYSTEM

Let us confine our discussion to systems in which the long
and short inplane molecular axes are principal magnetic axes
(i.e., of C_{2v}, D_{2h}, D_{3h} or D_{6h} symmetry). The free electron
model has proven successful in classifying and correlating the
π,π^* states of aromatic hydrocarbons. In all these molecules, the
lowest triplet state is the ^3L_a state (17), which indicates that
the electrons are on atoms, i.e., the nodes are across the bonds.
The L_a state is thus antisymmetric with respect to reflection in
a plane across the bonds. Let us then introduce the orthogonal
set of magnetic axes a, b, and N, where a is an axis that contains
the atoms, b is across the bonds and N is normal to the molecular
plane. The zf levels would then be τ_a, τ_b and τ_N respectively,
and will transform as rotations around the a, b, or the N axis,
respectively. τ_N is lowest in energy. For molecules of C_{2v} or
D_{2h} symmetry, the zf origin of the 0,0 band is very important
in determining the sign of E in a very simple manner. Its
radiation results from mixing with S_{σ,π^*} states via direct spin
orbit perturbation (H_{SO}) (17). For this to be true,
$<\tau_S S_{\sigma,\pi^*}|H_{SO}|\tau_i T_{\pi,\pi^*}>$ has to be totally symmetric (τ_S and τ_i
are the singlet and triplet spin functions, respectively, and
S and T are the singlet and triplet spacial functions,
respectively). Since H_{SO} and τ_S are of a_1 symmetry, the
symmetry of $S_{\sigma,\pi^*} \otimes T_{\pi,\pi^*}$ must be the same as that of τ_i. The
systems under consideration possess σ_{ab}, σ_{bN} reflection planes
if they possess C_{2v} symmetry, and σ_{aN}, in addition, if they
possess higher symmetry. The symmetry properties of S_{σ,π^*},
T_{π,π^*} and τ_i to these reflection planes are given in Table 2.
From this table, the coupling of S_{σ,π^*} with ^3L_a gives radiation
to the τ_b level whereas coupling it with the ^3L_b gives radiation
to the τ_a level. This conclusion assists in using PMDR methods
to assign the symmetry of the lowest triplet state of C_6H_6 in
C_6D_6. The 0,0 band is found to originate from τ_b, thus
confirming the ^3L_a assignment (18), as for the rest of aromatic
hydrocarbons.

Table 2

The Symmetry Properties of $S_{\sigma,\pi*}$ $T_{\pi,\pi*}$ and τ_i in Aromatic Compounds of C_{2v} or D_{2h} Symmetry

Function	σ_{ab}	σ_{aN}^{*}	σ_{bN}	Function	σ_{ab}	σ_{aN}^{*}	σ_{bN}
$S_{\sigma,\pi*}$	−	+	+	τ_a	−	−	+
L_a	+	+	−	τ_b	−	+	−
L_b	+	−	+	τ_N	+	−	−

*This operation is absent in C_{2v} type molecules.

V. GEOMETRY OF TRIPLET STATE FROM PMDR

For molecules of C_3 or higher (but not cubic) symmetry, τ_a and τ_b are degenerate and one zf transition is observed. For molecules of D_{2h}, C_{2v} or D_2 symmetry, one zf level is strongly radiative to the 0,0 band, τ_b for ^3L_a and τ_a for ^3L_b state. Since all the aromatic molecules have the ^3L_a lowest, let us confine our discussion to this state. For D_{2h} or C_{2v} type molecules, τ_b is highest in energy if saturating the $|D|+|E|$ transition, but not the $|D|-|E|$, affects the intensity of the 0,0 band. In this case, the b-axis is longer than the a-axis, as in the case of naphthalene, quinoxaline and the quinoidal structure of benzene and other distorted C_3 type molecules (i.e., contraction takes place along a). If saturating $|D|-|E|$, but not $|D|+|E|$, causes changes in the 0,0 band intensity, then τ_a is highest in energy and the a-axis is longer than the b-axis, as in the antiquinoidal structure of benzene and other distorted C_3 type molecules (expansion occurs along the a-axis).

For molecules of symmetry lower than C_{2h} and, in particular, those with no plane of symmetry, e.g., molecules with halogen over-crowding (19) or in a host lattice of low site symmetry and containing heavy atoms, the three zf levels radiate to the 0,0 band. The above simple rules can be used to determine the sign of E for substituted aromatic compounds as well as the nature of distortion of symmetric molecules, in particular, the Pseudo-Jahn Teller type distortion. Some of these results are given in Table 3.

VI. PMDR AND ENERGY TRANSFER

The use of PMDR in determining the mechanism of the ISC process in N-heterocyclics is discussed previously. It is found that the main route involves crossing from $S_{n,\pi*}$ to $T_{\pi,\pi*}$ or from $S_{\pi,\pi*}$ to $T_{n,\pi*}$ states (13), in agreement with previous proposals (20). In aromatic hydrocarbons, it is found that the

Table 3

Deviation of Triplet State Geometry
From D_{3h} Or D_{6h} Symmetry

Molecule (Solvent)	2E GH$_z$	Top zf	Structure
Benzene			
(in C_6D_6) (46)	0.4	τ_a	antiquinoidal
(in cyclohexane) (47)	0.7	τ_b	quinoidal
TCB			
(x-traps) (48)	2.58	τ_a	quinoidal
(in HMB) (48)	0	---	$>C_3$
HCB			
(x-traps) (49)	1.78		$<C_{2v}$
(in HMB) (49)	0.089		$<C_3$
Triphenylene (21)			
(in hexane)	0.070	τ_a	antiquinoidal
Corenene (21)			
(in hexane)	0.3	---	$<C_{2v}$

rate of ISC to the τ_N level is slower (21) than the sum of the rates to the τ_a and the τ_b levels by a factor of 1/5. Group theory predicts that τ_N is populated via spin orbit interaction between π,π^* states directly or indirectly involving inplane molecular distortion. τ_a and τ_b, on the other hand, are populated by second order interaction that involves spin orbit (s.o.) coupling between π,π^* and σ,π^* states and an out-of-plane distortion. Thus again the large s.o. matrix elements between π,π^* and σ,π^* states increase the importance of this coupling, even if it occurs in second order. This coupling occurs in first order in the radiative process to the 0,0 band of aromatic hydrocarbons and the corresponding coupling between π,π^* and n,π^* occurs in first order in both radiative and the ISC process of N-heterocyclics. Thus it seems that s.o. coupling between π,π^* and σ,π^* (or n,π^*), whether in first or second order, determines to a large extent the radiative and nonradiative properties of the triplet state of aromatic hydrocarbons and their N-heterocyclics.

PMDR has been used to confirm the exchange nature of triplet-triplet energy transfer (22). Recently, PMDR was used to determine the mechanism of the intermolecular energy transfer in mixed crystals in the following manner (23). In a mixed crystal, exciting the host to its singlet manifold (S_H) could lead to the following processes:

$$S_H \xrightarrow{S\text{-}S} S_G \xrightarrow{ISC_G} T_G \qquad \qquad 8a$$

$$S_H \xrightarrow{ISC_H} T_H \xrightarrow{T\text{-}T} T_G \qquad \qquad 8b$$

Mechanism 8a involves S-S transfer followed by ISC in the guest. This process leads to T_G with the same degree of spin alignment (and PMDR signals) as that obtained by direct excitation of S_G. Mechanism 8b, on the other hand, involves ISC in the host followed by T-T transfer and should give T_G with different degree of spin alignment than 8a. The former can be calculated from the relative rate constants of the ISC in the host and its crystal structure (11).

For systems of 10^{-3} M of 2,3 dichloroquinoxaline in 1,2,3,5 tetrachlorobenzene, mechanism 8a dominates whereas if tetrabromobenzene is used as a host, mechanism 8b takes over (23). This is consistent with the fact that the competition between S-S energy transfer and the ISC process in the host determines the important transfer mechanism. In hosts of fast ISC (those with heavy atoms), mechanism 8b should be important, whereas for hosts of large singlet Davydov splitting, large guest concentration or a slow ISC, mechanism 8a becomes important.

ACKNOWLEDGEMENTS

The author wishes to thank the fruitful and enjoyable collaboration of his research group, Drs. Tinti, Owens, Moomaw, Kallman, Hall and Chen, and Mssrs. Chodak, Lin, Leung, and Gwaiz and Miss Wilkerson. The support of the U. S. Office of Naval Research is greatly appreciated.

REFERENCES

1. (a) E. Fermi and F. Rasetti, Nature, 115, 764 (1925).
    (b) G. Breit and A. Ellet, Phys. Rev., 25, 888 (1925).

2. For a review see: R. A. Bernhein, "Optical Pumping: An Introduction", W. A. Benjamin, Inc., (1965).

3. M. Sharnoff, J. Chem. Phys., 46, 3263 (1967).

4. A. L. Kwiram, Chem. Phys. Letters, 1, 272 (1967).

5. J. Schmidt, I. A. Hesselmann, M. S. de Groot, and J. van der Waals, ibid, 1 434 (1967).

6. J. Schmidt and J. H. van der Waals, Chem. Phys. Letters, $\underline{2}$, 640 (1968).

7. J. H. van der Waals and M. S. de Groot, "The Triplet State" Zahlan, Editor. Cambridge U. Press, London, p. 101 (1967).

8. (a) J. Schmidt, W. S. Veeman and J. H. van der Waals, Chem. Phys. Letters, $\underline{4}$, 341 (1969).

    (b) D. A. Antheunis, J. Schmidt and J. H. van der Waals, ibid, $\underline{6}$, 255 (1970).

9. M. A. El-Sayed and J. Olmsted, Chem. Phys. Letters, in press.

10. (a) C. S. Harris, Proc. of 5th Molecular Crystal Symposium, Philadelphia, Pennsylvania.

    (b) C. S. Harris, J. Chem. Phys., $\underline{54}$, 972 (1971).

11. M. A. El-Sayed, J. Chem. Phys., $\underline{54}$, 680 (1971).

12. M. A. El-Sayed, "Advances in Electronic Excitation and Relaxation", E. Lim, Editor, Volume 1, Academic Press, Inc., in press (1972).

13. M. A. El-Sayed, Accts. Chem. Res., $\underline{4}$, 23 (1971).

14. M. A. El-Sayed, "Phosphorescence-Microwave Double Resonance Spectroscopy", International Review of Science, D. A. Ramsay, Editor, 1972.

15. J. Olmsted and M. A. El-Sayed, "Experimental Methods in Phosphorescence-Microwave Double Resonance", to be published in "The Creation and Detection of the Excited State", Marcel Dekker, Inc. (1972).

16. M. Sharnoff, Mol. Cryst., $\underline{9}$, 265 (1969).

17. S. P. McGlynn, T. Azumi and Kinoshita, "Molecular Spectroscopy of the Triplet State", Prentice-Hall (1969).

18. A. A. Gwaiz, M. A. El-Sayed and D. S. Tinti, Chem. Phys. Letters, $\underline{9}$, 454 (1971).

19. M. A. El-Sayed, M. Leung and C. T. Lin, Chem. Phys. Letters, $\underline{14}$, 329 (1972).

20. M. A. El-Sayed, J. Chem. Phys., $\underline{38}$, 2834 (1963).

21. W. R. Moomaw, M. A. El-Sayed, J. B. Chodak, in preparation.

22. M. A. El-Sayed, D. S. Tinti and E. M. Yee, J. Chem. Phys., 51, 5721 (1969).

23. A. K. Wilkerson, C. T. Lin and M. A. El-Sayed, Chem. Phys. Letters, in press.

24. M. Schwoerer and H. C. Wolf, in Proc. of the XIVth Collogue Ampere, 1966 ed., R. Blinc (North Holland, Amsterdam, 1967) p. 87.

25. I. Y. Chan, J. Schmidt and J. H. van der Waals, Chem. Phys. Letters, 4, 269 (1969).

26. C. B. Harris, D. S. Tinti, M. A. El-Sayed and A. H. Maki, Chem. Phys. Letters, 4, 409 (1969).

27. T. S. Kuan, D. S. Tinti and M. A. El-Sayed, Chem. Phys. Letters, 4, 507 (1970).

28. M. A. El-Sayed and O. F. Kalman, J. Chem. Phys., 52, 4903 (1970).

29. M. D. Fayer, C. B. Harris and D. A. Yuen, J. Chem. Phys., 53, 4719 (1970).

30. W. S. Veeman and J. H. van der Waals, Chem. Phys. Letters, 7, 65 (1970).

31. J. Schmidt, W. G. Van Dorp and J. H. van der Waals, ibid., 8, 345 (1971).

32. C. R. Chen and M. A. El-Sayed, Chem. Phys. Letters, 10, 307 (1971).

33. A. Shain, M. Sharnoff and M. Stombler, J. Chem. Phys., 56, 2475 (1972).

34. J. Schmidt, Chem. Phys. Letters, 14, 411 (1972).

35. M. Leung and M. A. El-Sayed, ibid., in press.

36. A. W. Hornig and J. S. Hyde, Mol. Phys., 6, 33 (1963).

37. M. S. deGroote, I. A. Hesselmann and J. H. van der Waals, Mol. Phys., 12, 259 (1967).

38. D. S. Tinti, M. A. El-Sayed, A. H. Maki and C. B. Harris,

Chem. Phys. Letters, $\underline{3}$, 343 (1969).

39. M. A. El-Sayed, D. V. Owens and D. S. Tinti, Chem. Phys. Letters, $\underline{6}$, 395 (1970).

40. D. S. Tinti and M. A. El-Sayed, J. Chem. Phys., $\underline{54}$, 2529 (1971).

41. M. A. El-Sayed, Molecular Luminescence, "Proceedings of International Conference on Molecular Luminescence", (Loyola U., Chicago, 1968), E. Lim, Editor, W. Benjamin, Inc., New York, 1969, p. 175.

42. M. A. El-Sayed, J. Chem. Phys., $\underline{52}$, 6438 (1970).

43. M. A. El-Sayed and J. Olmsted, Chem. Phys. Letters, in press.

44. M. Sharnoff and E. B. Iturbe, Phys. Rev. Letters, $\underline{27}$, 576 (1971).

45. A. H. Francis and C. B. Harris, Chem. Phys. Letters, $\underline{9}$, 188 (1971).

46. M. S. deGroot, I. A. Hesselman and J. H. van der Waals, Mol. Phys., $\underline{16}$, 45 (1969).

47. M. A. El-Sayed, W. R. Moomaw and J. B. Chodak, J. Chem. Phys., in press.

48. C. T. Lin and M. A. El-Sayed, unpublished results.

49. M. A. El-Sayed, A. A. Gwaiz and C. T. Lin, Chem. Phys. Letters, in press.

RECENT DEVELOPMENTS IN SENSITIZED LUMINESCENCE AND ENERGY TRANSFER PHENOMENA IN SOLIDS

D. L. Dexter

University of Rochester, Rochester, New York, USA

ABSTRACT

A brief review of traditional sensitized luminescence is given, followed by a discussion of cooperative and non-linear effects observed over the last few years.

In the 1920's it was found that electronic excitation energy could be transferred from one gaseous atom to another, e.g., excited Hg to Tl, the excess energy going into kinetic energy of both atoms. This was referred to as a "collision of the second kind". Then, around forty years ago, excitons were invented. In the thirties and forties much the same phenomena were observed in organic molecules in liquid solution, and Förster (1) correctly interpreted this effect with the Fermi Golden Rule:

$$P_{SA} = (4\pi^2/\hbar)|<H_{SA}>|^2 \rho_E \tag{1}$$

where P_{SA} is the probability of transfer from a sensitizer (S) to an activator (A). $<H_{SA}>$ is the matrix element of the interaction Hamiltonian between the molecules S and A -- in Förster's case the electric dipole-dipole interaction -- and ρ_E, the necessary density of states, was provided by the vibrational motion which contributed the line broadening of the optical transition of each molecule, just as the kinetic energy of the outgoing atoms did in the case of gases. Conservation of energy in the transfer process clearly requires that the emission band of S overlap an absorption band of A. This theory recognized that true transfer implied incoherence, in contrast, to two identical pendula coupled by a non-rigid mounting, such that the energy could flow periodically from one to the

other (2). Since the forties the effect has been widely studied
in solids also. In the fifties and sixties this theory was verified in detail (3), was extended to include higher multipole and
exchange interactions (4), and it was suggested that concentration
quenching should be dependent on transfer between like species, so
as to enhance the probability of reaching a "sink" (5). In this
extension to higher multipole interactions an aspect became clear
that will be important in much of what follows. The term "forbidden transition" in radiative transitions really means something
for an isolated atom, the transition probability decreasing in
powers of $(a_o/\lambda)^{2n}$ [here a_o is the Bohr radius, λ the wavelength,
n an integer], hence 10^{-6} - 10^{-7} for an electric quadrupole transition as compared with an electric dipole transition. In a solid,
odd parity vibrational modes can increase this figure. But in
energy transfer processes, depending as they do on near-zone interactions, the transition probabilities drop off not as $(a_o/\lambda)^{2n}$
but as $(a_o/\rho)^{2n}$, where ρ is the separation of the interacting ob-
jects. Hence energy transfer effects tend to be more pronounced
in systems with forbidden transitions (5).

     Nothing in the theory requires that S transfer <u>all</u> its ex-
citation energy. If there is an intermediate excited state, S
can drop to that and transfer the difference. (See Fig. la) This
has been observed (6). If there is not an intermediate excited
state it can transfer part of the excitation energy and emit a
photon of the energy difference. (Fig. lb) This cooperative effect has also recently been observed (7).

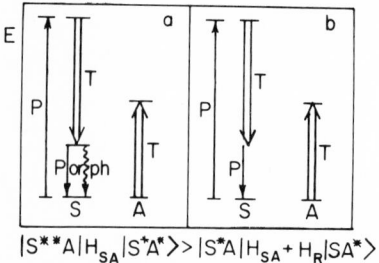 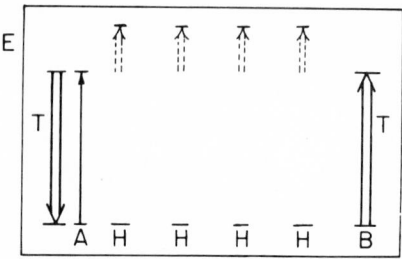

$|S^{**}A|H_{SA}|S^+A^*\rangle \gg |S^*A|H_{SA} + H_R|SA^*\rangle$

Fig. 1 a) Schematic representation of transfer of part of the elec-
            tronic excitation energy of S to A, utilizing a second
            real excited state in S. The double arrows indicate a
            transfer process. P stands for photon, T for transfer,
            ph for phonon processes.
        b) The same but with no real intermediate state in S.

Fig. 2 Transfer from A to B via virtual excitations of the
            intervening host atoms H.

The theory was also extended to include many-body effects, and it was shown that it may be easier to transfer energy from S to A via several virtual excitations of the host than by the direct near-zone multipole or exchange interaction between S and A (8). (See Fig. 2)

During the sixties several other new effects arose. The first, on which I shall dwell in some detail because it provides a framework for understanding some others, was the Varsanyi-Dieke experiment on cooperative absorption (9) among rare-earth ions; this was observed in a large variety of combinations. At first sight this seems impossible, since the perturbation radiation Hamiltonian is a sum of one-electron operators. Thus if one matrix element is non-zero, all others must be zero, and a two-electron excitation cannot occur. Yet beyond any doubt the photon's energy was being split to excite two atoms simultaneously. (A somewhat similar effect was seen in x-ray absorption spectroscopy by Parratt and Schnopper (10), in which two electrons on the same atom were excited simultaneously.) This suggests an interpretation of cooperative absorption as resulting from a two-center configuration interaction. Also observed, in the ultraviolet, was the inverse, the simultaneous two-photon decay from excited metastable states of atoms (11). Fig. 1 of Ref. 12 indicates schematically the Varsanyi-Dieke effect and its interpretation (12). It is seen that there are four routes to the first state described by $|ab>$, passing through virtual states $|a'b>$ or $|ab'>$ induced by the atomic interactions, or $|a'o>$ (or $|ab'>$) induced by the radiation Hamiltonian. The squared matrix element is of order

$$|M_{ab}|^2 \sim (ea_o\epsilon)^2 \ (e^2a_o{}^2/W\kappa\rho^3)^2$$

where the first factor would apply to an allowed atomic transition, W is some characteristic atomic energy denominator (0.1 - 1.0 ev), κ is a dielectric constant and ρ is the separation between the atoms. This applies to electric dipole-dipole transitions, in which case the second factor could be very large, $\sim 10^{-2}$. This was not the case with Varsanyi-Dieke experiments. There it involved forbidden transitions and was much smaller.

In parallel to this, Nakazawa and Shionoya (13) have observed emission in $YbPO_4$, in which two excited Yb^{+3} ions cooperate in producing green emission at approximately twice the energy of either ion. (See Fig. 3) (There are no Yb^{+3} states remotely near this energy.) Doubtless the ions interact, thus producing a virtual, intermediate state with one ion in the ground state, the other highly excited, which then emits a real photon at the sum of the excitation energy.

Much work has been done on infrared to visible light conversion. This can be accomplished in many ways, most of which are

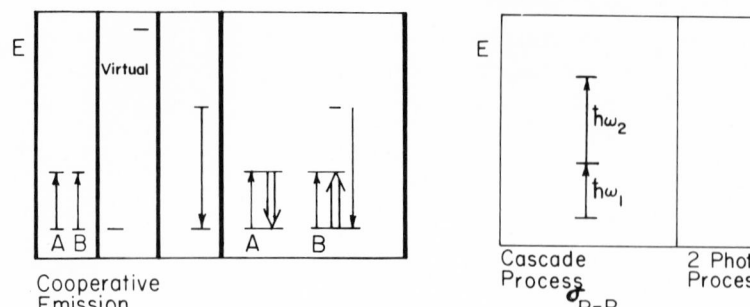

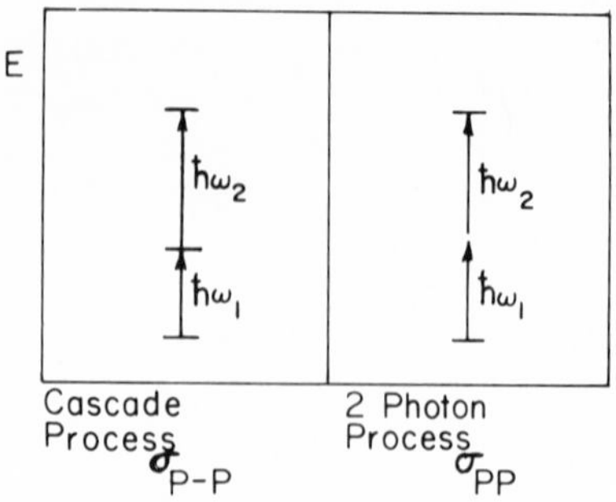

Fig. 3   Cooperative emission from a pair of ions, which need not
          be identical. This is the inverse of the Varsanyi-Dieke
          cooperative absorption. Only one of the four routes is
          shown here.

Fig. 4 a) An inverse cascade absorption process in which two pho-
          tons are successively absorbed on the same atom.
       b) A two-photon process in which two photons are absorbed
          simultaneously.

germane to this paper. One that is not is the inverse cascade process, (See Fig. 4a) in which an atom is excited by one photon and
then excited by another photon. Another possibility is that the
atom is excited by energy transfer and is then excited to a higher
state by a photon (or vice versa). (See Fig. 5) This is not really
anything new. It is encompassed by the traditional theory of
energy transfer as in sensitized luminescence.

      However a few years ago it was found that Er and Ho could be
induced to emit visible luminescence when Yb was also present in
the host, upon excitation in the IR Yb absorption band, the visible
luminescence being proportional to the square of the intensity of
IR excitation. This could result either from cooperative transfer
from two excited Yb ions, or from sequential transfer from Yb ions.
(There are other possibilities, discussed below.) One school of
thought was sequential (14), another cooperative (15). Miyakawa
and I did some calculations (16) on this and concluded that either
process could suffice, i.e., be faster than spontaneous emission
from Yb, but that the sequential process was much more favorable.
(See Fig. 4 of Ref. 16)

      It was found that Tm could also be excited by Yb excitation,
and this led to controversy. The Tm state could be excited in a
three-step sequential transfer process, or in principle, i.e.,
energetically, by a two-step cooperative process. (See Fig. 2 of
Ref. 16) Miyakawa and I concluded that except for very low excitation intensities, where luminescence would be too weak to be
observable, the three-step sequential process would dominate, and
the Tm luminescence should depend on the cube of Yb-excitation

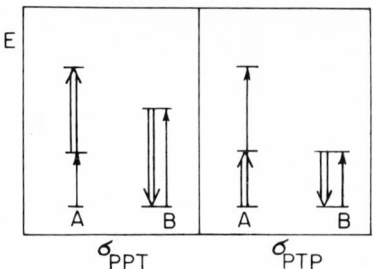

Fig. 5 a) Atom A absorbs a photon, as does B, which transfers its
          energy to A while A is still excited.
       b) Atom B absorbs a photon, transfers its energy to A,
          which then absorbs another photon.

intensity. (At sufficiently high intensity a saturation should
occur, and the dependence should gradually become quadratic, then
linear, then independent.) A careful experimental study by Hewes
and Sarver (17) confirms these conclusions.

The Yb plus Tb system, we suggested, should be an ideal one
to investigate, since sequential transfer is ruled out by the
energy mis-match. Obviously this also occurred to Livanova et al
independently, since they published a paper demonstrating the cooperative transfer simultaneously from two excited Yb to Tb (18).
This effect has been confirmed by Van Uitert et al in another host
(19).

During the sixties a great deal of work was done on fusion of
triplet excitons to produce singlet excitons in organic crystals
such as anthracene (20), and recently the fission of singlet excitons to produce two triplet excitons in tetracene (21), perhaps
along the lines suggested some years ago (22).

Another field that has been actively studied is two-photon absorption, wherein no excited state of the system B exists at either
photon energy (23). (See Fig. 4b) Clearly, since the angular momentum of each photon is unity, the angular momentum change of the
atom must be zero or two.

Altarelli and I studied (24) an alternative method of exciting system B by incorporating a sub-system A which does have a
state at one of the photon energies, $h\nu_1$, say, where the excitation
energy of B is $h\nu_1 + h\nu_2$. Here the energy $h\nu_1$ is transferred simultaneously from A to B while a photon $h\nu_2$ is absorbed by B.
(See Fig. 1 of Ref. 24) This process is several powers of ten more
probable, in typical cases than two-photon absorption. (Of course
it can be many powers of ten more probable when two-photon absorption is forbidden by symmetry.) This effect can be generalized,
of course; A (which could be the same as B) need not give up <u>all</u>

its energy, and this effect has been observed in CdS (25).

The question arises as to whether the cross-section for co-operative transfer + absorption (σ_{TP}) is more likely than coopera-tive transfer in Yb + Tb systems (σ_{TT}). The answer, according to a recent calculation by Altarelli, is no, not by several powers of ten under typical conditions. A basic principle seems to be that in general the most probable process is the one that uses the most real intermediate states. The physical interpretation of this is basically that photons travel fast, and their energies are not available in the medium for a very long time, in contrast to the excitation energy in a real excited state. Hence we would normally expect that in rare-earth combinations:

$$\sigma_{P-P} > \sigma_{T-T} \gg \sigma_{PT} \gg \sigma_{PP}.$$

See previous text and figure captions for definitions of these cross-sections.

It has not been possible in this time and space to cover all the specific systems investigated, nor to refer to all the hundreds of papers on these topics. I should acknowledge the fact that Prof. Toru Miyakawa, then in Rochester, now back in Japan, and Dr. Massimo Altarelli from Rome, now in Rochester, have helped a lot in understanding some of these processes.

REFERENCES

1. T. Forster, Ann. Phys. (Germany) 2, 55 (1948).

2. D. L. Dexter, T. Förster, and R. S. Knox, Phys. Stat. Solidi 34, K159 (1969).

3. See references in (2).

4. D. L. Dexter, J. Chem. Phys. 21, 836 (1953).

5. D. L. Dexter and J. H. Schulman, J. Chem. Phys. 22, 1063 (1954).

6. W. D. Partlow, Phys. Rev. Letters 21, 90 (1968).

7. P. V. Van der Ziel and L. C. Van Uitert, Solid State Communi-cations 7, 819 (1969).

8. J. D. Dow, Phys. Rev. 174, 962 (1968).

9. F. Varsanyi and G. H. Dieke, Phys. Rev. Letters 7, 442 (1961).

10. H. W. Schnopper, Phys. Rev. 131, 2558 (1963).

11. R. Novick, Science 177, 367 (1972) and references therein. See also W. D. Partlow and H. W. Moos, Phys. Rev. 157, 252 (1967).

12. D. L. Dexter, Phys. Rev. 126, 1962 (1962).

13. E. Nakazawa and S. Shionoya, Phys. Rev. Letters 25, 1710 (1970).

14. M. F. Auzel, Compt. Rend, 262B, 1016 (1966); 263B, 819 (1966).

15. V. V. Ovsyankin and P. P. Feofilov, Soviet Physics - JETP Letters 4, 317 (1966); Applied Optics 6, 1828 (1967).

16. T. Miyakawa and D. L. Dexter, Phys. Rev. B 1, 70 (1970).

17. R. A. Hewes and J. F. Sarver, Phys. Rev. 182, 427 (1969).

18. L. D. Livanova, I. G. Saitkulov and A. L. Stolov, Soviet Phys. Solid State 11, 750 (1969)

19. F. W. Ostermayer and L. G. Van Uitert, Phys. Rev. B 1, 4208 (1970).

20. P. Avakian and R. E. Merrifield, Molecular Crystals 5, 37 (1968).

21. N. Geacintov, M. Pope, and F. Vogel, Phys. Rev. Letters 22, 593 (1969). R. E. Merrifield, P. Avakian, and R. P. Groff, Chem. Phys. Letters 3, 155 (1969).

22. D. L. Dexter, Phys. Rev. 108, 630 (1957).

23. R. Braunstein and N. Oekmann, Phys. Rev. 134, A449 (1964).

24. M. Altarelli and D. L. Dexter, Optics Communications 2, 36 (1970).

25. D. W. Langer and T. Goto, Bull. Am. Phys. Soc. (Ser. 2) 16:1, p. 372 (March, 1971); Phys. Rev. Letters 27, 1004 (1971).

COOPERATIVE PROCESSES IN LUMINESCENT SYSTEMS

V. V. Ovsyankin and P. P. Feofilov

The State Optical Institute

Leningrad, USSR

ABSTRACT

This paper reviews various luminescence processes which occur in the systems of excited particles interacting with each other in the condensed state.

Typical contemporary luminescence investigations involve the study not only of isolated particles, but also the study of specific phenomena which are determined by the interaction between particles in an ensemble. The ensemble may no longer be described as a sum of one-particle Schrödinger equations. Rather the introduction of an operator describing the interparticle interaction, $V(r_i, r_j, p_i, p_j)$ is required. In the case of strong interactions the individuality of separate particles is lost and the necessity of accounting for these interactions is obvious. This was realized a long time ago and has led to the concepts of the luminescence for systems like crystalline phosphors. These concepts are based on the ideas of the zone theory of solids completed by the notions of quasiparticles and collective excitations (phonons, excitons).

When the interaction is weak (or, more exactly, when it may be accounted for within the limits of perturbation theory) and the particles retain their energetic individuality, the need to account for the interparticle interaction is not so obvious. Indeed, a great amount of fundamental information concerning the energy structure of real condensed systems was obtained by neglecting any interaction between the particles. In the meantime, it has become especially clear during the last few years that the existence even of weak interactions leads to a number of non-trivial consequences.

Some of these consequences will be considered in this paper.

Independent of their nature (Coulomb or exchange) all the interactions involved may be classified as <u>resonant</u> or <u>non-resonant</u>.

NON-RESONANT INTERACTIONS

Non-resonant interactions are the most general case of interactions wherein the degeneracy due to the identity of individual particles is not important. These interactions always exist in condensed media and the matrix element of the dipole transition between electronic states of various particles (which is equal to zero in the absence of the interaction) becomes finite:

$$M_{ab}^{a'b'} = \langle \psi_{ab} | d | \psi_{a'b'} \rangle = \sum_{m \neq a} \frac{\langle a'b' | V | mb \rangle \langle a | d | m \rangle}{E_A^{a'} + E_B^{b'} - E_A^m - E_B^b}$$

$$+ \sum_{n \neq b} \frac{\langle a'b' | V | an \rangle \langle b | d | n \rangle}{E_A^{a'} + E_B^{b'} - E_A^a - E_B^n} + (a'b' \longleftrightarrow ab)$$

(for notations see Fig. 1). It is thus possible to observe in absorption and luminescence <u>additional resonances</u>, the frequencies of which are simple combinations of frequencies of transitions in individual particles.

Thus, for example, the matrix element $\langle \psi_{00} | d | \psi_{a'b'} \rangle$ gives the probability of <u>cooperative excitation</u> of two interacting particles by a single photon (Fig. 1a) and of the inverse process of <u>cooperative emission</u> of one photon of summed frequency accompanying the simultaneous transition of the two excited particles into the ground state (Fig. 1b). The matrix element $\langle \psi_{a0} | d | \psi_{0b'} \rangle$ determines the probability of <u>combination excitation</u> (Fig. 1c), which consists of the transition from the state b' of one particle into the state a of another one initiated by the absorption of a photon the frequency of which is determined by the difference in energy between states a and b'. The same matrix element also determines the probability of the inverse process of <u>combination luminescence</u> (Fig. 1d) consisting of the transition from state a of one particle to the state b' of another particle which is accompanied by the emission of a quantum. Cooperative excitation was observed first in (1) and treated theoretically in (2). We predicted the existence of the remaining three processes in 1967 (3). Apart from the above-mentioned phenomena it is possible, in principle, to observe the <u>cooperative amplification of light</u> which propagates through a medium with an inverted population of levels and has a frequency equal to the sum of the energies of the interacting states. It is

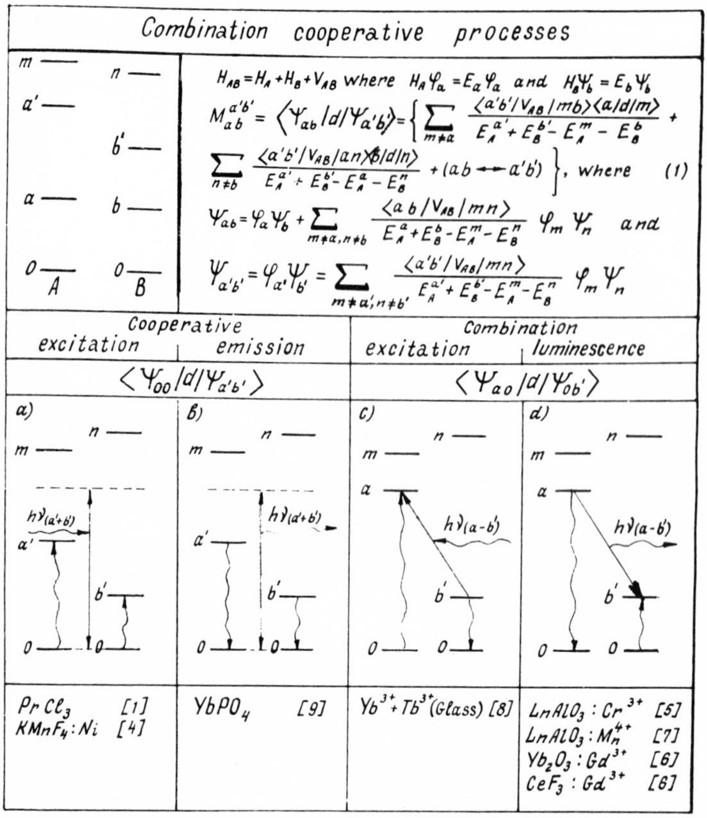

V. V. OVSYANKIN AND P. P. FEOFILOV

Combination cooperative processes

$H_{AB} = H_A + H_B + V_{AB}$ where $H_A\varphi_a = E_a\varphi_a$ and $H_B\Psi_b = E_b\Psi_b$

$$M_{ab}^{a'b'} = \langle \Psi_{ab}|d|\Psi_{a'b'}\rangle = \Big\{ \sum_{m\neq a} \frac{\langle a'b'|V_{AB}|mb\rangle\langle a|d|m\rangle}{E_A^{a'} + E_B^{b'} - E_A^m - E_B^b} +$$

$$\sum_{n\neq b} \frac{\langle a'b'|V_{AB}|an\rangle\langle b|d|n\rangle}{E_A^{a'} + E_B^{b'} - E_A^a - E_B^n} + (ab \leftrightarrow a'b') \Big\}, \text{ where } \quad (1)$$

$$\Psi_{ab} = \varphi_a\Psi_b + \sum_{m\neq a,\, n\neq b} \frac{\langle ab|V_{AB}|mn\rangle}{E_A^a + E_B^b - E_A^m - E_B^n}\, \varphi_m\Psi_n \quad \text{and}$$

$$\Psi_{a'b'} = \varphi_{a'}\Psi_{b'} = \sum_{m\neq a',\, n\neq b'} \frac{\langle a'b'|V_{AB}|mn\rangle}{E_A^{a'} + E_B^{b'} - E_A^m - E_B^n}\, \varphi_m\Psi_n$$

| Cooperative | | Combination | |
excitation	emission	excitation	luminescence
$\langle \Psi_{00}|d|\Psi_{a'b'}\rangle$		$\langle \Psi_{a0}|d|\Psi_{0b'}\rangle$	
a) $h\nu_{(a'+b')}$	b) $h\nu_{(a'+b')}$	c) $h\nu_{(a-b)}$	d) $h\nu_{(a-b)}$
$PrCl_3$ [1]	$YbPO_4$ [9]	$Yb^{3+}+Tb^{3+}$(Glass) [8]	$LnAlO_3:Cr^{3+}$ [5]
$KMnF_4:Ni$ [4]			$LnAlO_3:Mn^{4+}$ [7]
			$Yb_2O_3:Gd^{3+}$ [6]
			$CeF_3:Gd^{3+}$ [6]

Fig. 1   Combination cooperative processes.

also possible to imagine the enhancement of a one-particle two-photon absorption by the incorporation into the system of impurities which have real levels close to the energy of the single photon (3).

The phenomena described which we classify as combination cooperative phenomena may be observed in any condensed systems. However, the most favourable systems for the study of these processes are apparently the crystals activated by Ln^{3+} --or other transition metal ions with narrow spectral lines. The first-order corrections to the wave functions which are responsible for the processes under consideration correspond to the second-order correction to the energy of the combining states. This makes it possible to use a spectroscopic criterion for the identification of the phenomena: the frequencies of cooperative resonance have to be equal to the sum or difference of energies of individual states of interacting particles. The use of this criterion has permitted the

unambiguous identification of processes of cooperative excitation
(1,4) and combination luminescence (5-7). Recent publications
describe observations of combination excitation (8) and of cooperative emission (9). The summary of the data on systems in which
the combination cooperative processes were observed is given in
the lower part of Fig. 1.

RESONANT INTERACTIONS

Apart from non-resonant interactions in the luminescent
systems, interactions of resonant type are possible which are connected with an essential degeneracy in an ensemble of identical
particles, or with an accidental degeneracy in an ensemble of differing particles. When both interacting particles have a discrete
set of energy states, the interaction leads to a removal of the
degeneracy (energy splitting is $|V_{12}|$) -- which is spectroscopic
evidence for interaction -- and to the non-stationarity of wave
functions describing the individual states of particles. The
square of the amplitude coefficient of individual states gives the
probability that the excitation energy is localized on a definite
particle and varies periodically from 0 to 1 with a frequency
$\nu = |V_{12}|/2h$, thus providing kinetic evidence of interaction.

When one of the interacting particles has a continuous or
quasi-continuous spectrum of states in the region of discrete
states of another particle, the interaction leads to a shift,
broadening and spectral redistribution of transition intensity near
the discrete level. The latter effect is known as antiresonance
(10) and was invoked in the explanation of certain peculiarities
in absorption spectra of crystals (11-13). In some cases (14,15)
the observed resonances should be considered a result of intercon-
figuration interactions rather than interparticle ones. The
search for analogous effects in luminescence should be of great
interest.

The kinetic consequence of the interaction consists, in this
case, of the non-stationarity of the discrete states, which decay
with a rate proportional to the square of the matrix element of the
interaction operator. Non-stationarity of discrete state, mani-
festing itself in energy transfer from one particle to another, was
studied thoroughly in connection with the problems of luminescence
sensitization and quenching.

It is possible that, as a result of partial excitation energy
transfer (cross-relaxation), both interacting particles may be in
excited states (in particular radiative ones) and energy conversion
("exciton fission") takes place (1,16,17)).

NON-LINEAR COOPERATIVE PROCESSES

Both non-resonant and resonant interactions between excited particles may lead, in particular, to the "up-conversion" of light frequency, which is due to the population of higher energy states, which are unattainable by the absorption of single photons. These processes have recently attracted attention in connection with their practical use for the direct transformation of infrared radiation into visible, as well as with the problems of spectral sensitization mechanisms of various photophysical, photochemical and photobiolgical processes, the most important of them being sensitization of photographic processes and, of course, photosynthesis occurring in green plants.

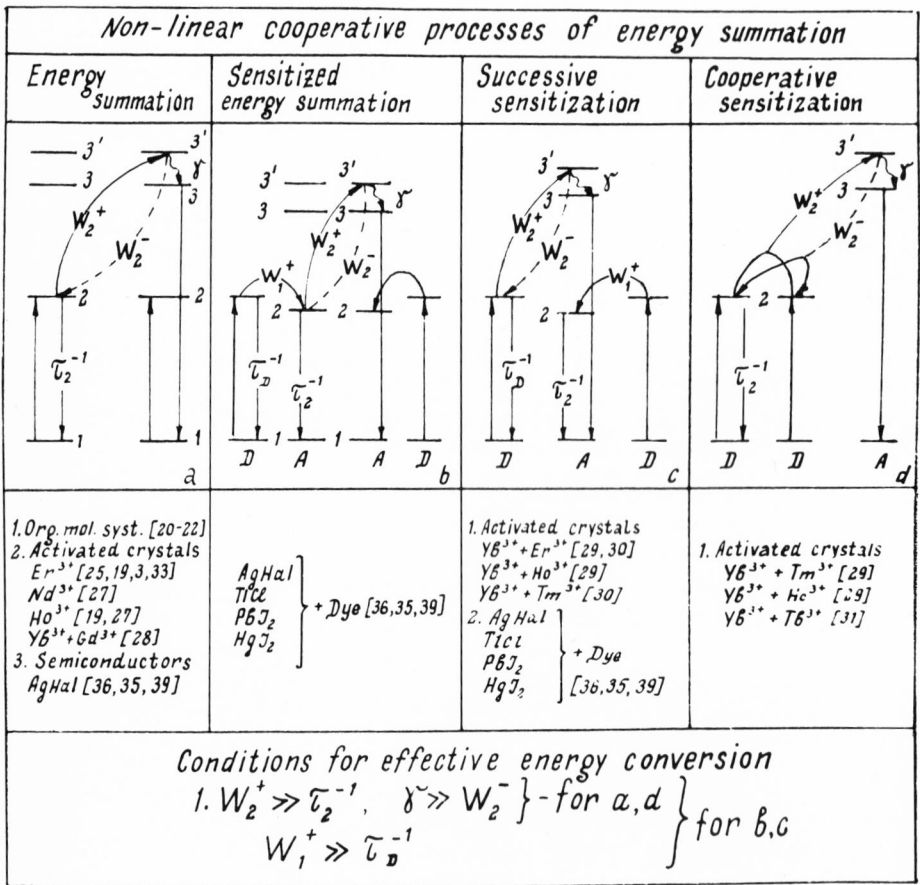

Fig. 2 Non-linear cooperative processes of energy summation.

The simplest examples of non-linear cooperative processes may be taken from the above-mentioned cooperative emission due to non-resonant interaction (Fig. 1) and the cumulation (summation) of energy of two excited particle to another particle already in an excited state (Fig. 2a). To non-linear cooperative processes should also be added resonant processes of sensitized summation of energy (Fig. 2b), in which the energy states which are summed are populated not by direct absorption of light but by energy transfer from other systems which are donors of energy; the processes of successive sensitization (Fig. 2c) which arise from successive energy transfer from donors to a single acceptor; and, finally, the process of cooperative sensitization, due to resonant as well as non-resonant interactions, which consists in simultaneous energy transfer from two donors to an acceptor.

The fate of the energy accumulated in higher excited state may be quite varied. First, a radiative transition to the ground state is possible--"anti-Stokes luminescence"; secondly, complete or partial transformation of electron excitation energy into other forms of energy which is not accompanied by radiative deactivation --"non-linear luminescence quenching" (18).

Kinetic and stationary characteristics of "anti-Stokes luminescence" may be obtained by the solution of simple rate equations for the population of the states involved in the process of absorbed energy transformation, and a series of experimental criteria (spectroscopy, kinetic, photometric and concentrational) may be formulated for the discrimination of these processes from each other as well as from one-particle processes of step-like excitation (19). In the Table an example is given of the criteria of this type for the differentiation of cases of energy cumulation and step-like excitation. Fig. 3 demonstrates the use of these criteria for the study of the mechanism of the green luminescence in BaF_2 - Er^{3+} excited by infrared radiation with $\lambda \underset{\sim}{\sim} 1\mu m$.

The use of criteria like those shown in the Table and in Fig. 3 has allowed the identification of the processes of energy summation in a variety of systems. There were, first of all, molecular systems in which it proved to be possible to interpret the so-called "delayed luminescence" of solutions and crystals which is due to the interaction of two triplet molecules followed by the formation of one excited singlet state (triplet-triplet annihilation) and by singlet luminescence (see reviews (20,21) and series of papers (22)).

Another group of systems with pronounced processes of energy summation is made up of crystals with rare-earth activators, where the presence of a great number of metastable levels is extremely favourable for observation of various phenomena connected with the

CHARACTERISTICS OF LUMINESCENCE EXCITED BY VARIOUS
TWO-PHOTON MECHANISMS

Two-step Excitation	Cooperative Excitation

$$W_2^+ = \alpha n_2$$

Rate Equations

$dn_1/dt = -\sigma_{12}En_1 + n_2\omega_{21} + n_3\omega_{31}$		$dn_1/dt = -\sigma_{12}En_1 + n_2(W_2^+ + \omega_{21})$	
$dn_2/dt = \sigma_{12}En_1 - (\sigma_{23}E + \omega_{21})n_2$		$+ n_3(\omega_{31} - W_2^-)$	
$+ n_3\omega_{32}$		$dn_2/dt = \sigma_{12}En_1 - n_2(2W_2^+ + \omega_{21})$	
$dn_3/dt = \sigma_{23}En_2 - n_3(\omega_{31} + \omega_{32})$		$+ n_3(2W_2^- + \omega_{32})$	
$N = n_1 + n_2 + n_3 \gg n_2, n_3$		$dn_3/dt = n_2W_2^+ - n_3(W_2^- + \omega_{32} + \omega_{31})'$	
When	When	When	When
$\sigma_{23}E \ll \omega_{21}$	$\sigma_{23}E \gg \omega_{21}, \omega_{31} \gg$	$W_2^+ \ll \omega_{21}$ or	$W_2^+ \gg \omega_{21}$ and
or $\omega_{31} \ll \omega_{32}$	ω_{32} and $\sigma_{23} \gg \sigma_{12}$	$W_2^- \gg (\omega_{31} + \omega_{32})$	$W_2^- \ll (\omega_{31} + \omega_{32})$
Dependence on Excitation Intensity			
E^2	E	E^2	E
Dependence on Activator Concentration			
C		C^2	C
Excitation Spectra			
$\sigma_{12}(\lambda) \cdot \sigma_{23}(\lambda)$	$\sigma_{12}(\lambda)$	$[\sigma_{12}(\lambda)]^2$	$\sigma_{12}(\lambda)$
Kinetics of Luminescence Decay			
$\exp(-\omega_{31} + \omega_{32})t$		$\omega_{31} \gg \omega_{21}$ and W_2^+	
		$\exp(-2\omega_{21}t)$	$(1 + 2\alpha n_2^{stat}t)^{-2}$

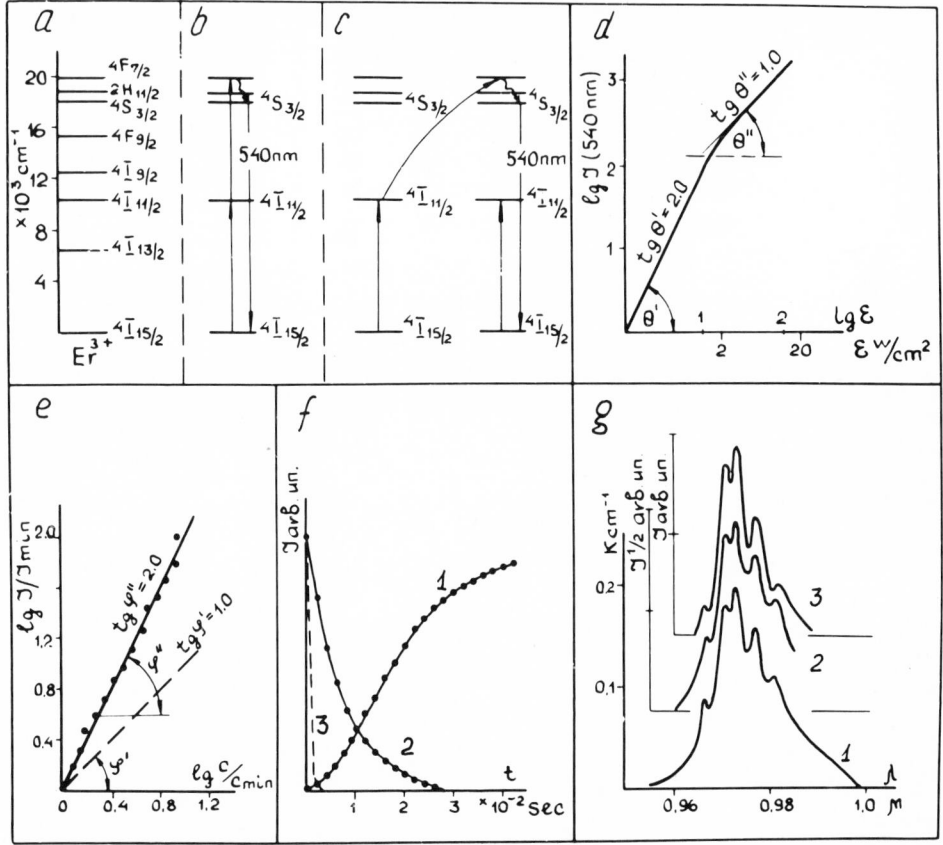

Fig. 3 Experimental study of energy summation mechanisms in
BaF$_2$:Er^{3+} crystals (after (3,19,25,26,33)). (a) term scheme
of Er^{3+} ion; (b) two-step excitation; (c) cooperative excitation; (d) dependence on excitation intensity (note dependence becomes linear at E $\approx$ 1W/cm^2); (e) dependence on Er^{3+}
concentration; (f) kinetics of anti-Stokes (green) luminescence. (1) Rise-time, (2) Decay, (3) Decay of Stokes-
excited luminescence; (g) absorption spectrum k(λ) (1), excitation spectrum I(λ) (3) and function [I(λ)]$^{\frac{1}{2}}$ (2) (these
experimental data prove the cooperative character of exci-
tation).

interaction of excited states (3). The study of "anti-Stokes
luminescence" of such crystals has shown that in some cases the
excitation proceeds not by the successive absorption of several
quanta by one activating ion, as was considered before (23) based
on the well-known scheme of Bloembergen's quantum counter (24), but
by a cooperative process of energy cumulation (25). The crystals

activated by Er^{3+} proved to be especially illustrative and allowed
the observation of energy summation in pairs of identical or differing excited states as well as the accumulation, in one ion, of the
energy of three primary excited ions. These processes are rather
effective and lead to the conversion of infrared radiation with
$\lambda \approx 1.0$ and 1.5 μm into visible (green or red) (3,19,26a,c).
Analogous processes of energy cumulation were observed also with
other ions, the interacting excited states being either identical
(27) or different (28).

The number of systems capable of cooperative energy summation
may be enlarged considerably when one uses crystals with activating
ions of two kinds (29,30). In these systems, besides energy summation and related resonant processes of sensitized summation and
successive sensitization, a process of cooperative sensitization
was detected which is conditioned by the joint action of resonant
and non-resonant interactions (29,31). This process is the inverse
of the process of cooperative deactivation--energy transfer from
one donor ion to two acceptor ions in one step--which was predicted
in (2) but was not observed until now.

Very high efficiency of infrared-to-visible conversion in
activated crystals stimulated a rapidly increasing number of investigations studying the transformation mechanism and the possibilities of the practical use of cooperative light converters. The
systems with ions Yb^{3+} as one of the activators were especially
investigated thoroughly since the absorption spectra of these ions
is fairly compatible with the emission spectrum of the GaAs:Si
diode (32). By combining the diodes with cooperative luminophors
one may design tiny light sources for various applications in
optoelectronic circuits.

Before considering the cooperative summation of energy in
other, more complex (and, in some cases, much more important)
systems, we shall discuss a question on the parameters which determine the efficiency of conversion of long-wave length radiation
into the energy of higher excited states.

EFFICIENCY OF COOPERATIVE PROCESSES OF ENERGY SUMMATION

The analysis of rate equations describing the processes of
two-quantum cooperative energy summation (see Fig. 2) shows that
the efficiency of these processes is, in general, a function of
the intensity of the exciting radiation (E) which is connected
with the superlinear dependence of the anti-Stokes luminescence
intensity $I_{31}(E)$. This function may, however, degenerate into a
linear dependence $I_{31}(E) = aE-b$ when the conditions

$$W_2^+ \gg \omega_{21} \text{ and } \gamma \gg W_2^- \tag{1}$$

are satisfied; γ is the rate of relaxation which takes the system out of resonance; the physical meaning of the phenomenological rates W_2^+, W_2^- and ω_{21} (averaged over the ensemble) differs to some extent for different processes and is clear from Fig. 2. For the systems operating according to 2a and 2b (Fig. 2), satisfying the conditions of (1) means that the quantum efficiency of cooperative summation is near its limiting value (0.5), and the energy efficiency (neglecting the relaxation losses $3' \rightsquigarrow 3$) is near unity. An analogous consequence for the systems 2b and 2c (Fig. 2) may be obtained when the conditions of (1) and an additional condition

$$W_1^+ \gg \tau_D^{-1} \tag{2}$$

are satisfied jointly. Therefore, apart from trival requirements (high absorption coefficient for process $1 \rightarrow 2$, and high quantum yield for transition $3 \rightarrow 1$), a two-quantum coverter may operate effectively only when the conditions of (1) and (2) are satisfied. The physical sense of this requirement is **very** simple: all the intermediate states have to relax via state 3.

In crystals with activating ions of one kind, operating according to scheme 2a (Fig. 2), the fulfillment of the conditions of (1) and, therefore, the realization of the utmost transformation efficiency may be attained when the intensity of the exciting radiation is of the order of a few watts/cm^2 (BaF_2-Er^{3+} crystals (33)). A further decrease by about 2 orders in the excitation intensity threshold (at which a linear dependence of $I(E)$ sets in) in this system is possible only with an increased activator concentration. In crystals with the activating ions of two kinds operating according to schemes 2c and 2d, the conditions of (1) may be satisfied at approximately the same intensities (34). However, further progress in increasing the efficiency of these systems depends on satisfying the condition of (2) and requires a search for specifically heterogeneous systems with an effective channel of energy transfer from ions--donors (D) to ions--acceptors (A).

These exists, however, a class of luminescent systems in which effective energy summation proceeds in a cooperative manner at considerably lower densities of converted radiation. This is the class of <u>semiconducting crystals with an adsorbed layer of dye molecules</u>. It is possible to observe in these systems the luminescence of semiconductors (acceptors) by exciting them in the absorption bands of adsorbed dye molecules (donors) which are considerably lower in energy (35,36). The linearization of these processes takes place in some cases (AgHal + Dye, sensitized photographic films) at intensities of incident radiation $\sim 10^{-11}$ watts/cm^2 (37). Satisfaction of condition (2) which follows from the absence in the spectrum of secondary radiation of any luminescence connected with the deactivation of intermediate states, allows us to assert that

these systems may be considered as unique converters of extremely
weak fluxes of low-energy quanta into the energy of higher
excited states.

Since the luminescence of silver halides and their photolysis
are alternative processes, the latter result allows us to assert
that the primary photolytic act in the formation of the latent
photographic image in the region of spectral sensitization has a
cooperative character.

COOPERATIVE LUMINESCENCE OF PHOTOSYNTHESIZING SYSTEMS

It is well known that for the realization of the elementary
act of photosynthesis--photochemical decomposition of water and
combination of liberated hydrogen with carbon dioxide--the energy
which is required is several times greater than that of single
quantum of red light absorbed by chlorophyll pigments. In connection with the discovery of cooperative luminescence in sensitized
semiconductors (which are similar in many aspects to photosynthesizing systems) we have looked for the analogous phenomena in green
parts of plants. It turned out, that, indeed, green leaves and
algae (chlorella), being excited in the long-wave length absorption
band of chlorophyll, luminesce much shorter wave lengths (38,39).
The study of this luminescense (which is rather weak due to the
efficiency of the known alternative pathways resulting in luminis-
cence and photosynthesis) have led us to propose a cooperative
mechanism for the population of higher excited states by energy
summation of two singlet or a singlet and a triplet chlorophyll
molecules. One can hope that a detailed study of cooperative
luminescence of natural photosynthesizing systems should bring us
nearer to the solution of one of the greatest mysteries of nature.

The above considered phenomena show the multiplicity and
variety of consequences which result from the fact that the atoms
and molecules in luminescent condensed media do not reveal themselves as a combination of isolated particles, but apart from their
individual characteristics the particles display their properties
as members of an ensemble. The investigation of these cooperative
phenomena opens up new ways to study the nature and mechanisms of
interparticle interactions, allows fabrication of new interesting
systems of optoelectronics and, what in our opinion is the most
important, gives the possibility to elucidate in a novel way the
mechanisms of the storage of electronic excitation energy in many
artificial and natural photolytic and photosynthetic processes which
are of great importance for practical applications and even for
human existence.

REFERENCES

1. F. Varsanyi and G. H. Dieke, Phys. Rev. Lett. $\underline{7}$, 442 (1961).

2. D. L. Dexter, Phys. Rev., $\underline{126}$, 1962).

3. V. V. Ovsyankin and P. P. Feofilov, in: Spectroscopy of Crystals, Ed. "Nauka" Press, Moscow, 1970, p. 135.

4. J. Ferguson, H. J. Guggenheim and Y. Tanabe, J. Chem. Phys. $\underline{45}$, 1134 (1966).

5. J. P. van der Ziel and L. G. Van Uitert, Phys. Rev. $\underline{21}$, 1334 (1968); Phys. Rev. $\underline{180}$, 343 (1969); $\underline{186}$, 332 (1969); Solid State Commun. $\underline{7}$, 819 (1969); J. P. van der Ziel, J. of Luminescence $\underline{1-2}$, 807 (1970).

6. P. P. Feofilov and A. K. Trofimov, Opt. i Spect. $\underline{27}$, 538 (1969).

7. P. P. Feofilov, Opt. i. Sectr. $\underline{31}$, 849 (1971).

8. V. I. Belak, G. M. Zverev, G. O. Karapetyan and A. M. Onishechenko, ZETP, Pisma $\underline{14}$, 301 (1971).

9. E. Nakazawa and Sh. Shionoya, Phys. Rev. Lett. $\underline{25}$, 1710 (1970).

10. U. Fano, Phys. Rev. $\underline{124}$ (1961) 1866; A. Shibatani and Y. Toyozawa, J. Phys. Soc. Japan $\underline{25}$, 335 (1968).

11. M. J. Taylor, Phys. Rev. Lett. $\underline{23}$, 405 (1969).

12. V. A. Arkhangelskaya and P. P. Feofilov, Opt. i Spectr. $\underline{28}$, 1219 (1970).

13. Y. Kato, C. I. Yu and T. Goto, J. Phys. Soc. Japan $\underline{28}$, 104 (1970).

14. M. D. Sturge, J. Chem. Phys. $\underline{51}$, 1254 (1969). M. D. Sturge, H. J. Guggenheim and M. H. L. Pryce, Phys. Rev. $\underline{B2}$, 2459 (1970).

15. G. A. Mokeeva, Opt. i. Spect. $\underline{32}$, 833 (1972).

16. J. F. Porter and H. W. Moos, Phys. Rev. $\underline{152}$, 300 (1966). H. W. Moos, J. of Luminescence $\underline{1-2}$, 106 (1970).

17. N. Geacintov, M. Pope , F. Vogel, Phys. Rev. Lett. $\underline{22}$, 593 (1969). R. E. Merrifield, P. Avakian, and R. P. Groff, Chem. Phys. Lett. $\underline{3}$, 155 (1969). R. P. Groff, P. Avakian, and R. E. Merrifield, J. of Luminescence $\underline{1-2}$, 218 (1970).

18. N. A. Tolstoi, A. P. Abramov, Opt. i Spect. $\underline{21}$, 171 (1966);
     $\underline{22}$, 501 (1967). N. A. Tolstoi, A. P. Abramov and I. N.
     Abramova, J. of Luminescence $\underline{1-2}$, 106 (1970).

19. V. V. Ovsyankin, P. P. Feofilov, in: Non-linear Optics, Ed.
     "Nauka" Press, Novosibirsk, 1968, p. 293.

20. C. A. Parker, Adv. in Photochemistry $\underline{2}$, 305 (1964).

21. S. K. Lower and M. El-Sayed, Chem. Rev. $\underline{66}$, 199 (1966).

22. S. P. McGlynn et al ., Photochem. & Photobiol. $\underline{8}$, 349 (1969).

23. J. F. Porter, J. Appl. Phys. $\underline{32}$, $\underline{825}$ (1961); Phys. Rev. Lett.
     $\underline{7}$,414 (1961). M. P. Brown and W. A. Shand, Phys. Rev. Lett.
     $\underline{11}$, 366 (1963); $\underline{12}$, 367 (1964); Phys. Lett. $\underline{8}$, 19 (1964); $\underline{11}$,
     218 (1964); IEEE J. Quant. Electron. $\underline{2}$, 251 (1966). V. L.
     Bakumenko et al. ZETP, Pisma $\underline{2}$, 27 (1965). L. Esterowitz and
     J. Noonan, Appl. Phys. Lett. $\underline{7}$, 281 (1965); Opt. Prop. of Ions
     in Solids, J. Wiley, N.Y., 1966, p. 485. M. P. Brown, W. A.
     Shand and J. S. S. Whiting, Brit. J. Appl. Phys. $\underline{16}$, 619 (1965).

24. N. Bloembergen, Phys. Rev. Lett. $\underline{2}$, 84 (1959).

25. V. V. Ovsyankin, P. P. Feofilov, ZETP, Pisma $\underline{3}$, 494 (1966).

26. V. V. Ovsyankin and P. P. Feofilov, Zhurn. Prikl. Spectr. $\underline{7}$,
     498 (1967). P. P. Feofilov and V. V. Ovsyankin, Appl. Opt. $\underline{6}$,
     1828 (1967).

27. W. D. Partlow, and H. W. Moos, Phys. Rev. $\underline{157}$,252 (1967). W.
     D. Gandrud and H. W. Moos, J. Chem. Phys. $\underline{49}$, 2170 (1968). C.
     M. Verber, D. R. Grieser and W. H. Jones, Jr., J. Appl. Phys.
     $\underline{42}$, 2767 (1971).

28. L. D. Livanova, I. G. Saitkulov and A. L. Stolov, Opt. i
     Spectr. $\underline{25}$, 609 (1968).

29. V. V. Ovsyankin and P. P. Feofilov, ZETP, Pisma $\underline{4}$, 471 (1966);
     Opt. i Spect. $\underline{31}$, 944 (1971).

30. F. Auzel, C. R. Acad. Sci. $\underline{262B}$, 1016 (1966); $\underline{263B}$, 819 (1966).
     F. Auzel and O. Deutschbein, Z. Naturforschung $\underline{24a}$, 1562 (1969).

31. L. D. Livanova, I. G. Saitkulov and A. L. Stolov, Fiz. Tv.
     Tela.$\underline{11}$, 918 (1969).

32. S. V. Galginaitis and G. E. Fenner, Proc. Intern. Conf. on
     GaAs. Dallas, USA, Oct. 1968, p. 131. J. D. Kingsley, G. E.

Fenner and S. V. Galginaitis, Appl. Phys. Lett. 15, 115 (1969).
L. F. Johnson, J. E. Geusic, H. J. Guggenheim, T. Kushida, S.
Singh and L. G. Van Uitert, Appl. Phys. Lett. 15, 48 (1969).
H. J. Guggenheim and L. F. Johnson, Appl. Phys. Lett. 15, 51
(1969). L. G. Van Uitert, H. J. Levinstein and W. H. Grod-
kiewicz, Mat. Res. Bull. 4, 381 (1969). L. G. Van Uitert, L.
Pictroski and W. H. Grodkiewicz, Mater. Res. Bull. 4, 777
(1969). M. Tamatani, K. Yokota and T. Nishimura, J. Phys. Soc.
Japan 29, 1099 (1970). J. L. Sommerdijk,W. L. Wanmaker and
J. G. Verriet, J. Luminescence 4, 404 (1971). J. L. Sommerdijk,
J. Luminescence 4, 441 (1971). Y. Seki and Y. Furukawa, Japan.
J. Appl. Phys. 10, 1293 (1971). P. N. Yocom, J. P. Wittke and
I. Ladany, Met. Trans. IAME 2, 763 (1971). J. E. Geusic, F.
W. Ostermayer, H. M. Marcos, L. G. Van Uitert, J. P. van der
Ziel, J. Appl. Phys. 42, 1958 (1971). M. P. Low and A. L.
Major, J. Luminescence 4, 357 (1971). E. Ya. Arapova et al.
Opt. i Spect. 32, 435 (1972).

33. V. V. Ovsyankin, Opt. i Spect. 28, 206 (1970).

34. F. W. Ostermayer, Jr. Metallurg.   Trans. 2, 747 (1971). P.
     N. Yocom, J. P. Wittke and I. Ladany, Metall. Trans., 2, 763
     (1971).

35. V. V. Ovsyankin and P. P. Feofilov, Doklady Acad. Nauk SSSR
     174, 787 (1967)

36. V. V. Ovsyankin and P. P. Feofilov, in: Proc. IX Intern. Conf.
     on Phys. of Semicond., Moscow, 1968, Ed. "Nauka" Press,
     Leningrad, 1968, Vol. 1, p. 237.

37. V. V. Ovsyankin and P. P. Feofilov, ZETP, Pisma 14, 548 (1971).

38. V. V. Ovsyankin and P. P. Feofilov, Biophysica 15, 589 (1970).

39. V. V. Ovsyankin and P. P. Feofilov, in: Molecular Photonics,
     Ed. "Nauka" Press, Leningrad, 1970, p. 86.

INTERMOLECULAR RELAXATION-FLUCTUATION PROCESSES

AND LUMINESCENCE SPECTRA OF SOLUTIONS

N. G. Bakhshiev

State Optical Institute, Leningrad, USSR

ABSTRACT

A brief review of the main experimental results and ideas
pertinent to the intermolecular relaxation-fluctuation processes
characteristic of the luminescence of solutions is given.

The last decade has witnessed notable achievements in the
study and interpretation of the general laws of solvent effects
on the luminescence spectra of organic molecules (1-7). These
achievements are the result of the analysis both of the
peculiarities of luminescence itself and of the mechanisms
underlying the appearance of various intermolecular processes in
radiation characteristics. Three main factors in this analysis
deserve detailed consideration.

First, valid concepts considering all fundamental types
of Van der Waals forces were developed. These concepts deal not
only with the classification of the forces mentioned (orientation,
induced, dispersion etc.) but also with the justification of the
methods for summation and averaging of dipole-dipole potentials
over configuration and space, i.e. the methods of derivation of
expressions for general "molecule-environment (solvate shell)"
potential. Wide application of the modern concepts of liquid
state physics (Onsager-Böttcher theory (7), for example) in
solution spectroscopy has played and still plays a special role.

Second, consideration of the spectroscopic aspect of the
problem consists of the understanding of the particular importance
of the Franck-Condon principle for intermolecular processes.

Unless this factor is taken into account it is practically impossible either to analyze in terms of the newest data the numerous experimental facts obtained while studying the absorption and emission spectra of condensed substances, or to develop the theory of these phenomena. Suffice it to say that it is in terms of the Franck-Condon principle that the analysis of the probabilities and rates for various intermolecular relaxations in "molecule solvate shell" systems after optical excitation can be carried out (7).

Third, a factor to which spectroscopists have paid special attention only recently is the necessity to account for the statistical nature of intermolecular forces. This factor and the fluctuation of all physical characteristics of the liquid resulting from it accounts for the spectral inhomogeneity inherent in experimentally observed absorption and emission spectra and represents the superposition of "elementary" spectra corresponding to molecules with different "molecule-environment" potentials.

Molecular rebuilding occurring in the "molecule-solvate shell" system after optical excitation is governed by the correlation between radiation lifetime τ_f and environment relaxation time τ_R. The main cases possible here are as follows: 1) $\tau_f \gg \tau_R$ - all molecular relaxations are completed within the time τ_f, i.e. the molecule at the moment of deexcitation is in thermodynamic equilibrium with all modes of the "molecule-solvate shell" system; 2) $\tau_f \sim \tau_R$ - molecular relaxations are realized only in part with the result that at the moment of emission the system is not in thermodynamic equilibrium; 3) $\tau_f \ll \tau_R$ - the case of extreme nonequilibrium, for the relaxations (orientational and translational in particular) do not proceed at all.

All of these statements are illustrated in Fig. 1, where the general pattern of electronic states and luminescence spectra is shown for the "molecule-solvate shell" system for all possible cases of correlation between τ_f and τ_R ($\overline{W}_i$ is the mean intermolecular potential values for corresponding energy states). Here E_g^o and E_e^o - are free molecule levels, E_g and E_e(F-C) - ground-state and Franck-Condon excited levels of the molecule in solution (absorption), E_e - excited equilibrium level (total relaxation, i.e. $\tau_f \gg \tau_R$), E_e' - excited nonequilibrium levels (for partial relaxation, $\tau_f \sim \tau_R$; for no relaxation, $\tau_f \ll \tau_R$), E_g(F-C) are ground-state Franck-Condon levels corresponding to initial levels E_e and E_e' (emission). It should be emphasized that in the systems considered the contributions of both enthalpy (interaction potential energy) and entropy (orientational effects of solvent molecules) are taken into account (7,9).

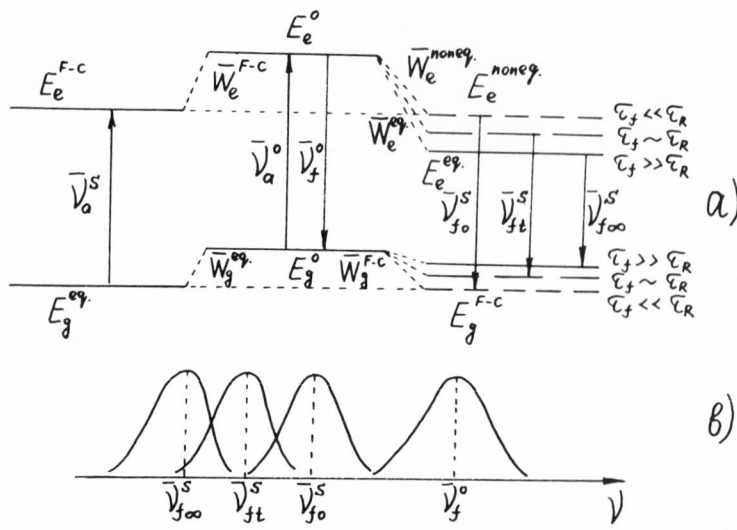

Fig. 1   The dependence of the electron states (a) and schematic
         pattern of the emission spectra (b) upon the relation
         between τ_f and τ_R (7,9) for a molecule in the gas phase
         or in solution.

     To give a complete picture of the influence of intermolecular
forces on spectra and positions of energy levels, it is necessary
to consider the fluctuation distribution of molecules in question
arising from the "molecule–solvate shell" interaction potential.
The case of total relaxation ($\tau_f \gg \tau_R$) is illustrated in Fig. 2,
where solid lines indicate the levels corresponding to the mean
values of intermolecular interaction potentials and dotted lines
show the levels corresponding to positive and negative fluctuations
of the potentials. The experimentally observed (superpositional)
and "elementary" (dotted line) spectra are schematically shown
in the same figure. The peculiarity of the pattern shown in
Fig. 2 lies in the fact that if $\tau_f \gg \tau_R$ the fluctuation distribution
functions for equilibrium ground $\rho_g(W_g^{eq})$ and excited $\rho_e(W_e^{eq})$ states
are different and independent. However, if $\tau_f \ll \tau_R$, the fluctuation
distribution is fixed and always corresponds to the ground state
of the molecule under investigation (7–9).

     The treatment of the data on solution luminescence shows
the approach developed to be in satisfactory agreement with
experiment. It enables us to interpret in terms of a unifying
theory many observations made on different systems under various
experimental conditions. It is enough to say that the validity of
these concepts has been confirmed for a wide range of molecules
differing not only in chemical nature and structure but also in
transition type, and that the concepts work well for various

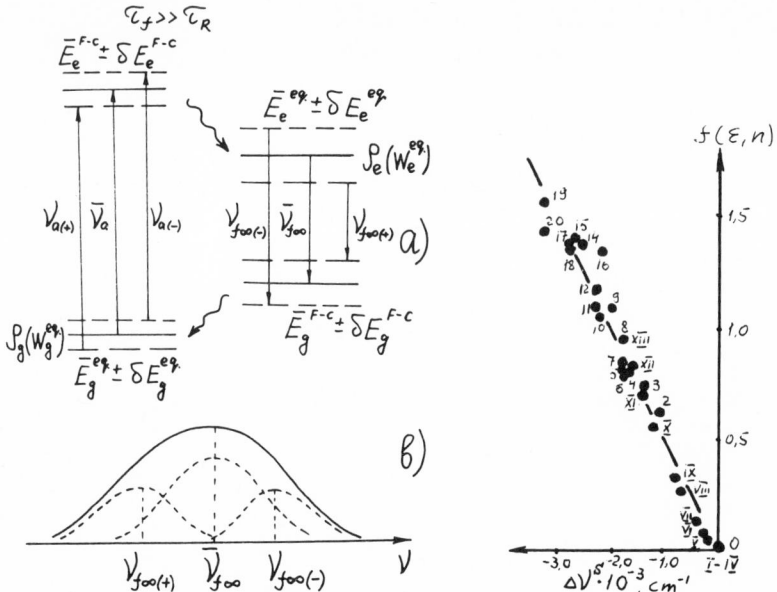

Fig. 2   The electron states (a) and emission spectra (b) of
         molecules in solution, showing the influence of
         fluctuation processes (the case of $\tau_f \gg \tau_R$) (7,9).

Fig. 3   The dependence of the shift of symmetry frequency ν^s
         on dielectric constant ε and refractive index n in the
         spectra of 3-monomethylaminophthalimide (7,10).

solvent media at widely varying temperatures and in any state of
aggregation ((7) and literature cited).

     The results obtained for the case of $\tau_f \gg \tau_R$ (liquid solutions)
explain the differing susceptibilities of absorption and
luminescence spectra due to solvent effects. The explanation,
which follows directly from the pattern in Fig. 1, is worth special
mention. The substantiation of the well known dependence of spectra
on solvent dielectric constant ε and refractive index n is also
of great importance. For example, Fig. 3 shows the typical results
obtained from the treatment of the experimental data using the
"universal-interaction-function" method (4,7) (Arabic numerals
designate the various solvents at room temperature; one of the
solvents (diethyl ether) at various temperatures and concentrations
(10) is designated by Roman numerals). The points are easily seen
to group around the theoretically predicted straight line
independently of the way in which the independent variables ε and
n are varied.

     From the analysis of correlations such as the one shown in

Fig. 3, effective methods of obtaining the characteristics of an
excited molecule from spectroscopic data are now derived. Some
results of these methods are shown in Table 1 (7).

Table 1

Dipole moment μ_e and polarizability α_e values
of some molecules in the first excited
electron state

Experimental technique / Molecule	μ_e (D)		
	I	II	III
4-amino-4-nitrodiphenyl	16.5	23	22.5
2-amino-7-nitrofluorene	19.5	23	19
4-dimethylamino-4-nitrostilbene	25	26.5	25
Experimental technique / Molecule	$\alpha_e \cdot 10^{24}$ cm3		
	I	IV	V
Benzene	12.5	–	12-13
Anthracene	41	47	–

(I) solvatochromic phenomena; (II) electric dichroism;
(III) polarized fluorescence; (IV) spectral electric field effect;
(V) theory.

Dipole moments μ_e and polarizabilities α_e determined from these
data are in good agreement with those determined by other methods.
Both the reliability of the μ_e and α_e values and the validity of
the physical premises underlying this method are confirmed
(including those based on solvatochromism theories).

The spectral peculiarities observed in the $\tau_f \sim \tau_R$ range
(viscous solutions) are also quite interesting. For example,
typical curves of the temperature dependence of solution
absorption and fluorescence spectra are shown in Fig. 4. It
is seen that the theory satisfactorily explains all of the main
features of this phenomenon (7). It should be emphasized that
this conclusion is confirmed by the direct investigation of the
kinetics of "instantaneous" solution luminescence spectra
obtained by means of nanosecond spectroscopy (11). The
interpretation of the changes in polarizational and temporal
luminescence spectra parameters (for the case of $\tau_f \sim \tau_R$) based on
the concepts developed is also worth mentioning (12).

Among the results for $\tau_f \ll \tau_R$ range (rigid solutions), the
quantitative interpretation of the failure of universal
correlation between absorption and fluorescence spectra should
be noted. This failure is illustrated by the observation that

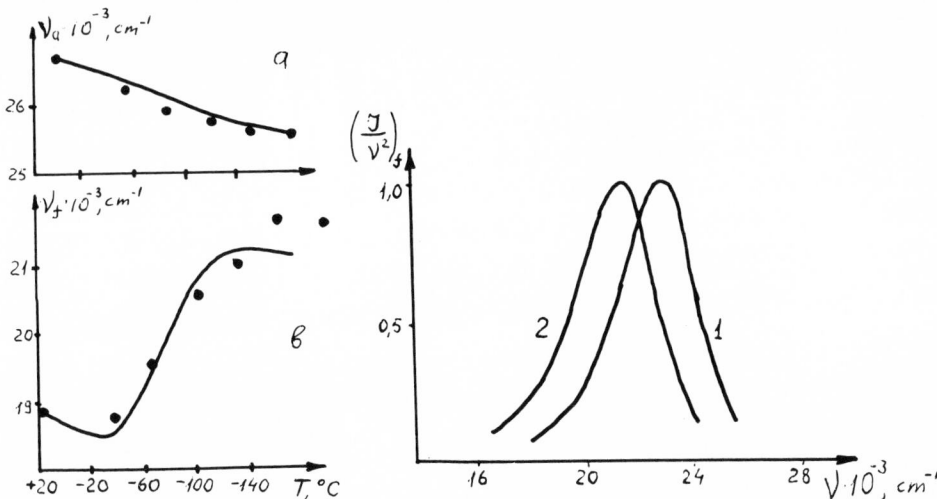

Fig. 4 Temperature dependence of the shift of the maxima
        of absorption (a) and fluorescence (b) spectra for
        4-aminophthalimide in propyl alcohol (7). Curves are
        calculated, points are from experimental data.

Fig. 5 The dependence of luminescence spectra of 3-amino-N-methyl-
        phthalimide in liquid binary solvent (hexane + alcohol)
        on the exciting light wavelength at the temperature -45°C.
        λ_{exc}(nm); 1, 365 nm; 2, 405 nm.

the temperature T_1 obtained from the universal correlation differs
markedly from the mean temperature of the whole solution T.
According to (13) this failure is caused by orientational
nonequilibrium between the excited molecule and its solvate shell.
In a rigid solution this nonequilibrium remains up to the moment
of emission. Another significant experimental fact based on
fluctuation processes is so called "bathochromic luminescence" (8),
which is the dependence of the emission spectra of rigid solutions
on excitation frequency. According to the principles stated above,
this phenomenon results from the fact that under anti-Stokes
excitation, if $\tau_f \ll \tau_R$, the molecules with longwave "elementary"
spectra are the ones that absorb, and hence radiate, light (see
Fig. 2).

    Let us outline briefly some of the newest data revealing the
predictive possibilities of the concepts developed. These data
show certain effects observed in the luminescence spectra of
solutions in liquid binary solvents. There are some reasons to
believe that in such systems the failure of universal correlation
is to be expected at a certain ratio of mixed solvent components.
Investigations have shown that the failure expected does take

place (see Table 2). It arises from the concentration gradient between an excited molecule and its solvate shell (14).

Table 2

T_1 Temperature Values, Obtained from Universal
Correlation, for 3-Aminophthalimide in Mixed Solvent
(Heptane (I) + Propyl Alcohol (II))

Temperature of experiment - 300°K

Binary solvent composition	T_1 (K)
I - 100 per cent	315
I + 0.1 vol per cent II	480
I + 0.5 vol per cent II	435
I + 8.0 vol per cent II	360
I + 15 vol per cent II	290
II - 100 per cent	290

In addition, the dependence of luminescence spectra on exciting light frequency is possible for multicomponent liquid solutions. Such a dependence, being the consequence of the universal correlation failure, is caused by the relatively long duration of the translational (diffusional) intermolecular processes involved in the formation of the solvent shell around the excited molecule; these processes sometimes do not equilibrate within the time τ_f. This phenomenon, predicted on the basis of relaxation-fluctuation concepts, turned out to be experimentally observed. Some results of these experiments for the temperature -45C are shown in Fig. 5. Note that the effect is largest in the very range of binary solvent concentrations at which the failure of universal correlation is most pronounced. The universal correlation accounts well for individual solvents at the same temperature.

In conclusion, it can be said that this general approach to the problems of liquid state spectroscopy, taking into account the principal features of a given state of the solvent is highly promising for investigation of both the spontaneous and the stimulated emission of solutions (9,15).

REFERENCES

1) N. Mataga, Y. Kaifu, M. Koizumi, Bull. Chem. Soc. Japan, 29, 465, (1956).

2) E. Lippert, Zs. Electrochem. 61, 962, (1957).

3) E. McRae, J. Phys. Chem. 61, 562, (1957).

4) N. G. Bakhshiev, Opt. i Spectr. 10, 717, (1961); 16, 821
    (1964).

5) W. Liptay, Zs. Naturforsch. 20A, 1441, (1965).

6) Yu. T. Mazurenko, N. G. Bakhshiev, Opt. i Spectr. 28, 905,
    (1970).

7) N. G. Bakhshiev, Spectroscopy of Intermolecular Interactions,
    Nauka, Leningrad (1972).

8) A. N. Rubinov, V. N. Tomin, Opt. i Spectr. 29, 1082, (1970).

9) N. G. Bakhshiev, Opt. i Spectr. 32, 1151, (1972).

10) B. S. Neporent, N. G. Bakhshiev, V. A. Lavrov, S. M. Korotkov,
    Opt. i Spectr. 13, 321, (1962).

11) W. Ware, P. Chow, S. Lee, Chem. Phys. Lett. 2, 356 (1968).

12) N. G. Bakhshiev, Yu. T. Mazurenko, I. V. Piterskaya,
    Izv. Acad. Nauka, USSR,ser.phys. 32, 1360, (1968).

13) N. G. Bakhshiev, I. V. Piterskaya, V. I. Studenov,
    A. V. Altaiskaya, Opt. i Spectr. 27, 349, (1969).

14) N. G. Bakhshiev, I. V. Piterskaya, A. V. Altaiskaya, Opt. i
    Spectr. 27, 1013, (1969).

15) N. G. Bakhshiev, V. I. Studenov, Opt. i Spectr. 33, 115,
    (1972).

VISIBLE LUMINESCENCE OF Yb^{3+}, Er^{3+} UNDER IR EXCITATION

J. L. Sommerdijk and A. Bril

Philips Research Laboratories

Eindhoven, The Netherlands

ABSTRACT

Possible mechanisms are considered for exciting green and red emission with infrared in crystals doped with Yb^{3+} and Er^{3+}. Excitation spectra and decay times are used in evaluating mechanisms. The analysis is applied to α-NaYF$_4$:Yb^{3+},Er^{3+}.

INTRODUCTION

Yb^{3+}, Er^{3+}-doped compounds are of practical interest since they can convert the IR emission of a Si-doped GaAs diode to green and/or red light. The present work deals with the excitation and emission processes in Yb^{3+}, Er^{3+} systems. Two aspects are considered: (i) the excitation routes for the green and red emission and (ii) the decay times of the IR and visible emissions. The latter are discussed for the phosphor α-NaYF$_4$:Yb^{3+}, Er^{3+} which shows a high IR-to-green conversion efficiency.

EXCITATION MECHANISMS

In Fig. 1 the relevant energy levels of Yb^{3+} and Er^{3+} are depicted together with possible excitation mechanisms for the green and red emission. The IR radiation (1 μm) is absorbed mainly by Yb^{3+}. Excitation of Er^{3+} occurs via transfer of the excitation energy from Yb^{3+}. The green emission is generated via one route (1), whereas for the red emission several routes are possible. The most obvious routes (2) are denoted by a, b and c in Fig. 1. An earlier paper (3) explains how one obtains information about the relative importance of these

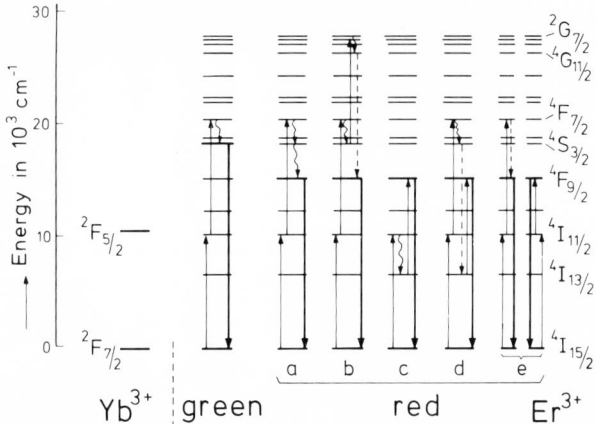

Fig. 1   Energy levels of Yb^{3+} and Er^{3+} and excitation routes of
the IR-excited green and red luminescence. Upward
pointing arrows: excitation of Er^{3+} via energy transfer
from Yb^{3+}, except for the transition $^4I_{11/2} \to ^4F_{9/2}$ in
route e, which occurs via transfer from another Er^{3+} ion
being in the $^4F_{7/2}$ state. Broken arrows: de-excitation
of Er^{3+} via energy transfer to Yb^{3+} (routes b and d) or
to Er^{3+} (route e). Wavy lines: non-radiative decay.
Downward pointing arrows: visible emissions.

routes by comparison of excitation and emission spectra under
IR and UV excitation and from the dependence of red intensity
on IR excitation density. We have found that in oxidic host
lattices route c is always dominant (>90%). The same applies
to fluoridic host lattices when the IR excitation density is
not higher than about 100 mW/cm^2. At higher densities, route
b also becomes important as is apparent from the super-quadratic
dependence of the red intensity on IR intensity. Routes d and
e were not considered in Ref. (3). In route d Er^{3+} is de-excited
from $^4S_{3/2}$ to $^4I_{13/2}$ due to coupling with the $^2F_{7/2} \to ^2F_{5/2}$
excitation of Yb^{3+}. A strong indication for this coupling is
that co-doping with Yb^{3+} results in a drastic decrease of the
decay time of the green Er^{3+} emission under cathode ray
excitation. If one assumes that the Yb^{3+} ion excited by the
transfer from Er^{3+}, transfers its energy back again to the
same Er^{3+} ion d_1, the contribution of d_1 cannot be distinguished
from that of a. From the above-indicated experiments it follows
then that the sum of the two contributions is less than 10% of
the total contribution to the red emission. If the excitation
energy of Yb^{3+} is not retransferred to the same Er^{3+} ion d_2, then
three IR photons are necessary for the population of $^4F_{9/2}$, leading
to a cubic dependence of this contribution on IR excitation just
as is the case of route b. Hence d_2 may be of importance when a

super-quadratic dependence of red intensity on IR power is observed.

Route e, suggested in Ref. (4), implies the interaction of two Er^{3+} ions, one in the $^4F_{7/2}$ state and the other in the $^4I_{11/2}$ state, resulting in two Er^{3+} ions both in the $^4F_{9/2}$ state and giving rise to two red photons. In our opinion e cannot be important. It would cause a sub-quadratic dependence of red intensity on IR density because three IR photons generate two red photons, and such a dependence has not been found experimentally. In addition, simultaneous excitation at both 0.5 μm (excitation to the $^4F_{7/2}$ level) and 1 μm did not result in a reduction of the green emission in comparison with the sum of the intensities measured at the separate excitations (4), as should be expected when route e is important.

In conclusion, we maintain that c is the dominant excitation route for the red emission. Only at high IR excitation densities in fluoridic lattices may routes b and/or d_2 become important.

DECAY TIMES OF α-NaYF$_4$:Yb^{3+},Er^{3+}

Information on the excitation and emission processes can also be obtained from the decay times of the IR and visible emissions. In this section we discuss the decay times measured for the system α-NaYF$_4$:Yb^{3+},Er^{3+}. The emission spectrum of this phosphor under IR excitation is shown in Fig. 2 together with that of the β-phase of NaYF$_4$:Yb^{3+},Er^{3+}. The α-phase emits mainly in the green and with high efficiency (5,6). The β-phase, on the other hand, shows a much lower (factor of 10) efficiency and emits mainly in the red. An explanation of this different behavior of the two phases is given elsewhere (7). In comparison with the best LaF$_3$ sample as described in Ref. (1) the green light output of the best α-NaYF$_4$ sample is about four times higher, when measured under the same IR excitation conditions. The decay times after IR and cathode ray (CR) excitation are summarized in Table 1.

Let us first consider the decay of the IR emission (1 μm). Due to the close match between the Yb^{3+} and Er^{3+} energy levels, the first Yb^{3+}, Er^{3+} energy transfer step is resonant, which means that the transfer rates in both directions are high with respect to the radiative rates of Yb^{3+} and Er^{3+}. If the Yb^{3+}→Er^{3+} and Er^{3+}→Yb^{3+} transfer rates are equal, the IR decay time τ_{IR} is given by Ref. (8) as: $\tau_{IR}=1/\{(f_y/\tau_y)+(f_e/\tau_e)\}$ where f_y and f_e are the relative fractions of Yb^{3+} and Er^{3+} ions and τ_y and τ_e are the IR decay times of the singly-doped Yb^{3+} and Er^{3+} samples, respectively. The above relation was indeed found for LaF$_3$ (8). Substitution of the values measured on α-NaYF$_4$ leads to $\tau_{IR}=2.2$ msec.

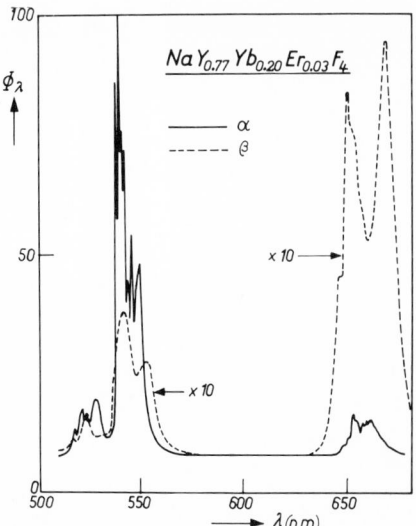

Fig. 2 Emission spectra of the α- and β-phases of NaYF₄:Yb³⁺,
Er³⁺ under IR excitation. Φ_λ denotes the spectral radiant power
in arbitrary units. IR excitation with a tungsten-iodine lamp
with appropriate filters; excitation density 50 mW/cm².

TABLE 1

Decay times after IR and CR excitation of α-NaYF₄ doped with 20%
Yb³⁺ and/or 3% Eu³⁺. IR excitation with a Si-doped GaAs diode
(2 msec. pulse of 50 mw/cm²). All decays were measured over
one decade and were found to be exponential. The values for the
Yb³⁺, Er³⁺ sample are averages over eight samples, the maximum
variations being ± 10%.

	Excitation	Emission	Decay Time
Yb³⁺	IR	IR	2.0 msec.
Er³⁺	IR	IR	12
Yb³⁺,Er³⁺	IR	IR	2.9
Er³⁺	CR	green	0.19
Yb³⁺,Er³⁺	CR	green	0.09
Yb³⁺,Er³⁺	IR	green	1.2
Yb³⁺,Er³⁺	CR	red	0.35
Yb³⁺,Er³⁺	IR	red	1.6

Experimentally, however, we find 2.9 msec. which indicates that
the transfer rates are unequal. Analysis of the rate equations
shows that this deviation (2.2 vs. 2.9 msec.) can be explained
if the ratio between the rates of the Yb³⁺→Er³⁺ and Er³⁺→Yb³⁺
transfers is about four. This difference between α-NaYF₄ and
LaF₃ can be explained as follows: The crystal structures of
α-NaYbF₄ and α-NaErF₄ are the same as that of α-NaYF₄ and the

ionic radii of Yb^{3+} (0.86 A) and Er^{3+} (0.88 A) are nearly equal
to that of Y^{3+} (0.92 A). The crystal structures of YbF_3 and ErF_3,
on the other hand, are quite different from that of LaF_3 and
the radius of La^{3+} (1.14A) is much larger than that of $Yb3+$ and
Er^{3+}. The incompatibility of the LaF_3 lattice with the YbF_3 and
ErF_3 lattices causes a broadening of the energy levels of 10,000 cm^{-1}
of both Yb^{3+} and Er^{3+}. This broadening results in a better overlap
of the Yb^{3+} and Er^{3+} levels so that the $Er^{3+} \rightarrow Yb^{3+}$ transfer is
promoted, in contrast with α-$NaYF_4$ where such a broadening does
not occur. A consequence of the weaker $Er^{3+} \rightarrow Yb^{3+}$ transfer in
α-$NaYF_4$ in comparison with that in LaF_3 is that the IR excitation
energy stays relatively long at the Er^{3+} ions. This may be a
qualitative explanation of the higher efficiency in α-$NaYF_4$ with
respect to that in LaF_3.

The decay times of the green and red emission after IR
excitation (τ_g', τ_r') are much longer than those after CR excitation
(τ_g, τ_r). The latter are determined by the de-population of
$^4S_{3/2}$ and $^4F_{9/2}$. τ_g' is determined not by the de-population of
$^4S_{3/2}$ but by the combined de-population of the $^2F_{5/2}$ level of Yb^{3+}
and the $^4I_{11/2}$ level of Er^{3+}, as governed by τ_{IR}. Since the green
emission intensity is proportional to the product of the populations
of the latter Yb^{3+} and Er^{3+} levels, τ_g' is given by 1/2 τ_{IR}, in
reasonable agreement with experiment. Furthermore, τ_r' is expected
to be longer than τ_g', since the long-living (12 msec.) $^4I_{13/2}$
level is involved in the main excitation route of the red emission
(route c). This is also found experimentally for α-$NaYF_4$:Yb^{3+},
Er^{3+} and for all other Yb^{3+},Er^{3+} systems studied. A quantitative
value for τ_r' is not feasible at the moment.

The reduction of τ_g due to co-doping with Yb^{3+} is attributed to
the $^4S_{3/2} \rightarrow ^4I_{13/2}$ transition coupled with the $^2F_{7/2} \rightarrow ^2F_{5/2}$ excitation
of Yb^{3+}. This coupling limits the efficiency of the green emission.
For the optimum Yb^{3+} concentration (20%), τ_g is reduced by a
factor of two, implying that the probability of the $Er^{3+} \rightarrow Yb^{3+}$ transfer is about equal to that of the decay within the Er^{3+} ion, which
is to be expected for the optimum Yb^{3+} concentration. At higher
Yb^{3+} concentrations the coupling becomes even stronger, resulting
in a lower efficiency of the green emission. As pointed out
above, this coupling is also involved in excitation route d of
the red emission. The contribution of d is, however, negligible
with respect to that of c because of the much lower population
density of $^4S_{3/2}$ compared to that of the IR reservoir. Only a
high IR excitation power (1 to 10 W/cm^2) may the two population
densities become comparable and then route d may compete with
route c.

ACKNOWLEDGEMENTS

Mr. R. E. Breemer and Mr. J. A. de Poorter are thanked for their experimental assistance.

REFERENCES

1. R. A. Hewes, J. Luminescence 1,2 778 (1970).

2. L. G. van Uitert, H. J. Levinstein and W. H. Grodkiewicz, Mater. Res. Bull. 4, 381 (1969).

3. J. L. Sommerdijk, J. Luminescence 4, 441 (1971).

4. J. P. Wittke, I. Ladany and P. N. Yocom, J. Appl. Phys. 43, 595 (1972).

5. G. Blasse and A.D.M. de Pauw, French Patent Specification Nr. 2107248 (application 3-9-1970).

6. T. Kano, H. Yamamoto and Y. Otomo, Spring Meeting Electrochemical Society, Abstract Nr. 82 (1972).

7. J. L. Sommerdijk, J. Luminescence, to be published.

8. J. D. Kingsley, J. Appl. Phys. 41, 175 (1970).

COOPERATIVE PROCESSES IN ACTIVATED GLASSES

B. M. Antipenko, A. V. Dmitruk, V. S. Zubkova, G. O.

Karapetyan and A. A. Mak

Institute of Exact Mechanics and Optics, Leningrad, USSR

ABSTRACT

We have investigated cooperative processes in activated
glasses, which are the result of the interaction of two or more
ions in excited states. It has been shown that in these glasses
(activators Yb-Tb, Yb-Eu) accumulation of energy occurs both by
cooperative sensitization and by combinative excitation. Quantitative characteristics of cooperative processes have been measured, i.e.: (a) dynamic-phenomenological constant for the rate
of cooperative interaction, (b) energetic-quantum yield of cooperative luminescence. The functional dependence of the efficiency
of cooperative processes on the power and energy of excitation,
structure and composition of the glass, temperature, and ratio of
the concentration of activators have also been investigated.

A visible glow can be excited by the IR Nd-laser irradiation
in glasses activated both with the $Tb^{3+} + Yb^{3+}$ and $Eu^{3+} + Yb^{3+}$
ions. The spectral and kinetic investigations of the excited
glow showed that it was the luminescence of the Tb^{3+} ions corresponding to transitions from the $^5D_4$ level to the $^7F_{6-0}$ level and
the luminescence of the Eu^{3+} ions related directly to transitions
from the $^5D_0$ level to $^7F_{0-6}$ level. In this paper we consider the
Tb^{3+} and Eu^{3+} luminescence excitation anti-Stokes mechanisms.

In Fig. 1 we show the oscillograph traces of the luminescence
signals of the Tb^{3+} (a, b, c) and Eu^{3+} (d,e,f) ions with the
samples being excited by the Q-switched Nd-laser ($t_r = 4x10^{-8}$ sec.,
$W_r = 0.35J$) (their compositions are given in the legend to Fig. 1).
Referring to Figure 1, the anti-Stokes excitation occurs at least

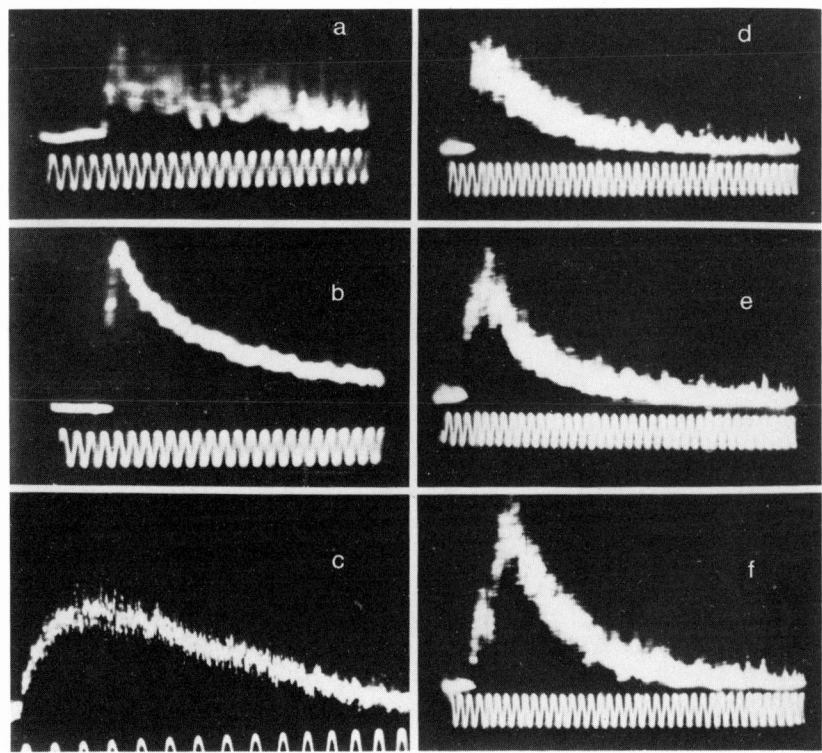

Fig. 1 Oscillograph traces of the luminescence of Tb^{3+} ions (a,b,
c) under monopulsed excitation (t_r = 4 x 10^{-8} sec, λ =
1.06μ) of the glasses: (a) 10 mole percent Yb^{3+} + 5 mole
percent Tb^{3+} in 35 percent SiO_2 + 25 percent B_2O_3 + 2 percent P_2O_5 + 1 percent K_2O + 12 percent BaO (matrix B) with
50 μsec marks (W_r = 0.07 J); (b) same glass as (a) but
W_r = 0.35 J; (c) 0.6 mole percent Yb^{3+} + 1.1 mole percent
Tb^{3+} in 15 percent Na_2O + 52.1 percent SiO_2 + 31.2 percent
B_2O_3 (matrix A) with 100 μsec marks (W_r = 0.35 J).   Traces
for Eu^{3+} ions (d,e,f) under monopulsed excitation (W_r =
0.35 J, t_r = 4 x 10^{-8} sec, λ = 1.06μ) of a glass with 2
mole percent Yb^{3+} and 4 mole percent of Eu^{3+} (matrix B);
(d) T = 293K; (e) T = 393K; (f) T = 493K with 50 μsec
marks.

by two mechanisms distinct in rates, efficiency, and induction
period. Excitation by the first mechanism (hereafter referred to
as "fast") occurs during the laser irradiation action, the photon

field of which in this manner is an immediate energy source for
the excitation. The length of the second mechanism (hereafter
called "slow") is more extended than the laser irradiation duration and approximately equal to the Yb^{3+} luminescence lifetime.
The last-mentioned circumstance points to the fact that in this
case the Yb^{3+} ions excited by the laser irradiation are the excitation energy source.

In a specific instance, when the experimental conditions are
varied, the distinct nature of the physical processes fundamental
to the slow and fast anti-Stokes excitation mechanisms enables one
of them to be prevailing and studied in its pure state. Thus a
differentiating factor for Eu^{3+} - Yb^{3+} glasses is the sample
temperature, in this case the activation energy of the fast process
(~ 400 cm^{-1}) being much lower than that of the slow one (~ 1200 cm^{-1}).
In the case of Tb^{3+} - Yb^{3+} glasses the ratio of the time required
for the encounter of two migrating Yb^{3+} excitations to the spontaneous Yb^{3+} decay time is the determining factor.

The attempted experimental investigation of the slow mechanism showed that the efficiency of the Tb^{3+} and Eu^{3+} ion excitation ($n'_{Tb,Eu}/W_r$) mechanism is determined by the energy of the
laser radiation rather than by its power, providing $t_r < \tau_{Tb}$.
Also it depends quadratically both on concentration of the excited
Yb^{3+} and the exciting light intensity, and linearly on $Tb^{3+}(Eu^{3+})$
concentration. On the basis of the data obtained the rate of the
excited Tb^{3+} and Eu^{3+} ion formation by means of this mechanism may
be phenomenologically written in the following form:

$$\alpha(N^o_{Tb,Eu}, N^o_{Yb}) \cdot (n'_{Yb})^2 \cdot N^o_{Tb,Eu} \, ,$$

where α is a rate constant weakly dependent on the chosen activator pair concentration (1) and the prime denotes an electronic
excited state. But such a rate expression is used for the description of cooperative sensitized luminescence (2) (Fig. 2b-I), the
process being thus fundamental for slow excitation mechanism. Of
particular interest was to evaluate some quantitative parameters
of the cooperative interaction, e.g. the phenomenological rate
constant α and quantum yield $\eta_{coop.} = \eta_{Tb,Eu} \cdot (n_{Tb,Eu}/n'_{Yb})$, which
are the features of its dynamics and efficiency. For the samples
investigated the parameters are as follows: for glasses containing 0.6 mole percent of Yb^{3+} and 1.1 mole percent of Tb^{3+} (matrix
A) $\alpha \simeq 10^{-39}$ cm^6/sec., $\eta \simeq 10^{-7}$, when $n'_{Yb} \simeq 10^{16}$; for glasses containing 2 mole percent of Yb^{3+} and 4 mole percent of Eu^{3+} (matrix
B) $\alpha \simeq 10^{-41}$ cm^6/sec., $\eta \simeq 10^{-8}$, when $n'_{Yb} \simeq 7 \times 10^{16}$. From these
data one can easily see that the cooperative energy transfer from
$2Yb^{3+}$ to Tb^{3+} is much more probable than to Eu^{3+}. It depends upon
the glass structure influence on the cooperative interactions
intensities (matrix A favors activator segregation much more than

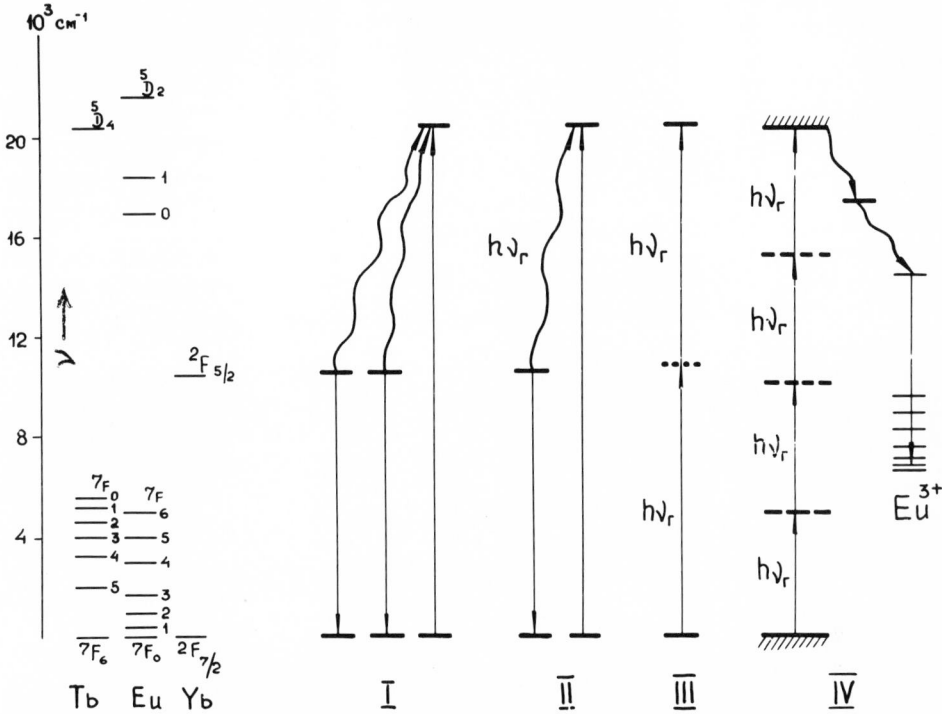

Fig. 2   (a) Empirical energy level scheme of the Tb^{3+}, Eu^{3+} and
          Yb^{3+} ions; (b) anti-Stokes excitation mechanism of the
          Tb^{3+} and Eu^{3+} ions in the glasses under Nd-laser irradia-
          tion: I. cooperative sensitization, II. combination exci-
          tation, III. two-photon absorption, IV. recombination
          mechanism of excitation of the Eu^{3+} ions with matrix zones.

matrix B (1)), and perhaps also on the resonance of initial and
final states of energy transfer, which is better in the case of
Tb^{3+}.

     The encounter of two excited Yb^{3+} ions is required for the
cooperative sensitization process to occur. The cooperative
sensitization efficiency may vary over a wide range by reducing
encounter yields both by lowering the density of the excited Yb^{3+}
(exciting light intensity), and by increasing their spontaneous
decay rate (a quencher of the Yb^{3+} excited states being injected
or Yb^{3+} concentration quenching being varied). This was the case
in Fig. 1, a,b,c.

     As illustrated in Fig. 1, different Tb^{3+} and Eu^{3+} excitation
luminescence processes activated by laser radiation compare
favorably with the cooperative sensitization process in efficiency.
Our investigation revealed that the efficiency of anti-Stokes

excitation activated by fast processes depends quadratically on
Nd-laser irradiation intensity. The two-photon transitions
(Fig. 2b-III) between the states of the same parity are allowed by
the selection rules, therefore at laser intensities they can play
a significant part in excitation of both the Tb^{3+} and Eu^{3+} ions.

Under our experimental conditions, i.e. at an exciting light
intensity of $\sim 3 \times 10^{15}$ quantum/cm^3) and a registration threshold of
$\sim 5 \times 10^7$ quantum/sec. it has been possible to observe only the Eu^{3+}
luminescence excited by the two-photon transition between the $^7F_1$
and $^5D_1$ states of Eu^{3+} in the glasses activated with Eu^{3+} or Tb^{3+}.
The energy gap of the $^7F_1$ and $^5D_1$ states is exactly equal to
$2h\nu_\Gamma$ and among other things the activation energy of the process
(400 cm^{-1}) coincides very closely with the $^7F_1$ state energy of the
Eu^{3+} ground multiplet. In the case of Tb^{3+} ions the probability
of the two-photon transition at $h\nu = 9400$ cm^{-1} is very small, however it may be abruptly increased to produce the cooperative process giving a real Yb^{3+} ($^2F_{5/2}$) level in the forbidden band of
Tb^{3+} energies. The two-photon process considered (Fig. 2b-II),
described as a combination excitation (2) is differentiated from
the usual one by the fact that in the process one of the intermediate transitions associates the states relative to dissimilar
ions (in this case $^5F_{5/2}(Yb)$——$^5D_4(Tb)$). Approximately 10^{10} Tb^{3+}
ions have been excited by two-photon processes with a 0.35 J mono-
pulse during 4×10^{-8} sec. in a glass with 1.2×10^{21} Tb^{3+} and
3.5×10^{21} Yb^{3+} ions (matrix B) and approximately 10^9 Eu^{3+} ions have
been excited in the glass with 1.4×10^{21} Eu^{3+} ions and 7×10^{20} Yb^{3+}
ions (matrix B). Yb^{3+} addition to the glasses activated with Eu^{3+}
does not significantly influence the efficiency of the two-photon
process.

In conclusion we shall consider another excitation mechanism
of anti-Stokes Eu^{3+} luminescence in the glasses coactivated with
Eu^{3+}, Yb^{3+} and Eu^{2+} ions (ca. 10 percent of Eu overall content).
Monopulsed irradiation of such glasses gave rise to effective (one
and a half orders of magnitude larger than in the case of two-
photon process) and non-inertial (with an accuracy of 3 μsec.)
Eu^{3+} luminescence excitation. Some preliminary experiments suggested the four-photon process to be the excitation mechanism.
Considering that four photons of laser irradiation are sufficient
for matrix excitation into the fundamental absorption band (3) and
that the effect revealed shows itself only in the presence of Eu^{2+},
the following working hypothesis may be supplied for its explanation: a pair of free carriers appears in the matrix zones under
laser irradiation which then recombines with the assistance of the
Eu^{2+}, luminescence being observed in the recombination process.

Thus the study carried out points to the fact that a dis-
ordered glass structure does not preclude cumulative processes

such as cooperative sensitization and combination excitation taking place therein, the cumulative process efficiency being comparable to the two-photon absorption efficiency with the participation of a virtual level.

REFERENCES

1. B. M. Antipenko, A. V. Dmitruk, V. S. Zubkova, G. O. Karapetyan, A. A. Mak, N. V. Mikhailova, Optika i Spectr. (in press).

2. V. V. Ovsyankin, P. P. Feofilov, in "Spectroscopy of Crystals" (1970).

3. G. O. Karapetyan, thesis to a Chem. Doctor's degree, Leningrad, 1967 (to be published).

MECHANISMS IN THE PRODUCTION OF THE DELAYED FLUORESCENCE OBSERVED IN AROMATIC AMINES TRAPPED IN RIGID GLASSES*

M. Ewald**, D. Muller*** and G. Durocher

Département de Chimie

Université de Montréal

Montréal, Canada

ABSTRACT

The delayed fluorescence of several aromatics has been studied in a series of rigid organic glasses at 77 K. Buildup and decay kinetics were investigated. All systems showed photoionization of the solute; the matrix quenches recombination delayed fluorescence and triplet-triplet annihilative delayed fluorescence is observed.

A systematic study of the different types of delayed fluorescences (DF) has been carried out in our laboratory. The triplet-triplet annihilation mechanism was shown to be responsible for the delayed fluorescence observed in the proflavine cation (1) and in the tryptophan molecule (2) trapped in an ethanol rigid glass matrix at 77 K, when excited in the first electronic absorption band system. On the other hand, consecutive biphotonic ionization has been studied in the TMPD (tetramethylparaphenyl-enediamine) molecule trapped in a 3-methyl pentane glass (3) by way of the observation of the isothermal recombination delayed fluorescence through a phosphoroscope.

The purpose of this paper is to point out a case where both the above mentioned mechanisms are effective in explaining the delayed fluorescence observed. Several aromatic amines including diphenylamine, carbazole, and tryptophan are studied but this paper will be concerned only with diphenylamine experimental results.

All emission spectra were obtained by the use of an Aminco-Bowman spectrophosphorimeter equipped with an ellipsoidal condensing system. Delayed fluorescence is observed through a Corning 7-60 filter. A photomultiplier 1P28 is used at the observation. The xy recorder has a response time of 0.25 sec. Details are given elsewhere (4).

A diphenylamine (DPA) solution in an EPA glass (1.5×10^{-3}M) at 77 K shows a delayed fluorescence (DF) with the following characteristics:

(1) The DF spectrum is identical in shape to the normal fluorescence (F). The intensity ratio I_{DF}/I_F is approximately 10^{-4}.

(2) The DF and the phosphorescence (P) intensities are related to the exciting light intensity (I) by the following equations:

$$I_{DF} \propto I^{2.0\pm0.1}$$

$$I_P \propto I^{1.0\pm0.1}$$

(3) While irradiating the matrix, a slow formation of DPA cation-radical is found and characterized by its absorption spectrum at 680 nm. (6).

(4) The same observations (1), (2) and (3) have also been made when using an ether or a 3-methyl pentane glass as the trapping matrix. The trapped electrons have been observed by direct absorption or by recording the spectra of the stimulated delayed fluorescence in the visible.

All these results show that at least a part of the DF is due to recombinations between cations and electrons initially produced by consecutive absorption of two photons by the molecule. These results agree with those of Lewis and Lipkin (6) who were the first to show that the DPA in EPA could be photoionized and could then give rise to an isothermal delayed luminescence at 77 K when an electron comes back into the positive hole. Linschitz, Berry and Schweitzer (7) in 1954 showed that the isothermal luminescence observed in the DPA molecule trapped in EPA is made of the phosphorescence of the solute molecule. We want here to show this luminescence is also composed of the solute fluorescence. A similar result has recently been obtained on other systems by Deroulede (8).

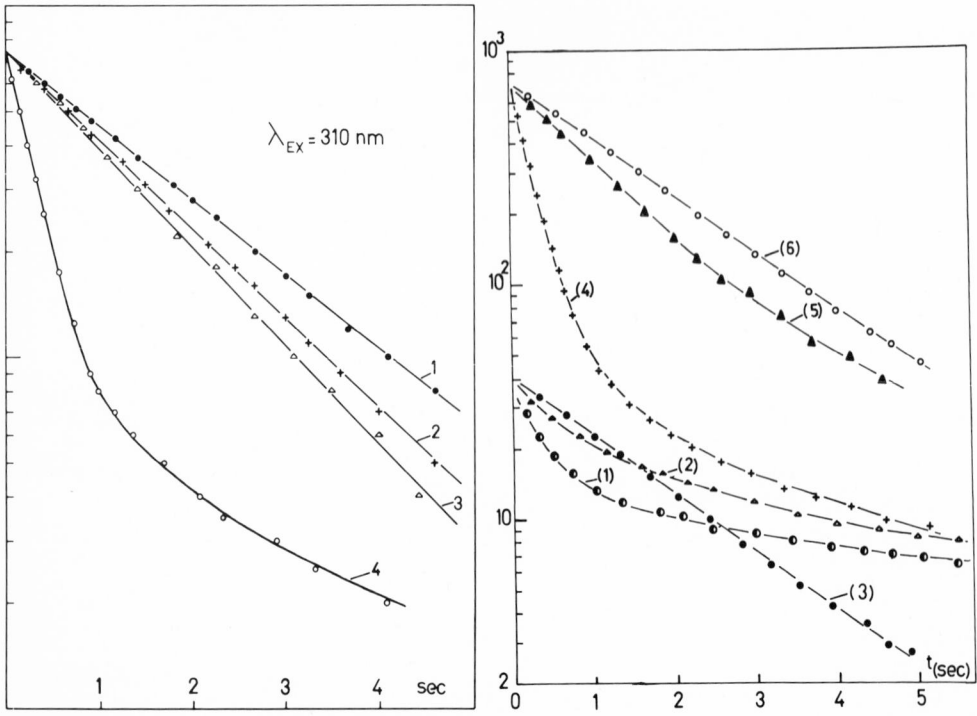

Fig. 1   DPA, 1.5 X 10^{-3}M, EPA, 77 K, λ_{ex} = 310 nm.
         Rise (R) and decay (D) of the luminescence observed
         through a phosphoroscope:
              1. D_P, observed at 400 nm;
              2. R_P, observed at 400 nm;
              3. R_{DF}, observed at 350 nm;
              4. D_{DF}, observed at 350 nm;
              arbitrary units.

Fig. 2   DPA, 1.5 X 10^{-3}M, 77 K, λ_{ex} = 300 nm:
         In 3MP alone
              1. D_{DF}, observed at 350 nm;
              2. R_{DF}, observed at 350 nm;
              3. D_P, observed at 390 nm.
         In 3MP + 10% ethanol
              4. D_{DF}, observed at 350 nm;
              5. R_{DF}, observed at 350 nm;
              6. D_P, observed at 390 nm
              arbitrary units.

(5) The rise (R) and decay (D) kinetics of the DF and the P for the DPA in EPA glass is presented in Fig. 1. The R_{DF}, R_P and D_P are all exponential with time constants in the following order: $\tau_{DF} < \tau_P < \tau_P'$. When the concentration of the solute is increased, the parameters τ_{DF} and τ_P spread out strongly. The first part of the DF decay is exponential and is followed by a non-exponential part. At the end of the observation the DF rate of decay is slower than the P rate of decay. This last point has been more clearly confirmed when ether or 3-methyl pentane (3MP) were used as solvent matrices (4). The slow part of the decay may be assigned to electron-cation recombinations. Similar results have been obtained in the TMPD-3MP system (4).

When the concentration of the solute increases, the exponential part becomes more and more important compared to the non-exponential part. On the other hand, in an ethanol glass the DF is observed even though it is quenched by a factor of about 3. The TMPD shows no DF in an ethanol glass in our limit of detectability, this corresponds then to a quenching factor of about 100.

The fact that $\tau_{DF} < \tau_P$ with an increase in the concentration of the solute cannot be interpreted with the use of the photo-ionization model where the triplet state acts as an intervening state in which case $\tau_{DF} \geq \tau_P$. At a concentration of 5×10^{-4}M of DPA in EPA, $\tau_{DF} \simeq \tau_P$ and the exponential part of the DF is practically absent (4). This shows that the second part of the decay (recombination DF) is really due to a photoionization produced by a consecutive absorption of two photons with the triplet state as intermediate. When the concentration increases from 5×10^{-4}M to 10^{-2}M, the DF would then be composed of a mixture of "recombination DF" and "annihilation DF" the last one being observed alone in an ethanol glass matrix where the electrons are definitely trapped (3,8,9). The exponential rise time τ_{DF} is faster than τ_P and its dependence upon the concentration might then be explained.

In a sense these results agree with the results obtained by different authors on solid solutions in which it was concluded that when a solution is frozen, some of the solute molecules are inhomogeneously dispersed in the matrix (15, 16). This matrix effect is remarkably shown in the study of the DF of DPA in the mixture of 3-MP + 10% ethanol and in 3 MP alone. The kinetic study in both solvents is presented in Fig. 2. The DF intensity is amplified by a factor of 20 when 10% ethanol is added to the 3 MP solution while the phosphorescence intensity remains approximately the same. Moreover this doping effect strongly affects the rise and the decay kinetic of the DF while the phosphorescence kinetics remains unchanged. The addition of ethanol strongly increases the "fast" part of the decay (D_{DF}) similar to the exponential part observed in EPA. The rise of the DF (R_{DF}) is faster in the mixture and τ_{DF} become less than τ_P just like in the EPA matrix. In the 3 MP alone the kinetics of rise and decay are polyexponential and very hard to analyze. This would imply that the electron traps in the matrix behave independently of each other and the trapping time is then the limiting step in the rise time kinetics. In the 3 MP doped with 10% ethanol the increase in the DF seems to be correlated with a considerable increase of the "annihilation DF" relative to the "recombination DF". The DPA molecules, thanks to hydrogen bond formation, are more soluble in ethanol and tend to concentrate in the ethanol regions of the matrix. Clusters of DPA molecules are then probably obtained and this greatly favors the triplet-triplet annihilation mechanism. These results on the doping of matrices follow similar observations on other systems as first published by Dupuis in 1964 (10).

These results show that the DF of the DPA in rigid matrices originates generally from two primary mechanisms: photo-ionization followed by recombination (α) and triplet-triplet annihilation (β). The relative importance of mechanisms α and β strictly depends upon the solvent. Similar results have been obtained on carbazole and tryptophan molecules and will be published later. The β mechanism is related to the heterogeneity of the solute molecules in the solvent after the freezing of the matrix. One of the requirements for exchange interactions to take place between molecules in the triplet states is that the solute molecules be separated by about 15A (11).

The question related to the minimum energy necessary to photoionize the DPA molecule in our rigid matrices has not yet been answered. Using photons at 310 nm (4.0 eV), one reaches, after $T_n \leftarrow T_1$ absorption, an energy of $E(T_1) + E(310 \text{ nm}) = 3.1 + 4.0 = 7.1 \text{eV}$. This energy is greater than the ionization potential of the rigid solution, I_s. Recently Bernas, Gauthier and Grand (12) have been able to measure in methylcyclohexane $I_s = 6.43 \pm 0.03$ eV and, when in the gas phase, $I_g = 7.25 \pm 0.03$ eV (13). As the difference $I_g - I_s = 0.82$ eV will certainly increase in polar solvents, it is realistic to think that the energy obtained after triplet-triplet annihilation $2 E(T_1) = 6.2$ eV is high enough to ionize the molecule in solvents like EPA, ether and ethanol. Such a route for the photoionization has been observed in fluid solutions at room temperature by the way of some photoconductivity experiments on aromatic hydrocarbons (14).

REFERENCES

* Taken in part from the "Thèse d'état ès Sciences Physiques" of M. Ewald, Paris-Orsay University, 1972.

** Present address: Département de Chimie-Physique Université de Bordeaux I 33-Talence, France.

*** Present address: Centre Universitaire, 93-St-Denis, France.

1. M. Ewald, B. Muel, G. Durocher, Chem. Phys. Lett., 5,83 (1970).

2. M. Ewald, D. Muller, G. Durocher, J.Luminescence, 5,69 (1972).

3. M. Ewald, B. Muel, Compt. Rend. (Paris), B268, 973 (1969).

4. M. Ewald, "Thèse d'état", Paris-Orsay, (1972).

5. D. Muller, M. Ewald, G. Durocher, Chem. Phys. Lett. in press.

6. G. N. Lewis, D. Lipkin, J. Am. Chem. Soc., 64,2801 (1942).

7. H. Linschitz, M. G. Berry, D. Schweitzer, J. Am. Chem. Soc., 76, 5833 (1954).

8. A. Deroulede, "Thèse d'état", Paris-Orsay, (1969).

9. A. Bernas, M. Gauthier-Bodard, D. Grand, D. Lonardi, Int. J. Rad. Phys. Chem., 1, 229 (1969).

10. F. Dupuy, R. Lochet, A. Rousset, Compt. Rend. (Paris), 258,4223 (1964).

11. K. R. Naqvi, Chem. Phys. Lett., 1,497 (1967).

12. A. Bernas, M. Gauthier, D. Grand, "Third International Conference on Photosensibilisation in Solids", Sarlat (France) Sept. (1971). J. Phys. Chem., in press.

13. A. Terenin, F. Vilessov, Adv. Photochemistry, 2,385 (1964).

14. R. C. Jarnagin, Accounts Chem. Res., 4, 420 (1971).

15. S. Leach, E. Migirdicyan, Actions Chimiques et Biologiques des Radiations, Ed. Masson, Paris, 9, 117 (1966).

16. G. Durocher, S. Leach, J. Chim. Phys., 66, 628 (1969).

EXCITON LUMINESCENCE AND PHOTOCONDUCTIVITY OF CdS UNDER HIGH INTENSITY EXCITATION

Hiroshi Saito and Shigeo Shionoya

Institute for Solid State Physics, University of Tokyo

Roppongi, Minato-ku, Tokyo 106, Japan

ABSTRACT

Luminescence and photoconductivity of CdS under high intensity excitation are studied at 1.8 K using a nitrogen laser and a nitrogen laser-pumped tunable dye laser, in order to investigate the dynamical behavior of high density excitons and excitonic molecules.

Three new luminescence lines, M, P_M and P, are reported. They are assigned to the radiative annihilation of an excitonic molecule, to the inelastic collision of excitonic molecules and to the inelastic collision of single excitons, respectively. The spectral shapes of the M and P_M lines are analyzed theoretically. Photoconductivity measurement shows that the magnitude of the photocurrent under exciton excitation tends to become nearly the same as that under band gap excitation with increasing excitation power.

I. EXPERIMENTAL PROCEDURES

A nitrogen laser Model C950 made by Avco-Everett Res. Lab. was used as the excitation source. This laser produces 337.1 nm pulsed light with peak power up to 100 kW, pulse duration of 10 nsec and repetition rate of up to 100 pps. Using this laser it is possible to obtain excitation densities of up to about 2 MW/cm^2 on the sample surface. A Dial-a-Line dye laser of Avco-Everett Res. Lab. was also used. This laser is operated by pumping with the nitrogen laser and is tunable in the range of 355 to 655 nm, producing a peak power of 4 to 8 kW, pulse duration of 2 to 8 nsec and spectral width of about 0.3 nm. The highest excitation density obtained by this laser is about one third of that by the nitrogen laser.

The luminescence spectra were measured by a combination of a Spex Model-1702 3/4 m grating monochromator, a photomultiplier EMI 6256S or HTV R453 and Princeton Appl. Res. Boxcar integrator Model 160. In the photoconductivity measurement, a 20 V d.c. voltage was applied across indium electrodes with a 0.5 mm gap, and the photo-current was detected by using the Boxcar integrator. The contacts were ohmic.

II. EXPERIMENTAL RESULTS AND DISCUSSION

Luminescence of CdS and some other II-VI compounds under high intensity excitation has been recently studied by several investi-gators (1-5). It was found that some new emission lines which do not exist under ordinary excitation are produced under high inten-sity excitation on the low energy side of the intrinsic exciton line. We show in Fig. 1 the luminescence spectra for a CdS cry-stal, obtained at 1.8 K under excitation by 337.1 nm light from the nitrogen laser at various power levels over a wide range. This crystal shows a strong I_2 emission line due to excitons bound to neutral donors under excitation by 365.0 nm mercury light. Under laser excitation, the M line located at 2.545 eV grows super-linearly. Fig. 2 shows a clearer M-line from another CdS crystal; the I_2 line is weak but, the I_1 line due to excitons bound to neutral acceptors is strong, under mercury light excitation.

As the M line becomes strong, the P_M line located at about 2.535 eV starts to grow very rapidly, and outstrips the M line in intensity. The P_M line is not seen in Fig. 2, since it is masked by a strong I_1 line. With further increasing excitation power, the P line located at about 2.53 eV increases in intensity rapidly and dominates the emission spectra. The changes with intensity of these three lines with excitation power are shown in Fig. 3, together with that of the A-2LO line, i.e. the emission line of the intrinsic A exciton accompanied by the emission of two longitudinal-optical phonons.

In the case of the A exciton excitation at about 2.55 eV with the dye laser, especially under high excitation density, the features of the emission spectra are very similar to those under nitrogen laser excitation. This indicates that the luminescence mechanisms of these lines are independent of whether the excitation is made by the band-to-band transition or exciton absorption.

We now show that the M line luminescence is due to the radia-tive annihilation of excitonic molecules (6). It seems that this line is the same as observed by Benoît à la Guillaume et al. (1) and called A by them. Some features of this line are that its in-tensity grows superlinearly with excitation power J according to $I \propto J^n$ with $n \approx 1.7$, as shown in Fig. 3, and that the line shape is

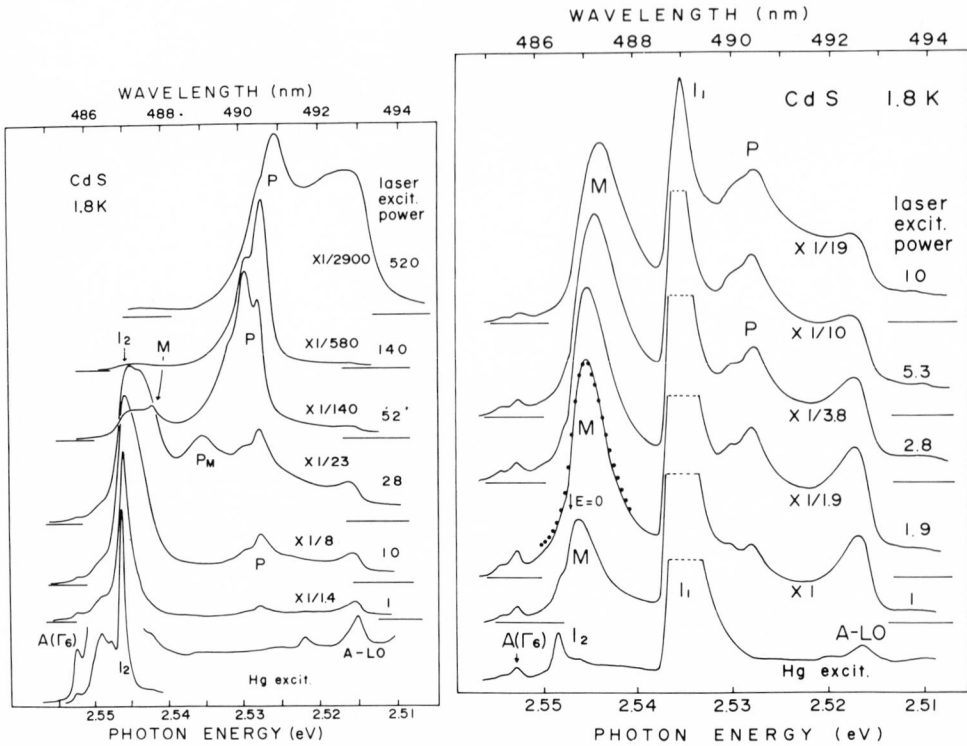

Fig. 1   Luminescence spectra of a CdS crystal at 1.8 K under the
         337.1 nm laser excitation. The highest excitation level
         corresponds to about 2MW/cm^2.

Fig. 2   Luminescence spectra of a CdS crystal at 1.8 K under 337.1
         nm laser excitation, showing the appearance of the M line.

asymmetric with a tail towards the low energy side as shown in
Fig. 2. The M line has been observed in all CdS crystals studied,
independent of the observed intensities of I_1 and I_2 bound exciton
lines. This indicates that the M line is intrinsic in nature.
This line has also been observed in CdSe at a similar position,
i.e. a little below the I_2 line, but such a line has not been found
in ZnSe.

Recently, luminescence of excitonic molecules has been clearly
demonstrated in CuCl (7). Further Akimoto and Hanamura (8) have
performed the theoretical calculation of the binding energy of an
excitonic molecule as a function of the ratio of electron and hole
effective masses $\sigma = m_e/m_h$, and have shown that excitonic molecules

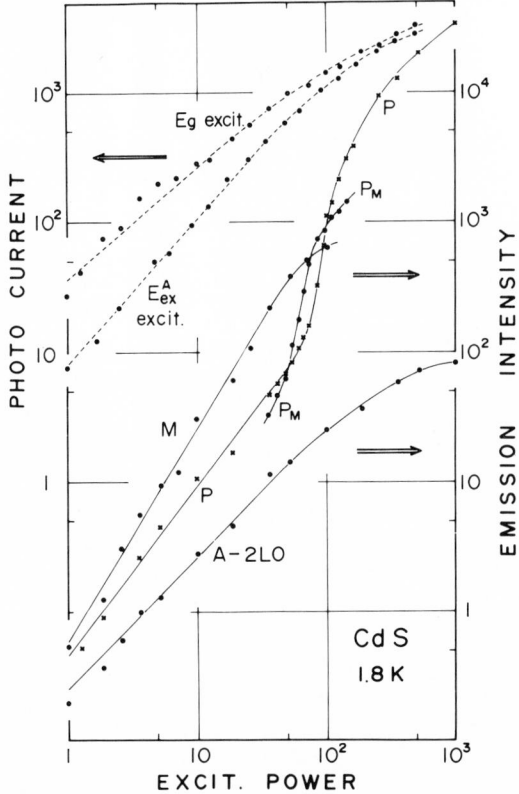

Fig. 3   The intensities of the M, P_M, P and A-2LO emission lines
and of the photocurrent under band gap and A exciton excitations versus the laser excitation power for a CdS crystal
at 1.8 K.

should be stable in the whole range of σ. The most probable process
of radiative annihilation of excitonic molecules is such that one
of the two excitons in a molecule is radiatively annihilated leaving
the other as a single free exciton. The energy of the emitted photon is

$$h\nu = E_{ex} - G_m - \hbar^2 k^2/4M, \tag{1}$$

where E_{ex} is the energy of single exciton, G_m is the binding energy
of excitonic molecule and M is the mass of a single exciton. This
indicates that the line shape of molecular luminescence is expressed
by the inverse Boltzmann distribution,

$$I(E) \propto E^{1/2} \exp(-E/kT), \tag{2}$$

where

$$E = E_{ex} - G_m - h\nu .$$ (3)

The asymmetric shape of the M line of CdS as observed implies, according to the above discussion, that this line is due to excitonic molecules. It is found that the observed line shape fits Eq. (2) fairly well on the low energy side but the high energy side is much broader than expected from Eq. (2). However, if one includes the effect of elastic collisions of excitonic molecules, it is possible to interpret the entire line shape. With an increase in density of excitonic molecules, the collisions between two excitonic molecules and between a molecule and a single exciton become considerable, so that their momentum states have some relaxation times. Taking this effect into account, the emission spectrum of an excitonic molecule is calculated (9) as a function of E defined by Eq. (3) as follows,

$$I(E) = A \int_0^\infty \frac{n(\varepsilon)\Gamma\varepsilon^{1/2} \, d\varepsilon}{(E - \Sigma - \varepsilon)^2 + \Gamma^2}$$

$$= \begin{cases} A'E^{1/2}\exp(-E/kT) & \text{for } \Gamma << kT \\[2mm] \dfrac{A\Gamma}{(E - \Sigma)^2 + \Gamma^2} & \text{for } \Gamma >> kT \end{cases}$$ (4)

Here, $n(\varepsilon)$ is the Boltzmann distribution function; Γ is the sum of the relaxation frequencies of a single exciton and an excitonic molecule; Σ is given by $N(w_0 - v_0)$, in which N is the number of excitonic molecules; and Nw_0 and Nv_0 are the shifts of the energy levels of an excitonic molecule and a single exciton, respectively, due to the collision with N excitonic molecules. This can be neglected in comparison with kT, i.e. $\Sigma \overset{\sim}{\sim} 0$. Eq. 4 indicates that the spectral shape changes from the inverse Boltzmann distribution in the low density limit to Lorentzian for high density.

The fitting of the M line shape to Eq. (4) has been made by taking Γ/kT and T as adjustable parameters. The fitting is satisfactory. An example is shown in Fig. 2. It is to be noted that the temperature of the exciton system is much higher than that of the lattice. The arrow in the figure indicates the points corresponding to E = 0. If one considers the dissociation process of an excitonic molecule to two single excitons, the energy required is the smallest when two triplet excitons, $A-\Gamma_6$ excitons in the present case, are produced. We take this energy as G_m. Further, if we assume that the position of E = 0 obtained above corresponds to the case where an $A-\Gamma_6$ exciton is left in the annihilation process of an excitonic molecule, then the separation between the position of E = 0 in the figure and the energy of $A-\Gamma_6$ exciton, 2.5524 eV, gives the value of G_m. This is estimated to be about 5.4 meV. With

increasing excitation power the peak of the M line shifts considerably to lower energies. If the annihilation process leaves a singlet A-Γ_5 exciton, the corresponding position of E = 0 should be
shifted to the low energy side by 1.3 meV, i.e. the difference between the Γ_5 and Γ_6 exciton energies. Therefore, the observed peak
shift implies that the contribution of the process leaving a Γ_5
exciton to the M line luminescence is increased with increase of
concentration of excitonic molecules.

In the case of CdSe, the detailed analysis of the M line shape
is impossible, since the M line overlaps the I_2 line. It is estimated that G_m may be roughly 3 meV.

It is to be noted that in CdS and CdSe the hole effective
masses are heavy and strongly anisotropic, while the electron
effective masses are light and nearly isotropic (11). We note that
the result of the calculation of G_m by Akimoto and Hanamura (8)
cannot by itself be applied to materials having anisotropic effective masses. We point out that in such cases as in CdS and CdSe
the anisotropy of the hole effective mass m_h can give a larger value
of G_m than calculated by the use of the geometrical mean of m_h.
This can be seen from the following discussion. Let us consider
the two configurations: one in which the molecular axis joining
the two holes coincides with the heavy direction of m_h, and the
other in which the molecular axis is perpendicular to it. If the
holes are sufficiently heavy so that their vibrational motion can
be treated in the adiabatic approximation, it is clear that the
former configuration gives smaller zero-point energy than the latter.
For the case in which the holes are not so heavy and move more
freely, the following considerations are appropriate. For the first
configuration, the two holes move in such a way that the two excitons can approach each other more closely without increasing the
overlap of the wave functions of the holes. Therefore, the attractive force between the two excitons is exerted more effectively, so
that the former configuration is more stable than the latter. Thus,
a molecule will be formed in such a specified, anisotropic configuration. The use of the geometrical mean of m_h gives only an average of the energies calculated for the above two configurations.

The anisotropic nature of m_h in CdS and CdSe is such that m_h
($\perp$c) = 0.7 and m_h($\parallel$ c)$\sim$5 for CdS, and 0.45 and $>$ 1.0 for CdSe (10).
The value of $\sigma = m_e/m_h$ in CdS calculated by using the geometrical
mean of m_h is $\sim$0.14. According to Akimoto and Hanamura's calculation, this corresponds to $G_m/G_{ex}\sim$0.086, where G_{ex} is the binding
energy of a single exciton. If one uses the binding energy of a
triplet A-Γ_6 exciton, 30.5 meV as the value of G_{ex}, the value of
G_m is estimated to be 2.6 meV. In the case of CdSe, the value of
G_m is estimated to be $>$0.94 meV. It is noted that these values of
G_m, estimated by using the geometrical means of m_h, are much smaller

than those experimentally obtained in accordance with the above argument. In this sense the excitonic molecules of CdS and CdSe are anisotropic.

It might be useful to attempt to evaluate G_m by taking $\sigma = m_e/m_h$ ($\parallel$) ~ 0.037 for CdS, as the effective σ value. According to Akimoto and Hanamura's calculation, $\sigma \sim 0.037$ corresponds to G_m/G_{ex} ~ 0.16. However, since their calculation is based on the variational method, the true G_m/G_{ex} value is expected to be higher than the calculated one. For $\sigma = 0$ their calculation gives $G_m/G_{ex} = 0.30$, which, however, should be 0.35. Taking this into account, we may assume that $\sigma \sim 0.037$ corresponds to $G_m/G_{ex} \sim 0.21$. Then we obtain $G_m \sim 6.4$ meV. It is noted that the G_m value obtained from the experiment, 5.4 meV, is close to this value rather than that calculated by using the geometrical mean of m_h, 2.6 meV. We wish to emphasize here that the anisotropy of the heavier particle in an exciton gives an important effect in the binding energy of an excitonic molecule.

In the case of ZnSe both the electron and hole masses are isotropic, and G_m is calculated to be 1.1 meV. As mentioned above, the temperature of the exciton system under high intensity excitation is estimated to be 17-20 K. In this temperature range, an excitonic molecule having G_m as small as 1.1 meV will be thermally dissociated. Therefore, it is quite reasonable that the M line is absent in ZnSe.

As described earlier, the temperature of excitonic molecules estimated from the luminescence line shape is much higher than the lattice temperature. If the spectral shape of the A-2LO line is analyzed according to the theory of Segall and Mahan (11), in the case of high intensity excitation one obtains T = 10-25 K (12).   The spectral shapes of the M and A-2LO lines under A exciton excitation are nearly the same as under nitrogen laser excitation, indicating that the temperature is raised even if high density excitons are created directly by optical absorption. Therefore, in the case of excitation of the band-to-band transition, it is assured that the phonon emission process following the optical absorption to produce excitons is not the cause for the observed rise in temperature.

It is clear that the collision of high density excitons and excitonic molecules brings about a rise in their temperature. In the following sections we will show that with increasing excitation power the inelastic collisions between excitonic molecules, as well as between single excitons, begin to take place frequently. Such collision processes cause the elevation in temperature of the exciton system.

With increasing excitation power the M line tends to become saturated. Just before this occurs, the P_M line, located at about 2.535 eV, starts to grow very rapidly with the superlinear relation $I \propto J^n$, $n \geqslant 4$, as shown in Fig. 3, and outstrips the M line in intensity. These facts strongly suggest that the P_M line is produced by the inelastic collision of two excitonic molecules (13). The shape of the P_M line is asymmetric with a tail towards the low energy side, although this is not clearly shown in Fig. 1. The apparent half-width is much broader than that of the M line.

Consider the radiative annihilation of an excitonic molecule during inelastic collision with another molecule. Then, the latter molecule is excited, which results in decomposition into two free excitons, since the excitonic molecule has no bound excited states. The energy of the emitted photon by this process is roughly given by $h\nu = 2E_m - 3E_{ex} = E_{ex} - 2G_m$, where E_m is the energy of excitonic molecule. Although the translational momentum must be conserved, the two excitons produced by the decomposition may have any value of momenta for their relative motion. Therefore, the shape of the luminescence line should show a rather long tail on the low energy side. The observed line shape is in agreement with this argument.

The calculation of the spectral shape of the P_M line, in which the relaxation times of the momentum states of excitonic molecules and excitons are taken into account, gives (9),

$$I(E) \propto (1/E^2)\{[(E-G_m)/2] + [(E-G_m)^2/4 + \Gamma^2/4]^{1/2}\}^{1/2} \quad (5)$$

where

$$E = E_{ex} - G_m - h\nu . \quad\quad\quad\quad\quad\quad (6)$$

and Γ is the sum of the relaxation frequencies due to the elastic collision of the excitonic molecules and single excitons involved in the emission process. Eq. 5 indicates that the rising point on the high energy side of the P_M line corresponds to the energy, $E_{ex}-2G_m$.

Experimentally, the rising point is at about 2.537 eV. If one assumes that the observed P_M line luminescence is due to the inelastic collision process of excitonic molecules in which three singlet Γ_5 excitons are produced, the observed position of 2.537 eV is reasonably understood. The fitting of the observed P_M line shape to Eq. (5) has been made by assuming that $E_{ex} = 2.5537$ eV, the energy of the Γ_5 exciton, and by taking Γ/G_m as an adjustable parameter. It has been found that if one assumes that $G_m = 8.5$ meV and chooses $\Gamma/G_m = 0.10$, the high energy side of the P_M line fits the equation fairly well. However, the fitting of the low energy side is rather poor. This may be because in the calculation of Eq. (5) the momen-

tum dependence of the matrix element for the collision of excitonic
molecules leading to their decomposition is not taken into account.

If one assumes that three Γ_5 excitons are produced in the collision process of excitonic molecules, the value of G_m in Eq. (5)
corresponds to the energy required to dissociate a molecule into
two singlet Γ_5 excitons. This energy is given by the sum of the
G_m value obtained from the M line analysis (5.4 meV), which gives
the binding energy of the excitonic molecule for dissociation into
two triplet Γ_6 excitons plus twice the energy separation between the
Γ_5 and Γ_6 excitons (1.3 meV), resulting in 8.0 meV. Therefore, the
assumption of $G_m = 8.5$ meV which gives a good fit of the P_M line
shape is reasonable. As was mentioned earlier, it seems that excitonic molecules tend to be dissociated into two singlet excitons,
but not into two triplet excitons, with an increase of their concentration.

We shall show now that the P line luminescence is caused by the
inelastic collision of single excitons, as was proposed earlier
(1, 4). As shown in Fig. 3, the P line is observed even under fairly low excitation power and grows with the superlinear relation:
$I \propto J^n$, $n \sim 1.5$. However, following the rapid growth of the P_M line,
the P line grows more rapidly with $n \geqslant 4$. After the P_M line becomes
saturated, the growth of the P line becomes similar to that for the
initial stage. The P line has a doublet structure, as shown in
Figs. 1 and 2, with peaks at 2.5296 and 2.5278 eV. It is noted that
the separation of these peaks from the A exciton line is nearly
equal to the energy difference between the $n = 2$ excited state and
the ground state of the A exciton.

These facts indicate that the P line is produced by the inelastic collision of two single excitons in which one is radiatively
annihilated while the other is raised to the $n = 2$ state. However,
the latter exciton can be raised to various excited states according
to the cross-sections for such collision processes (14). As will
be described shortly, it is experimentally verified for CdS that
free electron-hole pairs are produced as a result of the collision
of excitons. The observed position of the P line is interpreted
such that in the collision of two excitons the cross-section for
the process leaving an $n = 2$ exciton is the largest.

In this kind of collision, one may expect the appearance of
three emission lines, due to the collisions of two singlet excitons, one singlet and one triplet exciton and two triplet excitons.
These lines should be equally spaced with a separation of 1.3 meV,
the difference between the ground states of the singlet and triplet
excitons. The observed doublet structure of the P line has a
separation of 1.8 meV.

As mentioned above, the growth of the P line becomes very rapid just following the rapid growth of P_M line. Since in P_M line luminescence excitons are produced from excitonic molecules, the rapid growth of the P_M line brings about an abrupt increase of the number of single excitons. This effect is the cause of the observed rapid increase of the P line.

Previously Shionoya and coworkers (15) studied the photoconductivity of Cd(S,Se) crystals for the case in which high density excitons are photo-created directly. They used crystals for which the position of the A exciton coincides with the 488.0 nm argon ion laser line. It was found that the photocurrent increases as the square of the excitation intensity. This fact is evidence that free carriers are generated as a result of the inelastic collision of excitons.

In the present paper photoconductivity has been measured using a tunable dye laser corresponding to the band gap and the A exciton line. In Fig. 3 the magnitudes of the photocurrent under these two kinds of excitations are shown as functions of the excitation power. It is observed that the photocurrent under the A exciton excitation increases superlinearly, presenting further evidence for free carrier generation due to collision of excitons. Further, it is noticed that with increasing excitation the magnitude of the photocurrent with A exciton excitation tends to become nearly the same as that under band gap excitation. This indicates that under very high power excitation excitons and free carriers are in quasi-equilibrium, independent of whether the excitation is by exciton absorption or by the band-to-band transition.

ACKNOWLEDGMENTS

The authors would like to thank Prof. E. Hanamura and Dr. O. Akimoto for valuable discussions. The authors are grateful to Dr. C. H. Henry, Bell Telephone Laboratories for supplying some of the crystals used for this paper.

REFERENCES

1. C. Benoît à la Guillaume, J. M. Debever and F. Salvan, Phys. Rev. 177, 567 (1969).

2. K. Era and D. W. Langer, J. Luminescence 1,2, 514 (1970).

3. K. Era and D. W. Langer, J. Appl. Phys. 42, 1021 (1971).

4. D. Magde and H. Mahr, Phys. Rev. Lett. 24, 890 (1970); Phys. Rev. B 2, 4098 (1970).

5. T. Goto and D. W. Langer, J. Appl. Phys. $\underline{42}$, 5066 (1971).

6. S. Shionoya, H. Saito, E. Hanamura and O. Akimoto, Solid State Commun., to be published.

7. H. Souma, T. Goto, T. Ohta and M. Ueta, J. Phys. Soc. Japan $\underline{29}$, 697 (1970).

8. O. Akimoto and E. Hanamura, Solid State Commun. $\underline{10}$, 253 (1972).

9. E. Hanamura, this conference.

10. See, "Physics and Chemistry of II-VI Compounds", ed. M. Aven and J. S. Prener (North-Holland Pub. Co., 1967); and C. H. Henry and K. Nassau, Phys. Rev. B $\underline{2}$, 997 (1970).

11. B. Segall and G. D. Mahan, Phys. Rev. $\underline{171}$, 937 (1968).

12. R. F. Leheny, R. E. Nahory and K. L. Shaklee, Phys. Rev. Lett. $\underline{28}$, 437 (1972).

13. H. Saito, S. Shionoya and E. Hanamura, Solid State Commun., to be published.

14. H. Büttner, Phys. Stat. Sol. $\underline{42}$, 775 (1970).

15. H. Kukimoto, S. Shionoya and T. Kamejima, J. Phys. Soc. Japan $\underline{30}$, 1662 (1971).

LUMINESCENCE OF ANTHRACENE CRYSTALS UNDER HIGH INTENSITY OF

EXCITATION

Z. A. Chizhikova, M. D. Galanin and Sh.D. Khan-

Magometova

P. N. Lebedev Physical Institute

Moscow, USSR

ABSTRACT

Luminescence spectra of anthracene crystals have been studied under nitrogen laser excitation up to 10^{23} photons/cm^{2}sec at 4.2K. The sharp increase of the vibronic line at 23692 cm^{-1} has been observed at high intensity of excitation. The role of super-luminescence in a thin excited layer near the surface of the crystal is discussed.

The interaction of excitons to produce biexcitons or "exciton drops" has been observed in semiconductors, and attention has been attracted to a search for similar phenomena in molecular crystals. The interaction of excitons in molecular crystals results in several effects: singlet-singlet and singlet-triplet quenching, triplet-triplet annihilation with the excitation of singlet excitons, and some photoelectric effects. However, it is possible that the exciton interaction can result not only in nonradiative processes, but in changes in the luminescence spectrum.

In view of this possibility we have investigated the luminescence spectrum of the crystalline anthracene under high-intensity excitation at 4.2K. The excitation intensity was limited by the thermal destruction of the crystal.

To calculate the limiting excitation intensity and the maximum achievable exciton density one must take into account the "nonlinear" quenching due to exciton interaction. Under excitation with short pulses (10^{-8}sec) this quenching is determined by the dipole-dipole interaction of the singlet excitons (1,2). Because of the nonlinear quenching, the luminescence yield under high intensity excitation is low and we can assume that most of the excitation energy is dissipated into heat.

For a pulse energy of 5×10^{-3} joule/cm^2, or energy flux of 5×10^{23} photons/cm^2-sec, an energy of 150 joules/cm^3 is dissipated in the crystal (the extinction coefficient is about 3×10^4/cm). At room temperature the conductivity of molecular crystals is about $3 \times 10^{-3} cm^2$/sec, therefore under short-pulse excitation we may suppose that the heating is adiabatic. At room temperature the specific heat of anthracene is equal to 1.6 joules/cm^3deg. i.e. at the excitation energy mentioned the temperature reaches 100 K.

Thus if heating effects are to be negligible, the intensity of the excitation should not exceed 10^{23} photons/cm^2sec. At this excitation the nonlinear quenching diminishes the luminescence yield approximately ten times. Hence one can estimate the maximum density of excitations achieved at this excitation intensity.

The relative number of the excited molecules is determined by the value $F\sigma\tau$, where F is the excitation intensity (photons/cm^2sec.), σ, the absorption cross-section and τ, the life time. For anthracene $\sigma = 10^{-17} cm^2$ and $\tau = 5 \times 10^{-10}$sec, when the quenching is taken into account. It means that a fraction of 5×10^{-4} of all the molecules can be excited.

At helium temperature, the specific heat of the molecular crystals (benzene, naphthalene) decreases almost by three orders of magnitude. There are no data on the thermal conductivity at very low temperatures, but, probably it increases so much that the heating cannot be supposed to be adiabatic. Under these conditions it is best to obtain information about heating from the experimental data on the linewidth in the luminescence spectrum.

To excite the luminescence we used a nitrogen laser ($\lambda = 337$ nm) with a pulse duration of 15 nsec, repetition frequency of 10 c/sec. and pulse power of 1.8 kw. The laser radiation was focused with a lens to a spot of 0.1mm^2 on the surface of the crystal. The crystal was placed in liquid helium. We have studied anthracene crystals in the form of thin plates 1μm thick obtained by sublimation of zone refined anthracene. The crystals were fixed freely, i.e. without a contact with any surface.

The luminescence spectra were recorded with a double spectrometer DFS-12 (dispersion 5 A/mm). The resolution was about 1 cm^{-1}. The spectra were photoelectrically recorded with a photomultiplier, amplifier and recorder.

In Fig. 1 the spectra of the luminescence are shown in the region of 25000-23500 cm^{-1} obtained with excitation intensities differing by a factor of 7. The greatest intensity corresponds to an energy flux of about 10^{23} photons/cm^{2}sec. It is seen from the figure that there are considerable changes in the spectra: 1) the vibronic line at 23692 cm^{-1} increases in intensity, 2) the lines attributed to the impurities for example the line at 24902 cm^{-1}, decrease in intensity. In the spectral region of 23500 cm^{-1} there were no sharp changes in the spectrum. The observed changes are quite reversible.

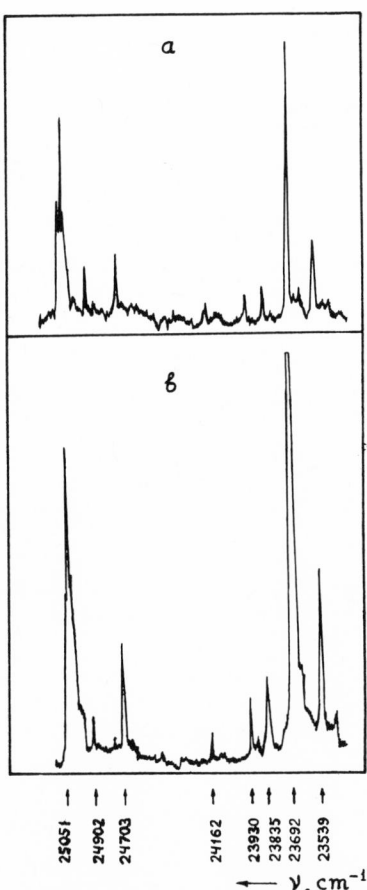

Fig. 1.   The luminescence spectrum of the anthracene crystal at 4.2K as a function of the excitation intensity: a) 1.4x10^{22} photons/cm^{2}sec, b) 1x10^{23} photons/cm^{2}sec. Positions of lines taken from references 4 and 5.

After recording the spectra at various intensities of excitation, the dependence of the intensity of the lines as a function of the excitation intensity was determined and is shown in Fig. 2.

The initial linear portions of the curves correspond to a constant luminescence yield. At increased excitation intensities the bands 25051, 24703 cm^{-1} (and also other vibronic bands) are quenched non-linearly. The impurity line at 24902 cm^{-1} even stops to grow with increasing excitation intensity (a saturation of the impurity bands was already observed earlier in (6,7)).

The sharpest effect consists in an intensive superlinear increase of the line at 23692 cm^{-1}. This effect is different in various samples, but was observed in quite a number of samples, and in some cases the increase in intensity of this line reached two orders of magnitude.

In Fig. 3 are shown (on a different scale) the lines 25051 and 23692 cm^{-1} at small and large excitation intensities. The halfwidth of the leading line at 25051 cm^{-1} grows with an increase

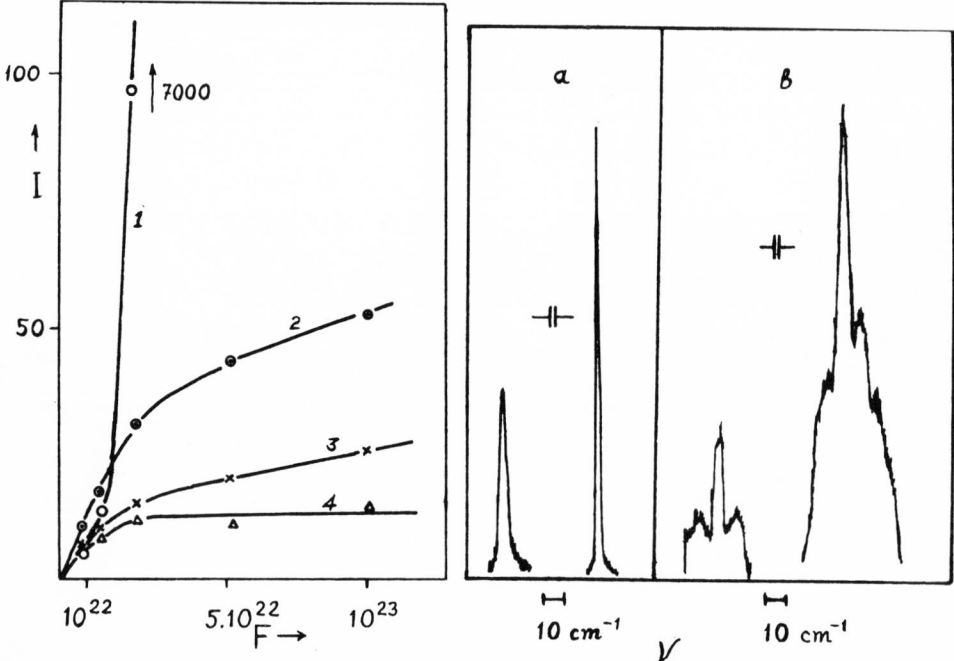

Fig. 2   The dependence of the luminescence intensity of an anthracene crystal on the excitation intensity (at 4.2K):
I - the 23692 cm^{-1} line (the vibronic transition to the vibrational level of 1400 cm^{-1}), 2 - the 25051 cm^{-1} line near the 0-0 transition, 3 - the 24703 cm^{-1} line (the vibronic transition to the vibrational level of 400 cm^{-1}), 4 - the impurity line at 24902 cm^{-1}.

Fig. 3   The lines in the luminescence spectrum at a low and high excitation intensity a) 23692 cm^{-1} and b) 25051 cm^{-1}.

in the excitation energy from 4 to 7 cm^{-1}. A similar broadening is detected with other lines. For instance, the line at 24703 cm^{-1} widens from 4 to 6.5 cm^{-1}; the 23539 cm^{-1} line from 6.5 to 9 cm^{-1}. According to the data on the dependence of the linewidth upon temperature (4) this broadening corresponds to a heating by several degrees only. On the contrary the line at 23692 cm^{-1} narrows from 5 cm^{-1} to at least 2 cm^{-1}.

It is natural to suppose that the increase in intensity and the narrowing of the line at 23692 cm^{-1} is due to stimulated emission. The role of the stimulated emission in the luminescence of anthracene crystals under high intensity of excitation was already discussed in (7) in connection with the differences in the increase of the luminescence intensity as a function of excitation intensity when the crystal surface or the edges of the crystals are observed.

However, in our case the difficulty is that the excited region of the crystal has the shape of a very thin layer near the surface of the platelet, and the effects are observed in a direction close to that of the normal to the plate. An estimate of the supposed gain across the layer shows that, in any case, it is too small to give rise to an observable superluminescence in the direction perpendicular to the crystal surface.

In fact, an estimate of the gain can be carried out with the known formula: $k = N\beta/8\pi\tau_0 c\nu^2\Delta\nu\mu^2$, where τ_0 is the radiative life time, N is the population inversion, ν and $\Delta\nu$ are the frequency and line width in wave numbers, μ is the refractive index, and β is the coefficient taking into account the part of the given line in the entire luminescence spectrum. Since we can neglect the population of the lower level which is 1400 cm^{-1} above the ground level, $N = n_2 = 2\times10^{18}$ cm^{-3}. The value of β was estimated roughly by the linewidth and the area of the luminescence spectrum, $\beta = 0.03\pm0.02$. By taking $\tau_0 = 5\times10^{-9}$sec, $\Delta\nu = 5$ cm^{-1}; $\mu^2 = 3$ we have $k = 1500\pm1000$ cm^{-1}.

In spite of such a large value of the gain it is still too small to give a considerable amplification across the excited layer. The thickness of the excited layer is determined by the absorption coefficient of the exciting light, i.e. it is equal to nearly 0.3μ. This value is possibly somewhat larger due to exciton diffusion (at room temperature the diffusion length is of the order of 0.1 μ; at helium temperature it is probably unknown).

In any case the thickness of the excited layer does not exceed that of the crystal, i.e. is of the order of 1μ. It gives for the amplification kx~0.15$\pm$0.1 which cannot explain the increase in intensity of the observed line.

At the same time the gain can be very large along the excited layer. The dimensions of the excited region is of the order of 0.1 mm, i.e. kx = 15±10 or $\exp(kx) \sim 10^{6\pm4}$. With such a large gain along the layer, one can observe not only the superluminescence, but also the generation, since a very small scattering is sufficient to create a feedback.

It is not yet clear why such an intense luminescence is observed not only along the excited layer, but in the direction near the normal to the surface as well (it is worth mentioning that we did not notice an essential difference in the spectrum when observing from the surface or from the edges of the platelet). We can assume that it is due to the small thickness of the excited layer and the large gain. With a layer thickness compared with the wavelength one should observe radiation diffraction for large angles, even exceeding that of the total reflection. On the other hand, with a large gain there can occur a considerable change in the refractive index of the excited layer, which can also influence the radiation propagation in the layer.

The stimulated emission reduces the lifetime of the excited state and stabilizes the population at a definite level. Thus, it can be considered as one more factor in limiting the exciton density in a crystal.

REFERENCES

1. N. A. Tolstoi, A. P. Abramov, Fizika tverdogo tela 9, 340 (1967).

2. S. D. Babenko, V. A. Benderskii, V. I. Goldanskii, A. I. Lavrushko and V. P. Tychinkii, Phy. stat. sol. 45, 91 (1971).

3. M. D. Galanin, Sh. D. Khan-Magometova, Z. A. Chizhikova, Short commun. in physics N 5, 34 (1972).

4. E. Glockner and H. C. Wolf, Z. Naturforsch. 34A, 94 (1969).

5. M. T. Shpak, N. I. Sheremet, Optica and spectroskopia 17, 694, (1967).

6. V. L. Broude, T. V. Klimusheva, V. S. Mashkevich, M. I. Cnopryenko, M. S. Soskin, E. F. Sheka, Journal Prikladnoy Spectroskopii; 1, 352 (1964).

7. V. L. Broude, E. F. Sheka, Quantum Electronics, Kiev, 1966, p. 188.

THEORY OF EMISSION SPECTRUM FROM HEAVILY EXCITED SEMICONDUCTORS

Eiichi Hanamura

Institute for Solid State Physics, University of Tokyo

Roppongi, Minatoku, Tokyo, Japan

ABSTRACT

After the important role of excitonic molecules in the emission spectrum from heavily excited semiconductors is pointed out, we calculate the emission spectra from the Bose condensed system of excitonic molecules and the system of interacting molecules. The characteristics of these spectra are listed and compared with the experimental results.

EXCITONIC MOLECULE AS A RADIATIVE CENTER

Recently, quasi-equilibrium of the many-exciton systems has become realized in a crystal and the optical and electrical properties of this system have been observed. In contrast to the electron gas in a uniform positive background, both attractive and repulsive interactions are of equal importance in our system and, as a consequence, various phenomena characteristic of these interactions are displayed with changes of excitation power. At low concentration of N electron-hole pairs, we could draw $N(N-1)$ lines of repulsive interactions among the particles with the same charge and N^2 attractive lines between pairs of electrons and holes. These N extra attractive interactions are taken into account most effectively by making N electron-hole exciton bound states. Furthermore, two excitons with antiparallel spins of both electrons and holes, were shown to be bound into an excitonic molecule for any mass ratio of electron and hole $\sigma = m_e/m_h$ (1). This excitonic molecule composed of four kinds of fermions--up and down spin electrons and holes--is considered to be the basic unit at low concentration because the repulsive interactions between molecules due to Pauli

121

effects overcome the exchange and van der Waals attractive interactions and the light translational mass of the excitonic molecules
prevents the formation of more complicated complexes of charged
particles.

Although the excitonic molecule may be considered to be formally
identical with the hydrogen molecule, the positive holes playing
the role of the protons, there exist some essential differences between the two. In the excitonic molecule, the mass ratio of the
two component particles of opposite charge is generally not so small
as that in the hydrogen molecule ($\sim$1/1840) and can have various
values for different crystals. As a result, the ratio of the molecular binding energy to that of an exciton decreases from 0.30 at
$\sigma = 0$ to 0.02 at $\sigma = 1$ depending on the mass ratio, and the average
spacing of heavier particles in the molecule increases from $1.44a_0$
at $\sigma = 0$ to $3.62a_0$ at $\sigma = 1$, as shown in Ref. (1), where a_0 is the
effective Bohr radius of the exciton. The anisotropy of the effective masses of the composite particles results in another characteristic of the excitonic molecule. For example, in CdS, the effec-
tive hole mass is considerably anisotropic and $m_h(||c)$ is very
heavy, i.e. $m_h = 0.7$ ($\perp$c), $5(||c)$, while the electron mass is light
and nearly isotropic. In such a case, the molecule forms a configuration such that the molecular axis joining the two holes coincides
with the heavy direction of the hole mass i.e. the c-axis. This
anisotropic excitonic molecule has a larger molecular binding energy
than one evaluated by the use of the geometrical mean of the hole
masses. This can be seen from the following discussion. In this
configuration, the two holes move in such a way that the two excitons can approach each other more closely without increasing the
overlap of the hole wave functions. Therefore, the attractive
force between the two excitons is exerted more effectively, and this
configuration is more stable than when the molecular axis perpendicular to the c-axis. The use of the geometrical means of m_h gives
only an average of the energies calculated for the above configurations.

In a direct band gap semiconductor, only when an exciton is
scattered on the dispersion line of a photon in the momentum space
is it radiative. On the other hand, an excitonic molecule with any
value of the translational momentum $\underline{K}$, is able to emit radiation
with momentum $\underline{K}_0$ by the process in which one electron-hole pair in
the molecule is annihilated through light emission while the other
pair remains as an exciton absorbing the momentum $\underline{K}-\underline{K}_0$ (3). In
this sense, the excitonic molecules work as the radiative centers
of the many exciton system. From the emission spectrum, much information on the electronic structure of this system and the interactions between the molecules is obtained. Therefore, the theoretical emission spectrum from the system of many excitonic molecules
is studied. How the boson-like character of excitonic molecules

is reflected in the emission spectrum and how the collisions between them modify the spectrum are discussed. It is then shown
that the inelastic collision between them induces a new line.
These theoretical results are compared with experimental results (4)
and from the comparison the binding energy of an excitonic molecule
and the value of its interactions is determined. More importantly,
the assignments of emission lines by the comparison of the emission
spectrum obtained theoretically with the experimental ones, with
respect to the position and the shape of spectral lines is confirmed.

EMISSION SPECTRUM FROM A BOSE CONDENSED SYSTEM OF EXCITONIC MOLECULES

An excitonic molecule, as well as an exciton, behaves like a
Bose particle at low density. As the repulsive interactions between the molecules due to the Pauli effects overcome the exchange
and van der Waals attractive interactions, some of the excitonic
molecules condense into the zero momentum state of the center-of-
mass motion at low enough temperature. The coherent state composed
of the condensed excitonic molecules is reflected in the emission
spectrum corresponding to that process in which an electron and a
hole of the excitonic molecule are annihilated, emitting light, and
the other pair remains as an exciton. The optical response of excitonic molecules is one of the best ways to study the coherent
state of the many exciton system. (5)

The emission spectrum from the system of N excitonic molecules
is written as follows:

$$I(\nu) = 2\pi \sum_f \mid <\psi_f \mid H' \mid \psi_i> \mid^2 \delta(E_f - E_i + h\nu) , \qquad (1)$$

where ψ_i is the wave function of N excitonic molecules with energy
E_i and ψ_f is the final state wave function with energy E_f of N-1
excitonic molecules and an exciton created by the emission process.
Here H' is the interaction Hamiltonian of the system with light and
is expanded in terms of the boson operators $C_{\nu K}$ and B_K corresponding
to an exciton and an excitonic molecule, respectively, as follows:

$$H' = \sum_{\nu K} G_\nu(\underline{K},\underline{P}_o) C^+_{\nu \underline{K}-\underline{P}_o} B_{\underline{K}} + h.c. \qquad (2)$$

where $G_\nu(\underline{K},\underline{P}_o)$ is proportional to the matrix element of the momentum
operator between the Bloch functions of the valence and conduction
bands and the envelope functions of an exciton and an excitonic
molecule. It is justified for Wannier excitons in semiconductors
to neglect the momentum dependence of $G_\nu(\underline{K},q)$ and approximate it
by a constant $G_0 \equiv G_{1s}(0,0)$. The molecules are considered to be in
the ground state for internal motion at the low temperature. We are
interested only in the 1s exciton state because transitions

involving the 1s state of an exciton make the largest contribution
to the emission spectrum. Here the deviations of the exciton and
the excitonic molecule from ideal bosons (5) due to Pauli effects,
as well as dynamical interactions, are taken into account as the
effective interaction between ideal bosons. Then the Hamiltonian
of our system is written as follows:

$$H = \sum_{\underline{K}} (2E_{1s} - E_m^b + K^2/4M)c_{\underline{K}}^+ c_{\underline{K}}$$

$$+ \tfrac{1}{2} \sum_{\underline{K},\underline{K}',\underline{q}} W(\underline{q};\underline{K},\underline{K}')c_{\underline{K}'+\underline{q}}^+ c_{\underline{K}-\underline{q}}^+ c_{\underline{K}} c_{\underline{K}'} \qquad (3)$$

$$+ \sum_{\underline{K}} (E_{1s}+\underline{K}^2/2M)B_{\underline{K}}^+ B_{\underline{K}} + \tfrac{1}{2} \sum_{\underline{K},\underline{K}',\underline{q}} V(\underline{q};\underline{K},\underline{K}')B_{\underline{K}'+\underline{q}}^+ c_{\underline{K}-\underline{q}}^+ c_{\underline{K}} B_{\underline{K}}$$

where $W(\underline{q};\underline{K},\underline{K}')$ denotes the effective interaction matrix element
describing the scattering of two excitonic molecules with the
translational momenta $\underline{K}$ and $\underline{K}'$ into the states with $\underline{K}-\underline{q}$ and $\underline{K}'+\underline{q}$,
due to the interaction between them.

As this paper is confined to low temperature case $kT<<E^b(ex)M/\mu$
and low concentration $Na_o^3 << \mu/2M$, we need only the values $V_o = V(0;0,0)$ and $W_o = W(0;0,0)$. The effects of multiple scattering
between molecules and between a molecule and an exciton are taken
into account by replacing W_o and V_o by $4\pi f_o a_o^2 E^b(ex)\mu/M$ and
$6\pi f_o' a_o^2 E^b(ex)\mu/M$, respectively. Here M is the mass of an exciton,
f_o is the scattering amplitude of two excitonic molecules in a
vacuum and f_o' is the scattering amplitude between an exciton and
and excitonic molecule.

Now let us discuss the emission spectrum from the condensed
system of excitonic molecules by assuming zero temperature. The
Green's functions of the condensed excitonic molecules are given
as follows:

$$G_m(\varepsilon,0) = \frac{N_o}{\varepsilon-\mu^*+i\delta} - \frac{N_o}{\varepsilon-\mu^*-i\delta} \text{ and}$$

$$G_m(\varepsilon,p) = \frac{u_p^2}{\varepsilon-\mu^*-E(p)+i\delta} - \frac{v_p^2}{\varepsilon-\mu^*+E(p)-i\delta}$$

with $\mu^*=E_m+NW_o$, $E^2(p) =(p^2/4M)[2NW_o+(p^2/4m)]$, $E_m = 2E_{1s}-E_m^b$,
$v_p^2 = (p^2/4M+N_oW_o-E(p))/2E_{(p)}$ and $u_p^2 + v_p^2 = 1$. Here N_o is the
number of condensed excitonic molecules with zero momentum. The
Green's function of an exciton created in the emission process is
obtained in the first order approximation as follows:

$$G_{ex}(\varepsilon,p) = 1/[\varepsilon-E_{1s} - p^2/2M - NV_o \pm i\delta] .$$

By the use of these Green's functions and the Hamiltonian eq. (2), we can evaluate the emission spectrum by neglecting the vertex correction in the emission process as follows:

$$I(\nu) = -G_o^2 \mathrm{Im} \int \frac{id\varepsilon}{2\pi} \int \frac{d\underline{K}}{(2\pi)^3} G_m(\varepsilon,k) G_{ex}(\varepsilon-h\nu,k)$$

$$= \pi G_o^2 N_o \delta(X) + \frac{2G_o^2}{3\pi a^3 E_{ex}^b} (M/2\mu)^{\frac{3}{2}} (\frac{NW_o}{E_{ex}^b})^{\frac{1}{2}} g(X/NW_o) , \quad (4)$$

where $X = -h\nu+(E_m-E_{1s})+N(W_o-V_o)$ and

$$g(y) = \frac{\left\{\frac{2+4y}{3} - \sqrt{(\frac{2+4y}{3})^2 - \frac{4y^2}{3}}\right\}^{\frac{1}{2}} \cdot \left\{y + 2 - \frac{3}{2}\sqrt{(\frac{2+4y}{3})^2 - \frac{4y^2}{3}}\right\}}{\sqrt{(\frac{2+4y}{3})^2 - \frac{4y2}{3}}}$$

The first term of eq. (4) describes the sharp emission line from the condensed molecules and the second term gives the side band on the low frequency side, coming from the interactions between the molecules. The emission spectrum eq. (4) is drawn in Fig. 1 for the numerical values of $NW_o = E^b(ex)/10$, $(V_o/W_o) = 1/2$ and $M/2\mu = 2$. We list the characteristics of the emission spectrum from the coherent system (5) as follows:

(1) The sharp emission line from the condensed state is expected at $N(W_o-V_o)$ above E_m-E_{1s}. The shift $N(W_o-V_o)$ is the difference between the level shifts of an excitonic molecule and an exciton due to the repulsive interactions with N-1 excitonic molecules and is negligibly small. It is noted here that the condensed <u>single excitons</u> with zero translational momentum can not be observed because the state with the zero momentum is not connected directly to the radiation.

(2) Below this line, the emission side band rises linearly with $E_m-E_{1s}-h\nu$ and the peak is formed with the tail at the low energy side decreasing in proportion to $(E_m-E_{1s}-h\nu)^{-3/2}$. This side band comes from the process in which a pair of condensed molecules are excited to the momentum states K and -K, and the emission process is induced in any of these molecules. Then the phonon-like collective mode of the excitonic molecules is left in the system and this process is reflected in the linear dependence of the side band on the energy difference X.

EMISSION SPECTRUM FROM THE SYSTEM OF MANY EXCITONIC MOLECULES AT FINITE TEMPERATURE

Under certain experimental conditions, the system of high density excitons is in thermal equilibrium with some definite

temperature higher than the lattice. It is observed from the
emission spectrum of excitonic molecules (4) in CdS at low density,
that they are in thermal equilibrium according to the Boltzmann
distribution at $\sim$ 20K. On the other hand, for such a high density
that the excitonic molecules collide with each other frequently,
the translational momentum of the molecule does not remain a good
quantum number. It is discussed in this section how this collision
modifies the line shape of the emission spectrum. In the same
approximations as in the previous section, the emission spectrum
is represented in terms of the temperature Green's functions
$g_{ex}(\varepsilon,p)$ and $g_m(\varepsilon,p)$ for an exciton and a molecule respectively,
as follows:

$$I(\nu) = -G_o^2 T \sum_\varepsilon \int \frac{dp}{(2\pi)^3} \, g_m(\varepsilon,p) g_{ex}(\varepsilon-h\nu,p) \; . \qquad (5)$$

Here when we take into account the level shifts $\Sigma_m(\varepsilon,p)$ and $\Sigma_{ex}(\varepsilon,p)$
and the level broadenings $\Gamma_m(\varepsilon,p)$ and $\Gamma_{ex}(\varepsilon,p)$ for a molecule and
and exciton, respectively, due to the interactions with N-1 exci-
tonic molecules, in the lowest order approximation, we obtain the
following expressions for the Green's functions:

$$g_m(\varepsilon,p) = 1/[i\varepsilon - E_m - p^2/4M - \Sigma_m(\varepsilon,p) \pm i\Gamma_m(\varepsilon,p)]$$

and

$$g_{ex}(\varepsilon,p) = 1/[i\varepsilon - E_{ex} - p^2/2M - \Sigma_{ex}(\varepsilon,p) \pm i\Gamma_{ex}(\varepsilon,p)]$$

where $\Sigma_m = NW_o$ and $\Sigma_{ex} = NV_o$,

$$\Gamma_m = (M/2\mu)^{3/2}(kT/E_{ex}^b)^{\frac{1}{2}}NW_o^2/8\pi^{7/2}E_{ex}^b a_o^3$$

and $\Gamma_{ex}=(M/2\mu)^{3/2}(kT/E_{ex}^b)^{\frac{1}{2}}NV_o^2/2\pi^{7/2}E_{ex}^b a_o^3$ were approximated by the
values at $\varepsilon = E_m+K^2/4M$ and $\varepsilon = E_{ex}+K^2/2M$, respectively, for small
value of K. The emission spectrum is evaluated as follows:

$$I(\nu) = \frac{G_o^2 M^{3/2}}{\sqrt{2}\pi^2} \int_{-\infty}^{\infty} \frac{\Gamma n(\varepsilon) \, d\varepsilon}{(\omega-\Sigma-\varepsilon)^2 + \Gamma^2} \qquad (6)$$

$$= (G_o^2 M^{3/2}/\sqrt{2}\pi^{3/2})\sqrt{X^2+\Gamma^2} \, \exp(-X/RT)\{\sin(\alpha-\Gamma/kT)E_i(X)$$

$$+ \cos(\alpha-\Gamma/kT)E_r(X)\}$$

where $\tan \alpha = (\sqrt{X^2 + \Gamma^2} + X)/\Gamma, \Gamma = \Gamma_{ex} + \Gamma_m$,

$$X = -h\nu + \Sigma = -h\nu + (E_m-E_{ls}) + N(W_o-V_o) \text{ and}$$

$$E_r(X) + iE_i(X) = \text{erfc}(y+ix) \text{ with } \binom{y}{x} = (\sqrt{X^2+\Gamma^2} \mp X)^{\frac{1}{2}}/\sqrt{2} \, .$$

As the values of W_0 and V_0 are of the same order, the difference $N(W_0-V_0)$ is much smaller than Γ and will be almost negligible. As Γ is also proportional to the concentration N, Γ is much smaller than kT at low concentrations. On the other hand, Γ becomes larger than kT at high concentrations. Therefore for both limiting cases, we obtain the following expression:

$$I(\nu) = \begin{cases} A'\sqrt{X} \exp(-X/kT) & \text{for } \Gamma \ll kT \\[2ex] \dfrac{A\Gamma}{(X - \Sigma_0)^2 + \Gamma^2} & \text{for } \Gamma \gg kT \end{cases} \qquad (6')$$

where $A=NG_0^2(M/\pi\mu)^{3/2}/(4E_{ex}^b a_0^3)$, $A'=\pi(kT)^{3/2}A$ and $\Sigma_0=\Sigma_m-\Sigma_{ex}$. This equation indicates that the spectral shape changes from the inverse Boltzmann distribution in the low density limit to the Lorentzian type in the case of high density. This change of the emission spectrum is in accordance with those observed in CdS (2,4). From this comparison of the theoretical spectrum with those observed in CdS, we can determine the point at $X=0$, i.e. $h\nu=E_{ex}-E_m^b$, from which we obtain the molecular binding energy $E_m^b = 5.4$ meV. In addition, Γ/kT is determined as $0.6 \sim 0.9$, depending on the excitation intensity, with $T = 17K \sim 20K$.

EFFECTS OF INELASTIC SCATTERING OF TWO MOLECULES

The excited state of the internal motion of an excitonic molecule is considered to be the unbound scattering state of two excitons due to the light translational exciton mass, at least for $\sigma=m_e/m_h > 0.1$. In this section, let us consider the effects on the emission spectrum of the inelastic scattering between molecules, in which one of the excitonic molecules is scattered into unbound two excitons and another excitonic molecule is converted to a single exciton with the emission of light. Some part of this effect is taken into account by adding the effect due to the inelastic scattering

$$\Gamma_{ex}^{inelastic}(\varepsilon,p) \text{ or } \Gamma_m^{inelastic}(\varepsilon,p) \text{ to } \Gamma \text{ in Eq. (6).}$$

As a result, we have a side band around $E_{ex}-2E_m^b$ as two molecules are converted into three excitons and radiation. In addition to these contributions illustrated by a and b of Fig. 3, the diagram c of Fig. 3 gives a side band with the same order of emission intensity. Summing up these three processes and denoting by Γ the sum of the relaxation frequencies due to the elastic scattering of the particles involved in the light emission, we obtain the emission

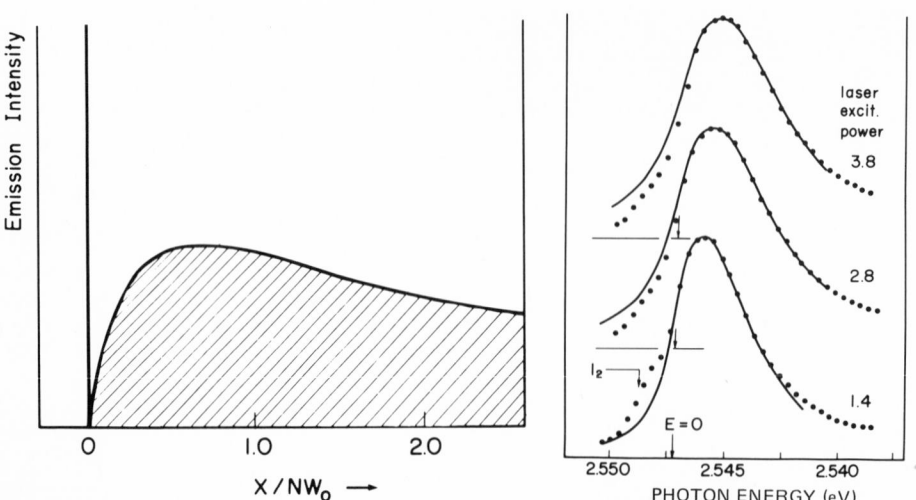

Fig. 1   Sharp emission line from the Bose condensed state of the
         excitonic molecules and the side band on the low energy
         side for $NW_O = E_m^b/10$, $V_O/W_O = 0.5$ and $m_e = m_h$.

         $X = E_{ex} - E_m^b + N(W_O - V_O) - h\nu$.

Fig. 2   The emission spectrum from interacting excitonic molecules
         changes from the Boltzmann shape to the Lorentzian shape at
         high excitation.   The points indicate the experimental data
         of Ref. 2.

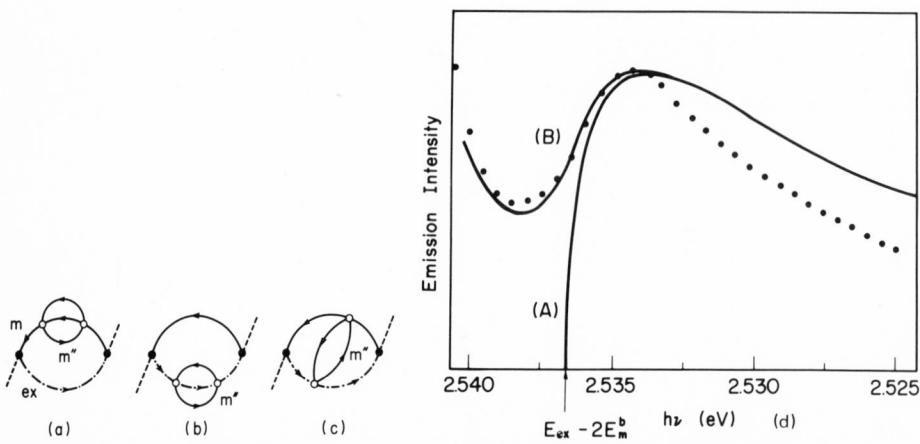

Fig. 3   The side band due to inelastic collision of excitonic mole-
         cules with $E_m^b = 8.5$ meV and $\Gamma = 0(A)$ and $\Gamma/E_m^b = 0.1(B)$.   The
         points indicate the experimental data of Ref. 9.   a,b, and
         c are the processes of photo-emission accompanied by in-
         elastic collision, where m" denotes the unbound state of
         two excitons.

spectrum of the side band which appears below $E_{ex}-2E_m^b$ as follows:

$$I(\omega) = \frac{N^2 G_o^2 V_o^2 (M/\mu)^3 [X+\sqrt{X^2 + \Gamma^2}\ ]^{\frac{1}{2}}}{64\pi^{5/2}(E_{ex}^b)^{5/2} a_o^6 (X + E_m^b)^2} \qquad (7)$$

where V_o, the matrix element of the inelastic scattering is approximated by a constant, and it is assumed that $kT \ll \Gamma$. According to this expression, the emission intensity of this side band is proportional to the square of the molecular concentration, i.e., the fourth power of the exciton concentration, in accordance with the experimental results of CdS (4). For small values of Γ, the emission spectrum rises as $\sqrt{X}$ from X=0, i.e., $h\nu=E_{ex}-2E_m^b$, where $X=E_{ex}-2E_m^b - h\nu$, and the peak consists of a tail on the low energy side decreasing in proportion to $X^{-3/2}$. As Γ increases, the sharp rise at X=0 becomes smeared. The long tail on the low energy side comes from the freedom in the relative motion of two excitons dissociated by the inelastic collision. Here the scattering potential was approximated by the value with the relative momentum equal to zero and as a result, the long tails are too much emphasized in comparison with the experimental data. However, the peak position and the characteristic long tail are taken into account by the theoretical curve drawn in Fig. 3d.

CONCLUSION

The emission from excitonic molecules dominates the emission spectrum from heavily excited semiconductors. Therefore, we have calculated the binding energy and the electronic structure of an excitonic molecule for any value of the isotropic effective masses (1) and discussed the case of anisotropic masses of the composite particles (2). In this paper, we have given the theory of the emission spectrum of a system of excitonic molecules. We have shown how the binding energy is estimated from the data by comparing the theoretical and experimental curves, how the boson-like character of the excitonic molecules--bose condensation--is reflected in the emission spectrum and how their interactions are reflected in the shape of the emission line and the side band. We believe these theoretical results are useful in analyzing experimental results.

The author is grateful to Professor S.Shionoya, Dr. H. Saito, O. Akimoto and M. Inoue for helpful discussions.

REFERENCES

1.   O. Akimoto and E. Hanamura: Solid State Commun. 10, 253 (1972) and J. Phys. Soc. Japn. 33 (1972) No. 6.

2.   S. Shionoya, H. Saito, E. Hanamura and O. Akimoto: to be
     published in Phys. Rev. Lett.

3.   H. Souma, T. Goto, T. Ohta and M. Ueta: J. Phys. Soc. Japan
     $\underline{29}$, 697 (1970). A. Mysyrewicz, J. B. Grun, R. Levy, A. Bivas
     and S. Nikitine: Phys. Lett. $\underline{26A}$, 615 (1968).

4.   H. Saito and S. Shionoya: to be published.

5.   E. Hanamura: Tech. Report of ISSP No. 520 (1972) and to be
     published in Solid State Commun. $\underline{11}$, 485 (1972).

SOME FEATURES OF INTERBAND LUMINESCENCE UNDER INTENSE LASER IRRADIATION

V. A. Kovarskii, E. Yu. Perlin and E. P. Sinyavskii

Institute of Applied Physics, Academy of Science of the

Moldavian SSR, Kishinev, USSR

ABSTRACT

A theoretical discussion of interband luminescence under conditions of laser irradiation is developed. A new luminescence band at frequencies corresponding to the gap in the quasi-energetic spectrum is predicted.

When the energy spectrum of a crystal is perturbed in the presence of radiation from a strong laser, some additional characteristics appear in the optical absorption and emission; these features are the topic of the present discussion.

Some features of interband absorption and luminescence in the intrinsic crystals corresponding to the gap in the renormalized quasi-energy band spectrum will be considered. The gap arises if the laser radiation frequency ω coincides with the distance between the first and second conduction band in a point $\underline{k}_0$ of the $\underline{k}$-space:

$$\varepsilon_{c2}(\underline{k}_0) - \varepsilon_{c1}(\underline{k}_0) = \hbar\omega \tag{1}$$

The following inequality is assumed to be satisfied: $\hbar\omega \ll \varepsilon_{c1}(\underline{k}) - \varepsilon_v(\underline{k})$, with $\varepsilon_v(\underline{k})$ being the upper valence band energy. Hence the laser irradiation does not generate single photon excitation of electron-hole pairs.

Equation (1) represents the degeneracy condition for the combined electron-photon system, consisting of the electron subsystem (the optical electrons of the intrinsic crystal) and the photon subsystem. The gap arises within the conduction band cl since the degeneracy is removed by the electron-photon interaction.

The renormalized quasi-energy spectrum of elementary electron-
hole excitations was obtained in the paper of Perlin and Kovarskii
(1), where the optical interband double resonance had been treated.
The following expression was derived for the electron-hole pair
energy under the resonant radiation:

$$\varepsilon^{(1,2)}(\underline{k}) = \varepsilon_{c1}(\underline{k}) - \varepsilon_v(\underline{k}) + \frac{a_{\underline{k}}}{2} \pm \frac{1}{2}\left[a_{\underline{k}}^2 + 4\lambda_{\underline{k}}^2\right]^{1/2}, \tag{2}$$

$$a_{\underline{k}} = \varepsilon_{c2}(\underline{k}) - \varepsilon_{c1}(\underline{k}) - \hbar\omega, \tag{3}$$

$$\lambda_{\underline{k}} = -\frac{ie\hbar}{m}\left(\frac{2\pi\hbar}{L^3}\right)^{1/2}\frac{<kc_1|F_{\underline{k}j}\nabla|kc_2>}{\sqrt{\omega_{\underline{k}}\kappa_{\underline{k}}}}\sqrt{n_{\underline{k}}} \tag{4}$$

Here $\underline{k}$ is the wave vector, j is the polarization of photons,
e and m are the free electron charge and mass, $\kappa_{\underline{k}}$ is the high
frequency dielectric constant, L^3 is the normalized volume, $|kc_i>$
are the Bloch wave functions of the electron in the i-th conduction band, $n_{\underline{k}}$ is the average photon number.

As it was shown in Ref. 1 the gap arises if the signs of the
effective masses μ_1 and μ_2 are opposite ($\mu_{1,2}^{-1} = m_{c1,2}^{-1}+m_v^{-1}$, m_{c1}, m_v
being the effective masses of the carriers in the conduction bands
and in the valance band, respectively). The relative positions of
the bands are shown in Fig. 1, where the bands in the absence of
laser irradiation are represented by solid lines. In the same
figure the broken lines schematically represent the quasi-energetic
band spectrum in the presence of the laser radiation field.

At the intersection of the resonance frequency,
$\Omega_0 = (1/\hbar)[\varepsilon_{c1}(\underline{k}_0)-\varepsilon_v(\underline{k}_0)]$, the absorption coefficient of the weak
light Ω goes to zero when

$$|\Omega-\Omega_0| < (2\lambda/\mu_1)[\mu_{12}(\mu_1-\mu_{12})]^{1/2}$$
$$(\mu_{12}^{-1} = \mu_1^{-1}-\mu_2^{-1}) \tag{5}$$

With the laser radiation intensity being of the order of 10MW/cm^2
and taking standard values for matrix elements of interband transitions both the width of the region of absorption "cut-out" and the
width of the gap reach values of $\sim 10^{-1}-10^{-2}$ eV.

Similar results were obtained independently by Yacoby. Yelesin
and co-workers (see, e.g. (3)) examined the gap in elementary crystal
excitation spectra, when the laser irradiation frequency ω is
resonant with the transitions between the valence and conduction
bands.

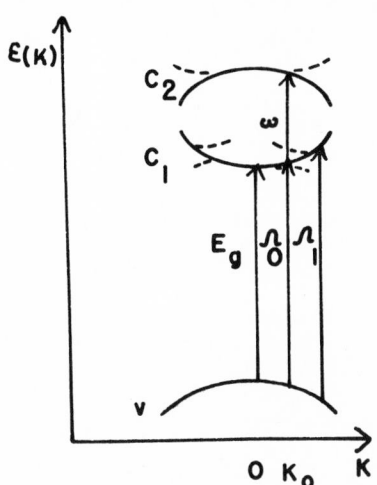

Fig. 1 Energy band spectrum and optical frequencies.

We shall consider a nondegenerate wide gap semiconductor. Let the electrons be pumped optically to the conduction band, c1, at pumping frequency $\Omega_1 > \Omega_0$. The intraband relaxation rate is supposed to be determined by the interaction with LO-phonons, their energy being less than the induced gap width.

The presence of the induced gap is similar to the rise of a new band edge, since the higher excitation energy branch has its minimum at $\underline{k} = \underline{k}_0$. Correspondingly the induced edge luminescence occurs at the frequencies $\Omega \sim \Omega_0$.

In general, if all intraband relaxation mechanisms are disregarded except the electron-phonon interaction, the interband recombination rate is expressed as:

$$W = \sum_{u,u',u''} \rho_u Sp\{\rho^0_{phon}\rho^0_{phot} \int_{-\infty}^{\infty} dt\ exp(+i\Omega t) \times$$

$$\times <u|\underline{d}|0><0|\underline{d}|u'>exp(i\ \varepsilon_{u'}t/\hbar)\ <u'|y(t)|u''><u''|Z(t)|u>\},\tag{6}$$

$$y(t) = T\ exp\{-\frac{i}{\hbar}\int_0^t dt_1\ exp[\frac{i}{\hbar}t_1(H^0_{el}+H^0_{phot})]\times$$

$$\times H_{ef}\ exp[-\frac{i}{\hbar}t_1(H^0_{el}+H^0_{phot})]\},\tag{7}$$

$$Z(t) = T \exp\{-\frac{i}{\hbar} \int_0^t \exp[\frac{i}{\hbar}(H^0_{e\ell}+H^0_{phon}+H_{ef})t_1] \times$$

$$\times H_{ep} \exp[-\frac{i}{\hbar}(H^0_{e\ell}+H^0_{phon}+H_{ef})t_1]dt_1\}$$

(8)

The notation in Eqs. (6)-(8) is as follows: $\underline{d}$ is the dipole momentum operator; $|0\rangle$ is the crystal ground state in the absence of electron-hole pairs; $|u\rangle$ are the crystal excited states with a single electron-hole pair; ρ_u is the initial excitation distribution; ρ^0_{phon} and ρ^0_{phot} are density operators of the phonon and photon subsystems; $H_{e\ell}$, H_{phon}, H_{phot} are Hamiltonians of free electron, phonon, photon subsystems; H_{ef} and H_{ep} denote the electron-photon and electron-phonon interaction; T is the time ordering symbol.

It is easy (1) to obtain the following expression for $\langle u|y|u'\rangle$:

$$\langle u|y(t)|u'\rangle \simeq \delta_{uu'}\langle u|y(t)|u\rangle \equiv \langle \underline{k}|y(t)|\underline{k}\rangle =$$

$$= \frac{1}{2}\{[1 - a_{\underline{k}}(a^2_{\underline{k}}+4\lambda_{\underline{k}})^{-1/2}] \exp[-\frac{i}{2}(a_{\underline{k}}+(a^2_{\underline{k}}+4\lambda^2_{\underline{k}})^{1/2})t] +$$

$$+ [1 + a_{\underline{k}}(a^2_{\underline{k}}+4\lambda^2_{\underline{k}})^{1/2}] \exp[-\frac{i}{2}(a_{\underline{k}}-(a^2_{\underline{k}}+4\lambda^2_{\underline{k}})^{1/2})t]\},$$

(9)

The methods of treating $\langle u'|Z|u\rangle$ were developed previously (4,5). For brevity we shall not present the expressions obtained, but rather go on to a qualitative analysis.

After optical pumping the electron goes to the higher subzone ε_1 of the split conduction band and relaxes to the minimum of this subzone (i.e. to the border of the gap). Then the electron can either recombine with a hole in the valence band or move to the lower subzone ε_2 with the subsequent relaxation to the edge of this band and emission of a photon $\hbar\Omega \sim E_g$ (E_g equals the forbidden band gap).

The most important mechanism generating the electronic transitions between ε_1 and ε_2 subzones is the interaction with LO-phonons. It should be born in mind, however, that single phonon transitions between the subzones are impossible, since the gap width is more than LO-phonon energy.

Two-phonon intraband scattering was investigated in the papers of Anselm and Lang (6,7). Following the formulae obtained in (3, 6,7), the ratio of the life times, τ_1 and τ_2, characterizing the single-phonon and two-phonon scattering is given by

$$\tau_1/\tau_2 \sim 10^{-4} - 10^{-5}$$

in the case of the low temperatures and crystals mentioned above. The analysis shows the presence of the gap to cause a five to tenfold decrease of two-phonon transition rates. It means that the lifetime of two-phonon transitions between the subzones can become more than the time of recombination between hot electrons and holes ($\tau_k \sim 10^{-7} - 10^{-8}$ sec).

Hence, in the presence of the strong laser irradiation, an additional interband luminescence band can be observed at the frequencies near Ω_0 corresponding to the gap in the quasi-energetic spectrum.

REFERENCES

1. E. Yu. Perlin and V. A. Kovarskii, Soviet Phys. Solid State, 12, 2512, (1972).

2. Y. Yacoby, Phys. Rev., B1, 1966 (1970).

3. V. F. Yelesin, Zh. Exper. Teor. Fiz. 59, 602 (1970).

4. E. Yu. Perlin and V. A. Kovarskii, Optika i Spektroskopiya 30, 323 (1971).

5. E. P. Sinyavaskii, Fizika Tverdogo Tela. 13, 2085 (1971).

6. A. I. Anselm and I. G. Lang, Fizika Tverdogo Tela. 1, 685 (1959).

7. I. G. Lang, in Fizika Tverdogo Tela. (2), edited by A. F. Ioffe, p. 186, Leningrad, (1959).

EXPERIMENTAL INVESTIGATIONS ON LUMINESCENCE AT HIGH

CONCENTRATIONS OF CARRIERS AND EXCITONS

R. Levy, A. Bivas and J. B. Grun

Laboratoire de spectroscopie et d'optique du corps

solide, Université Louis Pasteur, Strasbourg, France

ABSTRACT

The radiative recombination properties of some crystals (CuCl, CdS) excited by a powerful ultraviolet light source have been studied at low temperatures. New effects due to high concentrations of carriers and (or) excitons have been observed.

HIGHLY EXCITED COPPER CHLORIDE CRYSTALS

The luminescence of CuCl crystals has been measured for different intensities of the light excitation at 77K as shown in Figure 1. A strong emission line ν_B (25655 cm-1, 3.1802 eV) has been observed on the low energy side of the excitonic emission spectrum. This peak has been identified as the recomination of an excitonic molecule (1,2). One of the excitons of the molecule is scattered into the n=1 state of the free exciton while the other one is scattered into a photon-like state. This interpretation is based on two facts: the approximate square dependence on excitation intensity of the observed line intensity, and the emitted photon energy in agreement with the theory of the binding energy of the excitonic molecule (3,4,5,6).

The line ν_0 due to free exciton recombination is also observed. At low levels of excitation, this emission is strongly reabsorbed in the crystal. Two maxima (26040 cm-1, 25970 cm-1) separated by a minimum (26008 cm-1) corresponding to the n=1 excitonic absorption line are observed. When the light intensity is increased, the minimum due to the reabsorption disappears and

136

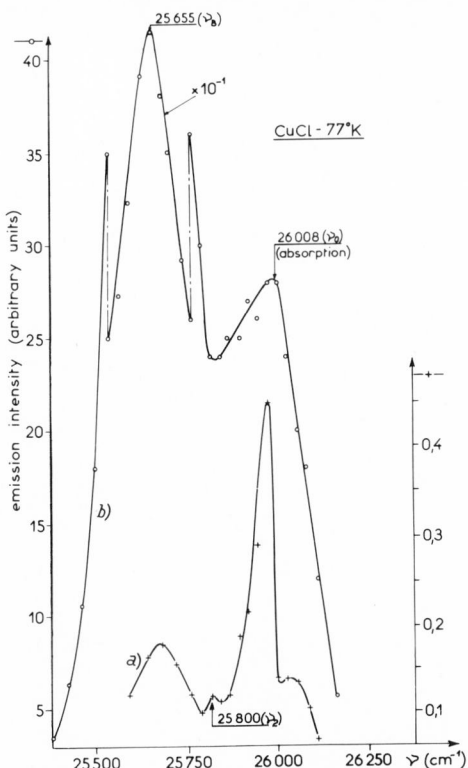

Fig. 1   Emission spectrum of CuCl at 77K excited by a powerful
ultraviolet laser light.   a) 0.7×10^{22} photons/cm^2 sec.
b) 10^{25} photons/cm^2 sec.

becomes a maximum of luminescence at exactly the value of the n=1
exciton state.

    In order to explain the disappearance of this reabsorption
in the resonant emission line ν_o, we have studied the excitonic
absorption of thin samples of CuCl during UV light excitation.   A
strong modification of the excitonic series has been observed as
the intensity of the UV light increased (7,8).

    The n=1 absorption line of the first excitonic series is
shifted towards higher energies (blue shift) while the resonance
n=1 emission line remains unshifted ruling out a trivial heating
of the sample.   As seen on Fig. 2 for two temperatures, this shift
is approximatively square-root dependent on the incident excitation light intensity.   Shifts of the order of 6 meV have been
obtained with 10^{25} UV photons/cm^2 s at liquid nitrogen temperature.

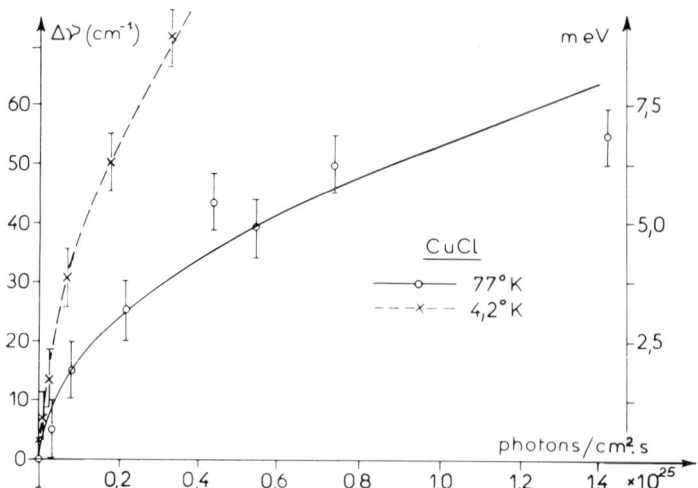

Fig. 2   Shift of the n=1 absorption line of CuCl versus the flux
density of ultraviolet light: o,x experimental points at
77K and 4.2K respectively, ———, ——— square-root dependence
on light flux density at 77K and 4.2K.

The n=1 absorption line of the second excitonic series has the
same behavior.

In these experiments, a large number of hot carriers are
created by absorption of the strong ultraviolet excitation. These
carriers thermalize very rapidly to the band extrema (10^{-11} sec).
These pairs of thermalized carriers will essentially form excitons
in this large band gap semiconductor as shown by Toyozawa (9).

A situation where electrons and holes are not forming excitons
very rapidly has been considered (7). A non-equilibrium situation
in which the excited electrons occupy the bottom of the conduction
band is assumed. When an exciton is then created by absorption,
its wave function cannot be formed out of all the conduction band
states because it must be orthogonal to the occupied states (Pauli
principle). The binding energy of such an exciton has been calculated by a variational principle. The energy of this exciton
increases with the carrier concentration N by the following
energy:

$$\Delta E = 32\pi \ a_x^3 \ N \ E_x$$

where E_x is the binding energy of the exciton, a_x the radius of
the exciton in the ground state. The carrier concentration N is
proportional to the square-root of the excitation intensity I.
The increase ΔE of the exciton energy is square-root dependent on

I. It must be emphasized that the above arguments remain valid
if the electrons and holes partially or completely form excitons.
In this case however the numerical value of the constant in the
above formula changes.

The excitonic absorption spectrum obtained corresponds to the
creation of new excitons in this non-equilibrium situation.   The
observed square-root dependence of the n=1 line shift on the
excitation intensity is predicted by this model. The theoretical
order of magnitude of the shift agrees well with the experimental
values for a carrier concentration of 10^{18}.

On the contrary, when the exciton recombines, on account of
the selection rules: $\Delta \underline{K}=o$, $\underline{K}=o$ where $\underline{K}$ is the exciton wave
vector, the "optical" excitons only will participate in this recom-
bination. These excitons are formed from all states so that their
binding energy is not changed.

The elementary excitation energy necessary to create a new
exciton in a high density of excitons has been also calculated by
Hanamura (10,11) and by Keldysh and Kozlov (12).

These two theories are almost equivalent and predict a shift
of the exciton level which is given by the formula:

$$\Delta E = f \, a_x^3 \, N_x \, E_x$$

where N_x is the exciton concentration and f a numerical factor
which is of the order of 9π for Hanamura and of the unity for
Keldysh and Kozlov.

However in these theories condensation and superfluity of
excitons are supposed. Further, the theory predicts that the
emission and the absorption involve the same energy level. Therefore the emission band should be shifted to higher energies in the
same way as the absorption band. This is, however, in contradiction with the experiment. One is tempted to conclude that the
experiment is in contradiction with Bose condensation of excitons.
It has been also suggested that the shift of the absorption line
could be due to a change of the dielectric constant (7). In this
case however the shift of both the absorption and the emission
bands is predicted, in opposition to the experiment.

In the present state of this research, it seems that the
above theory (7) is the only one in agreement with the experiment.

When a high density of excitons and (or) carriers is created
in CuCl, the exciton energy is then strongly changed. In a different type of crystal, such as CdS, this effect is not observed

but another phenomenon which may also be related to high concentration of carriers has been obtained as we shall now show.

HIGHLY EXCITED CADMIUM SULFIDE CRYSTALS

The luminescence of thick samples of CdS excited by a high density of ultraviolet photons has been studied at 4.2K. An intense emission line M (20385 cm^{-1}, 2.5269 eV) appears at high excitation levels on the low energy side of the excitonic spectrum. This peak has been identified by Benoît à la Guillaume et al. (13, 14) as an exciton - exciton collision recombination. One of the excitons scatters into a dissociation state while the other one scatters into a photon-like state. This interpretation is based on two facts: the approximate square dependence of the emission intensity on the excitation intensity and the emitted photon energy which is lower than the n=1 excitonic state energy by about the exciton binding energy.

We have observed a shift of the maximum of this line towards the low energy part of the spectrum, when the excitation intensity is increased (15). This shift has been measured for different excitation intensities. The experimental points obtained are shown on Fig. 3.

A global heating of the sample has been ruled out. The spectral position of the I_2 line due to the recombination of an exciton bound to a neutral donor has been studied under the same light excitations. This line shows no appreciable shift.

The anomalies of reflection corresponding to the excitonic absorption lines of these samples have been observed during the UV light excitation. No appreciable shift of the maxima and minima has been noticed when the UV light intensity is increased. This observation is confirmed by absorption induced measurements during the excitation, done by Goto and Langer (16).

As shown, the shift of the M line is not due to a change of the binding energy of the exciton. It can however be related to the recombination process involved in this emission. The energy balance of this process is the following: $\hbar\omega_M = E_g - 2E_x - T_{e,h}$ where $\hbar\omega_M$ is the energy of the emitted photon, E_g is the band gap energy, E_x the exciton binding energy, $T_{e,h}$ the kinetic energy of the unbound pairs of electrons and holes created during the recombination. The kinetic energy of the excitons is neglected.

When the excitation intensity is low, the bands are empty. The electrons and holes created during the exciton-exciton collision can have the lowest energies possible. Their kinetic

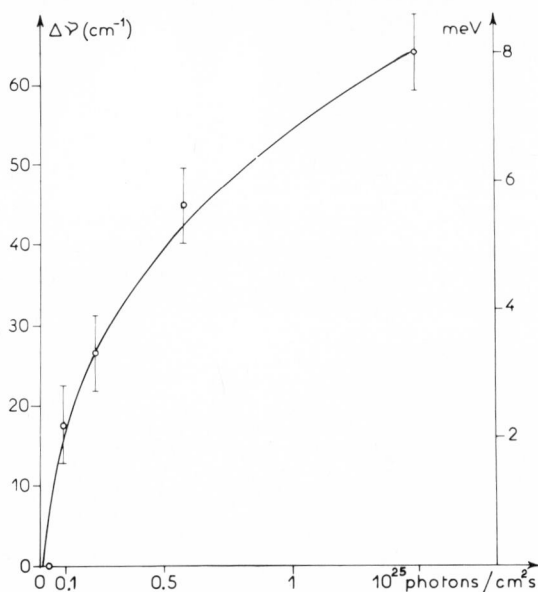

Fig. 3   Shift of the M emission line of CdS versus the flux
         density of ultraviolet photons at 4.2K. (We have taken
         for origin of the shift the position of the M line at the
         lowest flux density): o experimental points, - one-third
         power dependence on light flux density.

energy $T_{e,h}$ may be neglected.

When the excitation intensity is large, a high density of
free carriers is then created - The existence of a large number of
carriers has indeed been observed by Goto and Langer (16). They
have seen an induced absorption due to their carriers during the
UV light excitation of the crystals - Under these conditions, the
extrema of the bands are filled up to quasi Fermi levels in a new
stationary state different from that of thermal equilibrium. The
electrons and holes created during the exciton-exciton collision
must have higher energies. They need an additional kinetic energy
$T_{e,h}$. The energy of the emitted photon will then be decreased by:
$\hbar \cdot \Delta \omega = -T_{e,h}$. Assuming a band filled effect with elliptical bands
having only one extremum, Haken has calculated a line shift equal
to (15):

$$\hbar \cdot \Delta \omega = [m_e^{*-1} + m_h^{*-1}](h^2/8)(3/8\pi)^{2/3} N^{2/3}$$

m_e^* is equal to $(m_{x,e}^* m_{y,e}^* m_{z,e}^*)^{1/3}$ where $m_{i,e}^*$ are the elements of
the effective mass tensor, assumed diagonal, for the conduction
band. m_h^* is defined in an analogous way for the valence band. N
is the number of free carriers per unit volume and depends on the

square-root of the excitation intensity I in the steady state case.

The emission line shift depends on the one-third power of the excitation intensity. This theoretical dependence fits quite well with the experimental points as shown on the third figure.

The theoretical value of the shift is also in good agreement with the experimental value when 10^{18} carriers per unit volume are assumed in the excited crystals.

CONCLUSION

In highly excited crystals, intense new emission lines due to excitonic molecule recombination as in CuCl or to exciton-exciton collision recombination as in CdS have been observed. Other new effects related to the presence of high densities of excitons and (or) carriers have also been observed such as the "blue" shift of the free exciton absorption line in CuCl, the emission line being unshifted and the "red" shift of the exciton-exciton emission in CdS.

REFERENCES

1. A. Mysyrowicz, J. B. Grun, R. Levy, A. Bivas and S. Nikitine, Phys. Letters, 26 A, 615 (1968). S. Nikitine, A. Mysyrowicz and J. B. Grun, Helv. Phys. Acta, 41, 1058 (1968).

2. A. Bivas, R. Levy, S. Nikitine and J. B. Grun, J. Phys., 31, 227 (1970).

3. R. R. Sharma, Phys. Rev., 170, 770 (1968).

4. R. K. Wehner, Solid State Comm., 7, 457 (1969).

5. Adamovski, Bednarek and M. Suffczynski, Solid State Comm., 9, 2037 (1971).

6. O. Akimoto and E. Hanamura, Solid State Comm., 10, 253 (1972).

7. A. Bivas, R. Levy, J. B. Grun, C. Comte, H. Haken and S. Nikitine, Optics Comm., 2, 227 (1970).

8. R. Levy, A. Bivas and J. B. Grun, Physics Letters, 36A, 159 (1971).

9. Y. Toyozawa, Suppl. Progr. Theor. Phys., 12, 111 (1959).

10. E. Hanamura, Semiconductors Conference, Boston, 487 (1970).

11. E. Hanamura, J. Phys. Soc. Japan, 29, 50 (1970).

12. L. V. Keldysh and A. N. Kozlov, Soviet Phys. JETP 27, 521 (1968).

13. C. Benoit à la Guillaume, J. M. Debever and F. Salvan, Phys. Rev., 177, 567 (1969).

14. D. Magde and H. Mahr, Phys. Rev. Letters, 27, 890 (1970).

15. R. Levy, J. B. Grun, H. Haken and S. Nikitine, Solid State Comm., 10, 915 (1972).

16. T. Goto and D. W. Langer, Phys. Rev. Letters, 27, 1004 (1971).

EXCITON LUMINESCENCE AND PHOTOCONDUCTIVITY IN HIGHLY EXCITED GaP

A. Nakamura and K. Morigaki

Institute for Solid State Physics, University of Tokyo
Roppongi, Tokyo, Japan

ABSTRACT

This paper is concerned with excitonic processes at low temperatures in GaP. We present direct evidence for the non-radiative Auger process of bound exciton annihilation. Additionally, problems relating to the behavior of exciton luminescence and photoconductivity under intense excitation corresponding to the band-to-band transition in undoped or Te-doped GaP are examined.

1. INTRODUCTION

This paper is concerned with two problems related to excitonic processes at low temperatures in GaP. One of them is the annihilation of excitons bound to neutral donors. It has been believed (1) that the lifetime of bound excitons in GaP is determined by the non-radiative Auger process in which a bound exciton annihilates non-radiatively and its energy is put into exciting an electron bound to a donor into the conductive band. We have obtained direct evidence for the non-radiative Auger process of the bound exciton annihilation by observing a photocurrent peak at the position of the C-line (due to the radiative recombination of an exciton bound to a neutral donor (2))in the excitation spectrum of photoconductivity at 4.2K and 1.8K in Te-doped GaP (3).

The other problem is concerned with the behavior of exciton luminescence and photoconductivity under an intense excitation corresponding to the band to band transition in undoped or Te-doped GaP. The results obtained are presented and are discussed in Section 3.

2. EXPERIMENTAL PROCEDURE

The excitation source was a tunable-wavelength pulsed laser which consists of a pulsed N_2 laser and a dye cell placed in a wavelength-selectable cavity. The pulse width and repetition rate was 5 nsec and 10 Hz respectively. The maximum peak intensities of the laser light were $2.4 \times 10^5 W/cm^2$ at 4565A and $4 \times 10^4 W/cm^2$ at 5373A. The laser light was focused into a spot of 1.4 mm **diameter** covering the face with electrodes whose separation was 1 mm. Luminescence measurements were made using a single-type spectrometer (Spex model 1700).

The samples used in this experiment were undoped or Te-doped GaP grown epitaxially from the vapour phase. These samples contain sulfur donors with a concentration of $10^{15}/cm^3$ and tellurium donors with a concentration of $8.5 \times 10^{16}/cm^3$ respectively.

3. EXPERIMENTAL RESULTS AND DISCUSSION

As shown in Fig. 1 the relative photocurrent per incident photon was measured as a function of photon energy in Te-doped GaP at 1.8K. We can see a sharp peak at 2.308 eV and a steep increase in the relative photocurrent near 2.34 eV. In taking into account the fact that the band gap energy of GaP at 1.6 K is 2.339 eV, this steep increase is interpreted as due to the generation of conduction electrons by the band to band transition. A sharp peak at 2.308 eV which corresponds to the energy of the bound exciton is attributed to the non-radiative Auger process of the bound exciton annihilation. A quantitative analysis of the experimental results gives us the rate of the bound exciton annihilation by the Auger process to be $9 \times 10^6/sec$. This magnitude is almost consistent with the measured value of the lifetime of the bound exciton which was about 16 nsec

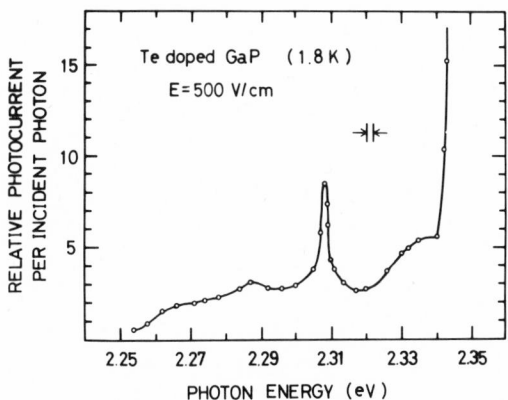

Fig. 1 Excitation spectrum of photoconductivity in Te-doped GaP at 1.8 K.

at 4.2 K. The detailed account of the Auger process has been described in Ref. (3).

    Figures 2(a) and (b) show the intensity of the emission due to
radiative recombination of excitons bound to neutral donors (the
C-line) and the photoconductivity as a function of the excitation
intensity of the laser light (4565A) at 4.2 K in Te-doped GaP
respectively. The resistivity of the sample decreased from $\rho > 10^{10}\Omega$
cm in the dark to $\rho \approx 0.05\ \Omega$cm at the maximum intensity of the laser
light at 4.2K. The resistivity under laser light excitation was
estimated by assuming that the carriers flow within the skin depth
of 2 μ. The emission intensity, I, varies linearly with the excitation intensity L for weak excitations and as the 0.6 power for
intense excitation at 4.2 K. The photoconductivity, σ, varies with
L in the same ways for the different excitation intensities at 4.2 K
and at 1.8 K. The excitation intensities at which the bend of the
slope of the I vs. L curve and of the σ vs. L curve occur are nearly
the same.

    The sample of undoped GaP exhibits similar behaviors in the
intensity of the emission (C-line) and the photoconductivity versus
the excitation intensity to that of Te-doped GaP. However, the
bend of the slope of the I vs. L and σ vs. L curves occur at an
excitation intensity one order of magnitude weaker than that for
Te-doped GaP.

    The laser light of 4565A corresponds to the band to band tran-
sition in the GaP. Therefore, free electron-free hole pairs are
created by this laser light, then some of them form free excitons
at low temperatures such as 4.2 K. In the present case, it is considered that most of free excitons are very quickly bound to neutral
donors. Furthermore, we can reasonably presume that the bound excitons, (N_{exc}), are annihilated by interactions with the neutral donors

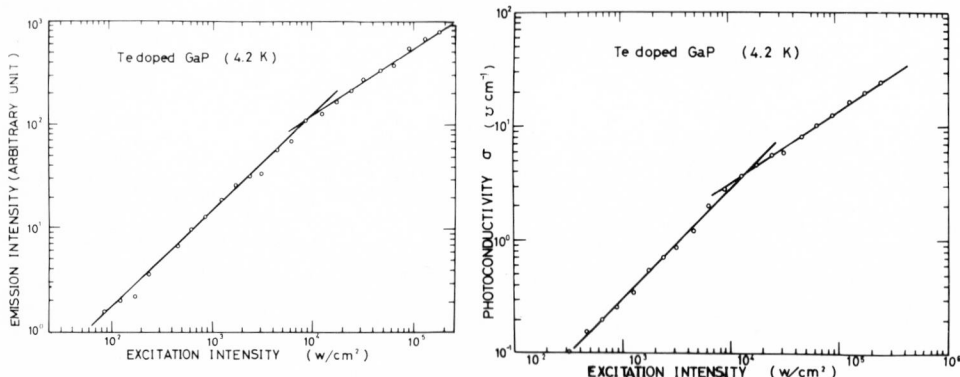

Fig. 2 (a) Emission intensity of the C-line and (b) photoconduc-
        tivity as a function of excitation intensity of the laser
        light (4564A).

and the conduction electrons, and the conduction electrons are trapped by ionized donors, (N_D^+), or combine with free holes to form excitons. Then the following rate equation is written:

$$dN_{exc}/dt = \gamma n^2 - (1/\tau_{nr})N_{exc} - \alpha n N_{exc} , \qquad (1)$$

$$dn/dt = L - \xi N_D^+ n - \gamma n^2 , \qquad (2)$$

where the concentration of free electrons, n, is assumed to be equal to that of free holes, and α, γ, and ξ are the reaction coefficients of the electron-bound exciton collision, the formation of excitons, and the trapping of electrons by ionized donors re- respectively, and τ_{nr} is the lifetime of bound exciton determined by the non-radiative Auger process, and L is the generation rate of free electron-hole pairs.

By considering the values of the parameters in Eqs. (1) and (2), we can conclude that the number of conduction electrons reaches the steady state value within the width of the laser pulse (~ 5 nsec); but, as for the number of bound excitons, the steady state condition is only satisfied in the range of intense excitation ($L \gtrsim 5 \times 10^3 W/cm^2$) where the collision of bound excitons with conduc- tion electrons predominates over the Auger process. From the com- parison of the steady solutions of Eqs. (1) and (2) with the experi- mental result, we conclude that the conduction electrons are anni- hilated by trapping by ionized donors in the region of weak excita- tion, and by exciton formation in the region of intense excitation. As to the bound exciton, it is found that the interaction of bound excitons with conduction electrons is a dominant process for the annihilation of bound excitons in the region of intense excitation. However, when the excitation is weak, it seems that the Auger pro- cess and the interaction of bound excitons with conduction electrons work to the same degree. In order to obtain a definite conclusion in this case, we would have to solve numerically Eqs. (1) and (2).

Accordingly, we can understand qualitatively the behavior of exciton luminescence and photoconductivity with variation of the excitation intensity.

ACKNOWLEDGEMENTS

The authors would like to thank Professor Y. Toyozawa and Professor E. Hanamura for helpful discussion. They also acknow- ledge Dr. I. Akasaki and Mr. I. Asao of Matsushita Research Insti- tute, Tokyo for providing them with samples used in this experi- ment. One of us (A.N.) would like to express his hearty thanks to Professor A. Odajima of Hokkaido University for his continuous encouragement.

REFERENCES

1.   D. F. Nelson, J. D. Cuthbert, P. J. Dean and D. G. Thomas
      Phys. Rev. Lett. 17, 1262 (1966).

2.   D. G. Thomas, M. Gershenzon and J. J. Hopfield, Phys. Rev.
      131, 2397 (1963).

3.   A. Nakamura and K. Morigaki, Tech. Rep. of ISSP, Ser. A
      No. 524, 1972.

INTERACTION OF SINGLET AND TRIPLET EXCITONS IN ORGANIC CRYSTALS

M. Trlifaj

Institute of Physics, Academy of Science

Prague, Czechoslovakia

ABSTRACT

The interaction of singlet and triplet excitons in organic crystals is examined theoretically in connection with the non-radiative annihilation of singlet excitons on triplet excitons and with the formation of incoherent singlet-triplet bi-excitons.

INTRODUCTION

The mechanism of the singlet exciton annihilation by triplet excitons is considered the non-radiative transfer of excitation energy from the singlet to the triplet exciton, and vice-versa, resulting in one triplet exciton with higher excitation energy. As the cause of the formation of the singlet-triplet bi-exciton we consider the dynamical interaction between two excited molecules in the crystal which is expressible in the form of a potential field.

MODEL OF TRIPLET AND SINGLET EXCITONS AND OF THE SINGLET EXCITON ANNIHILATION

For simplicity, we consider the above mentioned processes in organic crystal with one molecule per unit cell. We denote the sites of the molecules in the periodic arrangement in the crystal by lattice vectors $\underline{p}$. We assume that each molecule of the crystal may be divided into π-electrons and other electrons.

We describe the excited states of the electrons of the crystal, i.e. the π-electrons of the crystal, with the help of the neutral excited states of the isolated molecules in the one-electron approximation. In this approximation, we consider the neutral electron excitations of the molecule which are due to transitions of one π-electron from the state with the highest energy (which are π-electron states) in the ground electronic state of the molecule to all possible one-electron states with higher excitation energy. We characterize these electron excitations by the quantum index f and we denote their quasi-boson creation and annihilation operators with regard to the spin states at the p-th molecule of the crystal by $B_{pf}^{+}(S,M_S)$, $B_{pf}(S,M_S)$ where $S = 0,1$; $M_S = -S,...,+S$ are the spin quantum numbers. These neutral electron excitations have the same excitation energy $E_f(S)$ and in the presence of the intermolecular interaction migrate as a coherent or incoherent exciton through the crystal (1). In this description of the excited states of the electrons of the crystal we consider the triplet exciton in the incoherent random walk model as a neutral triplet electron excitation ($S=1$; $M_S=-1,0,1$) hopping randomly from one molecule to another (1). On the other hand, we consider the singlet exciton in the coherent model as a singlet electron excitation moving through the crystal in the form of a coherent excitation wave. These assumptions about the dynamics of triplet and singlet excitons allow us to introduce into the theory the quasi-stationary states of the singlet electron excitation in the field of the triplet electron excitation fixed on some molecule of the crystal. Then we can treat the non-radiative annihilation of the singlet exciton by the triplet exciton as a non-radiative spontaneous transition of the crystal from the initial state, with one singlet electron excitation in the field of the triplet electron excitation fixed on some molecule of the crystal, to the state with one triplet electron excitation with higher excitation energy. The operator H' describing this process of the singlet exciton annihilation is given by:

$$H' = \sum_{p'\neq p} \sum_{M} \sum_{f,f',f''} \{V_{1f_1f'f''}(\underline{p}-\underline{p}')B_{\underline{p}f}^{+}(1,M)B_{\underline{p}f''}(1,M)B_{\underline{p}'f'}(0,0) +$$

$$+ V_{2f_1f'f''}(\underline{p}-\underline{p}')B_{\underline{p}f}^{+}(1,M)B_{\underline{p}f'}(0,0)B_{\underline{p}'f''}(1,M)\}. \qquad (1)$$

In this expression $V_{if;f'f''}(\underline{p}-\underline{p}')$ ($i=1,2$) are the matrix elements of the non-radiative conversion of the f'-th singlet and f''-th triplet electron excitation pair situated at the p-th and p'-th molecule of the crystal into one triplet electron excitation localized on the p-th molecule of the crystal through electrostatic and exchange interaction ($i=1$), or through exchange interaction only ($i=2$).

INCOHERENT SINGLET-TRIPLET BI-EXCITON

Denote by $\underline{p}$ the vector of the singlet and triplet electron excitation separation. For the wave function $C_M(\underline{p})$ in the $\underline{p}$-representation and for the energy of the stationary states of the singlet electron excitation in the field of the triplet electron excitation f_2, M fixed on some molecule of the crystal we can derive the following set of different equations:

$$[E_{f_1}(0)+U_{f_1 f_2}(\underline{p})]C_M(\underline{p}) + \sum_{\underline{p}'\neq\underline{p}} L_{f_1}(\underline{p}-\underline{p}')C_M(\underline{p}') = 0. \qquad (2)$$

Here $U_{f_1 f_2}(\underline{p})$ is the potential energy of the dynamical interaction between f_1-th singlet and f_2-th triplet electron excitation pair at distance $\underline{p}$, and $L_{f_1}(\underline{p}-\underline{p}')$ is the transfer matrix element describing the non-radiative migration of the singlet electron excitation f_1 from one molecule to another.

Generally the set of Eqs. 2 has solutions with a continuous spectrum of energies and can also have solutions with a discrete spectrum of energies. Let λ denote the quantum indices of these solutions.

The solutions with a continuous spectrum of energies belong to the scattering states of the singlet-electron excitation in the field of the fixed triplet-electron excitation. In this case for a vanishing interaction between the singlet and triplet electron excitation, the wave function $C_M(\underline{p})$ of the singlet electron excitation passes over into the wave function of a coherent singlet exciton i.e.

$$C_M(\underline{p}) = (1/\sqrt{N})\ \exp[i(\underline{k}\cdot\underline{p})]. \qquad (3)$$

where $\underline{k}$ is the wave vector of the coherent singlet exciton and N is the number of the unit cells in the crystal.

The solutions with a discrete spectrum of energies belong to the bound states of the singlet electron excitation in the field of the fixed triplet electron excitation. In this case we have a bound pair of singlet and triplet electron excitation, that is, a singlet-triplet bi-exciton whose localization in the crystal determines the localization of the triplet electron exciation.

The transition probability of the triplet electron excitation with the bound singlet electron excitation from one molecule to another is given approximately by:

$$W_{\lambda_i f_1 f_2}(\underline{p}) = W_{f_2}(\underline{p})\ |G_{\lambda_i f_1}(\underline{p})|^2. \qquad (4)$$

Here $W_{f_2}(\underline{p})$ is the probability of the non-radiative transition of the triplet electron excitation from one molecule to another and $G_{\lambda_1 f_1}(\underline{p})$ is the corresponding overlap integral of the wave functions of the singlet electron excitation in the bound state λ. Because $|G_{\lambda_i f_1}(\underline{p})| < 1$ for all $\underline{p} \neq 0$ and for bound states λ, the transition probability of the triplet electron excitation with the bound singlet electron excitation from one molecule to another is always smaller than is the case for the incoherent triplet exciton alone. Therefore, in our model, the motion of the singlet-triplet bi-exciton as a whole must be analyzed in terms of an incoherent random walk model and we call this type of bi-exciton an incoherent singlet-triplet bi-exciton. In the diffusion approximation, the behavior of incoherent singlet-triplet bi-excitons in crystals is described by the diffusion tensor whose components $D_{ij}(S,T)$ $(i,j=1,2,3)$ can be written in the form:

$$D_{ij}(S,T) = \frac{1}{2} \sum_{\underline{p} \neq 0} W_{\lambda_i f_1 f_2}(\underline{p})(\underline{p} \cdot \underline{e}_i)(\underline{p} \cdot \underline{e}_j). \tag{5}$$

Here e_i $(i=1,2,3)$ are the unit vectors of the Cartesian axes.

PROBABILITY OF THE NON-RADIATIVE ANNIHILATION
OF SINGLET EXCITONS

Taking into consideration the broadening of the emission and absorption lines of the incoherent triplet exciton, we derive for the non-radiative annihilation of the singlet exciton, f_1, by the triplet exciton, f_2, with one resulting triplet exciton, f_2, in the final state, the following approximate expression:

$$\gamma = \gamma^1 + \gamma^2. \tag{6}$$

Here,

$$\gamma^1 = \frac{2\pi}{3\hbar} \sum_M \overline{\sum_\lambda} |\sum_{\underline{p} \neq 0} C_{\lambda M}(\underline{p}) V_{1f_2' if_1 f_2}(\underline{p})|^2 \sigma^a_{f_2 f_2'}(E_\lambda), \tag{7}$$

$$\gamma^2 = \frac{2\pi}{3\hbar} \sum_M \overline{\sum_\lambda} |C_{\lambda M}(\underline{p})|^2 |V_{2f_2' if_1 f_2}(\underline{p})|^2 \int \sigma^a_{f_2'}(E_\lambda + E) \sigma^e_{f_2}(E) dE, \tag{8}$$

where $\overline{\sum_\lambda}$ means average over all initial states of the singlet electron excitation in the field of the triplet electron excitation in thermodynamic equilibrium; $\sigma^a_f(E)$ and $\sigma^e_f(E)$ are generally the absorption and emission line shape factors of the f-th incoherent triplet exciton; and $\sigma^a_{f_2 f_2'}(E)$ is the absorption line shape factor belonging to the optical transition of the molecule in the crystal between triplet electron excited states f_2 and f_2'. The absorption and emission line shape factors fulfill the normalization conditions:

$$\int \sigma^a_{f'_2}(E)dE = \int \sigma^e_f(E)dE = \int \sigma^a_{f_2 f'_2}(E)dE = 1. \tag{9}$$

The first term in Eq. 6 is the probability of the non-radiative annihilation of the singlet exciton by the triplet exciton through non-radiative transfer of the excitation energy from the singlet exciton to the triplet exciton. This process takes place through electrostatic and exchange interactions. The second term in Eq. 6 is the probability of the non-radiative annihilation of the singlet exciton by the triplet exciton through non-radiative transfer of the excitation energy from the triplet exciton to the singlet exciton. This process takes place through exchange inter-action only.

Suppose first, that there are no bound states of the singlet electron excitation in the field of the fixed triplet electron excitation. In this case we use the dipole-dipole approximation for the interaction matrix element. Neglecting exchange interaction terms and the dynamical interaction between singlet and triplet electron excitations, and summing over all triplet excitons, we obtain for the probability of the non-radiative annihilation of the singlet exciton by the triplet exciton at low temperatures the approximate formula:

$$\gamma \simeq \frac{32\pi^3}{g\hbar v_0} |(\mu_{0f_1} \mu_{f_2 f'_2})|^2 \sigma^a_{f_2 f'_2}(F_{f_1 min}) n_T. \tag{10}$$

Here v_0 is the volume of the unit cell, n_T is the concentration of the triplet excitons in the crystal, $\mu_{ff'}$ is the transition dipole moment corresponding to the optical transition $f \to f'$ of the molecule in the crystal and $E_{f_1 min}$ is the energy of the coherent singlet exciton f_1 at the bottom of its energy band.

Assume, that $\mu_{0f_1} || \mu_{f_2 f_2}$ and that the energy $E_{f_1 min}$ is in resonance with the energy of the maximum of the absorption band $f_2 \to f'_2$. Then taking $|\mu_{0f_1}| \simeq |\mu_{f_2 f'_2}| = 10^{-1}e(3v_0/4\pi)^{1/3}$, $\sigma^a_{f_2 f'_2}(E_{f_1 min}) \simeq 1/(\Delta E_{f_2 f'_2})$ with $\Delta E_{f_2 f'_2} \simeq 10^{-1}$ eV as the bandwidth of the absorption band $f_2 f'_2$ and $v_0 \equiv 8 \times 10^{-22}$ cm^3, we get $\gamma \simeq 5 \times 10^{-8}$ n_T/sec. This value of the non-radiative annihilation of the singlet exciton by the triplet excitons is not too far from the experimental value $\gamma = (7 \pm 4) \times 10^{-9}$ n_T/sec, as determined for anthracene crystals (3).

When the bound states of a pair of a singlet and a triplet exciton exist, then the non-radiative annihilation of the singlet exciton by the triplet exciton, can go through formation of the incoherent singlet-triplet bi-exciton, as a monomolecular non-radiative decay of the incoherent singlet-triplet bi-exciton also. In this two-step process of non-radiative annihilation of the singlet exciton by the triplet exciton, we must add to the general kinetic scheme for the fluorescence of singlet excitons given in (3) the appropriate singlet-triplet bi-exciton rate equation.

REFERENCES

1. S. A. Rice and J. Jortner, Physics and Chemistry of the Organic Solid State, Vol. III, Interscience Publishers, New York, (1967), p. 199.

2. A. S. Davydov, Soviet Phys.-Uspekhi 82, 393 (1964).

3. S. D. Babenko, V. A. Benderskij, V. I. Goldanskij, A. G. Lavrushko and V. P. Tychinskij, Phys. Stat. Sol. (b) 45, 91 (1971).

ANTI-STOKES EXCITON EMISSION IN CdS CRYSTALS

E. Beckmann, I. Broser and R. Broser

Inst. für Elektronenmikroskopie der Max Planck Gesell-

schaft, Berlin-Dahlem and III. Physik. Inst. der Techn.

Univ., Berlin, West Germany

ABSTRACT

Two photon excited exciton emission induced by a giant pulse ruby laser is investigated in thick CdS platelets with laser intensity varying several orders of magnitude and the temperature between 2 and 300 K. The effect of an additional steady state excitation in the red and near infrared region is studied as well as the decay characteristics of the luminescence. A structured specimen is observed consisting mainly of one phonon replicas of free and bound exciton lines.

INTRODUCTION

Anti-Stokes luminescense with a large spectral separation between the exciting radiation (red) and the emitted light (green) was first observed in CdS by Halsted et al (1). Using different types of continuous light sources several authors (2,3) have studied the effect in greater detail. As a result it has been concluded that the excitation mechanism of steady-state anti-Stokes emission in CdS is a two-step absorption process via a real intermediate energy level in the forbidden gap.

Anti-Stokes exciton emission in CdS has been detected first by Braunstein and Ockman (4) with the aid of a pulsed ruby laser. From the fact that the intensity of this laser excited anti-Stokes emission was found to increase over several decades as the square of the corresponding steady state excited Stokes luminescence it was con-

cluded that in this case a two photon excitation process via virtual states of the ideal crystal is involved.

The absence of detailed structure in the reported anti-Stokes
exciton emission spectra (4,5) and the need of additional arguments
to rule out a two step absorption process for pulsed laser excitation have led us to study a variety of properties of anti-Stokes
exciton emission.

EXPERIMENTAL

Anti-Stokes luminescence and the corresponding non-linear
absorption were excited by a Q-switch ruby laser with a pulse duration of about 30 nsec and a maximum output energy of 1.5 wattsec
corresponding to 3×10^{18} photons with an energy $h\nu = 1.78$ eV. From
the diameter of the laser beam being 0.95 cm, there results a
photon flux of $1.4 \times 10^{26}/cm^2 sec$ and a photon density of $5 \times 10^{15}/cm^3$.
Stokes luminescence was excited with a high pressure mercury lamp
using the 3650 A Hg line with a photon flux of about $10^{17}/cm^2 sec$.
The emission spectra were analysed with the aid of a high resolution grating monochromator and recorded either via a photographic
plate or by a high speed photomultiplier. Intensity dependences
were measured using appropriate filter combinations in front of
and behind the crystal so that the signal at the detector was near
equal intensity for all incident laser fluxes. The induced absorption was measured either directly or by a double beam method, where
the laser was split into a constant probe beam with small intensity
and a high intensity beam which induced the non-linear absorption
effect.

The CdS crystals were grown from the vapor phase and have not
been doped intentionally. We selected platelets of some millimeters
thickness which showed edge emission and the I_1-line of the same
order of magnitude of intensity as the I_2-line at low temperatures
with UV light excitation. Crystals of this type have a low dark
conductivity; the Fermi level is situated in the lower half of the
forbidden gap.

EMISSION SPECTRA

Typical emission spectra of our crystals at 4 K under various
excitation conditions are shown in Fig. 1. Curve (a) and (b) both
stand for 'normal' UV light excitation. In (a) the detector is
mounted on the side of the irradiated surface ('reflected' luminescence), in (b) observation was made through the crystal (transmitted
luminescence). The difference in the shape of the spectra is due to
the fact that a crystal of some millimeters thickness absorbs most
of the no-phonon lines of the exciton emission, so that mainly the
phonon replicas are left. The anti-Stokes emission spectrum (c) was

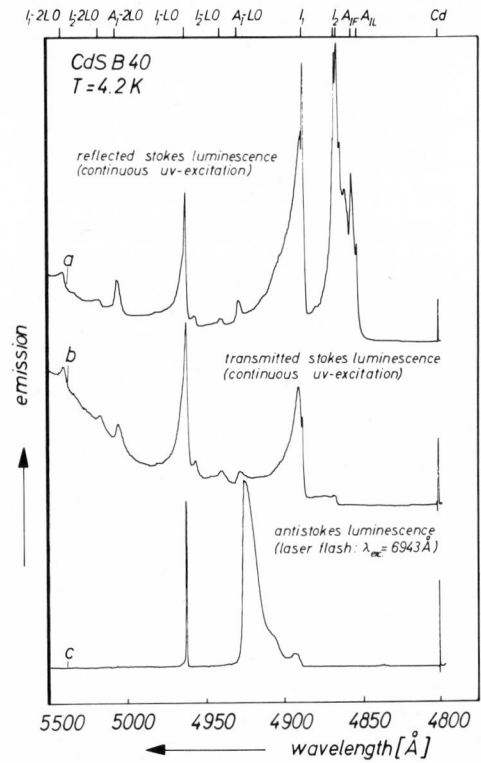

Fig. 1   Emission spectra of a CdS crystal under various excitation
         conditions

excited by a single laser flash. It is produced uniformly within
the crystal's volume and should therefore be compared only with
curve (b). In both cases the light has to pass through major parts
of the specimen. In contrast to earlier measurements (5) a pronounced structure has been found.

     The broad band consists of at least three lines. The smallest
at about 4895 A seems to be identical with the phonon side band of
the I_1-line, the biggest at 4920 A may be connected with the A-LO
phonon replica of the free exciton, however, there is a remarkable
shift of about 6A to shorter wavelength. The third line (λ=4908 A),
which is in some crystals the dominant line, could be a line described already as being due to an exciton-exciton interaction (6).
In any case, the shape and behavior of the broad band does not differ markedly from that of 'normal' excited exciton emission.

     The remaining line in spectrum (c) shows a more remarkable
effect. At the wavelength of the I_1-LO phonon line one observes a
very narrow line (limited by the slit width) which, in contrast to
the corresponding line in the Stokes case, has no detectable acoustic

phonon wings. We have intensified the laser flux to enlarge the
line by a factor of ten and found no other structure. However, at
77 K the I_1-LO line showed phonon broadening.

To explain this strange behavior and the total absence of all
the other phonon replicas, (of particular importance is the lack
of the bound phonon exciton line at the short wavelength part of
the I_1-LO line (7)) three differences between the continuous excited Stokes and the laser induced anti-Stokes luminescence have
to be considered: the excitation density is higher, the exposure
time shorter and the excited volume greater by several orders of
magnitude in the case of ruby laser excitation. The influence of
the first two points can be ruled out by the observation that
Q-switched UV laser excited crystals show just the opposite
effect, namely a strong broadening of all line structures (5).
Thus, there remains only the explanation that the different spatial
origin of the emission is responsible for the special shape of the
emission spectrum. In the interior of the crystal the exciton-
phonon coupling must then be entirely different from that of the
surface region; in particular two-phonon processes should not be
allowed.

EXCITATION MECHANISM

In order to decide whether a two step or a two photon process
creates the anti-Stokes emission the investigation of the intensity
dependence over several orders of magnitude is of great interest.
Fig. 2 shows our results for the broad band compared with edge
emission appearing at low intensities. Over three decades the
integrated emission of edge and exciton emission shows a slope of
two. Only if the two different emissions are separated does one
get deviations. At high laser fluxes the edge emission saturates
and exciton emission is, in accordance with Fig. 1c, the only sur-
viving luminescence. On the contrary, at low intensities mainly
edge emission is observable. The question arises, is the existence
of a quadratic law over many decades of intensity really sufficient
to favor a two photon process or can the same law also be explained
by a two step process? We discuss the two step process for the
case that the Fermi level is below the energetic position of the
level, which might, for example, be the excited state of an impurity
or a lattice defect. We assume further that the number of such
levels in the crystal is much greater than the number of absorbed
photons and that the time length of the pulse is small compared with
the lifetime of the center, e.g. that during the time of the flash
no recombination processes occur via the intermediate energy levels.
Both conditions seem to be fulfilled in our case. A concentration
M of about $10^{17}/cm^3$ centers is quite normal in CdS crystals. The
total number of photons absorbed during a single flash will reach
10^{17} only at the highest excitation level and thus a saturation of

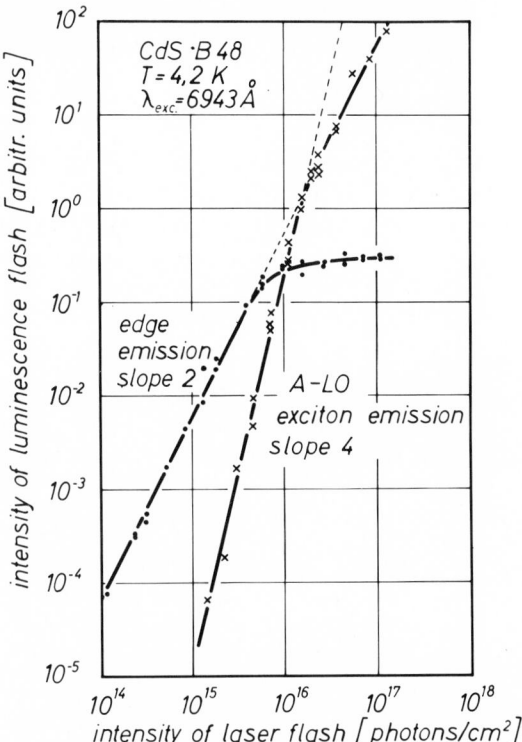

Fig. 2 Intensity dependence of the laser induced anti-Stokes
         emission.

centers is very unlikely. On the other hand the lifetime of most
known centers is much larger than a few nanoseconds. Solving a
simple differential equation, one can show that the number of
electrons Z(t) transported by the laser flash I(t) via the centers
into the conduction band is given by

$$Z_1(t)=[\alpha\beta/(\alpha+\beta)]MI(t)\{1-\exp[-(\alpha+\beta)\int_o^t I(t)dt]\}$$

$$\approx \alpha\beta MI(t)\int_o^t I(t)dt \tag{1}$$

where α,β are the cross sections for absorption of laser light from
the valence or into the conduction band respectively, while the
number of holes created in the valence band will normally ($\alpha \geq \beta$)
be much greater namely

$$Z_2(t) = [\alpha\beta/(\alpha+\beta)]MI(t)\{1+(\alpha/\beta)\exp[-(\alpha+\beta)\int_o^t I(t)dt]\}$$

$$\approx \alpha MI(t) \tag{2}$$

As the total luminescence will be proportional to the number of complete electron-hole-pairs, a slope of two over many orders of magnitude is expected according to Eq. 1. Thus the two step process can lead to the same law as the two photon process.

The details of Fig. 2 can be explained for both types of excitation processes in entirely the same way: for high intensities the luminescent centers responsible for edge emission can be saturated and all pairs recombine into excitons; for low intensities, however, the trapping of electrons and holes into donors and acceptors is the more efficient process and overcomes the exciton creation rate. A slope greater than two arises in any case and under certain assumptions the observed slope can be explained.

As a consequence of the preceding discussion we looked for additional information which might help to distinguish between the two possible excitation mechanisms. The most important are: (1) Increasing the intensity to very high values, several crystals showed not only saturation of anti-Stokes exciton emission, but even a marked reversible decrease. Other crystals of the same thickness did not show these effects. (2) Additional steady state red light increased, at low intensities of the laser flash, the efficiency of the emission by more than a factor of two; infrared light stimulated the opposite effect. (3) The induced absorption coefficient has been measured for a variety of different crystals and for different temperatures. We have found absorption coefficients between 10 and 10^{-1}/cm for the highest laser intensity. Some crystals showed no measurable effect. Variations of the temperature resulted in big changes in the absorption coefficient. (4) According to Eq. (1), the time-dependent non-linear absorption is proportional to the time dependent laser flash intensity times the integral over this intensity (two photon processes would give proportionality to the square of intensity). Thus the change of the shape of the transmitted light should be markedly sensitive to the excitation mechanism. With a relatively thick crystal we found an effect according to Eq. (1).

While all these observations clearly favor the two step excitation mechanism compared with the two photon process, the possibility that even in our experiments part of the absorption processes are due to a mixed process cannot be excluded. Using different crystals with less impurity content might very well show only the two photon effect. In this respect our measurements should not be used as an argument against the interpretation of the two photon spectroscopy in CdS (8), as long as the impurity and defect concentration of the crystals is not known.

ACKNOWLEDGEMENTS

We wish to thank Mrs. D. Maier-Hosch for technical help and the Deutsche Forschungsgemeinschaft for supplying us with some of the necessary equipment.

REFERENCES

1.   R. E. Halsted, E. F. Apple, J. S. Prener, Phys. Rev. Let. 2, 420 (1959).

2.   I. Broser, R. Broser-Warminsky, Luminescence of Organic and Inorganic Materials, p. 402 (1952).

3.   M. R. Brown, A. F. I. Cos, D. S. Orr, I. M. Williams, J. Woods J. Phys. C. Solid State Phys. 3, 1768 (1970).

4.   R. Braunstein, N. Ockman, Phys. Rev. 134, A 499 (1963).

5.   R. Levy, A. Bivas, I. B. Grun, J. de Physique 31, 507 (1970).

6.   C. Benoit à la Guillaume, J. M. Debever, F. Salvan, Phys. Rev. 177, 567 (1969).

7.   D. C. Reynolds, C. W. Litton, T. C. Collins, Phys. Rev. 4, 1868 (1971).

8.   F. Pradere, A. Mysyrowicz, Proc. 10th Int. Conf. Phys. of Semiconductors, Boston (1970).

EXCITON MECHANISMS OF EXCITATION OF IMPURITY CENTRE LUMINESCENCE IN IONIC CRYSTALS

Ch. Lushchik, G. Liidja, N. Lushchik, E. Vassil'chenko,

K. Kalder, R. Kink, T. Soovik

Institute of Physics and Astronomy, Estonian S.S.R.

Academy of Sciences, 202400 Tartu, U.S.S.R.

ABSTRACT

The exciton mechanisms of energy transfer to luminescence centers in alkali and alkali earth halides considered in the present paper are: (1) resonance transfer from self-trapped excitons to distant centers: (2) jump diffusion of axially relaxed excitons; and (3) migration of excitons before the axial vibrational relaxation. The last process is facilitated in iodides, because the relaxed state is achieved after surmounting an energy barrier.

INTRODUCTION

Our purpose is to consider the specific character of the excitation of the impurity center luminescence by optically created excitons in alkali and alkali earth halides.

It was shown in the 1950's that UV excitation in long-wavelength fundamental absorption bands leads to ionization and creation of color centers (1,2) and to the fast luminescence of impurity centers (2-4), but is not accompanied by delayed luminescence resulting from the motion of electrons and holes (2). It was concluded from these facts that long-wavelength absorption bands correspond to the formation of moving, uncharged electronic excitations, i.e. excitons.

It was shown in the 1960's that excitons become self-trapped and, consequently, immobile in alkali halides at low temperatures

(5,6).

Recently we have performed experiments to clarify the reasons for such a paradoxical situation, and the possible mechanisms of energy transfer by excitons to impurity centers.

EXCITON MECHANISMS OF ENERGY TRANSFER

The structure of fundamental absorption spectra of halide crystals gives evidence for the formation of excitons with the hole component localized on a single halide ion. Transient absorption spectra demonstrate that after the exciton has undergone vibrational relaxation, its hole component is located on two neighboring halide ions and forms a quasimolecular V_k center.

Fig. 1 represents simplified models and potential energy curves for axially non-relaxed and axially relaxed excitons. According to some hypotheses (7,10,12,13) there may exist an activation barrier E_r between relaxed and non-relaxed states. The radiationless annihilation of relaxed excitons is in many cases an internal quenching process passing over the potential barrier E_q.

Among the possible mechanisms of the exciton energy transfer to impurity centers one may distinguish the following four groups: (a) Resonance transfer of energy from self-trapped excitons to distant impurity centers. (b) Energy transfer by the jump diffusion of axially relaxed excitons. (c) Energy transfer as a result of the motion of non-relaxed or partially relaxed excitons. (d) Resonance transfer of energy to impurity centers from distant excitons before their complete vibronic relaxation.

Below we shall demonstrate that the first three mechanisms are realized in activated ionic crystals.

RESONANCE TRANSFER

Resonance transfer of energy or the reabsorption of the energy emitted by the self-trapped excitons is possible. The number of suitable systems is limited because self-trapped excitons emit comparatively low energy quanta.

Resonance transfer to impurity centers usually takes place after vibronic relaxation and reduces both the quantum efficiency and the decay time of the sensitizor emission.

Resonance transfer of energy from self-trapped excitons to impurity centers can undoubtedly be observed in CaF_2:Eu, SrF_2:Eu, BaF_2:Eu (11) KI:Sn (15) and KI:Eu crystals. Data for crystals KI:Eu and SrF_2:Eu are given in Fig. 2.

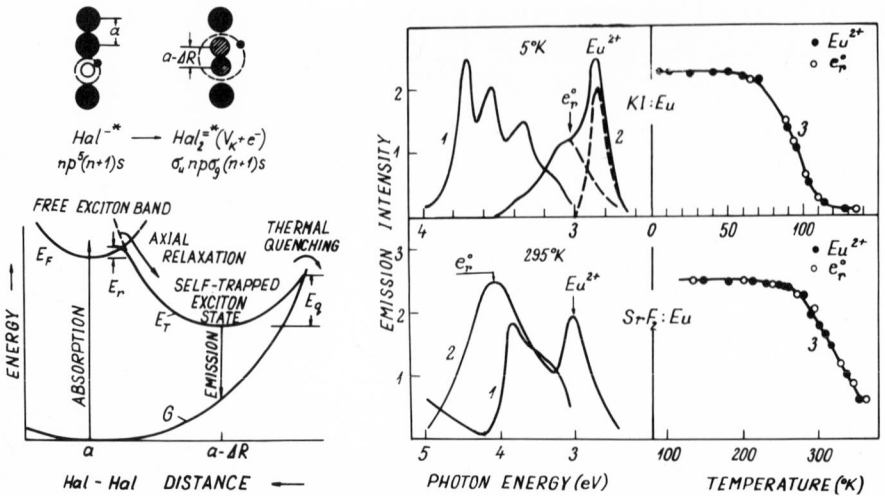

Fig. 1   Models and potential curves for a free and a self-trapped
         exciton state.

Fig. 2   Excitation spectra of the europium emission (1); emission
         spectra (2) and intensities of the exciton and activator
         emission excited by the photons of 6.0 eV for KI:Eu and
         10.7 eV for SrF₂:Eu as a function (3) of temperature.

The excitation spectrum of Eu^{2+} centers overlaps the emission
spectrum of the self-trapped exciton in SrF_2 (4 eV) and KI (3.3 eV).
The efficiency of the blue Eu^{2+} resonance emission and of the émis-
sion of self-trapped excitons have the same temperature dependence.
The decay time of excitons is also shorter after doping SrF_2 and KI
with europium.

JUMP DIFFUSION OF RELAXED EXCITONS

One can expect that the excitons which are self-trapped at low
temperatures start moving by jump diffusion as the temperature rises.

The efficiency of the exciton diffusion process depends on $w\tau$,
where $w = w_0 \exp(-E_j/kT)$ is the jump probability of the relaxed exciton
to the neighboring site over the energy barrier E_j, and $\tau =
(1/d_0)\exp(E_q/kT)$ is the life-time of excitons when their luminescence
is quenched as a result of non-radiative transitions. A sufficient
increase in the diffusion probability of excitons with temperature
requires $E_j > E_q$ and $w_0/d_0 \geq 1$.

We have found some crystals among alkali halides, e.g. NaCl
and CsBr (8), which satisfy these conditions. Data for CsBr:In with
various activator concentrations are represented in Fig. 3. The

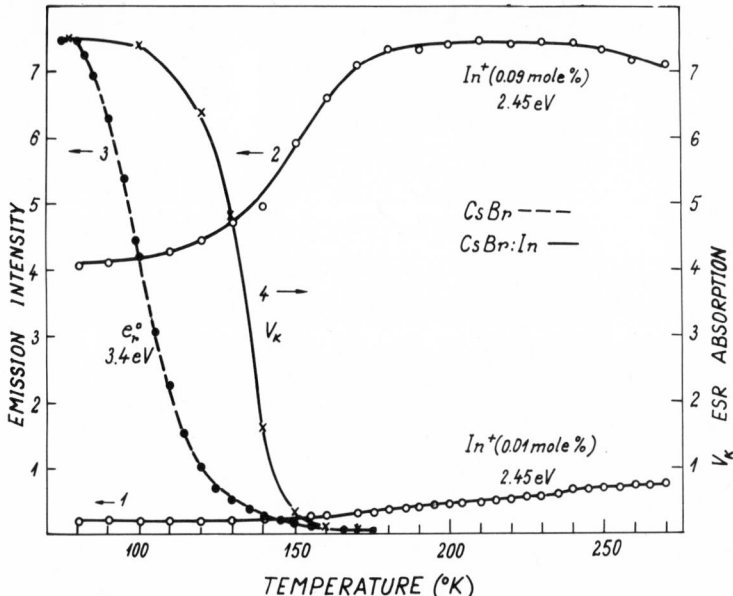

Fig. 3   Intensities of activator (1,2) and exciton (3) luminescence
         for CsBr:In, excited by the photons of 6.9 eV as a function
         of temperature. ESR signal of V_k centers (4) in a crystal
         x-irradiated at 80 K and measured at 80 K after preliminary
         annealing at various temperatures for 2 minutes.

efficiency of the excitation of In$^+$ centers by excitons (photons
of 6.9 eV were used) is nearly constant in the interval from 77 K
(and even from 5 K, as additional experiments have shown) to 140 K.
Subsequent heating leads to a significant increase of efficiency.
In spite of the short life-time of relaxed excitons (10^{-6}-10^{-9}sec)
their jump diffusion can be detected at higher temperatures than
is the case for long-living relaxed holes. The jump diffusion of
relaxed holes has been investigated by the ESR method (see Fig. 3).
Estimates for NaCl and CsBr (for details see (9))show that the
activation energy of jump diffusion for self-trapped excitons is
slightly smaller than for self-trapped holes.

     In KI:Tl and KBr:In the efficiency of energy transfer to im-
purity centers scarcely depends on temperature. Apparently, the
above mentioned conditions are not fulfilled in these systems.

MIGRATION OF NON-RELAXED EXCITONS

     The freezing of jump diffusion of excitons in NaCl and CsBr
at low temperatures does not completely stop the energy transfer
to impurity centers. The low-temperature exciton mechanism of
luminescence of NaCl:Ag and CsBr:In cannot be connected with

resonance transfer considered above. Actually, as is evident from
Fig. 3, the efficiency of the low temperature energy transfer is
nearly constant in the interval from 80 to 150 K, while the emission intensity of self-trapped excitons decreases by two orders of
magnitude. Moreover, in CsBr, activated by indium, the decay time
of exciton emission coincides essentially with that for CsBr,
whereas the efficiency of the exciton luminescence is sufficiently
diminished.

The low-temperature transfer of energy from excitons to impurity centers in this case occurs before the exciton becomes self-
trapped. One may explain this process as a result of the migration of axially non-relaxed excitons (6-11). Resonance transfer
from non-relaxed excitons is unlikely and, as in pure crystals, the
luminescence of non-relaxed excitons is not observed. An analogous
conclusion was made about the low-temperature diffusion of holes
in NaCl, KCl and CsBr (8,9,16).

It was estimated from the concentration dependence of the
efficiency of the low-temperature energy transfer in CsBr that
the volume from which excitons are collected to the In^+ ion, equals
about $400a^3$ (a = distance between nearest anions).

Fig. 4 demonstrates the low-temperature energy transfer by
excitons in CsBr:Rb. The excitation spectrum of the rubidium center
luminescence is separated from the exciton emission spectrum by 2eV,
and thus, resonance transfer is by no means possible. Excitation of
rubidium centers by polarized light produces polarized luminescence.
Rubidium emission excited by polarized light from exciton bands is
unpolarized. One expects that during the motion the system may
"forget" the anisotropy of excitation.

ACTIVATION BARRIER FOR RELAXATION

Exciton mechanisms of energy transfer to impurity centers in
alkali halides become complicated as temperature is decreased below
70 K (7,12,16).

In Fig. 5 are given results for KI (7) and KI:Tl. On the right,
emission spectra are given and on the left, temperature dependence
of integrated luminescence, all for the exciton band excitation
(6 eV). The emission of axially relaxed excitons (3.35 eV) decreases
approximately five fold as temperature is lowered from 60 to 30 K.
Simultaneously, a new emission band, peaked at 3.02 eV arises, and
is probably due to an unknown impurity. Each of these processes is
characterized by an energy activation barrier of 10 meV. In highly
purified crystals of KI doped with thallium the emission at 3.02 eV
diminishes at low temperature. In the range 60 to 30 K the quantum
yield of the exciton induced thallium luminescence rises three-fold

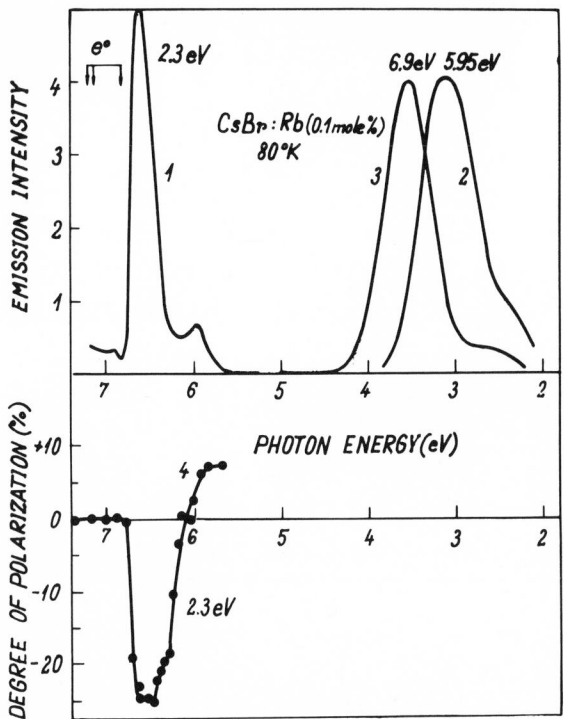

Fig. 4 The excitation (1) and degree of polarization (4) versus
photon energy of the exciting light for the 2.3 eV emission of a CsBr:Rb crystal. Emission spectra (2,3), excited by the photons of 5.95 and 6.90 eV, respectively.

(the integrated intensity of two prominent thallium bands at 3.7
and 2.85 eV having been measured). Analogous effects are observed
in NaI:Tl (12) and RbI:Tl (16) where the barrier heights are 15 and
18 meV, respectively.

At 5 K the decay time of exciton emission in KI (8 μsec) remains unchanged after doping thallium. The decay time of thallium
luminescence at 3.7 eV is remarkably shorter (2.3 μsec) than that
for excitons. These results suggest that energy is transferred to
Tl[+] centers before complete vibronic relaxation of excitons and
that the latter needs thermal activation. As assumed earlier (7,
10,12) the state of the axially relaxed exciton is separated from
the non-relaxed state by an energy barrier. Therefore, the exciton
lifetime before complete vibronic relaxation may increase up to
10^{-10} sec. This leads to a remarkable increase of the diffusion
path of excitons at low temperature. Our estimates for KI:Tl crystals with various thallium content show that at T < 30 K excitons
are converging from a volume of $1600a^3$ surrounding a Tl[+] center.

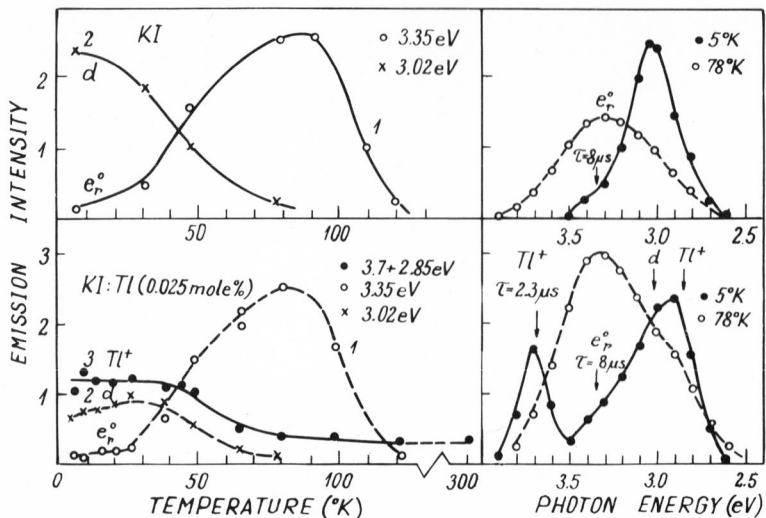

Fig. 5   Emission intensities of the self-trapped exciton (1), the
         unidentified (2) and thallium (3) impurity for KI and KI:Tl,
         excited by photons of 6.0 eV as a function of temperature.
         Emission spectra for the same excitation are given in the
         right-hand part.

Assuming the effective cross section of the interaction of the exciton with a Tl^+ center to be πa^2, the full diffusion length becomes
$500a = 2.5 \times 10^{-5}$ cm.

     Due to the activation barrier one may expect the appearance of
a weak emission from axially non-relaxed (free or partially relaxed)
excitons at low temperature. Up to the present we have not observed
this luminescence, perhaps because of the insufficient purity of our
crystals.

                        CONCLUDING REMARKS

     Finally, we give a table summarizing the various mechanisms of
exciton-induced luminescence which occur in different materials.

     The study of the exciton energy transfer to luminescence
centers leads to the conclusion that in A^1B^7 crystals there exist
not only "poor" (self-trapped) excitons but also, at least in iodides,
excitons capable of covering large distances. Both exciton types
may coexist in the same crystal.

     The detailed mechanism of exciton migration before their axial
relaxation to luminescence centers needs further investigation.
One has to determine effective cross sections for the exciton-
impurity interaction, to decide which excited states of the impurity

are immediately populated after the interaction, and to clarify the nature of the activation barrier. The investigation of an edge-type luminescence is of special interst because that would give some important information about the motion of excitons before axial relaxation.

TABLE

Crystal	L	R	D	M	M_b	Crystal	L	R	D	M	M_b
NaI:Tl	+	+		+	+	KBr:In	+		+		
KI:Tl	+			+	+	CsBr:In	+	+	+		
KI:Sn	+	+				CsBr:Tl	+	+	+		
KI:Eu	+	+	+			CsBr:Rb	+		+		
RbI:Tl	+			+	+	CaF$_2$:Eu	+	+			
NaCl:Ag	+		+	+		SrF$_2$:Eu	+	+			
NaBr:Tl	+	+		+	+						

Key:
L--luminescence of self-trapped excitons;
R--resonance transfer from self-trapped excitons;
D--diffusion of axially relaxed excitons;
M--migration of excitons prior to axial relaxation;
M_b--vibronic relaxation involving a barrier.

REFERENCES

1. L. Apker, E. Taft, Phys. Rev. 81 698 (1951); 82 814 (1951).

2. Ch. Lushchik, G. Liidja et al, Trudy Inst. Fiz. i Astron. Akad. Nauk Est. SSR No. 6, 63, 149 (1957); Proc. 7th Conf. on Luminescence, Tartu (1959) p. 101; Opt. i Spektroskopiya 9, 70 (1960).

3. K. Teegarden, Phys. Rev. 108, 660 (1957).

4. M. Tomura, Y. Kaifu, J. Phys. Soc. Japan 15, 1508 (1960).

5. M. Kabler, Phys. Rev. 136A, 1296 (1964). M. Kabler, D. Patterson, Phys. Rev. Lett. 19, 682 (1967).

6. R. Kink, G. Liidja, Ch. Lushchik, T. Soovik, Izv. Akad. Nauk SSSR, Sr. Fiz. 31, 1982 (1967); Trudy Inst. Fiz. i Astron. Akad. Nauk Est. SSR No. 36, 3 (1969).

7. R. Kink, G. Liidja, Fiz. Tverd. Tela, 11, 1641 (1969); Phys. Stat. Sol. 40, 379 (1970).

8. E. Vasil'chenko, N. Lushchik, Ch. Lushchik, Fiz. Tverd. Tela 12, 211 (1970); J. Luminescence 5, 117 (1972).

9.   E. Vasil'chenko, N. Lushchik, Ch. Lushchik, Trudy Inst. Fiz.i
     Astron. Akad. Nauk Est. SSR No. 39, 3 (1972).

10. Ch. Lushchik, J. Luminescence 1-2, 660 (1970).

11. K. Kalder, A. Malysheva, Opt. i Spektroskopiya 31, 252 (1971).

12. M. Fontana, N. Blume, W. van Sciver, Phys. Stat. Sol. 29, 159
     (1968); 31, 133 (1969).

13. H. Sumi, Y. Toyozawa, J. Phys. Soc. Japan 31, 342 (1971).
     Y. Toyozawa, J. Luminescence 1-2, 732 (1970).

14. T. Kamejima, S. Shionoya, A. Fukuda, J. Phys. Soc. Japan 30,
     1124 (1971).

15. E. Aluker, S. Chernov, J. Kalnin, I. Ushomirski, Izv. Akad.
     Nauk SSSR, Ser. Fiz. 35, 1352 (1971).

16. A. Hattori, M. Tomura, H. Nishimura, J. Phys. Soc. Japan 31,
     611 (1971).

MAGNETO-OPTIC EFFECTS IN RECOMBINATION LUMINESCENCE FROM

SELF TRAPPED EXCITONS

M. N. Kabler, M. J. Marrone and W. B. Fowler*

Naval Research Laboratory, Washington, D.C. and
*Lehigh University, Bethlehem, Penna., U.S.A.

ABSTRACT

Intrinsic recombination luminescence which originates in triplet states of the self trapped exciton in alkali halide crystals has been investigated experimentally and theoretically with regard to effects of applied magnetic fields. Lifetimes and circular polarizations are observed to vary anisotropically with field, and a level crossing resonance is identified in CsI and, tentatively, in CsBr. This provides evaluation of the zero-field splitting for the triplet state. Splittings and lifetimes for CsI and KI are compared with a theory based on a V_k-like hole and nonequilibrium level populations, and reasonable agreement is obtained. Electronic exchange energies which occur as parameters in the theory are in rough agreement with independent estimates.

I. INTRODUCTION

Exciton self trapping is a general phenomenon in halide lattices. It evidently occurs through the same mechanism which causes hole self trapping, a well known process resulting in a stable, intrinsic hole center, the V_k center. The self trapped exciton (STE) is, in fact, simply a V_k center which has trapped an electron. This metastable combination is responsible for the intrinsic luminescence exhibited by halide crystals at low temperatures. On the basis of lifetime and polarization measurements, luminescent transitions associated with both singlet and triplet STE states have been identified in a number of alkali halides (1). The allowed singlet-singlet transitions have lifetimes in the nanosecond range and generally occur at higher

energies than the forbidden triplet-singlet transitions, whose
lifetimes are longer and vary systematically from crystal to
crystal. Because of the strong spin-orbit coupling, one expects
the triplet state luminescence to be partially circular polarized
in an applied magnetic field. This expectation has been confirmed
by observation of magnetic circular polarization (MCP) in several
alkali chlorides, bromides, and iodides (2).

    In the present paper we are concerned primarily with magneto-
optic effects in KI and CsI. These include variation with field
of the triplet state luminescent lifetimes and intensities, and
MCP. The data provide additional insight concerning relationships
among the triplet states and give zero-field splittings which
compare favorably with a detailed theory. Significant magneto-
optic effects are not observed for the singlet-singlet transition.

    We begin by reviewing qualitatively the electronic states
relevant to the principal triplet-singlet emission. Figure 1
illustrates schematically the origin and splittings of these states.
In the ground state (not shown in Fig. 1) the halide ions have a
p^6 rare gas configuration, and the self trapped electronic
configurations are analogous to those of an excited diatomic rare
gas molecule. The relaxed states for the diatomic molecule
(symmetry $D_{\infty h}$) are at the center of Fig. 1. Correlations with
the p^5s configuration of a single excited halide ion, representing
the unrelaxed exciton, are indicated. The relevant angular
momentum quantum number for each state is also shown. Only odd
parity states derived from the p^5s configuration are considered,
since odd parity states initiate the luminescence and none of the
perturbations we shall consider break the parity selection rule.
The relaxed states of Fig. 1 are also those which result from
coupling the principal states of the V_k center with an electron
in a totally symmetric σ_g or a_g orbital. Although the V_k center
is actually a deep trap for the electron, the primary spacings
of the states shown will be assumed to approximate those for the
hole alone. Transient absorption spectra for the STE indicate that
this assumption is valid (3). We note also that the singlet state
responsible for the singlet-singlet luminescence transition at
higher energies is not $^1\Sigma_u^+$ of Fig. 1; the energies are such that
the singlet-singlet luminescence must originate in a configuration
involving a higher σ_g electron orbital.

    In a crystal of the NaCl structure, the D_{2h} crystal field
removes the remaining degeneracies, resulting in the states on the
right hand side of Fig. 1. These crystal field splittings are
relatively small. The triplet-singlet emission originates in the
lowest B_{2u} and B_{3u} states, as indicated. The spin forbiddenness
is broken by spin-orbit mixing of the remaining B_{2u} and B_{3u} states,
which brings some singlet character into the luminescent states.

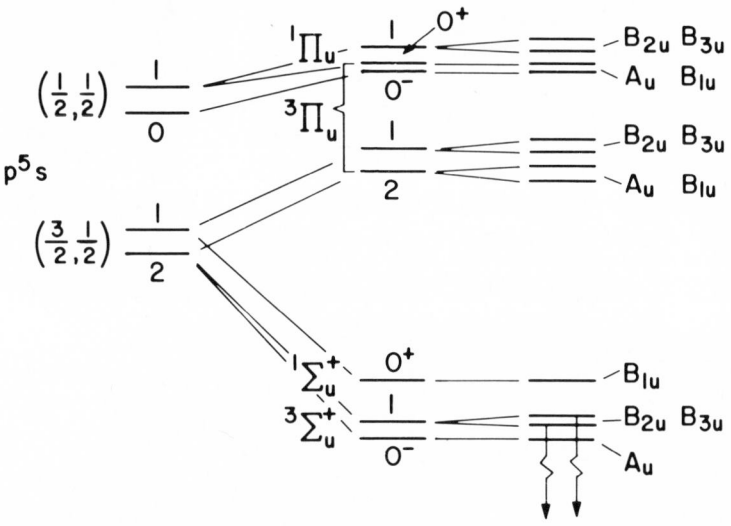

Fig. 1   Correlations among exciton states associated with
         (b_{1u}, a_g), (b_{2u}, a_g), and (b_{3u}, a_g) electronic
         configurations of the STE. Energies are not to scale.

II.   TRIPLET STATE ZERO-FIELD SPLITTINGS

For the present experiments, we are concerned with the
splittings of the lowest triplet states and with the actual
linear combinations of p-like hole orbitals, since these determine
the magnetic behavior. It is possible to compute these quantities
beforehand if one is willing to use input parameters obtained
from the V_k center alone and from the spectra of unrelaxed excitons.
A detailed exposition of the theory will be given in a subsequent
publication (4). One simply writes down a 12 x 12 Hamiltonian
matrix connecting the states on the right hand side of Fig. 1
using pure triplet and singlet states as a basis. Parameters
in this matrix are the molecular field (that is, the principal
$\Sigma_u - \Pi_u$ splitting), the crystal field, a halogen spin-orbit
coupling constant λ, and an exchange coupling parameter Δ. The
off-diagonal elements contain only λ (5). This Hamiltonian
matrix has been diagonalized numerically for several cases, and
results for two iodides are given in Table 1. The molecular and
crystal field parameters are taken from Schoemaker's recent
analysis of the anisotropic g factor for the V_k center (6). The
exchange parameter Δ is difficult to estimate for the STE; we have
taken a recent value for the unrelaxed exciton obtained from fine
structure in the KI absorption spectrum (7). These Δ and λ are
included in Table 1, together with the resulting splittings
between the three lowest levels of Fig. 1. Note that the

B_{2u} - B_{3u} splitting in Table 1[*]is extremely small, in fact negligible on the present scale of magnetic splittings. This appears to be a general theoretical result which obtains for chlorides and bromides as well; it is approximately proportional to the g anisotropy in the perpendicular plane, g_x - g_y.

Recent data of Fröhlich et al on the temperature dependence of triplet-singlet lifetimes provide a comparison with the theory, at least for KI (8). Below 20K two decay components were observed in the well known emission band at 3.3 eV, one with a constant lifetime of 0.6 microseconds, the other with lifetime decreasing exponentially with temperature. Excellent agreement was obtained with a simple kinetic theory based on nonequilibrium populations. The A_u state was assumed to empty only via thermal activation to B_{2u} or B_{3u}. Below 10K the slow component was shown to be given approximately by

$$\frac{1}{\tau} = \frac{1}{\tau_{ra}} + \frac{\tau_{ob} \, n}{\tau_{oa}(\tau_{ob}+\tau_{rb})} \qquad (1)$$

Here $1/\tau_{rb}$ is the radiative transition probability from B_{2u} or B_{3u} to the ground A_g state, n/τ_{oa} is the probability of a thermally activated transition from A_u to B_{2u} or B_{3u}, and $n = 1/[\exp(E_{ba}/kT)-1]$. Since there are two B levels, $\tau_{ob}/\tau_{oa} = 2$. The radiative probability $1/\tau_{ra}$ from A_u to the ground state is negligible for zero field but becomes significant in the magneto-optic experiments to be discussed. The energy E_{ba} derived from Fröhlich's data is just the A_u - B_{2u} or A_u - B_{3u} splitting in the lower right hand corner of Fig. 1 and in Table 1. The agreement is certainly fortuitous. In particular, it would be surprising if Δ actually had the same value for both the unrelaxed and self trapped excitons. Nevertheless, we feel that the parameters in Table 1 for KI give a reasonably good characterization of the state in which the 3.3 eV luminescence originates.

III. LIFETIME EFFECTS DUE TO MAGNETIC FIELD

The behavior of the lowest triplet states in an applied magnetic H is shown schematically in Fig. 2. Splittings are illustrated for H parallel to each of the three principal axes of the STE, the z axis being the line connecting the two halide ions. H_x and H_y each mix a B_u state into A_u, giving a finite transition moment for $A_u \rightarrow A_g$. The linear polarization of each transition is indicated. H_z mixes only B_{2u} and B_{3u} and produces partial circular polarization for any propagation direction not in the x, y plane. For both B_{2u} and B_{3u}, circular polarization is complete at high field for propagation along z and $H_x = H_y = 0$.

---

[*]Note: Table 1 is at the end of this paper.

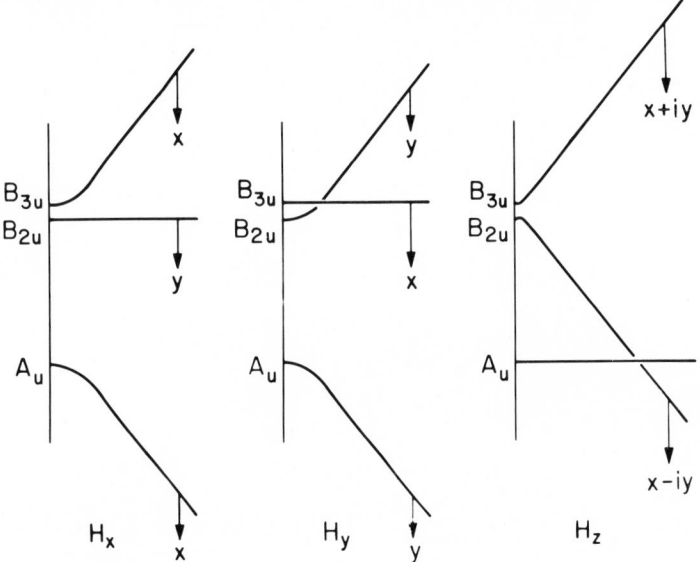

Fig. 2   Schematic level shifts and polarizations for luminescent
         triplet states with H along the principal axes of the STE.

For KI with $H\|[100]$, the six [110] axes will make an angle
of 45° or 90° with H; for the latter the H_x and H_y cases of
Fig. 2 apply. For the 45° orientation all three levels mix and
shift. These effects have been computed in the same way as
the zero-field splittings. The spin Hamiltonian describing the
field was taken to be $\beta \underline{H} \cdot \underline{L} + 2\beta \underline{H} \cdot \underline{S}$, and its matrix elements
were written down in the same basis. The 12 x 12 matrix was
then diagonalized numerically for the two orientations. Only
relative values for the dipole transition moments have been
calculated, since the exact spatial dependence of the a_g
electron orbital is not known.

Figure 3 gives the transition probability $1/\tau$ for the slow
component in KI as a function of H. The theoretical lifetimes
for 45° and 90° orientations were computed from Eq. (1), with
only τ_{oa} and τ_{ob} assumed to be independent of H. Although the
choice of initial parameters was not optimum, the trend in the
calculation for the 45° orientation agrees satisfactorily
with experiment. At high H the contribution from the $1/\tau_{ra}$ term
is substantial. The H dependence of intensity is also
anisotropic, and the calculations indicate that luminescence
propagating along H due to 90° orientations should be weak at
large H; probably as a consequence of this, we are unable to
identify this component in the data of Fig. 3. The effect
arises from the fact that for the 90° orientation, H mixes into

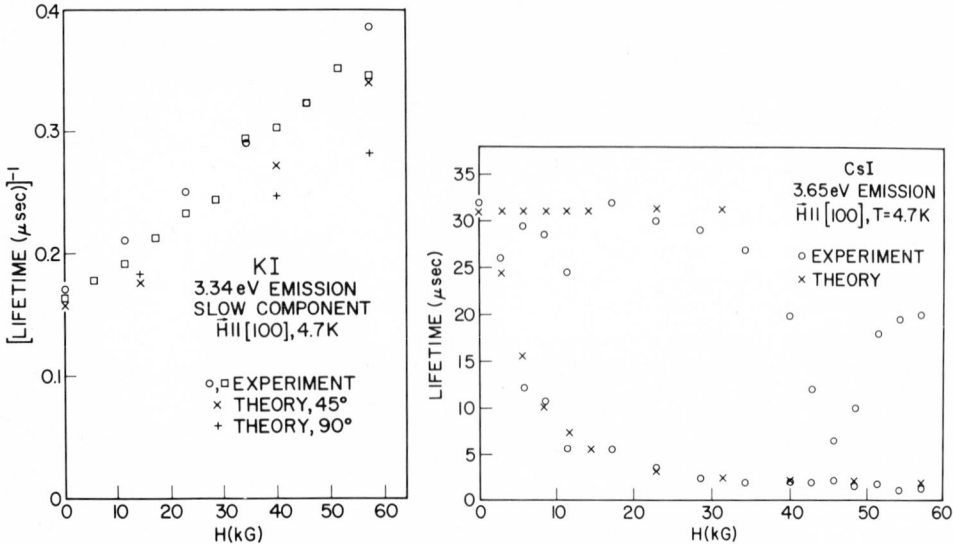

Fig. 3   Magnetic field dependence of $1/\tau$ for the slow
         luminescence component of KI. Square and
         circle data points are for H increasing and
         decreasing, respectively.

Fig. 4   Magnetic field dependence of lifetime τ of the
         slow component in the CsI 3.65 eV luminescence.

A_u only that p-like singlet component giving a transition moment
parallel to H. A large decrease with H of emitted light intensity
viewed along H has been observed also in CsI.

    The CsI structure produces STE's aligned along the three [100]
directions, and thus the splittings of Fig. 2 are applicable for
H||[100]. However, the B_u levels are degenerate for H = 0 and
become an E_u level in D4h symmetry. Data of Lamatsch et al (9) on
the temperature dependence of lifetime of the 3.65 eV band in
CsI show the same features as the KI data mentioned, and it has
been suggested that the cause is the same for both (8). We have
therefore computed lifetimes for the slow component in CsI on the
basis of a nonequilibrium formulation similar to that used in
deriving Eq. (1) (10). In this case the field dependences of all
parameters, including τ_{0a} and τ_{0b}, were taken into account.
The values of τ_{0a} and τ_{0b} were chosen to fit the lifetime at
H = 0 and at large H. The results are compared with experiment
in Fig. 4, where one notes a component arising from orientations
parallel to H and a faster component from orientations perpendicular
to H. The agreement for both is excellent. For H> 30kG, the
experimental data show a resonance-like feature in the component
we are identifying with orientations z||H. We attribute this to

level crossing, and take H_c = 45kG to be the field at which an
E_u level crosses A_{1u}. The zero-field splitting is consequently
$g\beta H_c$, the value of which is included as E_{ba} for CsI in Table 1.
This is, in fact, the value at H = 0 used in Eq. (1) to calculate
the lifetimes of Fig. 4. The resonance is evidently produced by
a rapid nonthermal discharge of A_{1u} population into E_u as the
levels cross. Thus all features of Fig. 4 are consistent with our
nonequilibrium formulation.

Lifetime data for the 3.65 eV band for H||[111] at 4.7K show
only one slow component, as all orientations are equivalent. The
behavior is similar to that of the faster component in Fig. 4,
except that the initial decrease with H is less steep. The theory
based on Eq. (1) with E_{ba} = 0.50 meV is in excellent agreement
with the data.

Table 1 also lists that value of Δ which, when used in the
above theory, gives an A_{1u} - E_u splitting equal to the measured
E_{ba} for CsI. There is no other estimate available for this Δ,
although our value is not far from that for KI.

IV. MAGNETIC CIRCULAR POLARIZATION

MCP has been measured for a number of alkali halides as a
function of temperature and field. When thermal equilibrium obtains,
when the B_{2u} - B_{3u} splitting is negligible, and when $g\beta H < E_{ba}$, a
simple two level model might be expected to be a reasonable
approximation. Thus the circular polarization would be proportional
to tanh $g\beta H/kT$. The data generally show this dependence in the
higher temperature range. In fact, in KI and KBr, this behavior
obtains qualitatively throughout the observed range down to 4.6K.
Evidently g $\sim$ 2 holds in all cases. Actually, at high fields,
$g\beta H \gtrsim E_{ba}$ and, at least for the iodides, thermal equilibrium fails
at low temperatures. Calculations taking these factors into
account will be described elsewhere (4).

Some of the experimental MCP data for CsI are shown in Fig. 5.
Here the circular polarization P = $(I_- - I_+)$/I is plotted as a
function of H. The data are normalized to P = 1.0; to reproduce
actual relationships the H||[100], 4.6K data should be multiplied
by 0.35 and the H||[100], 20K data should be multiplied by 0.59,
although this is unimportant for the present qualitative discussion.
The experimental method has been described (2). As for the lifetime measurements, the observed luminescence propagates along the
direction of H.

The data for H||[100] and 20K contain no obvious anomalies.
This is consistent with the foregoing discussion, since the
lifetime data imply thermal equilibrium near 20K. However, for

this same orientation at 4.5K, the behavior has changed markedly.
A resonance feature appears near 45kG, just as in the lifetime
data. It evidently arises because, at the level crossing,
essentially all the A_{1u} population is lost by radiation through
the <u>lower</u> E_u state. This is a nonequilibrium effect, and the 20K
data give no indication of the resonance.

An equally interesting feature of the H||[100] data is that
P = 0 up to 12kG at 4.6K. This implies equal populations in the
two E_u levels for those centers aligned parallel to H, as in the
right hand diagram of Fig. 2. However, according to Eq. (1), these
levels are being populated continuously through thermal excitation
from the A_{1u} level. The absence of circular polarization is
evidently a consequence of the fact that the nonradiative
transition probability peaks at an energy near E_{ba}, and thus the
populations of the two E_u levels can remain approximately
equal as the field splits them (11). The magnitude of this effect
should be a strong function of temperature, a premise substantiated
by additional low field data in the 4 - 10K range. The data of
Fig. 5 for H||[111] also show this decrease in slope at low H.
Otherwise, P increases smoothly with H as one might expect for
this orientation, regardless of nonequilibrium.

The MCP observed for CsBr is also anisotropic and, for
H||[100] and low temperature, shows a resonance. The peak is at
about 7.0kG and is substantially stronger than that of Fig. 5.

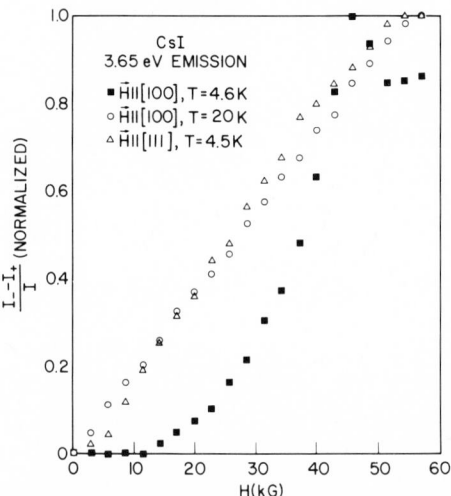

Fig. 5 Circular polarization P vs. H for CsI. All light emitted
        along H is included in I. Normalization constants are
        given in the text.

The relevant lifetime dependence has not yet been measured; however, if the MCP resonance is interpreted as due to level cross-ing, as for CsI, one can estimate Δ for CsBr: The relation $E_{ba} \propto \Delta\lambda^2$ is approximately valid (4) assuming the molecular field to be equivalent for CsBr and CsI. Taking $\lambda = 280$ meV for the bromide, (6) one obtains a value for Δ roughly 0.7 that of CsI. This small variation in Δ among CsBr, CsI, and KI is perhaps theoretically satisfying, and is consistent with the slight variations in Δ for unrelaxed excitons in rare gas crystals (12). However, spectra for the unrelaxed exciton in several alkali bromides (not CsBr) have been interpreted in terms of Δ values several times larger our estimate for the STE (13).

REFERENCES

1. M. N. Kabler and D. A. Patterson, Phys. Rev. Letters 19, 652 (1967).

2. M. J. Marrone and M. N. Kabler, Phys. Rev. Letters 27, 1283 (1971).

3. R. G. Fuller, R. T. Williams, and M. N. Kabler, Phys. Rev. Letters 25, 446 (1970).

4. W. B. Fowler, M. J. Marrone, and M. N. Kabler, to be published.

5. The Hamiltonian matrix reduces to four 3 x 3 matrices in the absence of an applied field.

6. D. Schoemaker, private communication.

7. Y. Petroff, R. Pinchaux, C. Chekroun, M. Balkanski, and H. Kamimura, Phys. Rev. Letters 27, 1377 (1971).

8. D. Fröhlich, U. U. Fischbach, and M. N. Kabler, J. Luminescence, to be published.

9. H. Lamatsch, J. Rossel, and E. Saurer, Phys. Stat. Sol (B) 48, 311 (1971).

10. As for KI, there is also a fast component ($\tau \sim 1\mu$sec) in the 3.65eV CsI band due to the initial depopulation of the E_u levels. Its lifetime does not appear to vary strongly with H, and it will not be discussed here.

11. The transition probability is proportional to $E^2 n$, where E is the field dependent level splitting. The factor E^2 comes from the phonon density of states. The field dependent

lifetimes of the E_u states must also be taken into account.

12. R. S. Knox, Phys. Rev. 110, 375 (1958).

13. Y. Onodera and Y. Toyozawa, J. Phys. Soc. Japan 22, 833 (1967).

Table 1

Zero-field energies (millivolts) for the luminescent triplet state. Calculated values are underlined. The exchange parameter Δ for KI is that measured for the unrelaxed exciton, Ref.(7). For CsI Δ has been chosen to give a splitting in agreement with our measured E_{ba}. The spin-orbit parameter λ, as well as other molecular and crystal field parameters not shown are from Ref.(6). The value of E_{ba} for KI is from Ref.(8), that for CsI from the present level-crossing data.

	Δ (exchg.)	λ (spin-orbit)	B_{2u}-A_u	B_{3u}-A_u	B_{3u}-B_{2u}	E_{ba} (expt.)
KI	43	584	0.704	0.706	0.002	0.69
CsI	33	584	0.50	0.50	0	0.50

THE EFFECTS OF RELAXATION PROCESSES ON THE SPECTRAL-LUMINESCENT

PROPERTIES OF ANTHRACENE VAPORS

V. P. Klochkov

The State Optical Institute, Leningrad, U.S.S.R.

ABSTRACT

A graphical representation of the relaxation processes from Franck-Condon to Born-Oppenheimer states is used to analyze the diffuse and the resonance fluorescence of anthracene vapor.

Anthracene vapors are known to exhibit fluorescent spectra of two kinds (1-6): discrete and diffuse (Fig. 1). Discrete spectra, observed first by Pringsheim (1) are obtained by excitation with photon energy equal to the interval between the zero vibrational levels 0",0' for two electronic states. It is well-known that the diffuse spectrum, in which vibronic peaks corresponding to the vibrations of ~ 400 and ~ 1400 cm^{-1} are observed, is obtained by excitation with radiation of different wavelengths. Variation of the exciting wavelength produces a shift in the spectrum and changes in its intensity distribution and diffuseness (7,8). Discrete and diffuse spectra of fluorescence have been extensively investigated, but the origin of the discrete spectrum is not well understood.

Consider the three following possible explanations: 1. discrete and diffuse spectra are emitted from two neighboring states of anthracene. 2. According to Terenin (9) both discrete and diffuse spectra are emitted by molecules in the same electronic state. The discrete spectrum is emitted from a narrow vibrational level; the diffuse, from a broad level. 3. Discrete and diffuse spectra are emitted after excitation of the same electronic state. The kind of spectrum observed depends on the type of relaxation process as a result of which the nuclear configuration may be changed (10,11). If there is no relaxation process the spectrum

181

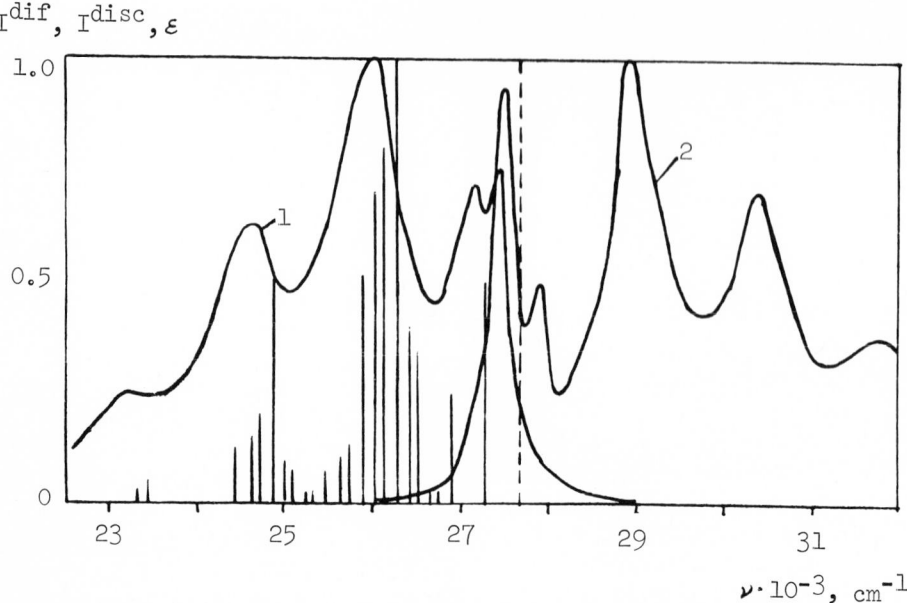

Fig. 1   The discrete and diffuse fluorescence spectrum (1) and
         absorption spectrum (2). The discrete spectrum is measured
         at T = 410 K, the diffuse spectra are measured at T = 413 K,
         $\lambda_{ex.}$ = 365 nm (8).

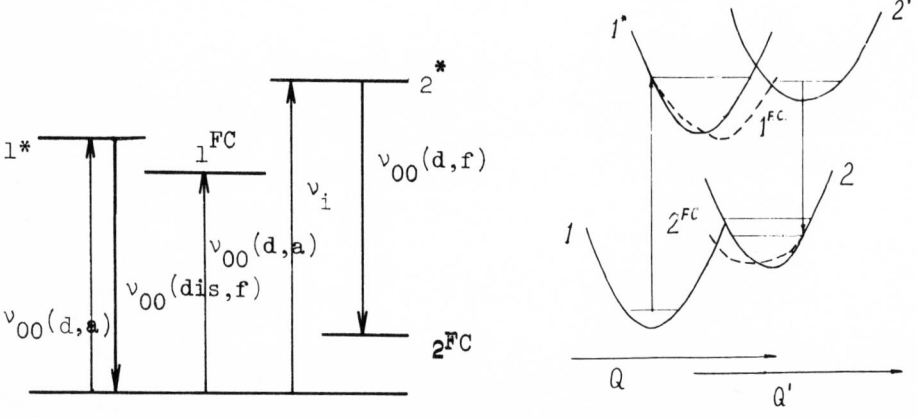

Fig. 2 Radiative transitions and configuration coordinate model.

is discrete (Fig. 2, transition $1 \leftrightarrow 1^*$). If there is a relaxation
process ($1FC \rightarrow 2^*$) (4,5), the diffuse spectrum is emitted (transition $2^* \rightarrow 2^{FC}$ between the level 2^* and the Franck–Condon level 2^{FC}).
Analysis of the experimental data (1-6,11,12) shows that possible
explanations 1 and 2 cannot explain the emission of discrete and

diffuse spectra.

Consider this question in detail: Anthracene is known to have
two nearby excited electronic states (B$^-$(3u) and B$^+$(2u) (14,15)).
If these states do not interact we may expect radiation from both
states. However the discrete and diffuse spectra of anthracene
must be considerably different with regard to both intensities and
lifetimes. According to (14) the oscillator strength of the transition into the B$^+$(2u) state is at least 200 times greater than that
into the B$^-$(3u) state. Experiments (4,5) on the influence of anthracene concentration and of dissolved gases show that the discrete
spectrum disappears at that concentration for which the time between
collisions becomes nearly equal to the lifetime of the excited state
responsible for the diffuse spectrum ($\sim$ 5x10^{-9} sec). This allows
one to suggest that the lifetimes of discrete and diffuse luminescence are approximately equal. If the transition into one of these
states occurs, then it is not understood why the discrete spectrum
due to anti-Stokes excitation is not observed (1,2,4,5). Assuming
that only the zero vibrational level of the excited state is narrow,
we may expect the appearance of a discrete spectrum as a result of
anti-Stokes excitation.

The objection to the second explanation is the absence of a
discrete spectrum of emission with anti-Stokes excitation. Besides,
it seems to be highly improbable that only the zero vibrational
level of the excited electronic state is of small width, with comparatively small interaction between vibrational levels of the
ground electronic state. This follows from the observation of the
narrow lines in the fluorescence spectrum involving high vibrational
levels (> 4000 cm^{-1} (3)). Small width of vibrational levels of the
excited state for complex molecules was obtained also by Bobovich
and Bortkevich (13) in an experiment where an inverted, induced
Raman spectrum was observed. The experimental data seems to agree
better with the suggestion that the type of luminescence spectrum
depends on the relaxation process. According to Neporent (10,11)
a change of the molecular nuclear configuration is possible. This
change arises by virtue of the absence of proper correlation between the electronic density distribution and the nuclear configuration in the newly formed state. After excitation, the molecule
relaxes from the nonstationary Franck-Condon state in which it was
formed (1^{FC}, Fig. 2) to the stationary state 2* (Fig. 2). We suppose that the relaxation process is initiated by the excitation of
the corresponding vibrational levels of the ground state or the
excited state. The state 1* is excited only by the 0''$\rightarrow$ 0' transition. A change in exciting photon energy by as little as 100 cm^{-1}
brings the relaxation process into play and leads to the disappearance of the discrete spectrum (4,5). We may suppose the potential
barrier of the relaxation processes in anthracene to be equal to or
smaller than this value. A necessary condition for the transition

to state 2^* is the participation of excited vibrational levels. The diffuse spectrum would appear not only with excitation at frequency $\nu_{ex.} \neq \nu_{oo}(dis)$ (where $\nu_{oo}(dis)$ is the frequency of the 0,0 band in the discrete spectrum), but also with excitation at frequency ν_{oo} (dis). It is clear that the latter case corresponds to transitions $1'' \to 1'$, $2'' \to 2'$, etc. In fact, discrete and diffuse spectra are always observed together. We note that if there is present a vibrational level which does not interact with any of the others, then excitation of this level would result in emission of the discrete spectrum alone.

Fig. 2a shows the relative energies of the 1^*, 1^{FC} and 2^* levels in accordance with the experimental data (see Table 1). The energy of the 2^* level is determined using the value of the inverse frequency, ν_i, measured by the method based on the effect of foreign gases on fluorescent spectral shape (8). With excitation by light of frequency, ν_i, the spectral shape does not depend on the pressure of foreign gases. This shows that the equilibrium distribution of molecular vibrational levels is maintained. The experimental data (16,20) show that for anthracene and its derivatives, ν_i is more than $\nu_{oo}(d,a)$ ($\nu_{oo}(d,a)$ is the frequency of the peak in the 0,0 band of diffuse absorption). Increase or decrease in the energy of the state 2^* relative to 1^{FC} is apparently determined by the change in the energy of the optically active electrons in the outermost shell interacting with electrons which are not active in this transition.

TABLE 1

Discrete and Diffuse Transitions in Anthracene and Related Compounds

Substance	ν_i (cm^{-1})	ν_{oo} (dis) (cm^{-1})	T K	ν_{oo} (d,a) (cm^{-1})	ν_{oo} (d,f) (cm^{-1})
Anthracene	28100c	27685	503	27490c	27470c
		27688.3 (12)	648	27360c	27320c
9-Methylanthracene	28000 (16)	26880 (17)	393	26950	26850a
			623	26700	26350a
β-Naphtylamin d	28300c	28900	423	29300b	28100b
Phenanthrene d		29070		29120 (19)	

(a) $\lambda_{ex.}$ = 365 nm
(b) is accounted for by the method in (18), $\lambda_{ex.}$ = 350 nm.
(c) is determined in (8).
(d) The discrete spectra for these substances have not been reported. Here the excitation frequencies are given for which the narrowest spectra of fluorescence are observed (18).

From Fig. 1 and Table 1 we can see that $\nu_{oo}(d,a) \neq \nu_{oo}(dis)$. The reason for this effect is probably the difference in the potential curves for 1^* and 1^{FC} states, since the potential curve of the 1^{FC} state must be dependent on relaxation process parameters which are probably influenced by environment, temperature and excitation wavelength.

In speaking of the potential curve of the 1^{FC} state it is necessary to remember that this function is a conditional concept. Precisely speaking this concept may be used only for stable states. Neporent (11) suggests that such potential curves be represented by lines of more than infinitesimal width. The potential curve of the Franck-Condon state is some average characteristic of a molecule which corresponds to the first semiperiod of molecular vibrations active in the electronic spectrum. In Fig. 3, the 1^* curve shows the dependence of electronic energy versus time if nuclear relaxation is absent (the normal coordinate Q is supposed to be $Q \sim \cos(\omega t)$). With the inclusion of relaxation, the form of the curve $E(t)$ is changed (curve 1^{FC}). Consequently, the potential curve $E(Q)$ of the state 1^{FC} must be changed. The intent of this kind of reasoning is a simple and graphic representation, within the framework of ordinary concepts concerning stationary states, of the influence of the relaxation process upon the spectral properties of molecules. In Fig. 2b potential curves of the states under consideration are shown. Here Q and Q' are normal coordinates of the ground (1) and excited (2) stable states, respectively. Table 1 shows that anthracene frequencies $\nu_{oo}(d,a)$ and $\nu_{oo}(d,f)$ are temperature dependent. This probably explains the experimentally observed difference between $\nu_{oo}(d,a)$ and $\nu_{oo}(dis)$. If the luminescence of anthracene vapor is measured at sufficiently low temperature, these values are expected to coincide. For 9-methylanthracene and for phenanthrene $\nu_{oo}(d,a)$ is nearly equal to $\nu_{oo}(dis)$ at low temperatures (see Table 1).

In conclusion let us make the following remarks. If diffuseness of spectra is determined only by relaxation processes the relaxation rate can be evaluated. For a relaxation time $t_r \gg \tau$ (τ is the excited state lifetime), the potential curves of states 1^* and 1^{FC} practically coincide and the spectrum is discrete (Shpolskii spectra (21)), and $\nu_{oo}(dis) = \nu_i$. For $T/2 \ll t_r < \tau$ (T is the period of the symmetric vibration active in the electronic spectrum) the influence of relaxation on the form of a potential curve is small. The spectrum is discrete, $\nu(dis) \neq \nu_i$. In the case $t_r \sim T/2$ the potential curves of states 1^{FC} and 1^* would be different. If T is the period of low-frequency vibrations, vibronic bands corresponding to such vibrations are completely diffuse. In such spectra we can observe vibronic bands of vibrations of high frequency (the diffuse spectrum of anthracene). If T is the period of high frequency vibrations the spectrum has no vibrational

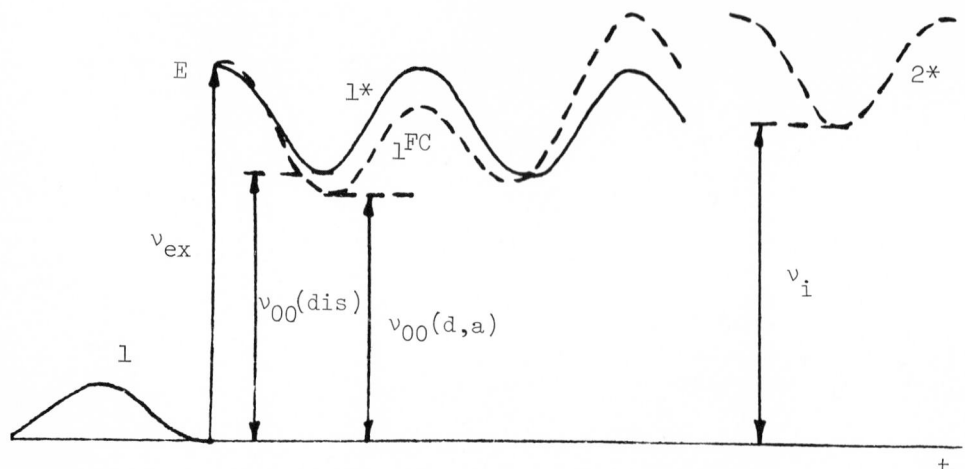

Fig. 3   Relaxation processes.

structure or has only a trace of it, $\nu_{oo}(d,a) \neq \nu_i \neq \nu(dis)$. For $t_r \ll T/2$ the change in electronic density is completed during the electronic transition. The rearrangement of nuclear configurations of the molecule is absent. The spectrum is discrete (binary molecules, benzene).

REFERENCES

1. P. Pringsheim, Ann. Acad. Warsaw, 5, 29 (1938).

2. K. H. Hardtl, A. Scharmann, Zs. Naturforsch., 12a, 715 (1957).

3. V. P. Klochkov, T. S. Smirnova, Optics i Spectry (USSR), 22, 851 (1967).

4. V. P. Klochkov, Optics i Spectry (USSR), 24, 40 (1968).

5. V. P. Klochkov, Izv. Akad. Nauk USSR, ser. fis., 32, 1525 (1968).

6. J. E. Haebig, J. Molec. Spectry, 25, 117 (1968).

7. N. A. Borisevich, V. V. Gruzinskii, Dokl. Akad. Nauk BSSR, 7, 309 (1963).

8. N. A. Borisevich, The Excited States of Complex Molecules in the Vapor Phase, Minsk (1967).

9. A. Terenin, Acta phys. Polon., 5, 229 (1936).

10. B. S. Neporent, Izv. Akad. Nauk USSR, ser. fis. 20, 455 (1956);
    Zh. Fis. Khim., 30, 1048 (1956); Tr. GOI, 25, 3 (1957).

11. B. S. Neporent, Optics i Spectry (USSR), 32, 252 (1972); 38,
    670 (1972).

12. J. P. Byrne, I. G. Ross, Canad. J. Chem., 43, 3253 (1965).

13. Ya. S. Bobovich, A. V. Bortkevich, Optics i Spectry (USSR),
    26, 1060 (1969).

14. M. Mestechkin, L. Gutyrja, V. Poltavec, Optics i Spectry
    (USSR), 28, 454 (1970).

15. R. Pariser, J. Chem. Phys., 24, 250 (1956).

16. V. P. Klochkov, A. M. Makushenko, Optics i Spectry (USSR),
    15, 237 (1963).

17. V. P. Klochkov, S. M. Korotkov, Optics i Spectry (USSR), 25,
    970 (1968).

18. V. P. Klochkov, Optics i Spectry (USSR), 19, 337 (1965).
    V. P. Klochkov, S. M. Korotkov, Optics i Spectry (USSR), 20,
    582 (1966).

19. O. P. Kharitonova, Optics i Spectry (USSR), 5, 29 (1958).

20. V. V. Gruzinskii, N. A. Borisevich, Optics i Spectry (USSR)
    15, 45 (1963).

21. E. V. Shpolskii, Usp. Fis. Nauk (USSR), 71, 215 (1960); 80,
    225 (1963).

ELECTRONIC ENERGY RADIATIONLESS TRANSITION AS RADIATIONLESS DIPOLE-DIPOLE ENERGY TRANSFER TO VIBRATIONS OF SOLVENT MOLECULES

E. B. Sveshnikova and I. B. Neporent

The State Optical Institute, Leningrad, U.S.S.R.

ABSTRACT

The main cause of the degradation of electronic excitation energy in rare-earth and transition metal ions is shown to be an inductive-resonance dipole-dipole energy transfer from excited ions to the solvent vibration overtones. The suggested mechanism can quantitatively explain such regularities as: 1. the linear dependence of log $k_{degr.}$ on ΔE, and the causes of deviation from linearity; 2. deuterium influence upon $k_{degr.}$; 3. existence of spin forbiddenness for radiationless transitions; and 4. dependence of $k_{degr.}$ on distance between the excited ions and the nearest high frequency vibration groups of the solvent. It is shown that this mechanism of energy degradation holds equally well for any aggregate state: liquid, glass and crystal.

The energy degradation of electronic excitation is the conversion of electronic excitation energy of the ion or molecule into vibrational energy. Present theories (1) give a qualitative explanation of the regularities observed experimentally. However, the nature of the perturbation causing radiationless transitions still remains to be determined.

We shall try to prove that the main cause of energy degradation of transition metals and rare-earth ions, independent of the phase of the ion matrix (glass, liquid or crystal), to be inductive-resonance dipole-dipole energy transfer from the excited ions to overtone vibrations of the solvent. As a result of such transfer, the excited ions return to the ground state and the excitation of overtone vibrations of the environment takes place simultaneously (Fig. 1).

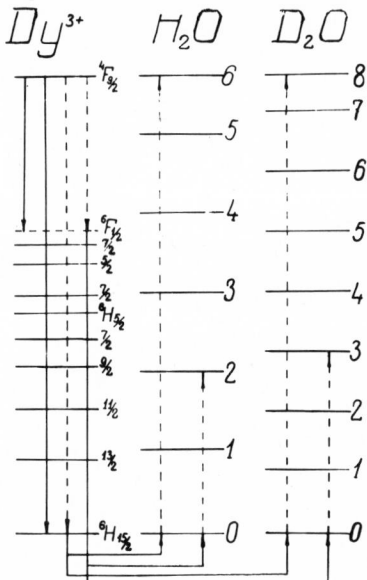

Fig. 1 Energy levels of the energy donor (e.g. a Dy^{3+} ion) and
        energy acceptors - high frequency vibration of H_2O and D_2O
        molecules, taking part in the phenomena of energy degra-
        dation; solid arrows - radiative transitions in the Dy^{3+}
        ion; broken arrows - energy transfer from the Dy^{3+} ion to
        isoenergetic levels of the solvent vibrations in the over-
        tone regions.

The frequency of the resulting vibration is equal to the
dissipated electronic energy quantum. The idea that the energy
degradation leads to excitation of high energy vibrations was
suggested by Heller (2).

In the work of Ermolaev and one of the present authors (3)
this approach was applied to the calculation of the probabilities
of energy degradation in solutions of the rare-earth ions. In contrast to the work of Ballard (4) and Heber (5) the calculations
herein were based on first order perturbation theory. The probabilities of energy degradation ($k_{degr.}$) were calculated via a dipole-dipole interaction approach, i.e. by Förster's formula

$$k_{degr.} = \sum_i 8.8 \times 10^{-25} n^{-4} \kappa^2 k_{rad} R_i^{-6} \int F(\nu) \varepsilon_s(\nu) \nu^{-4} d\nu$$

Since in this case the solvent molecules are the electronic energy
acceptors, $\varepsilon_s(\nu)$ is the molar decadic absorption coefficient of the
solvent in the overtone region (see Fig. 2b), and R_i is the distance

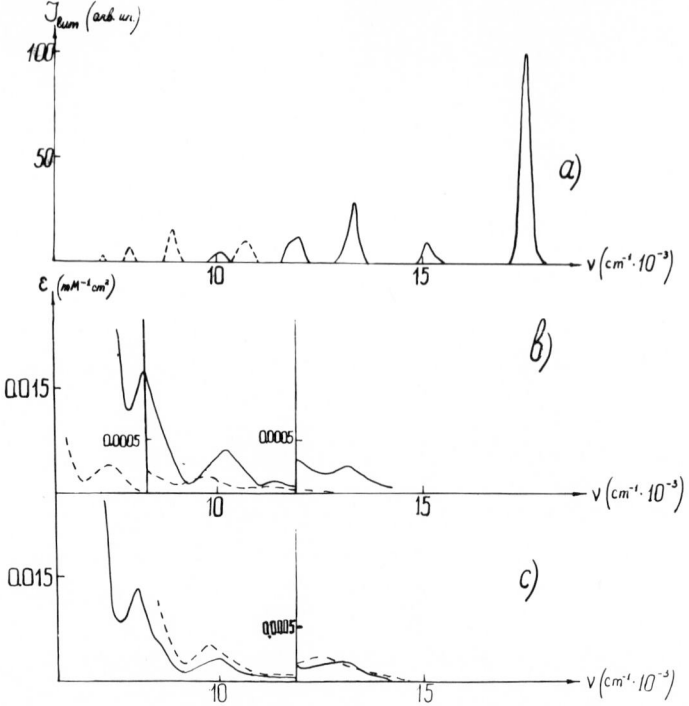

Fig.2 (a) Luminescence spectrum of the Dy^{3+} ion in D_2O; (b) H_2O
absorption spectrum - solid line; D_2O absorption spectrum -
broken line; (c) H_2O absorption spectrum in the $GdCl_3 \cdot 6H_2O$
crystal - solid line; ice absorption spectrum - broken line.
The last three infrared bands of the Dy^{3+} luminescence spectrum are taken from (6). Luminescence bands for which the
scale is increased by 10 times are shown with broken line.

from the excited ion to the i-th solvent molecule. The summation
is over all solvent molecules surrounding the excited ion.

Values of $k_{degr.}$ derived from experiment and from the above
formula are given in the table.

Comparison between columns 5 and 7 shows that the calculation
adequately predicts the isotope effect on $k_{degr.}$ upon deuteration
of the solvent. The calculation also correctly predicts the order
of the $k_{degr.}$ values. The discrepancy between experimental and
calculated values of $k_{degr.}$ is greatest for Eu^{3+} and Tb^{3+} (systems
for which we were unable to obtain sufficiently reliable $\varepsilon_s(\nu)$
values in the region of their luminescence).

Table

Degradation Constants

	:Medium	T K	Calculation $k^H_{degr.}$ sec^{-1}	Calculation $\dfrac{k^H_{degr.}}{k^D_{degr.}}$	Experiment $k^H_{degr.}$ sec^{-1}	Experiment $\dfrac{k^H_{degr.}}{k^D_{degr.}}$	$k_{rad.}$ sec^{-1}	R A
$Tb^{3+}(^5D_4)$	H_2O	293	5×10		2×10^3		97	
$Eu^{3+}(^5D_0)$	H_2O	-"-	1.1×10^2	$\Big\}$ 38	8×10^3	$\Big\}$ 48	80	2.5
-"-	D_2O	-"-	3		1.7×10^2		-"-	-"-
-"-	CH_3OH	-"-	4.4×10^2	$\Big\}$ 32	2.4×10^3	$\Big\}$ 21	130	2.5
-"-	CD_3OD	-"-	1.4		1.7×10^2		-"-	2.5
$Dy^{3+}(^4F_{3/2})$	H_2O	-"-	1.2×10^5	$\Big\}$ 18	3.3×10^5	$\Big\}$ 20	250	2.5
-"-	D_2O	-"-	6.5×10^3		1.6×10^4		-"-	2.5
-"-	$GdCl_3\cdot6H_2O$	-"-			2.5×10^5		250	7)
-"-	$GdCl_3\cdot6D_2O$	-"-			1.5×10^4			
-"-	CH_3OH	-"-	4.2×10^4	$\Big\}$ 17	1.1×10^5	$\Big\}$ 17	295	2.5
-"-	CD_3OD	-"-	2.5×10^3		6.4×10^3		295	2.5
-"-	$(CH_3)_2SO$	-"-	2.4×10^3		6×10^4		250	5
-"-	$(C_4H_9O)_3PO$	-"-	3×10^2		1.3×10^4		295	6.5 7.7
$Sm^{3+}(^4G_{5/2})$	H_2O	-"-	1.5×10^5	$\Big\}$ 32	3.3×10^5	$\Big\}$ 25	370	2.5
-"-	D_2O	-"-	4.7×10^3		1.3×10^4		370	2.5
$Pr^{3+}(^1D_2)$	CD_3OD	-"-	1.1×10^5		6.7×10^5		870	2.5
$Nd^{3+}(^4F_{3/2})$	H_2O	-"-	7.5×10^6	$\Big\}$ 9	1.2×10^8	$\Big\}$ 5	2500	2.5
-"-	D_2O	-"-	8.3×10^5		2.5×10^7		-"-	
$Cr^{3+}(^2E_g)$	$H_2O+NaCNS$	77	6.3×10					
			1.3×10^2		4.6×10^2		≤40	6.0 4.7

Table continued

continuation of Table

$C^{3+}(^2E_g)$	D$_2$O+NaCNS	77	5.3 1.2x10	7.1x10	≤40	6.0 4.7
-"-	CH$_3$OH	-"-	3.2x10 6.4x10	7.8x10	≤40	6.0 4.7
-"-	(C$_4$H$_9$O$_3$)PO	-"-	6.0 1.2x10	4.6x10	≤40	6.0 4.7

The dependence of the magnitude of $k_{degr.}$ upon the distance from the center of the electron excitation to the nearest high frequency vibration is established from experiment. The theoretical approach suggested above also predicts the existence of such a dependence.

We have made a comparison between the experimental and calculated values of $k_{degr.}$ for Dy^{3+} ions in a series of solvents: methyl alcohol, dimethylsulphoxide, tributylphosphate. These solvents were chosen so as to successively separate groups containing high frequency vibrations from rare-earth ions, i.e., increase R (see last column of the table). It is clear from the table that there is a qualitative correlation between the predictions of the theory and experimental data.

Many experimental and theoretical papers report the existence of one more regularity, i.e., the linear relation between log $k_{degr.}$ and ΔE, where ΔE is the energy gap between the ground and the nearest excited state. This relationship, however, is not universal. In organic chemistry it occurs only in a number of related compounds and it does not always hold for rare-earth ions.

Experimental data for various rare-earth ions in identical surroundings (methanol) are plotted in Fig. 3. The data for Yb^{3+} and Pr^{3+} do not satisfy the linear dependence. The dipole-dipole mechanism of energy degradation predicts a linear dependence between log $k_{degr.}$ and ΔE if similar intensity distribution of the ion luminescence spectra is observed, i.e., linearity in the case of the Tb^{3+}, Eu^{3+}, Sm^{3+} and Dy^{3+} ions, and deviations for Pr^{3+} and Yb^{3+} which exhibit intense long wavelength luminescence bands. Thus our approach enables us to explain this experimental feature as well.

A number of assumptions were made in our calculation: 1. we suppose a rigid structure of solvates of rare-earth ions in solution; 2. we do not take into consideration vibronic bands of the rare-earth spectra; 3. unlike Heber's work (5) we do not consider interactions such as electric dipole-quadrupole.

To answer the question what error is introduced by the assump-

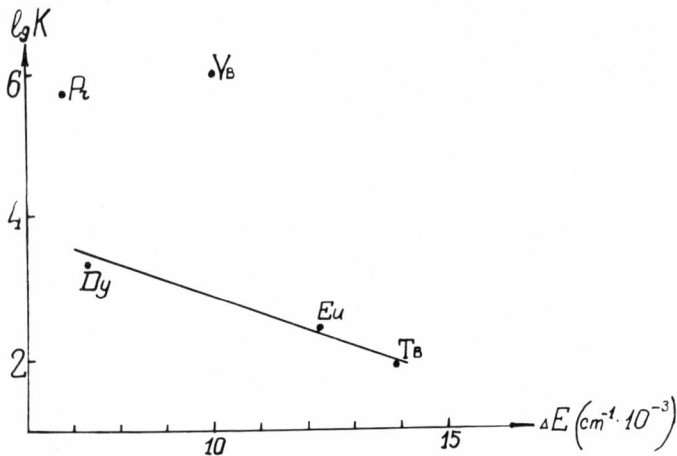

Fig. 3 Dependence of the logarithm of the energy degradation
        probability of the rare earth ions in methanol-d_4 at 293 K
        on the energy gap between the radiative level and nearest
        lower level.

tion of the rigid structure of solvates in solution, and to clear
up the question on the possibility of application of the proposed
mechanism to other aggregate states, we have measured τ of luminescence of the $GdCl_3.6H_2O$ and $GdCl_3.6D_2O$ crystals doped with
$DyCl_3$. To avoid concentration effects we have used crystals containing not more than one percent by weight of $DyCl_3$. The measurement of the water absorption spectra $\varepsilon_s(\nu)$ (Fig. 2c) have also been
made for $LaCl_3.7H_2O$ and $GdCl_3.6H_2O$ single crystals.

    The differences between the absorption spectra of distilled
water (Fig. 2b) and water in the rare-earth crystal are not critical. Therefore the differences in the rigidity of the structure
of the surroundings have no distinct effect on the value of $k_{degr.}$
in crystals and solutions; and τ of rare-earth ions should not
differ appreciably in crystals and solutions. Experimentally, this
is indeed the case.

    Thus, the model for the surroundings used by us for solutions
does not cause any appreciable errors. As all the parameters used
in the calculation of $k_{degr.}$ differ only slightly for crystals and
solutions, all the conclusions made for solutions are adequate for
crystals.

    To clear up the possibility of applying the suggested method
over a wide temperature range, we have measured the ice absorption
spectrum in the overtone region (Fig. 2c). The ice absorption

spectrum does not appear to differ much in absolute values from
the water spectrum at room temperature. Rare-earth luminescence
spectra and hence the integral

$$\int F(\nu)\epsilon_S(\nu)\nu^{-4}d\nu$$

also depend insignificantly on temperature (See Fig. 2). Thus the
theory predicts negligible temperature effect on k_{degr}..

If the change in the rare-earth ion solvate shell produced by
temperature variations is absent, the experiment also gives a weak
dependence on temperature for τ of rare-earth ion luminescence.

The energy degradation process in frozen solutions were studied by us in detail for other ions - the transition metal ions.
The anion $Cr(NCS)_6^{3-}$ was studied in different solvents. This ion
is of interest as the energy acceptors (the groups responsible for
the high-frequency vibrations of the solvent molecules) are separated from the Cr^{3+} ion by anions and are situated at a distance
ten times as large as the size of the excited ion. The application
of the dipole-dipole approach is more strict under such conditions
than in the case of rare-earth ions in water. The experimental and
calculated values of k_{degr}. obtained are in good agreement. Some
uncertainty in these calculated values of k_{degr}. is explained by
not taking into consideration the solvent libration of the S-H
hydrogen bond. We had to take two extreme possible magnitudes of R.

The investigation of the radiationless transitions of the
transition metal ions Cr^{3+} and Co^{3+} showed that the absence of the
isotope effect predicted by the calculations, and the discrepancy
between the order of experimental and calculated values of k_{degr}.,
could be a criterion for other processes dominating over degradation.

It was found (8) that concentration quenching was the main
cause of deactivation of Cr^{3+} ions in some solvents. Only by
eliminating this effect could we obtain agreement between theory
and experiment for such solvents as water, acetonitrile and dimethylsulphoxide.

Summing up the results, we can say that the mechanism suggested holds good for any aggregate state. Its application does
not depend on vibronic interaction of the ions. The mechanism
applies equally to rare-earth and transition metal ions even though
the vibronic band intensity of latter is comparable to the intensity of pure electronic transitions. Apparently, the energy degradation of simple molecules in the condensed state is effected by
the same mechanism. The solvent molecules take part in energy degradation as well. It is shown in the recent paper by Merkel,

Nilson and Kearns (9) that in a number of solvents $k_{degr.}$ is proportional to the overlap of the radiation spectrum and the solvent absorption spectrum.

We would like to emphasize that such a relation should be observed if the dipole-dipole energy degradation mechanism holds.

We shall not discuss the problem arising in attempts to apply this approach to the case of large organic molecules. This question has been discussed by Förster (10).

ACKNOWLEDGEMENT

The authors express their thanks to V.L. Ermolaev for the idea of this work and useful discussions.

REFERENCES

1. D.M. Burland, G.W. Robinson, J. Chem. Phys. <u>51</u>, 4548 (1968). J. Jortner "Transitions Nonradiatives dans les Molecules" 20ᵉ Reunion de la Societe de Chemie Physique, Paris, (1969) p. 9.

2. A. Heller, J. Am. Chem. Soc. <u>88</u>, 2058 (1966).

3. E.B. Sveshnikova, V.L. Ermolaev, Optika i Spectroscopia <u>30</u>, 379 (1971).

4. R.E. Ballard, Spectrochim. Acta <u>24A</u>, 65 (1968).

5. J. Heber, Phys. Kondens. Mat. <u>6</u>, 381 (1967).

6. G.A. Mokeeva, S.P. Lunkin, P.P. Feofilov, Jurnal Prikladnoi Spectroscopii, <u>4</u>, 245 (1966).

7. R.G. Wykoff, Crystal Structures Vol. 3, Interscience, New York (1965) p. 793.

8. E.B. Sveshnikova, I.B. Neporent, Izvest. Akad. Nauk SSSR (Ser. Fiz.) <u>36</u>, 1087 (1972).

9. P.B. Merkel R. Nilson, D.R. Kearns, J. Am. Chem. Soc. <u>94</u>, 1030 (1972).

10. Th. Förster, Chem. Phys. Lett. <u>12</u>, 422 (1971).

STUDY OF SPECTRAL AND TIME CHARACTERISTICS OF TETRAPYRROLE

MOLECULES IN THE TRIPLET STATE BY LASER PHOTOLYSIS

G. P. Gurinovich and B. M. Jagarov

Institute of Physics, Academy of Sciences of the

Byelorussian SSR, Minsk, USSR

ABSTRACT

The mechanism of radiationless intramolecular deactivation of the lowest singlet state has been ascertained for a number of free base porphyrins and their metal ion derivatives. The effect of the central metal ion on the intramolecular energetics is discussed.

The present paper is devoted to the elucidation of different deactivation pathway efficiencies of excited states. It is known that the dominant radiationless deactivation of the lowest singlet state occurs via the triplet state in the case of free base porphyrins and their complexes with closed shell metal ions (1-4). The experimental determinations of the triplet state formation intersystem crosing yield γ, the fluorescence yield ρ, and lifetimes τ have made it possible to calculate the probabilities of intersystem crossing, r, and fluorescence, f. The data obtained in our previous work have allowed us to come to the following conclusions: (a) the intramolecular probabilities are weakly influenced by side-chain substitution in the external position of pyrrol rings and methene bridges. (b) such structural changes, as aza-substitution and reduction of one pyrrol ring cause a simultaneous increase of the probabilities r and f. These changes are accompanied by the lifting of the quasi-prohibition of the longest wavelength electronic transition, which takes place in porphyrins. In particular, the aza-substitution causes the increase of the probabilities of non-radiative transition from the lowest triplet state which explains the weak phosphorescence in phthalocyanine molecules (5). The metal ions in the center of the parent porphyrins cause more fundamental changes in the intramolecular transitions. Our studies

of metalloporphyrins fall into three groups. A relatively simple
picture is envisioned for porphyrin complexes of diamagnetic
closed shell ions. The excited states of such ions lie much higher
than the lowest π^*-π singlet and triplet porphyrin states. The
tabulated data showed that central attachment of the diamagnetic
ion Zn^{2+} results in an increase of intersystem crossing probability
(see Table 1). The results obtained in studies of Mg-mesoporphyrin
in EPA at T=77°K (6) are also listed in Table 1, because the
measurements show that the yield for this compound is independent
of temperature and solvent. The value of r for mesoporphyrin and
its metalloporphyrins illustrates the qualitative picture of an
increase of intersystem crossing probability with increasing atomic
number of the central ion (Z=12 for Mg^{2+} and Z=30 for Zn^{2+}, which
is accompanied by a fluorescence yield drop as well as intersystem
crossing yield gain. Cd-complex (Z=48) exhibits still weaker
fluorescence (7). It is thus considered that an internal heavy
atom effect occurs in such compounds. Nevertheless, the detailed
analysis of the central ion effect is modified by other factors
such as symmetry.

All the above mentioned porphyrins show sufficient fluorescence. In degassed solutions of such compounds triplet-triplet
absorption with duration $\sim$ 500 µsec is observed. The investigation
of metalloporphyrins becomes more complicated when the parent
porphyrin is complexed with an ion with a partially filled d-shell.
These complexes do not exhibit fluorescence. Most of the deduc-
tions for such complexes have been based on phosphorescence studies,
but the recent work (8) proved that in many cases the phosphorescence was due to impurities. In the present situation T-T absorption data becomes of great importance. But all the attempts to
observe T-T absorption of Cu^{2+}, Co^{2+}, Fe^{2+} and other complexes by
traditional flash-photolysis with time resolution about 10 µsec
ended in failure. We have overcome this problem through a laser-
photolysis technique capable of a time resolution of 80 nsec with
an oscilloscope, and ultimately a resolution of 15-20 nsec using
a photographic recording apparatus. A detailed description of

Table 1

Quantum Yields of Triplet Formation, Fluorescence
Quantum Yields and Lifetimes, Transition Probabilities

Compound	γ	ρ	$\tau \times 10^9$ (sec)	$f \times 10^{-7}$ (sec^{-1})	$r \times 10^{-7}$ (sec^{-1})
Mesoporphyrin	0.80	0.12	19	0.63	4.1
Mg-mesoporphyrin	0.62	0.20	11	1.8	5.6
Zn-mesoporphyrin	0.86	0.05	2	2.5	43
Tetraphenylporphin	0.82	0.13	13.6	0.95	6.0
Zn-tetraphenylporphin	0.88	0.04	2.7	1.5	32.5

Table 2

Luminescence and Triplet-Triplet Absorption Data
of Metalloporphyrins

Compound	Configu-ration	Atomic number	Magnetic prop.	Fluor. yield	Phosph(a) yield	T-T absorp.	Triplet yield
Mg^{2+}-meso		12	diamag	0.20	weak	observed	0.62
Zn^{2+}-meso	d^{10}	30	diamag	0.05	0.07(b)	observed	0.86
Cu^{2+}-meso	d^9	29	paramag	none	0.60(c)	observed	∿1.0
Cu^{2+}-TPP	d^9	29	paramag	none	0.06(c)	observed	0.87
Co^{2+}-meso	d^7	27	paramag	none	none	---	---
Ag^{2+}-meso	d^9	47	paramag	none	none	none	---
Ni^{2+}-meso	d^8	28	diamag	none	none	none	---
Pd^{2+}-meso	d^8	46	diamag	10^{-4}(d)	0.50(c)	observed	∿1.0
Co^{3+}-meso	d^6	27	diamag	none	none	none	---

(a) All values of phosphorescence yields are found for etioporphy-
    rins complexes
(b) Ref. (14)
(c) Ref. (11)
(d) Ref. (12)

laser-photolysis apparatus, has been given previously (4,9). The
main results of our studies are listed in Table 2 and we shall now
briefly discuss them leaving a more detailed discussion for the
future. In our studies we have observed absorption of mesoporphyrins IX dimethylester complexed with Cu^{2+} and Pd^{2+} (Cu-meso and
Pd-meso, arising from the electronic phosphorescence state.

The lifetime of this phosphorescence state for Cu-meso and
Pd-meso were found to be 150 nsec and 250 nsec, respectively, in
undegassed toluene solutions at room temperature. The oxygen removal had little influence on the lifetime of the phosphorescence
state of Cu-meso. In degassed solutions of Pd-meso the triplet
state lifetime was 90 μsec. In this paper, we do not discuss the
nature of the electronic states in copper porphyrins. Gouterman
and his coworkers have recently published a systematic series of
experimental and theoretical works devoted to this question (10,11).

Using values of γ for tetraphenylporphin (TPP) and Zn^{2+}-tetra-
phenylporphin (Zn-TPP) as standard we have found the metastable
state formation yield of Cu-TPP in frozen glassy solution at 77°K.
Presuming that Zn-TPP and TPP triplet-triplet extinction coeffi-

cients do not differ markedly from absorption extinction coefficients, for the Cu-TPP phosphorescence state we have obtained the
value of γ for the copper complex, ~ 1.0 (4). We have not measured
the absolute value of γ for Pd-meso. But it is of interest to note
that under similar experimental conditions the optical absorption
attributed to the phosphorescence state is greater in the cases of
Pd-meso and Cu-meso than for freebase mesoporphyrins. This result
directly indicates a large value of γ, close to unity. Our result
is in a rather good agreement with the literature value of γ =
0.70 (12), especially if we take into account the accuracy of the
measurements. The large values of γ and the lack of fluorescence
for the two metallocomplexes provide evidence that the central
metal ion sufficiently increases the probability of nonradiative
processes which quench fluorescence. It is necessary to note that
the physical nature of the perturbation can differ. In Cu complexes the strong perturbation is possibly caused by the paramagnetic nature of the Cu ion. Another possible mechanism (10) assumes
the coupling of the unpaired electron with the porphyrin molecule
in singlet and triplet states. This mechanism explains the drastic
enhancement of nonradiative transition to the phosphorescence state.
In the case of Pd complexes the d-π interaction is responsible for
the perturbation.

The third group of experimental results were obtained for
Ag^{2+}, Co^{3+}, Ni^{2+}-mesoporphyrins. Entirely negative results were
obtained in our attempts to find the triplet-triplet absorption for
these transition metal complexes by the laser photolysis method.
The room temperature measurements were carried out in dilute solutions in the spectral region 400 nm-650 nm. The typical T-T spectra
of all the porphyrins, including Zn and Cu complexes, have been
placed in this region. It is difficult to imagine that the lack
of T-T absorption is due to an unusual energy level shift. At this
time, using the photographic variation of our experimental technique, we have found the bleaching of the singlet-singlet absorption bands which result from a depletion of the ground state caused
by the intense laser flash. Since the simple oscilloscope apparatus
was incapable of measuring the duration of the bleaching or even the
gross effect, it is obvious the duration of the bleaching is considerably shorter than 80 nsec. Our laser photolysis results agree
well with other experimental results in porphyrins phosphorescence
(11,13). Thus the lack of T-T absorption and the presence of ground
state depletion in Co, Ni, Ag complexes make it possible to suppose
the existence of a pathway for nonradiative deactivation of the
lowest excited singlet state other than the mechanism discussed
above for freebase porphyrins and their closed shell and Cu, Pd
complexes. The properties of ion complexes differ markedly depending on the nature of the ion. Recourse to analogies based on atomic
number and/or number of unpaired d-electrons cannot explain our results for Ag^{2+}, Ni^{2+} and Co^{3+}. Taking all factors into consider-

ation, we suggest that an intramolecular electron transfer to a low lying level centered on the metal ion takes place in Ag, Ni and Co porphyrin complexes.

REFERENCES

1. P. J. Bowers and G. Porter, Proc. Roy. Soc. 296A, 435 (1967).

2. A. T. Gradyushko, A. N. Sevchenko, K. N. Solovyov and M. P. Tsvirko, Izv. Akad. Nauk SSSR. Ser. Fiz. 34, 636 (1970).

3. B. M. Jagarov, Opt. i Spektr. 28, 66 (1970).

4. B. M. Jagarov, Izv. Akad. Nauk SSSR. Ser. Fiz. 36, 1093 (1972).

5. P. S. Vincett, E. M. Voigt and K. E. Rickhoff, J. Chem. Phys. 55, 4131 (1971).

6. G. P. Gurinovich, A. I. Patsko and A. N. Sevchenko, Dokl. Akad. Nauk. SSSR. 174, 837 (1967).

7. P. G. Seybold and M. Gouterman, J. Mol. Spectr. 31, 1 (1969).

8. D. Eastwood and M. Gouterman, J. Mol. Spectr. 35, 359 (1970).

9. B. M. Jagarov, Y. I. Kozlov, A. P. Simonov, G. P. Gurinovich, Opt. i Spektr. 32, 838 (1972).

10. R. L. Ake and M. Gouterman, Theoret. Chim. Acta. 15, 20 (1969).

11. D. Eastwood and M. Gouterman, J. Mol. Spectr. 30, 437 (1969).

12. J. B. Callis, M. Gouterman, Y. M. Jones, B. H. Henderson, J. Mol. Spectr. 39, 410 (1971).

13. R. S. Becker and J. B. Allison, J. Phys. Chem. 67, 2662 (1963).

14. A. T. Gradyushko, M. P. Tsvirko, Opt. i Spektr. 31, 548 (1971).

A NEW LASER TECHNIQUE FOR SIMULTANEOUS TWO WAVELENGTH STUDIES

Herschel S. Pilloff

Naval Research Laboratory

Washington, D.C. 20390, USA

ABSTRACT

A new technique has been developed which allows simultaneous two wavelength selection in both inhomogeneously and homogeneously broadened lasers. The two laser beams are collinear and collimated, have mutually orthogonal polarizations, and, in general, can be temporally synchronized. This technique has been applied to the pulsed nitrogen laser-pumped dye laser and features continuous tunability of the two wavelengths throughout the entire optical gain region of the particular dye solutions used. Following the description of this device, the extension of this technique to other lasers and a few potential applications will be briefly discussed.

The laser has already proven itself to be a very important tool for the study of a wide variety of photophysical and photochemical phenomena. Although the recent advent of continuously tunable laser sources has begun to extend our understanding of some of these processes, many future studies will require simultaneously two or more wavelengths. While it would, of course, be possible to use more than one laser source, there are many reasons why a tunable two-wavelength laser would be extremely useful, particularly if both output beams are perfectly collinear and collimated so that complete spatial overlap is maintained even when both beams are focused to the diffraction limited spot sizes. The technique to be described satisfies these requirements. Although it is applicable to other types of lasers, both inhomogeneously and homogeneously broadened, the N_2 laser-pumped dye laser was selected for consideration here primarily because it offers continuous wavelength tunability

201

throughout the visible spectrum (using different dye solutions) as well as comparatively large tuning ranges from a single dye solution - a maximum range of 1760 A (from 3910 to 5670 A) having already been reported (1).

Multiple wavelength operation of the homogeneously broadened dye lasers appeared to be possible, at least over small wavelength regions, based on previous reports of subsidiary intracavity etalon effects in which a number of closely spaced lines or channels appeared in the output spectra of dye lasers (2). While these experiments were in progress, Zalewski and Keller (3) reported two-wavelength operation in a flash lamp-pumped Rhodamine 6G dye laser using a different technique than will be described here. The problem thus seemed to be not whether it would be possible to operate a dye laser simultaneously at more than one wavelength, but rather, how to predetermine that each wavelength will interact only with those wavelength selective elements which are appropriate to that particular wavelength. The essential feature of the technique reported here is to effectively decouple the two wavelengths from each other by forcing the dye laser to produce the two wavelength beams with mutually orthogonal polarizations. This is accomplished by inserting a polarizing element in the laser cavity which both forces the laser to oscillate with perpendicular polarizations and at the same time spatially separates each polarization such that each can be directed to the appropriate wavelength selector. The two wavelength beams are then fed back into the polarizing element, spatially recombined, and ultimately output coupled from the laser.

The experimental arrangement is shown schematically in Fig. 1.

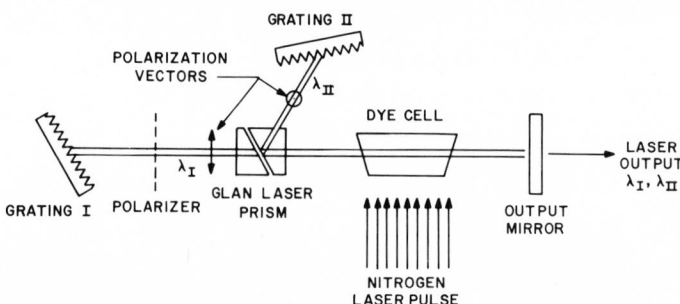

Fig. 1   Schematic diagram of two wavelength N_2 laser pumped dye laser. The symbol ⊙ represents optical polarization vector perpendicular to the plane of the diagram whereas ↕ represents the polarization parallel to the plane of the drawing. The polarizer, which is used to provide a variable loss, is arbitrarily shown in the laser cavity arm of grating I. The minimum total cavity lengths, distances between either grating I or II and the output mirror, are both approximately 20 cm.

The output from a commercially available N_2 laser, capable of
producing pulses of 100 kilowatts peak power and 10 nanoseconds
duration at 3371 A with a repetition rate of 100 pulses per second,
is focused with two cylindrical lenses into the dye cell. The dye
cell has a 2.5 cm. optical length and was designed to provide
transverse laminar flow of the dye solution with respect to the
longitudinal axis of the dye laser cavity. A Glan-laser prism, an
open escape window (all exterior surfaces broadband anti-reflection
coated) provides a low loss polarizing element which forces the two
wavelengths to have mutually orthogonal polarizations and also
spatially separates the two polarized beams such that each is
uniquely directed to the appropriate wavelength selector.
Initially, ruled (replica) gratings having 1800 grooves/mm. were
used in this experiment; however, the large spectral bandwidth
($\gtrsim$ 10 A) and, in some cases, complicated frequency sweeping effects
made their continued use undesirable. Instead, holographically
produced dielectric wavelength selectors were prepared in a manner
similar to that described by Kogelnik et al (4). Dichromated
gelatin plates were prepared from Kodak 649F spectroscopic plates
following both the sensitization and development procedures
described by Chang (5). The plates were exposed with the 4880 A
line from an argon ion laser and the beams were incident $\pm$ 45°,
which produced a spatial modulation frequency of 2900 lines/mm.
Those phase holograms having approximately 50% diffraction
efficiency as measured in the reconstructed beam were selected and
sandwiched with a totally reflecting mirror. The plane output
mirror is coated with a broad band dielectric coating having
approximately 70% transmission.

Although most of the experimental studies characterizing this
technique have been done with 4-Methylumbelliferone (4-MU)-ethanol
dye solutions of varying acidities (hydrochloric acid) because of
the very large wavelength tunability available from such single
solutions, other dye solutions have also been used and the results
in all cases were essentially the same. The two output beams are
collinear and collimated, and the two wavelengths are continuously
tunable anywhere within the optical gain region of the particular
dye solution used. In order to accommodate the components shown
in Fig. 1 it is necessary to physically increase the normal laser
cavity length, and this adversely affects both the spectral bandwidth and the output power characteristics of the N_2 laser-pumped
dye laser. The spectral bandwidths were measured with a scanning
Fabry-Perot interferometer having a free spectral range of 50 cm^{-1}.
The results for operation at maximum peak power and averaged over
1500 pulses (30 pulses/sec. for 50 sec.) showed a time averaged
spectral bandwidth of less than 2 A at 5000 A. Peak output powers
on the order of 10 kilowatts are obtained from this system with 4-MU.
A recent report that hydrochloric acid quenches both the spontaneous
and stimulated emission from the 4-MU proton exciplex whereas

perchloric does not, indicates that higher powers may be obtained from 4-MU with perchloric acid as the proton donor (6).

The time dependencies of the two-wavelength beams were measured with fast PIN photodiodes and displayed on a Tektronix 7904 oscilloscope. In general, a slightly different pulse shape is observed at each wavelength. To the extent that both wavelength beams have the same pulse shape, the two beams are not "a priori" automatically time synchronized because the wavelength having the higher gain will reach threshold first with the result that delay times of 4 nanoseconds or more have been observed. Unless the intensity of the first wavelength beam (having the shorter delay time) is attenuated in some way within the cavity, the intensity of the second beam will be strongly attenuated or even eliminated altogether as a result of extensive homogeneous broadening. This is usually accomplished by inserting a variable loss filter (polarizer) in the laser cavity arm of the first beam and in this way the relative intensities of the two beams can be controlled. In addition, the variable loss filter produces a slight increase in the delay of the first beam and a slight decrease in the delay of the second. In general, these changes in the delay times cannot be made large enough to synchronize the two beams because the very high gain of the N_2 laser pumped dye laser (7) and the comparatively short duration of the pumping pulse (relative to the round-trip optical cavity transit time) force the laser to operate in a nearly single-pass configuration. However, by increasing the laser cavity arm length of the first beam and/or decreasing the cavity arm length of the second beam, the two beams can be temporally synchronized over most of the tunable range of the dye solution.

After the two beams have been brought into synchronization, a completion effect of one beam on the other is often observed, i.e. when the cavity arm of one beam is blocked, the intensity of the other beam is increased. The extent of this competition is a function of the particular dye solution used, the choice of the two wavelengths, as well as the N_2 laser pump power. This effect results from the fact that the same volume region in the dye solution gives rise to both wavelengths and provides a direct demonstration of the high degree of homogeneous broadening common to dye lasers and possibly to a lesser extent, the rapid orientational relaxation often found in such systems. In fact, it should be possible to map both the extent and degree of homogeneous broadening in these solutions from a detailed study of this competition.

Some possible areas for application of this technique include infrared difference frequency generation in birefringent nonlinear crystals, two photon spectroscopy, optical double resonance, excited state spectroscopy, and photoselection, to mention only a few. The narrow banding techniques developed by Hänsch (8) should be applicable to the present system, although probably with some sacrifice of instrumental flexibility. Finally, it should be noted that the technique is rather general and has recently been applied to other lasers, e.g. the infrared HF chemical TEA laser (9). Because the calcite Glan-laser prism has high absorption losses at wavelengths longer than ∿2μ , it was replaced with a germanium flat having a small wedge angle and oriented at Brewster's angle. Germanium has an index of refraction ∿4.0 at HF wavelengths, and therefore, at Brewster's angle, reflects ∿ 76% of the polarization component perpendicular to the plane of the incident beam, which is adequate for many high gain systems.

The author acknowledges helpful discussions with Dr. L. S. Goldberg. He also thanks Drs. R. J. Schafer and J. A. Blodgett for their help in holographically exposing the dichromated gelatin plates.

REFERENCES

1. C. V. Shank, A. Dienes, A. M. Trozzolo, and J. A. Meyer, Appl. Phys. Lett. 16, 406 (1970).

2. M. Bass and T. F. Deutsch, Appl. Phys. Lett. 11, 89 (1967); and A. N. Rubinov and V. A. Mostovnikov, Bull. Acad. Sci. USSR, Phys. Ser. (English Transl.) 32, 1348 (1968).

3. E. F. Zalewski and R. A. Keller, Appl. Opt. 10, 2775 (1971).

4. H. Kogelnik, C. V. Shank, T. P. Sosnowski, and A. Dienes, Appl. Phys. Lett. 16, 499 (1970).

5. M. Chang, Appl. Opt. 10, 2550 (1971).

6. A. Bergman, R. David, and J. Jortner, Opt. Comm. 4, 431 (1972).

7. C. V. Shank, A. Dienes, and W. T. Silfvast, Appl. Phys. Lett. 17, 307 (1970); and T. W. Hansch, F. Varsanyi, and A. L. Schawlow, Appl. Phys. Lett. 18, 108 (1971).

8. T. W. Hänsch, Appl. Opt. 11, 895 (1972).

9. R. J. Cody and H. S. Pilloff, to be published elsewhere.

ON THE APPLICATION OF THE SINGLE PHOTON COUNTING TECHNIQUE FOR THE INVESTIGATION OF DYES WITH SHORT FLUORESCENCE DECAY TIME

P. E. Zinsli, H. P. Tschanz, O. Jenni, Th. Binkert

Institute of Applied Physics, University of Berne

Berne, Switzerland

ABSTRACT

The single photon method is used in fluorescence decay measurements with a excitation flash of small width and fast decay and an analog computer for evaluation of the rate parameters. With this equipment decay times down to 0.2 nanoseconds can be determined accurately and easily. This technique is applied to the investigation of the reaction kinetics in the excited state of 7-hydroxycoumarin and 4-methylumbelliferon in water-ethanol mixtures. Rate parameters for an adapted model and assignments of the fluorescent reaction products are given.

INTRODUCTION

The single photon method (1) is suited for measuring fluorescence decay time spectra in the nanosecond region. The small intensity required in this method is an advantage when monochromators are used and when a low excitation intensity is desired to avoid effects of high excitation density. When fluorescence is excited by a flash of the form $f(t)$, the measured intensity function $I(t)$ is proportional to the convolution

$$I(t) \sim \int_0^t g(t-t')i(t')dt' \tag{1}$$

of the fluorescence decay function $i(t)$ and the combined resolution function $g(t)$ of the flash $f(t)$ and the instrumental response $s(t)$, if linear fluorescence kinetics are obeyed. The interpretation of measurements (eq.(1)) is difficult for short decay times. Two improvements allow us exact statements on the function $i(t)$ in the

time region of one nanosecond: on the one hand we constructed a
flash-lamp of short pulse width (1 nsec FWHM) and fast decay time,
on the other hand we made use of an analog computer in evaluating
the data. The fluorescence kinetics of two coumarin dyes was so
investigated.

EXPERIMENTAL SETUP

The sample is excited by the optical pulses of a flash lamp,
while the electrical pulses serve as start-signals in a time to
pulseheight converter (TPC). The fluorescence emission of the
sample is filtered by a monochromator and detected by a photomultiplier with single electron resolution (RCA 8850). The single
electron events give the stop-signals in the TPC. The measured
time differences are stored in a multichannel analyzer (MCA). By
this arrangement the measured probability $p_1(t)dt$, that the first
photoelectron in the stop PM is released in the time interval $(t,
t+dt)$ after the beginning of the fluorescence, is proportional to
$I(t)dt$ (2) (3):

$$p_1(t)dt \sim I(t)dt \qquad\qquad\qquad (2)$$

The reconstructed function $p_1(t)$ in the MCA is now displayed on an
oscilloscope and evaluated with an analog computer.

The exciting flash lamp is a free-running spark discharge. A
steel tube with a quartz window is used, containing the electrodes
of the spark gap in a variable distance of about 0.1 mm, a induc-
tion-free highvoltage resistance (100 MΩ) and the attenuator for
the start-signal. The shortest pulsewidth (1 nsec FWHM) and
fastest decay (almost exponential over $2\frac{1}{2}$ decades with a time
constant of 0.5 ns) together with long term stability was obtained
in air with cylindrical tungsten electrodes. At atmospheric
pressure a line spectrum, mainly between 280 nm and 450 nm is
emitted, and at a voltage of 5kV the pulse frequency is 5 kHz.

The rate parameters of the model kinetics for $i(t)$ are evaluated from the measured curves $I(t)$ using a little analog computer.
The excitation function $g(t)$ is simulated by the convolution of a
Gaussian function with an exponential function. It is fed into
the input of the integrator corresponding to the initially excited
^1S state. The output voltages of the integrators now simulating
the population of the corresponding states are displayed alternatively with the measured curves on an oscilloscope. By adjusting
the settings of the coefficient-potentiometers (which correspond
to the values of the rate parameters) a fit of both curves is
attained if an applicable model was chosen. A storage oscilloscope
is advantageous to compare the curves. Advantages of this evaluation method are:

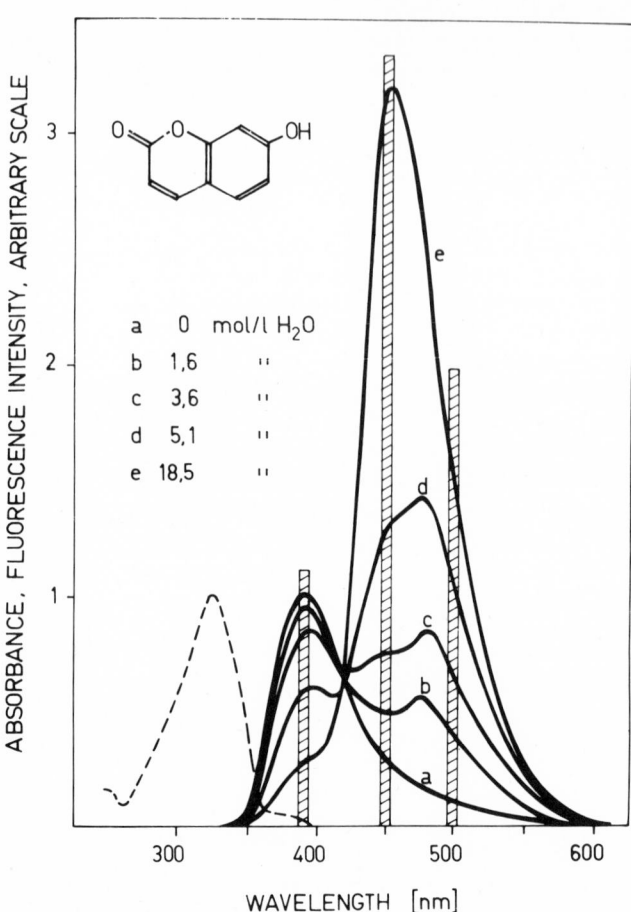

Fig. 1   Absorption and emission spectra (uncorrected) of 7-hydroxy-
         coumarin in ethanol with variable amounts of water added
         (in moles per liter solvent). The monochromator settings
         of the fluorescence decay measurements are marked with
         vertical bars.

    - Since the simulation of (1) requires only a few seconds,
      the results may be obtained just after the experiment and
      conclusions for further experiments may be drawn immediate-
      ly.

    - The explicit solution of intricate differential equations
      and the problem of deconvolution of spectra is avoided.

    - Additional effects in the experiment such as stray light,
      excitation of several ground-state molecular forms and
      superposition of the emission of different excited states

may be simply considered in the analog computer by addition
of the exciting pulse, exciting of several integrators and
addition of outputs.

- The parameters gained are excellent initial values for
  further handling of the data, e.g., by nonlinear regression
  in digital computers.

- In the fitting process one gets a sense of significant and
  unsignificant parameters of the model.

RESULTS

Using the solvent dependence of the fluorescence decay time
spectra the reaction kinetics of the excited states of 7-hydroxy-
coumarin (7-HC) and 4-methylumbelliferon (4-MU) is investigated.
These dyes (being stronger acids and stronger bases in the excited
state) show different fluorescence of neutral and ionic molecules,
while the absorption spectrum is constant over wide pH-ranges.
Overlapping emission peaks around 385 nm, 415 nm and 480 nm are
observed (4-7) depending on the acidity of the solvent. The
characteristic emission has found application in the operation of
a broadband tunable dye (4) laser. Fig. 1 shows the emission and
absorption spectra of 7-HC in ethanol water mixtures (similar
spectra are obtained for 4-MU): at low water concentration the
peak at 480 nm exceeds the peak at 450 nm, but it is smaller at
high concentrations. This behavior, being significant for the
kinetics, was investigated in the time spectra.

Approximate relative quantum yields ϕ_i/ϕ_k (i,k = A,B,C) of the
emitting molecular species A (emission maximum λ_e = 385 nm),
B (λ_e = 480 nm) and C (λ_e = 450 nm) are obtained from the emission
spectra (Fig. 1) by comparison with superpositions of pure spectra
of A, B and C obtained in ethanol, acid and water. Using the
relations between the relative quantum yields and the rate parameters we obtain conditions for the rate parameters which have to
be compared with the measured parameters of the time spectra.

Possible model kinetics were simulated on the analog computer
and the best fit was searched (Fig. 2). The overlap of emission
spectra was considered by summing integrator outputs, and very fast
nonstationary processes which do not influence the decay times are
taken into account by simultaneous excitation of several integrators. The choice of some parameters need not be definite. Measured decay times are listed in Table 1.

INTERPRETATION

A simultaneous interpretation of the fluorescence emission

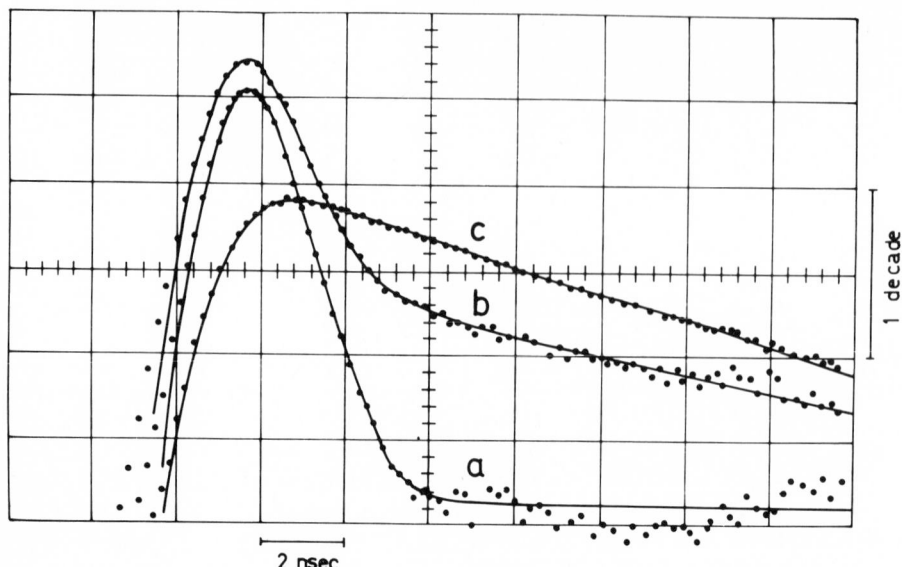

2 nsec

Fig. 2   Fluorescence decay time measurements (·····) of 7-hydroxy-
         coumarin in ethanol and fitted analog-computer curve
         (———) as displayed on a storage oscilloscope. a ex-
         citing pulse of the flash lamp, b fluorescence decay at
         390 nm (17.8 m H_2O), c fluorescence decay curve at 500 nm
         (17.8 m H_2O).

Table 1

Fluorescence decay rate parameters k [10^8 sec^{-1}]

Solvent	7-hydroxycoumarin			4-methylumbelliferone		
	390 nm(a)	450 nm	500 nm	390 nm(a)	450 nm	500 nm
ethanol	11.5+0.5	2.1+0.1	2.1+0.1	7.15+0.3	2.2+0.1	2.2+0.1
1 m water	11.6+0.5	2.1	2.1	7.2 +0.3	2.2	2.2
2 " "	11.7+0.5	2.1	2.1	7.2 +0.3	2.2	2.1
5 " "	12.5+0.8	2.0	2.1	8.5 +0.4	2.1	2.0
10 " "	17.5+1.5	1.9	2.1	12.0 +0.8	1.9	1.9
20 " "	> 30	1.8	2.1	24.5 +3	1.8	1.9
water	——	1.8	1.9	———	1.7	1.8
0.3 n HClO₄	——	——	2.2	———	——	2.1
0.25 n KOH	——	1.8	——	———	1.8	——

a) prompt decay (see fig. 2)

spectra and the decay times were only obtained in the kinetics

$$B^* \underset{k_{BA}}{\overset{k_{AB}}{\rightleftharpoons}} A^* \underset{k_{AC}}{\overset{k_{CA}}{\rightleftharpoons}} C^*$$
$$\underset{B}{\downarrow k_{MB}} \qquad \underset{A}{\downarrow k_{MA}} \qquad \underset{C}{\downarrow k_{MC}}$$

and with a rate parameter k_{CA} which is nonlinear in the concentration of water added. The following model is assumed: At the encounter of a neutral molecule A^* with a water molecule in ethanol solution the proton from the hydroxy group is transferred to the water molecule and a bound state B^* of the two ions is formed. For this first-order rate constant we obtain:

7-HC: $k_{BA} \lesssim 0.3 \times 10^8$ ℓ/mole·sec

4-MU: $k_{BA} \lesssim 0.4 \times 10^8$ ℓ/mole·sec

k_{BA} is much slower than an expected diffusion-determined reaction rate (8) $k_{diff} = 8RT/3000\eta = 6.6 \times 10^9$ ℓ/mol·sec which is explained by a stereospecific reaction site and a small reaction probability per encounter. By adding more water the dielectric constant of the solvent is raised, the ions separate more easily: the anion C^* is formed. In this case the stationary rate parameter is given by the product of the reaction rate constant $k'_{CA}[H_2O]_2$ and the probability for separation of the ions $\exp(-UkT)$ where $U = e^2/4\pi\epsilon_0\epsilon a$ and a = reaction impact parameter. If a linear dependence of the dielectric constant ϵ on the mole fraction X of water is taken, we obtain

$$k_{CA} = k'_{CA}[H_2O]\exp\{-\frac{575}{a}(\frac{1}{55.6X+24.6})\}:$$

7-HC: $k'_{CA} = 1.6 \times 10^9$ ℓ/mole·sec, a= 4A

4-MU: $k'_{CA} = 0.7 \times 10^9$ ℓ/mole·sec, a= 5A

The decay times of B^* and C^* being almost constant, we obtain as upper limits for first order back reaction rate constants:

7-HC: $k_{AB} \lesssim 0.05 \times 10^8$ ℓ/mole·sec , $k_{AC} \lesssim 0.1 \times 10^8$ ℓ/mole·sec

4-MU: $k_{AC} \lesssim 0.05 \times 10^8$ ℓ/mole·sec , $k_{AC} \lesssim 0.2 \times 10^8$ ℓ/mole·sec

CONCLUSION

Using a simple analog computer, fluorescence decay time measurements with the single photon method can be evaluated promptly and easily. With a very rapid excitation flash lamp, fluorescence kinetics and decay times in the region of one nanosecond may be investigated. The estimated accuracy lies at 0.05 to 0.1 ns below a decay time of 2 ns, at 5% above 2 ns and the lower limit of measurable decay times is at about 0.2 ns, given by the width of the flash, the time resolution of the setup and the subjective fitting process.

ACKNOWLEDGEMENTS

We wish to thank A. Dubied of the Medico-chemical Institute in Berne for leaving us the spectrometers.

We gratefully acknowledge the financial support of the "Kommission zur Förderung der wissenschaftlichen Forschung des eidgenössichen Volkswirtschaftsdepartementes".

REFERENCES

1. L. M. Bollinger, G. E. Thomas, Rev. Sci. Instr. $\underline{32}$ 1044 (1961).

2. J. A. Miehe, G. Ambard, J. Zampach, A. Coche, IEE-T-NS $\underline{17/3}$, 115 (1970).

3. Th. Binkert, H. P. Tschanz, P. E. Zinsli, J. of Lum., to be published.

4. A. Dienes, C. V. Shank, A. M. Trozzolo, Appl. Phys. Letters $\underline{17}$, 5, 189 (1970).

5. P. W. Fink, W. R. Koehler, Anal. Chem. $\underline{42}$, 9, 990 (1970).

6. G. J. Yakatan, R. J. Juneau, S. G. Schulman, Anal. Chem. $\underline{44}$, 6, 1044 (1972).

7. M. Nakashima, J. A. Sousa, R. D. Clapp, Nature Phys. Sci. $\underline{235}$, 16 (1972).

8. J. B. Birks in Photophysics of Aromatic Molecules Wiley-Interscience, London (1970).

THE RATIO OF QUANTUM YIELDS OF FLUORESCENCE AND PHOSPHORESCENCE OF

SOME N-HETEROCYCLIC COMPOUNDS IN SOLUTIONS AT 77 K

Ivan Janić

Institute of Physics and Mathematics, University of Novi

Sad, Novi   Sad, Yugoslavia

ABSTRACT

Measurements were made of the ratio of quantum yields of fluorescence and phosphorescence (η_f/η_p), the position of the 0-0 transition of the fluorescent and phosphorescent spectrum and the lifetime of phosphorescence in the molecules (free base) and cations (conjugate acid) of seven nitrogen containing heterocyclic compounds (5,6- and 7,8-benzoquinoline, 2-, 4-, 6-, 7- and 8-methylquinoline) dissolved in n-hexane and ethanol at different concentrations at 77 K. The n-hexane and ethanol solutions of benzoquinoline have a η_f/η_p between 0.5 and 6; the cation, between 2.5 and 17. In pure hydrocarbon solvents at 77 K, methylquinolines show only phosphorescence $(\eta_f/\eta_p < 0.01)$. In alcohol solvents at 77 K there occurs both fluorescence and phosphorescence (in ethanol $\eta_f/\eta_p = 0.8$ to 4.5).

INTRODUCTION

Some aza-heterocyclic compounds in hydrocarbon solvents exhibit very weak fluorescence or none at all. This quenching of fluorescence is a consequence of increased spin-orbit interaction, caused by the mutual action of n- and π-electrons in the fluorescent state (1,2,3). However, the same compounds show fluorescence in hydroxylic solvents such as alcohols, in which the hydrogen bond can be established with the nitrogen atom in the fluorescent molecule and the oxygen in the OH of alcohol. Hence, the quantum yields of fluorescence and phosphorescence largely depend on the solvent. The luminescent characteristics described above were found for quinoline (3-5). The luminescent spectra of 5,6- and 7,8-benzoquinoline (5-10) and of 2-, 4-, 6-, 7- and 8-methylquinoline (5) in solutions

213

have also been studied. In the present paper, further investigations
of their luminescent characteristics are reported.

EXPERIMENTAL RESULTS

The above benzo-and methylquinolines were examined in the following solvents: n-pentane, n-hexane, n-heptane, cyclohexane,
methylcyclohexane, n-propanol, methanol and ethanol. Since the
luminescent characteristics of the compounds examined are similar
in hydrocarbon solvents on the one hand, and in alcohol solvents on
the other, we shall limit the discussion to results obtained in
n-hexane and ethanol (Table I).

In the molecular state, 5,6- and 7,8-benzoquinoline show fluorescence (3450-4000 A) as well as phosphorescence (4550-5500 A),
in both polar and nonpolar solvents. The luminescence spectrum has
a clearly developed vibrational structure, though in n-hexane it
assumes a quasi-linear character (10). The lifetime of fluorescence
in aqueous solution (pH $\approx$ 13) at 293 K is of the order of 10^{-8} sec
(Table I, column 12). The η_f/η_p ratio largely depends on the solvent and the concentration (columns 6 and 7). For 5,6-benzoquino-
line in n-hexane at concentrations 10^{-2}, 10^{-3}, 10^{-4} and 10^{-5} M, the
η_f/η_p ratio amounts to 0.57, 0.52, 0.55 and 1.53, respectively; in
ethanol, 5.6, 4.3, 3.5 and 3.0, respectively. For 7,8-benzoquino-
line at the same concentrations in n-hexane the values of η_f/η_p are
0.82, 1.40, 1.55 and 2.20; in ethanol 11.0, 8.2, 6.8 and 5.6, respectively. The lifetime of phosphorescence does not depend on the
concentration.

Proton acceptance by benzoquinoline molecules produces a significant change in the luminescence spectrum. At a temperature of
77 K, the fluorescence spectrum of the cation has no vibrational
structure and is in the range of 3700-5200 A, with a maximum at
about 4000 A, while the phosphorescence spectrum has a significantly
reduced vibrational structure. The η_f/η_p ratio greatly depends on
the concentration, as can be clearly seen in the case of 5,6-
benzoquinoline in Fig. 1.

In pure hydrocarbon solvents solvents at 77 K methylquinoline
shows only phosphorescence (4550-5500 A). In the corresponding
n-paraffins, the phosphorescent spectrum has a typical quasi-linear
structure, differing in multiplicity with the individual methylquinolines (from singlet to quartet). By adding a small amount of
ethanol ($\sim$ 0.5 %) to the n-hexane solution an intense fluorescence
is induced at 77 K, which reaches saturation at an ethanol concentration of about 2%. It is evident that the hydrogen bond decreases the inner conversion from the fluorescent state to the
phosphorescent. The fluorescence spectrum in ethanol at 77 K is
further in the ultraviolet (3100-3700 A) than is the same spectrum

TABLE I

Experimental data for the position of the highest energy vibronic maxima of the fluorescence and phosphorescence spectra (ν_{oo}, in cm^{-1}), quantum yield rations of fluorescence and phosphorescence (η_f/η_p), lifetime of fluorescence and phosphorescence (τ) of the molecule (free base, B) and cation (conjugate acid, BH$^+$) in some benzo- and methylquinolines in n-hexane (n-hex.), ethanol (ET), ethanol+H$_2$SO$_4$ and water+H$_2$SO$_4$ solvents, at luminescent compound concentrations 10^{-4} M. The data in columns 12 and 13 were taken at 293 K, the rest at 77 K.

Compound	ν_{oo} (fluor.)		ν_{oo} (phosph.)			η_f/η_p			τ(sec) (phosph.)			τ(nsec) (fluor.)	
	n-hex.	ET	n-hex.	ET	ET + H$_2$SO$_4$	n-hex.	ET	ET + H$_2$SO$_4$	n-hex.	ET	ET + H$_2$SO$_4$	Water+H$_2$SO$_4$ pH≈13	Water+H$_2$SO$_4$ pH≈3
	(B)	(B)	(B)	(B)	(BH$^+$)	(B)	(B)	(BH$^+$)	(B)	(B)	(BH$^+$)	(B)	(BH$^+$)
	(1)	(2)	(3)	(4)	(5)	(6)	(7)	(8)	(9)	(10)	(11)	(12)	(13)
5,6-Benzoquinoline	28944	28736	21896	21867	21186	0.55	3.5	7.9	2.60	3.22	3.83	9	8
7,8-Benzoquinoline	28868	28612	21805	21782	20973	1.55	6.8	16.7	2.03	2.05	2.01	12	9
2-Methylquinoline		31626	21786	21877	21668	<0.01	3.3	8.5	1.10	1.42	2.50		14
4-Methylquinoline		31898	21777	21858	21626	<0.01	0.84	~50	1.07	1.52	2.60		13.5
6-Methylquinoline		31270	21290	21299	20661	<0.01	4.4	~70	0.82	1.05	1.26		6
7-Methylquinoline		31908	21753	21848	20921	<0.01	0.95	~70	0.84	1.35	2.06		23
8-Methylquinoline		(30796)	21345	21358	20121	<0.01	3.1	~70	0.82	1.20	1.08		38

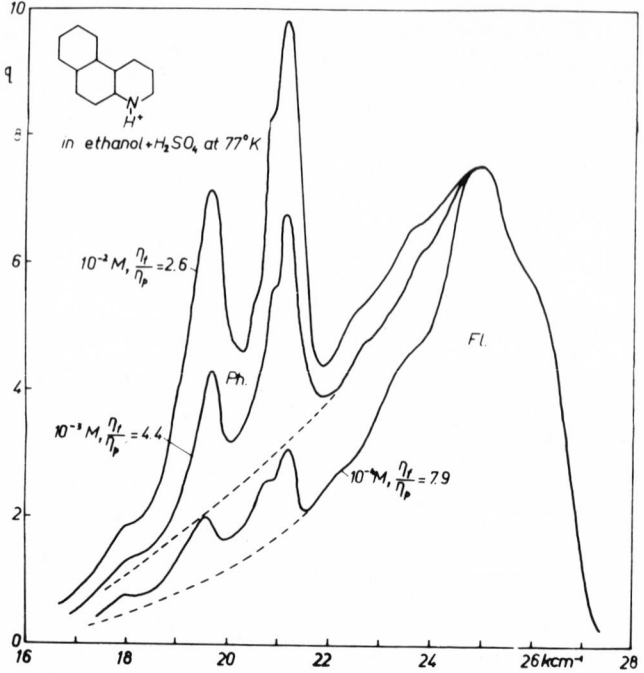

Fig. 1   Luminescence of 5,6-benzoquinoline cation.

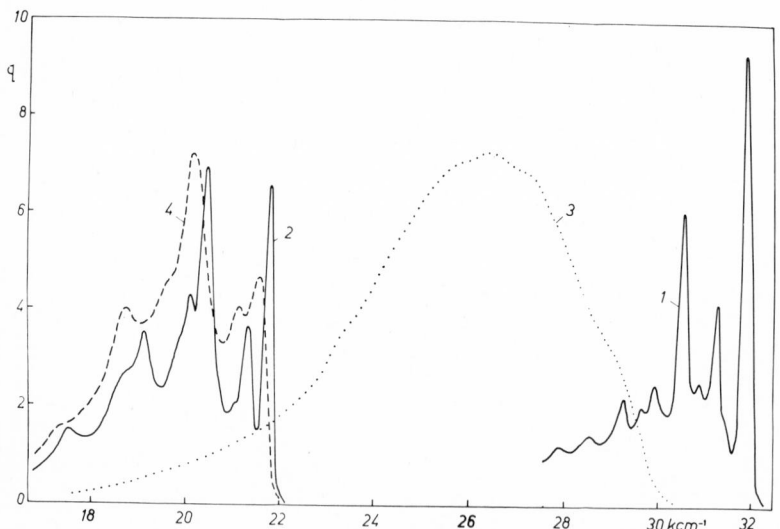

Fig. 2   Luminescence of 4-methylquinoline: 1 and 2 fluorescence and
         phosphorescence of base in EtOH; 3 and 4-fluorescence and
         phosphorescence of acid in EtOH+H_2SO_4 at 77 K.

for benzoquinolines. The phosphorescence spectra of methylquinolines are the same in ethanol and in n-hexane (columns 1, 2, 3 and
4 of Table I and Fig. 2). The luminescence spectra show a clear
vibrational structure, with the exception of 8-methylquinoline.
The intensity of luminescence decreases in a linear manner with increase in temperature: the phosphorescence disappearing at about
100 K and the fluorescence disappearing practically between 140-
160 K. But in some methylquinolines the fluorescence remains, with
a very low intensity, even at higher temperatures, e.g., in the case
of 8-methylquinoline up to as high as 293 K. The lifetime of phosphorescence and the η_f/η_p ratio do not depend on concentration
(columns 9 and 10).

The fluorescence spectrum (3400-5200 A) has a maximum at about
3800 A (Fig. 2) and no vibrational structure. The position of the
0-0 transition (3400-3500 A) is shifted considerably towards longer
wavelengths compared to the molecule (Fig. 2). It follows that the
first excited (singlet) electronic state of the molecule is certainly
of the (π,π^*) type, and that on the basis of Förster's cycle (11,12)
the molecule in this electronic state becomes a 5 to 7 pK-units
stronger base than when in the electronic ground state (7,13). The
cation phosphorescence spectrum has a considerably weaker vibrational
structure than the molecule. In the lowest triplet state the methylquinoline molecule is a 0.5 to 2 pK-units stronger base than when in
the electronic ground state (7,13) as can be estimated from the red
shift of the phosphorescence spectrum in the protonated molecule
(columns 4 and 5 of Table I). The η_f/η_p ratio in 2-methylquino-
line amounts to 8.5, while in the other methylquinolines $\eta_f/\eta_p \approx$
50 to 70.

ACKNOWLEDGEMENT

The lifetime measurements were made by Dr. M. Hauser (Physical
Chemistry Laboratory, University of Stuttgart) whose cooperation is
gratefully acknowledged.

REFERENCES

1. N. Mataga, S. Tsuno, Bull. Chem. Soc. Japan, 30, 368 (1957).

2. V. L. Ermolaev , I. P. Kotlar, Optika i Spektrosk., 9, 359
    (1960).

3. E. Bowen, F. Wokes, Fluorescence of Solutions, Longman, Green,
    London-New York (1953).

4. N. Mataga, Y. Kaifu, M. Koizumi, Bull. Chem. Soc. Japan, 29,
    373 (1956); 31, 459 (1958).

5.   I. Janić, Dissertation, Belgrade, 1963.

6.   Y. Kanda, R. Shimada, Spectrochim. Acta 13, 211 (1959).

7.   M. Nakamizo, Spectrochim. Acta 22, 2039 (1966).

8.   H. H. Perkampus, K. Kortüm, Z. Phys. Chem. 56, 73 (1967).

9.   A. Grabowska, B. Pakula, Photochem. Photobiol. 10, 415 (1969).

10.   I. Janić, A. Kawski, Bull. Acad. Polon. Sci., Ser. Sci. Math.
      Astr. Phys. 20, 253 (1972).

11.   Th. Förster, Z. Elektrochem., 54, 42, 531 (1950).

12.   A. Weller, Progress in Reaction Kinetics (G. Porter, Ed.) Vol.
      I, 189-214, Pergamon Press, Oxford 1961.

13.   I. Janić, Acta Acad. Paed. Civ., Pećs (Hungaria), 10, 31
      (1966).

FORMATION OF HETERO-EXCIMERS IN RADICAL ION RECOMBINATION REACTIONS

A. Weller and K. Zachariasse

Max-Planck-Institut für biophysikalische Chemie

Abt. Spektroskopie, Göttingen, Germany

ABSTRACT

Chemiluminescent reactions in ether solution between radical anions (A^-) of bitolyl and anthracene and different radical cation perchlorates ($D^+ClO_4^-$) are investigated. It is found that singlet hetero-excimers $^1(A^-D^+)$, and molecular triplet states, 3_A^* and 3_D^*, are the primary products of these reactions, the latter giving rise to molecular fluorescence by triplet-triplet annihilation and to hetero-excimer emission by mixed triplet-triplet annihilation. The results are interpreted on the basis of the energies involved and discussed in terms of the relative yields in which these excited species are produced from the radical ions.

INTRODUCTION

It is well known from fluorescence quenching and transformation studies (1,2) that hetero-excimers, $^1(A^-D^+)$, emitting a characteristic broad and structureless fluroescence band at longer wavelengths than the A or D fluorescence can be formed in the excited singlet state whenever suitable electron acceptors, A, and electron donors, D, are combined.

The question of whether or not the same type of hetero-excimers can also be formed from the radical ions (in their doublet ground states) according to:

$$2_A- \ + \ 2_D+ \longrightarrow {}^1(A\text{-}D^+)$$

has led to a search for chemiluminescent reactions (3-6) between the radical anions, A^-, of substituted and unsubstituted aromatic compounds and the stable radical cations, D^+, of tetramethyl-p-phenylenediamine (TMPD), tri-p-tolylamine (TPTA), tri-p-anisylamine (TPAA) and tri-(p-dimethylaminophenyl)-amine (TPDA) in a suitable flow system. The solvents used were dimethoxyethane (DME) and tetrahydrofuran (THF), which at room temperature have a dielectric constant of about 7.

In order to bring about chemiluminescence at all in these systems, the energy stored chemically in the radical ions must be transformed into excitation energy of the A,D system and, since this transformation cannot occur until the solvated radical ions have formed an encounter complex, $A^-..D^+$, it is the free enthalpy, $\Delta G(A^-..D^+)$, associated with the equilibrium: $A + D \longrightarrow A^-..D^+$ which under the prevailing isothermal conditions is available for subsequent reactions. The free enthalpy can be calculated according to Eq. 1 (derived earlier (3b):

$$\Delta G(A^-..D^+) = E_{1/2}(D/D^+) - E_{1/2}(A^-/A) + 0.20 \text{ eV} \qquad (1)$$

from the polarographic oxidation and reduction potentials of the donor and acceptor, respectively, measured in acetonitrile or dimethylformamide.

In view of its comparatively slow diffusional separation (10^6/sec in these low dielectric constant solvents) this encounter complex between the solvated radical ions represents a convenient origin of the competitive processes that eventually lead to the observed chemiluminescence. In particular, any excited configuration of the A,D system whose energy is greater than that calculated according to Eq. 1 for the radical ion pair will not be populated to any appreciable extent in the radical ion recombination reaction. In the case of the chemiluminescent systems described here the free energy deficiencies with respect to the molecular excited singlet states, $^1A^*$ and $^1D^*$, are greater than 0.38 eV so that direct production of these states from the radical ion pair can be ruled out.

An attempt to estimate the free enthalpy change, ΔG_{ipc}, associated with the equilibrium:

$$^2A^-..^2D^+ \rightleftharpoons {}^1(A^-D^+) \qquad (2)$$

on the basis of the thermodynamic and spectroscopic data of heteroexcimer systems in hexane (2) and taking into account (with the aid of the Onsager-Kirkwood continuum model) the increased heteroexcimer solvation free enthalpy in the ether solvents gave:

$$\Delta G_{ipc} \approx 0 \pm 0.12 \text{ eV} . \qquad (3)$$

Accordingly, the equilibrium constant K_{ipc} may vary between 0.01 and 100. This result indicates that direct hetero-excimer formation from the radical ions may, in principle, be possible, but because of the high uncertainties involved does not allow any general predictions as to the yields of such processes.

RESULTS AND DISCUSSION

Purification of the materials, the preparation of the solutions and the measurement of the chemiluminescence spectra were carried out as described previously (3) [cf. also (6)].

The upper part of Fig. 1 gives the chemiluminescence spectrum obtained with the system bitolyl[-]/TMPD[+]. The three bands observed can easily be identified as bitolyl fluorescence $^1A^*$, TMPD fluores-

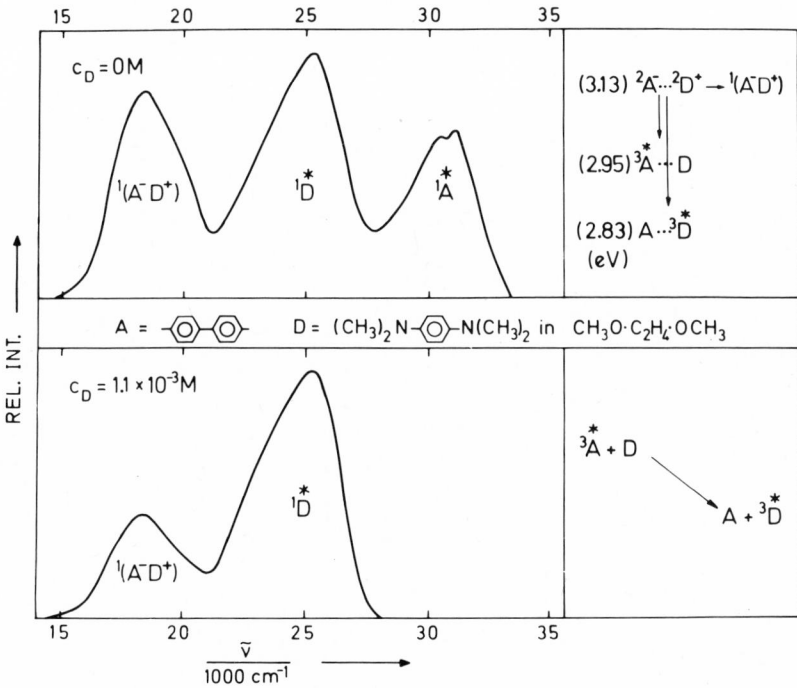

Fig. 1   Chemiluminescence spectra (uncorrected) obtained in flow experiments at room temperature with DME solutions of bitolyl radical anions (ca. 2 x 10[-4] M) and Wurster's Blue perchlorate (TMPD[+] ClO$_4$[-]). Upper spectrum: no TMPD added. Lower spectrum: 1.1 x 10[-3]M TMPD added to the radical anion solution.

cence $^1D^*$, and hetero-excimer fluorescence $^1(A^-D^+)$. For each band
the chemiluminescence yield determined as described in (4) is
roughly 10^{-4} Einstein/mole (anion).

The energy situation depicted on the right hand side of Fig. 1
shows that both molecular triplet states, $^3A^*$ and $^3D^*$, are energetically accessible from the radical ion pair so that the observation of both molecular fluorescences can be attributed to the
triplet-triplet annihilations

$$^3A^* + ^3A^* \longrightarrow ^1A^* + A$$

$$^3D^* + ^3D^* \longrightarrow ^1D^* + D$$

which follow triplet state production (via electron transfer in the
radical ion pair) by:

$$^2A^{\overline{\cdot}}\cdots^2D^+ \longrightarrow ^3A^* + D \qquad (4)$$

$$^2A^{\overline{\cdot}}\cdots^2D^+ \longrightarrow A + ^3D^* \qquad (5)$$

with yields γ_t^A and γ_t^D, respectively.

This interpretation of the experimental results implies that
the electron transfer processes (4) and (5) can, and evidently do,
occur without intermediate hetero-excimer formation. Furthermore,
they occur at encounter configurations that are unfavorable for
triplet energy transfer, since otherwise only the lowest triplet
state $^3D^*$ would result.

The occurrence of hetero-excimer fluorescence in this chemiluminescence experiment can be ascribed to a mixed triplet-triplet
annihilation process:

$$^3A^* + ^3D^* \longrightarrow ^1(A^-D^+)$$

and/or to direct hetero-excimer formation from the radical ion
pair:

$$^2A^{\overline{\cdot}}\cdots^2D^+ \longrightarrow ^1(A^-D^+) \qquad (6)$$

the yield of which is γ_c.

In order to decide which mechanism is operative in this system
chemiluminescence experiments were carried out with different
amounts of TMPD added to the radical anion solution. The chemiluminescence spectrum obtained with 1.1×10^{-3} M TMPD is depicted in
the lower part of Fig. 1. Although the bitolyl fluorescence is
completely absent, showing that virtually all bitolyl triplets pro-

duced initially by reaction (4) have been quenched (presumably via triplet energy transfer to TMPD), the hetero-excimer chemiluminescence has only decreased by a factor of 2. This means that about one half of the hetero-excimers are generated via mixed triplet-triplet annihilation while the other half (the unquenched ones) are produced directly from the radical ions according to reaction (6). The chemiluminescence yield of the latter, which amounts to about 5×10^{-5} Einstein/mole (anion), is then given by:

$$\chi' = \gamma_c \, \phi_o' \tag{7}$$

where ϕ_o' is the hetero-excimer fluorescence quantum yield. Since $\phi_o' < 1$, one has for the bitolyl$^-$/TMPD$^+$ system $\gamma_c > 5 \times 10^{-5}$.

The chemiluminescence spectra obtained with anthracene radical anions and three different radical cations in THF are shown in Fig. 2. The structured emission around 25,000 cm^{-1} is identical with the anthracene fluorescence, the structureless emissions with maxima below 20,000 cm^{-1} originate from hetero-excimers. The quantum yield rations, χ'/χ^A, (cf. Table 1) of hetero-excimer (χ') and anthracene (χ^A) chemiluminescence are identical with the intensity ratios obtained from the spectra after correcting for mutual overlap and for the non-linear spectral response of the detection system. The absolute values of χ' and χ^A have been determined according to the method described in (4) and are considerably less accurate than their ratios.

As shown by the energy data given in Table 1 the anthracene triplet is the only molecular excited state accessible from the radical ion pair. Therefore, mixed triplet-triplet annihilation can be ruled out as a source of hetero-excimer formation and the hetero-excimer chemiluminescence yield is given by Eq. 7.

While the anthracene chemiluminescence quantum yields, χ^A, are about equal in the three systems, showing that triplet state production and triplet-triplet annihilation are independent of the radical cation, the hetero-excimer chemiluminescence quantum yields differ by several orders of magnitude. Evidently γ_c (and, perhaps, also ϕ_o') are both decreasing as $\Delta G(A^-..D^+)$ becomes smaller, i.e. as the radical ion pair as well as the hetero-excimer (see $h\nu_{max}$ in Table 1) get closer to the ground state.

With this decreasing energy separation from the ground state two effects become operative which will both lead to a smaller value of γ_c:

(i) The amount of energy, U_{dest}, by which the hetero-excimer state is destabilized due to its interaction with the ground state increases as the energy difference between these two states

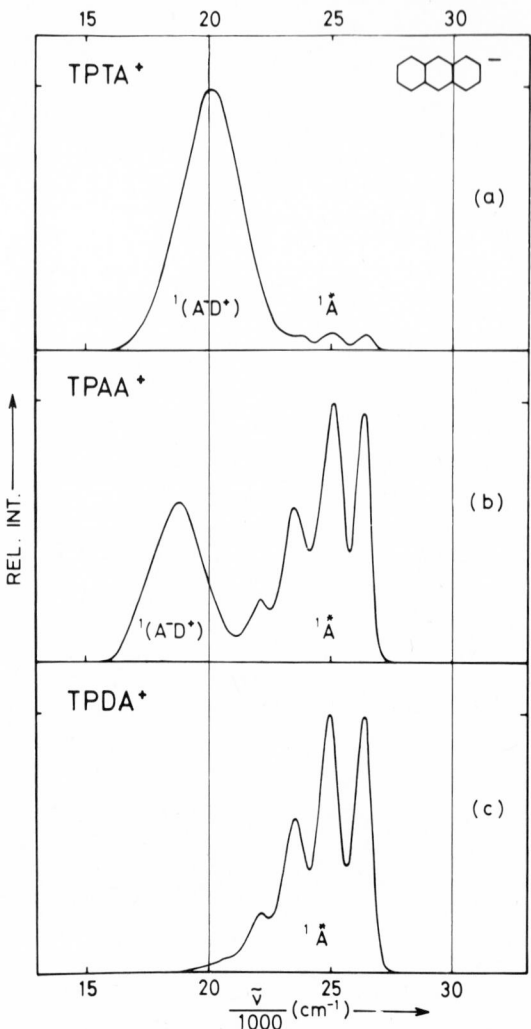

Fig. 2   Chemiluminescence spectra (uncorrected) obtained in flow
         experiments at room temperature with THF solutions of
         anthracene radical anions and the radical cation perchlor-
         ates of: tri-p-tolylamine (a); tri-p-anisylamine (b);
         tri-(p-dimethylaminophenyl)-amine (c).

decreases.   This makes ΔG_{ipc} (see Eqs. 2 and 3) more positive
(by U_{dest}) and, therefore leads to smaller γ_C values.

(ii)   The yield γ_g of the electron transfer process

$$2A^{\bar{\cdot}} \ldots 2D^+ \longrightarrow {}^1A + {}^1D \tag{8}$$

Table 1

Data for chemiluminescent radical ion recombinations

in tetrahydrofuran at room temperature with A = anthracene

$$E_{1/2}(A^-/A) = -1.96 eV \text{ (vs. SCE)};$$

$$\Delta E(^3A^*) = 1.82 \text{ eV}; \quad \Delta E(^1A^*) = 3.28 \text{ eV}$$

D	$E_{1/2}(D/D^+)$ (V)	$\Delta G(A^-.D^+)$ (eV)	$\Delta E(^3D^*)$ (eV)	$\Delta E(^1D^*)$ (eV)	$h\nu_{max}^{'a}$ (eV)	χ'/χ^A	χ'	$10^3\chi^A$
TPTA	0.74	2.90	2.96	3.51	2.39	25	0.02	0.8
TPAA	0.52	2.68	2.83	3.26	2.23	2.5	0.002	0.8
TPDA	0.08	2.24	2.63	2.99	(1.9)b	<0.03	<0.00002	0.7

(a) Hetero-excimer emission maximum.
(b) This estimated value corresponds to λ_{max} = 650 nm and shows
    that the hetero-excimer emission, if it occurred, would have
    been detected by our apparatus.
(c) Yields χ's are in Einstein/mole(anion).
(d) Half-wave potentials refer SCE.

leading directly to the (singlet) ground state is expected
to increase quite drastically as the amount of electronic
energy that has to be transformed into heat in this process
decreases.

Since for spin-statistical reasons 25% of the radical ion encounters may undergo spin-allowed reactions to give singlet states
and 75% to give triplet states, one can write:

$$\gamma_c + \gamma_g = 0.25 \qquad\qquad\qquad (9)$$

and $\qquad\qquad \gamma_t^A + \gamma_t^D = 0.75. \qquad\qquad\qquad (10)$

Thus if γ_g increases, γ_c becomes correspondingly smaller.
Furthermore, Eq. 10 (with $\gamma_t^D \approx 0$ for energetic reasons) according
to which γ_t^A should be independent of any effects which typically
affect γ_c or γ_g, provides an explanation for χ^A being constant in
this series.

It should be pointed out, however, that Eqs. 9 and 10 can
only be valid as long as the corresponding reaction rates fulfill
the conditions:

$$k_c + k_g > 1/\tau_{spin} \text{ and } k_t^A + k_t^D > 1/\tau_{spin}$$

where τ_{spin} is the spin relaxation time of the radical ion pair. If, for instance, both k_t^A and k_t^D are close to zero, as in systems where $\Delta G(A^-..D^+) < \Delta E(^3A^*)$, $\Delta E(^3D^*)$, all radical ion pairs will react to give singlet states and $\gamma_c + \gamma_g$ will approach unity.

ACKNOWLEDGEMENT

This work was supported by the Netherlands Foundation for Chemical Research (S.O.N.) with financial aid from the Netherlands Organization for the Advancement of Pure Research (Z. W. O.).

REFERENCES

(1)   H. Leonhardt and A. Weller, Ber. Bunsenges. physik. Chem. 67., 791 (1963); A. Weller, Nobel Symposium 5 (S. Claesson ed.) Almqvist and Wiksell, Stockholm (1967) p. 413 and literature cited therein.

(2)   H. Knibbe, D. Rehm and A. Weller, Ber. Bunsenges. physik. Chem. 73, 839 (1969); D. Rehm and A. Weller, Z. physik. Chem. NF, 69, 183 (1970).

(3)   A. Weller and K. Zachariasse, (a) J. Chem. Phys. 46, 4984 (1967); (b) Molecular Luminescence (E. C. Lim ed.) W. A. Benjamin Inc., New York (1969) p. 895; (c) Chem. Phys. Letters, 10, 197 (1971).

(4)   A. Weller and K. Zachariasse, Chem. Phys. Letters 10, 424 (1971).

(5)   A. Weller and K. Zachariasse, Chem. Phys. Letters 10, 590 (1971).

(6)   K. Zachariasse, thesis, Free University, Amsterdam (1972).

LUMINESCENCE PHENOMENA ASSOCIATED WITH POLAR LIQUIDS:

RELAXATION AND FLUCTUATION PROCESSES

Yu. T. Mazurenko

The University, Leningrad, USSR

ABSTRACT

It is shown how the complex kinetics of orientational relaxation in polar liquids, as described by the superposition of several exponential processes with radically different times, distinctly manifests itself in the characteristics of the luminescence of low temperature alcoholic solutions.

The intermolecular interactions usually cause an appreciable change of optical spectra when passing from the gaseous phase to the liquid. One of the interesting aspects of the formation of condensed phase spectra is the spectroscopic manifestation of molecular motion in liquids. In this paper we shall briefly consider the mechanism by which the orientational motion manifests itself in the electronic spectra of solutions, especially in their luminescence spectra.

The dipole orientation interaction in solution can be regarded as the interaction of the rigid dipole moment μ of the solute molecule with the field F produced by the rigid molecular dipoles of the solvent. The thermal motion of the solvent molecules causes the field F to depend randomly on time. Owing to the same reason, at every moment, F is different at different centers in the solution. On the other hand, since the field of the dipole moment μ induces an orientation of the dipoles of the surroundings, there must exist a dependence of F on μ. In order to find this dependence we should beforehand average F over a sufficiently long interval of time or over the ensemble of the centers in

227

solution. In that way we eliminate the thermal fluctuations.
We are interested not only in static but also in dynamic properties
of the dependence of $\underline{F}$ on $\underline{\mu}$ since the dynamic properties must manifest themselves when a sharp change of $\underline{\mu}$ accompanies the electronic
transition. Consequently, we shall present the relationship between
$\underline{\mu}$ and $\underline{F}$ in the following general form (in the linear approximation):

$$\overline{F}_\omega = \kappa_\omega \underline{\mu}_\omega, \tag{1}$$

where κ_ω is the complex susceptibility (including the average
electronic polarization) which adjusts the relationship between the
(complex) spectral components of changes of $\underline{F}$ and $\underline{\mu}$. The line
over $\underline{F}$ denotes the averaging of thermal fluctuations. The introduction of the susceptibility makes it at once possible to apply
the fluctuation dissipation relations of non-equilibrium statistical
mechanics (1,2). Thus we may consider from this one point of view
the broadening of spectra caused by fluctuations as well as the
kinetics connected with relaxation processes.

The average value of $\underline{F}$ under the condition of thermal equilibrium is equal to

$$\overline{F} = \kappa_0 \underline{\mu}, \tag{2}$$

where κ_0 is the static susceptibility. The average squared fluctuation of the projection of $\underline{F}$ is determined in the classical
approximation by

$$\overline{(\delta F)^2} = kT\kappa_0. \tag{3}$$

An instantaneous change of the dipole moment by $\Delta\underline{\mu}$ leads to
a violation of equilibrium and therefore must be accompanied
by relaxation of $\overline{F}$ according to: $\overline{\Delta F} = \kappa_0\Delta\underline{\mu}[1-\phi(t)]$, where $\phi(t)$
is the relaxation function which follows unambiguosly from κ_ω.

The spectral characteristics of solutions can now be
obtained on the basis of the following reasons: a) the field $\underline{F}$
is the random variable with the characteristics of Eqs. 1, 2 and 3;
b) during an electronic transition $\underline{F}$ does not change (as the
consequence of Frank-Condon principle).

The orientational spectral shift, η, corresponding to some
arbitrary value of the field $\underline{F}$ is determined by:

$$h\eta = -\Delta\underline{\mu}\underline{F}, \tag{4}$$

where $\Delta\underline{\mu}$ is the difference in the dipole moments of the upper
and lower states. After appropriate averaging it is possible

to find from Eq. 4 observable quantities such as the average shift η and its mean squared deviation $\overline{(\delta\eta)^2}$ which characterize the broadening of the spectrum:

$$h\overline{\eta} = -\Delta\underline{\mu} \cdot \overline{F} \equiv -\kappa_o \Delta\underline{\mu} \cdot \underline{\mu}, \qquad (5)$$

$$h_2\overline{(\delta\eta)^2} = (\Delta\mu)^2\overline{(\delta F)^2} \equiv kT\kappa_o \, (\Delta\mu)^2, \qquad (6)$$

where $\underline{\mu}$ is the dipole moment in the initial state of the quantum transition. We should point out that $\overline{\eta}$ does not represent the total spectral shift; but it is true that $\overline{(\delta\eta)^2}$ defines practically all of the broadening caused by the solution. Then it is not difficult to obtain

$$h\overline{(\delta\eta)^2} = kT(\overline{\eta}_a - \overline{\eta}_f), \qquad (7)$$

where a and f label absorption and emission (see Ref. (3)). Equations 5, 6, and 7 were obtained with the assumption of thermal equilibrium.

Such an equilibrium is not always realized. For example, immediately after the absorption of a quantum of light the condition of equilibrium Eq. 2 is violated. Therefore, the field F as well as the average frequency of emission $\overline{\nu}$ must undergo relaxation. Appropriate calculations lead to the expression:

$$[\overline{\nu}(t)-\overline{\nu}(\infty)]/[\overline{\nu}(o)-\overline{\nu}(\infty)] = \Phi(t) \qquad (8)$$

where t is the interval of time between the "moments" of excitation and observation of luminescence.

Up till now we have not made any assumptions concerning the explicit form of the susceptibility κ_ω. One of the ways of constructing κ_ω is founded on the Onsager model in which the environment is regarded as a dielectric continuum (4).

Let us consider some correlations between the derived results and experiment. The short time resolution spectroscopy (5) enables us, according to Eq. 8, to obtain the relaxation function $\Phi(t)$. On the other hand, $\Phi(t)$ can be calculated using the Onsager model and experimental data on dipolar relaxation in the solvent at hand. Calculations, performed for the case of n-propyl alcohol solutions, show that the high frequency dielectric relaxations τ_2 and τ_3 (6) have a significant effect on the character of $\Phi(t)$. In the dielectric behavior of alcohols these relaxations manifest themselves poorly, compared to the Debye relaxation τ_I. Nevertheless, owing to the nonlinear relationship between κ and ε (which is characteristic of the Onsager model) the contributions of τ_I, τ_2, τ_3 in $\Phi(t)$ appear to be roughly equal. Hence, $\Phi(t)$

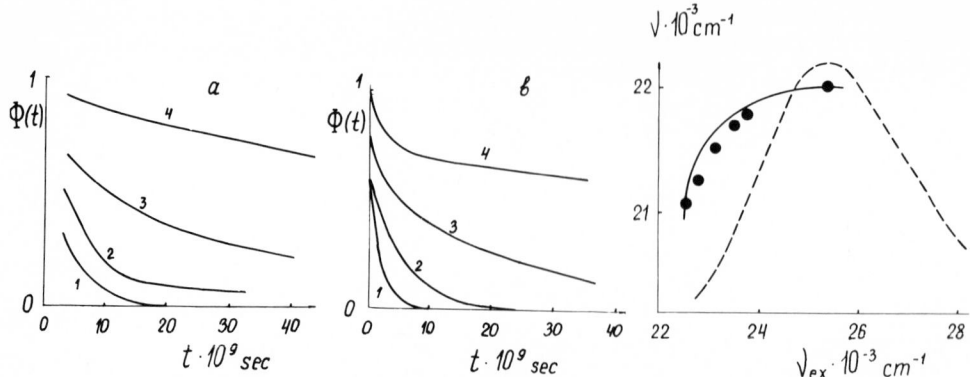

Fig. 1 a) Approximate character of relaxation function $\Phi(t)$ for
a solution of 4-aminophthalimide in n-propyl alcohol,
calculated from the data of (5). b) function $\Phi(t)$
calculated with the aid of Onsager model for the same
solution: 1 - 223K; 2 - 208K; 3 - 183K; 4 - 148K. For
the listed temperatures the magnitudes of τ_I are
1 x 10^{-8}, 2.7 x 10^{-8}, 1.8 x 10^{-7}, 1.4 x 10^{-5} sec.,
respectively.

Fig. 2 The dependence of position of luminescence spectrum
$\bar{\nu}$ (solution of 3-amino-N-methylphthalimide in isobutyl
alcohol at 128K) on the frequency of excitation ν_{ex}.
Points represent experiment (8), line represents
calculation. The absorption spectrum is shown as well.

becomes essentially nonexponential and the characteristic relaxation
times of the spectrum do not correlate with the Debye relaxation
time τ_I. This result is confirmed by experiment (5), Fig. 1.

Since the spectra of complex molecules are characterized by a
large intrinsic "intramolecular" broadening, the broadening
related to the influence of solvent appears to be latent to a
large degree. Nevertheless, there exists an interesting indirect
manifestation of solvent broadening (7,8). The spectral band of
absorption can be considered as a superposition of bands belonging
to the centers with different magnitudes of $\underline{F}$ and different shifts
η Eq. 4. On absorption of monochromatic light there must exist
some selectivity of excitation with respect to the magnitude of
$\underline{F}$, the degree of selectivity depending on the steepness of the
spectral curve in the region of the excitation line. If the
solution is sufficiently rigid, the set of the magnitudes of $\underline{F}$
formed on excitation, defines the position of the emission band.
Consequently, the spectrum of emission must depend on the frequency
of excitation. The universal formula, Eq. 7, makes it possible
to calculate this dependence in a simple way. If one proposes that

the intermolecular broadening is small compared to the intra-molecular broadening, then it is not difficult to obtain the relation:

$$h(\bar{\nu}_0 - \bar{\nu}) = kT(\bar{\nu}_0 - \bar{\nu}_\infty)[d\ln G(\nu)/d\nu]_{\nu = \nu_{ex}} \qquad (9)$$

where: $\bar{\nu}$ is the average frequency of emission, excited by a line at ν_{ex}; $\bar{\nu}_\infty$ is the average frequency of the completely relaxed spectrum ($\bar{\nu}_\infty$ can be determined on the basis of indirect experimental data); $\bar{\nu}_0$ is the average frequency of emission, while exciting at the absorption maximum; and $G(\nu)$ is the absorption coefficient. The correlation of Eq. 9 with experiment (8) is shown in Fig. 2.

REFERENCES

1. R. Kubo, J. Phys. Soc. Japan, 12, 570 (1957).

2. L. D. Landau, E. M. Lifshitz, Statistical Physics, Nauka Press Moscow, 1964.

3. R. A. Marcus, J. Chem. Phys. 43, 1261 (1965).

4. Yu. T. Mazurenko, N. G. Bakhshiev, Opt. i Spectr. 28, 905, (1970).

5. W. R. Ware, S. K. Lee, G. J. Brant, P. P. Chow, J. Chem. Phys. 54, 4729 (1971).

6. C. P. Smyth, Molecular Relaxation Processes, R. C. Cross, Ed., Academic Press, London and New York, 1966, p. 1.

7. A. N. Rubinov, V. I. Tomin, Opt. i Spectr. 29, 1082 (1970).

8. N. I. Rudik, L. G. Pikulik, Opt. i Spectr. 30, 275 (1971).

THE SPECTRAL INVESTIGATION OF THE LUMINESCENCE CENTERS

WITH THE GENERALIZED METHOD OF ALENTSEV

I. I. Antipova-Karataeva, N. P. Golubeva and M. V. Fock

Lebedev Phys. Inst., Vernadskii Inst. of Geo- and Anal.

Chem., Acad. of Sciences, Moscow, USSR

ABSTRACT

For crystals containing several kinds of centers with luminescent spectra having several overlapped bands, a method is given to find the shape and intensity of each component. The spectra are measured under different conditions in such a way that the contribution of each component is different in each case but the shape is constant. The method was applied to "undoped" self-activated ZnS crystals. The comparison of these bands with the ZnO band's shows that many are connected with the incorporation of oxygen. So e.g. the cathodoluminescent spectra excited by electrons with different energies show a different oxygen concentration of the surface and of inner parts (confirmed by successive chemical etching of the surface layer).

The luminescence kinetics of ZnS luminophors is very complicated due to the existence of many mechanisms of radiative and non-radiative recombination associated with a large variety of defects. Therefore, data based only on luminescence-kinetic investigations without taking into account physical-chemical considerations about the nature of these defects are hardly sufficient in the construction of models of luminescence centers. For example, a shift in the maximum of the blue luminescence band of self-activated ZnS into the long-wave regions during decay and the decrease of the excitation density can be explained by either of two models. Namely, luminescence is induced by the interimpurity recombination of a random set of donor-acceptor pairs (1,2) or by the luminescence of the discrete centers which are different in their nature and properties, but which have overlapping luminescence spectra. Both

preparatory and luminescent-kinetic considerations make us doubt
the first point of view for the self-activated luminescence of ZnS.
In fact, with cooling after annealing, association of the donor-
acceptor components arises and the recombination probability of
the donor-acceptor pairs of large separation is small because of
the small overlap of the Bohr orbits. The validity of the second
hypothesis is supported by the fact that the luminescence spectrum
can be decomposed into individual bands with explicit half-widths
and positions of the maxima. These bands are present in various
proportions in the spectra of different luminescent systems (powders,
single crystals, sublimated layers) (3). They are observed in
both the spectra of undoped and activated luminophors.

A decomposition of the luminescence spectrum into the individual bands was carried out using the generalized method of
Alentsev. This method is based on the fact that with a change in
the excitation conditions the luminescence spectrum is deformed by
the change in the relative intensity of the bands belonging to the
different luminescence centers. By using several luminescence
spectra one can calculate each band without making any assumptions
concerning its shape (4). For the present case, an algorithm has
been constructed and a program to carry out these calculations on
an electronic computer has been written. With this method it can
be shown that in the luminescence spectra of the various samples
of ZnS the bands with the maxima 1.82, 2.0, 2.25, 2.42, 2.50, 2.66,
2.91 and 3.07 ev (Fig. 1) (3,5) are present.

The fact that the decomposition into the individual bands of
the self-activated ZnS is possible for a comparatively small change
of excitation intensity or wavelength indicates differences in the
nature of the centers responsible for these bands. They may be
either native defects of ZnS or impurities. The concentrations of
all these defects in ZnS is great because of the high temperature
of thermal treatment and the impurity of the initial ZnS. Oxygen
is the impurity which is most widely observed in ZnS. The thermo-
dynamics of the system ZnS-ZnO shows that oxygen can enter the ZnS
lattice even with the small concentrations of oxygen in the "pure"
carrier gases used, except for H_2S. The conditions for the oxygen
entering are more favorable in a medium having a low sulphur
partial pressure over ZnS (ammonia, hydrogen, zinc). In ZnS
annealed with chlorides the presence of the oxygen formations as
solid solutions ZnS-ZnO and as oxychlorides is more probable. In
the "undoped" ZnS, ZnO can be also present. The luminescence of
the oxide formations in the undoped ZnS can be more intense than
in the activated one due to the low concentration of the competitive centers of luminescence.

An analysis of the preparation conditions allows one to make
assumptions concerning the oxygen formations responsible for not

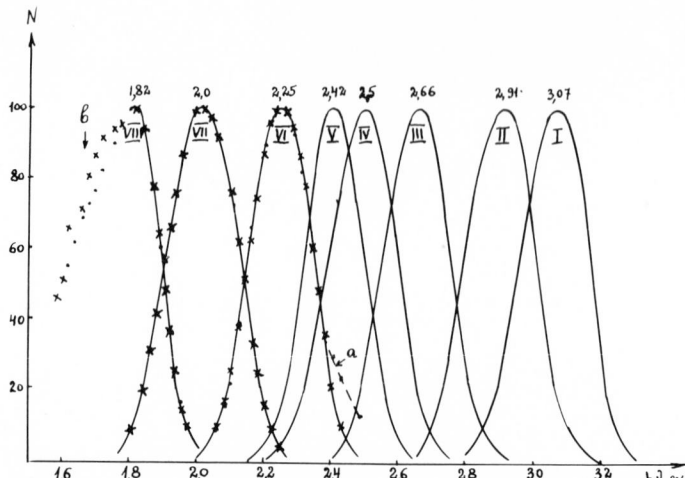

Fig. 1 Normalized individual bands in luminescence spectra of ZnS
        and ZnO (red). I-V: ZnS; VI-VIII: x=ZnS, ·=ZnO; a: green
        band of ZnO; b: spectral region of Hg.-lines.

only the green, but also for the less frequently studied luminescence in the yellow-red and dark blue-violet regions of the
spectrum. Moreover, by the Alentsev method it was found that the
bands VI, VII and VIII of ZnS are identical with the related bands
in the spectrum of ZnO red luminescence (Fig. 1). The intensity
of the yellow-red luminescence of air-annealed ZnS is weak due to
the presence of the intense green band. Its relative intensity is
highest in luminophors annealed in H_2S with subsequent surface
excitation (λ=313nm or electrons of 6 kev). From this, one can
assume that the red luminescence of the "undoped" ZnS annealed in
H_2S appears as a result of its surface oxidation, probably at the
cooling stage (5).

The dependence of the dark blue-violet luminescence of the
"undoped" ZnS on oxygen is evidently due to preparation, since it
is clearly seen in the luminophors obtained in a medium with low
sulphur partial pressure over ZnS (NH_3, Zn) and even upon annealing
ZnS in air (T=1150C). In the last case, the dark blue-violet
luminescence is primarily excited by λ=313nm and instead of the
band I, one detects that the band extends into the ultraviolet
region. λ=365nm excites the intense green luminescence where the
luminescence of the green ZnO is present.

A question arises about the nature of such an effective energy
transfer to the oxide centers, the concentration of which, at least
in ZnS annealed in H_2S, should be small. (The latter follows not

only from the preparation considerations, but from the superlinear
dependence of the red-band intensity upon the excitation energy.)
In order to make this question clear we have investigated in more
detail the ZnS luminescence in the region of 3.1 to 3.4 ev. In
this region there are bands of the edge and exciton luminescence
of ZnO and of the luminescence of a solid solution of ZnS-ZnO (6,
7). In this region, luminescence of the "undoped" ZnS is often
observed. It is significant in ZnS annealed in NH_3, practically
absent in powdered ZnS annealed in H_2S, appears again with further
oxidation and is also observed in the spectra of single crystals
grown in H_2S.

Two-layer luminophors are available for the investigation of
the luminescence of oxidized ZnS. The oxidized layer is produced
on the surface of the small crystals of ZnS already crystallized
at 1150C in H_2S. The ZnS oxidation layer was obtained both by a
second annealing in air at 980 and 1150C and by decomposition at
1150C of a $ZnSO_4$ layer placed on the surface of ZnS. In the
spectrum of the two-layered luminophors the edge luminescence of
the hexagonal ZnS and the bands at 3.1-3.4 ev weakly overlap the
intensity of the latter increasing under oxidation. These samples
allow one to detect ZnO-induced luminescence by comparing the
spectra before and after an etching of the sample by an acetic
acid solution which dissolves ZnO without affecting ZnS.

The ultraviolet luminescence was measured at 77K under an
electron-beam excitation, which allows an estimate (using the same
mechanism and excitation density) of the luminescence of both the
surface and interior of the crystal by changing the velocity of
the exciting electrons. The current density in the immobile beam
was $8 \times 10^{-7} a/cm^2$, the accelerating voltages were 6 and 21 kev (the
penetration depth of the electrons was 1.5 and 18μm respectively).

The character of the spectrum changes with oxidation depending
on the oxygen content in the atmosphere, temperature and duration
of the oxidative stage of annealing and the penetration depth of
the exciting electrons. The changes in the spectrum of the ZnS
edge luminescence are already evident at very low oxygen concentrations in the atmosphere. They are greater in ZnS oxidized at a
high temperature. In the spectrum of ZnS annealed at 1150C in non-
dry H_2S, one observes a relative increase of the intensity of the
3.7 ev band over the 3.66 ev band (Fig. 2, compare curves 1 and 2).
This implies an increase in the ZnS hexagonal phase contribution
which is stabilized by oxygen, i.e. these bands (especially 3.7 ev)
are most free from the superposition of the edge luminescence of
the cubic ZnS. With higher oxidation the maximum of the edge
luminescence spectrum is shifted into the shortwave region, and
luminescence of 3.1-3.4 ev appears. With a relatively low oxygen
content in the atmosphere, these changes are observed only with

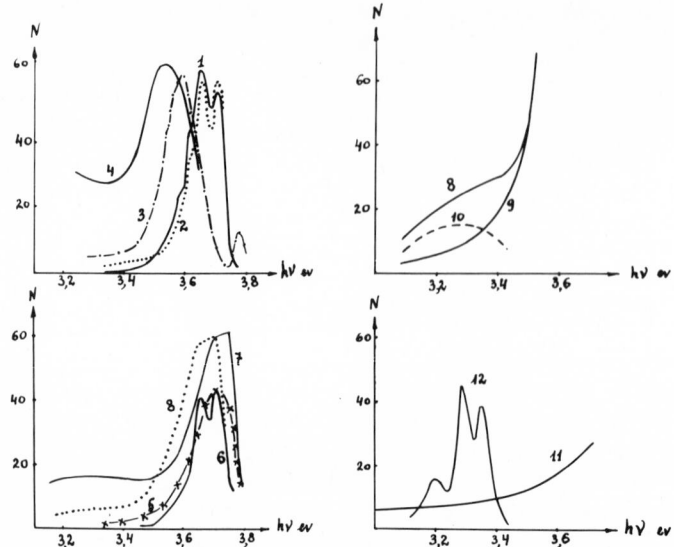

Fig. 2   Luminescence spectra of ZnS with different degrees of
oxidation. 1: ZnS annealed at 1150C in dry H_2S; 2: ZnS
annealed at 1150C in non-dry H_2S; 3: ZnS single crystal
grown in H_2S at 1350C; 4: ZnS annealed at 900C in NH3;
5 & 6: two-layer ZnS luminophor oxidized by 1150C with low
oxygen content in atmosphere (5-by 6 kev, 6-by 21 kev);
7: two-layer ZnS luminophor oxidized 1h. at 1150C; 8: two-
layer ZnS luminophor oxidized 1h. at 980C; 9: curve 8 after
etching in acetic acid; 10: difference between spectra 9'
and 8; 11: ZnO annealed with Zn at 1200C or with ZnS (1%)
at 900C; 12: ZnO annealed in oxygen at 1000C.
Note: All spectra (except 5) are measured at 21 kev.

6 kev excitation (Fig. 2 compares curves 5 and 6) and with an
increase in the oxygen concentration also at 21 kev (compares
curves 7 and 6).

The above-mentioned assumption that in the 3.1-3.4 ev band the
luminescence of ZnO is present (5) was supported by a noticeable
decrease of such luminescence after the etching in acetic acid of
samples annealed at 980C. Fig. 2 gives the difference of the
spectrum ordinates (curve 10) of the samples measured before (curve
8) and after etching (curve 9). The flat maximum of this curve
(of 3.2-3.4 ev) is placed in the region of the bands of the ZnO
edge and exciton luminescence (curve 12). The absence of distinct
structure of ZnO luminescence can be explained by the presence of
superstoichiometric zinc, which is produced under an annealing of
the system ZnS-ZnO. In the spectrum of ZnO with an intentional
excess of zinc present, this structure is also absent (Fig. 2,
curve 11).

The investigation of the dependence of the luminescence spectra on the penetration depth of the exciting electrons allows one to find the oxidation gradient toward the surface of the ZnS. For this, the ratio of the intensities of the luminescence band at 3.1-3.4 ev and the edge luminescence of ZnS measured at 6 and 21 kev was compared (Table 1). This comparison shows that in ZnS, oxidized at 980C, surface layers are oxidized. A reduction of the surface at 1150C can occur by the reaction:

$$ZnS + 2ZnO \rightarrow 3Zn + SO_2 \tag{1}$$

that should be taken into account when oxidizing ZnS at a high temperature.

Table 1

The dependence of the ratios of the intensities of the luminescence (I(3.1-3.4ev)/I(ZnS edge) on the penetration depth of the exciting electrons

	Temperature	1150C		980C	
Oxidation	Time	1h.	2.5h.	1h.	2.5h.
Excitation	6 kev	.17	.15	.13	.19
	21 kev	.27	.24	.11	.07

So, the oxidation at 1150C of the originally crystaline ZnS results in its recrystallization with an increase of the hexagonal phase and probably, that of the defect structure. The reduction process based on reaction (1), which competes with the oxidation, favors the oxygen to be fixed only in those places in ZnS similar in structure to ZnO. If the ZnS oxidation is carried out at a lower temperature (980C), the role of ZnO as a filter of ZnS edge luminescence becomes evident. Under a change from 1150C to 980C the velocity of the reaction (1) decreases, the equilibrium pressure of oxygen in the system ZnS-ZnO falls and, hence, ZnO can grow as a continuous layer in other than the most favorable places in the ZnS crystal. This can be detected in the luminescence spectrum measured at 6 kev as the non-structured band extending far into the ultraviolet region as in the ZnO spectrum with an intentional excess of Zn. This band disappears after the luminophor is etched with acetic acid, and the intensity of the edge luminescence of ZnS (at 6 kev) increases by 80-100 times. The oxide film can be found as well on the surface of the ZnS single crystals grown in H_2S. It is probably produced during the cooling process, and is one of the reasons for the lower intensity of the edge luminescence of the single crystals as compared with the powders. We observed a relative increase in the edge luminescence intensity of ZnS single crystals after etching by acetic acid. Thus, the ZnO acts also as a filter reabsorbing the edge luminescence of ZnS and transmitting

this excitation to its luminescence centers (Fig. 3). This results
in a high efficiency for the oxide luminescence in the "undoped"
ZnS which has few competing luminescence centers.

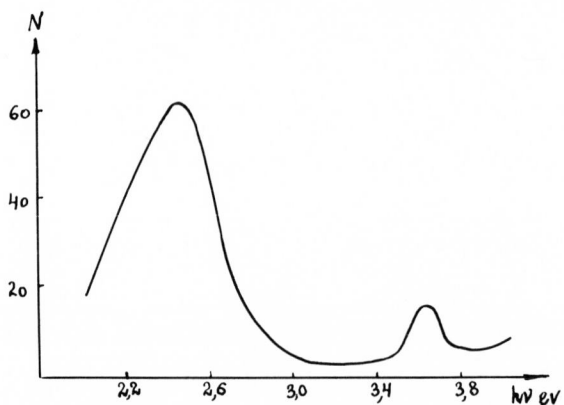

Fig. 3 Two-layer luminophor oxidized 2.5h. at 980C measured at
        6 kev.

REFERENCES

1. S. Shionoya, Y. Washizawa, and H. Ohmatsu, J. Phys. Chem.
    Solids, 29, 1843 (1968).

2. K. Era, S. Shionoya, and Y. Washizawa, J. Phys. Chem. Solids,
    29, 1827 (1968).

3. E. E. Bucke, T. I. Voznesenskaya, N. A. Gorbacheva, N. P.
    Golubeva, Z. P. Kaleeva, E. I. Panasyuk, and M. V. Fock, Zh.
    prikl. spektr. 12, 1047, 1970, Tr. FIAN 59, 25 (1972).

4. M. V. Fock, Zh. prikl. spektr., 11, 926, 1969, Tr. FIAN, 59,
    3 (1972).

5. N. P. Golubeva and M. V. Fock, Zh. prikl. spektr., 17, 376
    (1972).

6. J. Folinski and T. Skettrup, Solid State Commun, 6, 233 (1968).

7. F. A. Kröger and J. A. M. Dikhoff, J. Electrochem. Soc., 99,
    144 (1952).

AN INVESTIGATION OF THE COMPLEX STRUCTURE OF THE IMPURITY LUMINESCENCE SPECTRA OF ZnS SINGLE CRYSTALS

A. N. Georgobiani, A. I. Blazhevich, Yu. V. Ozerov,

E. I. Panasjuk, P. A. Todua and H. Friedrich*

Acad. of Sciences, Lebedev Phys. Inst., Moscow, USSR;

*ZIE, Deutschen Akad. der Wissenschaften, Berlin, DDR

ABSTRACT

The luminescence spectrum of ZnS crystals - both undoped and doped - in the visible consists of wide bands. The crystals cooled to liquid He temperature were excited by a mercury lamp or a nitrogen laser, and the spectra were measured by "modulation spectroscopy". Structure was found in the wide bands and attributed to donor-acceptor pair transitions and their phonon replicas. The dependences of emission spectra on temperature and on excitation intensity are explained in terms of changes in occupation of conduction and donor states.

The structure of the band scheme of zinc sulfide and the nature of its luminescence centers have for a long time been of interest to many investigators. However, until recently there has been no unified point of view on these questions because of an inconsistency in the experimental data, due to a great number of uncontrolled factors connected with the preparation of samples.

ZnS is a rather complicated system. Even in the simplest cases of "pure" and self-activated crystals (s.a.), i.e. crystals prepared without introduction of the compensating activator impurities, one observes blue or green emission.

For a long time the blue luminescence of s.a. ZnS was considered to be due to a single defect. In this region of the spectrum in addition to the recombination radiation of the s.a. ZnS band with $\lambda_{max} \approx 470$ nm, a second band with $\lambda_{max} \approx 415$ nm (1) was detected.

Shionoya et al. (2-5) have found some factors suggesting that the blue band of the s.a. ZnS is a result of the quasi-continuous set of donor-acceptor pairs, with various distances between the donor and acceptor, participating in the recombination transitions.

Tunitskaya et al. (6-8) have noticed that different parts of the blue band show different temperature behavior, and consequently the blue band of the s.a. ZnS consists of a small number of elementary bands overlapping each other. The different behavior of these components caused by variations in the excitation conditions, temperature, and so on can naturally characterize the behavior of the band as a whole.

Fock has suggested a method allowing one to determine the outlines of the elementary bands from the experimental spectrum of radiation (9-10). Using this method for an analysis of the spectrum of the blue luminescence of the s.a. ZnS, Fock et al. were able to separate four bands with maxima at 3.07; 2.91; 2.66; and 2.50 ev (11, 12).

We had in mind to find experimentally in a direct way the spectral structure of the recombination radiation of the s.a. ZnS single crystals. We have investigated ZnS:Cl, ZnS:I, ZnS:In single crystals, both with stochiometric composition and with deviation from stochiometry (treatment in the zinc melt to form ZnS:Cl, (Zn) and ZnS:I, (Zn) where (Zn) specifies excess Zn). The crystals studied were different both in the type of the coactivator impurity introduced, and in the method of preparation (gas transport reaction, sublimation method).

The monochromators used have resolution over the whole spectral region of 0.01 ev. With the purpose of making more precise the positions of features in the spectral distribution of radiation, a modulation spectroscopy method was used. Features of the radiation spectrum were also resolved by cooling the samples from room temperature to liquid nitrogen and helium temperatures.

To study the influence of excitation density on the spectral composition of the radiation, luminescence excitation was carried out with both a Hg lamp using the 313 nm line and pulsed nitrogen laser at 337 nm. The duration of the laser pulse was 15 nsec, the repetition frequency 25-30 c/s, and the pulse capacity (power) 10 vs.

Fig. 1 shows the luminescence spectrum of the ZnS:Cl single crystals (λ_{exc} = 313 nm) at T = 300, 77 and 4.2 K. At room temperature it is a rather wide band (with a half width of 0.5 ev) with the outlined structure in the region 2.8 - 2.9 ev. At liquid nitrogen temperature the spectrum half-width reduces to 0.3 ev and

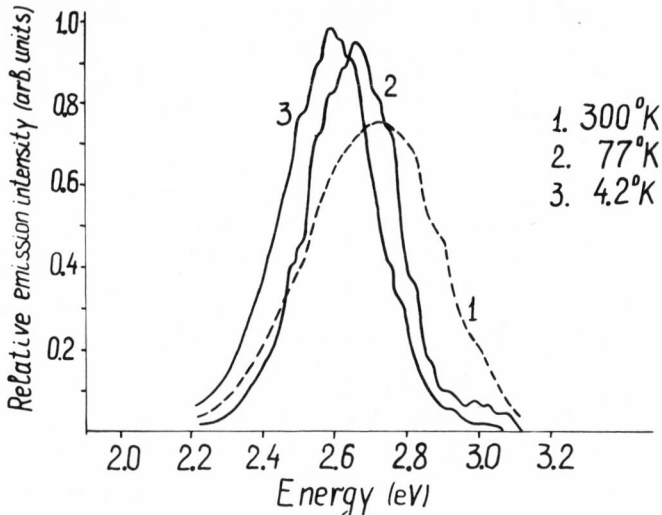

Fig. 1   Luminescence spectra of ZnS:Cl single crystals (λ_{exc} = 313 nm), measured at T_1 = 300 K, T_2 = 77 K, T_3 = 4,2 K.

maxima and shoulders appear near 3.06, 3.02, 2.98, 2.91, 2.87, 2.76, 2.72, 2.66, 2.62, 2.58, 2.56, 2.53, 2.50 and 2.48 ev. At the temperature of liquid helium the maximum of the total luminescence spectrum shifts towards ∿ 2.58 ev, and the halfwidth of the spectrum remains virtually unchanged compared with that at liquid nitrogen temperature. One observes some characteristic peculiarities. The positions of the maxima and shoulders in the spectra at liquid nitrogen and helium temperatures coincide. In other words, the shift in the maximum of the total spectrum towards smaller energies is apparently due to a relative change in the intensities of the components of the elementary bands, not their displacement.

In Fig. 2 the luminescence spectra of ZnS:Cl single crystals taken at the same temperatures under excitation by a nitrogen laser are shown. In this case a fine structure in the radiation spectra is also observed; however, the temperature shift of the maximum of the total luminescence spectrum is absent. Both at liquid nitrogen and liquid helium temperatures (at room temperature it is difficult to define the exact position of the maximum) the maximum is placed at hν = 2.58 ev. As it was shown above, the luminescence maximum upon excitation with a Hg lamp with the 313 nm line and at T = 4.2 K corresponds to this energy. The positions of the peaks in the radiation spectra at liquid nitrogen and helium temperatures coincide in this case too; only a relative redistribution of their intensities takes place.

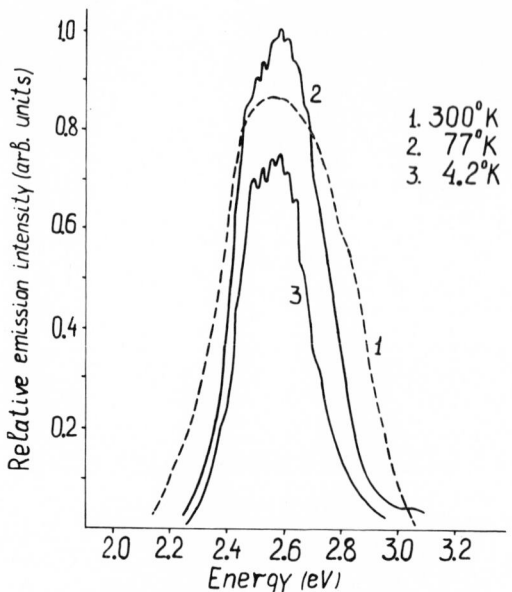

Fig. 2   Luminescence spectra of ZnS:Cl single crystals (λ_{exc} = 337 nm), measured at T_1 = 300 K, T_2 = 77 K, T_3 = 4.2 K.

In the radiation spectra of ZnS:I, ZnS:In, ZnS:Cl, (Zn), ZnS:I, (Zn) fine structure was also observed, and it appears that the positions of the peaks in the blue region of the spectrum do not depend, as a rule, upon the type of coactivator, the temperature of the sample (4.2 or 77 K) or the excitation density, but only on a redistribution of their intensities (up to the disappearance of either one or another band) occurs, resulting in a temperature shift of the total spectral maximum (at λ_{exc} = 313 nm).

In Table 1 are given the positions of the maxima in the luminescence spectra of the ZnS s.a. crystals at different temperatures and under different types of excitation.

The shift of the maximum of the total luminescence spectrum (λ_{exc} = 313 nm) towards small energies as the temperature decreases can probably be explained as follows: at room temperature the donor levels are ionized and radiation arises due to the electronic transition from the conduction band to the acceptor level, as the temperature decreases the occupation of the donor levels increases and the direct electron transition from the donor levels directly to the acceptor levels competes with the transition through the conduction band and thus causes the shift of the maximum of the total spectrum towards smaller energies. On the other hand, at high den-

Table 1

Maxima positions in luminescence spectra of s.a. ZnS crystals
at different temperatures
and using different excitation methods.

ZnS-In 313nm, 77°K	ZnS-I(Zn) 313nm, 77°K	ZnS-I 337nm, 4.2°K	ZnS-I 337nm, 77°K	ZnS-I 313nm, 77°K	ZnS-Cℓ(Zn) 313nm, 77°K	ZnS-Cℓ 337nm, 4.2°K	ZnS-Cℓ 337nm, 77°K	ZnS-Cℓ 313nm, 4.2°K	ZnS-Cℓ 313nm, 77°K
3.06	3.06			3.06	3.06				3.06
									3.02
2.98	2.98			2.98	2.98				2.98
2.94	2.94								
2.92	2.91	2.91	2.91	2.91	2.91			2.91	2.91
2.88				2.86	2.87				2.87
2.84									
2.82									
2.76				2.76	2.76	2.76	2.76	2.76	2.76
2.72	2.74					2.72	2.72	2.72	2.72
2.66	2.66	2.66	2.66	2.66	2.66	2.66	2.66	2.66	2.66
		2.64	2.64	2.64		2.64	2.64		
	2.62				2.62	2.62	2.62	2.62	2.62
2.58	2.58	2.58	2.58	2.58	2.58	2.58	2.58	2.58	2.58
			2.56		2.56	2.56	2.56	2.56	2.56
		2.54		2.54					
						2.52	2.52		2.53
2.50	2.50	2.50	2.50	2.50	2.50	2.50	2.50	2.50	2.50
		2.48							2.48
						2.46	2.46		

sities of optical excitation the quasi-fermi level for electrons approaches the bottom of the conduction band as the excitation density is increased, and consequently the filling of the donor levels increases and the transition from the donor to the acceptor can prevail over that from the conduction band to the acceptor. Thus, an increase in the excitation density acts in the same way as a decrease of temperature with weak excitation. From this fact one can probably explain the absence of the temperature shift of the maximum of the total luminescence spectrum under excitation of luminescence with a nitrogen laser.

From the regularity in the repetition of the bands it is possible to make an assumption that the bands at 3.06, 2.91, 2.76, 2.66, 2.58 and 2.50 ev which were measured by us and other authors (11, 12) are fundamental ones, and the others are their phonon replicas. For example, 3.06, 3.02 and 2.98 ev differ from each other by the energy of the optical phonon (LO and TO phonon combinations are possible, as well as their combinations with acoustic phonons).

REFERENCES

1. H. Samelson, A. Lempicki, Phys. Rev. $\underline{125}$, 901 (1962).

2. S. Shionoya, Proc. of the International Conf. on Luminescence Budapest, (1966), p. 962. Akademiai Kiado, Budapest (1968).

3. S. Shionoya, I. Washizawa, H. Ohmatsu, J. Phys. Chem. Solids, $\underline{29}$, 1843 (1968).

4. K. Era, S. Shionoya, I. Washizawa, J. Phys. Chem. Solids, $\underline{29}$, 1827 (1968).

5. S. Shionoya, J. of Luminescence, $\underline{1}$, $\underline{2}$, 17 (1970).

6. Z. P. Kaleeva, E. I. Panasyuk, V. F. Tunitskaya, T. F. Filina, J.P.S. $\underline{10}$, 819 (1969).

7. V. F. Tunitskaya, J. P. S. $\underline{10}$, 1004 (1969).

8. Z. P. Ilyuchina, E. I. Panasyuk, V. F. Tunitskaya, T. F. Filina, Trudi FIAN $\underline{59}$, 38 (1972).

9. M. V. Fock, J.P.S. $\underline{11}$, 926 (1969).

10. M. V. Fock, Trudi FIAN $\underline{59}$, 3 (1972).

11. E. E. Boukke, T. I. Vosnesenskaya, N. P. Golubeva, N. A. Gorbatcheva, Z. P. Kaleeva, E. I. Panasyuk, M. V. Fock, J.P.S. $\underline{12}$, 1047 (1970).

12. E. E. Boukke, T. I. Voshesenskaya, N. P. Golubeva, N. A. Gorbacheva, Z. P. Ilyuchina, E. I. Panasyuk, M. V. Fock, Trudi FIAN $\underline{59}$, 25 (1972).

AN INJECTION TYPE ELECTRO-LUMINESCENCE IN ZnSe-SnO$_2$ HETERO-JUNCTION

Kenji Ikeda,[*] Ken Uchida and Yoshihiro Hamakawa

Faculty of Engineering Science, Osaka University

Toyonaka, Osaka, Japan

Hiroshi Kimura, Hiroyoshi Komiya and Sumiake Ibuki

The Central Research Laboratories, Mitsubishi Electric

Corporation, Amagasaki, Hyogo, Japan

ABSTRACT

Injection type electro-luminescence has been observed in the forward biased ZnSe-SnO$_2$ heterostructure junction. The junction shows good rectification characteristics. Clearly separated four main emission bands are found at energies of 2.00, 2.35, 2.70 and 2.80 eV at 90 K. Emission light intensity has almost no frequency dependence from DC up to 10 KHz. Injection current and temperature dependence of the emission spectra are demonstrated. The C-V characteristic and the spectral response of photovoltaic effects are also measured to determine the potential profile of the junction.

INTRODUCTION

In recent years, strong attention has been paid to efficient light emitting devices for applications to opto-electronic information processing. Considerable success has been achieved with red and green light emitting diodes (LED) using p-n homojunctions of

[*] Present address: Semiconductor Research Department, Kita-Itami Works, Mitsubishi Electric Corporation, Kita-Itami, Hyogo, Japan.

III-V compounds. However, there is no efficient blue light solid
state LED at present. Semiconductors having an energy gap > 2.7 eV
such as ZnS, ZnSe, GaN, AlN and SiC are candidate materials for this
purpose. Among these materials, ZnSe is the most promising one from
the viewpoint of crystallographic symmetry, direct transition
material and level of the crystal growth technology. We have concentrated our attentions on these points, and a series of basic
experiments to get an injection-type luminescence has been conducted
on ZnSe single crystals. In this paper, experimental results
carried out on ZnSe-SnO$_2$ hetero-structure-junctions are presented.

SAMPLE PREPARATION AND EXPERIMENT

     The ZnSe single crystals used were grown by the high pressure
Bridgman method. In order to get a suitable conductance for making
a luminescent diode, the crystal was annealed in Al-vapor atmosphere
at 970 C for several hours, and then a Zn-firing treatment was
carried out on some crystals in molten-Zn solution at 920 C for
10-20 hours (1). After these treatments, the resistivities of the

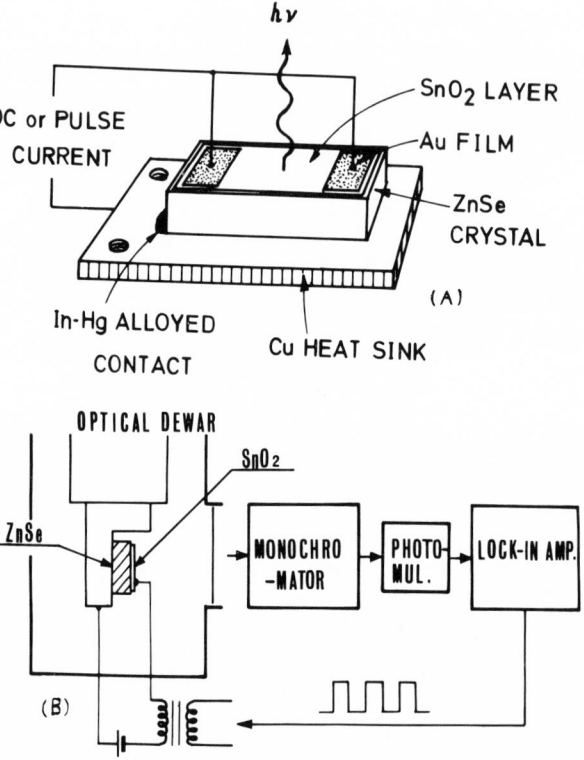

Fig. 1 A schematic diagram of the sample construction (A), the
        measurement system for the emission experiment (B).

crystals decreased to the order of 0.2-2.0 Ω-cm. The specimens
were prepared by cutting the crystal so as to have a 4x5 mm^2 area.
For making the ohmic contacts, two dots of In-Hg alloy were sintered
on the opposed faces of a block of ZnSe crystal and then the block
was cleaved into two pieces. The SnO$_2$ layer was grown on the cleaved
face by the evaporation of SnO$_2$ powder, and subsequent thermal oxidation at about 400 C in air for about 10 minutes. The grown SnO$_2$ is
transparent for photon energies up to about 3.5 eV. Au films having
a dimension of 1x3mm^2 were deposited on the SnO$_2$ layer for the
counter electrode, thereby leaving an optical window of 3x2 mm^2 as
shown in Fig. 1 (A). The sample was mounted on a Cu heat sink in
an optical dewar. The prepared hetero-junctions have good rectification characteristics, having rectification ratios of more than 10^3
at ± 2.0 V.

In measurements of emission spectra, square wave pulse voltages with a repetition frequency of 1 kHz having 50% duty cycle are
applied across the junction together with variable DC voltage. The
lock-in technique was employed for the detection of the electroluminescence signals (2). A schematic representation of the measurement system for the emission experiment is shown in Fig. 1 (B).

RESULTS AND DISCUSSION

Fig. 2 shows observed electro-luminescence spectra in a forward biased ZnSe-SnO$_2$ hetero-structure-junction measured at three
different temperatures. As can be seen in the figure, there are
four dominant structures A,B,C and D. These energies are 2.00,2.35,
2.70 and 2.80 eV which correspond to yellow, green, blue and violet.
Comparing these spectra with photo-luminescence data (3-5), the
first (A) and second (B) emission bands can be assigned to Cu impurity centers or to the self-activated center. The third band (C)
has fine structure at 23 K (1) whose energy separation (~ 30 meV)
corresponds to the energy of a LO-phonon measured in ZnSe crystals
(6). The main peak would be associated with the free-to-bound
transition at 105 K, and with the radiative recombination at a
donor-acceptor pair at 23 K (1). The remaining emission band (D)
is associated with the bound excitons (1).

The emission light intensity has almost no frequency dependence
from DC up to 10 kHz. It should be noted that ZnSe crystals have
an n-type conduction and that SnO$_2$ is considered to be a degenerate
n-type semiconductor. To determine the junction potential profile,
we have measured the spectral dependence of the photo-voltaic effect
upon irradiation with monochromatic light from the SnO$_2$ side. No
change in the sign of the voltage has been observed, and a very
sharp cut-off in the photo-voltage was seen at photon energies corresponding to the fundamental edge of ZnSe. The sign of the photo-
voltage shows that there is a depletion-type potential barrier on

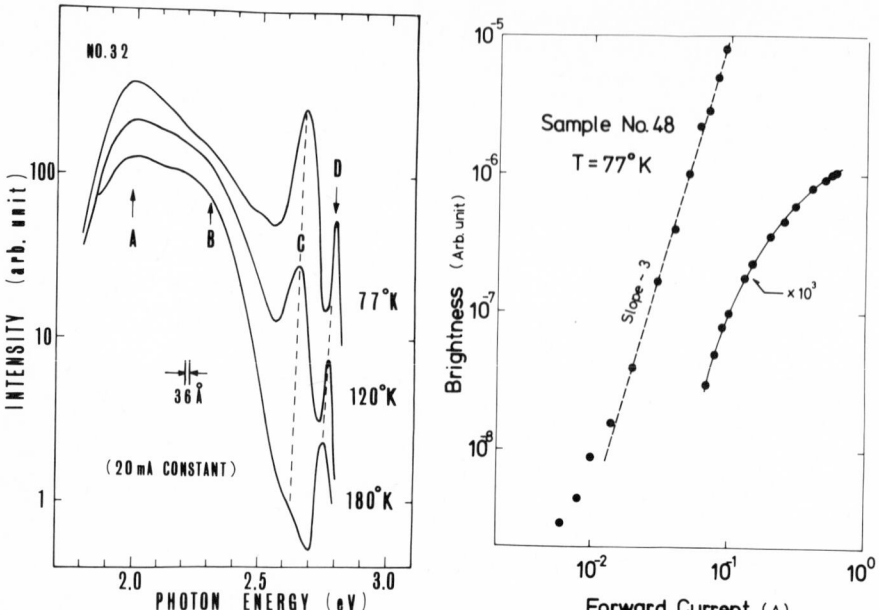

Fig. 2   Emission spectra of the forward biased ZnSe-SnO₂ heterostructure junction.

Fig. 3   The injection current dependence of the brightness.

the ZnSe side, and almost no dominant space charge layer on the
SnO$_2$ side. The diffusion voltage V_d is determined from the dependence of the open circuit photo-voltage on incident light intensity
and is estimated to be 0.98 eV at 300 K and 1.43 eV at 90 K. Fig. 3
shows the relation between the injection current and the emission
light intensity. As is seen in the figure, the brightness is proportional to the 3rd power of the current for a brightness range of
more than three decades. The junction capacitance C has been also
measured with a wide variety of bias voltages. The plot of $1/C^2$ vs
V does not follow a simple straight line. An analysis of the data
implies that there is a thin insulating layer in the interface region
of the junction.

By combining the results of the capacitance measurements with
the injection current dependence of the brightness, the carrier
transport mechanism in the forward biased condition must be mainly
due to tunneling via interface states in the hetero-junction. Although the detailed mechanism of this carrier transport across the
junction is not well-known at the present stage of experiments, the
brightness distribution in the specimen could be consistent with
the injection type luminescence. Since ZnSe has a large optical
dielectric constant, an optical confinement of the emitted light

in the crystal prevents a quantitative measurement of the distribution of emission in the vicinity of the junction. However, it has been confirmed by using a multi-electrode sample that the light emission is initiated from the ZnSe-SnO$_2$ junction side, and not from the side at the In-Hg alloyed contact, and also that no light emission is observed on the SnO$_2$ electrode side without applying the forward bias voltage.

The authors wish to express their thanks to Mr. T. Hamano for his assistance in the course of the experiments.

REFERENCES

1.   K. Ikeda, K. Uchida and Y. Hamakawa; to be published.

2.   K. Ikeda, K. Uchida and Y. Hamakawa; to be published.

3.   D. C. Reynolds, L. S. Petrotti and O. W. Larson; J. Appl. Phys. 32 (suppl.) 2250, (1961).

4.   M. Aven and H. H. Woodbury; Appl. Phys. Lett. 1 53, (1962).

5.   F. F. Morehead; J. Phys. Chem. Solids. 24 37, (1963).

6.   W. Taylor; Phys. Lett. 24A 556, (1967).

DONOR-DOUBLE ACCEPTOR LUMINESCENCE OF ZINC TELLURIDE

F. J. Bryant and A. T. J. Baker

Department of Physics

University of Hull, U.K.

ABSTRACT

Using a pulsed electron beam from a Van de Graaff accelerator, time resolved spectroscopy measurements at ~ 10K have been performed on the four broad band emissions occurring in zinc telluride at 531.4, 533.4, 551.9 and 553.9 nm and designated A_I, A_{II}, B_I and B_{II} respectively. These measurements have confirmed the assignment of the A_I and B_I emissions to free-to-bound transitions and the A_{II} and B_{II} emissions to bound-to-bound transitions. The data also permit an estimate of the donor depth as ~ 31 meV. Heat treatment studies and doping with various impurities have shown that enhancement of the bound-to-bound emissions is produced by phosphorus doping. The two acceptor levels are believed to result from the first and second ionized states of the zinc vacancy.

INTRODUCTION

At low temperatures the cathodoluminescence emission of zinc telluride near the band edge consists of sharp lines in the spectral region extending from 520 to 530 nm and broad bands in the 530 to 580 nm wavelength region. All these cathodoluminescence emissions are characterised by strong phonon coupling with a longitudinal optical phonon peak separation of 0.026 eV (1). The sharp lines are believed to be due to the radiative annihilation of excitons bound to defects in the lattice and are the subject of a parallel investigation using electron radiation damage. The tendency of zinc telluride to form complex defects as a result of

250

self-compensation has already been noted (2-5). Thus, the complex
nature of most defects is an important feature of the bound exciton
behaviour and also the broad band behaviour. The bound exciton
spectrum has been used to determine the threshold energies for
atomic displacement in zinc telluride (6,7).

The existence of a doubly ionizable acceptor centre involving
a zinc vacancy or zinc vacancy complex with acceptor levels at
0.048 meV and ~ 0.14 meV has been suggested by Aven and Segall (8)
on the basis of electrical measurements. Thomas and Sadowski (9)
have confirmed the existence of such a doubly ionizable zinc
vacancy acceptor by measuring the p-type conductivity between 700
and 950 C at various zinc pressures. Thus, there seems to be
considerable evidence for the existence of the doubly ionizable
acceptor involving the zinc vacancy.

The present paper is concerned with the four broad cathodoluminescence emission bands which occur at 531.4 nm (2.333 eV),
533.4 nm (2.324 eV), 551.9 nm (2.246 eV) and 553.9 nm (2.238 eV).
For convenience, these four emission bands are here referred to
as the A_I, A_{II}, B_I and B_{II} emission bands respectively. The
temperature dependence, excitation intensity dependence and time
resolved spectra of these four bands have been studied and lead
to the conclusion that they result from two free-to-bound transitions
(A_I and B_I) and two bound-to-bound transitions (A_{II} and B_{II})
involving a single donor centre and a doubly ionizable complex
acceptor centre (10).

EXPERIMENTAL PROCEDURE

The single crystals of zinc telluride were grown in this
department by recrystallization from the melt. The as-grown
crystals were tellurium rich and p-type. Mass spectrographic
analysis showed the major impurities to be, in p.p.m. atomic,
Si~100, Al~10, Fe~10, Mn~10, Cu~5. Phosphorus doped zinc telluride
samples were prepared by firing in phosphorus vapour for 72 hours
at 700 C. Other samples were fired in tellurium or zinc for 72
hours at 700 C. The crystal samples were typically 3mm x 3mm x 1mm
and were attached by thermally conducting silver paste to a copper
substrate which was mounted on the cold finger of a multisample
liquid helium cryostat (41).

The liquid helium cryostat shared a common vacuum (<5 μ torr)
with the drift tube of the Van de Graaff accelerator which was
used to provide the high energy electrons used for cathodoexcitation.
The exciting electrons were incident normally on the sample and the
emission was observed at 45° to this direction. During steady
cathodoexcitation the emission was chopped at 800 Hz before being
passed through a Hilger Monospek monochromator having a 1m grating

blazed at 1μm and then detected by a system consisting of a 9558
QB photomultiplier, low noise amplifier, phase sensitive detector
and potentiometric recorder. Steady cathodoluminescence was
excited by 3.3μA/cm² of 100 keV electrons. For pulsed cathodoexcitation the beam current was increased to 13.3μA/cm² and the
electron beam was electrically deflected off the sample at a
frequency of 10 kHz. The signal from the photomultiplier which
detected the analysed pulsed emission was fed into a boxcar
integrator together with a reference signal from the pulse generator.
The boxcar output was displayed on an oscilloscope and fed into
a pen recorder.

EXPERIMENTAL RESULTS

        A typical cathodoluminescence edge emission spectrum for zinc
telluride is shown in Fig. 1. Phosphorus doping, or indeed any
high temperature post-growth treatment, was found to destroy
the exciton emission which was present in as-grown samples. The
behaviour of the broad emissions was studied as the sample
temperature was varied and was studied as the excitation intensity
was varied. As the sample temperature was increased the intensity
of each of the A_{II} and B_{II} emission bands decreased whereas the
intensity of the A_I and B_I emission bands increased initially,
reached a maximum intensity at ~ 35 K, and then decreased. As the
excitation intensity was increased, by increasing the electron

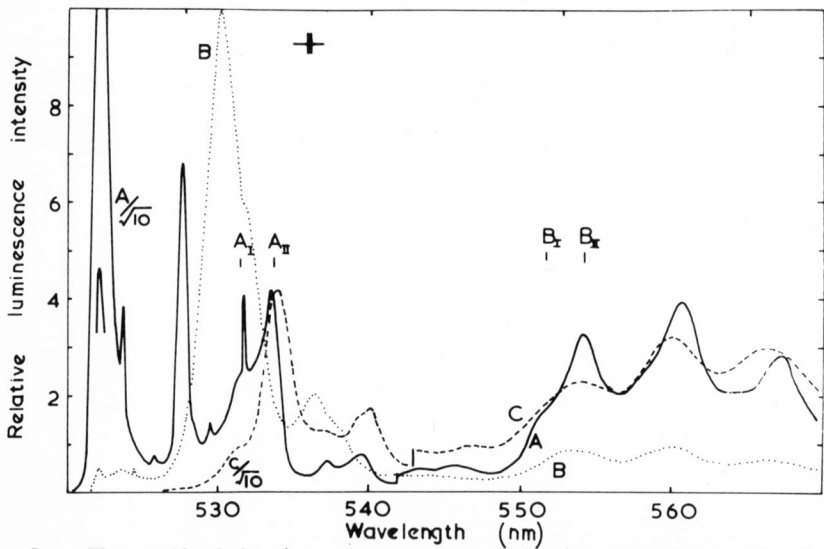

Fig. 1.   The cathodoluminescence edge emission spectrum of zinc
           telluride at ~ 10 K excited by 3.3μA/cm² of 100 keV
           electrons: <u>A</u>, as grown; <u>B</u>, sample fired at 700C in
           excess tellurium for 72 hours; <u>C</u>, sample fired at 700C
           in phosphorus vapour for 72 hours.

beam current, the A_{II} and B_{II} emission peak energies shifted to higher energies. Such peak position changes with excitation intensity were found to be significantly greater than the shifts due to variation in the sample temperature which also occurred. The temperature dependent and excitation intensity dependent behaviour has been previously interpreted as consistent with a model involving donor-acceptor pair recombination (10).

Further evidence for the donor-acceptor nature of the broad bands was sought from a study of the time resolved spectra of zinc telluride. Fig. 2 shows the spectra obtained at different times after the exciting electron beam was deflected from the sample. In all cases Fig. 2 shows that the peak energy positions of the bound-to-bound emissions (A_{II}, B_{II}) move to lower energies as the emission decays. From Fig. 2a it is clear that the lifetime of the free-to-bound emission (A_I) is much shorter than the lifetime of the bound-to-bound emission (A_{II}). From sample to sample there was some variation in the positions of the A_{II} and B_{II} emissions which we attribute to variation in the majority defect concentration. No free-to-bound emissions were observed for phosphorus doped samples (Figs. 2b and 2c). Fig. 2c includes

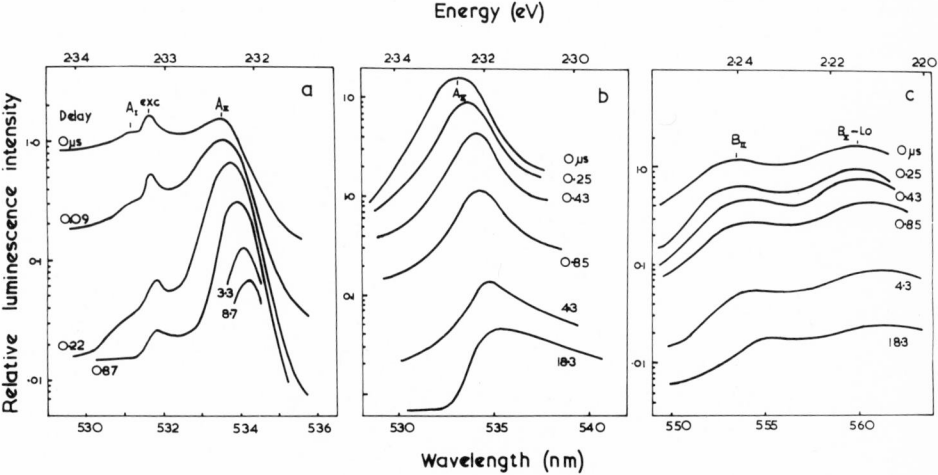

Fig. 2. Time resolved spectra of the edge emission of zinc telluride at ~ 10K; (a) A_I and A_{II} emissions for an as-grown sample; (b) A_{II} emission for a phosphorus doped sample; (c) B_{II} emission for a phosphorus doped sample. Intensities measured at different times after excitation by 13.3 $\mu A/cm^2$ of 100 keV electrons are plotted on a logarithmic scale.

a strong phonon replica of the B emission.

DISCUSSION

Study of the time resolved spectra of zinc telluride reveals
a shift of the peak energy to lower energies with increasing time
for each of the A_{II} and B_{II} emissions for both the as-grown and
phosphorus doped samples. Such behaviour is consistent with the
suggestion that each of the A_{II} and B_{II} emissions results from a
bound-to-bound donor-acceptor pair recombination process (10).
Concurrent with this peak energy shift there is observed a change
in the shape of each emission band. The change in shape occurs
because of the **faster** recombination of the closer donor-acceptor
pairs (i.e. those contributing to the emission band at higher
energies) and is clearly observed to be a narrowing of each band,
as is consistent with the theory of Thomas et al. (12).

The variation with time of the separation between the free-to-
bound emission peak energy, E_{fb}, and the bound-to-bound emission
peak energy, E_{bb}, can be used to estimate the donor depth, E_D, by
using the expression given by Colbow (13). For donor-acceptor
pair recombination with pair separation r, the emission energy,
E_{bb} is given by

$$E_{bb} = E_G - (E_A + E_D) + E_C \qquad (1)$$

where E_G is the band gap energy, E_A is the acceptor ionization
energy and the coulombic energy term, $E_C = e^2/\varepsilon r$. For free-to-
bound emission $E_{fb} = E_G - E_A$, thus at any time the separation
between corresponding free-to-bound and bound-to-bound emission
peak energies is given by $E_D - E_C$ (E_{fb} has no significant
coulombic term so does not change with time as the emission decays
after excitation). Values of $E_{fb} - E_{bb}$ have been measured for
various times after the cessation of excitation. By taking pairs
of values of $E_{fb} - E_{bb}$ at different times and using the expression
of Colbow (13) it has been possible, for each of the A and B
emissions, to obtain the values of E_D shown in Table 1. We note
that the shorter time intervals taken later in the decay yield
larger values of E_D which analysis shows to be more accurate (14).
We conclude that the donor depths are 31 meV for the A_{II} emission
and 32 meV for the B_{II} emission. This could be taken to indicate
that the donor involved in the A_{II} and B_{II} emissions is the same
donor centre, although of course in a particular material all
hydrogenic donor centres would be expected to lie at approximately
the same depth below the conduction band. For the steady state
excitation condition, t = 0, $E_D - E_C$ can be obtained directly
from Fig. 1 and is 8.53 meV. Since E_D has been shown to be ~ 31
meV then $E_C = e^2/\varepsilon r = 22.47$ meV and use of the value for ε of
10.4 given by Nahory and Fan (15) allows calculation of a value
of 62A for r.

TABLE 1

| | Time after excitation | | E_D |
	t_1	t_2	
Emission A	0.1 x 10^{-6}s 0.5 x 10^{-6} 1.0 x 10^{-6} 5.0 x 10^{-6} 10.0 x 10^{-6}	50 x 10^{-6} s " " " "	28.4 meV 28.9 29.1 29.8 30.6
Emission B	0.1 x 10^{-6} 0.5 x 10^{-6} 1.0 x 10^{-6} 5.0 x 10^{-6} 10.0 x 10^{-6}	20 x 10^{-6} " " " "	24.9 meV 27.3 27.5 30.9 32.0

Measurement of the emission energies of the two free-to-bound,
A_I and B_I, transitions (2.333 and 2.246 eV respectively) reported
here, allows estimates of the corresponding acceptor depths when
taken in conjunction with the value of the band gap energy of
2.391 eV (1). Thus the two acceptor levels corresponding to the
A and B emissions were found to be 0.058 and 0.145 eV from the
valence band. Such levels are in good agreement with the values
of 0.048 and ~ 0.14 eV reported by Aven and Segall (8) for the
singly and doubly ionized states of the zinc vacancy or zinc
vacancy complex in zinc telluride. Some support for the double
acceptor nature of the A_I, A_{II}, B_I and B_{II} bands is provided by
their behaviour following irradiation with electrons of sufficient
energy to produce atomic displacements. Meese (16) has observed
enhancement of two bands at 2.320 and 2.234 eV (following electron
irradiation above 185 keV and at 10 K) which is attributed to
displacement of zinc vacancies producing the doubly ionizable
centre (17). The damage sensitive emissions were believed to
be free-to-bound transitions with the bound states being the two
levels of the doubly ionizable centre. Enhancement of each of the
four emissions examined here might be expected to produce zinc
vacancy production in the lattice because their intensities would
depend on the concentration of acceptor levels present, and the
zinc vacancies produce singly ionized levels (for the A emissions)
and doubly ionized levels (for the B emissions). If the sample
is maintained at 10 K no enhancement of the four emissions examined
here has been observed. However enhancement of the B emission
only is produced, by electrons capable of displacing only zinc
atoms, if the sample is irradiated at room temperature or is
irradiated at 77 K and warmed to room temperature before examination at 10 K. This room temperature behaviour points to the

acceptor levels arising not from the presence of simple defects but
complex defects involving the zinc vacancy. Failure to observe
damage in some samples would then be due to the absence of the
impurity which associates with the zinc vacancy and in this way will
form the complex centre. The fact that electron damage at 300 K
produces enhancement of the B emission but not of the A emission
is taken to indicate movement of the Fermi level under room
temperature electron damage. Thus more acceptor centres for both
A and B emissions are produced but the enhancement is only observed
in the B emission because of the change in the Fermi level.

Evidence to support the involvement of the zinc vacancy in the
acceptor levels is also obtained from heat treatment of the samples.
Firing of ZnTe in a tellurium atmosphere is found to increase the
A and B emissions because of the greater concentration of zinc
vacancies produced, whereas firing in excess of zinc produces a
decrease of the emissions, and indeed causes complete disappearance
of the A emission.

The intensity of each bound-to-bound emission relative to the
corresponding free-to-bound emission is much greater in phosphorus
doped samples than in undoped samples (see Fig. 1) or samples
fired in dopants such as Cu, Fe, Mn or Al. Thus, although
phosphorus would be expected to enter the lattice substitutionally
for tellurium to give an acceptor centre, it is possible that in
the present case it is entering interstitially, or complexing,
and giving rise to a donor centre. This is not inconsistent with
the observations of Crowder and Pettit (4) who postulate a
phosphorus donor to explain the high degree of compensation
observed in their phosphorus doped samples which had been fired
subsequently in zinc (this does not exclude phosphorus being
involved in the acceptor levels also).

Watanabe and Usui (18) have studied the photoluminescence of
as-grown ZnTe and attributed a band at 2.323 eV to donor-acceptor
pair recombination with the acceptor level lying 31 meV above the
valence band. After firing in phosphorus they observed that the
2.323 eV band shifted 7 meV to higher energies. We note that
this is consistent with the donor-acceptor nature of the emission
because an increase in defect concentration would be expected to
lead to a decrease in donor-acceptor pair separation with a
consequent increase in the bound-to-bound emission peak energy.
They also observed a new emission at 2.370 eV which they attributed
to free-to-bound emission but they were unable to explain the
anomalous temperature dependence of the 2.370 eV emission intensity.
Their donor-acceptor band and the anomalous temperature dependence
of the 2.370 eV emission can both be explained on the model here,
which assigns the acceptor level to $E_V + 0.057$ eV and the donor
level to $E_c - 0.031$ eV. The 2.370 eV is then attributed to bound -
to-free emission (rather than free-to-bound emission) and its

intensity falls as increasing temperature thermalises the donor level and the intensity increases as the acceptor centres are emptied.

The authors thank W. E. Hagston for helpful discussions.

REFERENCES

1. A. C. Aten, C. Z. van Doorn and A. T. Vink, Proc. Int. Conf. Semiconductors Exeter (I.P.P.S. London) (1962) p. 696.

2. A. G. Fischer, J. N. Carides and J. Dresner, Solid State Comm. 2, 157 (1964).

3. R. S. Title, F. Morehead and G. Mandel, Phys. Rev. 136, 300 (1964).

4. B. L. Crowder and G. D. Pettit, Phys. Rev. 178, 1235 (1969).

5. T. C. Larsen, C. F. Varotto and D. A. Stevenson, J. Appl. Phys. 43, 172 (1972).

6. F. J. Bryant and A. T. J. Baker, Phys. Lett. 35A, 457 (1971).

7. F. J. Bryant and A. T. J. Baker, J. Phys. C. 5, 2283 (1972).

8. M. Aven and B. Segall, Phys. Rev. 130, 81 (1963).

9. D. G. Thomas and E. A. Sadowski, J. Phys. Chem. Solids 25, 395 (1964).

10. F. J. Bryant and A. T. J. Baker, Phys. Stat. Solidi (a) 11, 623 (1972).

11. F. J. Bryant and C. J. Radford, Cryogenics 10, 329 (1963).

12. D. G. Thomas, J. J. Hopfield and W. M. Augustyniak, Phys. Rev. 140, 202 (1965).

13. K. Colbow, Phys. Rev. 141, 742 (1966).

14. A. T. J. Baker, F. J. Bryant and W. E. Hagston, to be published.

15. R. E. Nahory and H. Y. Fan, Phys. Rev. 156, 825 (1965).

16. J. M. Meese, Appl. Phys. Lett. 19, 86 (1971).

17. W. R. Woody, R. A. House and J. M. Meese, Bull. Amer. Phys. Soc. II. 17, 859 (1971).

18. N. Watanabe and S. Usui, Jap. J. Appl. Phys. 6, 1253 (1967).

PHOTOLUMINESCENCE STUDY OF LATTICE DEFECTS IN CdTe BY MEANS OF HEAT TREATMENT AND ELECTRON IRRADIATION

T. Taguchi, J. Shirafuji and Y. Inuishi

Department of Electrical Engineering

Faculty of Engineering, Osaka University

Suita, Osaka, Japan

ABSTRACT

The effects of heat treatment in excess or dissociative Cd vapor pressure and the effects of electron or neutron irradiation on photoluminescence spectra at 4.2 K in CdTe single crystals have been investigated. Four groups of emission bands appeared. Heat treatment in dissociative Cd vapor suppressed all bands, but excess Cd vapor enhanced two of them, possibly due to the introduction of Cd interstitials or Te vacancies. One group was enhanced by electron or neutron irradiation, but the other group was enhanced only by neutron irradiation. Time resolved spectroscopy revealed that another group was mainly due to the electronic transitions between donor-acceptor pairs.

INTRODUCTION

The main purpose of the present paper is to report on a study of the photoluminescence properties of CdTe single crystals which was made in order to clarify the defects and the electronic transitions responsible for the edge emission and the emission band with a maximum at 1.42 eV (77 K), respectively.

CdTe single crystals were grown from the melt in an evacuated quartz ampoule by the usual Bridgman-Stockbarger method. The n-type samples were obtained by heat treatment under controlled Cd vapor pressure. Irradiation with 2 MeV electrons or neutrons including fast and thermal components was performed at 77 K with a dose of 3×10^{16} electrons/cm^2 and at 330 K with a dose of

258

5×10^{15} neutrons/cm^2, respectively. For time resolved measurements, a Xe flash lamp which generated high intensity pulsed light
of 0.5 μsec duration at a repetition rate of 40 Hz was employed.

EFFECTS OF HEAT TREATMENT AND IRRADIATION

     Fig. 1 shows the effects of heat treatment in excess or
dissociative Cd vapor pressure on photoluminescence spectra at
4.2 K in as-grown undoped p-type CdTe. The insert at the top of
Fig. 1 shows time resolved spectra of the 1.42 eV band at 4.2 K.
In the original p-type sample, there appeared three groups of
dominant emission bands resulting from native defects. The photon
energies of these groups were (i) 1.587 and 1.572 eV, (ii) 1.528 eV

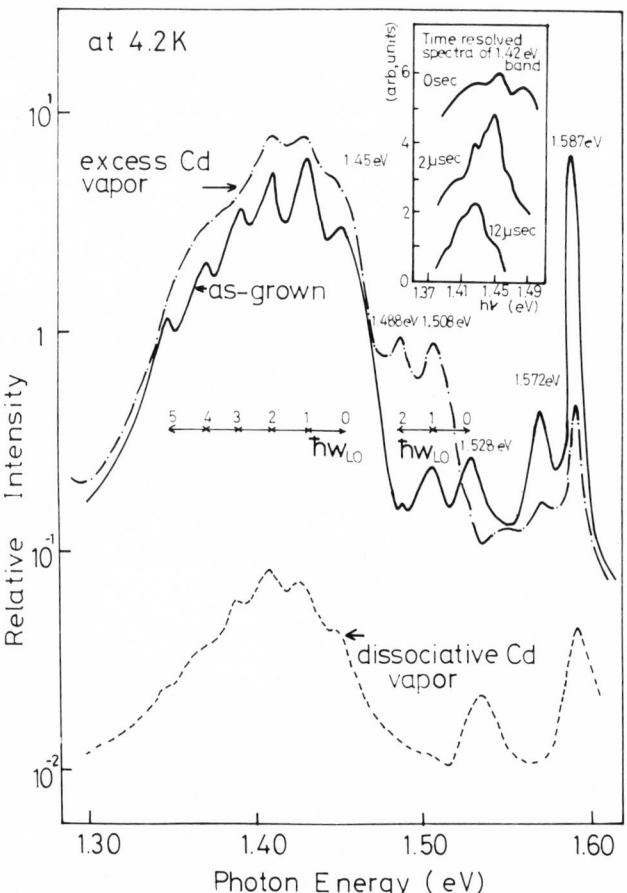

Fig. 1 Effect of heat treatment on the photoluminescence spectrum
        at 4.2 K in undoped p-type CdTe. An insert shows the time
        resolved spectra of the 1.42 eV emission band at 4.2 K in
        as-grown undoped p-type CdTe.

(and its LO phonon lines), and (iii) 1.45 eV (and its LO phonon
lines). After heat treatment for 200 hrs in excess Cd vapor at
850°C, the intensities of the (i) band, which may be related to a
bound exciton, decrease remarkably. In this case, the phonon
replicas of the (ii) band are enhanced in spite of the decrease in
the 1.528 eV band. The new emission bands centered at 1.55 and
1.2 eV (as shown in Fig. 2) appeared by these treatments. The
broad emission band at 1.45 eV shows little change in intensity
with some smearing out of the phonon lines. On the other hand,
heat treatment in a dissociative Cd vapor suppresses all the
emission bands and decreases the fine structure of the (ii) and
(iii) bands.

    Fig. 2 shows the effects of electron and neutron irradiation
on photoluminescence spectra at 4.2 K in as-grown undoped p-type
CdTe. Electron irradiation followed by room temperature annealing
reduces remarkably the emission intensities of the (i) band, while
those of the (ii) band are much enhanced. Band (iii) is little
influenced by these irradiations. By annealing at 100C for 15
minutes, the (i) and (ii) bands recover by about 10% of the
preirradiation value and a new band centered at 1.55 eV appears.
Neutron irradiation gives rise to enhancement of the (ii) band;
in addition, two new bands centered at 1.55 and 1.09 eV are

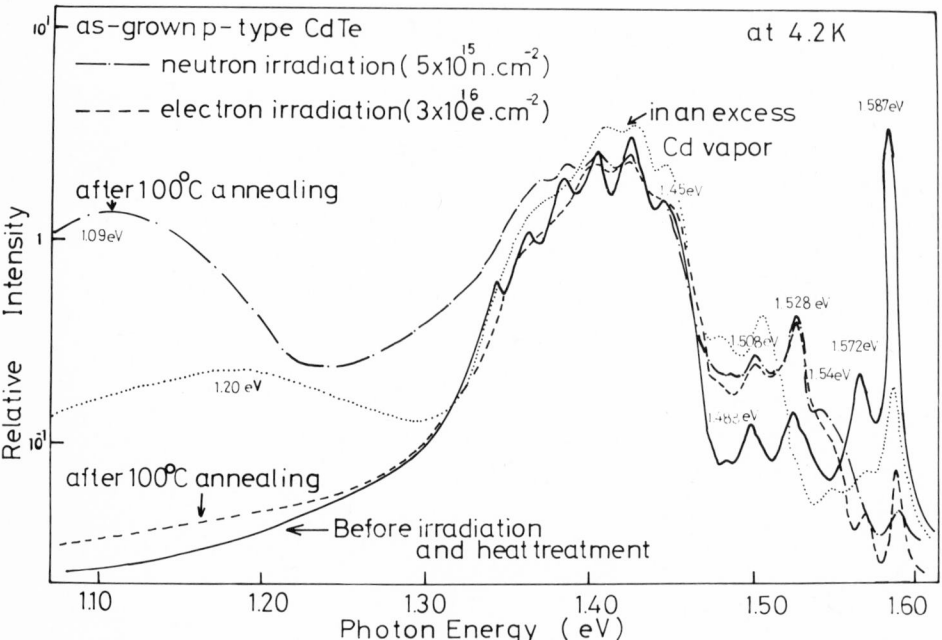

Fig. 2 Effects of heat treatment in an excess Cd vapor and electron
        or neutron irradiation on the photoluminescence spectrum
        at 4.2 K in undoped p-type CdTe.

observed. The 1.09 eV band appeared only in a neutron irradiated sample, possibly due to the defect aggregates. However, the 1.2 eV band was observed after firing in excess Cd vapor. In electron and neutron irradiated samples, the intensities of the (iii) band show only small changes. Enhancement of the (ii) band may be attributed to the introduction of Cd interstitials or Te vacancies, as in the case of heat treatment in Cd vapor (1).

TIME RESOLVED SPECTRA OF UNDOPED P-TYPE CdTe

In as-grown p-type CdTe, the (iii) band was observed as the band with a maximum at 1.42 eV at 77 K. This emission band has been interpreted to be either of the following electronic transitions: (a) the band is associated with the transition of electrons from the conduction band to the level of an acceptor center (2), (b) the band is attributed to electronic transitions from shallow donors to acceptors involving a Cd vacancy (3), or (c) transitions in donor-acceptor pairs formed by impurity ions and interstitial Cd atoms (4).

The characteristic properties of pair recombination are exhibited by long nonexponential time decay and the shift of the band maximum to lower energies with time (5). As shown in Fig. 1, from time resolved spectroscopy it is clear that the positions of the emission maximum shifts with time towards lower energies at 4.2 K. Using the equation expanded by Thomas et al. (6), the bound-to-bound constant is calculated. The reaction constant obtained is 3.5×10^7/sec. Calculating the average separation of pairs for various times after pulse excitation, we get r = 60 A at 1 μsec and r = 200 A at 12 μsec. The decay time constant of this band (8740 A) was approximately 5 μsec at 4.2 K.

REFERENCES

1. M. Caillot, Phys. Letts. 38A, 2 (1972).

2. D. de Nobel, Philips Res. Rep. 14, 361 (1959).

3. V. S. Vavilov, Int. Conf. II-VI Semiconducting Compounds (1967) p. 743.

4. Yu. V. Rud et al., Soviet Phys.-Semicond. 5, 573 (1971).

5. I. Filinski and B. Wojtowich-Natanson, J. Phys. Chem. Solids, 32, 2409 (1971).

6. D. G. Thomas et al., Phys. Rev. 140, A202 (1965).

LUMINESCENCE OF ZnTe AND ITS MODULATION BY INFRARED RADIATION

A. Maruani, J. P. Noblanc, G. Duraffourg

Centre National d'Etudes des Télécommunications

Bagneux, France

ABSTRACT

Measurements of reflectivity, photoluminescence and cathodo-luminescence for various electron beam voltages were performed on p-type single crystals at liquid nitrogen, hydrogen and helium temperatures. By an application of the microreversibility principle and consideration of the reflectivity data, the nature and the energies of the transitions were determined. In addition, we identify the lines due to the annihilation of exciton-neutral defect complexes, in ground or excited state. When the luminescence is modulated by a CO_2 laser ($h\nu_{IR} \sim 130$ meV) effects are observed at liquid helium temperature. A quenching effect is observed for the free-to-bound and pair recombinations. Quenching of the luminescence due to the annihilation of exciton isoelectronic oxygen complexes is a direct corroboration of the binding-in-two-step theory.

INTRODUCTION

Zinc telluride is a II-VI compound with a reported (1,2) direct band gap of 2.37 eV at 80 K and 2.39 eV at 4.2 K; as in other II-VI compounds the main absorption peak (several 10^5 cm^{-1}) is governed by free exciton formation with a binding energy in the range 10-13 meV according to several authors (3,4). The edge emission was first observed by Halsted et al (5) and then by Gross et al. (6).

We have studied the luminescence of ZnTe at liquid nitrogen,

hydrogen and helium temperatures and its modulation by CO_2 laser
radiation in three regions:
- The first region, 2.37 eV $<h\nu< 2.39$ eV, is that of annihilation
of free and "weakly" bound excitons.
- The second region, $2.00 <h\nu < 2.35$ eV at low temperatures, is the
analogue of the green edge emission of CdS (7).
- The third region, 1.85 eV $<h\nu < 1.99$ eV at low temperatures and
in crystals containing oxygen, was first pointed out by Dietz (8).
As studied by Merz (9) it involves the annihilation of an exciton
bound on isoelectronic oxygen with an intense participation of
phonons.

EXPERIMENTAL PROCEDURE

The experimental arrangement for luminescence or IR-modulated
photoluminescence was described elsewhere (10). In cathodoluminescence, the maximum voltage of the electron beam is 50 kV.
Photoluminescence is excited with a 4416 Å He-Cd laser with a
power of about 10 mW; the modulation of photoluminescence is
controlled by a Q-switched CO_2 laser with a peak power of around
100 W.

The single crystals were provided by the "Laboratoire de Magnétisme et de Physique du Solide" of CNRS. They were grown by a
Stockbarger method and are highly resistive p-type. The samples
were mechanically polished and etched in a bromine-ethanol solution.

LUMINESCENCE AT LIQUID NITROGEN TEMPERATURE

Figure 1(b) shows cathodoluminescence spectra in the energy
range of the free exciton transition for electron beam voltages
of 40 kV, 25 kV and 15 kV. The main peak undergoes a red shift
when the voltage increases; this is due to self-absorption which
increases with the exciting beam penetration depth, δ. This
effect must be taken into account, which can be done by considering
the function $a(h\nu) = I(h\nu) \exp (h\nu/kT)$, where $I(h\nu)$ is the
intensity of luminescence and T the temperature of the sample. It
can be shown that (11):
- If the self absorption is negligible, $a(h\nu)$ reproduces the shape
of the absorption coefficient α (which is the well known formula
resulting from the microreversibility principle).
- If the self absorption is important, $a(h\nu)$ - as far as the
excitation in the sample is a decreasing function of the distance
from the surface - exhibits a "plateau" centered on the maximum of
α.

The limit between these two domains is given by the relation
$\alpha\delta \sim 1$ ($\delta \sim 1\mu m$ in photoluminescence, $\sim 10\mu m$ in cathodoluminescence
for a beam voltage of 50 kV). Figure 1(c) shows the $a(h\nu)$ spectra

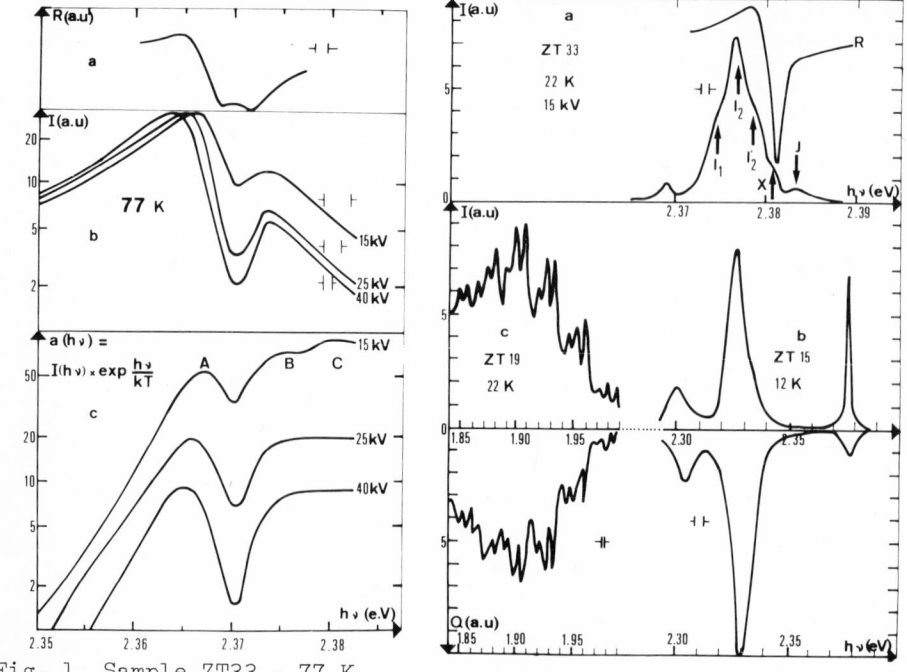

Fig. 1   Sample ZT33 - 77 K

        (a) R, reflectivity spectrum (artitrary linear scale);
        (b) I, cathodoluminescence spectra for different electron
            beam voltages (logarithmic scale);
        (c) a(hν) = I(hν) exp (hν/kT) from the above cathodo-
            luminescence spectra (logarithmic scale).

Fig. 2   (a) Sample ZT33 - 22 K

            Reflectivity spectrum (R) (arbitrary scale) and
            cathodoluminescence spectrum (I) for an electron
            beam voltage of 15 kV. (resolution: 1 meV).
        (b) Sample ZT15 - 12 K
            I, Photoluminescence spectrum (resolution: 0.6 meV);
            Q, Modulation spectrum (quenching), (resolution: 3.5
            meV).
        (c) Sample ZT19 - 22 K
            I, Photoluminescence spectrum (resolution: 0.8 meV);
            Q, Modulation spectrum (quenching), (resolution: 1.5
            meV).

computed from those of Fig. 1(b). We observe for each curve:
- a maximum A (ideally a plateau) slightly shifted towards low
  energies when the voltage increases, this is probably due to
  some heating of the sample,

- a plateau B, which is not followed by a decrease.
On the 15 kV spectrum, this plateau is followed by an other one,
C. Apparently B and C are blurred to form the same plateau on the
other two spectra. The position of the "maximum" A with respect
to the reflectivity curve (Fig. 1(a))enables us to interpret the
corresponding transition as the annihilation of the n=1 free
exciton, at 2.367 eV. We interpret then the plateaux B and C on the
15 kV spectrum as due to, respectively, the annihilation of the
n=2 free exciton (2.376 eV) and the band to band transition
(2.379 eV). This gives the binding energy of the exciton in ZnTe
as G = 12 meV $\pm$ 1 meV.

EXCITONIC EDGE EMISSION AT LIQUID HYDROGEN OR HELIUM TEMPERATURE

       Shown in Fig. 2(a) is the cathodoluminescence spectrum of one
sample at liquid hydrogen temperature for an electron beam voltage
of 15 kV. The main features of this spectrum are the "I" lines or
bumps and two "lines" referred to as X and J; as evidenced by
varying the beam voltage, the effect of reabsorption is the most
intense on X. A simultaneous examination of the a(hν) curve
computed from those data and the reflectivity spectrum enables us
to interpret X as the annihilation of the n=1 free exciton (2.381
eV). That provides a band gap energy of 2.393 eV at 22 K. The I
lines are then extrinsic and originate from the annihilation of
bound exciton complexes. The binding energies of the exciton are
6 meV, 4 meV, 3 meV for I_1, I_2, I_2'. Now, leaning upon the
so-called Halsted's rules (12) and the known orders of magnitude
of the ionization energies of the hydrogenic impurities in our
crystals, we interpret I_1, I_2 and I_2' as due to the annihilation
of excitons bound on neutral acceptors and donors respectively.
This procedure provides the ionization energies:

$$E_A = 60 \pm 5 \text{ meV} \quad E_D = 20 \pm 3 \text{ meV} \quad E_D' = 15 \pm 3 \text{ meV},$$

in fair agreement with what can be found in the literature (13,14).

       Located in energy between the n=1 and the n=2 free exciton
transitions, the J line correlates strongly with I_2 and is
interpreted (like the analogous line in CdTe (15)) as the
annihilation of an excited state of the bound exciton complex
giving rise to I_2. The energy of that transition, already seen
in absorption by Marple et al. (16) is 2.384 eV. At liquid
helium temperature the I_1 line is much greater than the other
ones; such a behavior expresses the variations of the statistical
occupation factors for the involved levels.

MODULATION OF THE EXCITONIC EDGE EMISSION BY AN INFRARED RADIATION

       Due to the neutralization of the ionized defects by the

exciting beam only neutral "impurities" are involved in the bound exciton luminescence. In order to confirm this point, one can try to ionize these defects during the excitation with an IR light. Such experiments were performed for absorption in CdS by Thomas and Hopfield (17). What is expected is a quenching of the I and J lines and perhaps some raising of a line due to an exciton on an ionized donor. Since, in luminescence, intense powers are needed (10) for that purpose, we made use of a CO_2 laser.

Fig. 2(b) shows the luminescence (I) and quenching (Q) spectra of one sample at liquid helium temperature (no modulation effect could be seen with our present apparatus at the temperature of liquid hydrogen). The poor resolution does not permit us to know whether the quenching peak is centered on I_1 or I_2. For some points of focusing on the sample, occasionally there could be seen the appearance of a line, interpreted to be the annihilation of an exciton bound to an ionized donor, at 4 meV below the X line (whereas a theoretical value for the binding energy of an exciton to an ionized donor is 3.2 meV (18)).

QUENCHING OF THE PAIR AND FREE-TO-BOUND RECOMBINATIONS

The luminescence in this region is affected by the CO_2 laser, as can be seen in Fig. 2(b). Such a quenching is interpreted in terms of the ionization of the defects. The slight shift between the luminescence and the quenching lines reflects the general observation that there are changes in the positions and the shapes of the different lines with the impact point of the beams on our inhomogeneous samples between two experiments.

LUMINESCENCE DUE TO ISOELECTRONIC OXYGEN

Fig. 2(c) shows the luminescence spectrum (I) of one sample and its quenching (Q) by IR radiation at liquid hydrogen temperature. The effect is more pronounced than in the other two regions of luminescence. The same uniform lowering is also seen at liquid helium temperature.

From a theoretical point of view, the binding energy of the exciton on the isoelectronic acceptor (0.4 eV) is explained in terms of a binding in two steps, the hole being much more loosely bound than the electron (19). Actually, Cuthbert (20) has indirectly estimated its binding energy to be $\varepsilon_h = 20$ meV. Interpreted as the freeing of the hole, such a quenching constitutes a direct corroboration of the binding-in-two-step theory; we get in fact $\varepsilon_h \lesssim h\nu_{IR} = 130$ meV.

ACKNOWLEDGEMENT

We feel very much indebted to Mrs. Rodot from the "Laboratoire de Magnétisme et de Physique du Solide" of CNRS (Meudon-Bellevue, France) for providing the samples. We are very grateful to Mr. Guglielmi for his aid in the laser measurements.

REFERENCES

1. R. Halsted, M. Aven, H. Coghill, J. Elec. Soc. 112, 177 (1965).

2. R. Nahory, H. Fan, Phys. Rev. 156, 825 (1967).

3. A. Aten, C. Van Doorn, A. Vink, Proc. Intern. Conf. on the Physics of Semiconductors, Exeter, July 1962, p. 696.

4. B. Segall, D. Marple, Physics and Chemistry of II-VI Compounds, edited by Aven and Prener, North Holland (1967), p. 335.

5. R. Halsted, M. Aven, Bull. Am. Phys. Soc. (1961) 312.

6. E. Gross, L. Suslina, A. Lifshitz, Sov. Phys. Sol. State 5, 582 (1963).

7. D. Thomas, Dingle, J. Cuthbert, II-VI Semiconducting Compounds, edited by D. G. Thomas, Benjamin, New York (1967), p. 863.

8. R. Dietz, D. Thomas, J. Hopfield, Phys. Rev. Letters 8, 391 (1962).

9. W. Merz, Intern. Conf. on Physics of Semiconductors, Cambridge, Mass. edited by S. P. Keller, J. C. Hensel, F. Stern, USAEC (1970), p. 251.

10. A. Maruani, Thèse de 3° Cycle, Paris VI, 1972.

11. J. Noblanc, J. Loudette, G. Duraffourg, Phys. Stat. Sol. 32 281 (1969).

12. R. Halsted, M. Aven, Phys. Rev. Letters 14 64 (1965).

13. M. Aven, J. Appl. Phys. 38, 444 (1967).

14. G. Neu, Y. Marfaing, C. R. Acad. Sci. Paris B 273 1112 (1971).

15. I. Noblanc, J. Loudette, G. Duraffourg, J. of Luminescence 1,2, 528 (1970).

16. D. Marple, M. Aven, Physics and Chemistry of II-VI Compounds, edited by D. G. Thomas, Benjamin (1967) p. 315.

17. D. Thomas, J. Hopfield, Phys. Rev. 128, 2135 (1962).

18. S. Elkomoss, Phys. Rev. B 4, 3411 (1971).

19. J. Hopfield, D. Thomas, R. Lynch, Phys. Rev. Letters 17, 312 (1966).

20. J. Cuthbert, J. Appl. Phys. 42, 739 (1971).

THE ELECTRONIC STRUCTURE AND LUMINESCENCE OF PHOSPHORS BASED ON

IIA-VIB COMPOUNDS

V. Mikhailin

State University, Moscow, USSR

ABSTRACT

For the understanding of the mechanism of luminescence excitation in the phosphors of IIA-VIB compounds the luminescence excitation and fundamental absorption spectra have been investigated. The measurements were performed using electron synchrotrons as radiation sources in the 5-45 eV region. The reflection spectra of evaporated thin films of alkaline earth sulphides CaS, SrS, BaS and single crystals of MgO, CaO, SrO are reported. The optical properties have been derived by Kramers-Kronig analysis of reflection data in the fundamental absorption region. Fine structure was interpreted as exciton and interband transitions from the valence band and from the cation core level of p-symmetry. For the investigation of energy migration the spectra of luminescence were studied. In the excitation spectra a sudden increase of the luminescence yield at photon energy more than 2 E_g was observed. The measurements at high energy show an increase of the luminescence yield which may be due to photon multiplication.

The present paper is aimed at the investigation and explanation of the optical properties of the alkaline earth sulphides which are often used as the basis of well-known phosphors. The knowledge of the band gap and fundamental absorption is necessary for understanding the mechanism of excitation of the luminescence in these phosphors and for studying the energy migration to the luminescence center (1,2).

Compounds of the group IIA-VIB, oxides and sulphides of Mg, Ca, Sr, Ba in particular, belong to the NaCl structure type and are divalent analogues of the alkali halide crystals.

It is very useful to know the theoretical optically important states of the band structure for understanding the electronic structure of the substance. These states can be determined on the basis of group theory analysis using the full symmetry of the crystal as the space group for obtaining the selection rules (3). Group-theoretical analysis of the electronic states helped us to identify the peaks as transitions between high-symmetry points in the reciprocal lattice by applying the calculated selection rules (4). These high symmetry points are critical points with different types of Van Hove's singularities in the behavior of the density of the state function near these points, which give the main contribution to the absorption and reflection spectra. The number of optical electrons is the sum of products of the number of critical points and the number of the different bands at each critical point. Let us analyse the critical points T, L, X, K (4). We may use for this purpose the two highest valence bands and two lowest conduction bands. Assuming the dipole approximation for transitions, we get the following selection rules from group theory analysis: $\Gamma_{15} \rightarrow \Gamma_1$, $\Gamma_{15} \rightarrow \Gamma'_{25}$, $X'_5 \rightarrow X_1$, $X'_4 \rightarrow X_3$, $X'_5 \rightarrow X_3$, $L_3 \rightarrow L_2$, $L_3 \rightarrow L_3$, $\Lambda_3 \rightarrow \Lambda_1$, $\Delta_5 \rightarrow \Delta_1$, $\Sigma_4 \rightarrow \Sigma_1$, $\Sigma_3 \rightarrow \Sigma_1$. In the region which is interesting to us the main contribution is due to the following transition: $\Gamma_{15} \rightarrow \Gamma_1$, $X'_5 \rightarrow X_1$, $L_3 \rightarrow L'_2$, $\Lambda_3 \rightarrow \Lambda_1$, $\Delta_5 \rightarrow \Delta_1$, $\Sigma_4 \rightarrow \Sigma_1$. The transition between the deeper bands gives a relatively small contribution (nearly 15% (3)).

We can use the results obtained for the interpretation of the MgO spectrum (5), taking into account the differences between Mg and Ca ions. The main features of the resulting CaS band structure are: the first conduction band is lower than in MgO and the X_3 point of the second conduction band is lower than that in MgO.

The measurements were performed by using the electron synchrotron DESY (Hamburg) and Lebedev Institute's synchrotron as vacuum ultraviolet radiation sources in the 6-60 eV region. The compressed samples were prepared from standard powders of "Soyuzchimexport" and from powders made by M. Alsallu in the University of Tartu, the latter containing 97% CaS. The excitation spectra of luminescence were measured on the samples CaS:Bi (2×10^{-4} g/g) also made in Tartu. The samples of sulphides activated with Mn were prepared in our laboratory.

The absorption spectra of the CaS, SrS, BaS show the doublet structure at the absorption edge which is due to the spin-orbit splitting of the Γ -exciton (1,2,6,7). Assuming that the absorption edge lies at the point of maximum slope of the curve we can estimate the positions of the absorption edge (7), which are for CaS 5.1 eV, for SrS ∿4.6 eV and for BaS ∿4 eV. The

shift of the absorption edge corresponds to the change in band gap
values and our estimation for CaS gives $E_g = (5.3 - 5.8)$ eV.
Spin-orbit splitting is estimated for CaS as 0.1 eV, for SrS and
BaS 0.12 and 0.17 eV, respectively. This change in spin-orbit
interaction of the singly-charged anion is due to the influence
of neighboring cations. The observed shift of the absorption
edge and the change of spin-orbit splitting values, both depending
on the cation weight for this sulphide series, are also confirmed
by our data on reflection of the powdered sulphides (7).

Near-normal-incidence reflectance spectra were obtained for
alkaline-earth sulphides by measuring the reflectance from thin
evaporated films with different thicknesses and from the samples
prepared from powder by compressing. These reflectance spectra of
different films and compressed samples were reproducible. Great
difficulties were encountered in the preparation and investigation
of thin alkaline-earth sulphide films because of the high evaporation
temperature and of the instability of the compounds during evaporation and measuring.

Fig. 1 shows the reflection spectrum in the fundamental
absorption region for the compressed sample of CaS. In our
measurement of the films of various thicknesses (600-2000 A) and
of compressed samples, the following maxima were reproducible:
10.6, 13.3, 15.1, ∿26, ∿29, 31.6, 34 and 43.8 eV.

The group of peaks in the range of 6-11 eV may be attributed
to the transition at Γ and X-points. The 10.6 peak may be due
to the transition between states of the type $X_4' {\rightarrow} X_3$ (4), and the
13.2 eV peak of the type $L_3 {\rightarrow} L_2'$. Other peaks may be interpreted
as optical structure corresponding to the symmetry points Λ, Δ
and Σ, and to the transitions from deeper bands. The structure
at energies more than 25 eV is ascribed to the transitions from
the 3p-state of Ca to the conduction band. Similar peaks are
observed in the CaO spectrum in the same spectral region.

For the investigation of the energy migration to the center
of luminescence we measured the excitation spectra of stationary
luminescence. Excitation spectra of manganese luminescence in
sulphides doped with manganese agree well in the long wavelength
region with the shift of the fundamental absorption (2). The
kinetics of manganese luminescence in CaS:Mn excited at the
absorption edge (6 min) indicates the probable annihilation of
excitons and the formation of electron-hole pairs. Beyond the
absorption edge at 6-10 eV the quantum yield for CaS:Mn is nearly
constant and low due to the migration losses. The excitation
spectra of SrS:Mn and BaS:Mn are shifted to lower energies but
their shape is quite similar. In the excitation spectra there
is a sharp increase of luminescence quantum yield for CaS at

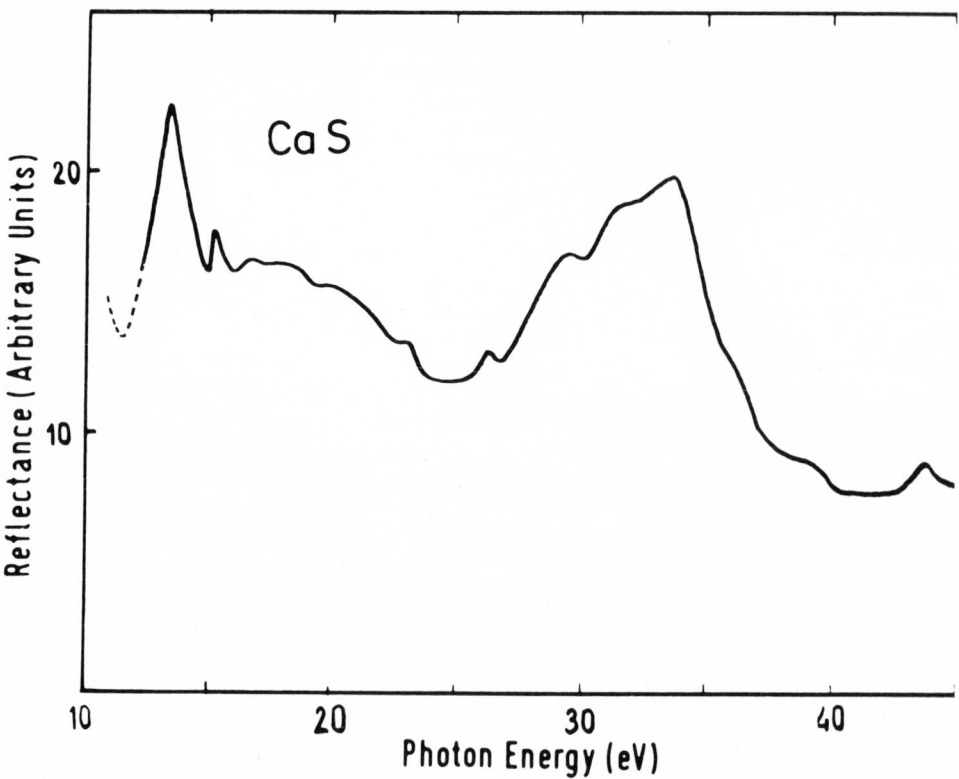

Fig. 1   The reflection spectrum for the compressed tablet of
         CaS (97%) irradiated with synchrotron radiation DESY
         from 13 to 45 eV.

10.8 eV, for SrS:Mn at 10.1 eV and for BaS:Mn at 9.4 eV, energies
corresponding to double the width of the band gap ($h\nu \geq E_g$). This
increase may be due to photon multiplication (8).

In the excitation spectrum of CaS:Bi beyond the absorption
edge, the quantum yield is also nearly constant and a sharp
increase in quantum efficiency occurs at 10.8 eV (Fig. 2).

In the excitation spectrum of CaS:Bi luminescence in the
14-17 eV region a new plateau is observed and the next increase
occurs at a photon energy corresponding to three times E_g. At
energies divisible by E_g some peculiarities in the spectra were
observed, a particularly sharp rise is observed at 58 eV. In
addition, the structure in the excitation spectrum correlates
with the structure in the reflection spectrum. Thus, the
excitation is ineffective in some reflection maxima, for example
at 10 eV. The kinetics of excitation changes as the energy
increases; namely, the contribution of the instantaneous component

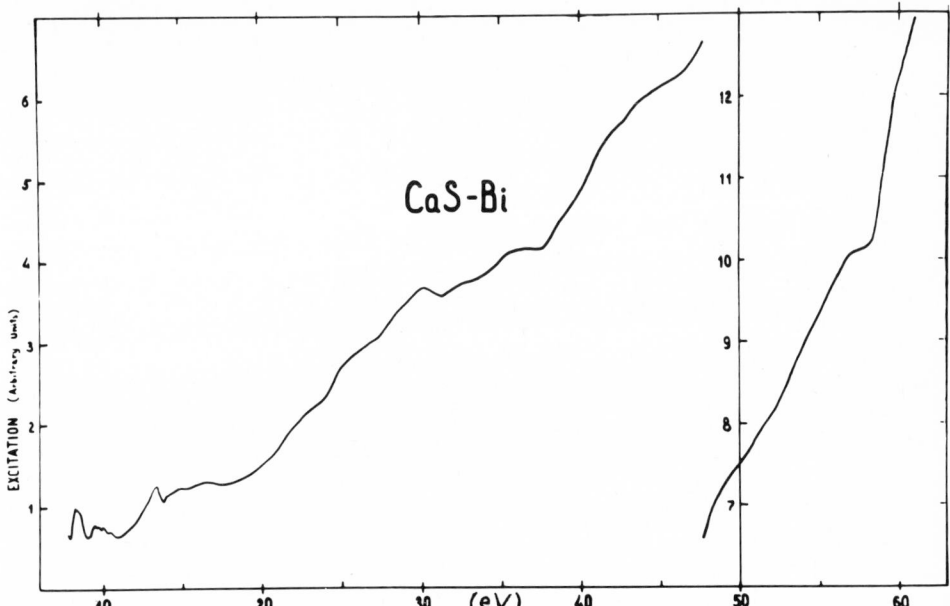

Fig. 2   Excitation spectrum of the steady-state luminescence
         (intensity of the luminescence/photon) for phosphor
         CaS:Bi (2×10^{-4} g/g), irradiated with synchrotron
         radiation DESY and Lebedev Institute's synchrotron.

increases while the time of the build-up of luminescence falls
to 60 sec at 20 eV.

ACKNOWLEDGEMENT

        The author is most thankful to Prof. R. Haensel, Prof. M.
Skibowski, Dr. E. Koch (Hamburg) for the opportunity of taking
measurements with the synchrotron radiation DESY and for their
helpful cooperation, to Prof. V. A. Petukhov (Moscow) for his
help and encouragement, to Dr. M. Alsallu (Tartu) for the samples
of CaS and to Dr. E. R. Ilmas (Tartu) for valuable discussion.

REFERENCES

1. V. L. Levshin, Proc. Intern. Conf. on Lum., 1966,
    Budapest, p. 39.

2. V. L. Levshin, V. V. Mikhailin, Proc. Intern. Conf. on Lum.,
    1966, Budapest, p. 1508.

3. B. K. Novosadov, L. K. Saulewitch (Kulman), D. T. Swiridow,
    Ju. F. Smirnov, Dokl. Akad. Nauk SSSR, 184, 82 (1969).

4. L. Kulman, V. Mikhailin, Tr. IFA Tartu (in press).

5. M. L. Cohen, P. J. Lin, D. M. Rossler, W. C. Walker, Phys. Rev.,
    155, 992 (1967).

6. G. A. Saum, E. B. Hensley, Phys. Rev., 113, 1019 (1959).

7. V. L. Levshin, V. V. Mikhailin, L. K. Saulevitch, Izv. Akad.
    Nauk SSSR, 33, 947 (1969).

8. E. R. Ilmas, Ch. B. Lushchik, Tr. IFA, 34, 5, 1966.

THEORY OF ENERGY MIGRATION IN A CRYSTAL

A. S. Davydov and A. A. Serikov

Institute of Physics, Kiev, USSR

ABSTRACT

Energy transfer phenomena in molecular systems and in solids
are investigated theoretically, including and not including relaxation processes. The dependences of energy transfer probability
on time and on the distance between impurity molecules of a crystal
are considered. Different methods of investigating energy transfer
processes are compared taking into account the effects of relaxation.

————————

At present energy migration in solids, polymers, bio-systems
are of considerable interest. The problem of energy migration at
the microscopic level is very complex as a consequence of the time
invertibility of the quantum mechanical equations. To simplify
the approach one often uses approximate methods, in particular the
treatment of quantum transitions using time-dependent perturbation
theory.

Förster's pioneer papers (1-3) were also based on the use
of the theory of quantum transitions. In these papers the excitation energy transfer between the donor and acceptor molecules
in the crystal matrix was investigated. In the Förster theory
the problem was considered to first order in perturbation theory.
The probability of excitation transfer per unit time was defined
via the probability of the excited donor state decay. Such a
definition holds true only in presence of one decay channel and
with an a priori assumption about the exponential character of
decay. As a rule, both these assumptions are not satisfied.

The Förster theory was further generalized by Dexter and

others (4,5) and by Galanin (6). However, the results of these
papers are limited by the use of first order perturbation theory
and have restricted range of applicability.

Another approach to a theoretical description of the energy
transfer phenomenon is based upon taking into account the relaxation processes which arise from the interaction of the donor and
acceptor molecules with the environment. This environment is
characterized by an infinite number of the degrees of freedom.
A phenomenological treatment of relaxation, in which terms describing the time variation of the probability amplitude of
various states of the system or the variations of the elements of
its density matrix were introduced, was carried out by some
authors (7-15). The main purpose of these papers was to define
the decay probabilities of the excited states of a donor molecule
and elucidate the range of applicability of perturbation theory.

For a three-level system (donor-acceptor) Burshtein (8) showed
that the damping law of donor excitation is different from exponential: $W_e(t) = W_e(0)\exp(-t/\tau)$. To characterize nonexponential damping
Burshtein introduced the value:

$$\tau_{eff} \equiv \int_0^\infty \frac{W(t)}{W(o)} dt$$

which defines the effective rate of decay $W(t)$. It was shown that
the approximate exponential decay should be observed at no resonance and at large distances between the donor and acceptor. With
strong interaction between the donor and acceptor the value $1/\tau_{eff}$
is essentially different from the decay rate calculated by means of
perturbation theory. Perlin and collaborators (9), Trifonov and
collaborators (10-12), Agabekyan (13), Kustov and Surogin (14)
obtained similar results.

Some papers [Zusman, Burshtein (15), Golubov, Konobeev (16),
Agabekyan (17)] have studied a more complicated problem in the
kinetics of the decay of the donor excited state which is surrounded by many acceptors. They used the approximation which corresponds
to the summation of probabilities of the excitation transfer to
each of them.

The above-mentioned papers have shown that the kinetics of
excitation transfer between the impurity molecules in a crystal
is rather complicated and requires further theoretical investigations, even in the simplest case of the system involving the
donor and one acceptor.

In our recent paper (18) a new method has been developed to
investigate the energy transfer process in the simplest dynamical

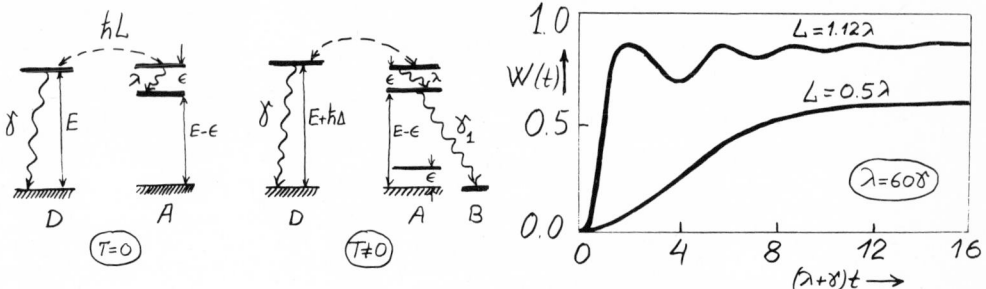

Fig. 1   Energy levels and transitions for donor-acceptor energy
         transfer.

Fig. 2   Time dependence of the excitation of the acceptor.

system (donor-acceptor) which interacts with a thermostat and a
field (Fig. 1, T = 0). This method is based on using the equations
for the density matrix $\rho_a(t)$ of the dynamical system which is
obtained when one averages the density matrix $\rho(t)$ of the whole
system over the equilibrium state of a thermostat and a field. In
this case the processes conditioned by the interaction with the
thermostat and the field are taken into account directly in the
Hamiltonian: $H = H_a + H_T + H_{rel}$ which defines the time evolution:

$$\frac{\partial \rho(t)}{\partial t} = \frac{1}{i\hbar}[H,\rho(t)] \tag{1}$$

of the density matrix $\rho(t)$ of the whole system. Where $H_a=H_0+H_{int}$
is the energy operator of the dynamical system "a"

$$H_0=ED^+D + (E-\epsilon)A^+A + \epsilon C^+C, \quad H_{int}=\hbar L(DA^+C^+ + D^+AC);$$

E, E-ϵ are the excitation energies of the donor and acceptor respectively; ϵ is the vibrational excitation energy of the acceptor
which can go into the thermostat. D^+, A^+ and C^+ are the creation
operators of the electronic excitation of the donor and acceptor
and the vibrational excitation of the acceptor. The operators
H_T and H_{rel} which characterize the state of the thermostat and the
field and their interaction with the dynamical system were defined
in (18) from considerations of the model.

     After averaging Eq. 1 over the states of the field and
thermostat at zero temperature, for the density matrix of the
dynamical system $\rho_a(t)=Sp_a\rho(t)$ the following equation was obtained.
Equations of this type were also used in the papers (19,20) to investigate the time evolution of the simplest quantum systems which
interact with the thermostat. The equation is:

$$\frac{\partial \rho_a(t)}{\partial t} = \frac{1}{i\hbar} [H_{int}, \rho_a] - \frac{\lambda}{2}\{[C^+C,\rho_a]_+ -2C\rho_a C^+\} - \frac{\gamma}{2}\{[D^+D,\rho_a]_+ -2D\rho_a D^+\},$$

$$(2)$$

where λ and γ are the parameters which define the vibrational acceptor energy transfer into thermostat and the emission of excitation donor energy: $[x,y]_+ \equiv xy + yx$.

If one introduces the operators

$$M(1)=D^+DAA^+CC^+; \ M(2) = \frac{i}{\sqrt{2}} (D^+AC-DAC^+); \ M(3)=DD^+A^+AC^+C; \ M(4) =$$

$$DD^+A^+ACC^+; \ M(5)=DD^+AA^+CC^+$$

which characterize various states of the dynamical system, then by means of Eq. 2 one can obtain the set of equations which define the time evolution of the probabilities

$$W_\ell(t) = Sp_a\{\rho_a(t)M(\ell)\}$$

of the corresponding states. When this system is solved with the initial condition $W_\ell(0) = 1$; $W_\ell(0) = 0$ ($\ell=2,3,4,5$) we obtain the explicit expression for the values $W_\ell(t)$ as the functions of the parameters L, λ and γ (see (18)).

Fig. 2 shows the time dependence of the probability $W(t) = W_3(t) + W_4(t)$ of the acceptor molecule excitation for two values of the set of parameters L, λ, γ. As it is seen from the plots, the energy transfer rate $dW(t)/dt$ has a constant value only in a small interval of time.

For the probabilities $W(\infty)$ which characterize efficiency of the energy transfer (quantum yield) the following expression was obtained:

$$\frac{W(\infty)}{K} = \frac{\alpha^2}{1+\alpha^2} \simeq \begin{cases} \alpha^2, & \text{if} & \alpha^2 \ll 1; \\ 0.5(\alpha-0.1), & \text{if} \ \ 0.2 \lesssim \alpha \lesssim 1.7; \\ 1-\alpha^{-2}, \text{if} & \alpha^2 \gg 1, \end{cases} \quad (3)$$

where

$$K \equiv \lambda/(\lambda+\gamma); \ \alpha \equiv 2L/(\lambda\gamma)^{1/2} \quad (4)$$

Thus, the energy transfer probability is proportional to the square of resonance interaction energy $\hbar L$ (the decay law R^{-6} for the dipole-dipole interaction) only under the condition:

$$|L| << (\lambda\gamma)^{1/2}$$

When the inverse inequality is satisfied the probability is practically independent of L. At the intermediate values , the probability of transfer is approximately proportional to the first degree of resonance interaction energy.

The decay law of the donor molecule excited state for the parameters $L=2\lambda$, $\lambda = 50\gamma$ is shown in Fig. 3. We see that in this case it is essentially nonexponential. If, according to Burshtein; we introduce the effective decay rate $1/\tau_{eff}$ we get

$$1/\tau_{eff} \equiv [\int_{o}^{\infty} \frac{W_1(t)}{W_1(0)} dt]^{-1} = \gamma + \frac{4\lambda L^2}{\lambda(\lambda+\gamma)+4L^2}$$

The theory developed in (18) can be generalized for the case of nonzero temperatures of the thermostat. Trofimov, Khomenko and the authors of the report have considered the dynamical system shown in Fig. 1 $(T\neq0)$, for temperatures at which only one-phonon vibrational excitations are essential. For the luminescence probability, from the electronic excited state of the acceptor molecule (transition to the state B), the following expression is obtained

$$W_B(\infty) = \frac{K\alpha^2}{1+\alpha^2+K\nu\frac{\gamma}{\gamma_1}\alpha^2+(2K\Delta/\lambda)^2}$$

K and α are defined according to (4); $\nu \equiv 1/[\exp(\epsilon/kT)-1]$; $\hbar\Delta$ is the difference of the excitation energies of the donor molecule electronic state and the vibronic state of acceptor molecule. Comparing (3) and (5) for $\Delta=0$ we see that with an increase in temperature the efficiency of the energy transfer decreases.

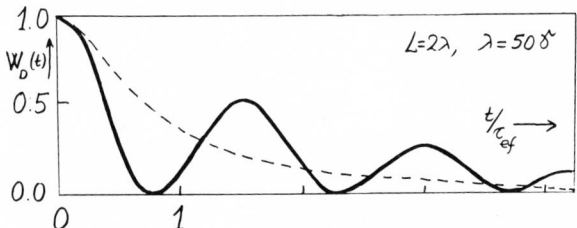

Fig. 3   Decay of the excited state of the donor molecule.

REFERENCES

1. Th. Förster, Naturwissenshaft, 33, 166 (1946).

2. Th. Förster, Ann. Phys., 2, 55 (1948).

3. Th. Förster, L. Naturforshung, 4a, 321 (1949).

4. D. L. Dexter, J. Chem. Phys., 21, 836 (1953).

5. D. L. Dexter, J. H. Schulman, J. Chem. Phys., 22, 1063 (1954).

6. M. D. Galanin: Resonance Energy Transfer in Luminescent Solutions, Proceedings of the Lebedev Physical Institute of the USSR Academy of Sciences, v.XII.

7. R. P. Frosh, J. W. Robinson: J. Chem. Phys., 37, 1962 (1962).

8. A. I. Burshtein: Teor. i Exper. Khim., I, 563 (1965).

9. V. Ya. Gamurar, Yu. E. Perlin, B.S. Tsukerblat: Fiz. Tverd. Tela, 11, 1193 (1969).

10. E. D. Trifonov, V. L. Shekhtman, Fiz. Tverd. Tela, 2984 (1969).

11. K. Poiker, E. D. Trifonov: Fiz. Tverd. Tela, 10, 1705 (1968).

12. E. D. Trifonov: Izv. Akad. Nauk SSSR, ser. Fiz., 35, 1330 (1971).

13. A. S. Agabekyan, A. O. Melikyan, Proceedings of the 1st International Seminar on Radiationless Energy Transfer in Condensed Media, Erevan, 1970, p. 5.

14. E. F. Kustov, L. I. Surogin, ibid, p. 61.

15. L. D. Zusman, A. I. Burshtein, Zh. Prikl. Spektr., XV, 124 (1971).

16. S. I. Golubov, Yu. V. Konobeev, Fiz. Tverd. Tela, 13, 3185 (1972).

17. A. S. Agabekyan: Opt. i Spektr., 30,449 (1971); 29, 71 (1970).

18. A. S. Davydov, A. A. Serikov, Phys. Stat. Sol., 51, 57 (1972).

19. J. R. Shen, Phys. Rev., _155_, 921 (1967).

20. B. Ya. Zeldovich, A. M. Perelomov, V. S. Popov: Zh. Exper. i
     Teor. Fiz., _55_, 589 (1968).

NEW TYPES OF LOCALIZED EXCITONS IN NAPHTHALENE CRYSTAL WITH IMPURITIES

N. I. Ostapenko*, V. I. Sugakov**, M. T. Shpak*

*Institute of Physics, Ukrainian SSR Academy of

Sciences, Kiev, USSR and

**Physics Department, Shevenko State University

Kiev, USSR

ABSTRACT

The effect of the impurity concentration of indole and thionaphthene in naphthalene crystals on localized excitons is studied both experimentally and theoretically through luminescence and absorption spectra. The observed structure is accounted for by the presence of clusters of perturbed host molecules.

THEORY

Molecules of indole and thionaphthene are similar to those of naphthalene. It is highly probable that as impurities in naphthalene they occupy lattice sites. For a random distribution of these impurities among the sites, groups (clusters) of differently positioned impurity molecules and crystal molecules perturbed by them are expected to exist in different regions of the crystal. Therefore, the spectrum of the impure crystal will appear as an overlap of cluster spectra for a wide range of impurity molecule separations and the spectra of the perturbed crystal molecules (Fig. 1).

The method of calculating the energy levels arising from doping the crystal with impurities which create pair centers has been given (8). This method is based on the following assumptions: (1) Energy levels of the impurity are far from the crystal exciton

282

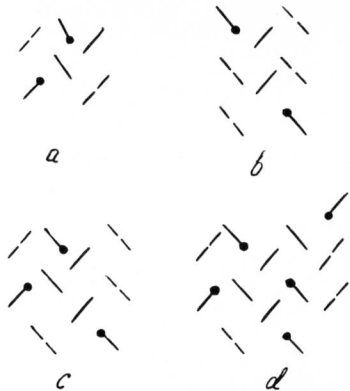

Fig. 1 Molecular arrangement in the ab-plane of naphthalene crystal
        for the $L_{2,n}$ center (a), $L_{2,e}$ center (b), L_3 center (c),
        L_4 center (d) (-- is naphthalene molecule, ——• is impurity
        molecule, —— is impurity perturbed molecule).

levels. (2) The impurity shifts the excitation energy of one nearest neighbor crystal molecule by the quantity Δ (the values of Δ
for thionaphthene, indole and benzofuran in naphthalene are determined in (6). (3) The values of the diagonal Green function $G_{0,0}$
were calculated from the spectral data of naphthalene in deuter-
onaphthalenes (9); the values of the nondiagonal Green functions
are smaller and the results are not sensitive to them, therefore,
they were calculated by an approximate dispersion law taking into
account only the resonance interaction between the nearest neighbor
molecules in non-equivalent positions.

     Application of this method for the energy calculation of local
excitons (6, 7) and pair centers (8) has confirmed its validity.
It can also be used in calculations of the excited states of more
complex clusters.

     Let us denote different configurations of impurity and dis-
torted molecules by L_m, where m is the number of distorted molecules
(Fig. 1). For a more detailed description of the cluster state we
shall use the notation:

$$L_m \binom{p_1\beta_1,\ p_2\beta_2 \ldots,\ p_m\beta_m}{n_1\alpha_1,\ n_2\alpha_2 \ldots,\ n_m\alpha_m}$$

where the upper elements correspond to the coordinates of the impurity molecules; the lower, the perturbed crystal molecules.
Thus, the center given in Fig. 1a is denoted as

$$L_2 \binom{0,\ (a+3b)/2}{(a+b)/2,\ a+b}$$

As mentioned above, the investigation of L_2 centers is done in (8). Omitting detailed calculations we write expressions from which the energy spectra of the more complex centers L_3 (Fig. 1c) and L_4 (Fig. 1d) are determined.

In the general case for the L_3 center there are three levels, the positions of which are found from the following three conditions. For the level $L_{3\alpha}$ (α, β, γ indicate excited states):

$$\Delta = 1/G_{0,0}(1 - G_{0,(a+b)/2}^{'2} \quad - G_{0,b}^{'}) \tag{1}$$

For the levels $L_{3\beta}$ and $L_{3\gamma}$

$$\Delta = 1/G_{0,0}\{1 - G_{0,(a+b)/2}^{'2} + (\tfrac{1}{2}) G_{0,b}^{'} \mp$$
$$[(1/4)G_{0,b}^{'2} + 2G_{0,(a+b)/2}^{'2}]^{1/2}\} \tag{2}$$

and, for the L_4 center (Fig. 1d) where the one doubly-degenerate level has a position found from an equation similar to Eq. 1, and two nondegenerate levels $L_{4\beta}$ and $L_{4\gamma}$ for which

$$\Delta = 1/G_{0,0}\{1 - G_{0,(a+b)/2}^{'2} \mp 2G_{0,(a+b)/2}\} \tag{3}$$

In Eqs (1-3) $G_{0,n_\alpha}^{'}$ is the Green function, $G_{0,n_\alpha}^{'} = G_{0,n_\alpha}^{'}/G_{0,0}$

$$G_{0,n_\alpha} = 1/N \sum_{\mu\kappa} \frac{C_\alpha^{\mu\kappa} C_{\alpha 0}^{\mu\kappa} \exp(i\kappa n_\alpha)}{E - E_\mu(\kappa)} \tag{4}$$

where $E_\mu(\kappa)$ is the exciton dispersion law in the μ-th band, $2N$ is the number of molecules in the crystal, the C's are the Fourier coefficients of the transformation from the coordinate to the momentum representation.

Eq. (1-3) are approximate. For their simplification, the terms in $G_{0,n_\alpha}^{'2}$ of order $n_\alpha > (a + b)/2$ are omitted. Since Green function values rapidly decrease as n_α increases, the terms omitted are much less than unity for levels more than 10 cm^{-1} from the band edge. Taking into account the omitted terms leads to corrections to the level position on the order of 1-2 cm^{-1}, within experimental and theoretical error.

Such a Green function calculation method is described in (8). Substituting their values into Eqs. 1-3, we obtain the dependence of the local level position on the shift of the molecule excitation energy Δ (Fig. 2)

EXPERIMENT AND DISCUSSION

Experimental investigations were made using monocrystalline

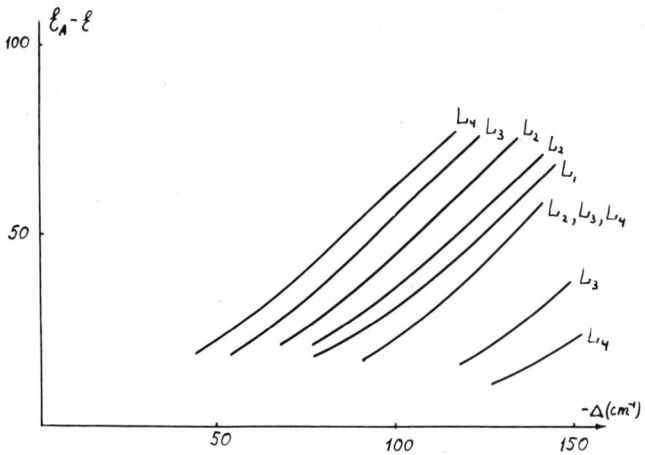

Fig. 2   Position of the 0-0 bands of L_2, L_3 and L_4 centers in the
         naphthalene crystal as a function of Δ. A similar plot
         for localized excitons is shown by the curve L_1. ε_A is
         the frequency of the A-band in the luminescence spectrum
         of the naphthalene crystal, ε is the frequency of the 0-0
         band in the luminescence spectrum of a particular center.

naphthalene films doped with indole and thionaphthene. Impurity
concentration, c, was varied in the range 1 to 10%.

     The luminescence spectra of naphthalene crystals with different
indole content are shown in Fig. 3. It is seen that the intensity
of the naphthalene luminescence (A-band) decreases with an increase
of indole concentration and it is practically gone at $c \sim 1\%$; in this
case on the long-wavelength side of the pure electronic transition
band of the local excitons ($\nu = 31462$ cm^{-1}) new bands appear with
maxima near 31458, 31449, 31444 and 31438 cm^{-1}. These bands are
the first bands of the series whose spacings coincide with the frequencies of the naphthalene intramolecular vibrations. Positions
of new band maxima do not depend on the impurity concentration as
only their relative intensities change with the concentration, the
intensities increasing more for bands further from the A-band.

     Polarization investigations of the luminescence spectra of
indole-doped naphthalene single crystals showed that the maximum
intensity ratio in the a and b-components is different for different
bands. For the above mentioned bands, the band at $\nu = 31449$ cm^{-1} is
the one most strongly polarized along the a-direction of the crystal;
the band at $\nu = 31444$ cm^{-1}, the most weakly polarized.

     With thionaphthene as an impurity, new bands in the lumines-
cence spectrum appear at the following frequencies: 31450, 31447,

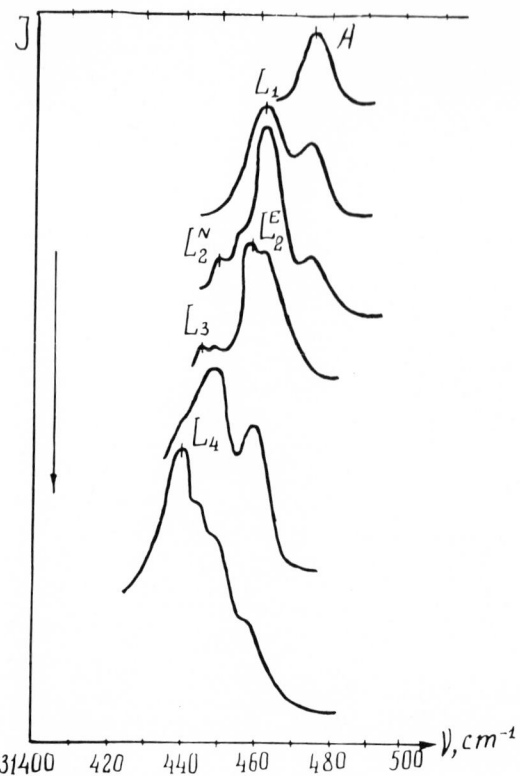

Fig. 3   Luminescence spectra for naphthalene crystals (region of
         pure electron transition) with different indole concentra-
         tion.   Increase of concentration is shown by the arrow.
         A indicates the 0-0 band of the naphthalene luminescence
         exciton series, L_1 indicates the 0-0 band of localized exci-
         tons, L_2^n shows the 0-0 band of pair centers with the per-
         turbed molecules in nonequivalent positions, L_2^e shows the
         0-0 band of pair centers with perturbed molecules in the
         equivalent positions, L_3 is the 0-0 band of centers consist-
         ing of three perturbed molecules, L_4 corresponds to 0-0
         band of centers consisting of four perturbed molecules.
         Intensity is plotted in arbitrary units.

31444, 31436, 31425 and 31418 cm^{-1}. Emergence of new naphthalene
bands with the impurity concentration increase can be explained on
the assumption that different type centers are formed in the crystal,
e.g., an isolated L_1, pair L_2, triplet L_3, etc. Complex centers can
also be formed both by equivalently positioned perturbed molecules
($L_{2,e}$; $L_{3,e}$; etc.) and by molecules in nonequivalent positions

($L_{2,n}$; $L_{3,n}$; etc.). Change of concentration changes the distribution of the different centers. At low concentrations ($c \ll 1\%$) mainly L_1-centers are formed. At higher concentration ($c \sim 1\%$) the number of L_2-centers increases, then L_3-centers appear, etc.

To assign the observed bands to the appropriate centers we compared theoretical and experimental values of the band positions and the magnitude of their polarization ratios. Theoretical values were determined from Eq. 1-4 for indole ($\Delta = -72$ cm^{-1}) and for thionaphthene ($\Delta = -97$ cm^{-1}). It follows from the calculations that complex centers lead to the splitting of several levels from the exciton bands. The polarization ratios calculated for them are in a good agreement with experiment. Comparison between the results of theory and experiment and interpretation of the local exciton bands are given in Table 1.

Table 1

Local Exciton Bands
A-Band Frequency minus Center's 0-0 Band Frequency

Center	Naphthalene with Indole		Naphthalene with Thionaphthene	
Model	Theory	:Experiment	Theory	:Experiment
$L_{2+,e}$, $L_{3\alpha}$, $L_{4\alpha}$	-10	-8,-9,-10	-20	-20,-23,-25
L_1	-13	-13	-28	-28
$L_{2-,e}$	-19	-17	-33	-31
$L_{2-,n}$	-23	-26	-42	-38
$L_{3\gamma}$	-30	-31	-50	-48
$L_{4\gamma}$	-37	-37	-59	-57

It can be seen from Table 1 that the theoretical values of the band positions show good agreement with experiment. It is worthwhile noting that agreement is good for L_2-centers with perturbed molecules in the equivalent positions $L_{2,e}$ even though in calculations the matrix element of the resonance interaction between such molecules was assumed small enough to neglect. It appears that in the resonance interaction between such molecules, nonequivalent molecules play an essential role. Such a process is possible for states close to the exciton bands. Agreement between theory and experiment indicates that this process of indirect interaction is more important in this case than the direct resonance interaction between molecules.

In the indole-doped naphthalene crystal, formation of L_2, L_3 and L_4 centers is observed in the luminescence spectra at lower concentrations of impurity than in the thionaphthene-doped naphtha-

lene. This could be explained by a lower solubility for thionaphthene than for indole. However, this is contradicted at low
thionaphthene concentrations by the intense luminescence due to
transitions from local excitons. Evidently the more effective migration of energy to the local exciton levels in the indole-doped
than in the thionaphthene-doped naphthalene is significant. According to Förster (10) and Dexter (11) the probability of energy transfer between two identical states is proportional to the fourth power
of the dipole transition matrix in the excited state. Since the
level created by indole is nearer to the exciton band bottom than
that created by thionaphthene, the dipole transition to this level
is larger. Therefore, the energy transfer over local levels of
indole-doped naphthalene is more intense than over local levels of
thionaphthene-doped naphthalene.

REFERENCES

(1) M. T. Shpak, N. I. Sheremet, sbornik Opt. i Spektroskopiya ,
vol. 1, Luminescence , Academy of Sciencies of the USSR,
Moscow, (1963) p. 110.

(2) A. Pröpstl and N. C. Wolf, Zeits. Naturforsch., 18a, 724 (1963).

(3) V. I. Sugakov, Opt. i Spektroskopiya , 21, 574 (1966).

(4) I. I. Osadko, Fiz. Tv. Tela, 11, 441 (1969).

(5) N. I. Ostapenko, M. T. Shpak, Phys. Stat. Solidi, 36, 515
(1969).

(6) V. I. Sugakov, Opt. i Spektroskopiya , 28, 695 (1970).

(7) N. I. Ostapenko, V. I. Sugakov, M. T. Shpak, Phys. Stat.
Sol., 45, 729 (1971).

(8) N. I. Ostapenko, V. I. Sugakov, M. T. Shpak, J. Luminescence
4, 261 (1971).

(9) E. F. Sheka, Fiz. Tv. T., 12, 1167 (1970).

(10) T. Förster, Ann. Physik, 2, 55 (1948); Zeits. Naturforsch.
4a, 321 (1949).

(11) D. L. Dexter, J. Chem. Phys., 21, 836 (1953).

LUMINESCENCE AND PHOTOCONDUCTIVITY OF NITRONAPHTHALENES

Frank Vogel[†] and Nicholas E. Geacintov

Department of Chemistry and Radiation and Solid State Laboratory[††]

New York University, New York, New York 10003

ABSTRACT

This work reports the first observation of phosphorescence from polycrystalline nitronaphthalenes. The phosphorescence spectrum is red-shifted from its molecular counterpart; its relative intensity decreases as the temperature is increased from 77°K, exhibiting a relatively weak activation energy. The emission is attributed to triplet ($\pi^* \leftarrow \pi$) states whose lifetimes at room temperature are 35µsec and 0.4 µsec for 1,5-dinitro-naphthalene and 1,8-dinitronaphthalene respectively. At room temperature in 1,5-dinitronaphthalene single crystal platelets, the photoconductivity observed is extrinsically generated, holes being more efficiently trapped than electrons, and is tentatively attributed to mobile excitons which migrate to the electrode and dissociate there producing holes and electrons.

INTRODUCTION

In recent years, electronic energy conversion processes in aromatic solids have been studied extensively (1,2). However, little is known about the effects of substituent groups on the luminescence, exciton motion, and carrier generation and transport mechanisms in these solids. Nitroaromatics such as the different nitro derivatives of naphthalene provide an interesting example of the effects of $-NO_2$ groups on electronic

properties of organic aromatic solids. In this paper we report
our investigations on the luminescence and photoconductivity of
two isomers of dinitronaphthalene, 1,8-dinitronaphthalene and in
particular 1,5-dinitronaphthalene.

LUMINESCENCE OF NITRONAPHTHALENES

The molecular electronic states of the nitronaphthalenes,
such as in solutions of EPA (Ether:Isopentane:Ethanol in 3:3:5
proportion) have been previously characterized (3). In contrast
to naphthalene (2), the nitronaphthalenes do not exhibit room
temperature emission. However, at 77°K in EPA, they exhibit
phosphorescence which has been characterized as a triplet to
singlet ground state ($\pi^* \leftarrow \pi$) transition(3). In EPA at 77°K, the
phosphorescence has a lifetime of approximately 10^{-2} seconds and
the emission spectra are similar to that of the naphthalene
triplet, but red-shifted by 2000 - 4000 cm^{-1}(3,4). Only
phosphorescence, and no fluorescence, is observed with molecular
nitronaphthalenes. This implies that the singlet-triplet
intersystem crossing rate is much greater than in the pure
aromatics, which is due to an enhanced spin-orbit coupling induced
by the nitro groups(4). As in the case of the pure aromatics, the
electronic structure of solid nitronaphthalenes is derived
from the molecular states. One may therefore conclude that,
in contrast to the aromatics, the solid nitronaphthalenes will
essentially have only triplet excited states.

We found that solid 1,5-dinitronaphthalene, 1,8-dinitro-
naphthalene, and 1-nitronaphthalene exhibit a luminescence at
low temperature and their lifetimes and spectra were determined.
In the lifetime measurements, the phosphorescence was produced
by a nanosecond nitrogen pumped dye laser tuned to approximately
3500 Å and recorded by an Intertechnique Didac 800 signal
averager. The lifetimes, calculated by employing a successive
approximations curve-fitting computer program, are summarized
in the Table. The temperature dependence of

TABLE

	1,5-dinitronaphthalene	1,8-dinitronaphthalene
Phosphorescence Activation Energy	600 ± 120 cm^{-1}	960 ± 160 cm^{-1}
Triplet Lifetime (EPA, 77°K)	110 msec (5)	80 msec (6)
Triplet Lifetime (Solid, 77°K)	60 ± 5 msec	26 ± 5 msec
Estimated Triplet Lifetime (Solid, 300°K)	35 μsec	0.4 μsec

the phosphorescence of the polycrystalline samples was also
measured in the range of 77°K to approximately 120°K. As the
temperature increases from 77°K, the luminescence decreases.
Above approximately 120°K, the phosphorescence intensity is lower
than the sensitivity threshold of the apparatus. The results
for 1,8-dinitronaphthalene are illustrated in Fig. 1 and
summarized for 1,8-dinitronaphthalene and 1,5-dinitronaphthalene
in the Table. At room temperature the triplets undergo
radiationless decay to the ground state, as evidenced by the
absence of room temperature luminescence. As the temperature
is decreased below 120°K, the phosphorescence appears and becomes
more intense (Figure 1), exhibiting an activation energy of
600 ± 120 cm^{-1} in 1,5-dinitronaphthalene and 960 ± 160 cm^{-1} in
1,8-dinitronaphthalene. Below approximately 88°K, the
phosphorescence intensity no longer increases as rapidly, indicating the dominance of other pathways for radiationless decay. At
77°K, the lifetimes of the triplet excited states in the solid
are 60, 26, and 17 msec ($\pm$ 5 msec) for 1,5-dinitronaphthalene,
1,8-dinitronaphthalene, and 1-nitronaphthalene, respectively.
These values are less than the reported values for the rigid matrix
samples, i.e., 110, 80, and 23 msec, respectively (5-7). Also,

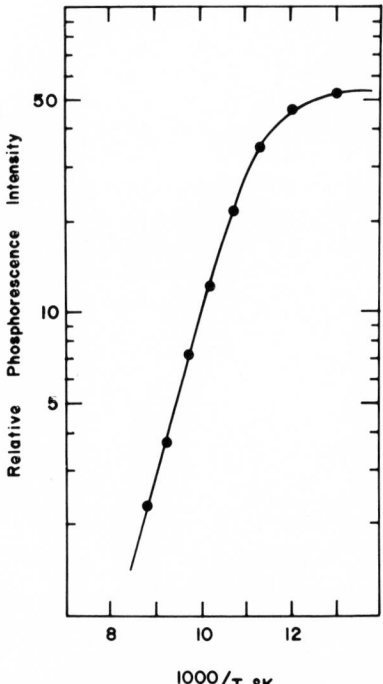

Fig. 1   Temperature Dependence of Phosphorescence Intensity for
         1,8-dinitronaphthalene

unlike with the pure aromatics, we have observed that the polycrystalline phosphorescence spectra for 1,5-dinitronaphthalene and
1,8-dinitronaphthalene are red-shifted approximately 800 cm^{-1} and
1200 cm^{-1}, respectively, from their EPA solution counterparts. This
red-shift is much larger in nitronaphthalenes than in the pure
aromatics such as anthracene (2) and may be attributed to greater
intermolecular interaction terms for triplet states in the nitronaphthalenes (4). From the lifetimes at 77°K and the temperature
dependences of the phosphorescence, we estimate that the triplet
excited state lifetimes at room temperature in 1,5-dinitro-
naphthalene and 1,8-dinitronaphthalene are 35µ sec and 0.4 µsec,
respectively. Because of the uncertainty in the phosphorescence
activation energy, the lifetimes reported are probably correct
to within a factor of three only. The lifetimes in the solid at
77°K are shorter than the EPA values at 77°K. As in anthracene,
this may be due to the quenching of the triplet excitons at
impurity sites (8). In order to determine if mobile triplet
excitons exist in nitronaphthalenes, we investigated the
photoconductive properties of 1,5-dinitronaphthalene. This
crystal was chosen because single crystal platelets can be grown.

PHOTOCONDUCTIVITY OF 1,5-DINITRONAPHTHALENE

Photoconductivity techniques have been utilized extensively
within the last decade to elucidate some of the excitonic
properties, e.g., the diffusion length, in organic crystals (9-12).
For example, in anthracene, light absorbed as deep as 2000 A
within the crystal can generate mobile species which can migrate
to the crystal-electrode interface and inject positive carriers
(13). In 1,5-dinitronaphthalene we have observed that light
absorbed deep within the crystal can extrinsically generate
both electrons and holes with nearly equal efficiency when water
electrodes are employed. This is in contrast to the behavior
observed in anthracene, wherein electrons are extrinsically
generated with appreciably smaller efficiency than holes (16).

It has been demonstrated that in organic molecular crystals,
the extrinsic photocurrent may be expressed as (9):

$$i = I_oeA \left(\frac{\varepsilon\lambda}{\varepsilon\lambda + 1} \right) \qquad (1)$$

where ε is the extinction coefficient, λ is the diffusion length,
A is the efficiency of charge production for $\varepsilon \gg (1/\lambda)$, I_O is the
incident light intensity, and e is the electronic charge. Hence,
for $\varepsilon\lambda \ll 1$, the extrinsic photocurrent should be linearly
dependent on ε. This is the case in anthracene wherein the
extrinsic photoconductivity has been attributed to singlet
excitons with diffusion length $\lambda \sim 400$ A (13). In 1,5-dinitro-
naphthalene however, singlet excited states rapidly decay by

intersystem crossing to the triplet manifold due to the strong
spin-orbit coupling (4). Hence any extrinsic photoconductivity
observed may be tentatively attributed to the migration of triplet
excited states to the crystal-electrode interface where the
charge injection process occurs. The dependence of the photocurrent on the extinction coefficient, and thus the possible role
of excitons in the photoconduction mechanisms, may be determined
by varying the incident photon energy. For energies up to 4 eV,
the excitation spectra of both positive and negative photocurrents
in 1,5-dinitronaphthalene are approximately linearly dependent
on the absorption coefficient of the crystal (14), see Fig. 2.
In addition, at a photon energy where the crystal absorption is
strongly polarized (i.e., 340 mμ in 1,5-dinitronaphthalene, see
Fig. 3), the absorption depth can be varied conveniently by
using polarized incident light. Hence, by changing θ, the angle
between the electric vector of the incident polarized light,
and a crystallographic axis, one can vary the absorption depth
of the light (10). In the case of 1,5-dinitronaphthalene, we
observed that the photocurrent was linearly dependent on the
polarization of the incident light (Fig. 3). These experiments
provide evidence for the existence of mobile excitons in 1,5-di-
nitronaphthalene. From the approximately linear dependence of
the photocurrent on extinction coefficient, one may conclude that
$\epsilon\lambda \ll 1$ (see Eqn. 1). From the published (14) absorption
spectrum of the crystal (Fig. 2) one obtains the approximate
optical density at 340 mμ. By assuming that the thickness of
the crystals used in the absorption measurements were of comparable
thickness to the ones employed in our photoconductivity experiments
we estimated the extinction coefficient at 340 mμ to be at least
10^3/cm. This leads to an upper limit to the diffusion length,
λ of the excitons of 10^{-4} cm, with 10^{-5} cm as the most probable
value. Furthermore, from the extrapolated lifetime (τ) of ~10^{-5}
seconds at room temperature, and from the relationship between the
diffusion coefficient, D, and the diffusion length, $\lambda = (D\tau)^{1/2}$,
we estimate the diffusion coefficient of the triplet excitons to be
of the order of $10^{-4} - 10^{-5}$ cm^2/sec. This value of D is of
similar magnitude to values of D in purely aromatic crystals such
as anthracene (15).

At photon energies greater than 4 eV, it is observed that the
photocurrent at first increases more rapidly than the absorption
coefficient (Fig. 3) and then drops off rapidly, whereas the
absorption continues to become stronger as the wavelength is
decreased. The rapid increase in the photocurrent relative to the
absorption is characteristic of the onset of autoionizing
transitions, as observed in purely aromatic crystals. The ensuing
decrease in the relative efficiency of charge carrier generation
at still shorter wavelengths is consistent with the hypothesis

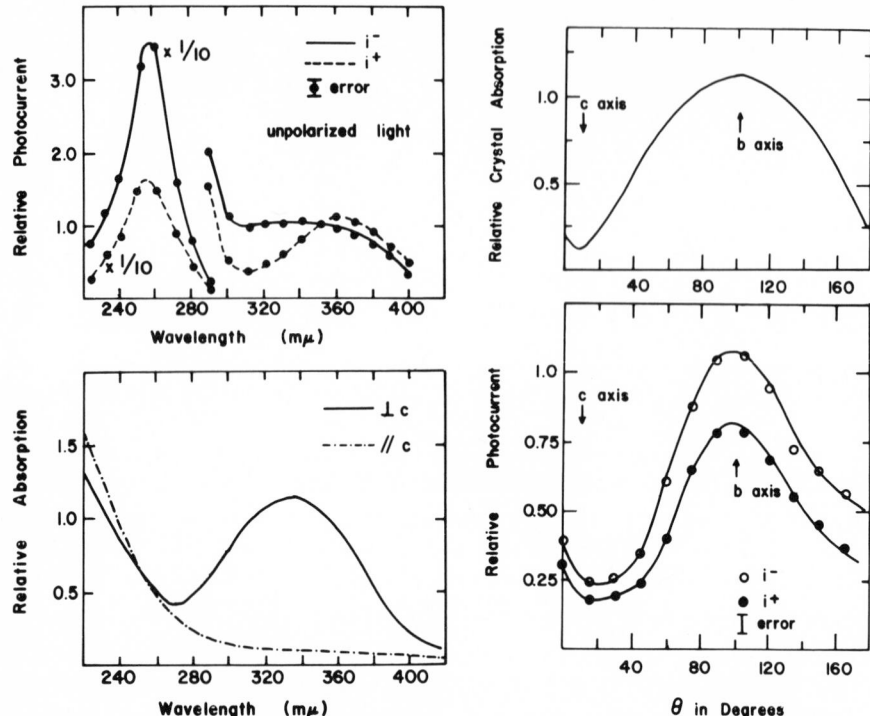

Fig. 2   The Relative Crystal Absorption (14) and Photocurrents
         as a Function of Wavelength in 1,5-dinitronaphthalene

Fig. 3   Relative Photocurrents and Crystal Absorption in
         1,5-dinitronaphthalene as a Function of Polarization
         of Exciting Light

         ($i^\pm$ is the photocurrent when the crystal is illuminated
         through the water electrode at a $\pm$ potential with respect
         to the dark electrode. Photocurrents ($\sim 10^{-12}$ amp)
         generated by 150 watt Xenon lamp; applied field $\sim$50 KV/cm.)

that the onset of lower-lying valence band transitions gives rise
to a lower ionization efficiency (12). If this interpretation
applies to 1,5-dinitronaphthalene, these results lead us to
conclude that the band gap in 1,5-dinitronaphthalene is approximately 4 eV.

SUMMARY AND CONCLUSION

    Our experiments indicate that in the solid nitronaphthalenes,
the excited electronic states are predominantly triplet excitons,
which are $(\pi^* \leftarrow \pi)$ in nature. At room temperature, the triplet
excitons, with a lifetime in the microsecond range, decay via
radiationless transitions to the ground state. Below approxi-

mately 120°K, the triplet excitons phosphoresce, the relative phosphorescence efficiency increasing as the temperature decreases, exhibiting an activation energy of approximately 800 cm^{-1}. At 77°K, the lifetimes are approximately 10^{-2} seconds. At room temperature, under an applied field of $\sim$50,000 V/cm, the triplet excitons in 1,5-dinitronaphthalene can migrate to a water-crystal interface and inject both electrons and holes. The triplet excitons have a diffusion length of approximately 10^{-4} - 10^{-5} cm and a diffusion coefficient of approximately 10^{-4} - 10^{-5} cm^2/sec at room temperature.

ACKNOWLEDGEMENT

This work was supported by the Army Research Office, Durham, N.C., grant number DA-ARO-D-31-124-72-G25.

The authors wish to thank Dr. Harry Fair and Professor Ferd Williams for interesting discussions and Mr. Charles Penn for his thoughtful technical assistance.

REFERENCES

† Also at Feltman Research Laboratories, Dover, N.J. 07834

†† Supported by the Atomic Energy Commission and the National
   Science Foundation

1. D. P. Craig in Physics and Chemistry of the Organic Solid
    State, eds. D. Fox, M. M. Labes, and A. Weissberger,
    Interscience Publishers, John Wiley and Sons, New York 1963,
    p. 585

2. D. P. Craig and S. H. Walmsley, Excitons in Molecular Crystals,
    W. A. Benjamin, Inc., New York 1968

3. R. Rusakowicz and A. C. Treata, Spectrochim. Acta 27A,
    787 (1971)

4. S. P. McGlynn, T. Azumi, and M. Kinoshita, Molecular
    Spectroscopy of the Triplet State, Prentice-Hall, Inc.,
    Englewood Cliffs 1969

5. D. S. McClure, J. Chem. Phys. 17, 905 (1949)

6. I. W. May, private communication

7. I. W. May and J. P. Kelso, Ballistic Research Laboratories
    Memorandum Report No. 2070 (1970)

8. P. Avakian and R. E. Merrifield, Mol. Cryst. and Liq. Cryst. $\underline{5}$, 37 (1968)

9. N. Geacintov, M. Pope, and H. Kallmann, J. Chem. Phys. $\underline{45}$, 2639 (1966)

10. N. Geacintov and M. Pope, J. Chem. Phys. $\underline{47}$, 1194 (1967)

11. N. Geacintov and M. Pope, J. Chem. Phys. $\underline{45}$, 3884 (1966)

12. N. Geacintov and M. Pope, J. Chem. Phys. $\underline{50}$, 814 (1969)

13. B. J. Mulder, Philips Research Report, Supplement 4 (1968)

14. M. Kojima, J. Tanaka, and S. Nagakura, Theoret. chim. Acta (Berl.) $\underline{3}$, 432 (1965)

15. F. Vogel, thesis, New York University (1971)

16. G. Castro and J. F. Hornig, J. Chem. Phys. $\underline{42}$, 1459 (1965)

A NEW METHOD FOR THE DIRECT DETERMINATION OF THE DIFFUSION

COEFFICIENT OF TRIPLET STATES: D OF TRIPLET ANTHRACENE IN

n-HEXADECANE AND n-HEXANE

B. Nickel and E. G. Meyer

Max-Planck-Institut für Biophysikalisch Chemie

Abteilung Spektroskopie

Göttingen-Nikolausberg, BRD

ABSTRACT

The diffusion coefficient D of triplet anthracene in n-hexa-decane (20 - 90 C) and n-hexane (25 C) was measured with the mod-ified Avakian-Merrifield method. For n-hexadecane large devia-tions from the Stokes-Einstein equation are found. With the experimental values of D the rate constants for diffusion-limited triplet-triplet annihilation are calculated for both solvents and compared with experimental values. In n-hexane triplet-triplet annihilation is slower than diffusion controlled.

INTRODUCTION

Avakian, Merrifield and Ern (1-3) have developed a method for the determination of diffusion coefficients of triplet excitons in molecular crystals. In a preceding paper (4) it has been pointed out that this method is not limited to the investigation of single crystals, but should be equally well suited to the determination of the diffusion coefficients of molecules in the triplet state in solution. A modification of the method has been proposed and, in order to test the applicability of the modified method, the diffusion coefficient of ^{3}pyrene* in glycerol was determined (5).

The aim of the present paper is to report on the determination

of the diffusion coefficient of ^{3}anthracene* in n-hexadecane and
n-hexane, to give an example for the application of the determined
diffusion coefficients, and to discuss the probable range of
application of the modified method.

EXPERIMENTAL METHOD

This method can, in principle, always be applied if a P-type
delayed fluorescence (DF) can be observed, i.e. if the interaction
of two triplet states T_1 leads to the formation of one excited
singlet state S_1 and if the fluorescence emission from this excited
singlet state can be observed:

$$T_1 + T_1 \xrightarrow{\gamma} S_0 + S_1$$

$$S_1 \longrightarrow S_0 + h\nu_{DF}$$

The intensity I_{DF} of the delayed fluorescence is proportional
to the square of the triplet concentration n:

$$I_{DF} \propto \gamma\, n^2 \tag{1}$$

If triplet-triplet annihilation can be neglected as a decay process, i.e. if

$$\gamma n \ll \beta \tag{2}$$

with $1/\beta$ the triplet lifetime, then the decay of the DF is exponential:

$$I_{DF} \propto \gamma n_0^2\, e^{-2\beta t} \tag{3}$$

If a sample is spatially inhomogeneously excited, and if during the
observation of a DF the relative distribution of triplet states
changes by diffusion, then the decay of the DF is no longer exponential. For a known initial distribution of triplet states the
diffusion coefficient D of the triplet states can be evaluated
from the nonexponential decay of the DF.

It has been shown that this evaluation is especially simple
if a sample is inhomogeneously excited with the aid of two interfering light beams. In the case for short excitation at t=0 the
time-dependence of I_{DF} is given by

$$I_{DF} \propto [1+(1/2)\exp(-2\delta t)]\exp(-2\beta t) \tag{4}$$

with

$$\delta = (4\pi^2/d^2)D \tag{5}$$

and d the spatial excitation period.

Let $2\phi_S$ be the angle between the two interfering light beams in the sample, λ the excitation wavelength in vacuo and n_S the refractive index of the sample at the wavelength of excitation. Then the spatial excitation period is

$$d = \lambda / (2 n_S \sin\phi_S) \tag{6}$$

If two interfering light beams are incident from the same side on the plane interface between two media 1 and 2, and if the angle of incidence is the same for both beams, then generally

$$n_1 \sin\phi_1 = n_2 \sin\phi_2 \tag{7}$$

i.e. it is sufficient to know the angle 2ϕ in air in order to calculate the excitation period d in the sample.

The experimental technique has been described in detail for excitation with a standing light wave, i.e. for $\phi = \pi/2$. The additional equipment needed for the general case $0 < \phi < \pi/2$ was a Mach-Zehnder interferometer and a device for the measurement of ϕ with an accuracy better than $\pm$ 1%.

RESULTS AND DISCUSSION

The diffusion coefficient D of ^{3}anthracene* in n-hexadecane was measured between 20 and 90 C. The concentration of anthracene was 1.3×10^{-5} M, the triplet lifetime $1/\beta$ was 23 ms at 20 C and 15 ms at 87 C. The angle 2ϕ between the two interfering light beams in air was ~ 2 °. The results are shown in Figure 1. The

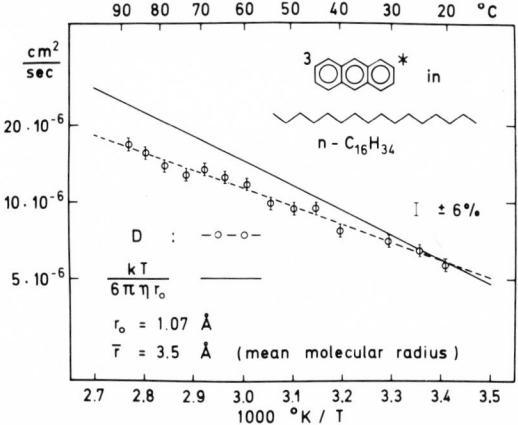

Fig. 1   Temperature dependence of the diffusion coefficient D of
^{3}anthracene* in n-hexadecane. D is plotted on a logarithmic scale.

diffusion coefficient D is plotted on a logarithmic scale against
1/T. Within an experimental accuracy of $\pm$ 6% lg D is a linear
function of 1/T.

The temperature dependence of D is considerably smaller than
expected from the Stokes-Einstein equation

$$D = kT/6\pi\eta r_o \tag{8}$$

and at 20 C the Stokes-radius (7) r_o = 1.07 A is much smaller than
the mean molecular radius $\bar{r}$ = 3.5 A. This results in agreement with
the fact that the Stokes-Einstein equation ceases to be a good
approximation for the diffusion coefficient D, if the solute
molecules are smaller than the solvent molecules. According to
Bowen and Metcalf (8) for anthracene in liquid paraffin the ratio
$\bar{r}/r_o$ increases with increasing viscosity and from their data one
should expect for anthracene in n-hexadecane at 20 C $\bar{r}/r_o \overset{\sim}{\sim}$ 3.
Osborne and Porter (9) have investigated the quenching of ^{3}naphtha-
lene* by α-iodonaphthalene in paraffins of different viscosity.
The experimental values of the bimolecular rate constant were up
to 7 times higher than expected for a diffusion-controlled reaction
from solvent viscosity, making use of eq. (8).

The concentration of anthracene was 3 x 10^{-5} M, the triplet
lifetime 1/β was 12 ms, the angle 2ϕ between the two interfering
light beams in air was $\sim$1°. It was found that:

$$D = (3.6 \pm 0.3) \cdot 10^{-5} \text{ cm}^2/\text{sec} .$$

With the viscosity of n-hexane (7) at 25 C: η = 0.294 cP one
obtains the Stokes-radius r_o = (2.06 $\pm$ 0.16) A. This value compares well with r_o = (1.59 $\pm$ 0.12) A for anthracene in liquid
paraffin (8) [of the viscosity η = 0.50 cP (this is approximately
the viscosity of n-octane)] and with r_o = (2.18 $\pm$ 0.15) A for
anthracene in acetonitrile (10).

The viscosity of n-hexane is lower than that of most other
commonly used solvents. Because of the condition (2), the determination of D is more difficult in less viscous solvents. Therefore, one may confidently state that diffusion coefficients of
molecules in the triplet state can be measured in a range of 6
orders of magnitude (the diffusion coefficient of ^{3}pyrene* in
glycerol (5) at -17 C is 8 x 10^{-11} cm^2 sec^{-1}).

If triplet-triplet annihilation is a diffusion controlled
reaction, then the bimolecular rate constant $\gamma_{diff.}$ can be calculated if the encounter distance ρ and the diffusion coefficient D
of the reacting molecules are known. According to Noyes (11), for
a diffusion controlled reaction of the type A + A $\longrightarrow$ products,

the rate constant $\gamma_{diff.}$ equals the expression:

$$\gamma_{diff.} = (4\pi \cdot 10^{-3} N_L \rho D)f \quad (0.5 \leq f \leq 1). \qquad (9)$$

This expression is smaller by a factor of 2 than the expression derived by Smoluchowski [12] for a diffusion-controlled reaction just of the type A + A $\longrightarrow$ products. We have not been able to understand Noyes' argument, nevertheless at the moment our data can be better understood on the basis of Noyes' expression. The factor f accounts for the fact that the result of a triplet-triplet interaction is on the average the annihilation of more than one and less than two triplets. For ^{3}anthracene* f $\approx$ f_{min} = 0.5 is probably a good estimate. The encounter distance ρ is assumed to be $\rho = 2 \bar{r} = 7.0$ A.

In table 1 the calculated minimum values of $\gamma_{diff.}$ in n-hexane and n-hexadecane are compared with experimental values $\gamma_{exptl.}$ obtained by flash-spectroscopy [13]. If eq. (9) is correct, then these data indicate that triplet-triplet annihilation is in n-hexadecane nearly diffusion-controlled, but in n-hexane considerably slower. It is desirable to extend the measurement of D and $\gamma_{exptl.}$ to paraffins of higher viscosity in order to see, whether a constant ratio of $\gamma_{exptl.}/\gamma_{diff.}$ is attained or not. If a constant ratio would be obtained, the lifetime of a triplet-triplet encounter complex (with respect to triplet-triplet annihilation) could be estimated.

Table 1

Comparison of experimental values and calculated values
of the rate constant of triplet-triplet annihilation.

	n-hexane 25 C	n-hexadecane 28 C
D (cm^2/sec)	$(3.6 \pm 0.3) \times 10^{-5}$	$(6.8 \pm 0.5) \cdot 10^{-6}$
$\gamma_{diff.}$(1/Msec)	$(9.5 \pm 0.7) \times 10^9$	$(1.8 \pm 0.1) \cdot 10^9$
$\gamma_{exptl.}$(1/Msec)	$(5.7 \pm 1.0) \times 10^9$	$(1.5 \pm 0.2) \cdot 10^9$
$\gamma_{exptl.}/\gamma_{diff.}$	0.60 ± 0.14	0.83 ± 0.16

$\gamma_{diff.}$ was calculated from equation (9) with f = 0.5 and ρ = 7.0 A. $\gamma_{exptl.}$ was calculated from the time dependence of triplet-triplet absorption with $(\varepsilon_{TT})_{max}$ = 6.3 $\cdot$ 10^4 M^{-1} cm^{-1} [14] in liquid paraffin. λ_{max} was 420.5 nm in n-hexane and 423.5 nm in n-hexadecane. The lower refractive index of n-hexane was not taken into account. Therefore the value of $\gamma_{exptl.}$ for n-hexane is probably ~10 % too high.

The usefulness of this method depends on the accuracy attainable and on the time necessary for the determination of one single diffusion coefficient. The maximum accuracy attainable is probably 1/3 of the accuracy with which the lifetime of a P-type delayed fluorescence can be measured. Thus, an absolute accuracy of at least ± 5% should be possible. The time typically needed for the measurement of two decay curves (one with homogeneous excitation and one with spatially periodic excitation) is ∿1 hour. In practice, with a single sample an unlimited number of determinations of D can be performed. Therefore, this method should be useful for systematic investigation of the temperature or pressure dependence of diffusion coefficients.

The specific advantage of this method is the fact that it allows the determination of diffusion coefficients of molecules in an electronically excited state. Thus, it should be possible to give at least a partial answer to the question of to what degree diffusion coefficients may depend on the state of electronic excitation.

For molecules with a maximum triplet lifetime of 1 sec or more it should be possible to measure translational diffusion as well as rotational diffusion via the time dependence of phosphorescence polarization. If both kinds of diffusion show the same temperature and viscosity dependence, then it should be possible to use measurements of the time dependence of phosphorescence polarization for the extrapolation of diffusion coefficients down to ∿ 10^{-15} cm^2/sec. Apart from this, it should be of interest to investigate experimentally the correlation between molecular shape, translational and rotational diffusion, and solvent structure.

ACKNOWLEDGEMENT

We wish to thank our colleague U. Suckow for the measurement of the triplet-triplet annihilation rate constants.

NOTE ADDED IN PROOF

If the rate constant $\gamma_{diff.}$ of a diffusion controlled reaction of the type A + A $\longrightarrow$ products is defined by $-dc_A/dt = \gamma_{diff.}$ x c_A^2, then $\gamma_{diff.} = 8\pi 10^{-3} N_L \rho D$. Therefore, the values of $\gamma_{diff.}$ in Table 1 are too small by a factor 2 and the values of the ratio $\gamma_{exptl.}/\gamma_{diff.}$ are too large by a factor 2. The conclusion to be drawn is now that even in n-hexadecane triplet-triplet annihilation is considerably slower than diffusion controlled.

REFERENCES

1. P. Avakian and R. E. Merrifield, Phys. Rev. Lett. $\underline{13}$, 541 (1964).

2. V. Ern, P. Avakian and R. E. Merrifield, Phys. Rev. $\underline{148}$, 862 (1966).

3. V. Ern, Phys. Rev. Lett. $\underline{22}$, 343 (1969).

4. B. Nickel, Ber. Bunsenges. Physik. Chem. $\underline{76}$, 582 (1972).

5. B. Nickel and U. Nickel, Ber. Bunsenges. Physik. Chem. $\underline{76}$, 584 (1972).

6. C. A. Parker and C. G. Hatchard, Proc. Chem. Soc., p. 147 (1962).

7. Viscosity data are taken from Landolt-Börnstein, 6. Edition, II. Band, 5. Teil, Transportphänomene I, Berlin (1969).

8. E. J. Bowen and W. S. Metcalf, Proc. Roy. Soc. A, $\underline{206}$, 437 (1951).

9. A. D. Osborne and G. Porter, Proc. Roy. Soc. A, $\underline{284}$, 9 (1965).

10. T. A. Miller, B. Prater, J. K. Lee and R. N. Adams, J. Am. Chem. Soc. $\underline{87}$, 121 (1965).

11. R. M. Noyes, "Effect of Diffusion Rates on Chemical Kinetics", in "Progress in Reaction Kinetics", ed. G. Porter, Vol. 1, Chapter 5, Pergamon Press, London (1961).

12. M. V. Smoluchowski, Z. Physik. Chemie $\underline{92}$, 129 (1917).

13. U. Suckow, unpublished results.

14. P. G. Bowers and G. Porter, Proc. Roy. Soc. A, $\underline{299}$, 348 (1967).

ELECTROPHOTOLUMINESCENCE IN A RIGID 3-METHYL PENTANE SOLUTION

OF N,N,N',N'-TETRAMETHYL-p-PHENYLENEDIAMINE AT 77K

R. Devonshire and A. C. Albrecht

Department of Chemistry

Cornell University, Ithaca, N. Y., USA

ABSTRACT

During studies to refine the model of the electrophoto-luminescence phenomenon in photoionized solutions of N,N,N',N'-Tetramethyl-p-phenylenediamine at 77K, a previously unreported contribution to the total recombination luminescence has been uncovered. Characteristics of the group of recombinations giving rise to this luminescence are described, and compared and con-trasted with those of the recombinations which involve the biphotonically produced trapped electrons previously reported in this system. The results suggest that a compact charge-pair is responsible for the new luminescence.

The possibility that coupling of the applied electric field with the polarizable cation is responsible for the electrophoto-luminescence found in this system is also discussed.

INTRODUCTION

A series of investigations in this laboratory (1) has shown that the tetramethyl-p-phenylenediamine (TMPD) molecule in rigid 3-methyl pentane (3-MP) at 77K is ionized by ultraviolet light to give the cation WB(TMPD$^+$) and a matrix trapped electron. The ionization step is a biphotonic one-electron process which normally proceeds via near-continuum states reached by the second photon from the lowest excited triplet state of TMPD.

After the sensitizing light is removed a slowly decaying

luminescence, called isothermal luminescence (ITL), is observed over a period of several hours. This luminescence is made up of the fluorescence and phosphorescence characteristic of TMPD and results from the recombination of the immobile cation and the electron which has been mobilized in some manner.

The rate of recombination can be greatly increased by shining near infra-red, visible or near-ultraviolet light on the sample thereby ionizing the trapped electron. A similar increase in rate is observed when the temperature of the sample is raised by a few degrees.

All of the above effects are similar to those found in a class of inorganic crystalline solids known as ZnS phosphors. In 1965 another effect was discovered in the organic system (2). Application of an electric field to a previously sensitized sample sharply increased the recombination rate. This is known in phosphor research as the Gudden-Pohl effect otherwise known as electro-photoluminescence (EPL). It was also observed that a second field application produced another marked increase in the recombination rate only if its polarity was opposite to that in the first field application, indicating a strong anisotropy of the field induced detrapping of the electron.

Based on a purely thermal model and a simple one-dimensional picture of electron detrapping, a one parameter equation was proposed (3) which relates the time course of the EPL to that of the ITL which would have been observed if the electric field had not been applied. The proposed relationship reads

$$EPL\ (t) = (\underline{a}/2)\ ITL\ (at) + (\underline{a}^{-1}/2)\ ITL\ (a^{-1}t) \tag{1}$$

where $\underline{a}$ is a parameter related to the electric field perturbation of the barrier heights which limit detrapping. Explicitly, $\underline{a} = \exp[\delta\varepsilon/kT]$ where $\delta\varepsilon$ is the perturbation of the trap depth. The one dimensional model asserts that all electron-cation pairs are geminate and the diffusion of the electron is governed by the Coulomb field of the cation giving a net inward flow. In this model the effect of an applied field is either to enhance the inward force of the Coulomb field or to reduce it, thus giving rise to two populations of recombining electrons corresponding to the two terms in Eq. 1. With $\underline{a} \geqslant 1$, the first term represents the group of electrons having enhanced inward flow while the second term represents the suppressed population. This equation appeared to meet with considerable success. A single $\underline{a}$ parameter at a given field strength could duplicate the time course of the EPL signal through Eq. 1 for long periods of time, although it failed conspicuously in the early part of the EPL decay. Furthermore, $\underline{a}$ proved to be exponential in the applied field confirming the expected linear dependence of $\delta\varepsilon$ on the field. That a one

dimensional picture should manage such success seemed very surprising so a more thorough study of the entire EPL phenomenon has
been undertaken. During these studies we have been able to
identify two quite distinct contributions to the total recombination luminescence as evident by their EPL behavior. In addition
to the usual trapped electron contribution, a new species,
probably a compact charge-pair has been discovered which is
especially responsive to the applied electric field. These are
the subject of this note.

EXPERIMENTAL

The experimental technique has already been described (3).
In the current studies care has been taken to view only the
fluorescence component of the recombination luminescence. For
this reason reference will be made to EPF (electrophotofluorescence)
and ITF (isothermalfluorescence).

RESULTS AND DISCUSSION

Accepting for the moment the validity of Eq. 1, then it is
possible from an ITF(t) curve and an EPF(t) curve to calculate the
value of $\underline{a}$ at each point in time. The resulting function, $\underline{a}(t)$,
generates the EPF at every time, t, from the experimental ITF(t)
through Eq. 1. To the extent that $\underline{a}(t)$ = constant the model underlying Eq. 1 may be considered a success. A pair of experimental
ITF(t) and EPF(t) curves and their corresponding $\underline{a}(t)$ function are
shown in Fig. 1.

It is seen that for long periods of time $\underline{a}$ does have a constant value and this corresponds to the values reported in earlier
papers. Let this plateau value for the parameter be symbolized as
$\underline{\bar{a}}$. Superimposed upon the observed EPF(t) curve is a calculated
EPF curve in which $\underline{a}(t)$ was taken as the constant, $\underline{\bar{a}}$. Deviations
from the experimental EPF(t) curve are evident at the shorter times
and they are especially conspicuous in the a(t) representation.
The $\underline{a}(t)$ curve provides a very sensitive measure of the failure of
Eq. 1 with a(t) = const. and is therefore useful in searching for
more subtle models. It turns out that the $\underline{a}(t)$ curve is double
valued whenever EPF>ITF. Only the smaller of the two roots is
plotted in Fig. 1, since, when an EPF(t) curve with $\underline{a}(t)$ = const.
is synthesized and from it the a(t) function regenerated, it is
the lesser root which has the same constant value on either side
of the EPF = ITF point. (The larger root rises rapidly in the
region EPF>ITF to a physically meaningless value at t=0.)

Figure 2 shows the $\underline{a}(t)$ functions derived from a series of

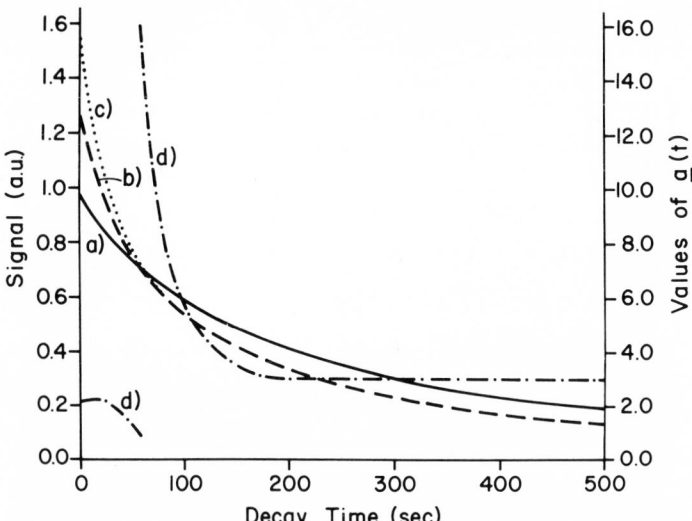

Fig. 1   Experimental ITF(t) and EPF(t) decay curves from the TMPD-
         3MP system at 77K: a) ITF(t) (——); b) EPF(t)＊ (---); c)
         EPF(t) (•••) computed from decay curve a) through Eq. 1
         with a = constant = 3.0 (the "plateau" value of a(t)); d)
         a(t) (-•-•-•) function derived from curves a) and b). The
         applied field strength is 5.08×10^5 V/cm (as it is for all
         the EPF data reported in this paper). The ordinate is in
         arbitrary units. ＊DWT 2 min. before field applied.

experiments in which the dark wait time (DWT, the period between
the end of sensitization and application of the field at time,
t=0) was varied from 30 secs up to 10 minutes. (The data in Fig.
1 refer to a 2 minute DWT.) There are several points to notice.
At long times all the a(t)'s converge to a constant value of a=3.0.
At intermediate times two extreme types of deviation from a are
evident. Short DWT's produce negative deviations from a, i.e.
a<ã, and long DWT's positive deviations with a>ã. These same
trends are also apparent at t=0, the instant the electric field is
applied.

      Values of ITF(0) and EPF(0) for the experiments reported in
Fig. 2 are shown in Table 1. At time, t=0, Eq. 1 takes the
simplified form

$$EPF(0)/ITF(0) = (1/2) [a + a^{-1}] \qquad (2)$$

Eq. 1 shows, as might be expected, that the ratio EPF(0)/ITF(0) is
a simple measure of the strength of the coupling of the field with
the charge recombination process. There is a steady increase in
this ratio with increasing DWT indicating that as the ITF decay
proceeds the recombination events which remain show an increasing
average sensitivity to the applied field. These remaining,

Table 1

The variation with dark wait time of the ratio EPF(0)/ITF(0) i.e.
the value of ITF just before the field is applied at time, t=0,
and the value of EPF just after the field is applied

Dark wait time (min)	EPF(0)[+]	ITF(0)[+]	EPF(0)/ITF(0)
0.5	4.02	3.50	1.15
1.0	3.17	2.37	1.34
2.0	2.37	1.45	1.63
4.0	1.45	0.82	1.77
10.0	0.73	0.29	2.52

[+] arbitrary units

hypersensitive recombinations evidently have a greater thermal
stability than the earlier less sensitive ones, although they, too,
eventually disappear as shown by the continual diminution of
EPF(0) with DWT. The new hypersensitive recombinations, best seen
after long DWT's, retain the same response with respect to field
reversal described earlier. Incidently, the quantitative behavior
of EPF with respect to field reversal is independent of the order
in which the polarities are applied.

The continuous climb in the ratio EPF(0) with DWT's of up to
an hour or more suggested that we try partial infra-red bleaching
of the sample as a substitute for a long dark-wait bleaching
process. If, indeed, the ratio EPF(0)/ITF(0) eventually reached
a steady value we might identify it solely with the hypersensitive
group. Fig. 3 shows the results of such experiments with varying
periods of IR bleaching in which the DWT is only 2 minutes in
every case.

With only a 5 sec. irradiation there is a precipitous fall in
the ITF(0) value. Further IR bleaching causes a steady, but much
slower, fall in the ITF(0) value. During the first 5 sec. of IR
bleaching the EPF(0) rises, despite the lower ITF(0) value, and
this is followed by a steady fall of EPF(0), paralleling the
behavior of ITF(0). (When these experiments were repeated several
times we found that the EPF(0) values after IR bleaching were
lower than the EPF(0) values with no IR treatment. We show however, the present set of results because of their mechanistic
significance.) The tabulation in Fig. 3 shows how IR bleaching
quickly establishes a constant and high value for the EPF(0)/ITF(0)
ratio. This strongly suggests that we have isolated the hypersensitive group by the IR treatment and that this is the ratio
that would have been reached after very long DWT's.

Having isolated one of these two groups by this post uv-

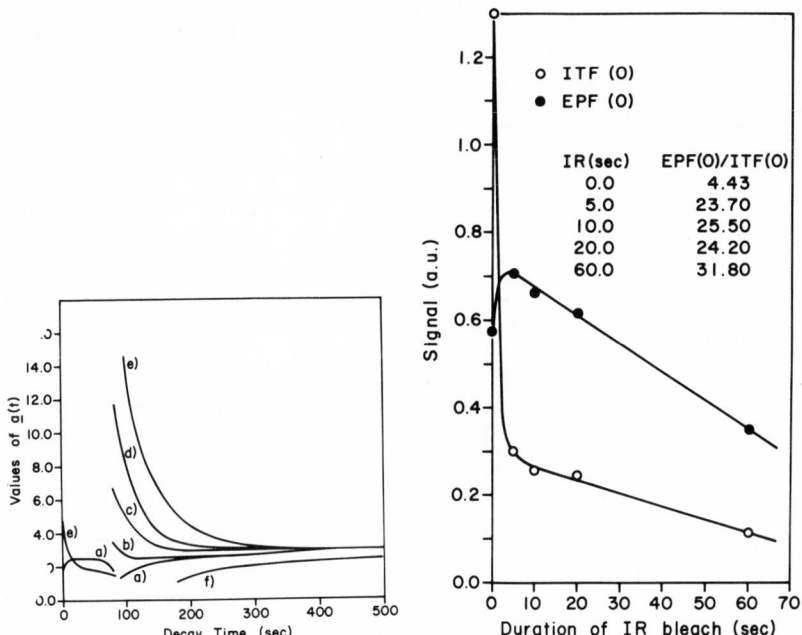

Fig. 2   The a(t) functions derived from a series of experiments
         with different dark wait periods (DWT) before application
         of the field at time, t=0: a) DWT = 0.5 min.; b) DWT =
         1.0 min.; c) DWT = 2.0 min.; d) DWT = 4.0 min.; e) DWT =
         10.0 min.; f) DWT = 2.0 min., data from a sample sensitized
         with a pulsed frequency-doubled ruby laser. We usually
         find different "plateau" a values for different samples.
         The trend exhibited by a(t) is, however, significant.

Fig. 3   The values of ITF(0) and EPF(0) from a series of experi-
         ments in which previously sensitized samples were given
         different periods of IR bleaching during the dark wait
         period before the field is applied. The dark wait time in
         each experiment is 2.0 min. The ordinate is in arbitrary
         units which are different for each of the two signals.

irradiation treatment we next considered the possibility of controlling their proportions by altering the conditions of the
initial uv-irradiation. We made a series of experiments in which
the uv-intensities were varied followed by a standard 2 minute DWT.
Then the EPF(0) and ITF(0) values were compared with and without a
5 second IR treatment. The results in Figure 3 suggest that we
might consider the normal EPF(t) and ITF(t) as being made up of
two components. Treatment with IR clearly gives us signals from
one of these components (the post-IR signal). We, therefore, sub-
tract the post-IR ITF(0) value from the ITF(0) value with no IR
treatment (the pre-IR ITF) and obtained hypothetical ITF(0) (the

pre-IR (corr.)) to be associated with the less sensitive group. The EPF(0) results can be treated in this same way. The results are summarized in Table 2 which shows the exponents obtained from log intensity vs. log signal plots.

Table 2

Exponents taken from log uv-intensity vs. log signal (ITF(0) and EPF(0)) plots. The data is taken from a series of experiments with and without a 5 sec. IR bleach during the dark wait period. The dark wait time is 2 minutes for each experiment. The pre-IR (corr.) signal is obtained from the pre-IR signal by subtracting from it the post-IR signal

| | Light Exponent | | F/P Ratio | |
	ITF	EPF	ITL	EPL
Post-IR	1.00	0.94	1.30	(Obscured by triplet lifetime)
Pre-IR (corr.)	1.60	1.25	0.70	

It is seen that the hypersensitive group (post-IR) is mono-photonically produced in the period of sensitization and that the less sensitive group displays a light exponent similar to that found in this system for the biphotonic process under photostationary conditions of uv-irradiation (4). Evidently by employing very intense uv-irradiation it should be possible to favor the less sensitive group. We therefore photoionized a sample with a frequency-doubled pulsed ruby laser giving light at 347 nm. In Fig. 2 the $\underline{a}(t)$ function for this experiment can be seen. Clearly it is of a form which we would expect at the shortest DWT's. This is just where we would expect the larger proportion of the less sensitive group.

During the measurements using partial IR bleaching of the sample we found that the fluorescence to phosphorescence ratio, F/P, was a variable. There is a pronounced enhancement of the F/P ratio associated with the hypersensitive group over that of the less sensitive group. This is also shown in Table 2. Comparing our results with previous work from this laboratory (5) we calculate that the fraction of electrons reaching the triplet manifold during the hypersensitive recombinations is 0.58 compared to the usual value of 3/4 which reflects purely statistical considerations.

In summary, we have been able to identify two distinct parts of the ITF(t) and EPF(t) decays seen in this system, one of which is associated with a previously unreported group of recombinations which has the following characteristics when compared with the normal biphotonically produced group: 1. Hypersensitivity to an

applied electric field, 2. Greater thermal stability, 3. Greater
resistance to IR bleaching, 4. Is monophotonically produced, 5.
Larger F/P ratio in recombination luminescence, 6. Can be formed
during the decay of the less sensitive group. (The rise of EPF(0)
in Fig. 3 seems to admit of no other interpretation.)

What is the nature of this new hypersensitive group? The
experimentally determined ionization threshold for the two photon
process found in this system is 5.9 ev (1). This group is monophotonically produced and the photon energy at 313 nm is only 4.0 ev
which is much below the energy required to reach the "plateau" of
the Coulomb potential in this low dielectric constant medium. This
argues strongly for a compact ion-pair. (There must be some charge
separation or there would be EPL activity.) What then is the
nature of this electron trap close to the cation and why are its
properties so different from those traps which are further away?
For example, how is the thermal stability of the anion to be understood given the very high coulomb field at the site of the trap?
The reasons might involve either different activation mechanisms
for identical traps, differentiated only by their distance from
the cation, or entirely different traps. Possible trapping sites
in the system are matrix cavities, oxygen, carbon dioxide, or a
partner TMPD molecule. On the other hand, the charge separation
might be fixed in some photochemical intermediate of TMPD itself.

If the slow isothermal recombination of this charge-pair is
thermally activated, then from the estimated decay rate a barrier
height of about 0.3 ev is suggested which is similar to that for
the biphotonic group. Then how is one to explain the contrasting
uv sensitivity to IR bleaching? Possibly the electron is highly
localized. This point, and others, suggest that tunneling may be
involved in the recombination of this group.

The observation of recombination fluorescence means that the
compact charge-pair lies at an energy not much below S_1. The
enhanced F/P ratio could result either from an enhancement of the
fluorescence or a quenching of the phosphorescence. If the electron
retained some spin correlation with the parent cation and, is made
directly from the singlet state, then recombination would show
enhancement of the fluorescence over that seen from the recombination of uncorrelated electrons. The identification of the new compact pair must await, at the very least, scavenging studies coupled
with EPL.

For the hypersensitive group application of an electric field
of 5×10^5 V/cm causes an approximately twentyfold increase in the
instantaneous rate of recombination (compared with, typically, a
twofold increase for the biphotonic group). We can interpret this
in terms of a thermal activation model or a tunnelling model. In

the former case the increase in rate is equivalent to a lowering of the activation barrier by 3 kT. In terms of the parameter d (used in the previous EPL work (3)), which is a measure of the coupling constant of the electron with the applied field, this is about 3A. (Corresponding figures for the biphotonic group are kT and d=1A.) The implication is a larger coupling distance for the hypersensitive group, an idea which for the electron, is difficult to believe because of the close proximity of the cation. We have, however, to consider also the effect of the applied field on the cation. For a compact charge-pair a field induced shift of the center of charge on the polarizable cation may significantly alter the barrier controlling its recombination (whether the mechanism proceeds by a thermal or by a tunneling route). For the biphotonic group the effect of such a shift of cationic charge must be much weaker. But then so is the observed coupling of this group with the applied field. This raises a new possibility as to how the applied field may couple with such charge pairs in general. If tunneling alone is responsible for the recombination of the hyper-sensitive group a simple estimate shows that a reduction of only 1/6 in the barrier width could account for the observed increase in rate.

Isotropic effects of the field such as electrostriction are unable to account for the anisotropic field behavior of the recombination luminescence.

Having now established that there is more than one group of electrons which contribute to the ITL and EPL decays observed in this system we can proceed with more confidence in refining the model of EPL behavior represented by Eq. 1. By suitable choice of experimental conditions we hope now to obtain the ITF(t) and EPF(t) decay curves associated with each of the groups. The present studies demonstrate the usefulness of EPL investigations of systems having charge separated intermediates.

ACKNOWLEDGMENTS

This work has been supported in part by a National Institutes of Health grant GM-10865 and by a grant from the National Science Foundation, and from the Material Science Center of Cornell University.

REFERENCES

1. K. Cadogan and A. C. Albrecht, J. Phys. Chem., 72, 929 (1968) (and earlier papers).

2. G. E. Johnson, W. M. McClain and A. C. Albrecht, J. Chem.
    Phys., 43, 2911 (1965).

3. J. Bullot and A. C. Albrecht, J. Chem. Phys., 51, 2220 (1969).

4. a) G. E. Johnson and A. C. Albrecht, J. Chem. Phys., 44, 3162
       (1966).
    b) Ibid, J. Chem. Phys., 44, 3179 (1966).

5. A. H. Kalantar and A. C. Albrecht, J. Phys. Chem., 66, 2279
    (1962).

HEAVY-ATOM EFFECT ON RADIATIVE AND RADIATIONLESS TRANSITIONS

Zbigniew R. Grabowski and Nina Sadlej*

Institute of Physical Chemistry, Polish Academy

of Sciences, Warsaw, Poland

ABSTRACT

Phosphorescence decay curves and quantum yields were determined for p-phenyl-benzophenone (PBz) and p-methoxyacetophenone (MAc) in rigid glasses containing ethyl bromide or ethyl iodide. Both radiative and radiationless deactivation are influenced to a similar extent by external heavy atoms.

The external heavy-atom effect on spin-forbidden radiative transitions is exerted (e.g., by ethyl iodide (1), by a variety of heavy metal alkyls (2) and by Xe) on the $T_1 \rightarrow S_0$ phosphorescence (3) and on $T_1 \leftarrow S_0$ absorption (4) of many aromatic compounds. The effect was found to be purely electronic in nature (5) but, in spite of numerous investigations, its mechanism is not yet fully understood (6).

An analogous heavy-atom effect on radiationless transitions (intersystem crossings) is less well-known. Even the experimental findings are controversial, in particular the quantitative comparisons of the effect of external heavy atoms on radiationless and radiative transitions. Most of the authors (but not all of them (7)) agree the effect to be most pronounced on $S_1 \rightsquigarrow T_1$ intersystem crossing between the states with a rather small electronic energy gap (8,9). As to the decay of the lowest excited triplet state, T_1, some authors find the radiative transition to be much more strongly enhanced by external heavy atoms than the radiationless deactivation $T_1 \rightsquigarrow S_0$, (3,6,7,10,11). If the last is catalyzed at all (12), some state that both transitions are catalyzed in a

314

comparable manner (13)while others find the radiationless deactivation $T_1 \leadsto S_0$ to be affected more than the $T_1 \to S_0$ phosphorescence
emission rate (8). Nearly all observations were made on the decay
of the triplet state population of aromatic hydrocarbons or on
their mono-halogen-derivatives.

In looking for the competition between radiative and radiationless deactivation rates (k_r and k_{nr}, respectively) of the triplet
state under the influence of external heavy atoms, we considered
the following three conditions to be important for the choice of
the compounds:

1. Triplet yield $\Phi_T \approx 1$, which allows the neglect of the
   fluorescence, internal conversion $S_1 \leadsto S_0$, or intersystem
   crossing $S_1 \leadsto T_1$, and the heavy atom effects on their
   rates.

2. Phosphorescence quantum yield $\Phi_P = k_r/(k_r + k_{nr}) \approx 0.5$
   in the absence of heavy atoms. This condition makes
   Φ_P measurements most suitable for the determination of
   relative changes of k_r and k_{nr} under the influence of
   heavy atoms.

3. The triplet state involved should be of the $^3(\pi,\pi^*)$
   nature, as the $^3(n,\pi^*) \leftrightarrow S_0$ radiative transitions are
   virtually not enhanced under the influence of external
   heavy atoms (14-16).

On this basis we have chosen two aromatic ketones, p-phenylbenzo-
phenone (PBz) (17) and p-methoxy-acetophenone (MAc) (15,18,19).

EXPERIMENTAL

MAc was synthesized from anizole and acetic anhydride in the
presence of $Mg(ClO_4)_2$ (20) and was recrystallized from hexane
(m.p. 36-36.5 C). PBz was prepared from benzoyl chloride and
biphenyl in the presence of $AlCl_3$ (21) purified by chromatography
on Al_2O_3 with C_6H_6 as solvent, and crystallized from CH_3OH
(m.p. 100.5-101.5 C). Absorption and phosphorescence spectra
of both compounds matched those previously published. The solvents
were carefully purified by chemical means and distillation. Three
solvents were used to form rigid glasses at 77 K: EPA (diethyl
ether + isopentane + ethanol, 5:5:2 by volume), EPB (diethyl ether +
isopentane + ethyl bromide, 2:2:1) and EPI (diethyl ether +
isopentane + ethyl iodide, 2:2:1).

For low-temperature measurements the solutions were placed in
a metal dewar with quartz windows at 90° for phosphorescence and
180° for absorption and refractive index measurements. Refractive

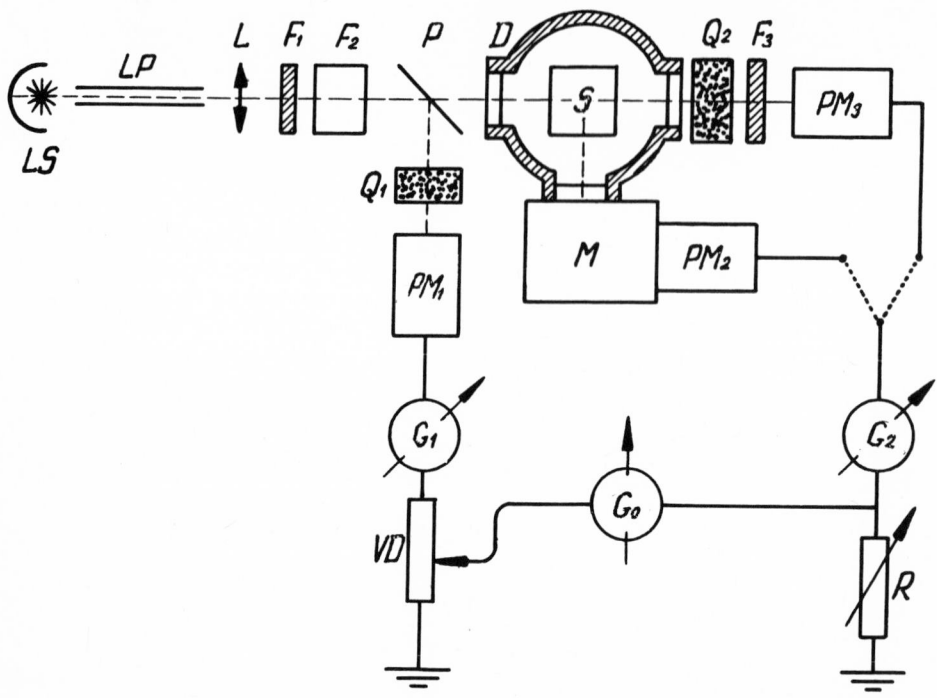

Fig. 1   Instrumental arrangement: LS - light source, Osram HBO 200
          highest pressure Hg lamp; LP - quartz capillary light-pipe;
          L - lens; F_1,F_2 - glass and liquid filters isolating the
          excitation wavelength (Hg 365 nm for PBz, or Hg 313 nm for
          MAc); P - partially reflecting quartz plate; D - metal
          dewar; S - sample; Q_1,Q_2 - light-scattering mat quartz
          blocks; M - monochromator; F_3 - filter absorbing the
          luminescence and transmitting the excitation wavelength;
          PM_1, PM_2, PM_3 - photomultipliers; G_1 - galvanometer
          indicating the reference signal; G_2 - galvanometer indicat-
          ing the signal to be measured; R - load resistance 0-100 kΩ;
          VD - voltage divider, total resistance 10 kΩ; G_o - zero
          indicating galvanometer.

indices were measured by means of a silica glass prism fitted into
the low temperature cell. At λ=436 nm we found n = 1.48 for EPA
glass at 77 K, and n=1.50 for the EPB glass.

    Phosphorescence spectra and absorption were measured by means
of a laboratory arrangement (Fig. 1) whereby the fluctuations of the
light source were eliminated by the use of a light-pipe and by
measuring always the ratio of luminescence signal (or transmitted
light) to the incoming light intensity. The ratios of the phosphorescence quantum yields were determined as:

$$\frac{\Phi_1}{\Phi_2} = \frac{n_2^2 \, (1 - T_1) \int_0^\infty [I_1(\nu)/S(\nu)]d\nu}{n_1^2 \, (1 - T_2) \int_0^\infty [I_2(\nu)/S(\nu)]d\nu} \qquad (1)$$

where T is the transmittance of the solution at the wavelength of excitation, $I(\nu)$ dν is the signal measured at a given spectral slit width and $S(\nu)$ is the spectral sensitivity factor of the instrument. To keep the geometry of the experiment unchanged, $T_1 \approx T_2$. It was assumed that the phosphorescence polarization in both solvents did not differ considerably. The spectral sensitivity curve, $S(\nu)$, was determined by the use of β-naphthol and quinine sulphate as luminescence standards (22) and by use of tungsten lamps calibrated in the Standard Office in Warsaw. The integrations were made over the whole phosphorescence spectrum.

The phosphorescence decay curves were photographed from the oscilloscope screen after the samples had been irradiated by a 200 J flash discharge between Fe electrodes in air, triggered by a shutter mechanism. The flash light was filtered by a UG2 Schott glass filter, the phosphorescence signal being isolated by an IF 425 nm Zeiss interference filter for MAc, or by a Hg 436 Zeiss Monochromat filter for PBz. The shutter screened the photo-multiplier for 2 ms during and after the flash. The output signal of the photomultiplier was supplied to the scope; the time constant RC of the load resistance and cables was 5 to 100 μsec.

The methods of measurement of both lifetimes τ and quantum yields were tested with known standards.

RESULTS

The replacement of EPA by a rigid glass solvent containing heavy atoms, EPB or especially EPI, accelerates the phosphorescence decay; if the effect is strong, the decay curves become nonexponent-ial (Fig. 2), as has already been known (7,10,11). Let us define the mean lifetime of phosphorescence

$$\langle\tau\rangle = [1/I(0)]\int_0^\infty I(t)dt, \qquad (2)$$

where $I(t)$ is the phosphorescence signal at time t. The same equation gives the true lifetimes in the simple exponential decay. If at a distance R from a heavy atom the phosphorescence decays exponentially,

$$I_R(t) = I_R(0)\exp(-t/\tau_R), \qquad (3)$$

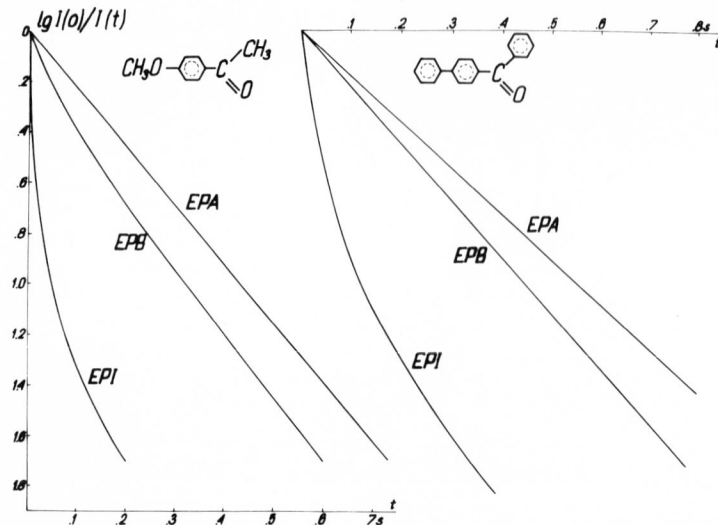

Fig. 2 Typical decay curves (semilogarithmic plots) of p-phenyl-
        benzophenone and p-methoxyacetophenone in EPA, EPB, and
        EPI rigid glasses at 77K.

the observed overall signal would be

$$I(t) = \int_R I_R(t) \, dR \qquad\qquad (4)$$

and then

$$\langle\tau\rangle = \frac{\int\int\limits_{0\ R}^{\infty} I_R(0)\exp(-t/\tau_R)\,dR\,dt}{\int\limits_R I_R(0)\,dR} = \frac{\int\limits_R I_R(0)\,\tau_R\,dR}{\int\limits_R I_R(0)\,dR} . \qquad (5)$$

The results are tabulated in Tables 1 and 2, along with the
results of the measurements of the quantum yield ratios. Each
value in the table is the mean of several measurements. The
indicated error limits are the standard deviations of the mean
value for the ratios Φ_{EPB}/Φ_{EPA}, and the standard deviations of
a single measurement for all other values, with the exception of
$\langle\tau\rangle$, to which an error of approximation was ascribed arbitrarily as
$\pm10\%$. The derived values of the rate constants, calculated as
$k_r=\Phi/\tau$, and $k_{nr} = (1/\tau)-k_r$, are associated with increasing errors
and have only an orientational value (see Tables 1 and 2).

Quantum yields Φ_{EPI} were not determined because of a marked,
probably sensitized, decomposition of ethyl iodide on longer
irradiations.

Table 1

Phosphorescence quantum yields Φ, lifetimes τ, overall heavy-atom catalysis factors τ_{EPA}/τ, and approximate radiative and radiation-less rate constants, k_r and k_{nr}, for p-phenyl-benzophenone triplet decay in EPA, EPB and EPI rigid solvents, 77K

Solvent	Φ/Φ_{EPA}	Φ	τ or $<\tau>$ [s]	τ_{EPA}/τ	k_r [s^{-1}]	k_{nr} [s^{-1}]
EPA	1	0.47(a) ±0.05	0.242(b) ±0.006	1	1.9 ±0.2	2.2 +0.4
EPB	0.89 ±0.06	0.42 ±0.07	0.196 ±0.004	1.23 ±0.06	2.1 ±0.4	2.9 ±0.5
EPI	.....	.....	0.053 ±0.005	4.6 ±0.5	$k_r + k_{nr}$ = 19 ± 2	

(a) ref. (17)
(b) for comparison: τ = 0.30s in ethanol + ether glass, 0.20 s in isopropanol, 0.15 s in cyclohexane, at 77K (24).

DISCUSSION

The most striking result seems to be a slight decrease of the phosphorescence yield, $\Phi_{EPB}/\Phi_{EPA} \leqslant 1$, which indicates a comparable heavy atom enhancement of both radiative and radiationless deactivation of the triplet state, or even a somewhat stronger effect in the latter case. Qualitatively, the same was found for both ketones, PBz and MAc.

Vibronic analysis of unperturbed phosphorescence of several carbonyl compounds, including MAc, led Lim, Li and Li (19) to the conclusion that both radiative and radiationless deactivations involve similar mixing of states, probably due to an out-of-plane distortion of the molecules in the triplet state. Discussing the mechanisms of the external heavy-atom effect (25-27) Giachino and Kearns (6) did not exclude the possibility of a comparably strong effect for both deactivation processes. This may be expected, especially taking into account the corrections (6) for the mixing in of $^3(n,\pi^*)$ and $^1(n,\pi^*)$ states lying close above the emitting $^3(\pi,\pi^*)$ state in both MAc and PBz molecules.

The nonexponential decay in heavy atom containing solvents

Table 2

Phosphorescence quantum yields Φ, lifetimes τ, overall heavy-atom catalysis factors τ_{EPA}/τ, and approximate radiative and radiationless rate constants, k_r and k_{nr}, for p-methoxy-acetophenone triplet decay in EPA, EPB and EPI rigid solvents, 77K.

Solvent	Φ/Φ_{EPA}	Φ	τ or $\langle\tau\rangle$ [s]	τ_{EPA}/τ	k_r [s^{-1}]	k_{nr} [s^{-1}]
EPA	1	0.72(a) $\pm$0.04	0.191(b) $\pm$0.003	1	3.8 $\pm$0.2	1.5 $\pm$0.3
EPB	0.93 $\pm$0.05	0.67 $\pm$0.08	0.12 $\pm$0.01	1.6 $\pm$0.2	5.5 $\pm$1.2	2.7 $\pm$1.0
EPI	.....	.....	0.022 $\pm$0.002	8.8 $\pm$1.0	k_r+k_{nr} = 46$\pm$5	

(a) ref. (18); error limit estimated by comparison to Φ_{EPA} = 0.68 in ref. (19).
(b) for comparison: τ_{EPA} is found to be 0.260 s (18); 0.190 s (19); for ether + ethanol 1:1 rigid solvent a complex decay curve is described (23) with τ_1 = 0.190 s and τ_2 = 0.530 s.

may give a key to understanding the distance-dependence of the perturbation. The basic theory of such decay curves has already been developed by Lin and Tweed (28). The final results of their work contain an error in their Eq. 28 and in subsequent conclusions, because they treated the heavy atoms only as quenchers (by application of the formalism of energy transfer by exchange mechanism (29,30)) and omitted the heavy atom effect on k_r. This led to the incorrect statement that Φ_P should decrease under the influence of external heavy atoms (loc. cit., Fig. 2).

In some cases the decay curves were resolved into two exponential decays of which the long τ was close to that of the unperturbed triplet and the short τ was ascribed to 1:1 complexes, e.g., of naphthalene with Xe (31) or with I$^-$ ion (32). The present decay curves for EPI glass (Fig. 2) also show, at high depletion of the triplet population, the decay rate approaches that in EPA. However, this may be consistent with any kind of short-range interactions leaving the more distant molecules virtually unperturbed. The apparent lack of influence of heavy cations like Cs^+ (32) or of a markedly lesser effect of $Pb(CH_3)_4$ than that of $Hg(CH_3)_2$ (9) may be due, in part at least, to the effect of the screening of the heavy atom by the solvent molecules or by the methyl groups, respectively.

Analysis of decay kinetics as a function of distance from the heavy nuclei, which hitherto has been unsuccessful (6), should finally form a bridge between the external and intramolecular heavy-atom effect, e.g., for the cases of heavy atoms being separated from the aromatic ring by an aliphatic chain such as the marked effect of Br across a — —$(CH_2)_3$-chain (33).

ACKNOWLEDGEMENT

We thank Dr. Jan Jasny for his advice and help in the optical measurements, in particular the measurement of the refractive index at low temperatures. Z.R.G. is deeply indebted to the Master and Fellows of Churchill College for electing him as an Overseas Fellow, thus allowing him to complete this work at Cambridge.

REFERENCES

*Present address: Institute of Physics, Technical University, Warsaw.

1. M. Kasha, J. Chem. Phys. 20, 7 (1952).

2. E. Vander Donckt and C. Vogels, Spectrochim. Acta 27A, 2157 (1971).

3. G. W. Robinson, J. Molec. Spectry 6, 58 (1961).

4. A. Grabowska, Spectrochim. Acta 19, 307 (1963).

5. G. G. Giachino and D. R. Kearns, J. Chem. Phys. 53, 3886 (1970).

6. G. G. Giachino and D. R. Kearns, J. Chem. Phys. 52, 2964 (1970), and the references therein; (a) erratum: ibid, 54, 3248 (1971).

7. S. Siegel and H. S. Judeikis, J. Chem. Phys. 42, 3060 (1965).

8. S. P. McGlynn, J. Daigre and F. J. Smith, J. Chem. Phys. 39, 675 (1963).

9. E. Vander Donckt and J. P. van Bellinghem, Chem. Physics Letters 7, 630 (1970); J. Chim. Phys. 68, 948 (1971).

10. S. P. Mc Glynn, M. J. Reynolds, G. W. Daigre and N. D. Christo-doyleas, J. Phys. Chem. 66,2499 (1962).

11. S. E. Webber, Chem. Physics Letters 5,466 (1970).

12. K. B. Eisenthal and M. A. El-Sayed, J. Chem. Phys. 42, 794 (1965).

13. M. S. de Groot and J. H. van der Waals, Molec. Physics 4, 189 (1961).

14. M. A. El-Sayed, J. Chem. Phys. 41,2462 (1964).

15. D. R. Kearns and W. A. Case, J. Amer. Chem. Soc. 88,5087 (1966).

16. R. F. Borkman and D. R. Kearns, J. Chem. Phys. 46,2333 (1967).

17. V. L. Ermolaev and A. N. Terenin, J. Chim. Phys. 55,698 (1958); Uspekhi Fiz. Nauk 71,137 (1960).

18. N. C. Yang, D. S. McClure, S. L. Murov, J. J. Houser and R. Dusenberg, J. Amer. Chem. Soc. 89,5466 (1967).

19. E. C. Lim, Y. H. Li and R. Li, J. Chem. Phys. 53,2443 (1970).

20. G. N. Dorofenko, V. J. Dulenko and L. M. Antonenko, Zh. Obshch. Khim. 32, 3047 (1962).

21. P. J. Montague, Rec. Trav. Chim. 27,327 (1908).

22. E. Lippert, W. Nägele, I. Seibold-Blankenstein, U. Steiger and W. Voss, Z. analyt. Chem. 170,1 (1959).

23. R. N. Griffin, Photochem. & Photobiol. 7,159 (1968).

24. G. Porter and P. Suppan, Trans. Faraday Soc. 61,1664 (1965).

25. G. J. Hoijtink, Molec. Physics 3,67 (1960).

26. J. N. Murrell, Molec. Physics 3,319 (1960).

27. H. Tsubomura and R. S. Mulliken, J. Amer. Chem. Soc. 82, 5966 (1960).

28. S. H. Lin and D. Tweed, Internat. J. Quantum Chem. 3S,315 (1969).

29. K. B. Eisenthal and S. Siegel, J. Chem. Phys. 41,652 (1964).

30.   M. Inokuti and F. Hirayama, J. Chem. Phys. <u>43</u>,1978 (1965).

31.   S. Siegel and H. S. Judeikis, J. Chem. Phys. <u>48</u>,1613 (1968).

32.   R. H. Hofeldt, R. Sahai and S. H. Lin, J. Chem. Phys. <u>53</u>, 4512 (1970).

33.   H. H. Perkampus and H. R. Vollbrecht, Spectrochim. Acta <u>27A</u>, 2178 (1971).

LUMINESCENCE KINETICS FOR ENERGY TRANSFER IN SOLID SOLUTIONS:

III. EXCITATION MIGRATION THROUGH THE DONOR

I. M. Rosman

Physico-Technical Institute, Sukhumi, USSR

ABSTRACT

The effect of excitation migration among donor molecules is included in a theoretical analysis of energy transfer from donor to acceptor molecules. Dipole-dipole interaction is assumed. The theory is applied to the fluorescence of rhodamine in water-glycerine solutions.

In the first and second papers of this series (1,2) we have considered solutions containing two "active" components: a donor (D) and acceptor (A) of the energy. Luminescence kinetics was studied as a function of C_A concentration for dipole-dipole and exchange transfer. In both cases the donor concentration was taken as very low and hence the excitation migration through donor molecules was not taken into account. In this paper such a migration has been taken into consideration.

For quantitative estimations the following model has been used. Energy migration is considered as a process of random walk and the corresponding diffusion coefficient D_{exc} is calcuated. Its effect on quantum yield of the transfer to the acceptor η_{DA} has been determined with the help of the known relationships for liquid solutions (3).

Energy migration due to dipole-dipole interaction has been studied earlier by Trlifaj (4). It will be shown that his expression for D_{exc} is suitable only for diluted solutions and weak interaction. A theory of luminescence in solid solutions taking into account energy migration has been recently proposed by Boyarsky and

Domsta (5). Their results differ from those obtained by us. The causes of this disagreement will also be studied.

Let us consider first a solution of one luminescent compound (D) in a non-absorbing solvent. Since we shall deal with appreciable migration in viscous solutions, let us assume that the transfer

$$D_k^* + D_j \longrightarrow D_k + D_j^* \tag{1}$$

is caused by dipole-dipole interaction and the rate constant is defined by the theory of Förster (6), i.e.

$$P_{DD} = (3/2)x^2(R_{oD}^6/r^6\tau_D^o) \tag{2}$$

where x^2 is a function of orientation of the transition dipole moments D_k and D_j, r is the intermolecular distance, τ_D^o is the lifetime of D^* in the absence of transfer, R_{oD}^6 is proportional to the quantum yield of the luminescence (η_D^o) and to the overlapping of absorption and luminescence of D.

Application of (2) for similar molecules is confirmed by the results of the fluorescence concentration depolarization studies of viscous solutions (7,8). To further simplify the calculations x^2 is substituted either by the mean square value of x^2 at $R_{oD} < 0.7$ R (R being the smallest intermolecular distance) or by the square of the mean, $\overline{x^2}$, at $R_{oD} > R$ (1):

$$P_{DD} = S^2R_{oD}^6/\tau_D^o r^6 \tag{3}$$

where S = 1 for the first case and 0.845 for the second one.

Now let us consider the energy migration as a random walk of a localized excitation (an excited state) through the molecules D and take an isotropic distribution of jump directions. Such a walk is equivalent to diffusion with the following coefficient

$$D_{ex} = (1/6)\nu\overline{\ell^2} \tag{4}$$

where ν is the frequency of jumps and $\overline{\ell^2}$ the mean square of the length. Let us calculate ν and ℓ^2.

The frequency ν is equal to the ratio of the total number of exciton jumps N to its lifetime τ_D^o, i.e. $\nu = N/\tau_D^o$. N can be expressed by the quantum yield of energy transfer through donor molecules η_{DD}. The probability that the exciton will decay without jumps is $(1-\eta_{DD})$, the probability that it will decay after the first jump is $\eta_{DD}(1-\eta_{DD})$, after the second jump $\eta_{DD}^2(1-\eta_{DD})$ and so on. Hence,

$$N = \sum_{k=1}^{\infty} k\eta_{DD}^k(1-\eta_{DD}) = \eta_{DD}/(1-\eta_{DD}).$$

To calculate ℓ^2 we express η_{DD} as a sum over spherical layers around D^*:

$$\eta_{DD} = C_D \int_R^\infty \eta_{DD}(r)4\pi r^2 dr .$$ (5)

Then

$$\eta_{DD}\overline{\ell^2} = 4\pi C_D \int_R^\infty \eta_{DD}(r)r^4 dr$$ (6)

According to Ref. (1) we have:

$$\eta_{DD} = \int_0^\infty (dH_D/du)\exp[-u-H_D(u)]du$$ (7)

where

$$H_D(u) = 4\pi C_D \int_R^\infty [1-\exp(-tP_{DD})]r^2 dr$$

and $u = t/\tau_D^0$. Comparing with (5) we obtain

$$\eta_{DD}(r) = \tau_D^0 P_{DD}(r) \int_0^\infty \exp[-u(1+\tau_D^0 P_{DD})-H_D(u)]du.$$ (8)

Thus

$$D_{exc} = (1/6\tau_D^0) \eta_{DD}\overline{\ell^2}/(1-\eta_{DD})$$ (9)

where η_{DD} is determined from (7) (See the tables in (1)) and $\overline{\ell^2}$ from (6) and (8). In the case of $q_D \ll 1$ we obtain

$$D_{exc} \approx (S^{5/3}R_{oD}^2/\pi^{\frac{1}{2}}\tau_D^0)q_D Z^{1/6}F(1/6,1/6,7/6,Z)$$ (10)

where

$$q_D = \frac{2}{3}\pi^{3/2}R_{oD}^3 C_D, \quad Z = S^2\gamma_D/(1+S^2\gamma_D), \quad \gamma_D = R_{oD}^6 R^{-6}$$

and F is a hypergeometric function equal to 1 at $\gamma_D = 0$ and $\pi/3$ at $\gamma_D = \infty$.

Further calculations will be done for a very weak ($\gamma_D \lesssim 0.1$) and very strong ($\gamma_D \gtrsim 10^3$) interaction using the approximate expressions of Ref. (1):

$$H_D \simeq 2q_D(\gamma_D/\pi)^{\frac{1}{2}}u \text{ and } H_D \simeq 2q_D Su^{\frac{1}{2}}$$

respectively. For the first case we obtain:

$$D_{exc} \simeq (R_{oD}^2/\pi^{\frac{1}{2}}\tau_D^0)q_D[1+2q_D(\gamma_D/\pi)^{\frac{1}{2}}]Z_1 F(1/6,1/6,7/6,Z_1)$$ (11)

where

$$Z_1 = \gamma_D/[1+\gamma_D+2q_D(\gamma_D/\pi)^{\frac{1}{2}}].$$

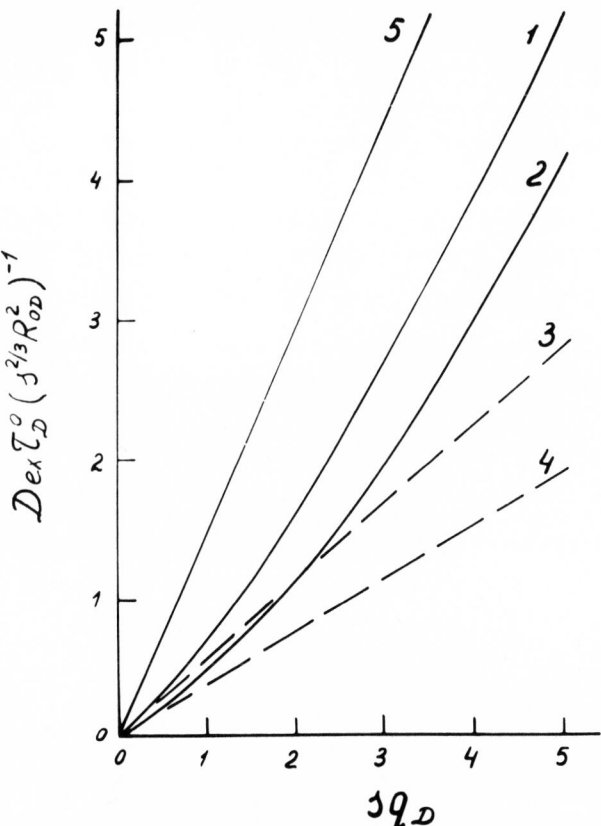

Fig. 1 The dependence of the excited state diffusion coefficient
on the concentration given. Curve 1--from (9) for $\gamma_D = \infty$;
curve 2--from (11) for $\gamma_D = 0.1$; curves 3,4--from (10) for
$\gamma_D = \infty$ and 0.1; curve 5--from (13) for $S^2\gamma_D = 10$.

For the second case numerical integration is needed. The results
are given in Fig. 1 (curve 1).

For the values $q_D \geq 3$ the following equation is obtained

$$D_{exc} \simeq 0.603(S^2 R_{oD}^2/\tau_D^o)q_D^{4/3} . \qquad (12)$$

It should be pointed out that for the given q_D the ratio D_{exc}/R_{oD}^2
is weakly dependent on R_{oD} (Fig. 1).

In the above mentioned paper (4) the following equation is
given (using our notation)

$$D_{exc} = (S^{5/3}R_{oD}^2/\pi^{\frac{1}{2}}\tau_D^o)\gamma_D^{1/6}q_D \qquad (13)$$

which is obtained averaging the transfer rate constant,

$$\overline{v\ell^2} = C_D \int P_{DD} r^2 dr \ ,$$

but not by averaging the probabilities of the process. It is clear that this is allowable only for the transfer being very slow (q_D and γ_D are small). It is easy to notice that under this condition Eq. (13) follows from (11) or (10). For $\gamma_D \gg 1$ Eq. (13) gives an enhanced value of D_{exc} (Fig. 1, curve 5).

Considering the energy migration as equivalent to the diffusion of molecules, the second part of our task can be solved on the basis of the general theory of transfer in liquid solutions (9,3). Generally speaking the analysis requires numerical methods. We have restricted ourselves to the case which allows analytical solution.

The D → A transfer is assumed to be due to dipole-dipole interaction with the parameter R_{oA} in Ref. (3). Then the H(u) function in the decay law of the donor excited state $P_D = \exp(-u \cdot H)$ is of the following form (3):

$$H(u) = 2Sq_A[u^{\frac{1}{2}} + (3/2S)^{\frac{1}{2}} \gamma_A^{-\frac{1}{4}} \sigma_{exc}^{3/4} u] \tag{14}$$

if we follow the conditions

$$\gamma_A \equiv R_{oA}^6 / R_1^6 > 10 \quad \text{and} \quad \sigma_{exc} \equiv D_{exc} \tau_D^0 / R_1^2 < 0.1\gamma_A \ ,$$

where R_1 is the sum of the D and A radii. Eq. (14) is the same as in theory of Sveshnikov (10) for the diffusion controlled transfer (the diffusion coefficient is D_{exc}, the radius of interaction is $R_1(\pi S/6)^{\frac{1}{2}}(\gamma_A/\sigma_{exc})^{\frac{1}{4}}$). The first condition limits the choice of donor-acceptor pairs while the second one limits the donor concentration. For example, if $\gamma_D \simeq \gamma_A$, $C_D \lesssim 0.5$ M. The quantum yield of the transfer to A is determined by (7) and (14) and is equal to:

$$\eta_{DA} = [\alpha Sq_A + QSq_A/(1+\alpha Sq_A)^{\frac{1}{2}}]/(1+\alpha Sq_A) \tag{15}$$

where

$$Q(x) = \pi^{\frac{1}{2}} x \exp(x^2) \mathrm{erfc}(x) \quad \text{and} \quad \alpha = (6/5)^{\frac{1}{2}} \gamma_A^{-\frac{1}{4}} \sigma_{exc}^{3/4} .$$

According to the theory of Bojarski (5) we have

$$\eta_{DA}' = q_A Q(Sq_A + Sq_D)/[q_A + q_D - q_D Q(Sq_A + Sq_D)] \tag{16}$$

It is seen from Fig. 2 that in the case of $R_{Do} \approx R_{Ao}$ equations (15) and (16) give close values of quantum yields. However, at $R_{Do} \ll R_{Ao}$, η_{DA} and η_{DA}' differ appreciably. This also occurs for

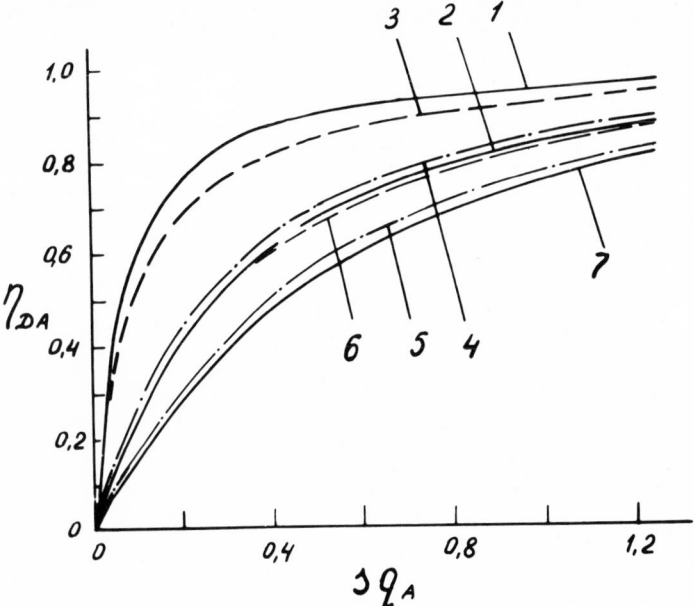

Fig. 2   Transfer quantum yield as a function of the acceptor con-
         centration. Curves 1,2,3--$Sq_D = 5$; 4,5,6--$Sq_D = 1$;
         7--$Sq_D = 0$. Curves 3,6--from (16); 1,2,4,5--from (15)
         $(R_{Do}/R_{Ao})^{3/2} = 1$ for 1 and 2 and equals 0.1 for 4 and 5.

$R_{oD} \gg R_{oA}$.

In Ref. (5) the quantum yield of the donor luminescence η_D is
calculated as a sum of yields for D^* of different generations
(directly excited by an outer source; excited by the first, second,
... transfer). At first the authors take into account the fact that
spatial distribution of the molecules was random and obtain an
expression for η_D, containing a solution for an infinite system of
interrelated differential equations. Note, however, that the orientation and the "forbidden volume" are not taken into account (S = 1,
$R = R_1 = 0$). They then assume that the D^* molecules of all generations experience similar conditions. We now show the limitations
of their analysis.

The total transfer quantum yield from D_j^* to D and A without
taking into account migration equals according to the Förster
theory

$$\eta_T = Q(q_A + q_D), \qquad (17)$$

η_T can also be written as

$$\eta_T = W_D + W_A$$

where W_D and W_A are transfer probabilities from D^* to the other D and to A, respectively. Then

$$\eta_{DA} = W_A(1 + W_D + W_D^2 + \ldots) = W_A/(1-W_D) \ .$$

Taking into account the relationship

$$W_A/W_D = q_A/q_D$$

we obtain

$$\eta_{DA} = q_A\eta_T/[q_A+q_D(1-\eta_T)] \tag{18}$$

which coincides with (16) if η_T is defined by Eq. (17). However, since all W_D and W_A are taken equal, the following expression should have been written:

$$\eta_T = K_A C_A + K_D C_D/[(1/\tau_D^o)+K_A C_A + K_D C_D] \ .$$

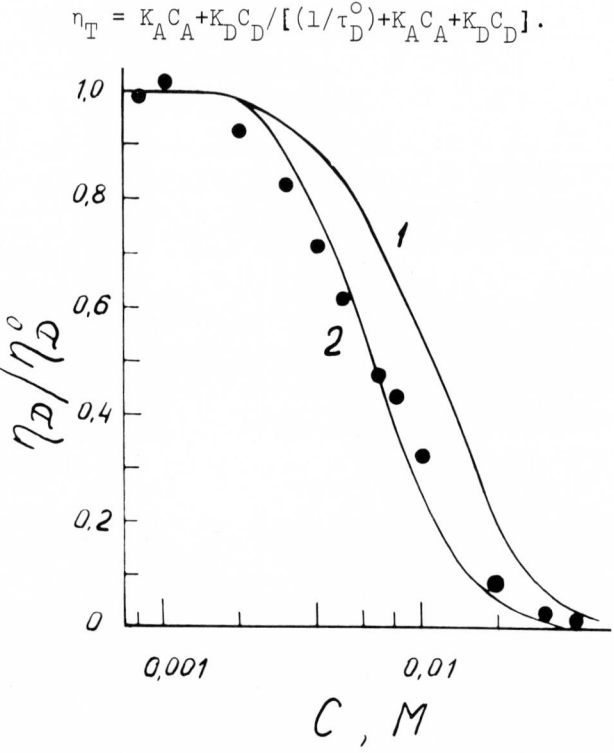

Fig. 3   Rhodamine 6G fluorescence yield as a function of concentra-
         tion: water-glycerine solutions (5.3 P) points are the
         results of measurements (12). Calculated curves:
         (1) without taking into account migration;
         (2) taking into account migration from (15).

Then from (18) we obtain

$$\eta_{DA} = K_A C_A / [(1/\tau_D^O) + K_A C_A],$$

i.e. η_{DA} does not depend on migration, which is incorrect. It is intuitively clear that the assumption W_A = const. is incomparable with migration of the excited state.

Fig. 3 presents the concentration dependence of rhodamine 6G fluorescence yield in water-glycerine solution (viscosity is 5.3 P) (11). Concentration quenching results from the energy transfer to non-fluorescent dimers, and the energy migration through monomer molecules should affect the quenching. Calculations by (15) have been carried out using the values of dimer concentrations and R_O for monomer-monomer and monomer-dimer transfer from Ref. (11). It is assumed that $\eta_D^O = 1$ (12). There is a good agreement of the theory with the experiment.

REFERENCES

1. M. M. Rikenglas, I. M. Rosman, Optica i Spectroscopia, in press.

2. M. M. Rikenglas and I. M. Rosman, Optica i Spectroscopia, in press.

3. I. M. Rosman, Izv. Akad. Nauk SSSR, ser. fiz. _36_, 922 (1972).

4. M. Trlifaj, Czechosl. J. Phys. _8_, 510 (1958).

5. C. Bojarski and J. Domsta, Acta Phys. Hung. _30_, 145 (1971).

6. Th. Förster, Ann. Physik _2_, 55 (1948).

7. M. D. Galanin, Trudy F.I.A.N. _12_, 3 (1960).

8. R. S. Knox, Physica _39_, 361 (1968).

9. M. M. Agrest, S. F. Kilin, M. M. Rikenglas and I. M. Rosman, Optica i Spectroscopia _27_, 946 (1969).

10. L. A. Kuznecowa, B. Ya. Sveshnikov and W. I. Shirokov, Optica i Spectroscopia _2_, 578 (1957).

11. C. Boyarski, J. Kusla and G. Obermüller, Zs. Naturforsch. _26a_ 255 (1971).

12. C. Bojarski, Z. Phys. Chem. N. F. _75_, 242 (1971).

ELECTRONIC ENERGY TRANSFER BETWEEN ORGANIC AND INORGANIC COMPOUNDS

IN SOLUTIONS

V. L. Ermolaev and V. S. Tachin

The State Optical Institute, Leningrad, USSR

ABSTRACT

Some new experimental facts concerning energy transfer in solutions between organic molecules, from organic molecules to rare earth ions and vice versa, and between different rare earth ions are considered and discussed. The possibilities for the application of energy transfer in the study of coordinate chemical processes in solutions are discussed.

INTRODUCTION

During recent years there has been an increasing interest in the study of energy transfer between organic and inorganic compounds. This is because of (1) the wide application of energy transfer as an instrument for the investigation of other problems of photophysics, photochemistry, photobiology and coordinate chemistry, and (2) the discovery of a number of new transfer phenomena, such as transfer forbidden by spin selection rules, transfer from higher excited states of organic molecules, and transfer between two excited molecules.

In the present report we shall briefly discuss a general classification of energy transfer processes between organic molecules from the viewpoint of spin selection rules. The mechanisms for radiationless energy transfer from organic molecules to rare earth (RE) ions and vice versa, and between different RE ions, will be discussed as well. These processes have been recently studied by us.

RESONANCE TRANSFER BETWEEN ORGANIC MOLECULES

The experimental situation regarding inductive-resonance and exchange-resonance energy transfer processes between organic molecules are summarized in Table 1. A crude estimate of the critical

Table 1

The Energy Transfer between Organic Molecules

Process	Medium	R_O (A) or k_t ($M^{-1}sec^{-1}$)	Ref.
1	2	3	4
1. Inductive-Resonant (Dipole-Dipole) Energy Transfer			
A. Spin allowed processes			
1. Singlet-singlet transfer $^1\Gamma_D{}^* + {}^1\Gamma_A \to {}^1\Gamma_D + {}^1\Gamma_A{}^*$	Rigid solutions Liquid solutions	$R_O{\sim}15\text{-}70$ A $k_t{\sim}10^{10}\text{-}10^{11}$	(7,8) (5,6)
2. Singlet-triplet annihilation $^1\Gamma_D{}^* + {}^3\Gamma_A \to {}^1\Gamma_D + {}^3\Gamma_A{}^*$	Rigid solutions	$R_O{\sim}30$	(9)
3. Singlet-singlet annihilation $^1\Gamma_D{}^* + {}^1\Gamma_A{}^* \to {}^1\Gamma_D + {}^1\Gamma_A{}^{**}$	Liquid solutions	$k_t{\sim}10^{10}$ $R_O{\sim}20$	(10,11)
B. Spin forbidden processes			
4. Triplet-singlet transfer $^3\Gamma_D + {}^1\Gamma_A \to {}^1\Gamma_D + {}^1\Gamma_A{}^*$	Rigid solutions	$R_O{\sim}20\text{-}60$	(1,12)
5. Triplet-triplet annihilation $^3\Gamma_D + {}^3\Gamma_A \to {}^1\Gamma_D + {}^3\Gamma_A{}^*$	Rigid solutions	$R_O{\sim}40$	(13)
II. Exchange-Resonant Energy Transfer			
A. Spin allowed processes			
6. Triplet-triplet transfer $^3\Gamma_D + {}^1\Gamma_A \to {}^1\Gamma_D + {}^3\Gamma_A$	Rigid solutions Liquid	$R_O{\sim}10\text{-}15$ $k_t{\sim}10^9\text{-}10^{10}$	(14,15)

cont.

Table 1 (continued)

7. Excited triplet-triplet transfer $^3\Gamma_D{}^* + {}^1\Gamma_A \rightarrow {}^1\Gamma_D + {}^3\Gamma_A$	Rigid solutions	$R_o \sim 6\text{-}8$	(16,17)
8. Triplet-triplet annihilation $^3\Gamma_D + {}^3\Gamma_A \rightarrow {}^1\Gamma_D + {}^1\Gamma_A{}^*$	Liquid solutions	$k_t \sim 10^9\text{-}10^{10}$ $R_o \sim 10\text{-}15$	(18)
9. Triplet-doublet transfer $^3\Gamma_D + {}^2\Gamma_A \rightarrow {}^1\Gamma_D + {}^2\Gamma_A{}^*$	Liquid solutions	$k_t \sim 10^9\text{-}10^{10}$	(19)
B. <u>Spin forbidden processes</u>			
10. Triplet-singlet transfer $^3\Gamma_D + {}^1\Gamma_A \rightarrow {}^1\Gamma_D + 1\Gamma_A$	Liquid solutions	$k_t \sim 10^6\text{-}10^8$	(3)
11. Singlet-triplet transfer $1\Gamma_D{}^* + {}^1\Gamma_A \rightarrow {}^1\Gamma_D + {}^3\Gamma_A$	Liquid solutions	$k_t \sim 10^6\text{-}10^8$	(4)

transfer radius (R_o in A) and (or) bimolecular rate constants (k_t in 1/Msec) are also given there. The rate constants of spin-forbidden transfers are some orders of magnitude less than those of the spin allowed transfer. The allowed transfer, for instance, singlet-singlet or triplet-triplet, does not depend on spin-orbit coupling in donor and acceptor molecules (1,2). The forbidden (triplet-singlet, singlet-triplet) transfer rate constants depend on the square of the spin-orbit factor for the donor or acceptor molecules (3,4).

Another problem arises in connection with the relation between energy transfer by exchange-resonance interaction and electron transfer. Fig. 1 shows the scheme of energy levels for the triplet-triplet transfer. The position of the levels is determined by the ionization potential of a molecule in the corresponding state for the vapor (20, 21). In all cases shown in Fig. 1, triplet energy transfer occurs but electron transfer does not.

However, electron donor-acceptor interactions can influence the energy transfer rate. For instance, in the case of singlet-triplet spin-forbidden transfer, the electron donor ($-OCH_3$) or electron acceptor ($-CN$) substitution in the naphthalene-energy acceptor increases the rate constant of the energy transfer in liquid solutions (4). In this case the transfer phenomenon cannot be described by the theory of weak interactions. Among other things, the transfer rate constant may not be proportional to the overlap integral of the donor emission spectrum and the acceptor absorption spectrum. Unfortunately no experimental proof of this hypothesis has been made.

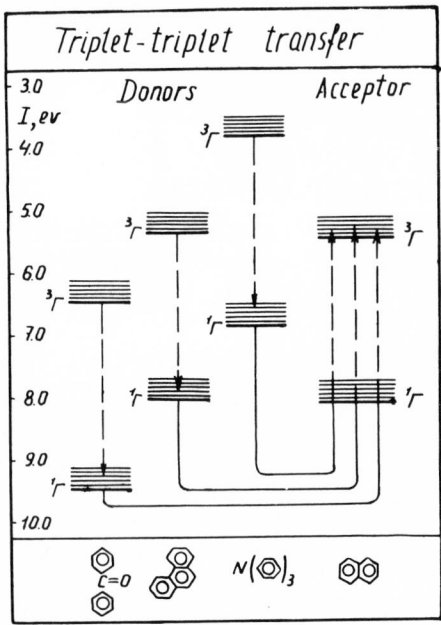

Fig. 1   Energy level scheme for triplet-triplet electronic energy
         transfer between organic molecules. The ordinate is the
         ionization potential (ev) of organic molecules in the ground
         and triplet states. The ionization potentials in the vapors
         are taken from (20, 21).

TRANSFER BETWEEN ORGANIC MOLECULES AND RARE EARTH IONS

     In the second part of our report we briefly review the works
on radiationless energy transfer between organic molecules and rare
earth (RE) ions. We have studied energy transfer in liquid solutions: 1. from organic molecules in the excited singlet or triplet
states to RE ions, 2. from excited RE ions to organic molecules
followed by a transition into singlet or triplet excited states,
and 3. between different RE ions. The mechanisms of the transfer
processes are given in Table 2.

     Radiationless transfer with the participation of RE ions is
distinct from that between organic molecules. Because the RE ions
have a positive charge, a repulsion exists between an ion-donor and
ion-acceptor in solutions. The repulsion results in a decrease in
the probability of an acceptor being found near the excited donor,
as compared with a random distribution. Alternatively, the attraction between RE ions and dye anions increases the probability of the
acceptor being found near the excited donor.

Table 2

The Energy Transfer with Participation
of Rare-Earth Ions (Ln^{3+}) in Solution

Process	Medium	R_0 (A) or k_t ($M^{-1}sec^{-1}$)	Ref.
1. Inductive-Resonant (Dipole-Dipole) Transfer			
1. $^1\Gamma_D(Org.^*) + {}^n\Gamma_j(Ln_A^{3+}) \rightarrow$ $\rightarrow {}^1\Gamma_D(Org.) + {}^m\Gamma_i^*(Ln_A^{3+})$	Liquid solutions	$k_t \sim 10^9 - 10^{10}$	(24)
2. $^m\Gamma_i^*(Ln_D^{3+}) + {}^1\Gamma_A(Org.) \rightarrow$ $\rightarrow {}^n\Gamma_j(Ln_D^{3+}) + {}^1\Gamma_A^*(Org.)$	Rigid solutions	$R_0 \sim 50 - 70$	(23)
	Liquid solutions	$k_t \sim 10^7 - 10^8$	(23)
3. $^m\Gamma_i^*(Ln_D^{3+}) + {}^n\Gamma_j(Ln_A^{3+}) \rightarrow$ $\rightarrow {}^1\Gamma_k(Ln_D^{3+}) + {}^m\Gamma_p^*(Ln_A^{3+})$	Liquid solutions	$k_t \sim 10^4$	(25)
	High DN**)	$R_0 \sim 6 - 10$	
II. Exchange-Resonant Transfer			
4. $^3\Gamma_D(Org.) + {}^n\Gamma_j(Ln_A^{3+}) \rightarrow$ $\rightarrow {}^1\Gamma_D(Org.) + {}^m\Gamma_i(Ln_A^{3+})$	Liquid solutions	$k_t \sim 10^6 - 10^9$	(28,19)
	Intra-complex	$k_t^{***} \sim 10^7 - 10^{11}$	(28,29)
5. $^m\Gamma_i^*(Ln_D^{3+}) + {}^1\Gamma_A(Org.) \rightarrow$ $\rightarrow {}^n\Gamma_j(Ln_D^{3+}) + {}^3\Gamma_A(Org.)$	Liquid solutions	$k_t \sim 10^4 - 10^6$	(30,25)
6. $^3\Gamma_D(Org.) + {}^3\Sigma_g^-(O_2) \rightarrow$ $\rightarrow {}^1\Gamma_D(Org.) + {}^1\Sigma_g^+$ or $^1\Delta_g(O_2)$	Liquid solutions	$k_t \sim 10^8 - 10^{10}$	(8,31)
III. Superexchange-Resonant Transfer			
7. $^m\Gamma_i^*(Ln_D^{3+}) + {}^n\Gamma_j(Ln_A^{3+}) \rightarrow$ $\rightarrow {}^1\Gamma_k(Ln_D^{3+}) + {}^m\Gamma_p(Ln_D^{3+})$	Liquid solutions Low DN	$k_t \sim 10^6 - 10^7$	(25)

*Org. is an organic molecule
**DN - donor number by Gutmann (26)
***k_t - rate constant of the first order (in sec.$^{-1}$)

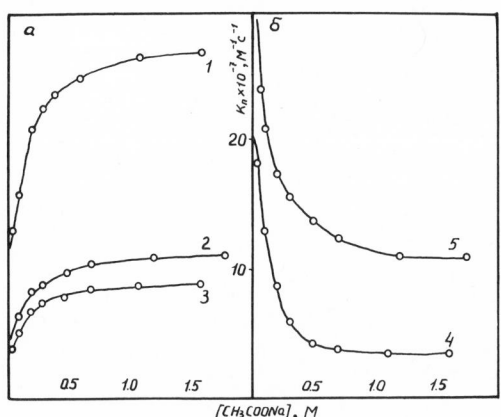

Fig. 2   The dependence of the rate constant, k_t, for energy transfer
        from Tb^{3+} ($^5D_4$) (c = 0.03 M) to different dyes in water at
        293 K, as a function of CH_3COONa concentration.   a - (3+
        ... 1+):(1) system + fuchsine, (2) + novel fuchsine,
        (3) + safranine T. b - (3+ ... 2-) system (4) - acid
        fuchsine, (5) - erythrosine.   From Ref. (23).

        The theory of the influence of donor-acceptor electrostatic
interactions on the dipole-dipole energy transfer in liquids and
viscous solutions was developed by Bodunov (22).   In our laboratory
Shakhverdov (23) showed the Bodunov theory to be in a good agreement
with experimental data on energy transfer between RE ions and anionic
and cationic dyes.   The influence of the addition of sodium acetate
on the rate constant for energy transfer from $Tb^{3+}(^5D_4)$ to the dyes
is shown in Fig. 2.   This addition gives rise to a neutral terbium
complex and leads to the disappearance of the electrostatic repulsion (Fig. 2) or attraction (Fig. 2b) between donor and acceptor.
However, Bodunov's theory does not quantitatively agree with the
experimental data in the case of energy transfer between RE ions.

        One of the authors and Antipenko found the energy transfer rate
constant between RE ions depends on solvent and alkaline metal salt
addition (32).   The transfer rate constants are small for those
solvents characterized by a high Gutmann donor number, e.g., dimethylsulfoxide, dimethylformamide, tributylphosphate, water, where
the bond strength of RE ion -solvent molecules increases with increase in the Gutmann donor number (26).   In these polar solvents
the first coordination sphere around the RE ions consists of neutral

solvent molecules and the ions are repelled from each other.

The transfer rate constants are much higher for those solvents characterized by a low Gutmann donor number, e.g., acetone, acetonitrile, ethylacetate. There the RE ions exist in neutral complexes containing anions (in our case NO_3^-) and the repulsion disappears.

In solvents with high donor number (water) the small transfer rate constants increase sharply upon the addition of alkaline salts $NaCOOCH_3$, $KCOOH$, $KCNS$, etc. to the solution. The salt anions neutralize the RE ion charge and decrease the electrostatic repulsion. The rate constant increase, however, is too great to be explained by the disappearance of the repulsion only.

The authors and Gruzdev (25) have suggested a model using unstable bridge complexes of the type:

$$(COOCH_3^-)_2 Tb^{3+} \ldots O = C(CH_3) = O \ldots Nd^{3+}(COOCH_3^-)_3.$$

The formation of these complexes leads to an encounter time delay between the excited donor (Tb^{3+}) and acceptor (Nd^{3+}) and to an increase in the observed transfer rate constant. The average lifetime of the complex depends on the properties of the solvents which are ordered as follows: water, methanol, ethanol, n-propanol. The maximal transfer rate constant increases in this way as well (see Table 3 and Fig. 3). Energy transfer between RE ion donor and RE

Table 3

The Rate Constants of the Energy Transfer
from Tb^{3+} to Nd^{3+} T = 293K

Solvent	R_o theor. (A)	k_t* theor. ($M^{-1}sec.^{-1}$)	k_t exp. ($M^{-1}sec.^{-1}$)	k_t max. with $NaCOOCH_3$	ε
Water	6-8	4×10^3	6×10^3	1.3×10^8	(81)
Methanol	10-11	2×10^4	2.4×10^4	4.4×10^6	(32,6)
Ethanol	10-11	2×10^4	3.8×10^4	13×10^6	(24,3)
Propanol	10-11	2×10^4	5.5×10^4	28×10^6	(20,1)

* $k_t = 4\pi R_o^6 / 3\tau_D^0 (r_D + r_A)^3$ was calculated by the Tupitskii-Bagdasaryan formulas for a discontinuous diffusion.

ion acceptor in bridge complexes occurs by the superexchange resonance mechanism. Energy transfer in those solvents with high donor numbers occurs due to inductive resonance dipole-dipole interactions. Thus, the transfer mechanism in the RE ion solutions depends on the nature of the solvents and on the addition of anions. Bodunov's

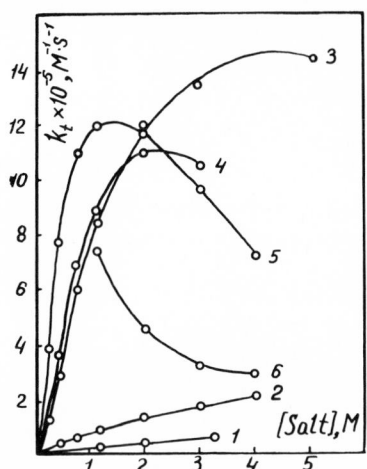

Fig. 3   The dependence of the rate constant, k_t, of energy transfer between Tb^{3+} ($^5D_4$) and Nd^{3+} ions as a function of concentration of alkali salts added to water at 293 K. 1, CCl_3COOK; 2, CCl_2HCOOK; 3, $HCOOK$; 4, $CClH_2COOK$; 5, CH_3COONa; 6, K_2CO_3. (From Ref. 32.)

theory developed for the case of dipole-dipole transfer cannot describe the data for systems involving exchange transfer.

Recently, one of the authors and Shakhverdov (23) have investigated the energy transfer from RE ions to the fluorescence state of dyes and showed it to occur through the dipole-dipole mechanism. The phenomenon is favorable for verification of theories of the influence of viscosity on dipole-dipole energy transfer. It is shown that the theories of Galanin (5), Tunitskii-Bagdasaryan (33) and Rosman, et al. (6) are in good agreement with experiment. On the contrary, the theories of Yokoto-Tanimoto (34) and Voltz, et al. (35) do not apply to transfer in solutions with low viscosity.

Energy transfer from organic molecules in the triplet state to RE ions and vice versa occurs in liquid solutions through the exchange resonance mechanism. This conclusion is confirmed by the dependence of the transfer rate constant on the presence of active substituents in the organic molecules. The most effective transfer takes place to and from aromatic ketones (28, 36) and aromatic acid anions (29). The oxygen non-bonding electrons of the group $\equiv C=O$ or $-COO^-$ form a coordination bond with the RE ion and energy transfer occurs in the complex. The observed transfer rate constant depends

on the lifetime of the unstable complex, as we reported earlier
(37).

As has been shown (38), the energy transfer rate constant from
ketones with an n,π^*-character for the lowest triplet state to RE
ions is more than one order of magnitude greater than that from
ketones with the π,π^*-triplet. It is clear, as concerns the case
of the n,π^*-triplet state, the overlap integral of the excited orbital localized in the region of C=O bond and the 4f-orbital of the
RE ions is much greater than the overlap integral of the π,π^*-
triplet state of ketones and the 4f-orbital.

Recently, Morina and Sveshnikova (38), have been shown that
both diffusion and intra-complex energy transfer from the sensitizer
(triplet ketone) to the acceptor (RE ion) can occur simultaneously
in some low donor number solvents (acetone, acetonitrile). The
stable complexes of RE ions with aromatic acid anions ($Ar-COO^-$) are
also formed in the solvents with higher donor number, e.g. methanol,
as compared to acetone or acetonitrile. The intra-complex transfer
rate constants are of the order of 10^9 to 10^{10}/sec (29). If an
isolated $-CH_2-$ group is put between the aromatic ring and the car-
boxyl group, the intra-complex transfer rate constant from the
ligand triplet state to RE ions should be one order of magnitude
lower than that without the $-CH_2-$ group (29).

To summarize this section, we can say that both inductive
resonance and exchange resonance energy transfer mechanisms occur
between organic molecules and RE ions. The transfer mechanism is
primarily determined by the electronic transition of the organic
molecules. If it is a spin-allowed transition (singlet-singlet),
the transfer mechanism is inductive resonance; if it is a spin-
forbidden transition, the transfer mechanism is exchange resonance.
The transfer between two different RE ions in solution can occur
both by the inductive resonance mechanism (in the case of electrostatic repulsion) and by the superexchange resonance mechanism (in
the case of bridge complexes). The novel factors that influence
the rate constant of energy transfer between RE ions are the electrostatic repulsion and the formation of unstable donor-acceptor complexes.

APPLICATIONS OF ENERGY TRANSFER
TO COORDINATE CHEMICAL PROCESSES

Energy transfer will certainly have extensive applications in
the research on coordinate chemical processes with RE ions in solutions (39). The complexing with anions results in an ion charge
neutralization, removal of electrostatic repulsion and formation
of bridge complexes. These processes are followed by a fast increase of the energy transfer rate constant between the RE ions.

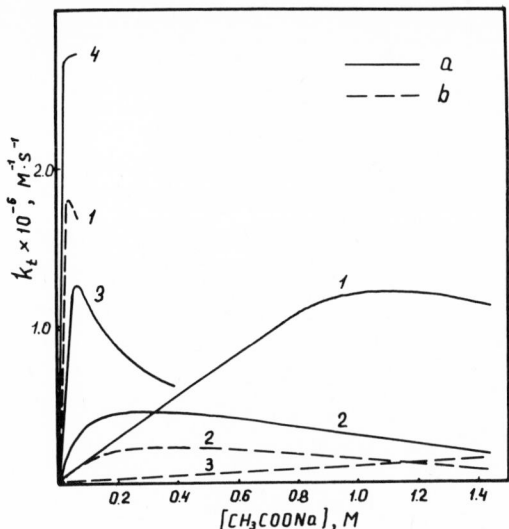

Fig. 4   The dependence of k_t on CH_3COONa concentration in different
          solvents at 293 K.   a - Energy transfer from Tb^{3+} ($^5D_4$) to
          Nd^{3+}; I,a water; II,a methanol; III,a ethanol; IV,a
          n-propanol.   b - The energy transfer from Eu^{3+} ($^5D_0$) to
          Nd^{3+}; I,b water; II,b methanol; IV,b n-propanol.

Therefore, the latter is a good indicator of complexing processes.
The dependence of k_t on Tb^{3+} ($^5D_4$) to Nd^{3+} in water as a function
of alkali acetate concentration is shown in Fig. 3, taken from the
work by one of the authors and Antipenko (32). The increase in
electron donating power in the series: CCl_3COO^-, CCl_2HCOO^-,
$CClH_2COO^-$, $HCOO^-$, CH_3COO^-, and Co_3^{2-}, gives rise to increase complexing with RE ions. The dependence of k_t on the CH_3COONa concentration in different solvents is shown in Fig. 4. The data of
Fig. 4 point out the role of the solvent in complexing.

     The study of the energy transfer rate between triplet organic
molecules and RE ions and vice versa is also a convenient indicator
of complexing processes. By this method the existence of both
stable and unstable aromatic ketone-RE complexes was shown (38) for
solvents with low donor number (acetonitrile, acetone). There are
no stable complexes in solvents with high donor number. Most likely
the energy transfer method suggested allows for the study of the
kinetics of complex formation. For example, it is shown that an
activation energy of about 1-3 kcal/mole is required for triplet
ketone-RE ion complex formation.

     The second approach to coordinate chemistry is based on the

idea suggested by one of the authors and Sveshnikova (40) relating
to the mechanism of electronic energy degradation (radiationless
transition) in RE ions and transition metal (TM) ions. In this
series of works (40, 41) it was shown that the degradation rate
constants of RE and TM ions could be calculated as arising from
dipole-dipole transfer to solvent molecules surrounding the ion, the
latter being excited to high vibrational states. The dependence of
the degradation rate constant (k_d) on the distance from the excited
ion to the high vibrational group makes it possible on the basis of
the luminescence decay time (or yield) to come to some conclusions
about the appearance or disappearance of high vibrational molecules
from the first ionic coordinational sphere. The use of solvents
with stable (deuterium) isotopes extends the possible range of this
method.

   For example, the dependence of the $Tb^{3+}(^5D_4)$ lifetime (τ_ℓ) on
temperature in various solvents is shown in Fig. 5. For the sol-

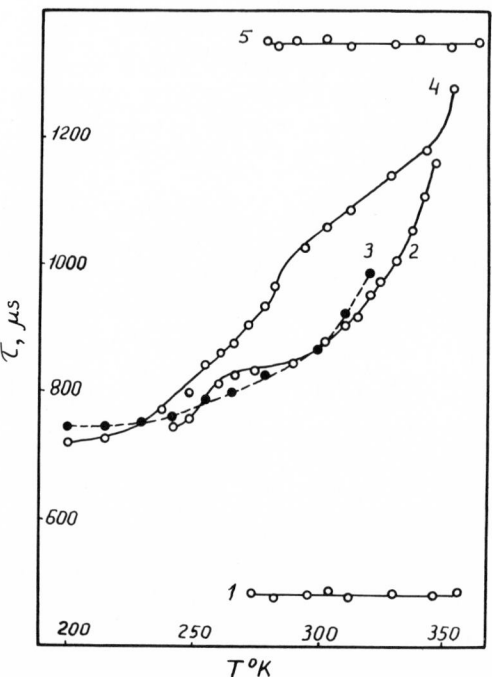

Fig. 5 The luminescence decay time (τ_ℓ) of Tb^{3+} $(^5D_4)$ in solutions
        as a function of the temperature: I water; II acetonitrile;
        III acetone-h_6; IV acetone-d_6; V dimethyl formamide.

vents with high donor numbers (methanol, water, dimethylsulfoxide, et al.), the $\tau(Tb^{3+})$ is independent of temperature. On the other hand, for solvents with low donor numbers (acetone, acetonitrile) there is an unusual dependence of $\tau(Tb^{3+})$ on temperature. Both τ and the quantum yield of the Tb^{3+} luminescence increase with increasing temperature (see Fig. 5). Such an anomalous behavior of $\tau(Tb^{3+})$ is evidence for a reorganization of the first Tb^{3+} coordinate sphere with temperature. It is well known (42), that at low temperature the RE ions in acetone are surrounded by three NO_3^- anions and four or five water molecules. It follows from the data obtained by us that some of the water molecules in the first coordination sphere are replaced by acetone molecules when the temperature increases. The experiment with acetone-d_6 confirms our conclusion. Many other conclusions on coordinate-chemistry processes in solutions with RE and TM ions have been made through the series of published works from our laboratory (25, 31, 32, 36-41).

We believe that energy transfer and electronic excitation degradation methods should be used in coordinate chemistry research because of their simplicity and possibilities. Moreover, the necessary luminescence apparatus is much simpler and inexpensive compared to those for nuclear magnetic resonance.

REFERENCES

(1) Th. Förster, Discus. Faraday Soc., 27, 7 (1959).

(2) V. L. Ermolaev, Opt. i Spectr., 6, 642 (1959).

(3) R. F. Vasilyev, Nature, 196, 668 (1962); 200, 773 (1963).

(4) V. L. Ermolaev, E. B. Sveshnikova, Opt. i Spectr. 28, 601 (1970).

(5) M. Galanin, Trudy Fiz. Inst. of Acd. of Sci. USSR, 12, 3 (1960).

(6) I. M. Rosman, Izv. Acad. Sci. USSR, ser. phys., 36, 922 (1972).

(7) Th. Förster, Ann. Phys., 2, 55 (1948).

(8) J. B. Birks, in: Photophysics of Aromatic Molecules, Wiley, N.Y., 1969.

(9) R. G. Bennett, J. Chem. Phys., 41, 3048 (1964).

(10) N. A. Tolstoi, A. P. Abramov, Fiz. Tverd. Tela, 9, 340 (1967).

(11) S. D. Babenko, V. A. Benderskii, A. G. Lavrushko, Izv. Acad.
      Sci. USSR, ser. Phys., $\underline{36}$, 1113 (1972).

(12) V. L. Ermolaev, E. B. Sveshnikova, Izv. Acad. Sci. USSR,
      ser. phys., $\underline{26}$, 29 (1962); Dokl. Acad. Sci. USSR $\underline{149}$, 1295
      (1963).

(13) R. E. Kellogg, J. Chem. Phys., $\underline{41}$, 3046 (1964).

(14) V. L. Ermolaev, A. N. Terenin, Memory S. I. Vavilov (Pamyati
      S. I. Vavilova), publ. Acad. Sci. USSR, 137 (1952).

(15) A. N. Terenin, V. L. Ermolaev, Trans. Farad. Soc., $\underline{52}$, 1042
      (1956).

(16) A. N. Terenin, V. V. Rylkov, V. E. Kholmogorov, Photochem.
      Photobiol., $\underline{5}$, 543 (1966).

(17) M. V. Alfimov, I. G. Batekha, Yu. B. Sheck, V. I. Gerko,
      Spectrochim. Acta, $\underline{27A}$, 329 (1971).

(18) C. A. Parker, C. G. Hatchard, Proc. Royal Soc. (L), $\underline{A269}$,
      574 (1962); A. I. Bogatyreva, A. G. Sklyarova, A. L.
      Bugachenko, Khim. Vys. Energii, $\underline{5}$, 37 (1971).

(19) R. B. Cundall, G. B. Evans, E. J. Land, J. Phys. Chem., $\underline{73}$,
      3982 (1969).

(20) V. I. Vedeneev, L. V. Gurvich, V. N. Kondratyev, V. A.
      Medvedev, E. L. Frankevich, in: Destruction Energies of
      Chemical Bonds. Ionization Potentials and Electron Affinity.
      Publ. Acad. Sci. USSR, M. (1962).

(21) F. I. Vilesov, Uspekhi Fotoniki, Publi. Leningrad State
      University, L., $\underline{1}$, 5 (1969).

(22) E. N. Bodunov, Izv. Acad. Sci. USSR, ser. phys., $\underline{36}$, 996
      (1972).

(23) T. A. Shakhverdov, Izv. Acad. Sci. USSR, ser. Phys., $\underline{36}$,
      1018 (1972).

(24) V. L. Ermolaev, T. A. Shakhverdov, Opt. i Spectr., $\underline{26}$,
      845 (1969).

(25) V. L. Ermolaev, V. P. Gruzdev, V. S. Tachin, Izv. Acad. Sci.
      USSR, ser. phys., $\underline{36}$, 984 (1972).

(26) V. Gutmann, in: Coordination Chemistry in Non-Aqueous Solu-
      tions, Springer-Verlag, N.Y. (1968).

(27) E. Matovich, C. K. Suzuki, J. Chem. Phys., $\underline{39}$, 1442 (1963).

(28) V. L. Ermolaev, B. M. Antipenko, E. B. Sveshnikova, V. S.
      Tachin, T. A. Shakhverdov, in: Molecular Photonic, Publishing
      "Nauka", L., 44 (1970).

(29) N. A. Kazanskaya, V. L. Ermolaev, A. V. Moshinskaya, A. A.
      Petrov, Yu. I. Kheruze, Opt. i Spectr., $\underline{28}$, 1150 (1970);
      $\underline{32}$, 82 (1972).

(30) A. P. Alexandrov, E. P. Volkova, V. N. Genkin, Opt. i
      Spectr., $\underline{27}$, 439 (1969).

(31) V. F. Morina, Izv. Acad. Sci. USSR, ser. phys., $\underline{36}$, 988
      (1972).

(32) B. M. Antipenko, V. L. Ermolaev, Opt. i Spectr., $\underline{28}$, 93
      (1970); $\underline{29}$, 90 (1970); $\underline{30}$, 75 (1971).

(33) N. N. Tunitskii, Ch. S. Bagdasaryan, Opt. i Spectr., $\underline{15}$,
      100 (1963).

(34) M. Yokota, O. Tanimoto, J. Phys. Soc. Japan, $\underline{22}$, 779 (1967).

(35) R. Voltz, G. Lanstriat, A. Coche, J. Chem. Phys. et Phys.-
      Chim. Biol., $\underline{63}$, 1253 (1966).

(36) V. L. Ermolaev, V. S. Tachin, Opt. i Spectr., $\underline{29}$, 93 (1970).

(37) V. L. Ermolaev, V. S. Tachin, Opt. i Spectr., $\underline{27}$, 1007
      (1969).

(38) V. F. Morina, E. B. Sveshnikova, Opt. i Spectr., $\underline{31}$, 599
      (1971).

(39) B. M. Antipenko, V. L. Ermolaev, T. A. Privalova, Jour.
      Inorg. Chem. (USSR), $\underline{17}$, 1252 (1972).

(40) E. B. Sveshnikova, V. L. Ermolaev, Opt. i Spectr., $\underline{30}$, 379
      (1971); Izv. Acad. Sci. USSR ser. phys., $\underline{35}$, 148 (1971).

(41) E. B. Sveshnikova, I. B. Neporent, Izv. Acad. Sci. USSR,
      ser. phys., $\underline{36}$, 1087 (1972); this journal.

(42) K. B. Yatsimirskii, V. A. Bidzilya, N. K. Davidenko, Doklady
      Acad. Sci. USSR, $\underline{202}$, 1379 (1972).

SPIN MEMORY IN TRIPLET-TRIPLET ENERGY TRANSFER[*]

Mark Sharnoff and Elsa B. Iturbe

Department of Physics, University of Delaware

Newark, Delaware 19711 USA

ABSTRACT

We have used optical methods to detect paramagnetic resonance signals from triplet excitons and from trapped triplet excitations in benzophenone, and we have shown that spin angular momentum is conserved in triplet-triplet energy transfer from the excitons to the traps.

In this communication we shall inquire into the nature of the interaction responsible for electronic energy transfer of the type $^3D^* + {}^1A \rightarrow {}^1D + {}^3A^*$. We know from the early work of Terenin and Ermolaev (1) that this interaction is of short range, of distance comparable with molecular diameters. The magnitudes of triplet exciton diffusion coefficients and of Davydov splittings in the $S_0 \rightarrow T_1$ absorption spectra of pure organic molecular crystals provide a very good idea of the interaction strength, which is found to lie between 0.01 and 50 cm^{-1}, depending upon molecular species and upon crystalline structure. These two pieces of information suggest that Heisenberg exchange is the dominant mechanism of triplet-triplet transfer in most systems, and this notion is supported by the calculations of Förster (2), of Dexter (3), and of Rice and Jortner (4). The most conspicuous property of triplet energy transfer via Heisenberg exchange has yet to be demonstrated, however; this is the conservation of spin angular momentum during its transfer from donor to acceptor. In this note we shall describe the experimental proof that the spin orientation is conserved during triplet energy transfer.

The donor species which we have used in our experiments are

346

free triplet excitons in pure crystalline benzophenone (orthorhombic phase). The acceptor species are benzophenone molecules which
are slightly mis-oriented in the lattice and which, because of
their distorted configurations, act as traps for the triplet excitons. The traps are shallow, and the temperature of the crystal
is accordingly maintained at 4.2 K. By resorting to an optical
method for the detection of electron paramagnetic resonance (EPR),
we have succeeded in observing the magnetic resonance of both
species. In this method, the EPR of the electronically excited
species of interest is alternately saturated by a microwave pulse,
applied at the resonance frequency, and then allowed to relax. The
train of intense microwave pulses thus produces periodic changes in
the populations of the spin sublevels, and these changes cause, in
turn, a periodic modulation of the phosphorescence. The modulation
of the phosphorescence can be detected photoelectrically and provides, when present, a very sensitive indicator that magnetic resonance has been excited. Details concerning this method and its
application to characterize the excitonic resonances and the
resonances of trapped excitations are given in earlier publications
(5,6,7,8).

Several optically detected EPR signals of excitonic and of
trap species are depicted in Fig. 1. The excitonic EPR has an
extremely short spin-lattice relaxation time ($\approx 10^{-7}$ sec. at 4.2 K),
and its pulsed EPR signals remain constant in intensity up to very
high microwave pulsation rates (Fig. 1 efg). The EPR of the traps
has a long spin-lattice relaxation time, and the populations of
the spin sublevels of the traps are not capable of following rapid
pulsations in microwave intensity. The amplitudes of the trap
signals therefore diminsh as the pulsation rate is increased. The
spin lattice relaxation times of the traps are also long in comparison with their phosphorescence lifetimes (6), which implies
that once an exciton has become trapped, it does not reemerge as
an exciton.[**] The magnetization which is imparted to the acceptor
at the moment in which it traps an exciton must persist throughout
the acceptor's lifetime. It should be noted (Fig. 1a), that the
trap signals are invariably positive. The low field exciton signal
is also positive, but the high field line is negative.

Since we can discriminate magnetically between excitonic EPR
and the EPR of traps, we can easily determine which portion of the
rather complicated crystalline phosphorescence spectrum (Fig. 2a)
arises from traps and which portion arises from excitons. For this
purpose the magnetic field is set at the value for which excitons
are in resonance, the pulsed monochromatic microwaves are applied,
and the pulsations which result in the phosphorescence are recorded
while the wavelength transmitted by the scanning spectrophotometer
is varied. The experiment is repeated with the magnetic field now
set to the center of the trap EPR. A set of typical results is

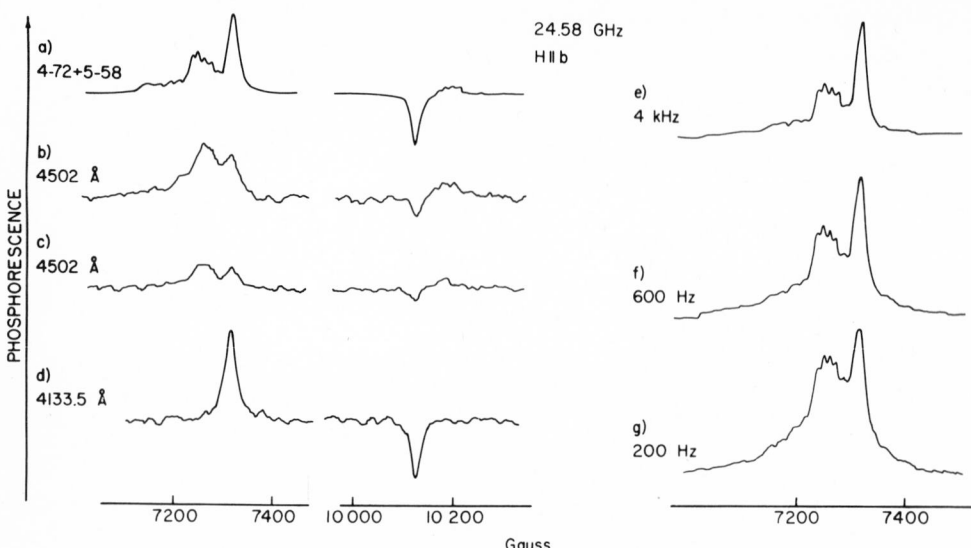

Fig. 1 Optically detected magnetic resonance signals from triplet
excitons (smooth peaks at 7330g and 10120g) and trapped
triplet excitations (positive shoulders) in crystalline
benzophenone at 4.2K. Ordinates represent the difference
between phosphorescence intensities with microwaves on and
off. In traces (a), (b), (c) and (d) the microwave repetition rates are 4 kHz, 200 Hz, 4 kHz, and 4 kHz, respectively.
(Trace (d) is unaffected by changes in the repetition rate.)
In traces (e), (f) and (g), CS 4-72 + 5-58 broadband filters
are again used and the traces scanned with the indicated
microwave, repetition rates. The exciton resonance is
nearly independent of microwave repetition rate, but the
trap signals diminish with increasing rate.

displayed in Fig. 2b. It is apparent that the excitonic emission
origin lies at 4133.5 A, in coincidence with the 0-0 band of the
crystalline $S_0 \rightarrow T_1$ absorption (9). The weak lines at 4210, 4232,
4524, and 4550 A and the strong lines at 4152, 4436, and 4457A are
also seen to be absent from the microwave-modulated phosphorescence
from the traps and are accordingly to be ascribed to excitonic
phosphorescence. Most of the remaining structure arises from trap
phosphorescence, as may be seen from its presence in both the trap
and the excitonic traces in Fig. 2b. (The appearance of some trap-
like structure in the excitonic trace is a consequence of spin-
selective energy transfer from the excitons to the traps, as we
shall presently show.) The separation and positive identification
of the components which have thus been achieved in this rather
complicated total emission spectrum represent an application of
microwave phosphorescence double resonance spectroscopy which has

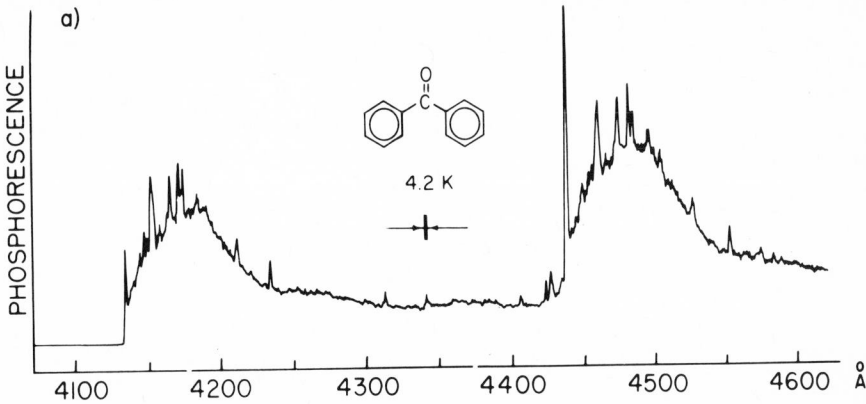

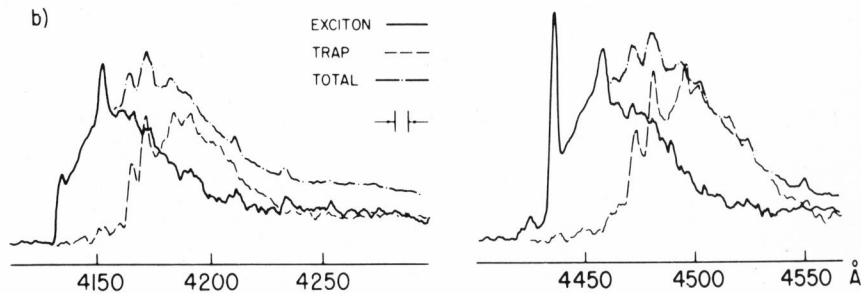

Fig. 2 (a) total phosphorescence of benzophenone at 4.2K, observed
        at 0.85A resolution; (b) comparison of total phosphorescence
        observed by DC methods with the amplitude modulated phos-
        phorescence observed under periodic satuaration of the ex-
        citon resonance or trap resonance. Resolution 3.3A. Cry-
        stal whose spectra appear is the same as the one used to
        obtain the data of Fig. 1.

not heretofore been described.

     Since the trap signals can be discriminated from the excitonic
EPR signals both optically and by magnetic means, it becomes possible to study changes which occur in the populations of the spin
sublevels of the traps when the excitonic EPR is saturated, and
vice versa. Such studies enable us to observe the excitons both
before and during the trapping process and to show that, as they
become trapped, their magnetization is transferred without change
in orientation. If we attempt optical detection of EPR by using
only light of wavelength close to the exciton 0-0 band at 4133.5A

(Fig. 1d), we find, as expected, that we observe only excitonic
EPR. Its amplitude is independent of the microwave pulsation rate.
If we carry out the detection using light near 4500A, which consists almost entirely of trap phosphorescence, we observe in addition to the usual trap EPR a pair of signals which occur at
precisely the magnetic fields at which exciton resonance normally
takes place. These extra signals have the same width and shape as
the exciton signals, and also the same signs. The negative sign
of the high field signal proves that the extra signals are indeed
excitonic resonances,for the direct excitation of trap EPR produces
only positive signals. We can be confident, furthermore, that
these extra exciton signals are conveyed via phosphorescence from
traps, rather than via small amounts of excitonic phosphorescence
present near 4500A, for the amplitudes diminish with increasing
microwave pulsation rate in just the same fashion as do the normal
trap resonance signals.

The conclusion just established implies that saturation of the
excitonic EPR causes a redistribution of the populations of the spin
sublevels of the traps, and we must next inquire into the mechanism
by which this redistribution takes place. Only two simple possibilities suggest themselves: (a) the populations of the trap spin
sublevels are continuously renewed by spin-selective trapping of
excitons; (b) the trapping of excitons is not spin-selective, but
the populations of the spin sublevels of the traps engage in magnetic cross-relaxation (10) with those of the excitons.† Possibility (b) can be quickly eliminated; for, in order to be effective
in communicating information to the trap spin sublevels, the cross
relaxation processes would have to occur rapidly in comparison to
phosphorescence from the traps. We have, however, already seen
that the spin-lattice relaxation times of the traps are long in
comparison to their phosphorescence lifetimes. We conclude that
cross-relaxation between traps and excitons, if present at all, is
far too week to give rise to extra exciton signals of amplitudes
comparable to those of the trap signals themselves, and that mechanism (a) must be responsible for their appearance.††

We will now consider the population dynamics in greater detail
and show that the spin-selective trapping mechanism is, in fact,
spin conservative. In Fig. 3 are shown Zeeman energy diagrams appropriate for a magnetic field directed along the crystalline b-
axis. The spin degeneracy is removed, even in zero field, by
magnetic dipolar interaction between the unpaired electrons. The
excitonic m=0 level is a pure spin state, regardless of field
strength, and this particular spin state is known to emit approximately 75% of the exciton phosphorescence (6). The m=0 state of
the trapped species is nearly a pure spin state because the characteristic z-axes of the traps are inclined by less than 20° to the
crystalline b-axis. The state to which it corresponds in zero field

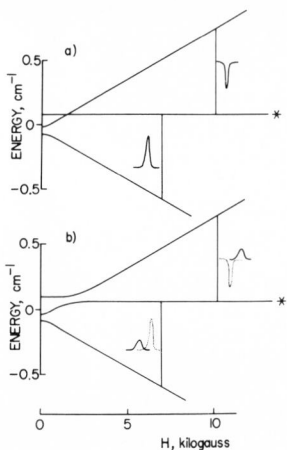

Fig. 3 Energy level and emission diagram for excitons (a) and
        trapped excitations (b) with magnetic field directed along
        crystalline b-axis. Emissive level is designated in each
        case by the asterisk. The allowed microwave transitions
        are represented by the vertical lines and produce the
        signal patterns indicated. In (b) the solid curves repre-
        sent the effects of saturating the trap EPR and the dotted
        lines the effects of saturating the exciton EPR.

is known to emit 90% of the phosphorescence from traps (11,12).

     Because the exciton spin-lattice relaxation time is orders of
magnitude shorter than the excitonic lifetime, the populations of
the excitonic spin sublevels are described by a Boltzmann distribution. When the low-field EPR (indicated by the vertical line near
7.75 kgauss) is suddenly saturated, the population of the m=0 level
increases. As this is the most radiative level, the excitonic phosphorescence intensity increases. Saturation of the high-field
excitonic EPR decreases the population of the most radiative, m=0
level, and the excitonic phosphorescence becomes weaker. The signs
of the excitonic EPR signals are thus explained.

     A different situation prevails in the case of the trap spin
levels. Here the spin-lattice relaxation times are long in comparison to the phosphorescence lifetimes, and the populations of
the less radiative levels will build up to values which are considerably larger than that of the most radiative level. Saturation
of either of the trap resonances will therefore cause an increase
in the population of the most radiative level, and hence both
resonances result in intensification of the phosphorescence.

     We now consider the effects which saturation of excitonic EPR

will have on the populations of the spin sublevels of the traps.
If we suppose that when an exciton in an m=0 spin state becomes
captured, m=0 trap states are exclusively formed as a result, and
that trapping of an exciton whose spin projection is +h/2π will
give rise exclusively to a trap with spin projection +h/2π, etc.,
we find that any process which diminishes the population of the
m=0 excitonic sublevel will result in a subsequent diminution of
the population of the m=0 levels of the traps. Saturation of the
high-field excitonic resonance reduces the population of the
excitonic m=0 level, and, in causing a consequent decrease in the
populations of the most radiative, m=0 levels of the traps, gives
rise to a trap-detected excitonic signal of negative sign. It is
thus clear why the low-field exciton resonance is positive, and the
high-field exciton resonance is negative, when detected via the
trap phosphorescence.

An elaboration of this kind of reasoning leads to the predic-
tion that when the magnetic field is directed along any of the
three crystallographic symmetry axes, the exciton resonances detected
via trap phosphorescence will have the same signs as when detected
directly via the excitonic phosphorescence. This is the state of
affairs which we find experimentally, and we offer it as proof that
spin orientation is conserved during the trapping of triplet excitons. We believe this result to be strong evidence that spin
orientation is normally conserved during any non-radiative[#] process,
resonant or non-resonant, of the type $3D^*+^1A \rightarrow {}^1D+^3A^*$, and that
Heisenberg exchange is the dominant mechanism.

In summary, we remark that we have been able to determine
separately the spin populations of the donor species and the
acceptor species and to observe by direct means the transfer of
angular momentum from one to the other. This method makes possible
a quantitative assessment of the degree to which spin non-conserving
interactions are involved, an assesment in which we are at present
actively engaged.

FOOTNOTES AND REFERENCES

*Research supported by the National Science Foundation.

**Re-emergence of trapped excitations would allow the trap signals
   to relax indirectly via the very rapid spin lattice relaxation
   of the excitons. The spin lattice relaxation times of the traps
   would then appear shorter than their phosphorescence lifetimes.

†Cross-relaxation is used here to mean interchange of Zeeman
   energy between two species without interchange of electronic
   excitation energy.

††Veeman and van der Waals (Ref. 13) have observed the occurrence of cross-relaxation between trapped species in benzophenone. The cross-relaxation occurred only under conditions which insured that a magnetic transition between spin sublevels of one species was in resonance with a magnetic transition in the other. We may note that no resonance between traps and excitons occurs under the conditions of our experiments.

#We must also exempt processes involving virtual photons (such as are responsible for the R^{-6} dependence in the Förster theory) from this conclusion.

1. A. N. Terenin & V. L. Ermolaev, Dokl, Akad. Nauk SSSR 85, 547 (1952).

2. Th. Förster, Ann. d. Physik 2, 55 (1948).

3. D. L. Dexter, J. Chem. Phys. 21, 836 (1953).

4. J. Jortner, S. A. Rice, J. L. Katz, and S. I. Choi, J. Chem. Phys. 42, 309 (1965). (See also R. H. Clarke and R. M. Hochstrasser, J. Chem. Phys. 46, 4532 (1967).)

5. M. Sharnoff, J. Chem. Phys. 46, 3263 (1967).

6. M. Sharnoff, Symposia Faraday Society, No. 3, 137 (1969).

7. M. Sharnoff and E. B. Iturbe, Phys. Rev. Lett., 27, 576 (1971).

8. M. Sharnoff, Chem. Phys. Lett. (in press).

9. S. Dym, R. M. Hochstrasser, and M. Schaefer, J. Chem. Phys. 48, 646 (1968).

10. N. Bloembergen, S. Shapiro, P. S. Pershan, and J. O. Artman, Phys. Rev. 114, 445 (1959).

11. I.-Y. Chan and J. Schmidt, Symposia Faraday Soc., No. 3, 156 (1969).

12. C. J. Winscom and A. H. Maki, Chem. Phys. Lett. 12, 264 (1971).

13. W. S. Veeman and J. H. van der Waals, Chem. Phys. Lett. 7, 65 (1970).

ENERGY TRANSFER FROM CHEMICALLY EXCITED CARBONYL COMPOUNDS TO

ANTHRACENE DERIVATIVES AND OXYGEN

V. A. Belyakov, G. F. Fedorova, and R. F. Vassil'ev

Inst.of Chem. Phys., Academy of Sciences of the USSR

Moscow, USSR

ABSTRACT

The fates of triplet carbonyls formed by disproportionation of peroxy radical are considered. The measured rate constants for T-T and T-S energy transfer, and quenching by O_2 are shown to be related to the probabilities of simple primary processes. The mechanisms of the latter are considered.

INTRODUCTION

The oxidation of organic compounds (RH_2) is accompanied by chemiluminescence of carbonyl compounds (R=O) which are the products of disproportionation of peroxy radicals ($HROO^\cdot$) (1,2):

$$2\ HROO^\cdot \rightarrow\ R = O + O_2 + HROH \qquad (1)$$

The enthalpy of reaction (1),$-\Delta H$ ($\simeq 100$ kcal/mole), is sufficient for electronic excitation of the O_2 ($^1\Delta_g$ or $^1\Sigma_g$ states with energies of 22.5 and 37.5 kcal/mole, respectively) or of the T_1 and S_1 states of R = O. A value of $\simeq 30\%$ has been obtained by a chemical method (3) for the yield of singlet oxygen, and 10^{-3} to 10^{-2} for the excitation yields, η_p^*, of triplet R = O (4). The η_p^* value has been shown (4) to depend on the energy distribution among three products of reaction (1) and on conversion of energy from a chemical (i.e. electronic) to a vibrational form.

The structure of R = O depends on the starting RH_2, i.e. cyclohexane, ethylbenzene, diphenylmethane, and methylethylketone yield cyclohexanone ($T_1 = 75$ kcal/mole), acetophenone ($T_1 = 74$),

benzophenone (T_1 = 69), and biacetyl (T_1 = 56), respectively.
Therefore, organic oxidation may be regarded as a method for production of triplet molecules. This method may be used as an instrument for the qualitative and quantitative study of the fates
of triplet states in solution (2,4-6). This paper deals with the
elementary processes of intermolecular interaction of carbonyl
triplets with molecular oxygen and luminescent acceptors of energy.

Usually the rate of reaction (1), w, is 10^{-9} to 10^{-8} mole/liter
sec. This yields excitation rates, $\eta_P^* w$, of 10^9 to 10^{11} excited
molecules/cm^{3}sec and (allowing for the variations in T_1-S_0 emission
efficiency, η_P, from 10^{-6} to 10^{-3}) chemiluminescence emission rates,
$I = \eta_F \eta_P^* w$, of 10^3 to 10^8 quanta/cm^{3}sec. The chemiluminescent mixture consists mostly of a 5 to 10% solution of RH$_2$ in benzene, small

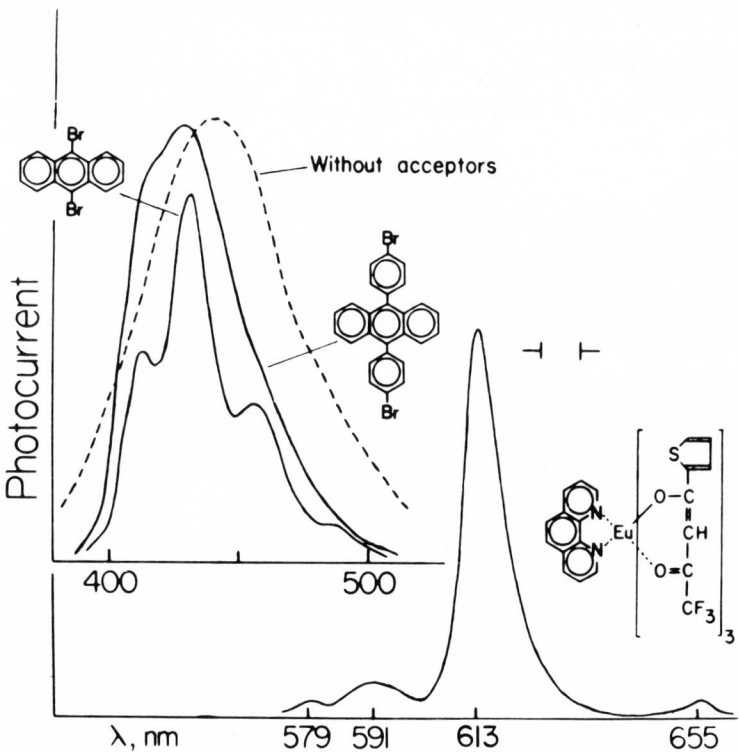

Fig. 1   Spectra of chemiluminescence in oxidation of ethylbenzene
         at 60°C with and without energy acceptors:
         9,10-dibromoanthracene, 9,10-di(p-bromophenyl)antracene,
         and europium tris-thenoyltrifluoroacetonate with
         1,10-phenanthroline (Eu(TTA)$_3$Phen).

amounts (10^{-5} to 10^{-3} mole/liter) of the initiator (peroxide or an azo compound), and not less than 10^{-7} mole/liter of dissolved O_2, the total volume being some ten milliliters.

ENERGY TRANSFER TO ACCEPTORS WITH STRONG
SPIN-ORBIT COUPLING

The chemiluminescence spectra for ethylbenzene oxidation are shown in Fig. 1. They coincide with fluorescence spectra of additives when the latter are introduced into the solution. Only the shapes of spectra are shown in Fig. 1, but the absolute chemiluminescence intensity in the case of di-bromoanthracene is several times higher than that for di-(p-bromophenyl) anthracene in spite of the fact that the fluorescence quantum yield for the former ($\eta_A = 0.11$) is seven times lower than that for the latter ($\eta_A = 0.79$). Europium chelate ($\eta_A = 0.2$) enhances chemiluminescence much more than does any substituted anthracene.

Fig. 2 gives a qualitative explanation of the results obtained. The most efficient path of intermolecular deactivation of the triplet product is the triplet-triplet energy transfer that is known to be diffusion-controlled. In the case of chelate this leads to excitation of the triplet state(s) of the ligand followed by efficient energy transfer to the ion, and by strong radiative transitions between $^5D_0$ and $^7F_{0,1,2,3}$ states.

With anthracences, the triplet-singlet energy transfer results in fluorescence excitation. This process must compete with the triplet-triplet transfer, i.e. with quenching. The rate constants

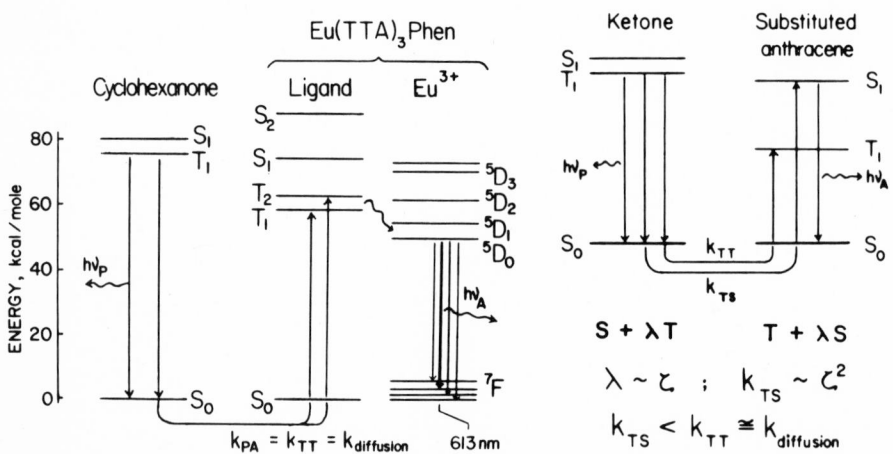

Fig. 2   Illustrating energy transfer from cyclohexanone to europium chelate and from a typical ketone to a typical anthracene derivative.

for triplet-singlet energy transfer, k_{TS}, can be found from chemi-luminescence intensities as a function of acceptor concentrations, [A] (2,5). For A ≡ europium chelate, τ_P is the experimental life-time of the triplet donor and k_{TT} is the triplet-triplet rate constant, one obtains:

$$1/[(I/I_o)-1] = 1/[(\eta_A/\eta_P)-1] + 1/[(\eta_A/\eta_P)-1] \cdot 1/[(k_{TT}\tau_P[A]].$$

Then with A ≡ anthracene derivative and measurement of k_{TS}:

$$\frac{1}{\frac{I}{I_o}-1} = \frac{1}{\frac{\eta_A k_{TS}}{\eta_P(k_{TT}+k_{TS})}-1} + \frac{1}{\frac{\eta_A k_{TS}}{\eta_P(k_{TT}+k_{TS})}-1} \cdot \frac{1}{(k_{TT}+k_{TS})\tau_P[A]}.$$

To what extent are the rate constants k_{TS} related to the elementary process of interaction between excited donor, P^* (i.e., $P(T_1)$) and the ground state acceptor, $A(S_o)$? Consider the fate of the collision complex

$$P(T_1) + A(S_o) \underset{P_{oo}}{\overset{}{\rightleftharpoons}} [P(T_1)...A(S_o)] \overset{P_{TT}}{\underset{P_{TS}}{\longrightarrow}} \begin{matrix} P(S_o)+A(T_1) \\ P(S_o)+A(S_1). \end{matrix} \quad (2)$$

The above probabilities (P_{TT}, P_{TS}) are the rate constants in reciprocal seconds. P_{TT} can be calculated from the equation given by Dexter (7) for the exchange transfer mechanism:

$$P_{TT} \sim Z_1^2 \int F_P(\nu)\epsilon_A^T(\nu) \, d\nu .$$

The triplet-singlet exchange transfer is spin-forbidden when the states in question are "pure". However, in heavy-atom sub-stituted molecules the spin-orbit coupling is enhanced and the mixing of "pure" states relaxes the selection rule (8): P_{TS} can be derived from a modification of the Dexter expression applied to the triplet admixture to the singlet state:

$$P_{TS} \sim \frac{\sum_i \zeta_i^2}{(E_S - E_T)^2} Z_2^2 \int F_P(\nu)\epsilon_A^S(\nu)d\nu. \quad (3)$$

The quantity P_{oo} may be represented, say, as the frequency of vibrations in the cage.

The processes in Eq. (2) constitute a complete system of events for decomposition of the collision complex. Their relative (dimen-sionless) probabilities are evidently the following:

$$P_T = P_{TT}/(P_{TT} + P_{TS} + P_{oo}) ,$$

$$P_S = P_{TS}/(P_{TT} + P_{TS} + P_{OO}) \; ,$$

and

$$P_O = P_{OO}/(P_{TT} + P_{TS} + P_{OO})$$

$$= 1 - P_S - P_T \; .$$

At each encounter the partners suffer a certain number, n, of collisions and each collision results, depending on the corresponding probability, in one of the events of Eq. (2). We assume the probabilities are independent and therefore we may write for bimolecular rate constants:

$$k_{TS} = k_{diff.}[1 - (1 - P_T - P_S)^n] \frac{P_S}{P_T + P_S} \; ;$$

$$\tag{4}$$

$$k_{TT} = k_{diff.}[1 - (1 - P_T - P_S)^n] \frac{P_T}{P_T + P_S} \; .$$

Therefore, the experimental rate constants for competitive energy transfer are proportional to the probabilities of elementary transfer processes and should be related to the transfer mechanisms. A scale for comparing rate constants has been suggested (2,8). It is based on a calculation of the quantity $\Sigma \zeta_i^2$ for each acceptor,

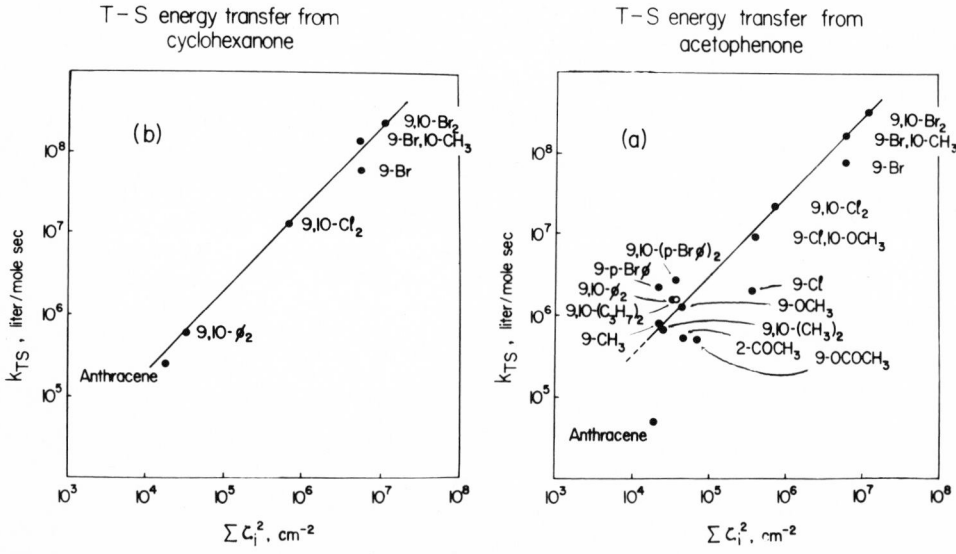

Fig. 3   T-S energy transfer rate constants as a function of spin-orbit coupling effects in anthracene acceptor
(a) transfer from acetophenone
(b) transfer from cyclohexanone
$[RH_2]$ = 10% v/v in benzene at 40 C.

where ζ_i is the spin-orbit coupling constant for each of the atoms constituting the acceptor molecule. In Fig. 3 this scale is used for plotting k_{TS} for anthracences studied earlier (2,8) and also in this paper. Only substituents in the conjugated ring system have been taken into account in calculating the sum $\Sigma\zeta_i^2$. It will be noted that the other parameters in Eq. (3) are not essentially different for various anthracene derivatives. A satisfactory fit of the re-- sults obtained from Eq. (3) (Fig. 3) confirms the mixing mechanism. Lower rate constants for anthracene, 9-chloro- and 9-bromo anthra-- cenes may be reasonably explained by higher singlet excited levels (i.e. by poorer spectral overlapping), as well as by higher values of k_{TS} for p-bromo phenyl derivatives which are probably due to a certain degree of conjugation between anthracene and phenyl moieties. These results allow rejecting such alternative mechanisms as: (a) Triplet-triplet transfer to the T_2 level followed by inter-- system T_2-S_1 crossing; this must strongly depend on T_2-S_1 splitting which is rather sensitive to substituents. (The triplet-triplet absorption data show that T_2 is close to S_1 for anthracenes; S_1 is rather sensitive to the substituents whereas T_2 is not (9).) How-- ever, our experiments on the temperature and concentration effects in the mixed benzene-cyclohexane solvents do not confirm these ex-- pectations. (b) Spin-independent long-range resonance transfer (see discussion in (2)). (c) Energy transfer from dimers of singlet oxygen-Kasha mechanism (10).

Gas-phase measurements (11) yield values of 10^{-3} to 10^{-1} for P_T (and, presumably, for P_S) but because of hundreds of repetitive collisions (n in Eq. (4)) almost every encounter results in either T-T or T-S transfer and k_{TT} is close to k_{diff}.

QUENCHING OF THE TRIPLET PRODUCT BY MOLECULAR OXYGEN

The same approach allows one to conclude that the rate con-- stants for oxygen quenching are strictly relevant to elementary

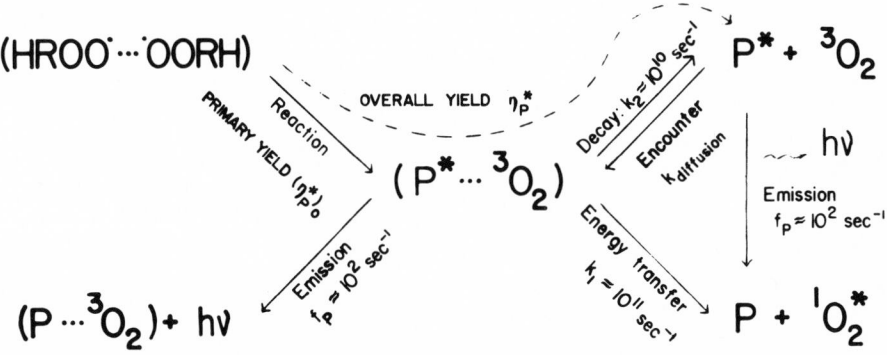

Fig. 4 Mechanism of chemical excitation and quenching.

interaction. This aspect is of particular importance for organic chemiluminescence, since quenching may occur not only upon the encounter of P^* and O_2 but just after formation of P^* in reaction (1) when the newly formed P^* and O_2 are in the same cage. Both types of cages may be reasonably considered as similar, and chemical excitation and quenching may be represented by Fig. 4.

Assume that the rate constant f_p is of the order 10^2/sec (2) and shows only a small dependence on the presence of O_2, that the lifetime for energy transfer, $1/k_1$, is about 10^{-11} sec (12), and the lifetime of the cage, $1/k_2$, is several times higher (12). Then it can be concluded that: (1) Escape of P^* into the bulk of the solution is much more probable than its emission from the cage. The ratio of bulk to cage emission can easily be shown to be $k_2\tau_p$, where τ_p is the actual lifetime of triplet P^* in solution. (2) The experimental excitation yield, η_P^*, is smaller than the primary excitation yield, $(\eta_P^*)_0$: $(\eta_P^*)_0/\eta_P^* = (k_1 + k_2)/k_2$. However, for fluid solutions this ratio is only several units. We disagree with the conclusion of Kellogg (13) that this ratio may reach 10^8 and that the primary excitation yield is close to unity. This author compared the probabilities of emission and deactivation in the cage, rather than the excitation yields. Allowing for the above values, the ratio of the former must be low: $f_p/k_1 \approx 10^{-9}$. (3) The actual rate constant for quenching of the triplet by O_2, $k_{oxy.}$, equals $k_{diff.} \cdot k_1/(k_1+k_2)$ or $k_{diff.} \cdot [1 - \eta_P^*/(\eta_P^*)_0]$. The known rate constants are indeed lower than those in diffusion-controlled quenching (14-16), in spite of the abnormally high diffusion coefficients of O_2 (17). If $k_{oxy.}$ were equal to $k_{diff.}$, then the measured chemiluminescence excitation yield would be zero: the triplet ketones formed by reaction (1) could not survive quenching by O_2 formed in the same reaction; therefore, there would be no chemiluminescence at all. Multiplicity statistics (14) seems to be responsible for successful competing of k_2 with k_1 in Fig. 4.

REFERENCES

1. R. F. Vassil'ev, Progr. Reaction Kinetics, 4, 305 (1967).

2. V. A. Belyakov and R. F. Vassil'ev, Photochem. Photobiol., 11, 179 (1970).

3. J. A. Howard and K. U. Ingold, J. Am. Chem. Soc., 90, 1056 (1968).

4. V. A. Belyakov, R. F. Vassil'ev and G. F. Fedorova, Izv. Akad. Nauk SSSR, Ser. Fiz., 32, 1325 (1968).

5. V. A. Belyakov and R. F. Vassil'ev, in: Molekulyarnaya Fotonika, Nauka, Moscow, 70 (1970).

6. R. F. Vassil'ev and G. F. Fedorova, Optika i Spektroskopiya,
    24, 419 (1968).

7. D. L. Dexter, J. Chem. Phys., 21,836 (1953).

8. R. F. Vassil'ev, Nature, 200, 773 (1963).

9. P. G. Bennett, P. J. McCartin, J. Chem. Phys., 44, 1969 (1966).

10. A. U. Khan and M. Kasha, J. Am. Chem. Soc., 88, 1574 (1966);
    92, 3293 (1970).

11. R. E. Rebbert, P. Ausloos, J. Am. Chem. Soc., 87, 1847 (1965);
    87, 5569 (1965). N. A. Borisevich, A. A. Kotov, G. B. Tol-
    storozhev, this conference.

12. P. J. Wagner and I. Kochevar, J. Am. Chem. Soc., 90, 2232
    (1968).

13. R. E. Kellogg, J. Am. Chem. Soc., 91, 5432 (1969).

14. G. Porter and M. R. Windsor, Proc. Roy. Soc., A245, 238 (1958).

15. H. L. J. Bäkström and K. Sandros, Acta Chem. Scand., 12, 832
    (1958).

16. G. Jackson, R. Livingston and A. C. Pugh, Trans. Faraday Soc.,
    56, 1635 (1960).

17. W. R. Ware, J. Phys. Chem., 66, 455 (1962).

RADIATIVE AND RADIATIONLESS TRANSITIONS IN RARE EARTH CHELATES

V. V. Kuznetsova, R. A. Puko, V. S. Khomenko,

T. I. Razvina

Institute of Physics, Minsk, USSR

ABSTRACT

The luminescence kinetics and relative yields of crystalline tetrakis-ligand isomers of rare earth β-diketonates were measured in the temperature range from 77 K to 293 K. The isomers differ in their luminescence spectra, yields and lifetimes. It is found that the Eu^{3+} $^5D_1$ luminescence level is excited by means of the triplet state of the ligand. The $^5D_0$ level is populated as a result of the radiationless $^5D_1 \rightarrow ^5D_0$ transition and the probability of this process increases with temperature.

INTRODUCTION

The mechanisms of radiationless processes taking place in the rare earth (RE) chelates are not clear in spite of a series of publications (1-5) concerning this problem. In order to elucidate the probabilities of relaxation processes in the RE chelates we studied the luminescence kinetics and relative yields of crystalline piperidinium tetrakis-dibenzoylmethanato europate (III) and gadolinate (III) isomers (α, β, γ - forms of Eu(DBM) HP and Gd(DBM)$_4$HP, where DBM = dibenzoylmethane anion, HP = piperidinium). The isomers have the same chemical composition but different structures of coordination polyhedra (6,7).

The syntheses of complexes were made according to Ref. (7). Fluorescence spectra and relative luminescence yields were obtained as described previously (8). A nitrogen laser with pulse duration 2×10^{-8} sec. was employed as an excitation source for investigation of luminescence kinetics (λ=337.1 nm). Time resolution of recording system was 1.5 X 10^{-7} sec.

RESULTS AND DISCUSSION

The fine structure of the Eu^{3+} luminescence and absorption spectra (7) indicates the presence of a two-fold axis of symmetry in the molecules of the $Eu(DBM)_4HP$ isomers. This is in agreement with the X-ray diffraction data (9). The α-isomer of the complex is the most symmetric and is nearly cubic. The molecules of the proton acceptors do not enter the internal coordination sphere of the complex and their influence on the $Eu(DEM)_4^-$ anion structure is negligible. A stronger influence is exerted by structural factors on the location of the ligand energy levels, and on the intensity and kinetics of chelate luminescence.

Relative fluorescence and phosphorescence yields and the phosphorescence duration of $Gd(DBM)_4HP$ isomers are given in Table 1. It is obvious from comparison of relative fluorescence and phosphorescence yields that interconversion probability is about two orders of magnitude higher than the S_1-S_0 transition probability. The α-isomer phosphorescence yield is lower than those of the β- and γ-isomers. Based on analysis of the results of phosphorescence duration and relative fluorescence yield measurements we conclude that degradation of excitation energy in α-isomer occurs at some intermediate state rather than at the fluorescence or phosphorescence states.

Table 1

The luminescence duration and relative yield of $Gd(DBM)_4HP$ isomers.

Compound	t_e X 10^4sec	B_{ph}	B_{fl}	B_{fl}/B_{ph}
$\alpha Gd(DBM)_4HP$	2.8 $\pm$ 0.3	0.28	0.011	0.039
$\beta Gd(DBM)_4HP$	1.9 $\pm$ 0.2	1.00	0.012	0.012
$\gamma Gd(DBM)_4HP$	3.9 $\pm$ 0.4	0.87	0.013	0.015

B_{ph} and B_{fl} are relative phosphorescence and fluorescence yields, respectively.

A scheme of energy levels and transitions between them for $Eu(DEM)_4HP$ isomers is given in Fig. 1. The triplet level energies were determined from the phosphorescence spectra of Gd^{3+} chelates. Transitions from excited ion $^5D_1$ and $^5D_0$ levels were observed in Eu^{3+} ion emission. The luminescence decay corresponding to transitions from $^5D_0$ levels in the measured range is exponential, to a good approximation. The maximum value of luminescence intensity for β- and γ-isomers is achieved at t = (8 $\pm$ 1) X 10^{-6} sec. (77 K). The luminescence rise reflects the processes of populating the luminescence level. The luminescence rise curves $^5D_0 \rightarrow {}^7F_j$ for all isomers obey an exponential law, moreover, it is necessary to stress

that at T = 77 K and 293 K, the rise time constant is in satisfactory agreement with luminescence decay time from the $^5D_1$
level of corresponding isomers. The population of $^5D_0$ levels
proved to occur as a result of a radiationless $^5D_1 \rightarrow ^5D_0$ transition,
that is in agreement with Refs. (1, 3). Taking into account
the high luminescence yield at T = 77 K, it can be stated that
this process plays the main role in the de-activation of the $^5D_1$
level, that is $\tau_3 \approx 1/d_{32}$. The duration of the $^5D_0 \rightarrow ^7F_j$ luminescence at 77 K for all isomers is practically the same as can be
seen in Table 2. The rise time for the $^5D_0 \rightarrow ^7F_j$ transition and the
decay time for the $^5D_1 \rightarrow ^7F_j$ transition increase in order of
$\alpha \rightarrow \beta \rightarrow \gamma$. Whereas the $^5D_0 \rightarrow ^7F_j$ luminescence yields for β- and γ-isomers
are equal, the luminescence yield for α-isomer is three times less:
the same yield behavior can be observed for the luminescence
from the $^5D_1$ levels. The decrease of Eu^{3+} luminescence yield in
α-isomer in comparison with β- and γ-isomers cannot be explained
solely by an increase of radiationless transition probabilities in
the ion. It follows from a comparison B and τ (see Table 1) that
the luminescence yield decrease for α-Eu(DBM)₄HP is caused by
excitation energy losses in the organic part of the complex, which
exceed the analogous losses in the β- and γ-isomers. Comparatively
low α-Eu(DBM)₄HP luminescence and α-Gd(DBM)₄HP phosphorescence
yields (Table 1) are likely to have the same cause, that is a low
yield of triplet state formation. These results confirm the
supposition that excitation of the RE ion luminescence levels
occurs as a result of intramolecular energy transfer from the
triplet level of the organic ligand.

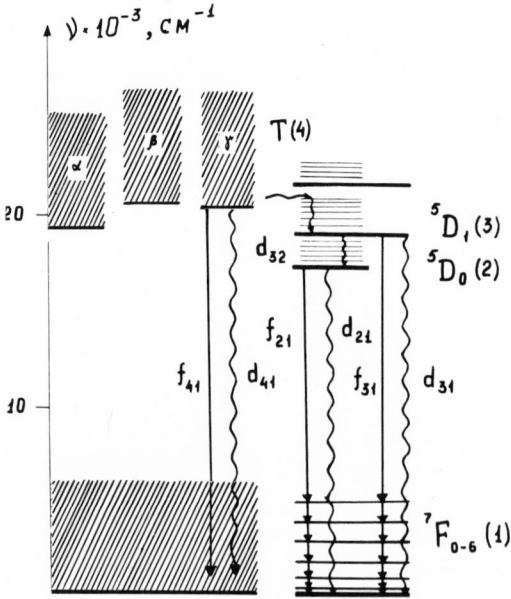

Fig. 1   Energy-level scheme of Eu(DBM)₄HP isomers.

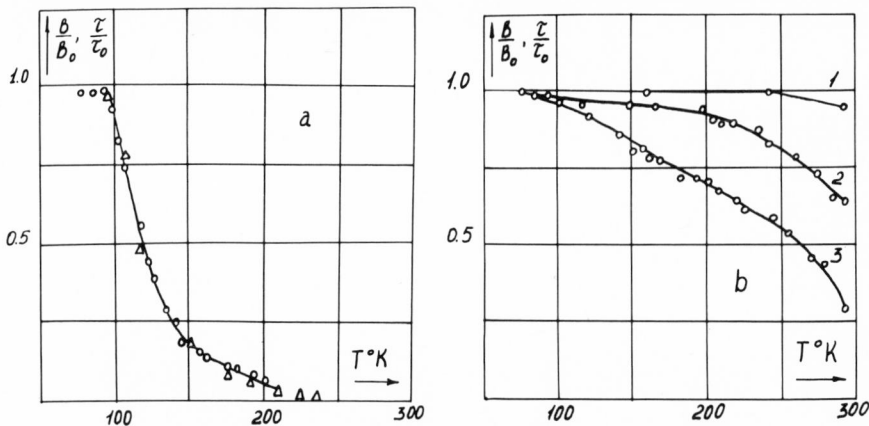

Fig. 2 (a) Temperature dependences of phosphorescence duration
        -Δ- and yield -o- of αGd(DBM)₄HP. (b) Temperature
        dependences of fluorescence duration and yield of γ-
        Eu(DBM)₄HP isomer: 1 - luminescence duration ($^5D_0 \to {}^7F_j$);
        2 - luminescence yield ($^5D_0 \to {}^7F_j$); and 3 - luminescence
        duration ($^5D_1 \to {}^7F_j$).

Table 2
The luminescence duration and relative yield of Eu(DBM)₄HP isomers.

Compound	T°K	$\tau_3 \times 10^6$ sec decay	B_3	$^5D_1 - {}^7F_j$ $\dfrac{\tau_3^{293K}}{\tau_3^{77K}}$	$\dfrac{B_3^{293K}}{B_3^{77K}}$	$\tau_2 \times 10^6$ sec rise	$^5D_0 - {}^7F_j$ $\tau_2 \times 10^6$ sec decay	B_2	$\dfrac{\tau_2\tau_3^{293K}}{\tau_2\tau_3^{77K}}$	$\dfrac{B_2^{293K}}{B_2^{77K}}$
αEu(DBM)₄HP	293	0.14		0.1	0.08		290		0.06	0.37
	77	1.35	0.29			1.27	479	0.30		
βEu(DBM)₄HP	293	0.83		0.48	0.34	0.68	486		0.47	0.58
	77	1.72	0.99			1.59	499	0.96		
γEu(DBM)₄HP	293	0.63		0.29	0.30	0.57	472		0.28	0.65
	77	2.12	1.00			1.83	496	1.00		

        Fig. 2a shows the temperature dependences of duration and
yield of Gd(DBM)₄HP phosphorescence. Phosphorescence is practically absent in the Eu(DMB)₄HP complex. Triplet state lifetimes
in these chelates appear to be completely defined by a process of
intramolecular excitation energy transfer to the ion, whose
probability is greater than $f_{41} + d_{41}$. Temperature dependence of
intensity and duration of Eu^{3+} luminescence for transitions from
$^5D_1$ and $^5D_0$ levels differ markedly (Fig. 2b). Luminescence yield
from the $^5D_1$ level is expressed by

$$B_3 = \rho f_{31}/(f_{31} + d_{31} + d_{32}) = \rho f_{31}\tau_3,$$

where ρ is the yield of $^5D_1$ level excitation.

As is seen from Table 2, there exists a good correlation between $B_3(293K)/B_3(77K)$ and $\tau_3(293K)/\tau_3(77K)$. The value of ρ proves to be temperature independent, i.e., the probability of intramolecular excitation energy transfer to the ion surpasses the probability of ligand triplet level de-activation (d_{41}), both at 77 K and at room temperature.

The change from 77 K to 293 K has the greatest influence on the luminescence duration and yield from the $^5D_1$ level. The temperature dependence of the lifetime of this state was explained in Ref. (3) by an increase in the probabilities of radiationless transitions $(^5D_1 \rightarrow ^7F_j)$ while the probability of the $^5D_1 \rightarrow ^5D_0$ transition was assumed to be temperature independent. For the latter to be correct, $B_2(293K)/B_2(77K)$ should coincide with $\tau_3(273K) \cdot \tau_2(273K)/\tau_3(77K) \cdot \tau_2(77K)$. It follows that in the temperature range where $\tau_2 = $ const, the $^5D_0 \rightarrow ^7F_j$ luminescence yield should vary similarly with changes of the $^5D_1 \rightarrow ^7F_j$ luminescence yield and duration. No such correlation is evident from Fig. 2 and Table 2. The luminescence yield decreases less than does the time parameter for the same temperature increase. This is true for all isomers and indicates an increase in probability of the $^5D_1 \rightarrow ^5D_0$ radiationless transition with temperature.

REFERENCES

1. E. Nardy, S. Yatsive, J. Chem. Phys., **37**, 2333 (1962).

2. V. V. Kuznetsova, A. N. Sevchenko, Phizicheskie problemy spectroscopy, **1**, M., 236 (1962).

3. M. L. Bhaumik, L. J. Nugent, J. Chem. Phys., **43**, 1680 (1965).

4. G. A. Crosby, Mol. Crystals, **1**, 37 (1966).

5. M. Kleinerman, J. Chem. Phys., **51**, 2370 (1969).

6. V. S. Khomenko, V. V. Kuznetsova, Z. pry. spectr., **7**, 850 (1967).

7. V. V. Kuznetsova, V. S. Khomenko, T. I. Razvina, R. A. Puko, Z. pry. spectr., **16**, 279 (1972).

8. V. V. Kuznetsova, Kand. dissertation, Minsk, (1961).

9. A. V. Aristov, et al., Teor. Eksper. chim., **6**, 61 (1970).

ENERGY TRANSFER FROM HIGHER TRIPLET STATES

M. V. Alfimov, I. G. Batekha, V. I. Gerko, Yu. B. Sheck

Inst. of Chem. Phys., Academy of Sciences of the USSR

Chernogolovka, Moscow, USSR

ABSTRACT

The observation of the effect of energy transfer from higher triplet states by the decrease in phosphorescence of aromatic compounds dissolved in transparent glassy solvents under excitation of solute molecules in their triplet-triplet absorption is reported. The quantum efficiency of the energy transfer was shown to increase with the energy gap between the excited state T^* and the lowest triplet state T_1 of the solute and with the multiple chemical bonds of the solvent molecules. The quantum efficiency for transfer from higher triplet states has a value of 0.01-0.2.

It has long been assumed that the high values of intramolecular conversion rates within a singlet or triplet manifold of organic molecules prevent the realization of various processes involving the participation of highly excited molecules. The first indications of the possible importance of processes involving highly excited triplet molecules were stated in the papers devoted to photosensitized biphotonic reactions in glassy solutions (1,2). But since similar results were interpreted differently by other authors (3,4) and the statement of the energy transfer from higher triplet states ($T^* \rightarrow T$ energy transfer, Fig. 1) itself was based on indirect results, the feasiblity of the $T^* \rightarrow T$ energy transfer was still questionable.

In view of this we have undertaken a search for direct experimental evidence of the occurrence of $T^* \rightarrow T$ energy transfer (5). In planning the experiments we proceeded from the assumption that if the energy transfer from higher triplet states of solute organic

367

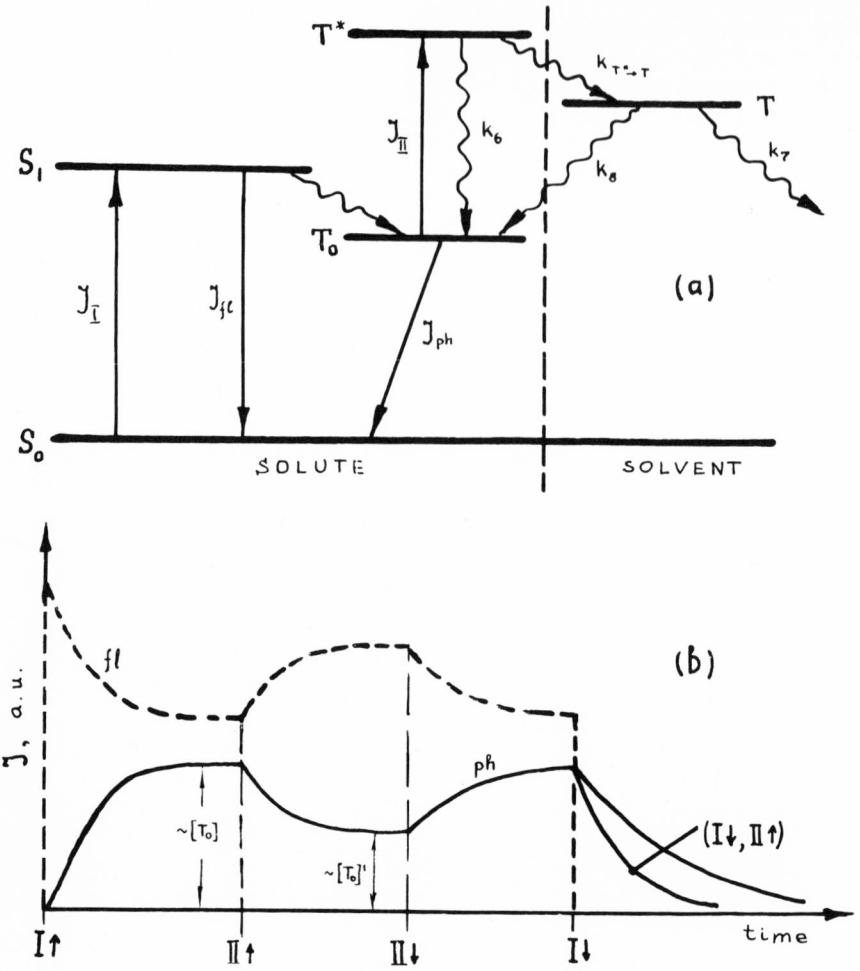

Fig. 1 (a) Energy diagram and transitions between the levels and
(b) effect of $T^* \to T$ energy transfer. The straight arrows
refer to processes involving the absorption or emission of
a photon, the wavy arrows denote nonradiative transitions.
In Fig. b the moments when light sources I and II are
switched on (↑) and off (↓) are indicated.

molecules to the neighboring molecules of the glassy solvent took
place, then the excitation of solute molecules in their triplet-to-
triplet absorption band ($T^* \leftarrow T_1$ absorption) would have led: (a) to
the depopulation of the lowest riplet state T_1 (due to the appearance of an additional pathway of T_1 deactivation) and (b) to the
appearance of triplet excitations in the matrix.

The solutes used were aromatic molecules displaying fluorescence and phosphorescence in rigid media: naphthalene-D_8, diphenylamine, carbazole, phenanthrene, etc. The matrix used in most experiments was glassy toluene at 77 K, the diffusion of triplet excitations in which had already been established previously (6). The experiments were carried out with standard luminescent equipment with two sources of excitation (5,7). One of them (I, λ = 300-340 nm) populated the T_1 level of solute molecules while the second (II), giving the monochromatic light of one of the resonant Hg-lines (λ = 366, 436, 546 or 579 nm), served for the excitation of the level T^* (Fig. 1a).

After source I was switched on (Fig. 1b) the solute fluorescence intensity I_{fl} (in the case of Fig. 1b the solute is carbazole) immediately grows and then decreases with a phosphorescence rise time τ_r to some stationary value which is a consequence of the ground state, S_o, depletion. By this time the phosphorescence intensity I_{ph} is also reaching its stationary state. Switching source II on results in a considerable decrease of I_{ph} and an increase of the I_{fl} value. Therefore, illumination of the solution by light II which is absorbed in the solute $T^* \leftarrow T_1$ transition actually causes T_1 depopulation and growth in the concentration of solute molecules in their ground state. When source II is switched off the system returns to its initial state (Fig. 1b).

The appearance of triplet excitations in the matrix has been demonstrated in the experiments with acceptors of triplet energy embedded in the solution. Biphenyl proved to be a very convenient and suitable acceptor in the system toluene-chrysene (5). In this system, after switching source II on, one may observe the appearance of biphenyl phosphorescence; this fact points out the occurrence of triplet energy transfer from the solute to the acceptor via toluene, i.e. the appearance of triplet excitations in the matrix. Hence, all experimental data lead to the conclusion that during the lifetime of the higher triplet state T^* there exists a finite probability of energy escape from a highly excited molecule to the neighboring matrix molecules.

The processes occurring with $T^* \rightarrow T$ energy transfer are schematically illustrated in Fig. 1a. The kinetic treatment of the scheme (7) leads to the expression for the quantum efficiency of the $T^* \rightarrow T$ energy transfer:

$$\beta = [k_{T^* \rightarrow T}/(k_{T^* \rightarrow T} + k_6)] \cdot [k_7/(k_7 + k_8)]$$

The β value has a clear physical meaning, namely, it indicates which part of the quanta absorbed in the $T^* \leftarrow T_1$ absorption band results in irreversible energy escape from the higher triplets of solute molecules to the solvent. The β value may be obtained on

TABLE 1

Quantum efficiency of energy transfer from higher triplet states (8).

Solute / Solvent	Naphthalene-D_8 *	Naphthalene-D_8 β	Chrysene *	Chrysene β	Diphenyl-amine *	Diphenyl-amine β	Phenanth-rene *	Phenanth-rene β
Toluene	366 2370	0.13	366 2500	0.20	546 26000	0.03	436 3100	0.02
	405 5500	0.13	546 9000	0.02	579 14000	0.03		
1-Hexene	366 1650	0.08						
Polymethyl-methacrylate	366 2200	0.05 77,293 K			546 26000	0.08	436 3100	0.01 (293 K)
Methylcyclo-hexane**		$<3\times10^{-4}$		$<3\times10^{-4}$		$<3\times10^{-4}$		$<3\times10^{-4}$

* The upper number denotes the wavelength of the source II in nm, the lower is the ε_T^λ value in $1/M \cdot cm$.

** The effect of energy transfer was undetectable. The figure of 3×10^{-4} reflects the limit of the method sensitivity only.

the basis of experimentally attainable parameters according to the following equation (7):

$$[T_1]/[T_1]' = 1 + [2.3\varepsilon_T(\lambda)/6.02\times10^{20}]\tau_r J_{II}\beta$$

where $[T_1]/[T_1]'$ is the ratio of the phosphorescence intensities in the absence of and in the presence of source II (see Fig. 1b); $\varepsilon_T(\lambda)$ is the extinction coefficient for $T^* \leftarrow T_1$ absorption at the wavelength of source II in $1/M \cdot cm$; J_{II} is the intensity of the mono-chromatic light flux from source II in quanta/sec·cm^2. The results of the measurements (8) are given in Table 1 from which it may be seen that the escape efficiency, β, depends on the type of solute molecules embedded in the solvent. Under the assumption that $k_{T^* \to T}$ and k_8 values depend weakly on the kind of solute molecules, the observed difference in β values may be attributed to the varia-tion in the lifetimes ($1/k_6$) of higher triplet states, T^*. This conclusion seems to be consistent with the well-known model for nonradiative transitions (9), according to which $1/k_6$ is propor-tional to the energy gap between states T^* and T_1.

As may be seen from Table 1 (naphthalene-D_8) excitation II within the same electronic-vibrational band for $T^* \leftarrow T_1$ absorption does not cause variations in β values, but excitation II in dif-ferent bands may (chrysene). This effect also seems to be due to the different lifetimes of states T_n^* and T_{n+1}^*.

The β values depend also on the type of solvent molecules (Table 1). There is (8) qualitative correlation between the β values and double bonds in the solvent molecules; the effect of $T^* \to$ T energy transfer is absent when the glassy solvent is formed by saturated hydrocarbons. These facts lead to the conclusion that the presence of multiple bonds in solvent molecules is one of the important requirements for the feasibility of effective $T^* \to$ T energy transfer (8). The fulfillment of this requirement eventually leads to the presence of a relatively low-lying triplet level in the molecule forming the glassy matrix. The lowest triplet state in saturated hydrocarbons is probably above the T^* level of the solute molecules used and this accounts for the absence of the energy transfer effect in this case.

It should be noted in conclusion that the feasibility of $T^* \to$ T energy transfer had been independently documented by Liu and Kellogg (10), and that this process was recently shown to be a primary act in some photochemical reactions (11,12).

REFERENCES

1. V. V. Rylkov, V. E. Kholmogorov, A. N. Terenin, Dokl. Akad. Nauk SSSR 165, 356 (1965). A. N. Terenin, V. V. Rylkov, V. E. Kholmogorov, Photochem. Photobiol. 5, 543 (1966).

2. S. Siegel, K. Eisenthal, J. Chem. Phys. 42, 2494 (1965).

3. Kh. S. Bagdasaryan, Kinetika Kataliz 8, 1073 (1967).

4. M. V. Alfimov, I. G. Batekha, V. A. Smirnov, ibid., 7, 766 (1966); Khim. High Energies 2, 123 (1968).

5. M. V. Alfimov, I. G. Batekha, V. A. Smirnov, Dokl. Acad. Nauk SSSR 185, 626 (1969).

6. M. V. Alfimov, I. G. Batekha, Yu. B. Sheck, Izv. Akad. Nauk SSSR. Ser. Phys. 32, 1488 (1968); Khim. High Energies 2, 215 (1968).

7. M. V. Alfimov, I. G. Batekha, Yu. B. Sheck, V. I. Gerko, Spectrochimica Acta, 27A, 329 (1971).

8. V. I. Gerko, Yu. B. Sheck, I. G. Batekha, M. V. Alfimov, Opt. i Spectra. 30, 456 (1971).

9. G. W. Robinson, R. P. Frosch, J. Chem. Phys. 38, 1187 (1963); 37, 1962 (1962).

10. R. S. H. Liu, R. E. Kellogg, J. Am. Chem. Soc. 91, 250 (1969).

11. I. G. Batekha, Yu. B. Sheck, S. A. Krysanov, M. V. Alfimov,
     Dokl. Akad. Nauk SSSR 197, 614 (1971).

12. Yu. B. Sheck, I. G. Batekha, M. V. Alfimov, Khim. High Energies
     5, 297 (1971).

ON THE CONCENTRATION QUENCHING OF FLUORESCENCE POLARIZATION

A. Kawski and J. Kamiński

Luminescence Research Group, Institute of Physics

University of Gdańsk, Gdańsk, Poland

ABSTRACT

Applying Förster's "excitation master equations" which describe the motion of the excitation energy among the molecules of a solution, an expression is found for the probability $\bar{p}_{1k}$ that the excitation terminates on the same molecule which served as the absorber of the incident photon. With the quantity $\bar{p}_{1k}$ we obtain for the emission anisotropy as a function of concentration, the same expression obtained previously for the multilayer model.

The well-known phenomenon of concentration quenching of the fluorescence polarization (self-depolarization, concentration depolarization) of dyes in rigid or viscous isotropic solutions has recently been re-examined theoretically (1-14). Concentration quenching of fluorescence polarization (CQFP) results from an excitation transfer from molecules initially excited by absorption of the primary light to molecules initially unexcited and having orientations different from those initially excited. When a fluorescent solution consists of molecules of the same kind, energy migration due to inductive resonance causes re-excitation of initially excited molecules. Hence, the calculation of the anisotropy of photoluminescence emission (EA), r, as a function of concentration is a complicated one and simplifying assumptions are unavoidable. (The emission anisotropy, r, is related to the degree of polarization P by r = 2P/(3-P).) A number of authors (1,2,3,5,6,15-18) have studied the problem of CQFP utilizing various simplifying assumptions.

Critical reviews of the earlier theories on CQFP are given in the papers of Ore and Eriksen (4,8,19) and Knox (5). In previous publications by Bojarski (3) and the present authors (6) modified versions of the Jablonski shell model of a luminescent center were worked out. The luminescent center in a solution consists of an excited luminescent molecule surrounded by concentric spherical shells of equal thickness which may contain unexcited molecules of the same kind. Different centers in the same photoluminescent solution differ in the occupancy of their shells by unexcited molecules. Thus we have a group of centers without unexcited molecules, a group with one single unexcited molecule in the first shell and so on. Their configuration may also be characterized by the numbers k_1, $k_2, \ldots, k_z$ of the unexcited molecules in the particular shells of the luminescence center. The distribution of the unexcited luminescent molecules in successive layers in the environment of the initially excited luminescent molecule is given by the generalized Smoluchowski (20) distribution:

$$P(k_1, \ldots, k_z) = \prod_{\ell=1}^{z} \{\exp(-\nu_\ell)\}(\nu_\ell^{k_\ell}/k_\ell!) \tag{1}$$

Here z notes the assumed number of shells, $\nu_\ell = V_\ell \cdot n = A_\ell \cdot 4\pi R_1^3 \cdot n/3$; V_ℓ, the volume of the ℓ-th shell; k_1, the population of the ℓ-th shell by unexcited molecules; R_1, the "effective radius"; n, the number density of unexcited molecules which is almost equal to the number density of the luminescent molecules in the solution (we restrict the analysis to low intensities of excitation); and, $A_\ell = [1+(\ell-1)\delta]^3 - [1+(\ell-2)\delta]^3$, where δ characterizes the thickness of the layers. It is assumed that the probability for energy transfer from an excited to an unexcited molecule is the same for all molecules in one layer, and is different for different layers. By an appropriate selection of the layer thickness, there is the possibility of making the rate constant of the transfer $\mu_\ell = (R_1/R_\ell)^6/\tau_0$ dependent on the distance R_ℓ to the unexcited molecules in the ℓ-th layer.

With the assumption that only molecules initially excited by light absorption are responsible for the polarized fluorescence emission, the emission anisotropy, r, can be computed from the relation given in our previous paper (6):

$$\frac{r}{r_o} = \frac{<\eta_{1k}>}{<\eta_k>} = \frac{\tau_n}{\tau_o} \sum_{k_1=1}^{\infty} \ldots \sum_{k_z=0}^{\infty} P(k_1, \ldots, k_z)\eta_{1k}, \tag{2}$$

where $<\eta_{1k}>$ is the mean value of the quantum yield of fluorescence of molecules excited directly by absorption of exciting radiation, $<\eta_k>$ is the mean value of the overall quantum yield, r_o is the fundamental emission anisotropy, τ_n is the intrinsic lifetime of the excited molecular state and τ_o is the actual mean lifetime (it is assumed that the lifetime for all groups of centers is the same, $\tau_k = \tau_o$).

According to the definition of quantum yield we have:

$$\eta_{lk} = \bar{\rho}_{lk}\, \tau_o/\tau_n ,\qquad (3)$$

where

$$\bar{\rho}_{lk} = \int_o^\infty \rho_{lk}(t)dt/\tau_o \qquad (4)$$

is the probability that the fluorescent radiation is emitted by an initially excited molecule (where the symbol $\bar{\rho}_{lk}$ has been taken from the Knox paper (5)). The quantity $\bar{\rho}_{lk}$ is not the time average of $\rho_{lk}(t)$ which is zero. Substituting (3) into (2) yields

$$\frac{r}{r_o} = \sum_{k_1=1}^\infty \cdots \sum_{k_z=0}^\infty P(k_1,\ldots,k_z)\bar{\rho}_{lk} . \qquad (5)$$

According to Eq. (5), the calculation of EA leads to the determination of the quantity $\bar{\rho}_{lk}$. A formal expression for $\bar{\rho}_{lk}$ can be obtained using the Förster (16) "master equations" which describe the motion of the excitation energy among the molecules of a solution. Applying the pair-wise transfer rate: $\mu_{ij} = (R_o/R_{ij})^6/\tau_o$, where R_{ij} is the distance from molecule i to molecule j, and R_o is a parameter whose value may be computed from absorption and emission data (16,22), to a system of k interacting molecules in which ρ_{ik} is the probability that the i-th molecule is excited at time t, Förster (16,21) obtained a coupled system of equations that describes the motion of the excitation:

$$d\rho_{ik}(t)/dt = -(1/\tau_o)\rho_{ik}(t) + \sum_{j=1}^k [\mu_{ji}\rho_{jk}(t)-\mu_{ij}\rho_{ik}(t)], \qquad (6)$$

where i = 1,2,...,k and the following conditions are fulfilled: $\rho_{ik}(0) = \delta_{li}$, $\rho_{ik}(\infty) = 0$, where δ_{li} is the Kronecker delta. By the aid of Eq. 4, and $r_{ij} \equiv \tau_o\mu_{ij}$, $r_{ii} \equiv 0$, the system of Eqs. (6), in accordance with Ore and Eriksen (8), becomes a simple set of ordinary linear equations for $\bar{\rho}_{ik}$:

$$\delta_{li} = (1 + \sum_{j=2}^k r_{ij})\bar{\rho}_{ik} - \sum_{j=2}^k r_{ij}\bar{\rho}_{jk} \qquad (7)$$

with i = 1,2,...,k. Since we are only interested in the probability of emission of the initially excited molecules, we can reduce Eq. (7) to a system of two equations. We have assumed that the probability of emission of the system of k-1 initially unexcited molecules is given by the sum of probabilities of the emission of individual molecules:

$$\bar{\rho}_{IIk} = \sum_{j=2}^k \bar{\rho}_{jk} \qquad (8)$$

The system of k Eqs. (7) can then be represented in the following way:

$$1 = (1 + \sum_{j=2}^{k} r_{1j}) \bar{\rho}_{1k} - < r_{j1} > \bar{\rho}_{IIk} , \qquad (9)$$

$$0 = (\sum_{j=2}^{k} r_{1j}) \bar{\rho}_{1k} - (1 + <r_{j1}>) \bar{\rho}_{IIk} , \qquad (10)$$

in which it is assumed for the multi-shell-model that

$$< r_{j1} > = r_{21} = r_{31} = \ldots = r_{k1} = \tau_o < \mu_{j1} > , \qquad (11)$$

where

$$< \mu_{j1} > = (\sum_{\ell=1}^{z} k_\ell \mu_\ell - \mu_1)/(\sum_{\ell=1}^{z} k_\ell - 1) \qquad (12)$$

is the mean value of the rate constant for energy transfer back to the initially excited molecule, being different for different center groups. In previous papers (3,6) it was assumed that after a sufficiently long time for migration of the excitation energy the probability of finding it on any of the molecules belonging to a given center is the same for all of them

$$w_{1k}(\infty) = 1/ \sum_{\ell=1}^{z} k_\ell . \qquad (13)$$

Solution of Eqs. (10) and (11) yields

$$\bar{\rho}_{1k} = (1+<r_{j1}>)/(1+<r_{j1}> + \sum_{j=2}^{k} r_{1j}) \qquad (14)$$

where

$$\sum_{j=2}^{k} r_{1j} = \tau_o (\sum_{\ell=1}^{z} k_\ell \mu_\ell - \mu_1) .$$

Eq. (14) does not contain any of the rate constants for the energy transfer μ_{1j} other than those $\mu_{\ell j}$ involving the initially excited molecules. The expression

$$\sum_{\ell=1}^{z} k_\ell \mu_\ell$$

is decreased by μ_1, because in the first layer of the center there are only $k_1 - 1$ unexcited molecules to which the energy may be transferred. Putting $\mu_1 = 1/\tau_0$ and taking into account $\mu_\ell = \mu_1/[1+(\ell-1)\delta]^6$ (see (6)) we obtain from Eqs. (5) and (14) with Eq. 1, the expression for the emission anisotropy, which has been obtained previously (6).

It should be noted that for $\langle r_{j1} \rangle = 0$ (re-excitation by energy migration of each initially excited molecule is neglected), we obtain from Eq. 14 our expression (23) for the quenching of photoluminescence of solutions by foreign absorbing substances resulting from intermolecular transfer.

REFERENCES

1. A. Jablonski, Acta Phys. Polon. <u>14</u>, 295 (1955); <u>17</u>, 481 (1958).

2. A. Ore, J. Chem. Phys. <u>31</u>, 442 (1959).

3. C. Bojarski, Acta Phys. Polon. <u>22</u>, 211 (1962); <u>34</u>, 853 (1968).

4. E. L. Eriksen and A. Ore, Phys. Norveg. <u>2</u>, 159 (1967).

5. R. S. Knox, Physica <u>39</u>, 361 (1968).

6. A. Kawski and J. Kaminski, Acta Phys. Polon. <u>A37</u>, 591 (1970); <u>A41</u> (1972).

7. A. Jablonski, Acta Phys. Polon. <u>A38</u>, 453 (1970); <u>A39</u>, 87 (1971).

8. A. Ore and E. L. Eriksen, Phys. Norveg. <u>5</u>, 57 (1971).

9. F. W. Craver and R. S. Knox, Molec. Phys. <u>22</u>, 385 (1971).

10. F. W. Craver, Mol. Phys. <u>22</u>, 403 (1971).

11. C. Bojarski and J. Domsta, Acta Phys. Hung. <u>30</u>, 145 (1971).

12. R. E. Dale and R. K. Bauer, Acta Phys. Polon. <u>A40</u>, 853 (1971).

13. A. Jablonski, Acta Phys. Polon. <u>A41</u>, 85 (1972).

14. A. Jablonski, Bull. Acad. Polon. Sci., Ser. sci. math. astr. et phys. <u>20</u>, 243 (1972).

15. G. Weber, Fluorescence and Phosphorescence Analysis, Editor: D. M. Hercules, Interscience Publishers, New York, 1966.

16. Th. Förster, Ann. Phys. <u>2</u>, 55 (1948); Fluoreszenz organischer Verbindungen, Vandenhoeck und Ruprecht, Göttingen (1951).

17. S. I. Vavilov, J. Phys. SSSR $\underline{7}$, 141 (1943); Mikrostruktur des Lichtes, Akademie-Verlag, Berlin, 1954.

18. M. D. Galanin, J. Exper. Theor. Phys. SSSR $\underline{28}$, 485 (1955).

19. A. Ore, Physica $\underline{54}$, 237 (1971).

20. M. Smoluchowski, Festschrift-L. Boltzmann, Editor: S. Meyer, Leipzig, 626 (1904); Bull. Internat. Acad. Polon. Sci., 1057 (1907).

21. M. Trlifaj, Czech. J. Phys. $\underline{6}$, 533 (1956); $\underline{8}$, 510 (1958).

22. D. L. Dexter, J. Chem. Phys. $\underline{21}$, 836 (1953).

23. A. Kawski, J. Kaminski and E. Kuten, J. Phys. B: Atom. Mol. Phys. $\underline{4}$, 609 (1971).

INTERACTIONS AND ENERGY TRANSFER BETWEEN EUROPIUM IONS AT DIFFERENT LATTICE SITES IN Y_2O_3

J. Heber and U. Köbler

Inst. für Tech. Physik der Tech. Hochschule

Darmstadt, BRD (Germany)

ABSTRACT

In Y_2O_3 the Eu^{3+} ions can occupy two different lattice sites with the site symmetries C_{3i} and C_2. We have studied stationary and time-resolved energy transfer from the $Eu^{3+}(C_{3i})$ ions to the $Eu^{3+}(C_2)$ ions and the spectrum of Eu^{3+} ion pairs at identical lattice sites. The experimental results can be explained with a fixed range of interaction $R_0 \approx 9$ A between the Eu^{3+} ions. The results show some evidence for coherency effects in the energy transfer from $Eu^{3+}(C_{3i})$ ions to $Eu^{3+}(C_2)$ ions.

INTRODUCTION

The crystal Y_2O_3 belongs to the cubic space group T_h^7 (1, 2). The unit cell contains 16 formula units. Eight of the 32 Y^{3+} lattice sites possess the site symmetry C_{3i}; the remaining 24, symmetry C_2. The Eu^{3+} ions substitute for Y^{3+} ions on both lattice sites with nearly the same probability (3). At the two different lattice sites the Eu^{3+} ions have different spectra and energy level schemes. Fig. 1 shows the energy level diagrams for both ion types, $Eu^{3+}(C_2)$ and $Eu^{3+}(C_{3i})$, together with the optical transitions from which the diagrams have been constructed. The characteristic fluorescence of the Eu^{3+} ions originates from the $^5D_0$ level. The fluorescence from the $^5D_1$ levels is small because of the relatively fast radiationless transition to the $^5D_0$ levels. At the C_{3i} sites the crystal field has a center of symmetry. Therefore the 4f eigenstates of the Eu^{3+} ions in the crystal all have the same well-defined parity and only magnetic dipole transitions are possible between them. At C_2 sites the crystal field has no center of

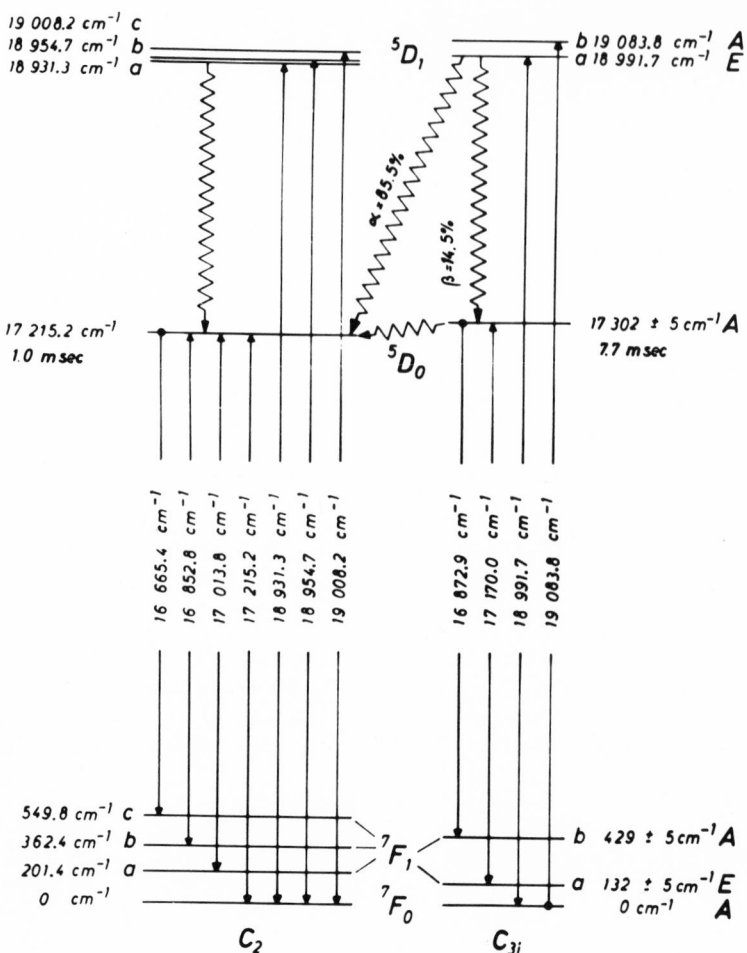

Fig. 1   Energy transfer in Y_2O_3:Eu^{3+} showing the paths of the energy
          transfer for the energy level scheme of both Eu^{3+} ion types.

symmetry so that electric dipole transitions induced by the crystal
field are also possible. Therefore the lifetime of the fluorescence
level $^5D_0$ is expected to be shorter for ions residing at C_2 sites
than for ions at C_{3i} sites. The measured lifetimes are 1.00 $\pm$ 0.01
msec and 7.7 $\pm$ 0.1 msec, respectively, both values being for crystals at 77 K with low Eu^{3+} concentration.

As can be seen in Fig. 1, the metastable level $^5D_0$ has a somewhat higher energy for the C_{3i} ions than for the C_2 ions. Therefore, energy transfer from the C_{3i} ions to the C_2 ions is expected,
as indicated by the twisted arrow. Energy transfer is also found

between the $^5D_1$ level of the C_{3i} ions and the $^5D_0$ level of the C_2 ions. According to our experimental results, this transfer is realized by two paths: directly and indirectly via the $^5D_0$ (C_{3i}) level. The probabilities for direct energy transfer and for internal conversion, leading to indirect transfer, are found to be 85.5% and 14.5%, respectively. A third path via the $^5D_1(C_2)$ levels is ruled out by the experiment as it was not possible to detect any $^5D_1(C_2)$ fluorescence by exciting into the $^5D_1(C_{3i})$ levels. It is surprising that the intersystem process $^5D_1(C_{3i}) \rightarrow {}^5D_0(C_2)$ with a high mismatch of the electronic energies is very active, whereas the intersystem process $^5D_1(C_{3i}) \rightarrow {}^5D_1(C_2)$ with small energy mismatch is not observed. The reason may be that for the latter process, acoustic phonons of low energy and long wavelength and having a small ion-phonon interaction and a small density of states are needed; but in the former, optical phonons with a strong ion-phonon interaction and a high density of states can be active. Furthermore it is of some interest that the direct intersystem process $^5D_1(C_{3i}) \rightarrow {}^5D_0(C_2)$ is preferred over the intrasystem process $^5D_1(C_{3i}) \rightarrow {}^5D_0(C_{3i})$ with nearly the same energy mismatch. From this we can conclude that the pairs have a stronger ion-phonon coupling than do the single ions.

In addition to the energy transfer measurements between Eu^{3+} ions at different lattice sites, we investigated the pair spectra of Eu^{3+} ions at identical lattice sites, especially those with C_{3i} symmetry. Altogether we used three different methods to study the interaction between Eu^{3+} ions in Y_2O_3: stationary energy transfer measurements; time-resolved energy transfer measurements; and spectroscopy of coherent pair states. All experimental results could be consistently explained using a fixed-range-of-interaction model. For distances smaller than the interaction range R_0 the interaction between the ions is approximately constant whereas for greater distances it becomes negligible. Let us now discuss the experimental results using the proposed model.

STATIONARY ENERGY TRANSFER

In our model the macroscopic probability w_T for the energy transfer from C_{3i} ions to C_2 ions is given by the product of the probability that the next C_2 ion is within the interaction range R_0 around a C_{3i} ion and a distance independent microscopic probability W_T for the energy transfer from the C_{3i} ion to the C_2 ion. Using Poisson statistics, we find

$$w_T = W_T \{1 - \exp[-(4/3)\pi R_0^3 n(C_2)]\}$$

where $n(C_2)$ is the density of the $Eu^{3+}(C_2)$ ions. Expressing the interaction range R_0 for an equivalent $Eu^{3+}(C_2)$ concentration by the relation:

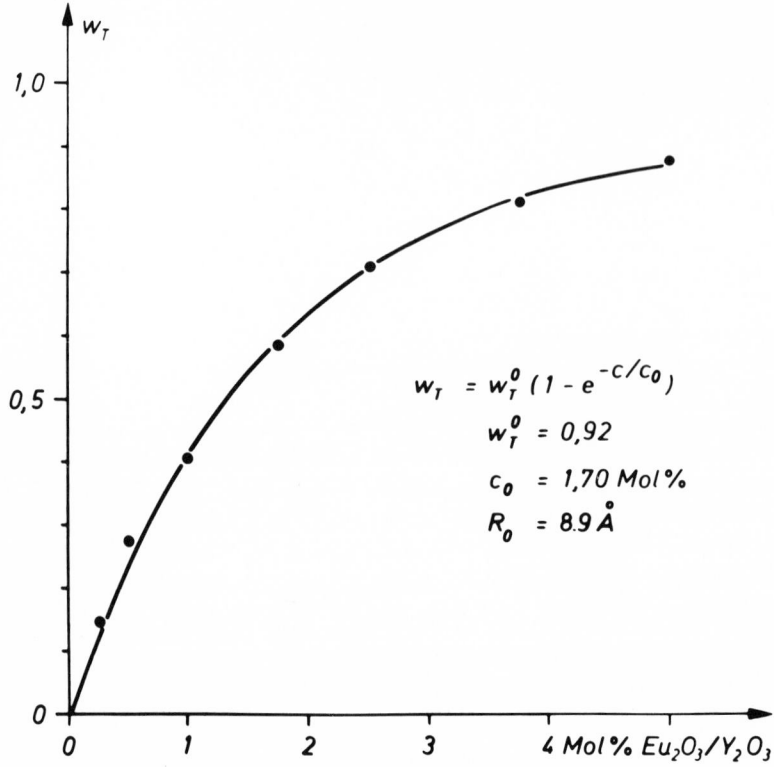

Fig. 2   Energy transfer in $Y_2O_3:Eu^{3+}$ comparing the experimental
results (dots) with the model (full line). $T = 77K$. The
parameters of the model are $w_T = 0.92$, $R_O = 8.9A$.

$$\mu_O(C_2) = 1/(4/3)\pi R_O^3$$

we find

$$w_T = W_T \{1-\exp(-n/n_O)\} = W_T \{1-\exp(-c/c_O)\}$$

where c is the molar concentration of the Eu_2O_3 in Y_2O_3. R_O is
related to c_O by the equation

$$R_O = (\pi c_O N_Y)^{-1/3}$$

where N_Y is the density of the Y^{3+} sites.

In Fig. 2 we show the theoretical curve fitted to experimental
results with the model parameters $W_T = 0.92$ and $R_O = 8.9A$. Experi-

mental values were found at 77K by comparing the excitation and the absorption spectra of the Eu^{3+} ions (4). The experimental results at other temperatures, T = 4.2 and 293K, can be fitted to the model as well, with only slightly different values of the parameters W_T and R_O.

TIME-RESOLVED ENERGY TRANSFER

The model is also able to describe time-resolved energy transfer experiments using a value for the interaction range R_O which

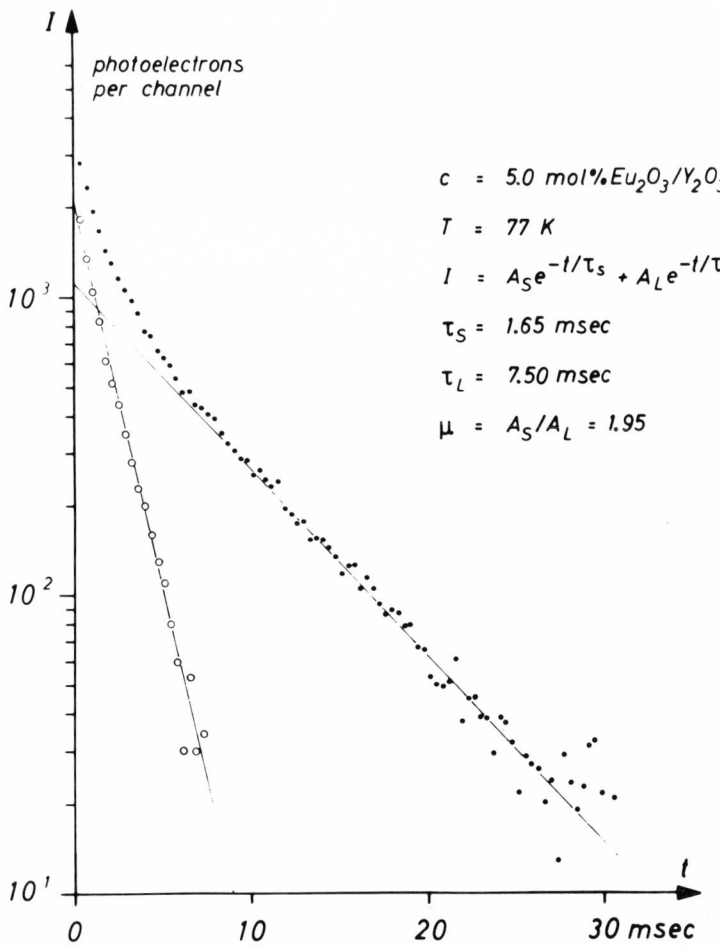

Fig. 3 Time-resolved energy transfer in Y_2O_3:Eu^{3+}. The left side shows a typical fluorescence decay curve of the ^5D_O level of the $Eu^{3+}(C_{3i})$ ions, which becomes nonexponential due to the energy transfer to the $Eu^{3+}(C_2)$ ions. The temperature is T = 77K.

is equal (within experimental error) to that found from the stationary experiments. Fig. 3 shows a typical decay curve of the $^5D_0$
fluorescence of the $Eu^{3+}(C_{3i})$ ions after a pulsed optical excitation
into the $^5D_1(C_{3i})$ levels. This curve can be decomposed into two exponential decay functions with decay times τ_L and τ_S such that

$$I(t) = A_S exp(-t/\tau_S) + A_L exp(-t/\tau_L)$$

The two time constants τ_S and τ_L are concentration independent and
only the ratio, $\mu = A_S/A_L = \mu(c)$, is a function of the concentration.
If we identify the time constant $\tau_L = 7.5$ msec with the fluorescence
lifetime of the unperturbed $Eu^{3+}(C_{3i})$ ions, and the time constant
$\tau_S = 1.65$ msec with that of the perturbed ions having an $Eu^{3+}(C_2)$
ion within its interaction range R_O, the experimental results can
very easily be explained by our model. The amplitude A_L must be
proportional to the probability that within the interaction range
R_O of the $Eu^{3+}(C_{3i})$ ion there is no C_2 ion. Using Poisson statistics, we have $A \sim exp(-c/c_O)$. The amplitude A_S must be proportional
to the probability, $1-exp(-c/c_O)$, that the next C_2 ion is within the
interaction range, times the probability β (Fig. 1) that the excitation of the $^5D_1$ level decays to the $^5D_0$ level: $A_S \sim \beta\{1-exp(-c/c_O)\}$.
Thus we find for the ratio $\mu(c) = A_S/A_L = \beta\{exp(-c/c_O)-1\}$.

In Fig. 4 the theoretical dependence is fitted to the experimental data for two different excitation conditions. As can be
seen, we get in both cases the same value for c_O and therefore for
the interaction range R_O (within experimental error), but different
values for the probability β. For $^5D_1$ excitation we get $\beta = 0.145$.
That means that only 14.5% of the perturbed $Eu^{3+}(C_{3i})$ ions undergo
internal conversion from the $^5D_1$ level to the $^5D_0$ level, the remaining 85.5% transferring their excitation energy directly to the $^5D_0$
level of the C_2 ions. From the value $c_O = 1.8\%$ we get an interaction range $R_O = 8.7$ A in good agreement with the results from the
stationary energy transfer measurements.

PAIR SPECTRA OF THE $Eu^{3+}(C_{3i})$ IONS

Until now we have considered the interaction between Eu^{3+} ions
at different lattice sites giving rise to a transfer of optical excitation energy. Let us now consider the interaction between two
identical Eu^{3+} ions, especially those on C_{3i} sites. If the interaction is weak and symmetric to the permutation of the two ions the
eigenfunctions should have the form

$$\psi = \psi_a \cdot \psi_b$$

for the ground state, and

$$\psi^*_{1,2} = (1/2)^{1/2}(\psi^*_a\psi_b \pm \psi_a\psi^*_b)$$

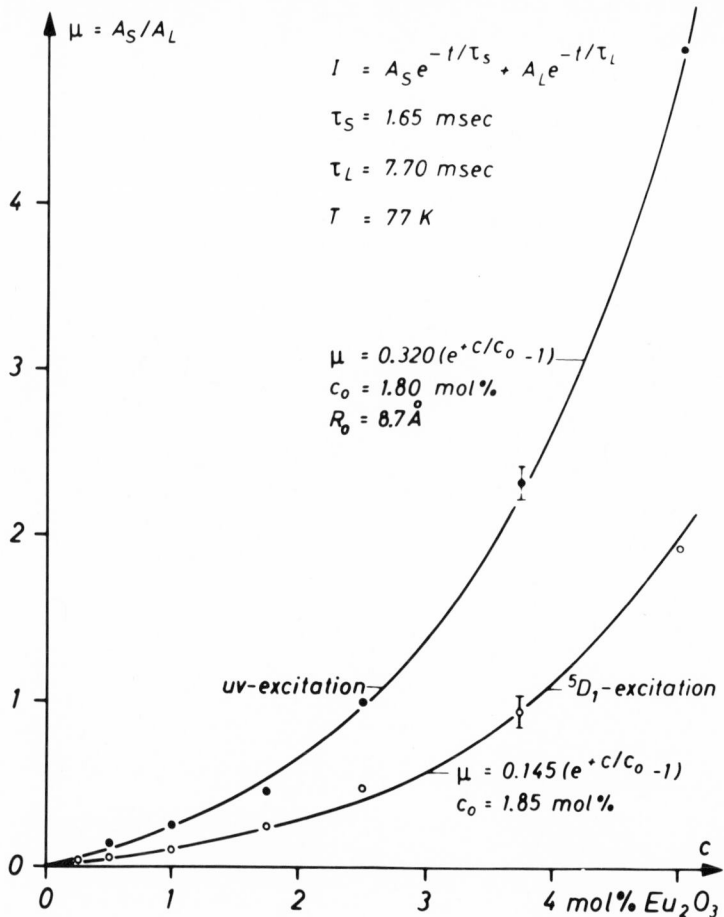

Fig. 4   The concentration dependence of the ratio $\mu = A_S/A_L$ fitted
to the model with an interaction range $R_O = 8.7A$.

for the excited state which consists of a symmetric and an antisymmetric state. Because optical transitions are only possible
between states of the same symmetry for each pair type only one
transition is expected from the symmetric ground state to the
symmetric excited state. The energy of this transition should be
shifted from the energy of the single ion line by half the resonance
splitting.

The distances for the first four neighbor pair types of the
$Eu^{3+}(C_{3i})$ ions are 5.30, 7.50, 9.18, and 10.60A. Remembering that
the interaction radius R_O for the energy transfer was $R_O \approx 9A$, we

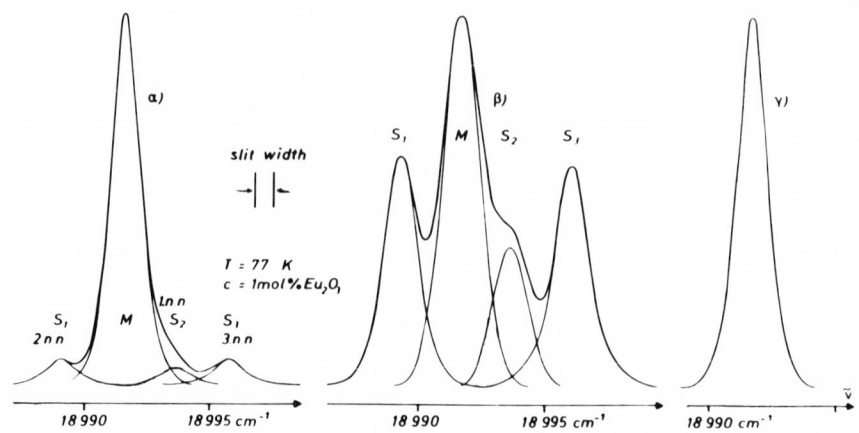

Fig. 5   The $^7F_o \rightarrow a^5D_1$ transitions of $Eu^{3+}(C_{3i})$ ions seen:
in the absorption spectrum (α); in the excitation spectrum
of the strongest $Eu^{3+}(C_2)$ fluorescence transition $^5D_o \rightarrow {}^7F_2$
at $\nu = 16\ 356\ cm^{-1}$ (β); and in the fluorescence excitation
spectrum of the $Eu^{3+}(C_{3i})$ transition $^5D_o \rightarrow a^7F_1$ at $\nu = 17170$
$cm^{-1}(\gamma)$.   T = 77 K, c = 1.0 mole%.

can expect up to three satellite lines in the $Eu^{3+}(C_{3i})$ spectrum if
the interaction is of the same kind. This is indeed the case, as
can be seen from Fig. 5 α,β. From the concentration dependence of
the intensity of the satellite lines S_1, S_2, and S_3 relative to the
intensity of the main line in the absorption spectra (Fig. 5α) it
was possible to identify these lines as belonging to the second,
first, and third nearest neighbor pair types, respectively (5). In
Fig. 5β the same $Eu^{3+}(C_{3i})$ transition as in Fig. 5α is given but
taken in the excitation spectrum of the $Eu^{3+}(C_2)$ fluorescence. As
can be seen, the intensity of the satellites is much enhanced over
the intensity of the single ion line. This means that the pairs
show a stronger energy transfer as compared with the single ions.
Finally, Fig. 5γ shows the same transition in the excitation spectrum of the $Eu^{3+}(C_{3i})$ fluorescence itself. Within experimental
error no satellites could be detected. Thus, all excitation energy
absorbed in the $Eu^{3+}(C_{3i})$ pairs is quantitatively transferred to
the C_2 ions. This cannot be explained by a simple doubling of the
interaction volume of the pairs compared with the single ions. In
this case an energy transfer rate from the C_{3i} pairs to the C_2 ions
of less than 60% is expected for c = 1 mol%. So, we must conclude
that the formation of coherent pair states has a great influence on
the energy transfer process from the $Eu^{3+}(C_{3i})$ ions to the $Eu^{3+}(C_2)$
ions.

In Table 1 the energy shifts of the observed pair transitions

Table 1

Energy shifts for the transitions of the first, second, and third
nearest $Eu^{3+}(C_{3i})$ pairs with respect to the single ion transitions.

Single ion transition	$\Delta\nu$		
	1.n.n.	2.n.n.	3.n.n.
$7_{F_O} \rightarrow b^5D_1$ $\nu = 19\ 083.8\ cm^{-1}$	+8.2 cm^{-1}	-6.5 cm^{-1}	+3.5 cm^{-1}
$7_{F_O} \rightarrow a^5D_1$ $\nu = 18\ 991.7\ cm^{-1}$	+1.9 cm^{-1}	-2.4 cm^{-1}	+4.4 cm^{-1}
$a^7F_1 \rightarrow {}^5D_O$ $\nu = 17\ 170.0\ cm^{-1}$	+3.3 cm^{-1}	-4.8 cm^{-1}	+8.7 cm^{-1}
$\Sigma\ \Delta\nu$	+13.4 cm^{-1}	-13.7 cm^{-1}	+16.6 cm^{-1}

are summarized. As can be seen, all shifts are of approximately the
same magnitude independent of the order of the neighbors, especially
if we take the sums over different transitions of the same pair.
This means that the interaction between the $Eu^{3+}(C_{3i})$ ions is nearly
constant up to the third neighbors, that is up to an interaction
range of $R_O \simeq 9A$, and then becomes negligible. So we find justification for the proposed energy transfer model. Analysis of the pair
spectra clearly shows that the interaction between Eu^{3+} ions in Y_2O_3
has more the character of a short range step function than of the
continuous analytical potential as used by Förster (6) and Dexter
(7). Thus, we find that superexchange over the intervening oxygen
ions explains the observed energy transfer and pair spectra.

REFERENCES

(1) K. A. Gschneider, Rare Earth Alloys, D. Van Nostrand, Princeton,
    (1961) p. 242.

(2) L. Pauling, M.D. Shappel, Z. Kristallographie 75, 128 (1930).

(3) G. Schäfer, Phys. kond. Materie 9, 359 (1969).

(4) J. Heber, E. H. Hellwege, U. Köbler, and H. Murmann, Z. Physik
    237, 189 (1970).

(5) U. Köbler, Z. Physik 247, 289 (1971).

(6) T. Förster, Ann. Phys. 2, 55 (1958).

(7) D. L. Dexter, J. Chem. Phys. 21, 836 (1953).

ENERGY TRANSFER BETWEEN MANGANESE AND ERBIUM IONS IN MnF$_2$

J. M. Flaherty and B. Di Bartolo

Department of Physics, Boston College

Chestnut Hill, Massachuetts 02167, USA

ABSTRACT

We have studied the fluorescence characteristics of the system
MnF$_2$:Er and found evidence of a Mn$^{2+}\to$Er^{3+} energy transfer process.
The temperature dependence of this process reveals itself through
the spectral behavior of the Er system.

We have recently discovered the presence of energy transfer
from divalent manganese to trivalent erbium in the cyrstal MnF$_2$:Er
(1%). The host lattice, manganese fluoride, has attracted some
interest in the past because it is a material which is both optically active and antiferromagnetic (1,2).

The energy level scheme of Mn^{2+} and Er^{3+} in MnF$_2$ is presented
in Fig. 1, as derived from absorption, fluorescence and excitation
measurements. We measured the fluorescence spectrum of a pure MnF$_2$
sample in the temperature region 20-110 K. This range includes the
Neél temperature, $T_N \simeq 67$ K and the temperature $\sim \frac{1}{2}T_N$ corresponding,
according to Prohofsky (1), to the ordering of the excited Mn^{2+} ions.
In this sample, the Mn emission is centered at ~ 6300 A and is
~ 600 A wide at 80 K, and at 26 K it is centered at ~ 5820 A and is
~ 525 A wide. Sharp fluorescence lines appear on the short wavelength side at low temperature. Going down in temperature the Mn
fluorescence shows a sharp increase at $\sim \frac{1}{2}T_N$, consistent with some
previous data (2). The structure of the fluorescence emission suggests the presence of two bands centered at 5820 A and 6300 A at low
temperatures; for this reasons the Mn emission is symbolized in
Fig. 1 as due to two transitions originating from two overlapping
wide levels. The same behavior is presented by the emission of Mn

388

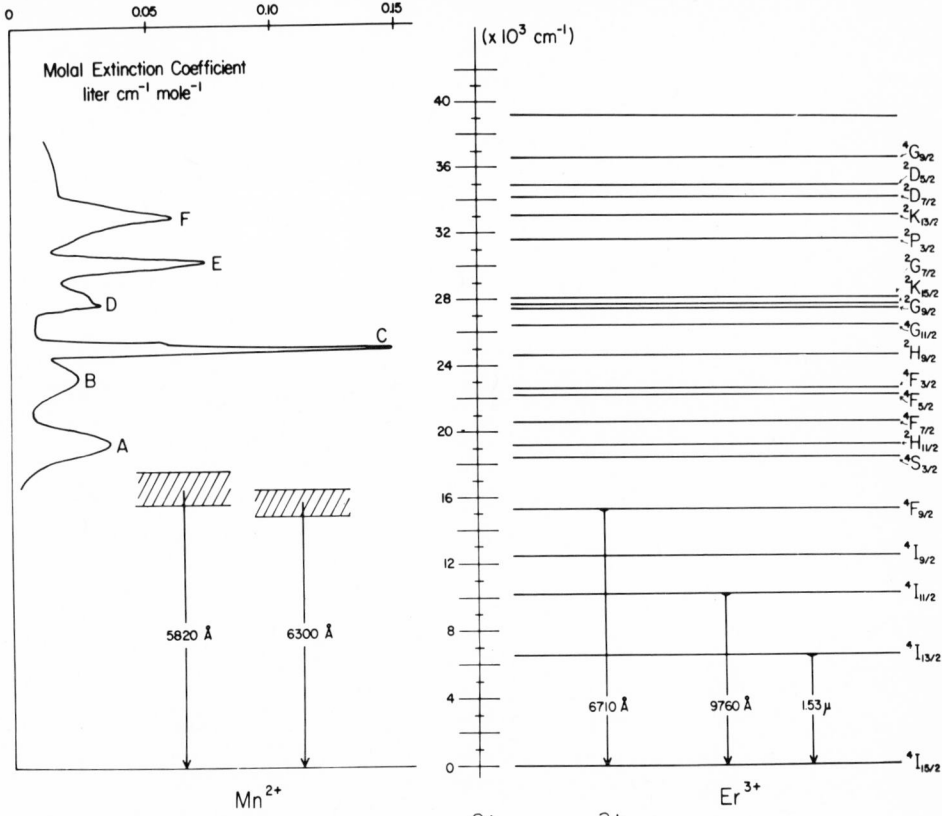

Fig. 1   Energy Level Scheme of Mn^{2+} and Er^{3+} in MnF_2.

in a sample MnF_2:Er(1%). Also in Fig. 1 the Mn absorption spectrum is presented with the bands labeled as A, B, C, etc.

The fluorescence spectrum of Er^{3+} in MnF_2:Er consists of three groups of lines centered at 6710 A, 9760 A and 1.53 μ which we assigned to transitions from the $^4F_{9/2}$, $^4I_{11/2}$ and $^4I_{13/2}$ levels to the $^4I_{15/2}$ ground level, respectively. In what follows we shall refer to these three groups of lines either as to the three "lines" at 6710 A, 9760 A and 1.53 μ, or as to the $^4F_{9/2}$, $^4I_{11/2}$ and $^4I_{13/2}$ "fluorescence".

We measured the thermal dependence of the integrated intensity of each of the three Er lines. As the temperature is raised from 4.2 to 77 K, the integrated intensity of all the three Er lines increases; the increase factor is 1.7, 2.0, and 2.6 for the fluorescence originating at the $^4F_{9/2}$, $^4I_{11/2}$, and $^4I_{13/2}$ levels respectively. Under the same temperature change, the intensity of the Mn emission in both samples MnF_2 and MnF_2:Er was seen to decrease; this is contrary to the behavior expected for a radiative transfer from

Mn to Er. Going from 77 to 300 K, the integrated intensity of the $^4F_{9/2}$ fluorescence decreases by a factor of 30, the integrated intensity of the $^4I_{11/2}$ fluorescence decreases by a factor of 10.6, and the integrated intensity of the $^4I_{13/2}$ fluorescence increases by a factor of 1.7. These measurements were made by optically exciting the sample through a $CuSO_4$ filter (band pass 3500-5200 A) resulting in the pumping of the system through the excitation bands

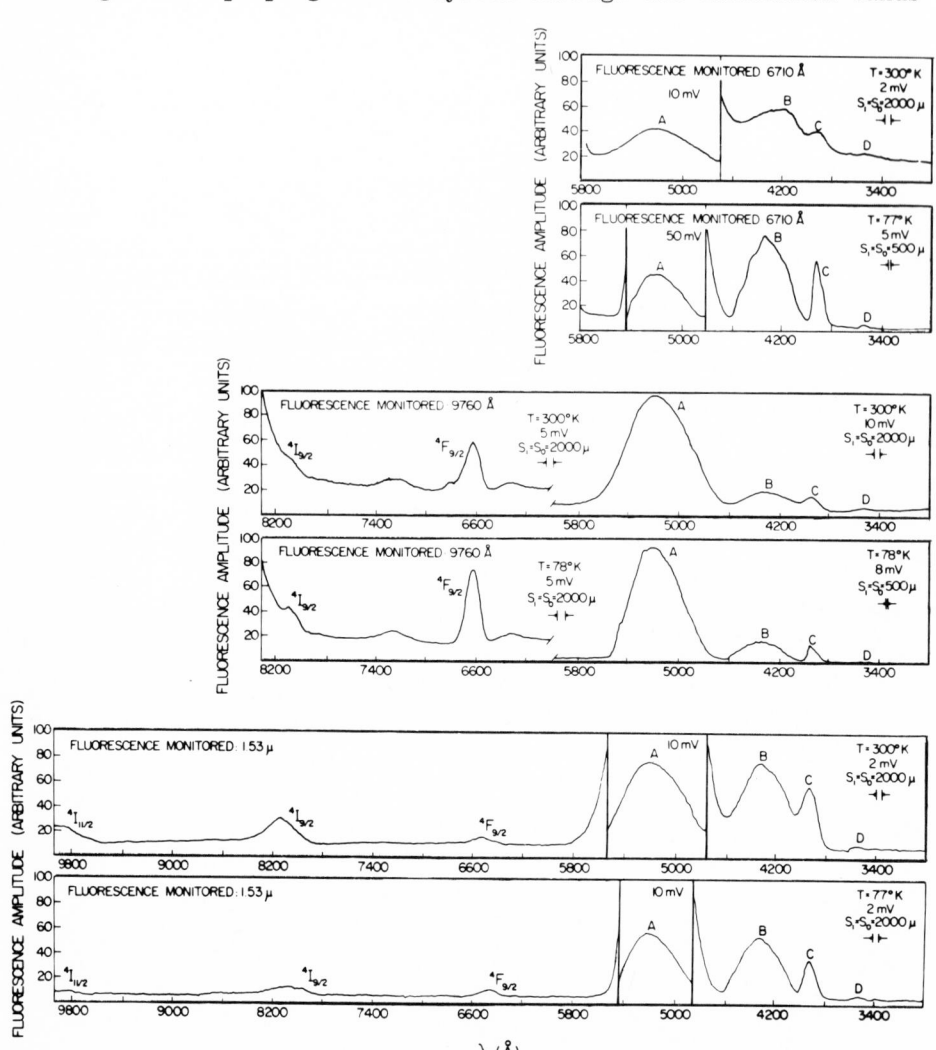

Fig. 2   Excitation Spectra of Er^{3+} in MnF_2:$Er(1\%)$. S_I and S_o indicate the widths of the input and output slits, respectively, of the Jarrell Ash Model 82-410 monochromator. The mV indicate the sensitivity scale of the P.A.R. Model 122 lock-in amplifier.

above the $^4F_{9/2}$ level.

The excitation spectra of each of the three Er transitions at 77 and 300 K are presented in Fig. 2. We can recognize the appearance of the Mn absorption bands A, B, C and D in each excitation spectrum at both temperatures. This gives direct evidence of the existence of the Mn$\longrightarrow$Er energy transfer process. The lower Er absorption bands are also present, but most of the Er excitation energy is a result of the Mn$\longrightarrow$Er transfer. An additional proof of the nonradiative nature of the transfer is the fact that even when no Mn fluorescence is detectable (say, at room temperature), Mn$\longrightarrow$Er energy transfer is still present as is shown by the excitation spectra.

Consider now the excitation spectra of the 9760 A line; the contribution of the Mn bands substantially decreases with respect to that of the Er bands when the temperature is raised from 77 to 300 K. (Notice the smaller slit widths used on the monochromator through which the exciting light is passed for the Mn bands at the lower temperature.) This indicates that the transfer process is more efficient at 77 than at 300 K.

We measured the decay pattern of the MnF$_2$ emission at 5820 and 6300 A by subjecting the sample to pulsed optical excitation. The lifetime at 5820 A is 38 msec at 4.2 K, 34 msec at 24 K and decreases abruptly to 3 msec at $\sim$ 32 K, the same temperature at which the shift in the peak of the Mn fluorescence occurs. The lifetime at 6300 A is 35 msec at 4.2 K, 28 msec at 32 K, 28 msec at 59 K, decreases abruptly to a value of 5.8 msec at 68 K and continues to decrease smoothly at higher temperatures. The fact that the decay patterns at 5820 and 6300 A differ in the temperature region 4.2 to 32 K, implies the existence of at least two real metastable levels for Mn^{2+}. In the attempt to interpret the thermal dependence of the fluorescence lifetimes of Mn, we fitted the curve corresponding to the 5820 A emission to a theoretical expression which implies a thermal depletion of the level associated with this emission into the lowest Mn absorption band. The same behavior was shown by the decay pattern of Mn in MnF$_2$:Er.

We also measured the decay patterns of the three Er lines in MnF$_2$. The lifetimes of the $^4F_{9/2}$, $^4I_{11/2}$ and $^4I_{13/2}$ levels at 4.2 K are 279 μsec, 9.8 msec and 20 msec, at 77 K are 245 μsec, 9.4 msec, and 17 msec, and reduce at 300 K to 128 μsec, 6.2 msec and 15.7 msec, respectively. The decay pattern of the $^4F_{9/2}$ level is essentially exponential with a deviation to longer time constants at the end of the decay. The decay patterns of the lower two levels present a fluorescence rise at the end of the pulse followed by an exponential decay. In any case, the overall behavior of the Er decay patterns reveals a mode of relaxation for the Er ion whereby a fluorescent level is excited only by nonradiative multiphonon decay of the next higher level. Since no fluorescence was observed from Er levels higher in energy than the $^4F_{9/2}$ level, every initial excitation of these levels results in the excitation of the $^4F_{9/2}$ level.

Under these conditions of stepwise relaxation, if the sample was optically pumped only into the Er levels above $^4F_{9/2}$, the intensity of the $^4F_{9/2}$ fluorescence and the $^4F_{9/2}$ lifetime should present the same thermal dependence. We observed experimentally that from 4.2 to 77 K this fluorescence increases by a factor of 1.7 while the lifetime decreases by a factor of 1.14; from 77 to 300 K the fluorescence decreases by $\sim$ 30 and the lifetime decreases only by a factor of 1.5. We explain the difference in the thermal dependences of the $^4F_{9/2}$ fluorescence and lifetime as due to the fact that the Er system is also excited via the Mn absorption bands and that the efficiency of the Mn$\longrightarrow$Er energy transfer process is temperature dependent. On the basis of the above results it can be said that the efficiency of this process increases from 4.2 to 77 K and then decreases from 77 to 300 K.

On the basis of the experimental results we have established the following model. The Mn emission originates from at least two real metastable levels which are the result of nonequivalent localized traps. Both of these traps are excited via the Mn absorption bands. Because of the high concentration of Mn and the relatively strong interaction between the Mn ions in MnF_2, the optical absorption may be expected to be an exciton type process, in which the excitation energy is delocalized and moves freely throughout the crystal. As the temperature is increased from 4.2 K, the fluorescence from the 5820 A trap, being energetically closer to the first absorption band, becomes thermally quenched. This quenching provides an increase in the amount of excitation of the lowest Mn absorption band. Considering the energy level scheme we can see that there are several Er levels which are energetically in resonance with the Mn absorption bands. We may assume that the Mn$\longrightarrow$Er energy transfer takes place via these resonances. The observed thermal dependence in the energy transfer process is explained in terms of the amount of excitation energy present in the Mn A band. Specifically, at very low temperatures, the excitons are relatively free to move throughout the lattice except for their interaction with defects and the Er impurities. The number of excitons maintained in the A band actually increases as the temperature is raised above 4.2 K, since this band is replenished by the thermal quenching of the higher Mn fluorescence traps; this allows a greater number of excitons to interact with Er. As the temperature is raised above 77 K, exciton-phonon interactions become increasingly competitive with the impurity interaction; this results in a decreasing amount of energy transferred to Er as the temperature is further raised.

REFERENCES

1. E. W. Prohofsky, Phys. Rev. Lett. $\underline{14}$, 302 (1965).

2. W. W. Holloway, E. W. Prohofsky, and M. Kestigian, Phys. Rev. $\underline{A139}$, 954 (1965).

VIBRONIC STRUCTURE IN THE LUMINESCENT SPECTRA OF IONS IN SOLIDS

Toru Miyakawa

Defense Academy

Yokosuka, Japan

ABSTRACT

A unified treatment of the vibronic structure taking into account both diagonal and off-diagonal parts of Hamiltonian for the electron-phonon coupling is given. It is shown that both the ratio of integrated intensities of one-phonon component to that of zero-phonon component and the profile of one-phonon side band for an almost forbidden transition can be different from those for an allowed transition. Also the different temperature dependence of the integrated intensities is pointed out. These results are discussed in the light of recent experimental observations.

Vibronic structures in absorption and emission spectra are important in various phases of luminescence. Two causes of the appearance of these structures are well known and arises from diagonal and off-diagonal parts of the Hamiltonian for electron-phonon coupling. They are usually called the relative shift of the equilibrium positions of lattice coordinates and the forced electric dipole transition, respectively. Let us call them, for short, Δ- and M-processes, respectively.

Although one or the other of these mechanisms has been stressed in the literature (1,2), a unified treatment of them seems to be lacking. Here we derive an expression for the vibronic structure which we hope will give a unified view of the mechanisms governing this structure and point out that the temperature dependence of integrated intensities is different according to which one of the two mechanisms is operative. Also, depending upon the nature of

the electronic transition we may expect different weight factors
for the phonon density of states, and accordingly, the ratio of
integrated intensities of the one-phonon band to the zero-phonon
line R can vary from one line to the other.

The absorption coefficient by an electron localized on an ion
in a host crystal with the refractive index n is;

$$A(\omega) = \frac{n}{2\hbar^2 c} \sum_\sigma \int_{-\infty}^{\infty} dt\, f_{T\sigma}(t)\exp(-i\omega t),$$ (1)

$$f_{T\sigma}(t) = \mathrm{trace}_L \{ <T^*_{\sigma k} e^{iHt/\hbar} T_{\sigma k} > e^{-(\beta+it/\hbar)H_g} \}/\mathrm{trace}_L (e^{-\beta H_g}).$$ (2)

Here $T_{\sigma k}$ denotes the transition operator by which a photon with
wave vector k and polarization σ is either created or annihilated
with an electronic transition $1 \rightarrow 1'$. Trace_L means taking the
diagonal sum of the quantity in the lattice eigenstates. H is the
Hamiltonian for the electron phonon system, H_g is the same for the
ground state of the electron.

The generating function $f_{T\sigma}(t)$ contains all the information
we need. If we assume that the lattice relaxation time is much
shorter than the lifetime of the excited electronic state by
spontaneous emission and ignore the complications arising from
multiphonon relaxation; it also gives the emission probability
after appropriate changes in notation and multiplicative constants.
Although Eq. 1 can be used in the static approximation; we use
here the adiabatic approximation and follow the method of Kubo and
Toyozawa (3). The electron-phonon coupling Hamiltonian is assumed
to be linear in the lattice coordinates.

$$H_{eL}(\underline{r},\underline{Q}) = \sum_j v_j(\underline{r}) \cdot \underline{Q}_j.$$ (3)

This is because we are not interested in the shift of lattice
frequency or effects associated with Jahn-Teller distortions.
Formally, however, the effect of quadratic terms may be incorporated
into matrix elements M_2, S_2 or S_3.

Expanding the wave functions for the electron up to second
order in v, we have the following forms for the matrix elements of
optical transition operator M and the non-adiabatic part of the
Hamiltonian H_{NA}

$$M_{\ell'\ell} = (M_0)_{\ell'\ell} + (M_1)_{\ell'\ell} \cdot \underline{Q} + \underline{Q} \cdot (M_2)_{\ell'\ell} \cdot \underline{Q},$$ (4)

$$H_{NA} = -i\underline{S}^{(1)} \cdot \partial/\partial\underline{Q} - i\underline{S}^{(2)}\underline{Q} \cdot \partial/\partial\underline{Q} - S^{(3)}.$$ (5)

Expanding $\exp(iHt/\hbar)$ into powers of H_{NA} up to second order the generating function may be put into the following form:

$$f_{T\sigma}(t) = g_M(t)\, f_0(t)\, f_S(t),\qquad\qquad\qquad (6)$$

$$g_M(t) = |M_0|^2 + \{(M_1^* M_0 + c.c.)/2\}\{1 + ne^{-i\omega t} - (n+1)e^{i\omega t}\}\Delta$$

$$+ |M_1|^2[(\hbar/2\omega)\{ne^{-i\omega t} + (n+1)e^{i\omega t}\} + \{1 + ne^{-i\omega t} - (n+1)e^{i\omega t}\}^2 \Delta^2/4]$$

$$+ [(M_2^* M_0 + c.c.)/2][(\hbar/2\omega)(2n+1)$$

$$+ \{1 + ne^{-i\omega t} - (n+1)e^{i\omega t}\}^2 \Delta^2/4],\qquad\qquad (7)$$

$$f_0(t) = \exp\{i\varepsilon_{\ell'\ell} t/\hbar - g + g_+(t) + g_-(t)\},\qquad\qquad (8)$$

$$g_{\underline{+}}(t) = \int_0^\infty d\omega D(\omega) \binom{n+1}{n} \exp(\underline{+}i\omega t),$$

$$g = g_+(0) + g_-(0),\qquad\qquad\qquad (9)$$

$$D(\omega) = \sum_j (\omega_j \Delta_j^2/2\hbar)\delta(\omega - \omega_j).$$

$$\Delta_j = (1/\omega_j^2) \int v_j(\underline{r})[|\phi_{\ell'}(\underline{r})|^2 - |\phi_\ell(r)|^2]dv .\qquad\qquad (10)$$

We shall not go into discussion of $f_S(t)$ here because it gives only small correction terms unless there are closely spaced excited electronic states such that $\hbar\omega/\varepsilon \sim 1$. Apart from the last line in Eq. 7, Eqs. 7-10 were derived by Mulazzi et al (4). However, they did not go into the consequences of their results.

If we put $\Delta = 0$ for all modes and neglect M_2 in Eq. 7 the usual expression for phonon-induced transitions results. The intensity of the zero-phonon line is proportional to $|M_0|^2$ while that of the one-phonon side band is proportional to $|M_1|^2(\hbar/2\omega)(n+1)$. Thus the ratio R can take almost any value depending on the ratio of the two matrix elements. For an almost forbidden transition with a nearby level (or a group of levels) to which the transition is allowed from the ground or excited state this ratio can be quite large, and in the extreme case the zero-phonon line may be missing. However, this does not imply that the two-phonon side band is strong. On the contrary we should expect a sharp cut-off unless Δ_j is non-zero for some of the modes.

The profile of the phonon side band is determined by

$$\sum_j (\hbar/2\omega_j)|v_{j\ell n} M_{n\ell}/\varepsilon_{\ell n}|^2.\qquad\qquad (11)$$

It should be noted that the sum of integrated intensities of zero- and one-phonon components increases with temperature in this simplified approximation unless the condition

$$|M_1|^2 + (M_2^* M_0 + c.c.)/2 = 0,$$

is satisfied and provided no complication arises from the relaxation processes via nearby levels.

The phonon side bands observed in $CaWO_4$:Eu,Na system (5) seem to exemplify the different ratios of one-phonon vs. zero-phonon intensities. Namely it is almost 10 for the $^7F_0 \rightarrow ^5D_0$ transition while it is 0.03 and 0.25 for the $^7F_0 \rightarrow ^5D_1$ and $^7F_0 \rightarrow ^5D_2$ transitions, respectively. The weak side band observed for the magnetic dipole transition was explained by Yamada et al. (5). As the phonon side band is allowed in first order in both transitions terminating in the $^5D_0$ and $^5D_2$ states states but the electronic transition is allowed only in the second order in the former transition, this small M_0 seems to be responsible for the relatively strong side band. Unfortunately, there seems to be no report on the detailed analysis of the temperature dependence of integrated intensities.

On the other hand, if, neglecting M_1 and M_2, we expand $f_0(t)$ into powers of g, g_+ and g_- we find a series of multiphonon side bands. The ratio of the integrated intensities of successive multiphonon components is determined by the coupling constant g in this case and their profile is given by the n-th convolution of the spectral density of electron-phonon coupling $D(\omega)$ (1).

It should be noted that in this case the sum of all the phonon side band intensities including zero-phonon line is conserved at any temperature. The temperature dependence of the zero-phonon line is determined by the temperature dependent coupling constant g. The $4f^{n-1}5d \rightarrow 4f^n$ spectrum observed for Eu^{2+} in SrF_2 (6) or in alkali-halides (7) seems to fall into this category.

If both Δ- and M-processes are operative the one-phonon component will be a superposition of profiles discussed above. As the weight-factor as well as the symmetry of interacting modes are different for the two processes, there may be additional peaks in the spectra than those expected from Δ- or M-process alone.

The temperature dependence of the zero-phonon line may be put into an approximate form neglecting for the moment the effect of M_2:

$$|M_0|^2 e^{-\bar{g}} (1+\bar{r}\bar{g}_m)\{1 - 2\bar{n}(\bar{n}+1)\bar{r}\bar{g}_m/(1+\bar{r}\bar{g}_m) - 2\bar{n}\bar{g}/(1+\bar{r}\bar{g}_m)\}, \tag{12}$$

where $\bar{r} = (\hbar/2\bar{\omega})|M_1/M_0|^2$, and $\bar{g}_m = \sum_j (\omega_{jm}/2\hbar)\Delta_{jm}^2$ is an average coupling constant for modes coupling through M_1.

In this approximation the ratio R is given by $R = \bar{r}/(1+\bar{r}\bar{g}_m)$ while the ratio of two-phonon to one-phonon component is determined by $\bar{g}_m$. If M_0 is small, $\bar{r}$ can be quite large, and accordingly R can be large. This large value of $\bar{r}$ enhances the effect of the small Δ_{jm} shift and at least for small values of $\bar{n}$ equation 12 may be approximated by

$$|M_0|^2 e^{-\bar{g}}(1+\mathrm{rg}_m)\ \exp(-2\bar{n}g_a) \tag{12a}$$

with an apparent coupling constant $g_a = (\bar{g}+\bar{r}\bar{g}_m)/(1+\bar{r}\bar{g}_m)$. For large $\bar{r}$ and small $\bar{g}$ and $\bar{g}_m$ values one may expect a strong one-phonon side band but a negligible two-phonon ecomponent and still an appreciable temperature dependence of the zero-phonon line with an apparent Huang-Rhys factor of the order of unity.

This might explain a rather puzzling fact that from the temperature dependence of the zero-phonon line the Δ-process seems to give a dominant contribution to the vibronic structure even in systems like Sm^{2+} in alkali-halides (7) or Cr^{3+} in YAG (8).

In conclusion we wish to stress that detailed analysis of the temperature dependence of the vibronic structure may yield valuable information regarding the electron-phonon coupling in these systems.

REFERENCES

1. For instance, Y. Toyozawa: Dynamical Processes in Optical Spectra of Solids (Syokabo, Tokyo and W. A. Benjamin, New York, (1967), p. 90.

2. M. Wagner: Z. Phys. 214, 78 (1968).

3. R. Kubo and Y. Toyozawa: Prog. Theor. Phys. 13, 161 (1955).

4. E. Mulazzi, G. F. Nardelli and N. Terzi: Phys. Rev. 172, 847 (1968).

5. N. Yamada and S. Shionoya: J. Phys. Soc. Japan 31, 841 (1971).

6. E. Cohen and H. J. Guggenheim: Phys. Rev. 175, 354 (1968).

7. G. Baldini and M. Guzzi: Phys. Stat. Solid 30, 601 (1968).

8. W. A. Wall, J. T. Karpick and B. di Bartolo: J. Phys. C4, 3258 (1971).

LUMINESCENCE OF SIMPLE POLYATOMIC ANIONS*

S. P. McGlynn

Coates Chemical Laboratories, The Louisiana State

University, Baton Rouge, Louisiana, 70803, USA

ABSTRACT

The luminescence of simple anions such as nitrite, nitrate, oxalate, formate, cyanide, azide, etc. is discussed and interpreted. The effects of associated metal cations are elaborated, and a theory of color associated with heavy-metal salts is developed.

INTRODUCTION

It is our intention to discuss the colors commonly exhibited by heavy-metal salts of which neither the anion nor cation components are normally colored. We will, for simplicity, restrict discussion to non-transition metal cations and so avoid the prolixity that any consideration of $d \leftrightarrow d$ and $f \leftrightarrow f$ excitations would require. Similarly, we will only consider polyatomic anions; this tactic enables us to evade the difficult area of halide salt crystals and to focus primary concern on the topics of triplet states of the anion and cation $\leftarrow$ anion inter-ion charge transfer. We emphasize, however, that these exclusions are only matters of convenience and that effects identical to those to be discussed here are observed in both transition metal salts and in halide crystals and that the interpretation of these latter effects follows logically along the same lines as are broached here.

The specific phenomenon of interest is illustrated in Table 1 for a series of silver salts and in Tables 2 and 3 for series of nitrite and chlorite salts, respectively. It is quite clear from these tables that heavy-metal, non-transition-metal cations induce considerable degrees of color. This phenomenon has been interpreted

399

Table 1

Colors of Ag^+ Salts

Anion	Color	Ag-O distance (Angstroms)	$S_1 \leftarrow S_0$ Transition Energy (mμ)
ClO_3^-	None	2.51	190
$SO_4^=$	None	2.34	< 200
$MoO_4^=$	Yellow	2.42	210
(KAg) CO_3	None	2.42	210
$PO_4^=$	Yellow	2.34	< 200
$AsO_4^{\equiv}$	Red	2.34	< 200
$CO_3^=$	Yellow	2.30	210
NO_2^-	Yellow	2.04	355
ClO_2^-	Green-yellow	2.20	290
$N_2O_2^=$	Yellow	----	248
NO_3^-	None	----	300
$SO_3^=$	None	----	< 200

Table 2

Colors of NO_2^- Salts (from Ref. 2)

Cation	Color	ε(mole^{-1}cm^2)
Na(I)	None Yellow Tinge	3×10^{-3}
Cd (II)	Yellowish	2.3×10^{-2}
Ag (I)	Pale Yellow	0.5
Hg (I)	Sulfur Yellow	---
Pb (II)	Orange Yellow	11.7
Tl (I)	Orange	0.64

Table 3

Colors of chlorite salts (from Ref. 6)

Cation	Comments	Color
Na (I)	Crystal	None
NH_4 (I)	Crystal	None
Ba (II)	Crystal	None
Ag (I)	Crystal	Green-Yellow
Pb (II)	Crystal	Orange-Yellow
Tl (I)	Unstable Solution	Yellow
Hg (II)	Probably $3Hg(ClO_2)_2 \cdot HgO$	Red

by us previously (1-4) and it is with the details of this interpretation and its validation that this paper is concerned.

The gist of our thesis is as follows:

(i) Heavy-metal salts in solution and, particularly, in the crystal exhibit considerable degrees of cation-anion covalency. (This effect is particularly evident (5) in $AgMnO_4$ which is electrovalent in aqueous solution and which possesses the standard red-purple color of the aquated MnO_4^- ion, but which is clearly covalent in the crystal, the Ag-O distance being 2.19 A, and which possesses a green color characteristic of an $AgMnO_4$ covalent unit.)

(ii) As a result of this covalency, inter-ion charge transfer (CT) processes of type cation ← anion become possible and may be of low energy. (In the specific case of $AgNO_2$ in the crystal, such CT transitions are predicted (4) to occur at 3.04 ev and to be coincident with, or even at lower energy than, the first $S_1 \leftarrow S_0$ transition characteristic of the NO_2^- ion. In the case of NO_3^- salts such CT processes are known (7) to occur at considerably lower energy than the 3000 A band which is characteristic of the NO_3^- ion.)

(iii) As a result of the covalency between anion and cation, the orbitals of the anion delocalize onto the heavy-metal center and become subject to the large spin-orbit coupling available at that center. Thus, the incidence of spin-orbit coupling to which the electrons of the formerly-isolated anion are subjected undergoes a large increase, an increase which is further augmented because of the energetic proximity of high-intensity CT transitions to formerly spin-forbidden excitations of the "isolated" anion. It is our contention that it is just such an increase of spin-orbit coupling and its enhancement of triplet ← singlet transitions

Table 4

Phosphorescence lifetimes of NO_2^- salts (sec)[a]

Cation	$\tau_p(EM)$	$\tau_p(ABS)$	$\tau_p(CALC)$
Na (I)	3.3×10^{-3}, Y	3.3×10^{-2}	2.3×10^{-2}, Y 1.76, Z 0, X
Ag (I)	1×10^{-4} 6×10^{-4} 3.5×10^{-3}	---	$2. \times 10^{-4}$, X 6×10^{-4}, Y 2×10^{-2}, Z
Tl (I)	9×10^{-5}	6×10^{-4}	---
Pb (II)	7×10^{-5}	5×10^{-5}	---

[a]The X-axis is perpendicular to the NO_2^- plane. The Z-axis lies along the principal axis of the C_{2v} ion. The $\tau_p(EM)$ data were obtained by direct decay measurement at temperatures in the range 77 - 1.8 K. The $\tau_p(ABS)$ data were calculated using the extinction data of Table 2. The $\tau_p(CALC)$ data were obtained by MO and 1st order spin-orbit coupling computations methods.

characteristic of the "isolated" anion which is responsible for the colors of heavy-metal salts. In accord with this thesis, we can provide a ready interpretation (4) of the colors of the heavy-metal salts of NO_2^- and ClO_2^- and the absence of color in the heavy-metal salts of OCN^-, NCO^-, N_3^-, $Ox^=$, Benzoate$^-$, $HCOO^-$, CH_3COO^-, NO_3^-, ClO_3^- $SO_4^=$. We can make no comment on the heavy-metal salts of the other anions for the simple reason that we have not subjected any of them to either experimental or theoretical investigation.

(iv) As a result of the increased spin-orbital coupling, the luminescence of heavy-metal salts is usually a phosphorescence of short duration (i.e., $\sim 10^{-4}$ sec).

(v) In some instances, cation $\leftarrow$ anion CT processes may be sufficiently low in energy to dominate the colors and emissive characteristics of certain heavy-metal salts. The fact that we cannot identify any such salts presently merely attests to the lack of investigative effort in this area and not to the non-existence of such salt types.

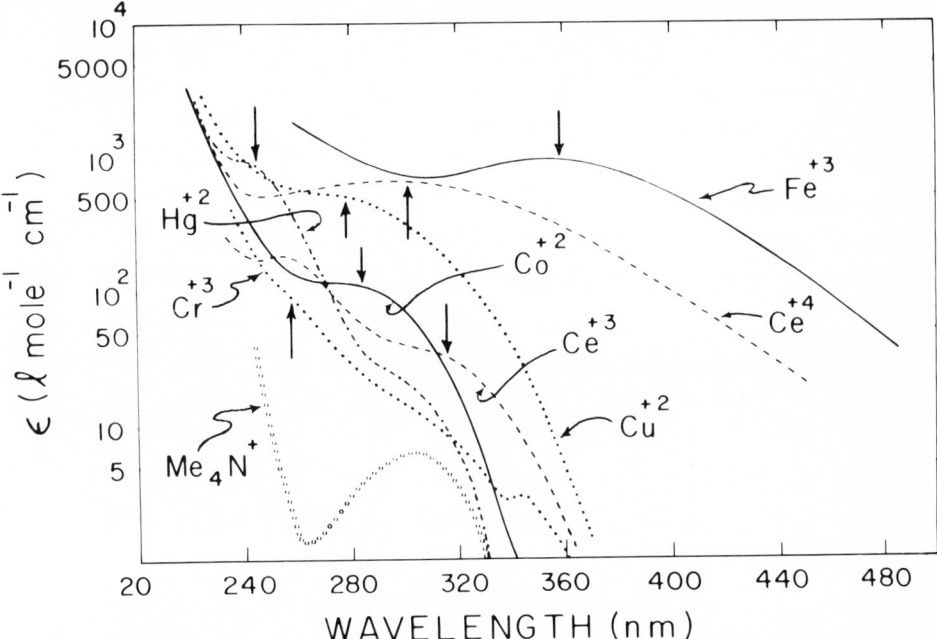

Fig. 1   Absorption spectra of nitrate salts in anhydrous methanol.
In the case of cobaltous nitrate, the solvent is t-
butylalcohol.   Solutions of ferric nitrate were acidified
with nitric acid to prevent hydrolysis.   Charge transfer
bands are indicated by a vertical arrow.   The band at 256 nm
in cerous nitrate is due to an internal transition of the
cerous ion.   The weak band ($\varepsilon \approx 3$) at 344 nm in $Cr(NO_3)_3$
is not present in the spectrum of an aqueous solution of
this salt; it is probably due to an enhanced singlet-
triplet transition of the nitrate ion.

REPRESENTATIVE SYSTEMS

The nitrite ion, NO_2^-, has been discussed by us previously
(1-4,8).   It shall suffice here to point out that covalency of
anion and cation does occur, that cation ← anion CT bands have been
identified, that spin-orbit coupling undergoes enhancement by a few
orders of magnitude and that the colors and luminescences of these
salts are fully interpretable along the lines of postulates (iii)
and (iv), respectively.   A representative set of data illustrating
the concordance of luminescence, absorption and theoretical results
is given in Table 4.   It is clear that the concordance evident in
this Table is excellent.

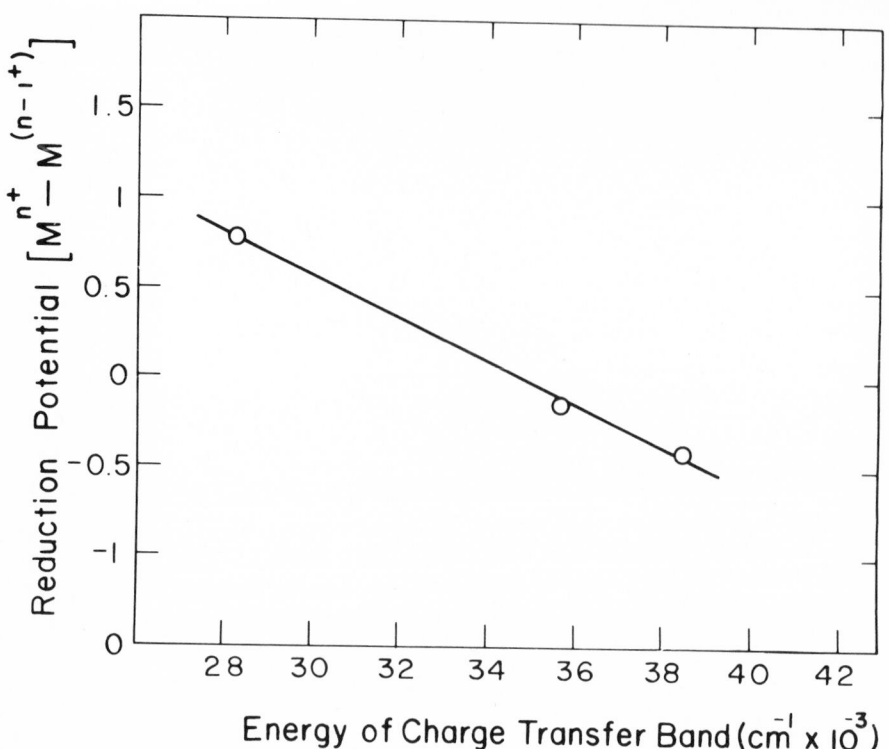

Fig. 2   Energy of the charge-transfer band of nitrate salts in
         methanol plotted against the reduction potential of the
         positive ion.

        The nitrate ion, NO_3^-, has been discussed by us previously
(4,7). It exhibits all the effects discussed above for NO_2^-, though
not to quite as large a degree. The reason for this is rather
straightforward, being associated with the highly-acidic nature of
the nitrate ion and its reluctance to form complexes with other
than transition-metal ions, as well as with its ease of decomposition to nitrite ion and the consequent complication of emission
data. Nonetheless, the reason why NO_3^- does not form colored heavy-
metal salts is perfectly clear: The lowest triplet state of NO_3^-
ion lies at $\lambda < 3800$ A and the enhancement of the $T_1 \leftarrow S_0$ transition,
which is weak anyway because of small covalency effects, does not
produce color because of its spectral location in the ultraviolet.
The spectra of transition metal nitrates in non-aqueous solvents
are of interest, however, for another reason: They exhibit low-
energy absorption bands which are identifiable, with a high degree
of certainty as cation $\leftarrow$ anion charge transfer bands. These bands
are illustrated in Fig. 1 and are denoted by vertical arrows. They
are further abstracted in Fig. 2 where they are plotted against the
reduction potential of the counter-ion. The paucity of points in

Table 5

Luminescence data for cyanate and isocyanate salte (9,10)

Compound	Matrix	λ_{max} (mµ)	τ_p (sec)
NaOCN	Crystal	406	0.21
	Glass	406	0.17
Cd(NCO)$_2$	Crystal	425	$10^{-1} - 10^{-2}$
Hg(NCO)$_2$	Crystal	460	3×10^{-2}
Pb(NCO)$_2$	Crystal	490	7×10^{-3}
CH$_3$NCO	Glass	418	2.05
φ-NCO	Glass	387	3.15
HNCO	Glass	430	---

Fig. 2 exhausts all the available information on reduction potentials and is not, consequently, entirely satisfactory. Nonetheless, Fig. 2 is adequate to point up the CT nature of these new absorption bands; to show that they are not, as had been inferred previously, the enhanced 3000 A absorption band of NO_3^- (shown for tetramethyl-ammonium nitrate in Fig. 1); and to suggest that such spectra might better serve the inverse purpose of generating reduction potential data.

Cyanate and isocyanate ion spectra have been discussed previously (9,10). A synopsis of this luminescence behavior is given in Table 5. They exhibit all the characteristics required by postulates (iii) and (iv). Nonetheless since the $T_1 \leftarrow S_0$ absorption lies in the region $3300 \geq \lambda \geq 2500$ A, heavy-atom salts are not expected to exhibit color--a conclusion which accords with observation.

CONCLUSION

This discussion could be prolonged indefinitely by merely referring to different anions and their various salts. Rather than do so, we now list those anion systems which have been studied: azides (11); chlorites (6); oxalates (12,13); benzoates (14); phenylcarboxylates (14); thiocyanates (15); hydroxides (16). The validity of postulates (i)-(v) is upheld in all instances.

Finally, a large number of other anion categories, among them some of the more interesting and important systems, remain as yet wholly uninvestigated. There is every reason to believe that their investigation will induce order in a very untidy area and that this order will lie much along the lines of postulates (i) through (v) outlined above.

REFERENCES

*This research was supported by contract between the United States Atomic Energy Commission - Biology Branch and The Louisiana State University.

1. H. J. Maria, A. T. Armstrong and S. P. McGlynn, J. Chem. Phys. 48, 4694 (1968).

2. H. J. Maria, A. Wahlborg and S. P. McGlynn, J. Chem. Phys. 49, 4925 (1968).

3. H. J. Maria, B. N. Srinivasan and S. P. McGlynn, Molecular Luminescence, Benjamin, Inc., New York, (1969) (edited by E. C. Lim), p. 787.

4. L. E. Harris, H. J. Maria and S. P. McGlynn, Czech. J. Phys. B20, 1007 (1970); S. P. McGlynn, ibid., B5, 654 (1970).

5. Observations by L. W. Johnson and S. P. McGlynn, to appear in Chem. Phys. Letters.

6. Observations by H. J. Maria and S. P. McGlynn, in preparation for publication.

7. H. J. Maria, J. R. McDonald and S. P. McGlynn, J. Am. Chem. Soc., in press.

8. H. J. Maria, A. T. Armstrong and S. P. McGlynn, J. Chem. Phys. 50, 2777 (1969).

9. J. W. Rabalais, J. R. McDonald and S. P. McGlynn, J. Chem. Phys. 51, 5095 (1969).

10. J. W. Rabalais, J. R. McDonald and S. P. McGlynn, J. Chem. Phys. 51, 5103 (1969).

11. J. R. McDonald, J. W. Rabalais and S. P. McGlynn, J. Chem. Phys. 52, 1332 (1970).

12. H. J. Maria and S. P. McGlynn, J. Mol. Spectroscop. 42, 177 (1972).

13. H. J. Maria and S. P. McGlynn, J. Mol. Spectroscop. 42, 296 (1972).

14. H. J. Maria and S. P. McGlynn, J. Chem. Phys. 52, 3399 (1970).

15. J. R. McDonald, V. M. Scherr and S. P. McGlynn, J. Chem. Phys. 51, 1723 (1969).

16. H. J. Maria and S. P. McGlynn, J. Chem. Phys. 52, 3402 (1970).

MULTIPLICITY OF DIELECTRIC LOCAL MODES: PHONON-IMPURITY CENTER

BOUND STATES

E. I. Rashba

L. D. Landau Institute for Theoretical Physics

Academy of Sciences of the USSR, Moscow

ABSTRACT

The general theory of binding of optical phonons to electronic impurity centers is presented. The structure of the bound state spectrum is investigated. A convenient method for the calculation of binding energies is developed.

INTRODUCTION

The phonon spectrum is perturbed in the neighborhood of impurity centers due to electron-phonon coupling and, under favorable conditions, local modes of some special type may arise. Their formation is caused solely by the coupling of impurity electrons to optical phonons and has nothing to do with the usual mechanisms of the formation of the local modes. These new modes, predicted by Kogan and Suris (1), are known as dielectric modes. They were found by Dean and Manchon (2) in GaP doped with donor impurities. In (3) they were considered as quantum states of phonons bound to impurity centers.

In (1,3) the dielectric mode formation was considered as some type of resonance effect which arises, when one of the excitation energies of an impurity center is close to the optical phonon frequency ω_0. Therefore, the two-level model was used in which, of the whole electronic spectrum of the center, the wave functions corresponding to the ground level E_0 and to the resonance level E_1 were the only functions taken into account. Such an approach is justified only when the distance from the resonance $|E_1-E_0-\omega_0|$ is much less than the separation of the successive electronic levels. It can be seen that, even in the two-level model, the dielectric mode formation, by itself, is not connected with the resonance: when the phonon dispersion is neglected these modes always arise. In this connection some important questions arise: How general is this conclusion and is it correct for the many-level centers? How many dielectric modes arise and how can they be classified? How can their frequencies be calculated? These questions are discussed below. They are of great importance because the centers, for which the dielectric modes were found in the impurity exciton phonon-assisted emission spectra, were far from resonance.

THE BASIC EQUATIONS

The electron-phonon coupling is supposed to be weak and phonon dispersion to be neglectable. Furthermore we take a quite general model of the center with its ground state non-degenerate. When the coupling is weak the dielectric mode frequencies must be near ω_0, and they may be found as poles of the electronic Green's function $G(\omega)$ which are close to ω_0; thus, close to the phonon emission threshold. Here and below we count the frequency parameter ω_0 in G from E_0. At the threshold, the mass operator $M(\omega)$ calculated to second order of the perturbation theory diverges as

$$M(\omega) \propto 1/|\omega - \omega_0| , \qquad (1)$$

and, thus, a series of "dangerous" diagrams must be summed. The series of diagrams divergent at threshold singled out by Pitayevsky (5) is shown in Fig. 1 for the weak coupling case. The summing of

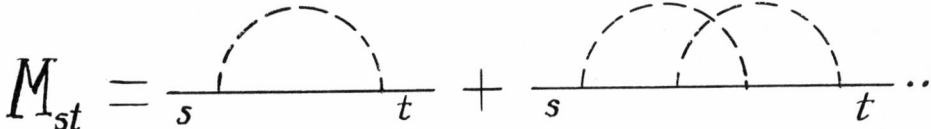

Fig. 1 Diagrams summed for the mass operator.

such a series was already performed in some related problems concerning the binding of the electron with the optical phonons. Mel'nikov and Rashba (6) showed that for polarons with weak coupling conditions, bound states do not arise. On the contrary, Levinson (7) showed that for magnetopolarons a rich spectrum of bound states arises.

The solution of the impurity center problem is complicated by the fact that G and M are nondiagonal in quantum numbers of the electronic levels. Eventually, however, the frequencies may be gotten as solutions of the equation found by Rashba (8):

$$\text{Det} \, ||A_{st} - \delta_{st}[(\omega - \omega_0)(\omega^2 - \varepsilon_s^2)/2\varepsilon_s]|| = 0, \quad s,t = 1,2,\ldots\infty. \qquad (2)$$

Here s,t indicate all excited electronic states of the center,
$\varepsilon_s = E_s - E_o$ are excitation energies,

$$A_{st} = \sum_q \gamma_{so}(\underline{q}) \gamma_{ot}(-\underline{q}), \qquad \gamma_{st}(\underline{q}) = c_q <s|\exp(i\underline{q}\cdot\underline{r})|t> , \qquad (3)$$

where c_q are the coefficients in the electron-phonon interaction
Hamiltonian, and index "o" in γ_{so} and γ_{ot} corresponds to the
ground state.

PARTICULAR CASES

Let one of ε_s be close to ω_o, i.e., the separation between the
electronic levels $\Delta\varepsilon >> |\varepsilon_s - \omega_o|$. Then it is possible to use the two-
level approach; this can be done by setting the corresponding ma-
trix element in Eq. (2) equal to zero:

$$A_{ss} = (\omega - \omega_o)(\omega - \varepsilon_s).$$

If the system is extremely close to the resonance, then

$$\omega - \omega_o \approx \pm A_{ss}^{\frac{1}{2}} \qquad \text{at} \qquad |\varepsilon_s - \omega_o| << A_{ss}^{\frac{1}{2}} . \qquad (4)$$

When the system moves away from resonance, then

$$\omega - \omega_o \approx A_{ss}/(\omega_o - \varepsilon_s) \quad \text{at} \quad A_{ss}^{\frac{1}{2}} << |\varepsilon_s - \omega_o| << \Delta\varepsilon. \qquad (5)$$

The mean number of phonons at resonance is about ½ for both
states, i.e., zero-phonon and one-phonon states are mixed with
comparable weights. Such states may be designated as hybrid ones.

When the system moves away from resonance, the wave function
is dominated by the one-phonon contributions since the electronic
system is mainly in the ground state. Such states may be considered
as the bound states of a phonon to the impurity center where the
binding energy is $\lambda = \omega_o - \omega$.

GENERAL STRUCTURE OF THE SPECTRUM

When resonance levels satisfying the criterion (4) are excluded
from consideration, equation (2) may be simplified

$$\text{Det} \left| \left| A_{st} - \delta_{st} \frac{\omega_o^2 - \varepsilon_s}{2\varepsilon_o} (\omega - \omega_o) \right| \right| = 0, \quad s,t = 1,2,\ldots\infty . \qquad (6)$$

It may be shown that, if all $\varepsilon_s > \omega_o$, then all the frequencies
of the dielectric modes defined in Eq. (6) are less than ω_o, i.e.
all the phonon binding energies are positive. When the center is a
spherically symmetric one, bound states arise that correspond to all

the values of the angular momentum. It is very significant that an infinite number of frequencies corresponds to every value of the angular momentum. If several $\varepsilon_S < \omega_O$, then the same number of frequencies larger than ω_O arises.

VARIATIONAL PRINCIPLE

Eq. (6) is not convenient for calculations; thus, we return to a configurational representation. It follows from Eq. (3) that in the r-representation:

$$A(\underline{r},\underline{r}')=\psi_O(\underline{r})V(\underline{r}-\underline{r}')\psi_O(\underline{r}'), \quad V(\underline{r})= \sum_{\underline{q}}|c_{\underline{q}}|^2\exp(i\underline{q}\cdot\underline{r}), \tag{7}$$

where ψ_O is the ground state electronic wavefunction. When an electron interacts with polarizational phonons:

$$V(r) = (e^2\omega_O/2r)(\frac{1}{\kappa_\infty} - \frac{1}{\kappa_O}) . \tag{8}$$

It is convenient now to replace Eq. (6) by the variational problem

$$\omega_O-\omega=\lambda=2\max_\phi\{ [\phi,(H-E_O)A\phi]/[\phi,((H-E_O)^2-\omega_O^2)^2\phi]\} , \tag{9}$$

in the class of functions $(\psi_O,\phi)=0$. Here H is the Hamiltonian of the center without electron-phonon coupling. It is easy to prove the equivalence of Eq. (6) and (9) by expanding ϕ over the functions ψ_S with $s\neq0$, and writing the condition (9) for the expansion coefficients.

The significant limiting case corresponds to the deep center with the ionization potential $R \gg \omega_O$; under such conditions (9) may be replaced by:

$$\lambda=2\max_\phi\{ [\phi,A\phi]/[\phi,(H-E_O)\phi]\}, \quad (\psi_O,\phi) = 0 . \tag{10}$$

For states with angular momenta $\ell \geq 1$ the condition (10) is equivalent to the equation

$$\lambda(H-E_O)\phi - 2A\phi = 0 . \tag{11}$$

Using (9) or (10) it is easy to find the approximate value of the binding energy. For example, for the hydrogen-like model of the center with $R \gg \omega_O$ we get:

$$\lambda_{2p} \approx 0.15(\kappa_O/\kappa_\infty - 1)\omega_O ; \tag{12}$$

this is three times larger than in the two-level scheme.

CONCLUDING REMARKS

Formulas (9) and (10) allow us to find the scale of binding energy. For polarizational phonons

$$\lambda_k = b_k(\kappa_o/\kappa_\infty - 1)\omega_o = b_k \alpha \omega_o (\omega_o/R)^{\frac{1}{2}} , \qquad (13)$$

where b_k are numerical factors. It follows from (12) that for the first levels $b_k \sim 0.1$. It is seen from (13) that for fixed κ_o/κ_∞ the binding energy does not depend on R, i.e., the bound states should arise even for relatively deep centers. Their radius, a, must remain macroscopic for phonon dispersion to occur for small momenta $q \sim \pi/a$.

Reynolds et al. (4) discovered a whole series of bound states in the luminescence spectrum of CdS. Their classification of the symmetry is a tentative one. It is likely that there are some levels with the same symmetry.

REFERENCES

1. Sh. M. Kogan and R. A. Suris, Zh. Eksp. Teor. Fiz. 50, 1279 (1966).

2. D. D. Manchon, Jr. and P. J. Dean, Proc. X th Intern. Conf. on the Physics of Semiconductors, Cambridge Mass. 1970, (Published by the US Atomic Energy Commission, 1970), p. 760.

3. P. J. Dean, D. D. Manchon, Jr. and J. J. Hopfield. Phys. Rev. Lett., 25, 1027 (1970).

4. D. C. Reynolds, C. W. Litton and T. C. Collins, Phys. Rev., 4, B1868 (1971).

5. L. P. Pitayevsky, Zh. Eksp. Teor. Fiz. 36, 1168 (1959).

6. V. I. Mel'nikov and E. I. Rashba. ZhETF Pis. Red. 10, 95 (1969).

7. Y. B. Levinson, ZhETF Pis. Red. 12, 496 (1970).

8. E. I. Rashba, ZhETF Pis. Red. 15, 577 (1972).

STIMULATION OF NONRADIATIVE DECAY UNDER INTENSIVE LIGHT EXCITATION

A. M. Tkachuk and A. A. Fedorov

The State Optical Institute, Leningrad, USSR

ABSTRACT

The light emission by impurity ions in crystals under Stokes excitation is preceded by a nonradiative transition which populates the radiative level. Information about the probability of non-radiative transition may be obtained by studying the emission rate after ultra-short exciting pulses of high intensity. The emission rate of CaF_2 crystals activated by Sm^{2+} ions was investigated. Sub-nanosecond pulses of the second harmonic of a Nd-glass mode-locked laser were used for the excitation. The probabilities of nonradiative decay were determined. It was found that in $CaF_2:Sm^{2+}$ the non-radiative decay probability depends on the exciting light intensity. Results are interpreted in terms of the induced nonradiative transitions by "hot" phonons.

The emission of light by impurity ions in crystals under Stokes excitation is preceded by nonradiative relaxation from the level initially excited to the radiative level. The probability of the nonradiative decay is determined by the selection rules for electron-phonon transition, and depends on the temperature. The probability usually does not depends on the exciting light intensity.

In investigating the nonradiative relaxation in samarium-activated fluorite, we have observed stimulated radiationless relaxation under conditions of high intensity excitation.

We have studied the emission rate of $CaF_2:Sm^{2+}$ crystals ($\nu = 14118$ cm^{-1}) at T = 77 K. Subnanosecond pulses from a Nd-glass mode-locked laser were used for the excitation. The series of 10-14 short exciting pulses is shown in Fig. la. The energy of each

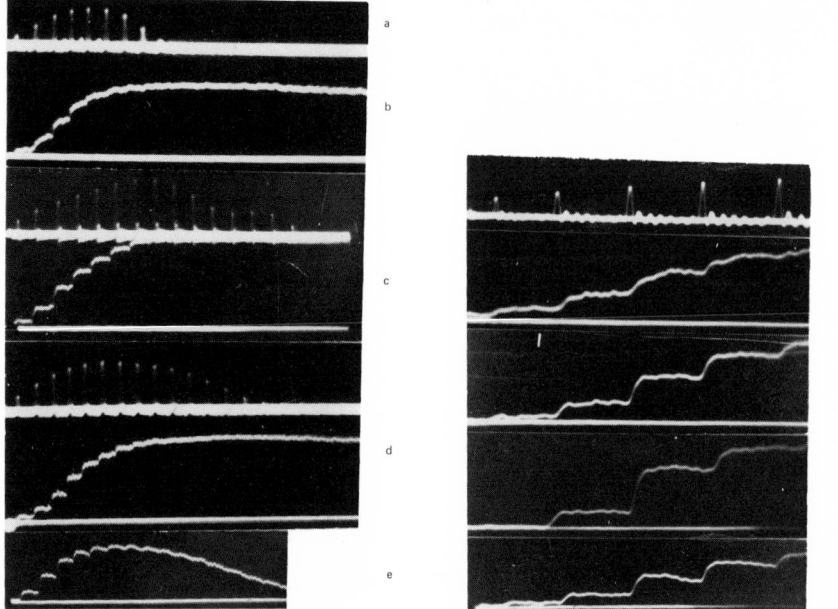

Fig. 1   Oscillograms of the excited pulses (a) (ν = 18888 cm^{-1}) and
         luminescence signals (b)-i(ν = 14118 cm^{-1}) at 77 and 300 K.
         Excited light density is: (b) P/S $\sim$ 400 W/cm^2, S = 40 mm^2,
         T = 77 K; (c) P/S $\sim$ 2 kW/cm^2, S = 33 mm^2, T = 77 K; (d)
         P/S $\sim$ 360 MW/cm^2, S = 1 mm^2, T = 77 K; (e) P/S $\sim$ 4.5 kW/cm^2,
         S = 13 mm^2, T = 300 K; (f) P/S $\sim$ 2 kW/cm^2, S = 1 mm^2, T =
         77 K; (g) P/S $\sim$ 2kW/cm^2, S = 33 mm^2, T = 77 K; (h) P/S $\sim$
         260 kW/cm^2, S = 1 mm^2, T = 77 K; (i) P/S $\sim$ 4.5 kW/cm^2, S =
         13 mm^2, T = 300 K. The distances between short excited
         pulses are $\sim$ 16.5 nsec.

short pulse was about 0.1-0.2 J (at λ = 530 nm), with pulse duration
$\leq$ 1-1.15 nsec (time resolution limited by the registration system).
The exciting light density varied from some hundreds W/cm^2 to tens
GW/cm^2, the lowest value being limited by the sensitivity of the
photo-multiplier and oscilloscope.

    Upon excitation of CaF$_2$:Sm^{2+} crystals at λ = 530 nm the transi-
tion 1 $\rightarrow$ 3 (Fig. 2) corresponds to transitions between the levels
of 4f and 4f5d electron configurations in Sm^{2+}: oscillator strength
$\sim$ 0.6x10^{-4} (1,2). The luminescence (2 $\rightarrow$ 1, Fig. 2) is interpreted
as a transition between the lowest level of 4f5d configuration and
the level $^{7}F_1$ of 4f configuration (3-7). The nonradiative decay
probability is designated by W$_{32}$ (transition 3 $\rightarrow$ 2 in Fig. 2). There
are no other nonradiative transitions at 77 K, because the quantum
yield for luminescence of the system is equal to 1 if the spectral
region of the exciting light is 300-600 nm (1). The lifetime of the

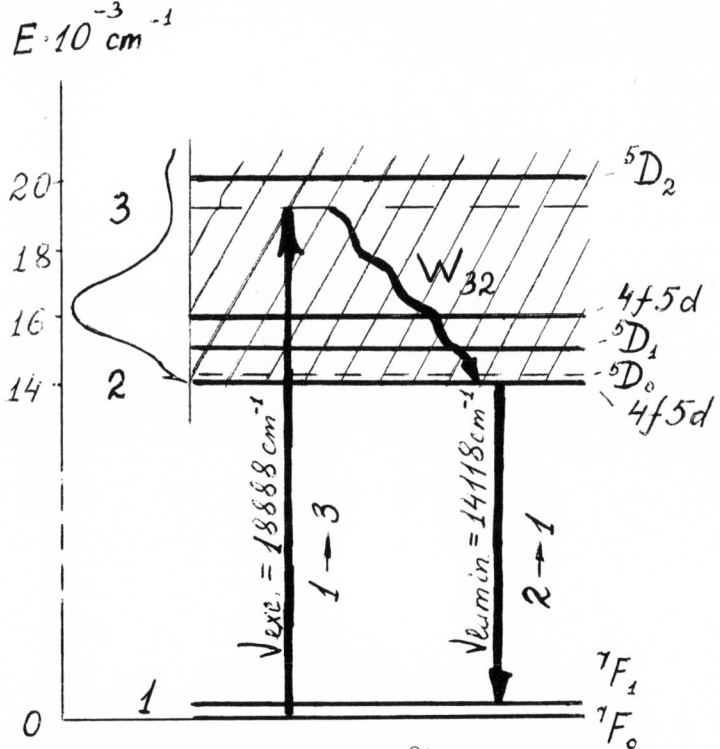

$E \cdot 10^{-3} cm^{-1}$

Fig. 2   Energy levels of the $CaF_2:Sm^{2+}$ crystals.

emission level "2" is $\tau_2 \sim 2 \times 10^{-6}$ sec (1,3). If the exciting light pulse is $F(t)$, $\tau_2 \gg 1/W_{32}$ and $\tau_2 \gg t_p$ (time duration of excited pulse), the emission rate from level "2" is:

$$I(t) = const \int_o^t \exp(-W_{32})[\int_o^t F(t)\exp(W_{32}t)dt]dt \qquad (1)$$

Expression (1) becomes simplier, if either

$$t_p \gg 1/W_{32}, \text{ in which case } I(t) = const \int_o^t F(t)dt; \text{ or} \qquad (2)$$

$$t_p \ll 1/W_{32}, \text{ in which case } I(t) = const[1-\exp(-W_{32})] \qquad (3)$$

Hence, we can determine W_{32} from $I(t)$. See (8,9) where W_{32} for ruby has been determined.

In Fig. 1 are shown some luminescence signal oscillograms of $CaF_2:Sm^{2+}$ crystals at 77 K or 300 K and under different light excitation density, i.e. from $P/S \sim 400$ W/cm^2 to 360 MW/cm^2. From Fig. 1 one can see that the rise time of the luminescence signal is long in comparison to the duration of one short excitation pulse. It follows from Eq. 3 that the time of the nonradiative decay is $1/W_{32} \sim 10$ nsec. This time becomes shorter at higher excitation density, and at $P/S = 2kW/cm^2$ (Fig. 1c) $1/W_{32}$ is about 3 nsec. At $P/S > 100 \ kW/cm^2$ (Fig. 1d) the rise time of luminescence is already no longer than the excitation pulse duration, thus $1/W_{32} < t_p = 1.5$ nsec. Therefore changing the excitation density from 400 W/cm2 to $P/S > 10 \ kW/cm^2$, we can increase the nonradiative decay probability W_{32} from 10^8 to $> 5x10^{10} \ sec^{-1}$.

The observed effect can be connected with: (1) temperature dependence of W_{32}, it being envisioned that the local temperature of the activated center increases with increasing excitation density; (2) interaction between activated center and excitation light, which alters the nonradiative decay probability.

In our experiments, even at $P/S = 200 \ MW/cm^2$, $S = 0.01 \ cm^2$ the illuminated volume of the crystal could be heated only by $\Delta t \sim 1.5$ degree, and at $P/S = 200 \ kW/cm^2$, $\Delta t \sim 1.5x10^{-3}$ degree. It is clear that the usual heating does not need to be taken into account. The degree of elevation of the local temperature can be checked by studying the τ_2-lifetime of the emission level. As follows from the temperature dependence $\tau_2(T)$ (3) and Fig. 2, τ_2 falls at 300 K to $\leq 4x10^{-8}$ sec. There is no remarkable decreasing of the luminescence signal at 77 K for the time $\sim 10^{-7}$ sec (Fig. 1); which means $\tau > 10^{-6}$ sec. and, hence, the local temperature could not be more than 100 K (2). The interaction between the activated center and excitation light field can be revealed through the increase in W_{32} which arises from the mixing of excited levels. This effect is connected with the transitions within the center and depends only on the excitation light intensity. The luminescence rate I(t) from Eq. 2 depends only on the value of W_{32} in the cases of temperature dependence and mixing the levels by light.

However, it follows from the experimental data that at excitation density $P/S \leq 10 \ kW/cm^2$ one can easily distinguish two components in the build-up of the luminescence. The component one is fast, corresponding to an intensity jump in the oscillograms with duration ≤ 1.5 nsec. The second component is slower with a lifetime ≥ 3 nsec. The contribution of the rapid component diminishes, in comparison to that of the slow component, when the ordinal number of the short excitation pulses in the series (and hence the integrated intensity of excitation) increase, Fig. 1.

The experimental facts could be explained assuming that there is a stimulated emission of nonequilibrium "hot" phonons under excitation by high intensity light. The possibility of such a process was discussed (10). The stimulated nonradiative transitions would occur only if there were a population inversion in the levels and if, in addition, the impurity centers interact via the photon or phonon field. The inversion of the levels 3 and 2 (See Fig. 2) takes place for the first, second and third short excitation pulses, during which the greater part of the luminescence buildup occurs (Fig. 2). For the next short pulses in the excitation train, the population of the third level will remain larger than the second level population only under the condition that the number of centers excited by each succeeding short pulse is not smaller than the total number of centers excited by all previous short pulses. That is why the amplitude of the rapid component of the luminescence diminishes as the number of pulses in the excitation train rises beyond three; after the seventh pulse, the rapid component has disappeared completely. The nonlinear rise of the luminescence signal and its saturation during the course of the excitation may be connected with the progressive rise of absorption from the second level, whose population rises with each successive pulse. The concentration of the excited centers even at $P/S = 200$ MW/cm^2 (which corresponds to $N = 2 \times 10^{15}$ quanta/cm^2) is smaller than the usual concentration of the activated centers, $C = 3 \times 10^{18}$ cm^{-3}. It should be noted that the rapid components and overall nonlinear rise of the luminescence intensity are observed at excitation densities of $P/S \geq (1-10)$kW/cm^2 $= (2.6-26) \times 10^{12}$ quanta/cm^2. Under these conditions the distances between excited centers are about 0.6×10^{-4} cm; i.e., they are comparable with the wave length of exciting light. Phonon-stimulated radiationless decay, should it occur under such conditions, would imply a phonon lifetime exceeding 10^{-10} sec., which appears improbable for phonons in CaF$_2$. Therefore, we suppose that for CaF$_2$:Sm^{2+} stimulation of the nonradiative decay is caused by the coherent excitation light field.

ACKNOWLEDGEMENT

The authors are grateful to Professor A. M. Bonch-Bruevich for his continued interest and encouragement and Professor P. P. Feofilov for useful discussion of the experimental data.

We are thankful to L. I. Andreeva, S. M. Semichastnova and B. M. Stepanov for the use of the photomultiplier.

REFERENCES

1. D. L. Wood and W. Kaiser, Phys. Rev. 126, 2079 (1962).

2. V. A. Archangelskaya, M. N. Kiseleva and V. M. Shriber, Opt. i Spect. 23, 509 (1967).

3. P. P. Feofilov, Opt. i Spectr. 1, 992 (1956).

4. A. A. Kaplyanskii and P. P. Feofilov, Opt. i Spectr. 13, 493 (1962).

5. A. A. Kaplyanskii and A. N. Przevusskii, Opt. i Spectr. 13, 882 (1962).

6. B. P. Zakharchenya and A. J. Ryskin, Opt. i Spect. 13, 875 (1962).

7. P. P. Feofilov, Acta Physica Polonica 26, 331 (1964).

8. M. Anson and R. C. Smith, IEEE J. Quant. Electron. 6, 268 (1970).

9. P. N. Everett, J. Appl. Phys. 42, 2106 (1971).

10. V. R. Nagibarov and U. H. Kopvillem, JETF 54, 312 (1970).

THE ENERGY-SHIFT OF DONOR-ACCEPTOR EMISSION ON VARYING THE EXCITATION INTENSITY

E. Zacks and A. Halperin

The Racah Institute of Physics, The Hebrew University of Jerusalem, Jerusalem, Israel

ABSTRACT

An analytic expression for the dependence on the excitation intensity of the emission due to donor-acceptor pair recombination is given. Experimental values taken from literature were fitted by a non-linear-least-square method to the theoretical expression. Values for the limiting phonon energy for distant pairs ($h\nu_\infty$), and for the Bohr-radius (R_B) of the impurity were then obtained. These parameters were calculated for GaP with C and S impurities, for ZnSe, for the blue emission of ZnS:Ag,Al and for self-activated ZnS:Cl. The method offered in the present work is most useful in cases where the line-structure of the D-A pairs is difficult to resolve. It should also be useful in cases in which the D-A line spectrum is resolved, when the determined parameters ($h\nu_\infty$ and R_B) may help in the classification of the line spectrum. An expression for the energy-shift of the emission with the time of the phosphorescence decay is given.

INTRODUCTION

The model of donor-acceptor pair emission was established in recent years and more and more of the recently observed emission spectra in crystals are interpreted by this concept. The most direct evidence for pair-emission was the observation of sharp emission lines fitting the relation:

$$E_n = E_G - (E_D + E_A) + e^2/\varepsilon r_n, \qquad (1)$$

where E_n is the photon energy of the emitted line, E_G is the gap

energy, E_D and E_A the binding energies for the donor and acceptor
respectively, e is the electronic charge, ε is the low frequency
dielectric constant and r_n is the donor-acceptor separation. This
relation should hold for pair-separations not too small so that
van der Waals interactions, and other short range effects may be
neglected. The first line series fitting the above relation has
been observed in GaP by Thomas et al. (1). Since then, many other
crystals have been shown to exhibit such an emission.

The sharp line emission is not always observable (2). Still,
there are more effects typical of pair-recombination. We shall
mention here only two of them, namely the shift towards lower
energies of the unresolved emission band during the decay of the
luminescence (the so called "t shift") and a similar shift with
decreasing excitation intensity ("j - shift") (3). These shifts
follow directly from the concept of pair-recombination emission.
It is, however, of importance to derive the quantitative relation
between the energy of the peak of the unresolved emission band, and
the intensity of excitation J or the decay time t. Comparison with
the experimentally observed shifts can then give values for the
parameters involved, and thus provide better evidence that the
emission indeed involves pair-recombination.

In the present work we deal mainly with the j-shift. The
derived expression was compared with experimental data taken from
literature using a non-linear best-fit computer program (4).

                                    THEORY

The derivation of the j-shift expression was described previously (5), and will therefore be given here only briefly. Let us
assume a p-type crystal ($N_D \ll N_A$) at a temperature low enough so
that ionization of the donors or acceptors is negligible. The
intensity of the emission due to recombination of pairs with
separations between r and r + dr can then be expressed by:

$$I(r) \propto r^2 P(r)/\tau(r), \qquad (2)$$

where $P(r)$ is the fraction of neutral donors, and $\tau(r)$ is the lifetime of the pairs having the separation r. A given intensity of
exciting light will produce in the steady state a flux of free
carriers M, and with a capture cross-section $\sigma(r)$ for an electron
(or hole) the rate of excitation will be given by:

$$W_{exc} = 1/T(r) = M\sigma(r), \qquad (3)$$

where $T(r)$ is the lifetime of the unexcited pair under the given
excitation.

At steady state the generation and annihilation by recombination are just equal which gives $P(r)/\tau(r) = \{1 - P(r)\}/T(r)$, or $P(r) = \tau(r)/\{\tau(r) + T(r)\}$. Inserting in (2) and expressing I, τ and T as function of the energy-shift E, $\{E(r) = -e^2/\varepsilon r$, $dE = (e^2/\varepsilon r^2)dr\}$ we have:

$$I(E) \propto (e^2/\varepsilon)^3/E^4\{\tau(E) + T(E)\} \qquad (4)$$

We now assume that one impurity, let us say the donor, gives a shallow hydrogenic level compared to the much deeper level of the other impurity. We then may write (6) $1/\tau(E) = W_o\exp\{-2r/R_B\}$ with W_o a constant and R_B the donor Bohr radius. Setting $E_B = e^2/\varepsilon R_B$ we get:

$$1/\tau(E) = W_o\exp\{-2E_B/E\} \qquad (5)$$

Putting $\sigma(r) \propto r^2$, (see References 6, 7, 8) and equating to zero the derivative of (4) with respect to E we obtain the expression:

$$J = D\{E_m^3/(E_B - 2E_m)\}\exp\{-2E_B/E_m\} \qquad (6)$$

where D is a proportionality factor, J is the excitation intensity (and so $M \propto J$), and E_m is the peak energy of the emission band.

COMPARISON WITH EXPERIMENT

For comparison with experiment it is more convenient to use the measured photon energies instead of the shifts $E(r)$. Putting $E_m = h\nu_m - h\nu_\infty$ and $E_B = h\nu_B - h\nu_\infty$ gives:

$$J = D \frac{(h\nu_m - h\nu_\infty)^3}{h\nu_B + h\nu_\infty - 2h\nu_m} \exp\left\{- \frac{2(h\nu_B - h\nu_\infty)}{h\nu_m - h\nu_\infty}\right\} \qquad (7)$$

Expression (7) was used for a few crystals for which experimental data for the peak energy $(h\nu_m)$ over a wide range of excitation energies was available. Results are given below.

Thomas et al. (1) give the peak energies $(h\nu_m)$ of the pair-emission band for GaP:C,S for excitation intensities varying over five orders of magnitude. Using their data (Reference 1, Fig. 5) the best fit computer program for Eq. (7) gave the values for $h\nu_\infty$ and $h\nu_B$ as listed in the upper line of Table I. Values for the Bohr radius as obtained from $h\nu_B - h\nu_\infty = e^2/\varepsilon R_B$ are also given in Table I.

An independent set of data for GaP:C,S was taken from Maeda (9). Fig. 1 shows the experimental values given by Maeda, and the solid curve gives the best fit of Eq. (7) to these points. The computed parameters are given in the second line of Table I. The

Table I

Computed parameters, sums of $E_D + E_A$ and Bohr radii for GaP, ZnSe and ZnS. References to paper from which experimental points were taken are given in column 1

Crystal	$h\nu_\infty$(eV)	E_G(eV)	(E_D+E_A)(meV)	$h\nu_B$(eV)	R_B(A)
GaP:C,S (1)	2.188	2.339 (15)	151	2.242	24.8
GaP:C,S (9)	2.190	2.339 (15)	149	2.247	23.5
ZnSe (10)	2.691	2.822 (10)	131	2.750	28.1
ZnS:Ag,Al (13)	2.670	3.83 (16)	1160	2.92	6.0
ZnS:Cl (13)	2.490	3.83 (16)	1340	2.663	8.7

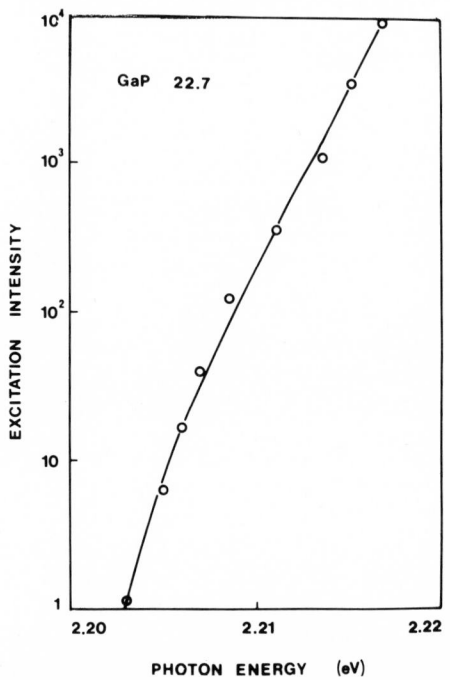

Fig. 1   Least-square fit of Eq. (7) to experimental points for GaP at 22.7 K (after Maeda (9)). Parameters in theoretical curve (full curve): $h\nu_\infty$ = 2.190 eV, $h\nu_B$ = 2.247 eV and D = 1.32 x 10^6.

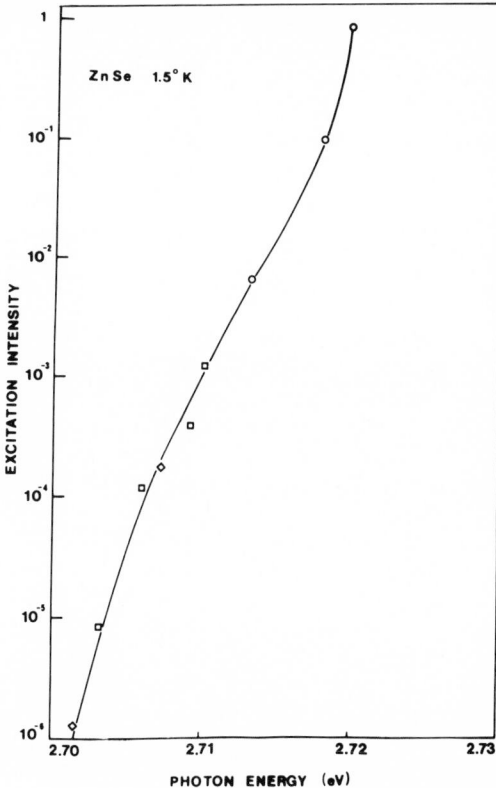

Fig. 2   As Fig. 1 but for ZnSe at 1.5 K (after Dean and Merz (10)).
Parameters: $h\nu_\infty$ = 2.691 eV, $h\nu_B$ = 2.750 eV and D = 16.18.

agreement between the two sets of parameters for GaP:C,S is quite
good and the values for R_B fit well the value R_B = 24 A obtained
from decay curves (6).

Data for peak-energies for excitation intensities varying over
six orders of magnitude were reported for ZnSe by Dean and Merz (10),
who seem to be the first to recognize that the points in a plot of
log J versus $h\nu_m$ do not fall on a straight line. Their experimental
points and the best fit to Eq. (7) are shown in Fig. 2, and the
parameters are listed in Table I. In this case our value $h\nu_\infty$ =
2.691 eV, deviates from that given by Dean and Merz (2.681 eV).
The latter was obtained from the assignment of the shell-numbers
for the various lines in the discrete line spectrum. The discrepancy
between our results and those obtained by Dean and Merz (10) may
have originated from a misclassification of the line spectrum in
Reference 10. This notion seems to be supported by an examination
of the high-energy threshold of the discrete line spectrum (see

Ref. 10, Fig. 5), which theoretically should occur at about $0.64~R_B$ (11,12). In addition, comparison of the resulting values for $E_D + E_A$ also seems to fit better our value for $h\nu_\infty$ (see Reference 5).

Peak-energies for various excitation intensities for ZnS were reported by Shionoya and Washizawa (13). The range of intensities given is, however, not wide enough to give high accuracy in the computed parameters. In addition the reported data are for powder samples, and may contain a mixture of particles having different structures.

The results obtained by best fit of the data to Eq. (7) are summarized in lines 4 and 5 of Table I for the blue band of ZnS:Ag,Al and ZnS:Cl respectively. The computed values for $E_D + E_A$ seem to be reasonable. They were obtained assuming the ZnS powder to be purely cubic, and taking 3.83 eV for the gap energy (16). We have not found in the literature any value for the Bohr radius of the Al donor level in ZnS. That for Cl in ZnS may be compared with the value of 8.4 ± 0.5 A reported by Riehl for ZnS:Cu:Cl, which agrees very well with our value of 8.7 A.

In conclusion, using Eq. (7) with experimental data for the peak energies of a pair-recombination band over a wide range of excitation intensities, one can get accurate values for the parameters appearing in the theoretical expression; namely, for the sum of $E_A + E_D$ and for the Bohr radius R_B. The method can also be of help as a guide in the classification of the lines in the resolved pair-spectrum.

The "t-shift" can be used in a similar way to that of the j-shift. An expression for the intensity of the pair emission as function of time $I_E(t)$ was given by Thomas et al. (Ref. 6, Eq. 21). Taking derivatives with respect to E and equating to zero one obtains:

$$W_0 t = (E_B - E_m)/E_B \exp\{2E_B/E_m\}, \tag{8}$$

with all the symbols as defined above, and using the photon energies one gets:

$$W_0 t = \frac{h\nu_B + h\nu_\infty - 2h\nu_m}{h\nu_B - h\nu_\infty} \exp\left\{\frac{2(h\nu_B - h\nu_\infty)}{h\nu_m - h\nu_\infty}\right\}$$

It should be noted that the formula in this form is correct only for saturated excitation.

Preliminary work shows that the "t-shift" formula can be used in a similar way to that of the j-shift formula. Thus, one can obtain two independent sets of values for $h\nu_\infty$ and $h\nu_B$. In addition the last equation can also give the value of W_0, which can then be compared with the one obtained from phosphorescence decay curves.

REFERENCES

1. D. G. Thomas, M. Gershenzon, and F. A. Trumbore, Phys. Rev.
    133, A269 (1964) (also J. J. Hopfield, D. G. Thomas, and M.
    Gershenzon Phys. Rev. Letters 10, 162 (1963).

2. C. H. Henry, R. A. Faulkner and K. Nassau, Phys. Rev. 183,798
    (1969).

3. K. Era, Sh. Shionoya, Y. Washizawa and H. Ohwatsu, J. Phys.
    Chem. Solids, 29,1843 (1968).

4. BMDx85, Biomedical Computer Programs, x-Series Supplement,
    edited by W. J. Dixon (University of California Press,
    Berkeley, Calif. 1969).

5. E. Zacks and A. Halperin, Phys. Rev. B15, (1972), in print.

6. D. G. Thomas, J. J. Hopfield, and W. M. Augustiniak, Phys.
    Rev. 140,A202 (1965).

7. P. J. Dean and L. Patrick, Phys. Rev. B2,4959 (1970).

8. M. Lax, Phys. Rev. 119,1502 (1960).

9. K. Maeda, J. Phys. Chem. Solids 26 595 (1965).

10. P. J. Dean and J. L. Merz, Phys. Rev. 178,1310 (1969).

11. M. R. Lorenz, T. N. Morgan, G. D. Pettit, and W. Y. Turner,
    Phys. Rev. 168,902 (1968).

12. W. B. Brown and R. E. Roberts, J. Chem. Phys. 16,2006 (1967).

13. K. Era, Sh. Shionoya and Y. Washizawa, J. Phys. Chem. Solids
    29,1827 (1968).

14. N. Riehl, J. Luminescence 1,2 1-16 (1970).

15. P. J. Dean and D. G. Thomas, Phys. Rev. 150, 690 (1966).

16. J. Nahir, M. Sc. Thesis, The Hebrew University of Jerusalem
    (1967).

ELECTROLUMINESCENCE IN GaN

J. I. Pankove, E. A. Miller and J. E. Berkeyheiser

RCA Laboratories

Princeton, New Jersey, USA

ABSTRACT

Light-emitting diodes of GaN have been made, which can generate CW light over any portion of the visible spectrum and in the near ultraviolet. External power efficiencies of the order of 10^{-4} can be obtained at room temperature. Annealing treatments of Zn-doped insulating GaN greatly increase the emission efficiency.

MATERIAL AND DEVICE PREPARATION

The GaN was synthesized by a vapor-transport technique (1) which generates a single crystal layer of this compound on a sapphire substrate. The undoped material is n-type with an electron concentration of about $10^{18}/cm^3$ and a mobility of about 130 $cm^2/Vsec$. The donors are believed to be nitrogen vacancies. The addition of Zn vapor to the gaseous ambient in the furnace during growth causes a compensation of the native donors and the material becomes insulating. In this fashion, it is possible to grow i-n structures which can be processed into electroluminescent diodes.

To make the diodes, the crystal was cut into 1 mm diameter discs. Indium was soldered around the periphery of the disk to form an ohmic contact to the n-type conducting layer. An indium dot was soldered to the insulating layer at the center of the disc to complete the diode. Gold connections were made to the two electrodes.

ELECTROLUMINESCENCE

When current is passed through the diode, light is emitted in

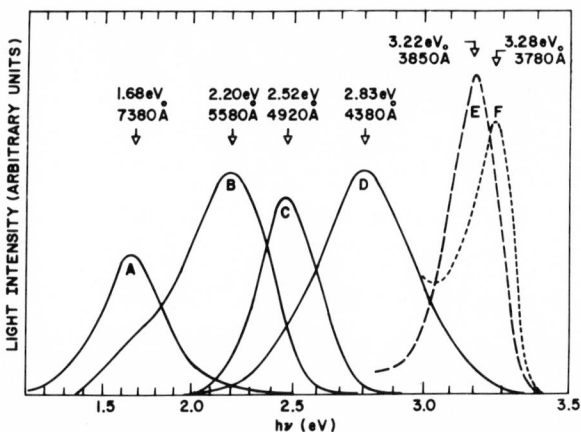

Fig. 1   Examples of electroluminescence spectra from GaN diodes at
          room temperature. The solid lines are for different Zn-
          doped i-n diodes: (A) -80V, 0.5 mA; (B) +34V, 130 μA;
          (C) +78V, 1 mA; and (D) -10V, 9.3 mA. The non-solid lines
          are for undoped GaN: (E) "Aquadag" electrode +12.5V, 10 mA;
          and (F) "MIS" diode biased 50 nsec at -30V then 50 nsec at
          +25V.

the insulating layer. The luminescence mechanism is believed to be
due to the radiative recombination of electrons at deep centers
which are ionized by the high electric field in the insulating
region. This process can occur with either polarity of bias,
generating CW light at room temperature.

      The electroluminescent emission spectrum falls in the visible
range. Red, yellow (2), green (3), and blue (4) light have been
observed, as shown by curves A, B, C, and D, respectively, in Fig.
1. These spectra originate from several different diodes fabricated
from different wafers. The spectral position of the emission peak
is determined by growth conditions, such as the Zn vapor pressure
during growth, the rate of growth and the duration of the growth.
The spectral width at half maximum is at least 350 meV. As the
current through the diode is increased, the emission intensity
grows, but does not shift in position. In some diodes, the shape
and position of the emission spectrum depend on the polarity of the
bias. Thus, blue light is emitted when the indium dot is biased
negatively; green and yellow light can be generated with either
polarity; red is usually found only with a negative bias.

                    ELECTRICAL CHARACTERISTICS

      The $I(V)$ characteristics of the diode exhibit a power depen-
dence: $I \propto V^n$, where n can be in the range of 3 to 6 at low currents (I smaller than about 10^{-5} A), becoming approximately 2 at

higher currents. The quadratic dependence of the I(V) characteristics (Fig. 2) is suggestive of transport by single-carrier charge-limited current in the presence of traps (5). The dependence of current on voltage becomes stronger when the electric power input is of the order of 0.5 W causing the diode to become warm. Light is emitted mostly over the quadratic portion of the I(V) characteristic and the emission continues at the higher bias where heating occurs.

EFFICIENCY

The radiated power is proportional to the electric power input, but tends to saturate when the power input exceeds 0.5 W. If, however, the diode is pulsed, the linear dependence of power output on driving power can be extended by at least one order of magnitude. Although occasionally external power efficiencies of 10^{-3} have been obtained with yellow light emitting diodes, power efficienies in the range 10^{-4} to 10^{-5} are more common.

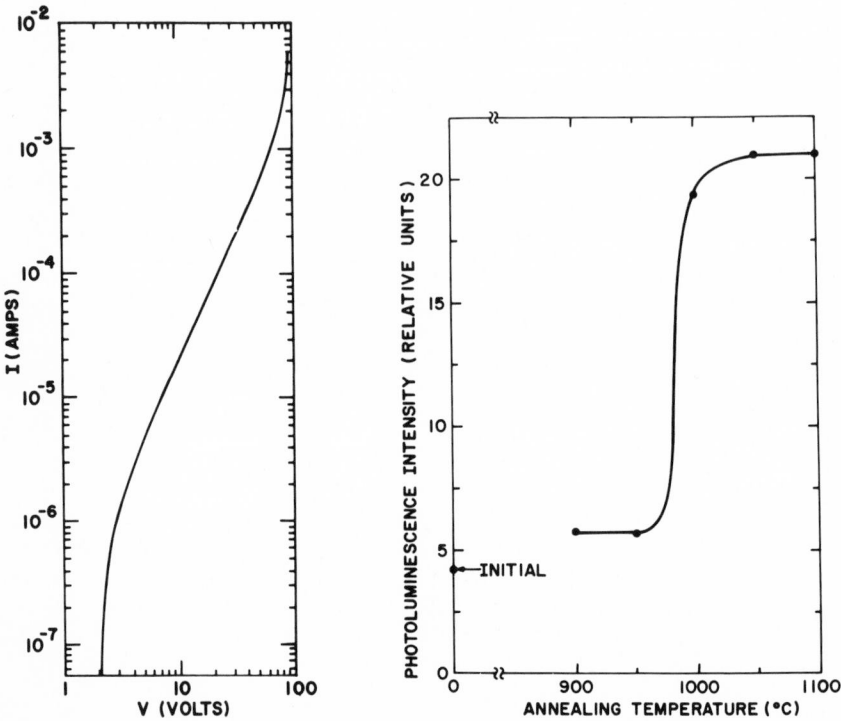

Fig. 2 I(V) characteristic of i-n diode. T = 300 K.

Fig. 3 Photoluminescence intensity of self-supporting Zn-doped GaN after annealing in ammonia for ½ hour at the indicated temperatures.

ULTRAVIOLET ELECTROLUMINESCENCE

UV electroluminescence has been obtained in undoped GaN by the following two techniques. In the first method, a surface barrier (perhaps a Schottky barrier) is made under a dot of colloidal carbon ("Aquadag"). In this case, dc UV electroluminescence peaking at 3.22 eV has been obtained-curve E in Fig. 1. In the second approach (6), a metal-insulator-semiconductor ("MIS") structure is fabricated; first a negative pulse is applied across the device to produce a surface inversion layer (rendering the surface quasi-p-type), then a positive pulse is used to inject the surface holes into the n-type bulk where radiative recombination occurs emitting ultraviolet radiation at about 3800 A-curve F in Fig. 1.

EFFECT OF ANNEALING

Because of a mismatch between the lattices of the GaN and of the sapphire substrate, and also because of the differences in the coefficients of thermal expansion for these two materials, severe strains are generated at the sapphire-GaN interface. These strains create dislocations and other crystal imperfections which can be efficient centers for nonradiative recombination. Some of these imperfections can be annealed if the GaN is separated from the substrate. Fortunately, we have developed an etching technique (7) which removes preferentially the n-type layer, thus freeing the insulating layer. This zinc-doped ($\sim$ 100 μm thick) insulating layer which has become self-supporting can now be annealed. Annealing is done in ammonia to prevent the decomposition of GaN. Photoluminescence measurements were made on such a specimen at 78 K at various stages of the annealing treatment. The results shown in Fig. 3 indicate that the photoluminescence efficiency increases by a factor of about four when the specimen is annealed above 1000 C. No further improvement is obtained beyond 1050 C. Initially and after each annealing step, the 78 K photoluminescence spectrum peaked in the blue at 2.88 eV; there was no appreciable change in the shape of the emission spectrum.

CONCLUSION

The incorporation of Zn in GaN produces deep luminescent centers which permit the generation of visible light of several different colors under optical or electrical excitation. The luminescence efficiency can be increased substantially by annealing. In undoped GaN, UV electroluminescence can be obtained.

REFERENCES

1. H. P. Maruska and J. J. Tietjen, Appl. Phys. Lett. 15, 327 (1969).

2.   J. I. Pankove, E. A. Miller, and J. E. Berkeyheiser, to be
     published.

3.   J. I. Pankove, E. A. Miller, and J. E. Berkeyheiser, RCA
     Review 32, 383 (1971).

4.   J. I. Pankove, E. A. Miller, and J. E. Berkeyheiser, J.
     Luminescence 5, 84 (1972).

5.   M. A. Lampert and P. Mark, "Current Injection in Solids,"
     Academic Press (1970).

6.   J. I. Pankove and P. E. Norris, RCA Review (in press).

7.   J. I. Pankove, J. Electrochemical Soc. (in press).

LUMINESCENCE STUDY OF THE ELECTRONIC BAND STRUCTURE OF $In_{1-x}Ga_xAs_{1-y}P_y$

A. Onton and R. J. Chicotka

IBM T. J. Watson Research Center

Yorktown Heights, New York 10598, USA

ABSTRACT

A luminescence study of $In_{1-x}Ga_xAs_{1-y}P_y$ has been performed to determine the electronic band structure of this quaternary alloy. Electron microprobe cathodoluminescence at 300 K was used to determine the direct $\Gamma_{8v}-\Gamma_{1c}$ gap and photoluminescence at 2 K for the indirect $\Gamma_{8v}-X_{1c}$ gap. Comparison of the experimentally determined gaps is made with the theory of Van Vechten and Bergstresser and with a simple interpolation procedure employing the experimentally determined gaps of the ternary alloy systems which constitute the boundaries of this quaternary.

INTRODUCTION

There has been considerable interest recently in alloys of III-V compounds. Stimulation of this interest comes largely from the prospect that through an understanding of the alloys one may at some time be able to produce to order materials with prescribed electronic and optical properties. To a large extent most of the effort has been concentrated on alloys of the type represented by $In_{1-x}Ga_xP$ and $GaAs_{1-y}P_y$, where substitution occurs on either the group III or group V sublattice – but not both (1). There does not seem to be any experimental study of a III-V quaternary alloy in the literature where both the final composition of the alloy and electronic band structure were measured.

Previous studies of quaternary III-V alloy systems were made by Müller and Richards (2) who showed that Vegard's law appears to be satisfied in a number of them, and by Sirota et al. (3) who

431

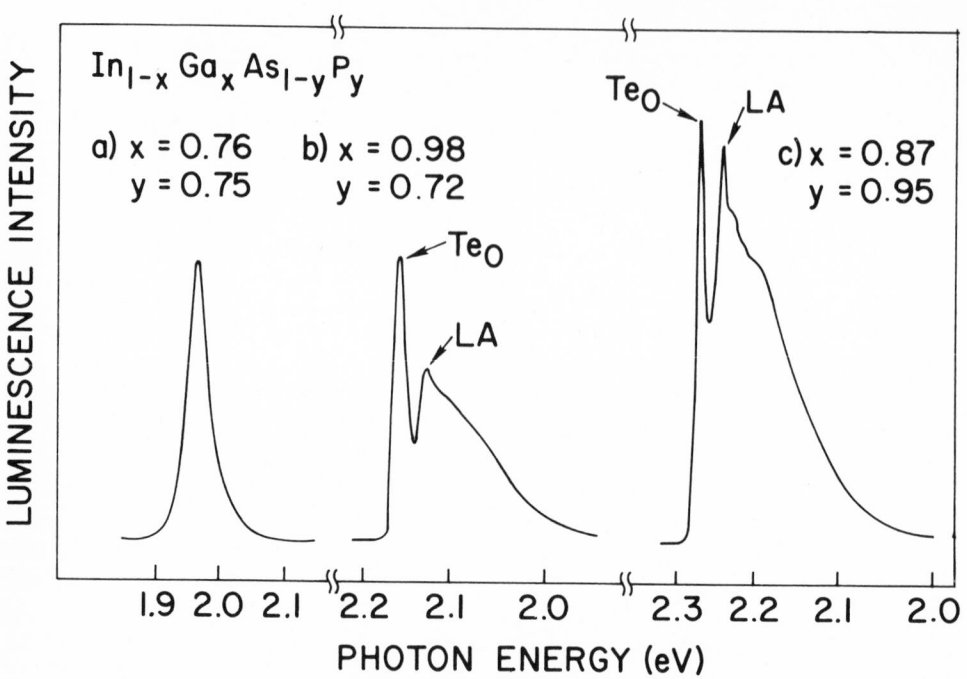

Fig. 1   (a) Cathodoluminescence spectrum of $In_{1-x}Ga_xAs_{1-y}P_y$ at
300 K; and (b) and (c), Photoluminescence spectra of the
quaternary alloy at 2 K excited by the 4880 A line of an
argon ion laser. The peaks Te_0 and LA are exciton
emission, the underlying band at low energy in donor-
acceptor pair emission.

showed that alloys formed of InP-GaAs mixtures existed with a
larger bandgap than that of either of the substituent compounds.
Recently $Ga_{1-x}Al_xAs_{1-y}P_y$ has been employed in double heterojunction
lasers by Burnham et al. (4), although no study of the electronic
band structure of this alloy appears in the literature.

Our interest in $In_{1-x}Ga_xAs_{1-y}P_y$ has been as a prototype of the
general class of quaternary III-V alloys. It is of crucial importance to test experimentally the usual assumption that to first
order the alloy band structure varies "linearly" with alloy composition as is the case in the ternary alloys. Next one may test
the "intrinsic" dielectric two-band theory of Van Vechten and
Bergstresser (5) (VVB) as to how it describes the quaternary alloy
band structure in its deviations from "linearity." Finally, a
quaternary alloy is an excellent test medium for the theoretically
predicted "extrinsic" deviations from the "linear" band structure
caused by disorder effects in the alloy (5) since the quaternary

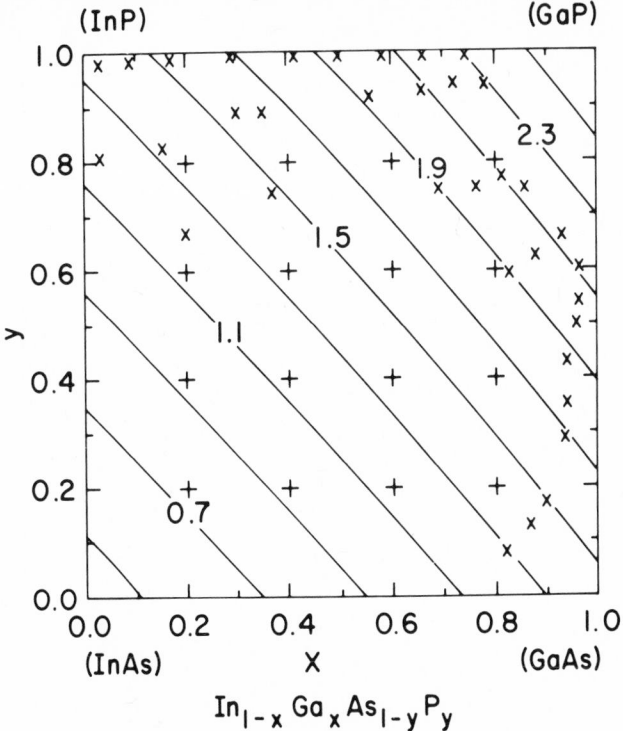

Fig. 2 The theoretical $\Gamma_{8v}-\Gamma_{1c}$ band gap of $In_{1-x}Ga_xAs_{1-y}P_y$ as
computed according to Van Vechten and Bergstresser (Ref. 5).
The positions of the samples whose data is given in Table I
is shown by x's.

has disorder on both lattice sites whereas the ternary systems have
it on only one sublattice.

We have measured the direct bandgap in the quaternary
$In_{1-x}Ga_xAs_{1-y}P_y$ by electron microprobe cathodoluminescence (6). In
this technique which appears to have been used first by Wittry et
al. (7), a volume with linear dimensions of the order of 1 μm is
excited by an electron beam. The resulting X-ray emission is
analyzed to give alloy composition; band gap luminescence to give
the band gap. The region of the alloy with an indirect band gap
was measured in photoluminescence at 2 K. Shallow bound exciton
recombination is used to estimate the gap there (8), the composition
being determined later by electron microprobe. Our samples were
prepared by a vertical Bridgman technique, consisted of large
polycrystalline grains, and had an intentional gradual alloy composition gradient to yield a number of samples from each ingot.
The alloy was doped n-type with Te to give n~10^{17}/cm^3. The accuracy
of the alloy composition determinations is estimated at ±0.02 in

Table 1

The direct Γ_{8v}-Γ_{1c} energy gap of $In_{1-x}Ga_xAs_{1-y}P_y$ at 300 K.
E_o^{exp} is the experimentally measured gap. Δ_L ($\equiv E_o^{exp} - E_o^{int}$)
is the deviation from the value obtained by the Laplace
interpolation procedure; and Δ_{VVB} ($\equiv E_o^{exp} - E_o^{VVB}$) from
Van Vechten and Bergstresser's "intrinsic"
dielectric two-band model (Ref. 5)

x	y	E_o^{exp}(eV)	Δ_L(meV)	Δ_{VVB}(meV)
0.74	0.996	2.222	−20	−80
0.66	0.994	2.124	+10	−50
0.58	0.994	1.993	0	−60
0.5	0.991	1.881	+20	−60
0.41	0.992	1.738	0	−80
0.29	0.99	1.592	0	−80
0.17	0.988	1.475	0	−50
0.09	0.983	1.351	−40	−80
0.03	0.985	1.326	−30	−40
0.78	0.94	2.203	−70	−80
0.72	0.94	2.135	−20	−60
0.66	0.93	2.038	−10	−50
0.56	0.92	1.869	−20	−60
0.35	0.89	1.535	−40	−90
0.3	0.89	1.472	−50	−90
0.15	0.83	1.236	−60	−90
0.025	0.81	1.165	+20	−10
0.86	0.75	2.101	−70	−50
0.81	0.77	2.041	−80	−60
0.76	0.75	1.953	−70	−50
0.69	0.75	1.809	−100	−90
0.37	0.74	1.388	−90	−90
0.2	0.67	1.17	−30	−40
0.91	0.66	2.077	−40	−30
0.88	0.63	1.978	−60	−50
0.83	0.59	1.85	−80	−50
0.97	0.6	2.086	−40	−40
0.96	0.54	2.019	−20	−10
0.96	0.5	1.953	−30	−30
0.94	0.43	1.854	−30	−10
0.94	0.35	1.75	−20	−10
0.94	0.29	1.686	−20	0
0.9	0.17	1.47	−30	−30
0.87	0.13	1.389	−20	−20
0.81	0.08	1.268	−20	−10

the full range 0 to 1 for each of the alloy composition parameters x and y (6). The precision is somewhat better.

LUMINESCENCE MEASUREMENTS

Typical results of the luminescence measurements are shown in Fig. 1. The cathodoluminescence spectrum of a sample at room temperature is shown in Fig. 1 (a). The precise experimental conditions for such a measurement have been described in detail previously (6). It was shown there that the energy at the cathodoluminescence peak corresponds to within 10 meV to the band gap in direct band gap semiconductors. The results, both alloy composition and direct band gap, for thirty-five (of 88) such samples are listed in Table I and their alloy compositions are plotted in Fig. 2. Comparison with theory will be made in the next section. To the eye the luminescence efficiency of the quaternary appeared equal to efficiencies encountered in ternary alloys (6). There has been criticism of the cathodoluminescence technique and Bridgman material with respect to previous results (6) obtained on $In_{1-x}Ga_xP$ (9,10). Those earlier $In_{1-x}Ga_xP$ results have been confirmed (11) on material prepared by epitaxial growth from solution and vapor grown material from a number of sources. The results obtained with the quaternary near the $In_{1-x}Ga_xP$ and $GaAs_{1-y}P_y$ boundaries are in excellent agreement with the previous work on these ternary alloys (6). The $InAs_{1-y}P_y$ and $In_{1-x}Ga_xAs$ systems have also been studied previously (12).

Also shown in Fig. 1 (b) and (c), are photoluminescence spectra obtained at 2 K with alloy material with an indirect band gap. The sharp structure at the high energy edge of these spectra may be assigned to recombination of excitons bound to neutral donors, Te_o, by comparison with similar studies in $In_{1-x}Ga_xP$ (8) and $GaAs_{1-y}P_y$ (13). By adding 30 meV to Te_o (10 meV free exciton correlation energy +20 meV binding energy for the exciton) one can estimate the indirect energy gap, E_g. These quantities are listed in Table II. The values of Te_o near $In_{1-x}Ga_xP$ range between 20 and 80 meV lower than those of $In_{1-x}Ga_xP$ (8), but the differences are consistent with a monotonic decrease toward $GaAs_{1-y}P_y$. Similarly the values of Te_o near $GaAs_{1-y}P_y$ are slightly higher than those found for $GaAs_{1-y}P_y$ (13).

COMPARISON WITH INTERPOLATION AND THEORY

We propose here that the simplest interpolation scheme for the energy gap in a quaternary is the solution of Laplace's equation in the interior of the square whose sides are the ternary compounds as shown in Fig. 2. In our case the problem reduces to a solution of:

Table II

Indirect Γ_{8v}-X_{1c} band gap of $In_{1-x}Ga_xAs_{1-y}P_y$ at 2 K as determined from the energy of the exciton bound to neutral Te donors, Te_o.

x	y	Te_o(eV)	E_g(eV)
0.89	0.95	2.282	2.31
0.87	0.95	2.280	2.31
0.79	0.94	2.253	2.28
0.74	0.93	2.258	2.29
0.70	0.91	2.215	2.25
0.98	0.74	2.172	2.20
0.98	0.72	2.158	2.19
0.97	0.62	2.127	2.16
0.95	0.54	2.094	2.12

$$\nabla^2 E_o(x,y) = 0, \tag{1}$$

with the boundary conditions:

$$x = 0, \quad E_o(x,y) = 0.36 + 0.705y + 0.275y^2 \tag{2}$$

$$x = 1, \quad E_o(x,y) = 1.441 + 1.091y + 0.210y^2 \tag{3}$$

$$y = 0, \quad E_o(x,y) = 0.36 + 0.790x + 0.280x^2 \tag{4}$$

$$y = 1, \quad E_o(x,y) = 1.34 + 0.668x + 0.758x^2 \tag{5}$$

where Eqs. (2-5) are the experimentally determined energy gaps of the ternary alloys (Refs. 12, 14, 12, and 6, respectively). Comparison of the experimental values of E_o with the results of the solution of Eq. 1 is given in Table I where Δ_L is the difference between the interpolated and experimental energy gaps. As it should be, the agreement is good near $In_{1-x}Ga_xP$ and $GaAs_{1-y}P$. Toward the center of the alloy system (third group of points in Table I), however, the experimental points lie at least 70 to 80 meV below the interpolated ones. This additional "bowing" of the energy gap in the quaternary can be viewed as the analog of the bowing in ternary systems with respect to linear interpolation.

We have also computed the value of the direct Γ_{8v}-Γ_{1c} gap of $In_{1-x}Ga_xAs_{1-y}P_y$ according to the prescription of Van Vechten and Bergstresser (5). We have hypothesized in setting up this calculation that Vegard's Law in a quaternary system implies that the lattice constant must satisfy Laplace's equation

$$\nabla^2 d(x,y) = 0 \tag{6}$$

with $d(x,y)$ varying linearly along the boundaries of Fig. 2. Under these conditions a solution of Laplace's equation is:

$$d(x,y) = (6.037 - 0.384x - 0.168y - 0.034xy) \times 10^{-10} \text{ m.} \qquad (7)$$

The other parameters prescribed by VVB to vary linearly, as well as the spin-orbit splitting, were similarly taken to satisfy Laplace's equation. The results of this computation are given in Fig. 2 as a contour map for the E_0 gap and comparison with experimental values is made in Table I. The "extrinsic" bowing of VVB is not included in these results. The calculated E_0 values are in excellent agreement with experiment near $GaAs_{1-y}P_y$, but are over 60 meV higher than the experimental values of $In_{1-x}Ga_xP$. The VVB theory seems to describe the E_0 gap better near the center of the quaternary than does the interpolation procedure. There seems to be no clear evidence in our data that the "extrinsic" disorder effects (see VVB) produce an excess bowing in the quaternary due to disorder on both the group III and V sublattices. However, the effect could be lost in the combined errors of the "intrinsic" VVB calculation and our experimental data. Application of the "extrinsic" correction for disorder effects proposed by VVB (not done explicitly here) improves agreement near $In_{1-x}Ga_xP$, but affects the $GaAs_{1-y}P_y$ agreement badly. A refinement of theory appears necessary, perhaps including a more detailed treatment of disorder effects as proposed by Stroud (15).

CONCLUSION

We have determined experimentally the direct $\Gamma_{8v}-\Gamma_{1c}$ and indirect $\Gamma_{8v}-X_{1c}$ minimum band gaps of $In_{1-x}Ga_xAs_{1-y}P_y$ for a large range of alloy compositions through luminescence measurements. A representative number of alloy band gaps are tabulated together with deviations from the dielectric two-band theory and from an interpolation procedure. With the present results the E_0 gap can be estimated to ±30 meV anywhere in the alloy system.

ACKNOWLEDGMENTS

We wish to thank R. E. Fern for technical assistance in the optical measurements and J. D. Kuptsis, F. Cardone, and B. M. Chider for microprobe analysis. Programming of numerical solutions of Laplace's equation was done with the advice of Dr. R. A. Toupin.

REFERENCES

1. For a survey of various III-V alloy band structure studies see for example: M. R. Lorenz and A. Onton, Proceedings of the Tenth International Conference on the Physics of Semiconductors, Cambridge, Mass., 1970, eds. S. P. Keller, J. C. Hensel, and F. Stern (U.S. Atomic Energy Commission, Oak Ridge, Tennessee, 1970), p. 444.

2. E. K. Müller and J. L. Richards, J. Appl. Phys. $\underline{35}$, 1233 (1964).

3. N. N. Sirota, E. I. Bolvanaovich, L. A. Makovetskaya, V. V. Rosov, V. P. Shipilo, V. I. Osinsky, and G. G. Shiyonok, Proceedings of the Ninth International Conference on the Physics of Semiconductors, Moscow, 1968 (Nauka, Leningrad, 1968), p. 1217.

4. R. D. Burnham, N. Holonyak, Jr., H. W. Korb, H. M. Macksay, D. R. Scifres, J. B. Woodhouse, and Zh. I. Alferov, Appl. Phys. Letters $\underline{19}$, 25 (1971).

5. J. A. Van Vechten and T. K. Bergstresser, Phys. Rev. $\underline{B1}$, 3351 (1970).

6. A. Onton, M. R. Lorenz, and W. Reuter, J. Appl. Phys. $\underline{42}$, 3420 (1971).

7. D. B. Wittry, D. F. Kyser, E. L. Miller, T. Rao-Sahib, J. McCoy, and A. Van Couvering, Consolidated Semiannual Progress Report No. 4, Electronic Sciences Laboratory of the School of Engineering, University of Southern California, Sec. 1.1A (1966).

8. A. Onton and R. J. Chicotka, Phys. Rev. $\underline{B4}$, 1847 (1971).

9. A. W. Mabbitt, Solid State Commun. $\underline{9}$, 245 (1971).

10. A. Laugier and J. Chevallier, Solid State Commun. $\underline{10}$, 353 (1972).

11. A. Onton, M. R. Lorenz, and W. Reuter, unpublished data.

12. A. G. Thompson, J. E. Rowe, and M. Rubenstein, J. Appl. Phys. $\underline{40}$, 3280 (1969); Yu. M. Burdukov, N. V. Zotova, and Kh. A. Khalilov, Sov. Phys.-Semicond. $\underline{4}$, 138 (1970); A. G. Thompson and J. C. Woolley, Can. J. Phys. $\underline{45}$, 255 (1967).

13. A. Onton and L. M. Foster, to be published; M. G. Craford, private communication.

14. A. G. Thompson, M. Cardona, K. L. Shaklee, and J. C. Woolley, Phys. Rev. $\underline{146}$, 601 (1966).

15. D. Stroud, Phys. Rev. $\underline{B5}$, 3366 (1972).

CHARACTERIZATION OF DEFECTS IN GaAs BY PHOTOLUMINESCENCE

MEASUREMENTS

E. Fabre

Laboratories d'Electronique et de Physique Appliquee

Limeil-Brévannes, France

ABSTRACT

Photoluminescence measurements at 4.2 K have been applied to the characterization of defects in gallium arsenide, either in bulk or in epitaxial material. This paper deals with a particular acceptor center: the gallium vacancy-donor complex. The interactions of this defect with dislocations on the one hand, and with copper on the other, are investigated and discussed.

INTRODUCTION

Low temperature photoluminescence studies have been used for quality assessment in the fabrication of semiconductors for many years. These luminescence measurements appear as a very interesting characterization method, i.e. they can show electronic transitions involving defects or impurities which allow, in some cases, one to construct a simple model of the energy levels of the impurities inside the band gap. Such measurements are widely applied in the characterization of gallium arsenide, either bulk or epitaxial (1). For instance, in bulk material, the influence of heat treatment on some centers has been found (2), and in epitaxial material a higher concentration of silicon on an arsenic site has been observed at the epilayer-substrate interface than in the layer (3,4).

This paper deals with a particular acceptor center in gallium arsenide, the gallium vacancy-donor complex, and its interaction with different defects or impurities in the case of highly n[+] doped materials (n > 2x10[17]/cm[3]). The experimental apparatus has been previously described (3). The sample is immersed in liquid helium and excited with an He/Ne laser (5 mW) through a fluorine window.

The luminescent light is analyzed by a grating monochromator and detected with a photomultiplier having a Sl photocathode. Measurements were carried out with a lock-in amplifier.

THE GALLIUM VACANCY-DONOR COMPLEX

A typical low temperature photoluminescence spectrum of highly doped gallium arsenide is shown in Fig. 1. In addition to the known lines due to band to band recombination or conduction band-shallow acceptor transitions, one can observe the presence of a center at 1.22 eV which is assigned to a gallium vacancy-donor complex (5). The energy position of this peak depends weakly on the nature of the donor, which can be located either on an arsenic or a gallium site. The dopants used in this study were Te_{As}, Se_{As}, S_{As}, Si_{Ga}, Sn_{Ga}. This center is almost always observed in bulk Bridgeman materials, but seldomly in epitaxial materials in which case its photoluminescence intensity is always very low.

INTERACTION WITH THE DISLOCATIONS

Fig. 1 shows three photoluminescence spectra (1a, 1b, 1c) obtained on the bulk material highly doped with tellurium. These three spectra are measured under the same conditions of excitation and detection and differ only in the intensity of the 1.22 eV band. On these samples, the densities of dislocations have been determined by X-ray reflection topography and are, respectively, $5 \times 10^4/cm^2$ (1a) and $5 \times 10^3/cm^2$ (1 b). X-ray reflection topography does not detect any observable dislocation line in case 1c, and it can be seen that the contrast is lower in case 1b than in case 1a. However, other measurements (especially etching figures and cathodoluminescence performed with a scanning electron microscope) always reveal the presence of dislocations, even in case 1c. This correlation between the intensity of the 1.22 eV luminescence band and the density of dislocations detected by X-ray reflection topography has been checked with a great number of samples.

We are led, therefore, to attribute this to the influence of the gallium vacancy-donor complexes (detected by photoluminescence) on the strain field around a dislocation line. Indeed, it is known that the contrast in X-ray reflection topography is dependent on the extension of the lattice strain field around a dislocation line.

INTERACTION WITH COPPER

The study of this interaction has been carried out by following copper diffusion: (i) in highly doped n^+ materials, the photoluminescence spectrum of which reveals the presence of the gallium vacancy-donor complex; and (ii) in semi-insulating materials in which this complex is never observed. In fact, copper is not purposely intro-

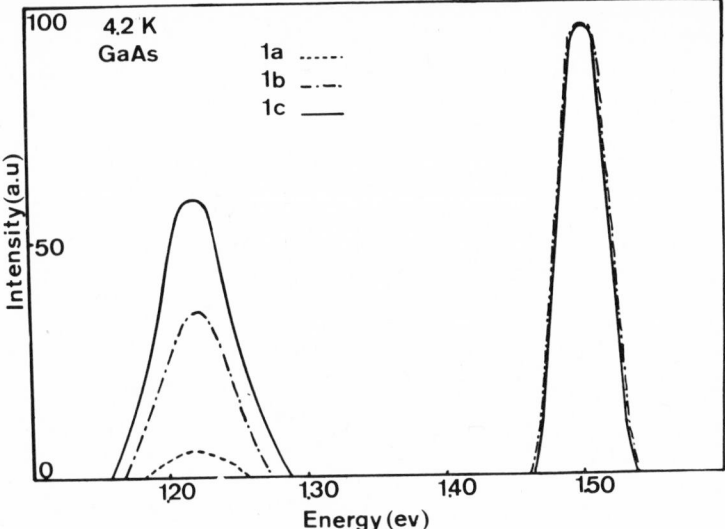

Fig. 1   Photoluminescence spectra at 4.2 K of n⁺ Te-doped GaAs.

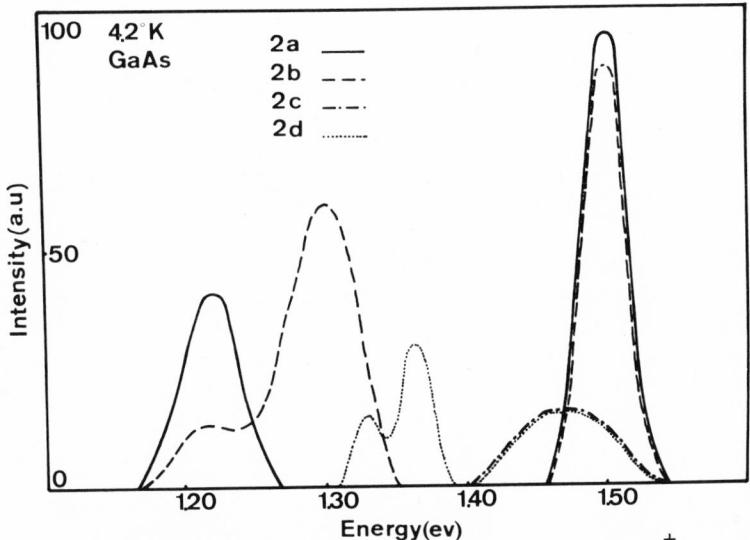

Fig. 2   Photoluminescence spectra at 4.2 K of: (2a) n⁺ Te-doped
material before copper diffusion, (2b) n⁺ Te-doped material
after copper diffusion, (2c) semi-insulating material before
copper diffusion, (2d) semi-insulating material after copper
diffusion.

duced into the material, but enters by contamination during vapor
epitaxial growth at 750 C on the material used as a substrate.
Copper diffuses very fast at this temperature through the epitaxial
layer into the substrate. The photoluminescence spectra before and

after epitaxial growth (after the epilayer has been chemically removed) are shown in Fig. 2. In a semi-insulating material the appearance of a peak at nearly 1.36 eV related to the acceptor center Cu_{Ga} (7) may be observed. This is evidence for the introduction of copper during the growth since the substrate has been initially chosen to be free of copper. On the other hand, the center at 1.36 eV is not detected in the n^+ material. In this case, the 1.22 eV band has almost completely disappeared and a peak which is due to a $Cu_{Ga}-Te_{As}$ complex (7,8) appears at 1.31 eV. A similar result is observed with silicon doped bulk material and the corresponding center at 1.29 eV is attributed to the $Cu_{Ga}-Si_{Ga}$ complex (9,10,11).

It appears that copper introduced by diffusion interacts with the gallium vacancy-donor complex to give another complex Cu_{Ga}-donor, the copper being fixed on the gallium vacancy in the following manner:

$$Cu + V_{Ga}/Donor(\sim 1.22eV) \rightarrow Cu_{Ga}/Donor(\sim 1.30eV)$$

This interaction has also been observed with other donors in the case of n^+ epitaxial layers highly doped with S_{As}, Se_{As}, Sn_{Ga}. H. Nakashima (11) has reported the same effect with Ge_{Ga} and has determined that the change of the energy position of the peak related to the $Cu_{Ga}/Donor$ complex as a function of the donor was of the same order as that of the change of the energy position of the peak related to the $V_{Ga}/Donor$ complex.

When copper contamination during the vapor epitaxial growth occurs only at the beginning of the epitaxial process ("in situ" etching of the substrates, purge, and the first moments of the growth), this study is helpful to explain the differences that we saw in the photoluminescence spectra of two epitaxial layers deposited on a semi-insulating and a n^+ substrates, respectively. In the first case, we have seen the presence of Cu_{Ga} (1.36 eV band with a phonon replica at 1.33 eV) but no peak was observed in the second case, neither at 1.36 eV, nor at 1.30 eV. Indeed, in the case of the epilayer deposited on the semi-insulating substrate, the copper, which is introduced during a certain time at the beginning of the epitaxial process, diffuses first into the substrate and then into the epilayer. In the case of the n^+ substrate, the copper is fixed in the $Cu_{Ga}/Donor$ complex in the first microns of the substrate and cannot diffuse further.

CONCLUSIONS

It appears that photoluminescence is a method well adapted to the characterization of GaAs. We restricted this study to one center only but photoluminescence also gives interesting information on other centers (especially the acceptor level related to the presence of Si_{As}). It has also defined the limits of validity of other characterization techniques, for instance X-ray reflection topography. A strong correlation between the results obtained by different techniques appears to be more and more necessary.

REFERENCES

1. E. W. Williams and D. M. Blacknall, Transactions of the Metallurgical Society of AIME 239, 387 (1967).

2. C. J. Hwang, J. Appl. Phys. 40, 4584 (1969); 40, 4591 (1969); 39, 1654 (1968); 38, 4811, (1967).

3. E. Fabre, Solid State Commun. 9, 635 (1971).

4. H. Iwasaki and K. Sigibuchi, Appl. Phys. Lett. 18, 420 (1971).

5. E. W. Williams, Phys. Rev. 168, 922 (1968).

6. E. Fabre and C. Schiller, Solid State Commun. 10, 81 (1972).

7. H. J. Queisser, C. S. Fuller, J. Appl. Phys. 37, 4895 (1966).

8. T. D. Dzhafarov, Soviet Phys. Solid State 12, 2259 (1971).

9. H. Nakashima and M. Hirao, Japan J. Appl. Phys. 9, 1495 (1970).

10. E. Fabre, Phys. Stat. Solidi 9, 259 (1972).

11. H. Nakashima, Japan J. Appl. Phys. 10, 1737 (1970).

RADIATIVE RECOMBINATION INVOLVING ACCEPTORS IN GaSb

W. Rühle, W. Jakowetz and M. Pilkuhn

Stuttgart University, Physics Institute

Stuttgart, BRD

ABSTRACT

The 2 K photoluminescence of n- and p-type GaSb is compared. It is shown that in both cases several emission lines can be explained by the existence of a doubly ionizable acceptor (ionization energies of 34.5 and 102 meV). This acceptor was found to be always present in GaSb. Evidence is presented that donors participate in a transition involving the first ionization state of this acceptor (even in undoped samples). The ionization energies of two other acceptors are determined to be 57 and 13 meV. The $(LO)_\Gamma$-phonon replicas ($\hbar\omega_{LO} \overset{\sim}{\sim} 30$ meV) of several transitions have been observed.

INTRODUCTION

The photoluminescence spectra of GaSb differ drastically, if one compares undoped p-samples and Te-doped n-samples: While p-samples show a pronounced structure (Fig. 1), n-samples generally show only one very broad emission band (Fig. 2). The reason for this different behavior is a recombination involving a doubly ionizable acceptor which is always present in GaSb and which seems to be caused by a native lattice defect (1,2). This doubly ionizable acceptor was postulated from Hall-measurements (1). Recently Kastalskii and coworkers (3) have studied and discussed its influence on the luminescence of n-type samples.

EXPERIMENTS

We made photoluminescence experiments of n-type (Te-doped) as well as p-type (undoped) GaSb, and in both cases we were able to identify luminescence involving this doubly ionizable acceptor.

444

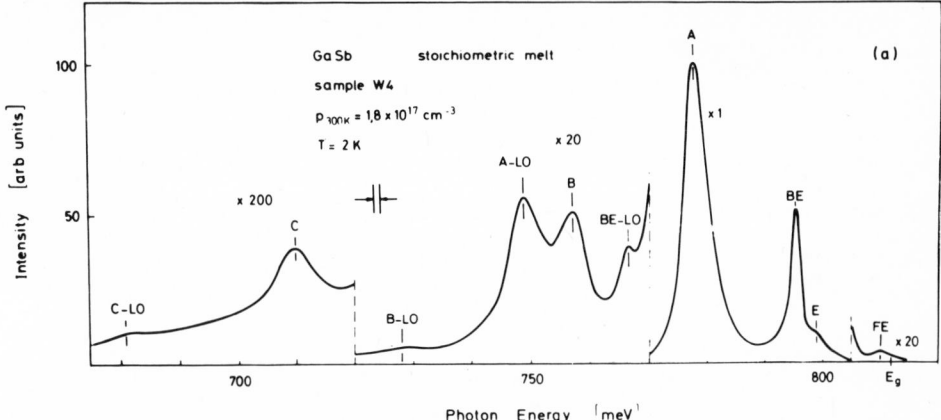

Fig. 1   Photoluminescence spectrum of an undoped GaSb—sample,
         grown from stoichiometric melt.

The values for its ionization energies, especially for the second
ionization state, were determined accurately.   In addition, we
could prove that in p-samples localized donors can be involved in
the recombination via the first ionization state of this doubly
ionizable acceptor (D-A transitions).

     A krypton laser with a maximum power output of about 300 mW
was used for excitation and the luminescence light was dispersed
by either a 0.25 m Jarrell Ash or 1 m Spex monochromator.   The
light was detected by a PbS photoconductive cell. Fig. 1 shows
the spectrum at 2 K of an undoped sample grown from a nearly
stoichiometric melt.   In this paper only the lines A, B, C, and E

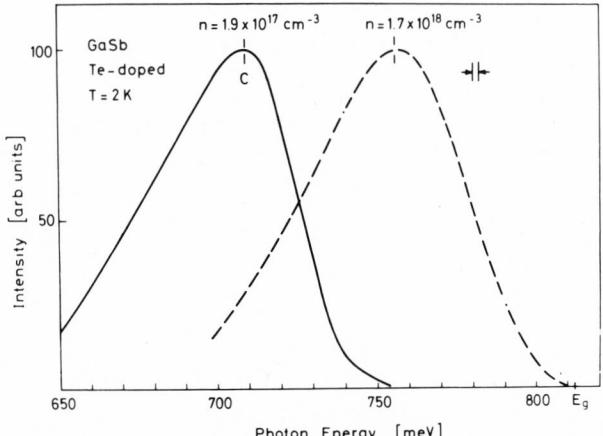

Fig. 2 Photoluminescence spectra of Te-doped GaSb-samples with
       different doping levels.

are discussed: they are all transitions involving acceptor levels.

DISCUSSION OF RESULTS

We first discuss line A. Although the small binding energy (about 2.5 meV) of the effective mass donors associated with the Γ – minimum makes it very difficult to distinguish between donor-acceptor and conduction band-acceptor transitions, we were able to identify line A as a donor-acceptor pair recombination. We did the following three experiments in order to prove this: 1. The energy shift of line A in a magnetic field was measured. A quadratic shift at low magnetic fields was clearly observed. This diamagnetic shift cannot be explained as being due to tightly bound holes, which move on a small Bohr radius. The magnitude of the shift is so large that it must be caused by a bound particle with a large Bohr radius i.e. a donor electron associated with the Γ -minimum. 2. The temperature dependence of the peak of line A was measured. Up to about 15 K, line A follows the energy gap. At higher temperatures it moves to higher energies. This temperature behavior can be explained by the ionization of the shallow donors. At higher temperatures the conduction band-acceptor transition with linear temperature dependence is observed. 3. The energy of line A shifts to higher energy with increasing excitation power. The shift is 1.1 meV when the excitation power is altered by three orders of magnitude. This can be explained by an increase of the Coulomb term which appears in the energy balance of D-A transitions. All these three experiments show clearly, that line A is a donor-acceptor pair transition.

Line C at 710 meV is always correlated in intensity with line A. Quite accurately, a constant intensity ratio of the two lines was found in all samples. The energetic distance between the two lines is too large to explain line C as a phonon replica of line A. It is concluded that line C must be attributed to the second ionization state of the acceptor involved in line A. Lines B and E are transitions involving acceptor levels which are 55 and 13 meV deep. Both acceptor levels were also seen in photo-conductivity either in n-type or p-type samples (2). Lines A, B, and C are accompanied by $(LO)_\Gamma$ -phonon replicas. These replicas are labeled A - LO, B - LO, and C - LO. The energetic distance between the zero phonon line and the phonon satellite is about 30 meV corresponding to the $(LO)_\Gamma$ -phonon energy.

In Fig. 2, the spectra of two Te-doped n-samples with different doping levels are shown. The spectra consist of only one extremely broad line. In these transitions, the same acceptor level, namely the second ionization state of the doubly ionizable acceptor, is involved. When the Fermi level lies in the conduction band, the acceptor levels are both occupied by electrons.

At low excitation level, the holes generated can occupy only the second ionization state; the occupation of the first ionization state corresponds to a successive capture of two holes. The large shift of the line with doping level is caused by the shift of the Fermi level in the conduction band. The effect is large because of the small effective mass m_e. In the case of a recombination involving the second ionization state of a doubly ionizable acceptor, the electron must tunnel through a repulsive Coulomb barrier before it can recombine with the hole. This Coulomb barrier is caused by the remaining negative charge of the acceptor. Electrons with larger kinetic energy can tunnel easier. Therefore, even in emission, a large "Burstein-Moss-shift" is observed. This model also explains the large linewidth of this recombination (3).

In Table 1, the different acceptor levels are listed. In the last column, the intensity ratios of zero phonon line to phonon replica are given. The expected increase of phonon participation with increasing binding energy of the acceptor is indeed observed.

TABLE 1

Intensity ratios of zero phonon lines to phonon replicas in dependence on acceptor ionization energies

Line	Energy Level (meV)	Acceptor Ionization Energy (meV)	$I_X:I_{X-LO}$
A	777.5	34.5	1:0.02$\pm$0.01
B	755	54	1:0.03$\pm$0.01
C	710	102	1:0.1 $\pm$0.05
E	799	13	-------

CONCLUSION

In conclusion, it was demonstrated that the recombination in both n- and p-type GaSb is dominated by one doubly ionizable acceptor. In this model the origin and behavior of several lines are easily understood. Although the donor binding energy is very small, it was possible to prove donor participation in the transition involving the first ionization state of this acceptor.

ACKNOWLEDGMENT

We thank Dr. D. Bimberg, who gave us the opportunity to carry out the magnetic field measurements on his apparatus.

448   W. RUHLE, W. JAKOWETZ, M. PILKUHN

REFERENCES

1.   Y. J. van der Meulen, J. Phys. Chem. Sol. <u>28</u>, 25 (1967).

2.   W. Jakowetz, W. Rühle, K. Breuninger, M. Pilkuhn, Phys.
     Stat. Sol. <u>12</u>, July (1972).

3.   A. A. Kastal'skii, T. Risbaev, I. M. Fishman, Yu. G. Shreter,
     Sov. Phys. - Semic. <u>5</u>, 1391 (1972); (Fiz. Tekhn. Polupr. <u>5</u>,
     1596 (1971)).

PHOTOLUMINESCENCE OF GaSe

A. Cingolani, F. Evangelisti, A. Minafra, A. Rizzo

Istituto di Fisica, CSATA, Bari, Italy

ABSTRACT

Systematic photoluminescence measurements of the p- and n-type GaSe have been carried out in the temperature range between 4.2 and 300 K. The main results, reported here, are: 1) direct free exciton emission centered at 2.109 eV; 2) a structured emission localized between 2.103 and 2.060 eV and attributed to excitons bound to neutral acceptors; 3) impurity states emission in the energy range between 2.057 and 1.904 eV.

INTRODUCTION

Gallium selenide has been extensively studied, because of its interesting layer structure, principally from the point of view of band structure and optical properties (1,2). In particular, final evidence of an indirect absorption edge at 53 meV below the direct gap has been given by absorption measurements (1). On the other hand, some authors (3) have reported stimulated emission at 2 K at an energy corresponding to the indirect gap. Very few works on spontaneous luminescence of this compound are available (4) and only at quite high temperature.

This work has been undertaken in order to systematically investigate, in the temperature range between 4.2 and 300 K, a variety of samples, both p- and n-type, grown by different techniques and with a high degree of monocrystallinity.

449

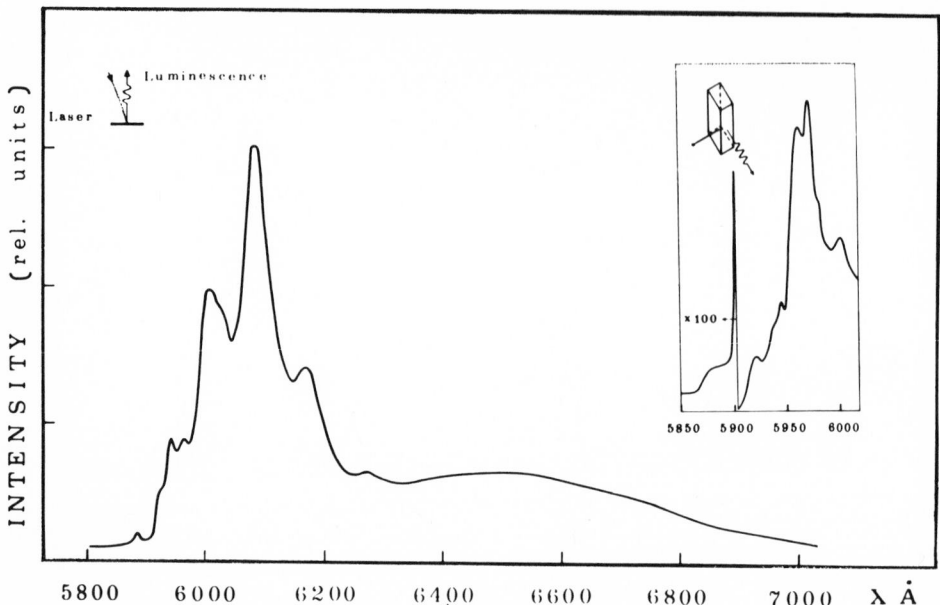

Fig. 1 Photoluminescence spectrum of GaSe at 4.2 K. The insert
        shows details of the B region.

EXPERIMENTAL

Most of the GaSe single crystals, used in this work, were
grown in our laboratory. The p-type samples have been obtained
from the melt by the Bridgman-Stockbarger method and from the
vapor by closed tube sublimation, while the n-type ones have been
grown by iodine transport. The growth conditions and the structural and electrical properties have been reported elsewhere (5,6).
Some other melt grown p-type samples, obtained from "Semi-Elements
Inc.",have also been used.

The samples were mounted in a continuum flow liquid helium
cryostat and excited by a d.c. Ar$^+$ laser with maximum flux of
10^{19} quanta/cm^2 sec. Several excitation arrangements were utilized
in order to reduce self-absorption effects for radiation close to
the fundamental optical edge. A diffraction grating monochromator
of 600 mm. focal length (Monospek D 600) with a phase-locked photoelectric detection system was used to record the luminescence spectra in the temperature range between 4.2 K and 300 K.

In the same range of temperature photoconductivity measurements were carried out by exciting the sample with a 1000 W tungsten lamp.

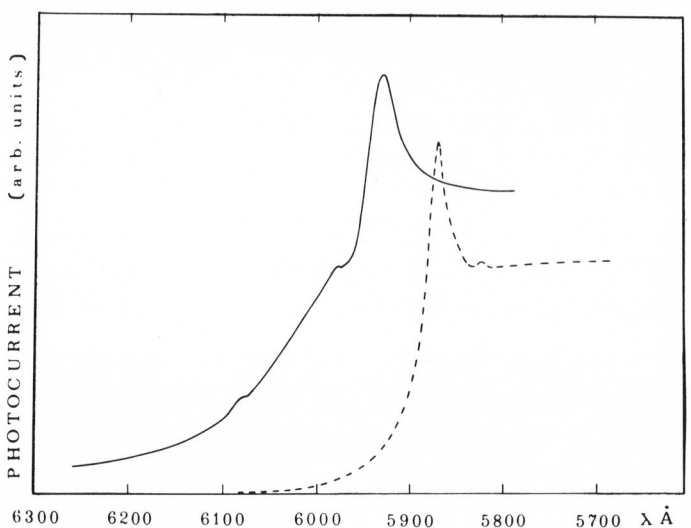

Fig. 2 Photoconductivity spectra of GaSe at 4.2 K (dashed line)
and 140 K (full line).

RESULTS AND DISCUSSION

In Fig. 1 is shown a typical emission spectrum of a p-type
GaSe at 4.2 K. The emitted light has been collected from the
excited surface. We may divide the spectrum into three regions:
(A) free exciton emission, (B) a structured emission localized
between the direct and indirect edge, (C) emission below the indirect gap, including four narrow bands at 2.057, 2.038, 2.011,
1.984 eV, and a broad band centered at 1.904 eV.

The identification of peak A with the direct free exciton is
energetically consistent with absorption data (1). This is confirmed by the thermal shift of this peak which shows a temperature
dependence of the form $E = E_0 - aT^{2.3}$, which is consistent with
band gap transitions (7). Moreover, the linear dependence between the intensities of the emitted and the exciting light (J),
observed for the whole available range of J, may be theoretically
derived from a simple model of the free exciton, by using the
"method of the rate equations" (8).

The region B is shown with more details in the insert of Fig.1
for an excitation level of about 10^{17} quanta/cm^2 sec. This region
consists of many overlapping structures covering the energy range
between 2.103 and 2.060 eV. The relative intensities of the peaks
are different for different samples, while the transition energies
remain constant. The main features are: a) the emission is localized between the direct and the indirect gap. This suggests the
possibility, for GaSe, of radiative transitions from resonant

states, similar to that reported recently in some III-V mixed compounds (9). b) In the p-type samples, the emission intensity, measured as a function of J, shows a linear behavior followed by saturation, while in the thinner ones, a quenching of the luminescence is observed at the higher values of J. On the other hand, the luminescence of n-type samples is linear also at the higher excitations. c) The temperature dependence of the transition energy is similar to that of the free exciton; the thermal quenching of intensity increases sharply at 150 K (corresponds to $kT \stackrel{\sim}{\sim} 12$ meV).

The photoconductivity spectra, measured at 4.2 K and at 140 K, are shown in Fig. 2. At very low temperature only the free exciton contributes to photoconductivity, while with increasing temperature structures become evident below the excitonic peak. These, allowing for the thermal shift, are consistent with the peaks of the region B.

Recently Mooser et al (10) have reported electroluminescence measurements at 2 K, which show emission in the energy range corresponding to our B region. Such an emission is attributed to the recombination of excitons bound to neutral acceptors. In particular, they divide the B region into two zones, separated by 20 meV which they justify by the hypothesis of an acceptor level splitting due to the polytype structure of GaSe. The same hypothesis explains the fine structure of the two zones. The acceptor level, responsible for the B emission, has been determined by electric (10) and photovoltaic (11) measurements at about 100 meV from the valence band. This leads, on the basis of the Sharma and Rodriguez theory (12), to a binding energy of the bound exciton equal to about 12 meV, which results by assuming, according to Mooser, an effective mass ratio of five ($m_p/m_n = 5$) for the GaSe.

Our experimental results, mainly the thermal quenching of the luminescence and the photoconductivity, confirm the attribution of the B emission to excitons bound to neutral acceptors. Moreover, the different dependence on excitation intensity for p- and n-type samples is also in agreement. In fact, the saturation and quenching, shown by p-samples are typical of the emission of excitons bound to neutral acceptors in p-type materials at low temperatures (8,13).

In the n-type samples the iodine, acting as donor, compensates acceptor centers, and, according to D. Bimberg et al (14), a linear behavior of the emission intensity is expected at low temperature.

Finally, we wish to note that the B emission, when analyzed for polarization, may also be divided into two zones: the former B_1, ranging from 2.103 to 2.085 eV, is unpolarized; the latter B_2, ranging from 2.084 to 2.060 eV, favors $E||c$. The energy separation between B_1 and B_2 is 20 meV and corresponds to Mooser's energy separation.

As far as the C region is concerned, i.e. the emission below the indirect gap, the bands, C_1, C_2, C_3, C_4 at 2.038, 2.011, 1.984, 1.094 eV, respectively, are observed only in p-type samples. This region can be identified as due to transitions involving impurity states. Such an assumption is supported by the J dependence of the intensities of the C bands, which is sublinear with a trend to saturation.

We want finally to note that Raman effect measurements on GaSe (15) give a phonon of 27 meV which corresponds to the energy separation of C_1, C_2, C_3. This suggests that the C_2 and C_3 structures are phonon replicas of the C_1 transition.

ACKNOWLEDGEMENTS

We wish to thank Professor F. Ferrero for his interest and Mr. G. Savoia for useful help during the measurements.

REFERENCES

1. E. Aulich, J.L. Brebner and E. Mooser, Phys. Stat. Sol. **31**, 129 (1969) and references cited herein.

2. H. Kaminura and K. Nakao, J. of Phys. Soc. of Japan **24**, 1313, (1968) and references cited herein.

3. R.E. Nahory, K.L. Shaklee, R.F. Leheny, J.C. De Winter, Solid State Comm. **9**, 1107, (1971).

4. M.I. Karaman and V.P. Mushinskii, Soviet Phys. Semicond. **4**, 662 (1970) and references cited herein.

5. V.L. Cardetta, A.M. Mancini, A. Rizzo, J. of Crystal Growth, to be published.

6. V.L. Cardetta, A.M. Mancini, C. Manfredotti and A. Rizzo, Proceeding of the II Int. Conference on "Vapor Growth and Epitaxy" Jerusalem, Israel (1972).

7. C.D. Mobsby, E.C. Lightowlers and G. Davies, J. of Luminescence **4**, 29, (1971).

8. C. Benoit, à la Guillaume, Debever, F. Salvan, Phys. Rev. **117**, 567 (1969).

9. D.R. Scrifres et al, Phys. Rev. B, $\underline{5}$, 2206 (1972).

10. J.P. Voitchovsky, E. Mooser - preprint.

11. F. Adduci et al, to be published.

12. R.R. Sharma, S. Rodriguez, Phys. Rev. $\underline{159}$, 649 (1967).

13. Koh Era, D. Langer, J. of Luminescence $\underline{1,2}$,525 (1970).

14. D. Bimberg, W. Scairer, M. Sindergeld, T.O. Yep, J. of Luminescence $\underline{3}$, 175 (1970).

15. T.J. Wieting, J.L. Verble, Phys. Rev. $\underline{5}$, 1473 (1972).

THE RESONANT LUMINESCENCE OF FREE EXCITONS IN CRYSTALS

E. Gross,[*] S. Permogrov, V. Travnikov and A. Selkin

A. F. Ioffe Physical Technical Institute

Leningrad, USSR

ABSTRACT

The emission spectrum of the zero-phonon n = 1 A exciton line in CdS crystals has been investigated for temperatures in the range 2 to 25 K with high resolution and compared with the polariton dispersion curve which has been measured experimentally. It has been found that the resonant exciton luminescence arises from the emission of polaritons of the lower branch of the polariton dispersion curve. The emission line consists of two maxima, one being connected with the "knee" of the polariton dispersion curve and the other arising from the anomaly in the crystal boundary transparency.

Luminescence as a method for the study of electronic properties of matter readily gives information concerning the positions of energy levels and the probabilities of optical transitions. Studies in recent years have shown that luminescence methods can also provide more refined information, e.g., information about the dynamics of electron excitations. However, up to the present little attention has been given to the question of how the electron excitation, i.e., the motion of the electrons, is transformed into light, i.e., into electromagnetic waves.

This problem is most easily understood for the intrinsic electronic excitations of crystals, i.e., excitons, because of the wave nature of these excitations. The wave nature of excitons has been well understood from the very beginning of the development of exciton theory. It has been shown that the transverse exciton waves

---

[*] Deceased.

are strongly coupled to the transverse electromagnetic waves (1).
Because of this fact it is not possible to separate light and excitons inside the crystal. Instead a new kind of crystal excitation
should be considered. This mixed exciton-photon excitation is now
well known as the polariton (2) and can be treated as "exciton-like"
or "photon-like" depending on the energy region under consideration.

The dispersion curve of the polariton consists of two branches
with a peculiar region of "anticrossing" for which the exciton-
photon mixing is largest. This region gives rise to the processes
of resonant "exciton" absorption and emission of light by the crystal.

In the process of exciton zero-phonon luminescence the polariton wave, which is propagating inside the crystal, reaches the crystal boundary and there transforms into a light wave which can be
detected as a crystal emission. There is a rather wide energy
region near the resonance where such transfer can take place. The
resulting exciton emission line is "inhomogeneously" broadened,
since different polariton states contribute to it.

It is very important to specify the polariton states which
take part in such a process so we have undertaken the experimental
study of the polariton emission spectrum in CdS crystals. The exciton emission in II-VI semiconductors from the polariton point of
view has been investigated by Tait and Weiher (3) and Benoit a la
Guillaume et al (4). In this paper we have studied the resonant
emission spectrum of the lowest exciton state n = 1 A in CdS crystals (5) with high resolution and have compared it with the polariton dispersion curve determined experimentally in our previous
work (6).

All measurements were carried out with the light beams directed
perpendicular to the surface of the crystal platelets with the
light, polarized in a direction which is an allowed polarization
for the n = 1 A optical transition. This geometry is the most
simple for the analysis since it provides the possibility of avoid-
ing the difficulties connected with crystal anisotropy.

In Fig. 1 the original tracing of the n = 1 A emission line
at T = 2 K is presented. The solid line is the polariton dispersion curve as given by (7)

$$E = E_T + E_{LT}/(1 - c^2K^2\hbar^2/\kappa E_T^2) + \hbar^2K^2/2M \qquad (1)$$

where E and K are the polariton energy and wave vector, respectively
M is the exciton effective mass, E_T and E_L are the energies of the
transverse and longitudinal excitons at the zone center, E_{LT} =
E_L - E_T, and κ is the "background" dielectric constant in the region
of exciton resonance.

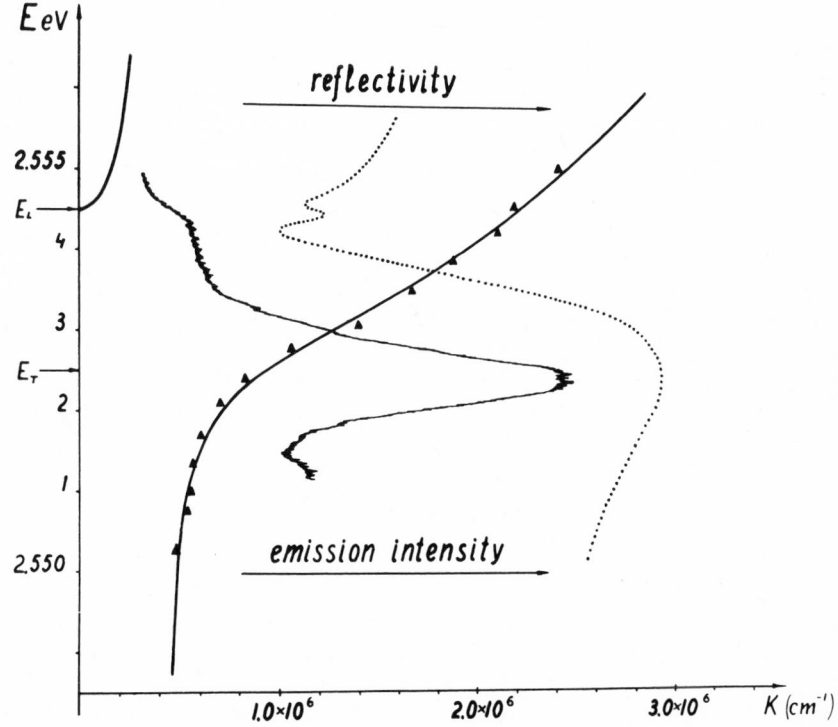

Fig. 1   Original tracing of the n = 1 A exciton emission line in
          CdS crystals at T = 2 K. The solid line is a plot of the
          polariton dispersion curve and the filled triangles are the
          experimental points (6). The dotted line is the reflectance
          spectrum taken under the same conditions as the emission
          spectrum.

      The parameter values E_T = 2.5525 eV, E_{LT} = 0.002 eV, M = 0.9m
and κ = 8.3 were used to get the best fit to the experimental points
(filled triangles) obtained from the inelastic polariton scattering
as described in our previous papers (6). The experimental points
in Fig. 1 were taken with improved resolution which lead us to a
somewhat different set of parameters as compared with reference (6).

      The dotted line in Fig. 1 is the reflection spectrum taken at
normal incidence under the same conditions as the emission spectrum.
The characteristic feature of this spectrum is the "spike" at the
energy E_L. This spike arises from the spatial dispersion of polari-

ton waves in the crystal. The detailed analysis of the reflection spectra in the region of polariton resonance will be given elsewhere (12).

As can be seen from Fig. 1, the emission line of n = 1 A excitons has a doublet structure. Both emission maxima are situated below the onset of the upper polariton branch. Thus, it may be concluded that all the resonant crystal emission is connected with the polaritons of the lower branch. In order to understand the occurrence of two maxima in the emission spectrum the thermal distribution of polaritons in the crystal and the transmission properties of the crystal boundary should be taken into account.

At low temperatures the polaritons, which are generated in the crystal by the ultraviolet excitation, interact with the phonons and populate the narrow energy interval just above the "knee" of the lower polariton branch. Below the knee the density of states is small. Moreover, the relaxation of polaritons below the knee is slowed down due to the small value of the phonon wave vectors in such a process (8). In the energy region under consideration the share of photon properties in the polariton state increases as the energy of the polariton decreases. The product of the two above mentioned factors will give rise to a sharp low energy emission maximum at an energy close to the knee of the polariton dispersion curve.

The high energy maximum of the exciton emission line is connected with the maximum of the crystal boundary transparency. In the energy region below E_L there exists in the crystal only one polariton wave with a real wave vector. It can be shown (3) that in this case the transmittance at the crystal vacuum boundary T is given by: T = 1-R where R is the reflection coefficient on the vacuum-crystal boundary. It can be seen that the overall minimum of the reflection spectrum, and hence the maximum of the boundary transparency, is situated just below the energy E_L.

At elevated temperatures, the maximum of the polariton energy distribution in the crystal shifts toward higher energies. As can be seen on Fig. 2, at T = 25 K this increases the growth of the high energy maximum of the exciton emission line with respect to the low energy one. At higher temperatures, the complex structure of the exciton emission line is masked by phonon broadening.

The doublet structure of the exciton emission lines have also been observed in other crystals such as CdTe (9), GaAs (10) and CuCl (11). It seems likely that our qualitative arguments should be applicable to the analysis of the experimental data on these materials.

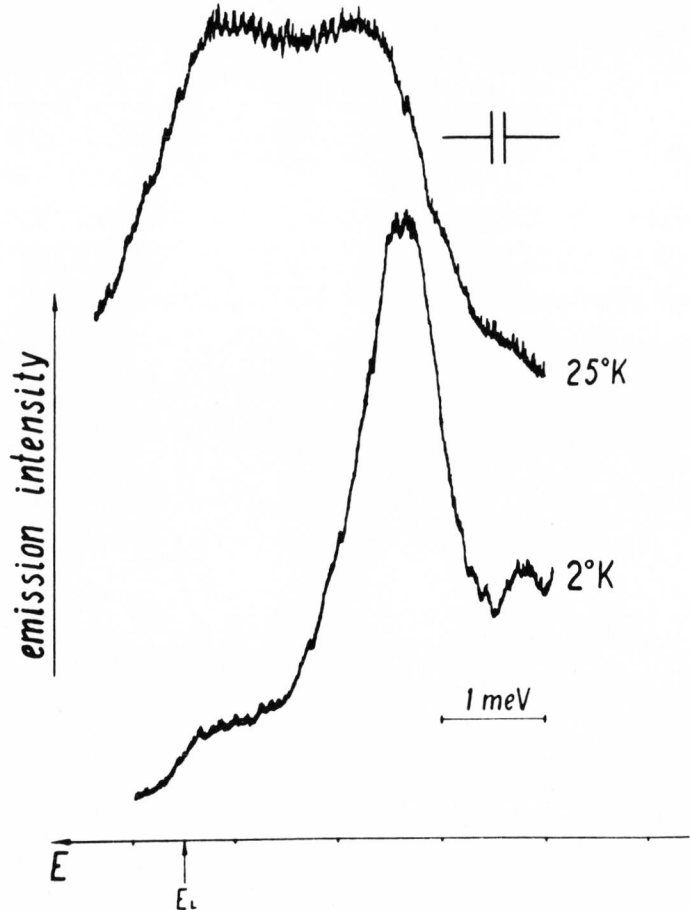

Fig. 2   The emission spectrum of n = 1 A exciton line in CdS
         crystals at T = 2 and 25 K.   The energy scales for the two
         spectra are shifted so that the energy E_L coincides for
         both temperatures.

REFERENCES

1.   M. Born, K. Huang, Dynamical Theory of Crystal Lattice, p. 82,
      Oxford University Press, London (1954).

2.   J. J. Hopfield, Phys. Rev., 112, 1555 (1958).

3.   W. C. Tait, R. L. Weiher, Phys. Rev. 178, 1404 (1969).

4.   C. Benoit a la Guillaume, A. Bonnot, J. M. Debever, Phys.

Rev. Lett., 24, 1235 (1970).

5. D. G. Thomas, J. J. Hopfield, Phys. Rev., 116, 573 (1959).

6. E. F. Gross, S. A. Permogorov, V. V. Travnikov, A. V. Selkin, Soviet Physics, Solid State, 13, 578, 585 (1971).

7. J. J. Hopfield, Phys. Rev., 182, 945, (1969).

8. Y. Toyazawa, Progr. Theor. Phys. (Kyoto) Suppl., 12, 111 (1959).

9. A. Gippius, V. Vavilov, J. Panosian, V. Uschakov, Kratkie Sovobschenia po Fizike 7, 8 (1970) (in Russian).

10. D. D. Dell, R. Dingle, S. E. Stokowski, J. V. Dilorenzo, Phys. Rev. Lett. 27, 1644, (1971).

11. A. Bivas, R. Levy, J. B. Grun, C. Comte, H. Haken, S. Nikitine, Opt. Commun. 2, 227, (1970).

12. S. Permogorov, V. Travnikov, A. Selkin, Soviet Physics, Solid State (to be published).

RESONANCE INTERACTION BETWEEN ORTHO- AND PARA-EXCITON LEVELS

DUE TO PHONONS IN A CUPROUS OXIDE CRYSTAL

F. I. Kreingold and V. L. Makarov

Institute of Physics, Leningrad State University

Leningrad, USSR

ABSTRACT

The exciton emission in Cu_2O single crystals has been investigated under uniaxial stress. The splittings of ortho-exciton
emission lines and of their phonon replicas have been observed.
An anomalous line does not split and is attributed to the phonon
replica of the para-exciton transition. Above a threshold uniaxial stress a new narrow line appears and is attributed to
resonance involving the para-exciton. The temperature dependences
of these spectra are reported and explained.

INTRODUCTION

In spite of the fact that a comprehensive study has for some
years been made of the cuprous oxide crystal, it remains an important object of investigation in studying exciton properties.
Investigation of absorption spectra made it possible to find the
exciton energy structure, while the influence of uniaxial deformation and of electric and magnetic fields on absorption spectra
allowed the determination of the symmetry of exciton states and
allowed an adequate theoretical interpretation.

It was found that the ground state of the exciton n=1 of the
"yellow" series has a symmetry Γ_{25}^+, and the transition to this
state is quadrupolar (1, 2, 3). Apart from this triply degenerate
state (ortho-exciton) the existence of a singlet exciton state
with a symmetry Γ_2^+ (para-exciton) was predicted. However, up to
the present there has been no confirmation of the existence of a
para-exciton level in Cu_2O, which is not accidental, since an

immediate observation of the para-exciton in absorption and emission was impossible because the corresponding transitions are forbidden both in dipole and quadrupole approximations.

It is well known that in Cu_2O crystals, apart from the direct transitions to the exciton states, the indirect transitions with participation of phonons have great probability. Such transitions give rise to the well known absorption "steps" near the exciton line and to the phonon satellites in the luminescence.

Knowing the exciton symmetry, it is possible to determine the symmetry of the phonons participating in indirect transitions. In the dipole approximation near the Γ point of the Brillouin zone the transitions to the exciton state Γ_{25}^+ are possible with the participation of phonons having symmetries: Γ_2^-, Γ_{12}^-, Γ_{25}^-, Γ_{15}^-, (1). Since the indirect transitions into the para-exciton state are allowed, it is possible to identify the Γ_2^+ level in the indirect absorption or luminescence spectra.

PHONON REPLICAS OF EXCITON LUMINESCENCE IN CUPROUS OXIDE

In the Cu_2O luminescence spectrum, apart from the no-phonon line Γ_{25}^+ of the exciton luminescence, a number of bands were observed which result from exciton annihilation with a simultaneous creation (or destruction) of phonons (Fig. 1a) (4). Investigating the spectra of Cu_2O exciton luminescence we were able to determine the energies of all optical phonons (Table 1).

Apart from the phonon replicas given in the table, a luminescence band is observed at a distance of $180 cm^{-1}$ from the ortho-exciton line having the characteristic Maxwell contour of the phonon replica of a free exciton (band B_1 in Fig. 1a). This band does not fit in with the general diagram of the cuprous oxide phonon spectrum (5, 6) and appears to be "superfluous". Investi-

Table 1

The energies of optical phonons in Cu_2O crystal measured from the exciton luminescence spectra.

Symmetry of phonons	Γ_{25}^-	Γ_{12}^-	Γ_2^-	Γ_{25}^+	Γ_{15}^-		Γ_{15}^-	
					long.	trans.	long.	trans.
Energy, cm^{-1}	84	110	350	515	150	–	640	610

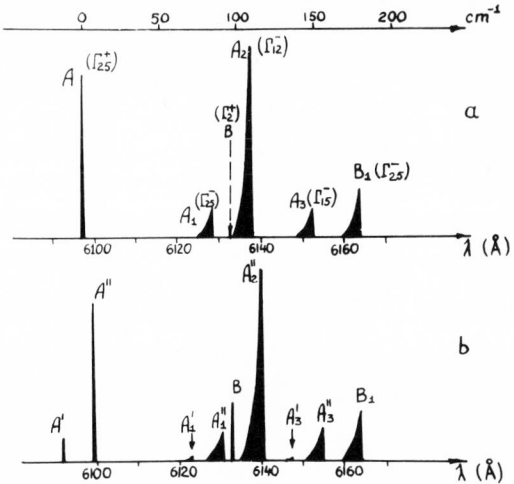

Fig. 1   Spectrum of exciton luminescence of a Cu$_2$O crystal: a –
undeformed crystal; b – crystal subjected to stress along
the symmetry axis of fourth order C$_4$; A and B – no-phonon
lines of ortho- and para-excitons, A$_{1,2,3}$ and B$_1$ – phonon
replicas of ortho- and para-excitons and A',A'' – components of the ortho-exciton no-phonon line appearing at
deformation.   In Fig. 1a there is no no-phonon line of
the para-exciton.   The place where the para-exciton line
must be is marked by a dotted arrow.

gation of a great number of Cu$_2$O samples shows that, while the
relative intensities of the phonon replicas of line Γ_{25}^+ were constant in all crystals, the intensity of the phonon replica, $h\nu$
=180cm^{-1}, was changed from sample to sample.   It was observed
that in crystals with a smaller amount of impurities the relative
intensity of this band was noticeably greater.   Investigation
shows that the temperature dependence of this band differs very
markedly from that of the other bands due to indirect transitions
from Γ_{25}^+ level.   Experiments on uniaxial compression reveal that
while the no-phonon luminescence line Γ_{25}^+ and its phonon replicas
split, phonon replica B$_1$ (Fig. 1b) does not undergo any splitting
and its position in the spectrum remains unchanged (7).   Finally,
the phonon replica with an energy of 180 cm^{-1} never occurs in the
luminescence spectra of cuprous oxide bound excitons.

All this leads us to suppose that the observed band results
from the interaction of a phonon with another (not Γ_{25}^+) exciton
state.   It is natural to suppose that the exciton state responsible
for the "superfluous" peak observed is the para-exciton state Γ_2^+.
From the position of the para-exciton phonon satellite it is

possible to position level Γ_2^+ itself. As a matter of fact, luminescence from the para-exciton level in the dipole approximation is
possible with the participation of only one phonon with a symmetry
Γ_{25}^- where: $\Gamma_2^+ \times \Gamma_{15}^- \times \Gamma_1^+ = \Gamma_{25}^-$. We have identified such a phonon,
its energy (see Table 1) being 84 cm^{-1}. The para-exciton level
must, consequently, be sought on the long-wavelength side of the
ortho-exciton level at a distance $\Delta = (180 \text{ cm}^{-1} - 84 \text{ cm}^{-1}) =$
96 cm^{-1}, i.e., the wavelength of the para-exciton line must be
6135A (at T=4.2 K).

Although it was thus clear where to seek the level Γ_2^+, it was
not possible to find it either in the absorption or in the luminescence spectra and only the experiments on uniaxial deformation
confirmed the existence of the para-exciton level in the calculated
position.

EXCITON LUMINESCENCE OF Cu$_2$O CRYSTALS
UNDER THE UNIAXIAL STRESS

The influence of uniaxial stress on the exciton spectra in
Cu$_2$O is widely discussed in (2, 3, 8). It has been shown that
under deformation the degeneracy of the Γ_{25}^+ level is lifted, which
leads to the splitting of the n=1 exciton line into two or three
components. With the deformation along C$_4$ we have observed the
doublet splitting of the Γ_{25}^+ emission line. This splitting corresponds to the splitting of the absorption line. The phonon replicas
of the ortho-exciton line show the same splitting. In contrast,
the phonon peak B$_1$ does not split or shift under the stress. At
pressures exceeding a critical value we have observed the abrupt
appearance of a new sharp emission line with a halfwidth much less
than that of the phonon replicas. Further increase of the pressure
does not change the intensity of this line appreciably. The position of this emission line corresponds to the calculated position
of para-exciton level 96 cm^{-1} below the ortho-exciton level. It
has been found that the appearance of this line in the emission
spectrum takes place for pressures at which the phonon replica A_2'
of the high energy A' exciton component passes through the calculated position of the para-exciton level, i.e., is of resonant
character. Such a resonance generation of the para-exciton line
may be conditioned by the mixing of Γ_2^+ and Γ_{25}^+ states through
phonon interaction. However, further experiments are necessary
to explain this phenomenon in detail.

THE TEMPERATURE DEPENDENCE OF THE EXCITON LUMINESCENCE

As a result of our study of exciton luminescence in samples
of Cu$_2$O with different impurity concentration, it was shown that
the temperature dependence of the luminescence in all samples is
quite similar. Fig. 2a shows the temperature dependence of the

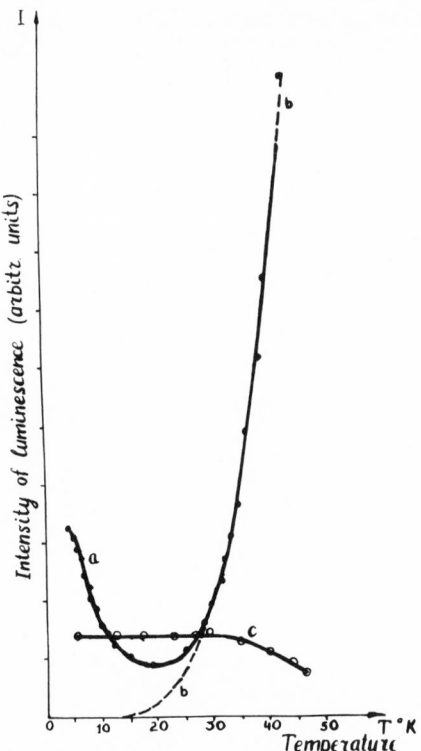

Fig. 2   Temperature dependence of exciton luminescence: a - for
         ortho-exciton; b - theoretical curve for the case of
         equilibrium exciton distribution between levels Γ_{25}^+ and
         Γ_2^+; and c - for phonon replica Γ_{25}^- of the para-exciton.

integral intensity of the luminescence of the ortho-exciton.   The
main feature of this dependence is the pronounced minimum always
in the same temperature interval, namely at 20 K.   The temperature
dependence of the intensity of the para-exciton phonon replica
(B_1-peak) is quite different (Fig. 2c).   The existence of a para-
exciton level accounts for these experimental data.

     Since the para-exciton level is deeper than the ortho-exciton,
the density of ortho-excitons must increase with temperature.   The
integral intensity of the ortho-exciton luminescence should, consequently, increase with the temperature as follows: $I = I_0 \exp(-\Delta/kT)$,
where $\Delta = 96$ cm^{-1} is the energy separation between Γ_{25}^+ and Γ_2^+
levels.   The density of the para-excitons and the intensity of the
phonon repetition B_1 should decrease.   It is seen from experiment
that the increase in the intensity of the ortho-exciton luminescence

and simultaneous decrease in the intensity of the para-exciton
phonon replica B_1 occur only above 20 K. At the same time, the
increase in the intensity of ortho-exciton luminescence follows
the exponential law, and the value of Δ obtained from the experimental curve proved to be equal to 95 cm^{-1}. With an equilibrium
distribution of the excitons between Γ_2^+ and Γ_{25}^+ the temperature
dependence of ortho-exciton luminescence would be exponential
(Fig. 2b). Comparison with the experimental dependence in the interval 4.2 K to 20 K makes it possible to assert that in this temperature range there is a deviation from equilbrium distribution.
The existence of the minimum is explained by the fact that the
degree of non-equilibrium depends on temperature: with increasing
temperature of the crystal the non-equilibrium in filling the
ortho- and para-exciton levels decreases.

ACKNOWLEDGEMENT

Concluding, we feel it a pleasant duty to express our deep
gratitude to Prof. S. A. Moskalenko for discussion of the results
and interest shown in this publication.

REFERENCES

(1) R. I. Elliott, Phys. Rev., 124, 340 (1961).

(2) E. F. Gross, A. A. Kaplyanski, Fiz. Tverd. Tela, 2, 379
    (1960).

(3) E. F. Gross, A. A. Kaplyanski, Fiz. Tverd. Tela, 2, 2968
    (1960).

(4) E. F. Gross, F. I. Kreingold, ZhETF Pis. Red., 7, 281 (1968).

(5) C. Carabatos, Phys. Stat. Sol., 37, 773 (1970).

(6) C. Carabatos, B. Prevot, Phys. Stat. Sol., 44, 701 (1971).

(7) E. F. Gross, F. I. Kreingold, V. L. Makarov, ZhETF Pis. Red.,
    15, 383 (1972).

(8) E. F. Gross, A. A. Kaplyanski, V. T. Agekyan, Fiz. Tverd.
    Tela, 4, 1009 (1962).

THE EFFECT OF DISLOCATIONS ON THE LUMINESCENCE SPECTRA OF CdS AND CdSe SINGLE CRYSTALS

Yu. A. Osipian and E. A. Steinman

Institute of Solid State Physics, Academy of Sciences

Chernogolovka, USSR

ABSTRACT

The effect of the dislocation density on the relative quantum efficiency of the luminescence of excitons, bound to neutral and ionized centers in CdS and CdSe, was studied. Dislocations were found to be effective centers for inelastic scattering of excitons, causing a considerable decrease in the exciton life time. As dislocation densities exceeding $10^7/cm^2$ the exciton life time becomes shorter than the relaxation time of the excitons causing a change in both spontaneous and stimulated emission spectra.

It has been shown (1) that the plastic deformation of CdS single crystals leads to a marked change of the luminescence spectra. The changes observed have been assumed to be due to dislocations, the density of which show a sharp increase under plastic deformation. It is known that the presence of slowly decreasing elastic fields near the dislocation as well as translational symmetry along the dislocation line may substantially increase the effectiveness of their interaction with quasiparticles (2-6). On the other hand the growth conditions and the following heat treatment of II-VI compounds considerably affect their electric properties. The influence is usually explained by a change in the stoichiometric composition (in cases of no diffusion of foreign impurities). Still, in a number of papers (7-11) it has been reported that growth conditions greatly change the concentration and distribution of dislocations in crystals. Thus, one may believe that at least some part of the observed changes of electrical and optical properties is caused by dislocations.

The purpose of the present investigation was to study optical
properties of kinetics of excitons in CdS and CdSe crystals with a
directed and controlled change in the dislocation structure in
these crystals. Dislocations were introduced through the uniaxial
compression of the specimens at temperatures between 150 and 300 C.
The specimens were cut from large single crystals grown from the
gas phase. The specimen resistivity ranged from 10^9 to 10^{11} $\Omega \cdot cm$.
The geometrical orientation was chosen so that the slip could be
performed in the basal plane, and edge dislocations were easily
introduced. Optical spectra were measured at 1.5 K with a diffraction spectrometer with a linear dispersion of 5 A/mm.

Fig. 1(a) shows the luminescence spectra of CdS crystals before and after the plastic deformation. It should be noted that

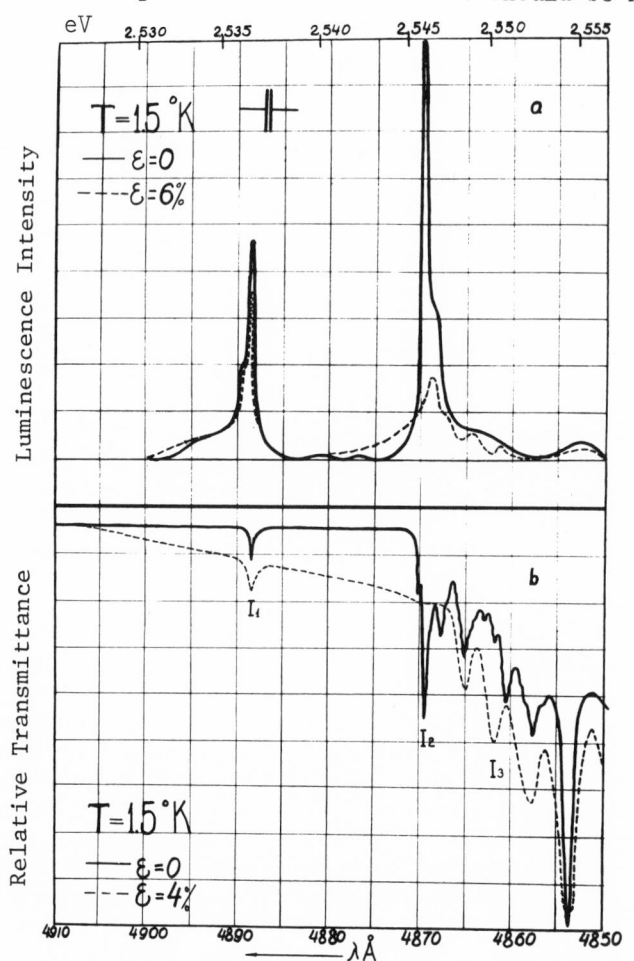

Fig. 1   The luminescence spectra (a) and absorption spectra (b)
         of non-deformed and deformed crystals.

two lines I_1 and I_2 associated with excitons localized at neutral
acceptors and donors, respectively, dominate in the luminescence
spectrum at 1.5 K. It is seen that the plastic deformation markedly alters the intensity relation between the two lines. Also, the
weak structure on the short-wavelength side of I_2 is observed in
the spectrum. Similar changes can be observed in the absorption
spectra of deformed crystals. In the spectrum given in Fig. l(b),
one can clearly see that simultaneously with a decrease in the I_2
line intensity, that of the I_3 line increases. Redistribution of
line intensities due to dislocations introduced in the crystal is
in many ways similar to the effects observed in experiments on
infrared quenching of the line I_2 (13), in which the concentration
of neutral and ionized donors was changed. Therefore, we believe
that the plastic deformation gives rise to some new levels in the
crystals. These levels are in a position to capture donor electrons
thus decreasing the concentration of neutral donors and increasing
that of ionized donors. It is necessary to add that when introducing dislocations, the change of intensities of some lines occurs
against the background of a total decrease of quantum yield. Taking
into account the fact that the luminescence is excited by ultraviolet radiation, i.e. the electrons are excited deep into the band,
one may suppose that between the absorption of high-energy photons
and the radiation the following chain of transitions takes place:
electron-hole pairs-excitons-localized excitons-radiation (14).

   As stated above at low temperatures the whole crystal radiation
is concentrated mainly in the region of the lines of bound excitons.
Hence, a decrease of the quantum yield of the luminescence implies
the appearance of an additional channel of non-radiative energy
consumption on one or several links of the chain. To find out the
nature of the channel, we conducted experiments to study the exciton
luminescence at 77 K through excitation with an argon laser,
$\lambda = 4880A$ (13). In Fig. 2(a,b) the luminescence spectra of a non-
deformed specimen and a specimen containing about $10^7/cm^2$ dislocations are shown. Qualitatively the observed change could be explained as follows: when exciting with an argon laser monochromatic
excitons with a kinetic energy of 26 meV and the wave vector
$\sim 10^7 cm^{-1}$ are generated. The distribution in the spectrum of spontaneous luminescence corresponds to the equilibrium distribution of
excitons in the former case and to the substantially non-equilibrium
one in the latter (15). This indicates that in the non-deformed
crystal the exciton system during its lifetime comes to equilibrium
with the lattice while in the deformed crystal it does not, i.e.
when introducing dislocations the exciton lifetime in the band becomes comparable to the time of exciton relaxation along the energy
axis or becomes less. This assumption is also qualitatively confirmed by the observation of the induced luminescence in CdSe crystals where an increase in the dislocation density leads to a marked
decrease in the luminescence intensity (15).

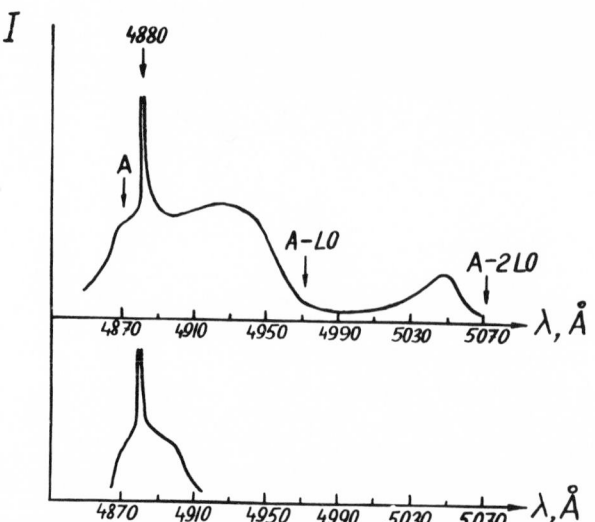

Fig. 2 The spectrum of spontaneous exciton-phonon luminescence
under argon laser excitation for a non-deformed crystal
(a), and a deformed crystal (b).

Aiming at determining the exciton lifetime dependence in the
band on the dislocation density, we have conducted experimental
investigations on the absorption coefficient dependence in the
exciton-phonon region of the spectrum on the power density of the
incident radiation in CdSe crystals. It was shown earlier that
the coefficient of exciton-phonon absorption α in the region of
transitions followed by phonon absorption is determined by the
difference of $n(k)$- exciton and $\nu(-k)$- phonon occupation numbers,
which in turn depends on the power radiation density (16).

$$\alpha = \hbar\omega \sum_k W(k,\omega)(\nu-n) \tag{1}$$

where $W(k,\omega)$ is the function characterising the optical transition
probability. The obtained results are given in Table 1. From the
data presented in the Table it follows that the exciton lifetime
in the crystals with a high dislocation density is decreased markedly. This is in good agreement with the quantum yield drop of the
exciton-phonon luminescence as well as with the disturbance of
equilibrium distribution of excitons in the band. Thus, dislocations are efficient centers of non-elastic interactions with excitons. Considering that with a high dislocation density the process
of exciton destruction during their interaction with dislocations
becomes predominant, the radius of exciton interaction with dislocations can be roughly estimated from the formula:

Table 1

$\varepsilon = 0$		$\varepsilon = 3\%$	
$T(k)$ $E_O(eV)$	$t(sec)$	$T(k)$ $E_O(eV)$	$t(sec)$
86°	2×10^{-8}	84°	8.8×10^{-11}
1.24×10^{-3}		1.24×10^{-3}	
93°	3.5×10^{-9}	108°	2.7×10^{-11}
3.7×10^{-3}		9.6×10^{-3}	

$$R = 1/N_d v \tau \qquad (2)$$

where N_d is the dislocation density and v, τ are the velocity and lifetime of excitons, respectively. Substituting the exciton thermal velocity and lifetime from the Table we obtain $R_{max.} = 10^{-3}$ cm.

We now consider the problem connected with the mechanism of exciton scattering by dislocations. There are two possibilities available here: either the exciton dissociates on collision with the dislocation or its energy is consumed for dislocation ionization (the latter process is similar to Auger recombination); to accomplish the former process, an exciton must obtain the additional energy necessary for its dissociation. This energy might be obtained from the kinetic energy, however, in our experiments the kinetic energy of excitons is always less than that of the binding energy, or the additional energy might be obtained from the excited dislocation, but this process is even less probable. In view of what has been said above, we think that the most probable process is the one involving dislocation ionization.

Thus, dislocations in CdS and CdSe crystals may capture electrons from neutral donors which gives rise to the redistribution of line intensities in the luminescence and absorption spectra. At the same time dislocations are efficient centers of non-elastic scattering of excitons and heavily reduce the exciton lifetime in the band. Under these conditions the thermal equilibrium has not occurred for the exciton band and it is likely to be the main cause for a decrease in the luminescence quantum yield. In this case, the spectra of spontaneous and induced luminescence correspond to a non-equilibrium state.

REFERENCES

1. Yu. A. Osipian, E. A. Steinman, V. B. Timofeev, Phys. St. Sol., 32, K 121 (1969).

2. W. T. Read, Phil. Mag., 45, 775 (1954); 45, 1119 (1954).

3. V. L. Bonch-Bruevich and V. B. Glasko. Fiz. tverd. tela., 1, 36 (1961).

4. P. R. Emtage, Phys. Rev., 163, 865 (1967).

5. I. M. Lifshitz, H. I. Pushkarov, Letters Zh. eksper. teor. Fiz., 11, 456 (1970).

6. V. A. Grazhulis, Yu. A. Osipian, Fiz. tverd. Tela, 1, 36 (1961).

7. Chikawa Jun-ichi, Nakayama, T., J. Appl. Phys., 35, 2493 (1964).

8. Chikawa Jun-ichi, Japan, J. Appl. Phys., 3, 229 (1964).

9. M. Ja. Skorochod, L. I. Dacenko, Kristallographija, 11, N2, 300 (1966).

10. E. V. Markov, A. A. Davidov. Zh neorg. mat., 7, 575 (1971).

11. M. Rubenstein and F. M. Ryan, II-VI Semiconductors Compounds 1967, International Conference, W. A. Benjamin, Inc., N.Y. Amsterdam, p. 402 (1967).

12. R. R. Holmes and C. Elbaum., Phys. Rev., 173, 803 (1968).

13. D. G. Thomas, J. J. Hopfield, Phys. Rev., 128, 2135 (1962).

14. V. L. Broude, I. I. Tartakovskii, V. B. Timofeev, Fiz. tverd. Tela, to be published.

15. Yu. A. Osipian, V. B. Timofeev, E. A. Steinman, Zh. exper. teor. Fiz., 62, 272 (1972).

16. A. F. Dite, V. B. Timofeev, V. M. Fain, E. G. Jashin, Zh. exper. teor. Fiz., 58, 450 (1970).

EFFECTS OF THERMOLUMINESCENCE EXCITATION IN SEMICONDUCTING DIAMONDS

S. A. A. Winer, N. Kristianpoller and R. Chen

Department of Physics and Astronomy

Tel-Aviv University, Israel

ABSTRACT

Results are presented which indicate a clear connection between the excitation mechanisms of the 260 K and 150 K glow peaks in semiconducting diamond when excited at 80 K by $\lambda=360$ nm. Excitations in this wavelength region and with $\lambda=225$ nm at $T{\geq}150$ K, however, seem to indicate that under these conditions, different mechanisms are involved. A model is discussed which takes into account the new observations.

INTRODUCTION

Thermoluminescence (TL) of semiconducting diamonds was first investigated by Halperin and Nahum (1). After excitation at 80 K (LNT) with uv light of 225 nm (5.5 eV), corresponding to the band to band transition, they found an intense blue glow peak at 250-260 K and a blue peak at ∿150 K, both depending linearly on the radiation dose. The activation energies of these peaks were found to be 0.2 eV and 0.35 eV for the lower and higher temperature peak respectively. The latter value was in good agreement with 0.37 eV, the activation energy for release of holes in semiconducting diamonds as measured by electrical methods (2).

Halperin and Chen (3) showed that the two TL peaks could be excited also in the 300-400 nm region by employing higher light doses for excitation. For this wavelength range, the dependence of the 260 K peak on the radiation dose (D) was found to be super-linear at low doses. The maximum intensity of this peak increased with D^n, where n depended on the excitation wavelength λ and showed values of up to 3. The 150 K peak depended linearly on the excitation dose.

The superlinear excitation was explained by assuming a multi-stage transition of electrons excited from the valence to the conduction band before being trapped at the center. This model was supported by the results of pre-excitation experiments (4). Emission spectra of the band to band excited TL in semiconducting diamonds have recently been measured by Walsh, Lightowlers, and Collins (5).

In the present work, a connection is established between the excitation of the 150 and 260 K peaks at different exciting wavelengths. The processes involved are discussed.

EXPERIMENTAL

The samples used were the C_2, C_3 and C_4 semiconducting diamonds (II_b) mentioned in the previous work (3). The crystal was mounted in a metal vacuum cryostat and was excited at LNT or at higher temperatures up to 300 K. A 900 Watt xenon arc lamp in conjunction with a Hilger and Watts D285 quartz monochromator was used for uv excitation. The emission was measured with a thermoelectrically cooled EMI9558 QB (S 20) photomultiplier. The signal was amplified by a Keithley 414S picoammeter and fed into the Y axis of a YEW 3073 X-Y recorder. The reading of a copper-constantan thermocouple attached to the crystal holder was fed into the X channel. The intensity of the exciting light was kept constant in each series of measurements. The time of irradiation was 1 minute for $\lambda=225$ nm and 10 minutes for 360 nm.

RESULTS AND DISCUSSION

The experiments consisted mainly of exciting the sample at various temperatures. Curve a of Fig. 1 is a typical blue glow curve excited with 360 nm light at LNT. The TL peaks appear at 155 K and 260 K, the height of the latter depending superlinearly on the excitation dose (3). Normally, the intensity of a peak is reduced substantially when excited at temperatures close to its maximum. This is indeed how the low temperature peak behaved as the temperature of excitation approached 150 K. This is shown in curve b, c and d which were obtained after excitation at 130, 140 and 150 K respectively. Excitation at temperatures below 110 K resulted in the same glow curve as 1 a. The behavior of the 260 K peak was surprising: it fell rather sharply as the temperature of excitation approached 150 K. When the temperature of excitation was above 150 K, the 260 K peak did not appear at all.

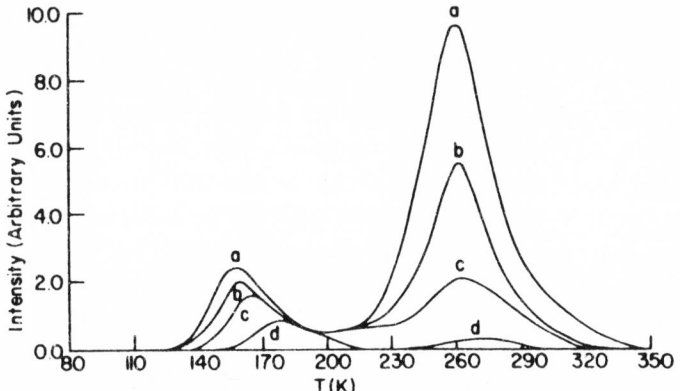

Fig. 1   Glow curves of semiconducting diamonds excited by 360 nm
         at LNT (curve a), 130 K (curve b), 140 K (curve c) and
         150 K (curve d).

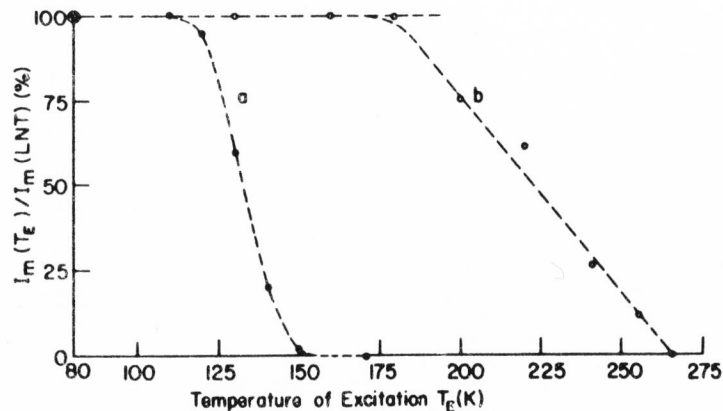

Fig. 2   The maximum intensity of the 260 K glow peak obtained at
         temperatures T_E divided by the maximum intensity found
         after excitation at LNT.   Curve a gives the results for
         excitation with 360 nm and curve b for 225 nm.

         The dependence of the maximum intensity of the 260 K peak on
the temperature of excitation, (T_E) between 80 and 170 K is shown
in curve a of Fig. 2 for excitation with 360 nm.   In this curve,
the Y axis represents the ratio (in percents) of the maximum
intensity obtained at T_E, to that obtained after excitation at LNT.
This ratio reduces sharply from 100% at 110 K to practically zero
at 155 K.

The results of similar series of measurements but with excitation wavelength of 225 nm are shown in curve b of Fig. 2. In this case, the maximum intensity I_m remains constant up to 180 K and decreases rather slowly as the temperature of excitation approaches the temperature of the peak, 260 K.

The difference between the behavior of the 260 K peak excited by 360 nm and 225 nm is readily seen in Fig. 2. As explained before (1,3), the 225 nm light raises electrons from the valence band directly to the conduction band. These electrons are trapped at a level which acts later as a recombination center for thermally released holes. If T_E is relatively high, recombination may occur during the excitation due to the thermal release of trapped holes from the 0.35 eV trap. The occupation of the recombination center is thus expected to be lower than after excitation at LNT, resulting in a less intense glow peak.

The situation is different after excitations with 360 nm. It seems that this is due to the fact that one of the levels involved in the superlinear excitation (3,4) is unstable above 150 K. This may result from the relatively large number of free holes in the valence band at temperatures approaching 150 K. These holes are probably released from the 0.2 eV trap which is the one responsible for the 150 K peak. Since this intermediate state is unstable above 150 K, electrons cannot accumulate at this level and therefore electrons cannot reach the recombination center. This explains the connection between the change in the 150 K peak and the impossibility of exciting the 260 K peak above 150 K.

Although the 260 K peak could not be excited by 360 nm at $T_E > 150$ K, irradiation at $150 < T_E < 200$ K prior to LNT excitation was found to substantially enhance the TL peak, compared to that produced by LNT excitation alone. This indicates that it is probably the second intermediate level of the multistage excitation proposed by Halperin and Chen (3,4) which is temperature unstable. This point is being further investigated presently.

In order to verify that it is one of the intermediate states which is unstable above 150 K and not the recombination center itself, the following measurements have been performed. The sample was irradiated at LNT with 360 nm for a certain period of time, then heated to 180 K and held at this temperature for 15 minutes, then cooled to LNT and heated as usual. The 260 K peak had then the same intensity as without intermediate heating. This supports the assumption that the instability above 150 K is related only to an intermediate state involved in the excitation and not to the recombination center.

The results described and discussed here give an additional

proof to the fact that excitation of the blue glow peak in semi-conducting diamonds by light in the 300-400 nm range is due to an entirely different mechanism than the band to band excitation, with 225 nm.

ACKNOWLEDGMENT

We would like to thank Prof. A. Halperin for providing the samples used in this work.

REFERENCES

1. A. Halperin and J. Nahum, J. Phys. Chem. Solids 18, 297 (1961).

2. A. T. Collins and E. C. Lightowlers, Phys. Rev. 171, 843 (1968).

3. A. Halperin and R. Chen, Phys. Rev. 148, 839 (1966).

4. R. Chen and A. Halperin, Proc. Int. Conf. Lumin., Budapest, 1414, (1966).

5. P. S. Walsh, E. C. Lightowlers and A. T. Collins, J. Lumin. 4, 369 (1971).

RELAXED EXCITED STATES OF KI:Tl$^+$-TYPE PHOSPHORS

A. Fukuda,[*] K. Cho,[*] and H. J. Paus

Physikalisches Institut der Universität

7 Stuttgart 1, BRD (Germany)

ABSTRACT

Adiabatic potential energy surfaces, especially in the ε_g sub-space, have been calculated by assuming the linear Jahn-Teller effect (JTE) within the $a_{1g}t_{1u}$ excited states and by using values of coupling constants estimated experimentally by several investigators. The calculation can explain one (the A_T band) of the two emission bands excitable by the A band but not the other (the A_X band).

1. INTRODUCTION

This paper will consider the question of whether or not the relaxed excited states responsible for the emission bands (1-5) produced by A-band excitation can be described at least qualitatively by the linear electron-lattice interaction, i.e. the linear Jahn-Teller effect (JTE), within the $a_{1g}t_{1u}$ excited states. The linear JTE was studied in detail by Öpik and Pryce (6) and the result was applied to KCl:Tl$^+$ by Kamimura and Sugano (7). After some further work was done by Trinkler and Plyavin (8) and by Kristoffel (9), Fukuda (4) made a rather systematic investigation of the emission bands excitable in the A band and qualitatively explained most of the observed features of the emission bands on the basis of the model which was originally proposed by Kamimura and Sugano (7). This model assumes the existence of two kinds of minima on the Γ_4^-(A) adiabatic potential energy surface (APES) in the ε_g subspace. (Hereafter, the APES which continuously connects with the Γ_1^- unrelaxed excited state etc. will be called the Γ_1^- APES, etc.) Quite recently, however, Ranfagni (10) showed that two kinds of minima cannot coexist when the spin-orbit interaction is small, as in Ga$^+$. In this way, so far, we cannot explain even qualitatively the characteristic features of the emission bands using the linear JTE within the $a_{1g}t_{1u}$ electron configuration.

In order to see the extent to which the linear JTE can explain the observed features of emission bands, we will make a rather general calculation in Section 2 to obtain the APES's, especially in the ε_g subspace. The observed features will be correlated with the calculated results obtained by using values of coupling constants estimated experimentally by several investigators (5,11-17), and problems will be pointed out in Section 3. One essential problem is whether all the emission bands produced by A-band excitation are explainable by assuming the $a_{1g}t_{1u}$ electron configuration. Although we could not solve the problem definitively, we will show some circumstantial evidence for the assumption and suggest in Section 4 how to improve the above calculation.

2. CALCULATION OF ADIABATIC POTENTIAL ENERGY SURFACES

In order to make a general discussion on various KI:Tl$^+$-type phosphors we normalize the energy by the separation between the C and A bands, $\Delta = W_C - W_A$, and describe the $a_{1g}t_{1u}$ excited states in terms of the parameters, Δ, $\varepsilon = (W_B - W_A)/\Delta$, and $\lambda = 1$. We know experimentally that Δ is about 1 eV (18-20). The position of the A band is taken as energy zero. Then, without taking into consideration the electron-lattice interaction, the eigenvalues of all the eigenstates will change as a function of ε as shown in Fig. 1; in particular, the relative position of Γ_1^- is given by

$$\{W(\Gamma_1^-)-W_A\}/\Delta = (\tfrac{1}{2})\{1-[(3/2)-2(\varepsilon-\tfrac{1}{2})^2]^{\tfrac{1}{2}}\}.$$

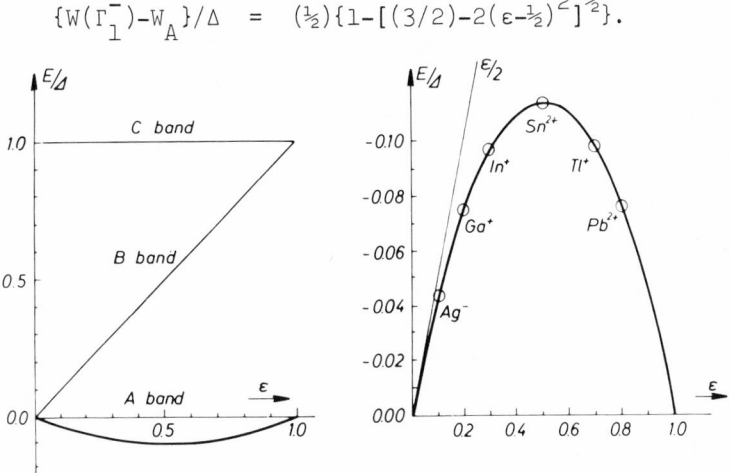

Fig. 1 Eigenvalues of Γ_4^-(C), Γ_5^-(B), Γ_3^-(B), Γ_4^-(A), Γ_1^- as a function of $\varepsilon = (W_B - W_A)/\Delta$, where $\Delta = W_C - W_A$.

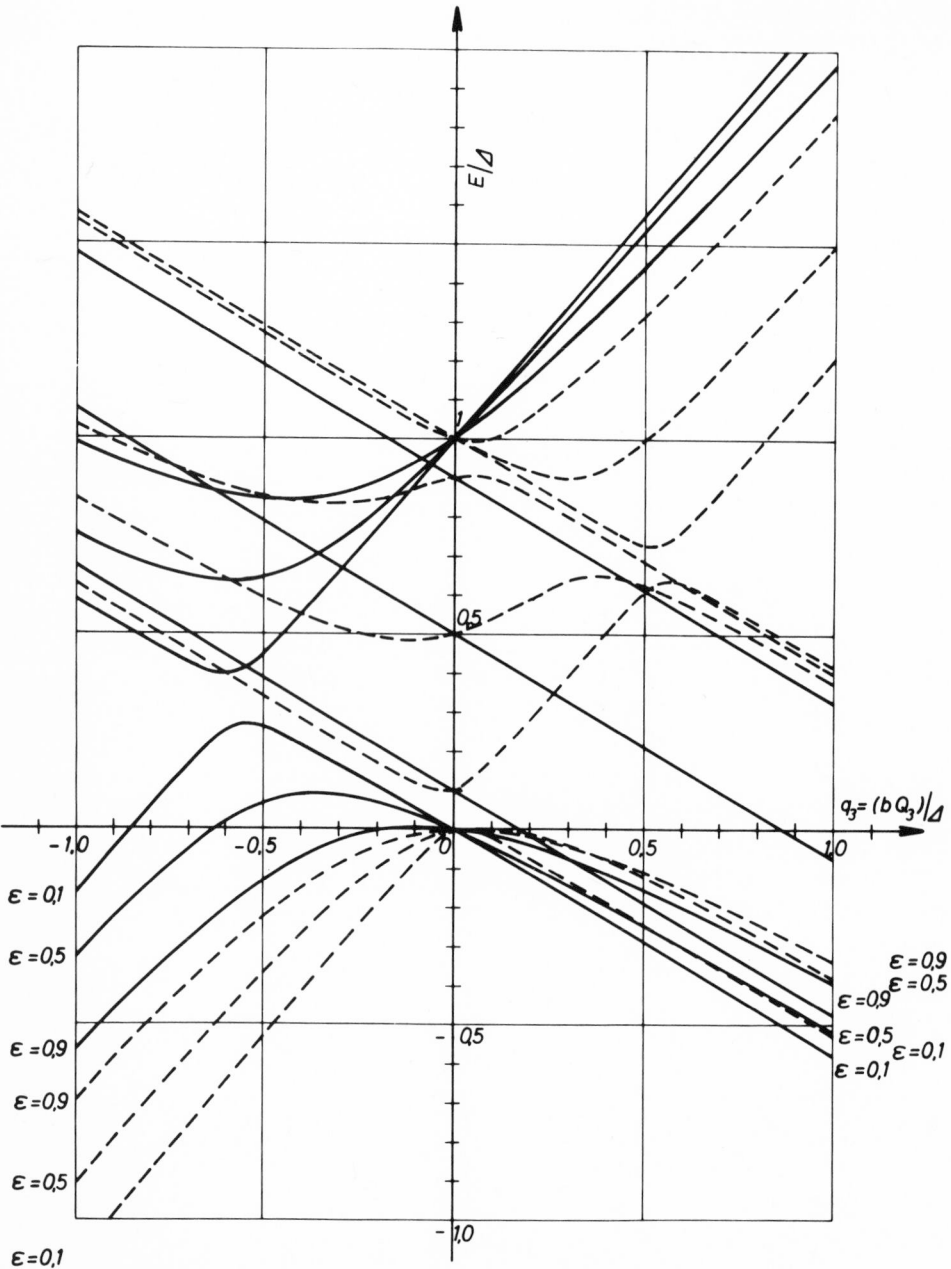

Fig. 2 Cross section of the APES's containing the q_3 axis (purely
        tetragonal distortions). For clarity only the $\Gamma_4^-(C)$, $\Gamma_5^-(B)$,
        and $\Gamma_1^-(A)$ APES's are shown and lattice potential energies
        are omitted.

We also normalize the interaction-mode coordinates (11,12) as follows:

$$q_1 = aQ_1/\Delta, \; q_2 = bQ_2/\Delta, \; \ldots \; , \; q_6 = cQ_6/\Delta \; .$$

Then the lattice potential energies are given by

$$Q_1^2/\Delta = (\Delta/a^2)q_1^2 = Aq_1^2, \; \ldots, \; Q_6^2/\Delta = (\Delta/c^2)q_6^2 = Cq_6^2 \; .$$

APES's are obtained as the eigenvalues of the interaction matrix as given in (4) and (21) plus the lattice potential energies given above.

It is difficult to diagonalize the matrix except for the special cases of $\varepsilon = 0$ and $\varepsilon \ll 1$ (6). Some cross sections of the APES's are, however, easily obtained. For example, cross sections containing the q_3 axis (purely tetragonal distortions) are obtained by solving cubic equations and are shown in Fig. 2 where the lattice potential energies are omitted. Adding the lattice potential energies, we will generally obtain on the $\Gamma_4^-(A)$ APES, 4 points at which E/Δ is stationary if regarded as a function of q_3 as shown in Fig. 3. When we calculate cross sections containing both the q_3 and q_2 axes (APES's in the ε_g subspace) and E/Δ is calculated as a function of q_3 and q_2, it becomes clear that: 1. When R is deeper than T, R is related to two of the six rhombic minima (one of three nearly degenerate, almost tetragonal minima) and T is not a minimum. When T is deeper than R, R is not related to rhombic minima and T is one of the three tetragonal minima. In contrast to previous speculation (4), these two kinds of minima, tetragonal

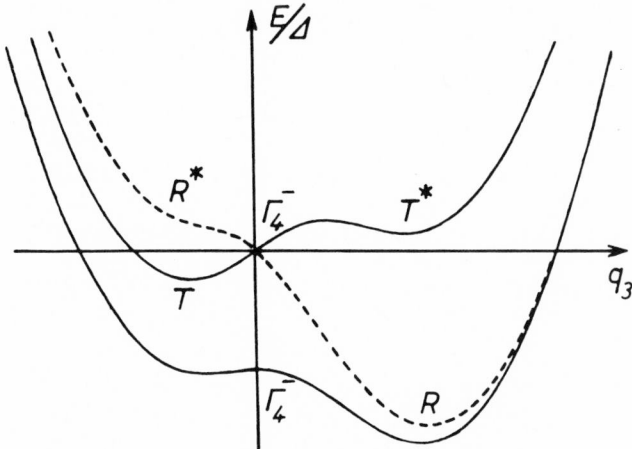

Fig. 3   Possible 4 points on the $\Gamma_4^-(A)$ APES at which E/Δ is stationary if regarded as a function of q_3.

and rhombic, do not coexist. 2. When R is related to the rhombic
minima, T^* may also be one of the other three tetragonal minima.
These tetragonal minima are very unstable in the sense that they
may exist only when ε and B have some restricted values. 3. R is
not related to any minima. 4. When R is related to the rhombic
minima, three tetragonal minima may also exist on the Γ_1^- APES and
may be very close to the rhombic minima on the $\Gamma_4^-(A)$ APES.
5. When T is one of the three tetragonal minima, there exists only
one totally symmetric minimum on the Γ_1^- APES.

It has not been proven that these rhombic or tetragonal minima
calculated as a function of q_3 and q_2 are real minima when treated
as a function of all interaction-mode coordinates. These are, however, the only possible minima from which arises emission polarized
in such a way that the degree of linear polarization as a function
of azimuthal angle in the (100) plane becomes zero in the <011>
direction. It has also not been proven that there are no minima
other than these rhombic or tetragonal minima. However, it is very
difficult to imagine the existence of minima which are deeper than
the rhombic minima, at least when ε is as small as in Ga^+. For
example, defining coordinate q_ζ of purely trigonal distortions along
the [111] axis such that: $q_4 = q_5 = q_6 = (1/\sqrt{3})q_\zeta$, we obtain the
cross sections containing the q_3 axis and find that such distortions
do not produce any minima deeper than the rhombic minima. Similar
discussion can be made for intermediate distortions.

3. CORRELATION OF THE CALCULATED RESULTS WITH THE OBSERVED DATA

At temperatures as low as 4.2 K A-band excitation produces
only one dominant emission band (1-5). Except for Ag^- in some
alkali halides and Au^-, this emission band is polarized in such a
way that the degree of linear polarization as a function of azimuthal angle in the (100) plane becomes zero in the <011> direction
(22-27). This linearly polarized emission band for Ag^-, Ga^+, In^+,
Sn^{2+}, Tl^+, and Pb^{2+} will be called A_T (T means tetragonal). According to the preceeding section, the only way is to correlate the A_T
band with the rhombic or tetragonal minima on the $\Gamma_4^-(A)$ APES and/or
with the tetragonal minima on the Γ_1^- APES. The values of ε and B
listed in Table I indicate that the rhombic minima are the only
possible minima on the $\Gamma_4^-(A)$ APES when ε is small as in Ag^-, Ga^+
and In^+ or when ε is not small but B is very small as in Sn^{2+} and
Pb^{2+}. On the other hand, in Tl^+ where both ε and B are not small,
the tetragonal minima are the only possible minima. For the former
case the tetragonal minima may also exist on the Γ_1^- APES and may be
very close to the rhombic minima on the $\Gamma_4^-(A)$ APES; for the latter
case, there is only one totally symmetric minimum on the Γ_1^- APES.
The calculation is, however, too approximate to tell which are
the actual minima, rhombic or tetragonal. Thus, the rhombic
minima on the $\Gamma_4^-(A)$ APES may actually be responsible for the A_T

TABLE I

List of $\Delta = W_C - W_A$, $\varepsilon = (W_B - W_A)/\Delta$, a^2, b^2, c^2, $A = \Delta/a^2$, $B = \Delta/b^2$, and $C = \Delta/c^2$ estimated experimentally by several investigators. The moment analysis of absorption bands cannot determine b^2 and c^2 separately but determine $(3b^2 + 2c^2)/5$, which are shown in parenthesis.

		Δ	ε	a^2	b^2 c^2 $(2b^2+3c^2)/5$		A	B	C	Refs.
Ag⁻	KCl	1.26	0.129	0.29	0.11	0.22	4.29	11.0	5.96	5
					0.10	0.15		12.6	8.40	17
	CsBr	1.08	0.093		1.4			0.75		5
Ga⁺	KCl	1.05	0.200	0.40	0.59	0.48	2.62	1.78	2.19	12
					0.98				1.07	11
	KI	0.67	0.164	0.30	0.77	0.30	2.23	0.87	2.23	12
In⁺	NaCl	1.04	0.317	0.94	2.00	0.69	1.11	0.52	1.51	12
	KCl	1.07	0.280	0.40	0.62	0.52	2.68	1.73	2.06	12
					1.17				0.92	11
				1.06	(0.55)		1.00	(1.94)		13
	CsBr	0.90	0.289	0.21	2.19	0.46	4.28	0.41	1.95	12
	KI	0.72	0.264	0.22	0.61	0.24	3.27	1.18	3.00	12
				0.88	(0.23)		0.82	(2.90)		13
Sn²⁺	KCl	1.00	0.580	0.25	4.70	1.82	4.00	0.21	0.55	12
					2.46				0.41	11
	KI	0.66	0.500	0.15	2.25	0.52	4.40	0.29	1.27	12
					1.18				0.35	11
Tl⁺	KCl	1.33	0.676	0.24	0.50	0.46	5.54	2.66	2.89	12
					0.83				1.60	11
				0.29	(0.62)		4.59	(2.15)		13
				0.26	(0.72)		5.11	(1.85)		14,15
	KBr	1.14	0.693	0.14	(0.58)		0.81	(1.97)		14,15
				0.41	(0.20)		0.28	(5.70)		14
	KI	0.92	0.783	0.08	0.36	0.14	11.5	2.56	6.56	12
					0.90	0.03		1.02	3.06	16
				0.09	(0.24)		10.2	(3.84)		14,15
				0.44	(0.07)		2.09	(13.1)		14
	RbI	0.87	0.783	0.44	(0.08)		1.98	(10.9)		14
Pb²⁺	NaCl	1.70	0.800		2.49				0.68	11
	KCl	1.66	0.777		1.82				0.91	11
	RbCl				2.90					11

band even in Tl⁺ and the A_T band in Sn²⁺ may actually arise from the tetragonal minima on the $\Gamma_4^-(A)$ APES as will be explained in connection with the higher order JTE and the effect of a charge-compensating vacancy. Experimentally it is not easy to determine which minima, rhombic or tetragonal, are responsible for the A_T band. In other words, most of the observed data can be explained in terms of the rhombic minima as well as in terms of the tetragonal minima. As far as the present authors are aware, the negative

magnetic circular polarization observed in KI:Sn^{2+} is the only
observation that can be explained by the tetragonal minima and not
by the rhombic minima (29). The remarkably large circular polarization observed in KI:Ga$^+$ and KI:In$^+$ seems to suggest the rhombic
minima, but not conclusively (29). Although Shimada and Ishiguro
(16) concluded that the A_T band in KI:Tl$^+$ arises from the tetragonal minima, the results of their uniaxial stress experiment can
also be explained in terms of the rhombic minima. In this way it
is an interesting problem to decide experimentally which are the
actual minima, rhombic or tetragonal.

One success of the calculation based on linear JTE is that it
predicts the tetragonal minima on the Γ_1^- APES as a trapping state
and that the tetragonal minima may disappear when both ϵ and B are
not small, as in Tl$^+$. When there are no tetragonal minima on the
Γ_1^- APES, there is no region between the origin and minima where
both of the $\Gamma_4^-(A)$ and Γ_1^- APES's go down parallel to each other. In
such a case the non-radiative transition from the $\Gamma_4^-(A)$ to Γ_1^- APES's
may barely occur, and thus we understand qualitatively why in KI:Tl$^+$
and KBr:Tl$^+$ the A_T band arises mainly from the minima on the $\Gamma_4^-(A)$
APES and not from the minima on the Γ_1^- APES. In connection with the
non-radiative transition from the $\Gamma_4^-(A)$ to Γ_1^- APES's two observed
facts will be mentioned as follows: First, Kleenmann (5) found
that at low temperatures, C-band excitation also produces an emission band on the low energy side of the A band though the efficiency
is very low, and that this emission band is slightly shifted to low
energy compared to the A_T band. Although he ascribed this emission
to the A_X band, which will be discussed in the following, we believe
that this emission band must also be the A_T band and arises from
the minima on the Γ_1^- APES. The ratio in population between the
minima on the Γ_1^- APES and those on the $\Gamma_4^-(A)$ APES is larger in C-
band excitation than in A-band excitation. Since the separation
between these minima is rather large in Ag$^-$ because of the rather
large value of B, these two excitations may produce slightly different A_T bands. Second, Gerhardt (28) observed a similar slight
difference in the A_T band of KCl:Tl$^+$ by time resolved spectroscopy
and found the negative linear polarization of the A_T band from the
tetragonal minima on the Γ_1^- APES. The negative linear polarization
is very interesting and must be related to the non-radiative transition from the $\Gamma_4^-(A)$ to Γ_1^- APES's before relaxation is completed.

Now let us consider the emission spectrum at high temperatures.
A-band excitation produces two emission bands in some KI:Tl$^+$-type
phosphors and only one emission band in others. There are two
simple views regarding the interpretation of the case where only
one emission band appears. First, the one band is regarded as two
bands accidentally overlapping each other. Second, the one band
is regarded as one of two bands and the other of the two is considered, for some reason, not to appear. The second view seems

more reasonable than the first, because there are at least three
phosphors which have only one emission band and hence we have to
expect accidental overlap frequently. Moreover, the observed temperature variation of the one band does not indicate any overlap
of two bands. In other words, the accidental overlap must be very
exact if we interpret the one band as the overlap of the two bands.
Hereafter let us denote the emission band which appears at high
temperatures as A_X (X means "unknown").

     In order to explain the A_X band we need another kind of re-
laxed excited state than that for the A_T band. An old and still
debated question is whether the relaxed excited state responsible
for the A_X band is described by the $a_{1g}t_{1u}$ electron configuration
or not. So far only one relaxed excited state with an electron
configuration other than $a_{1g}t_{1u}$ has been proposed, i.e. the
localized exciton or V_K center plus an electron perturbed by a Tl$^+$-
type impurity ion. However, the difficulty of this proposal has
already been pointed out by Fukuda (4). Moreover, recent magnetic
circular polarization experiments on the emission bands (29) suggest
that the relaxed state responsible for the A_X band is very similar
to that for the A_T band, in particular the relaxed excited state
for the A_X band, as well as that for the A_T band, has a trapping
state which can be considered to be related to the Γ_1^- unrelaxed
excited state. Thus it is worth-while to try to obtain the APES's
more accurately than in Section 2.

4. POSSIBILITIES TO IMPROVE THE CALCULATION

     Obviously the calculations made in Section 2 are not satis-
factory in the following respects: 1. It is not difficult to take
into account a difference in the radial wave functions between $^3T_{1u}$
and $^1T_{1u}$ by assuming $\lambda \neq 1$. However, we do not think this assumption could explain the additional minima for the A_X band on the
$\Gamma_4^-(A)$ APES. 2. We have not proven that there are no minima outside
the ε_g subspace which might be responsible for the A_X band. However, we do not think such an exact calculation could explain the
observed facts because the A_X band does not appear in KCl:In$^+$ but
does appear in KI:In$^+$ although both the phosphors have nearly the
same values of A, B, C, and ε. We encounter a similar situation
in KCl:Tl$^+$ and KI:Tl$^+$. 3. Higher order JTE must be important because the linear JTE is very strong and hence the Stokes shift is
as large as 1 eV or more. The lattice potential energy in the
$a_{1g}t_{1u}$ excited state may be smaller than that in the a_{1g}^2 ground
state as assumed in an ordinary 1-dimensional configuration coordinate scheme. The difference in lattice potential energy is the
diagonal part of the quadratic electron-lattice interaction and has
already been taken into account by Honma (30) to explain the asymmetry of the C band. It is worth-while to consider the quadratic
JTE (31) more generally, including off-diagonal elements. However,

it is not easy to do so because of the following: first, we cannot
use the interaction-mode coordinates; second, even if we assume
that only the six coordinates, $Q_1, Q_2, \ldots, Q_6$, which show the large
linear interaction, have also the quadratic interaction, we need
eight coupling constants to represent the quadratic interaction.
4. In Sn^{2+} a charge-compensating vacancy must play an important
role (26); the calculation based on the linear JTE shows that the
rhombic minima are responsible for the A_T band but the magnetic
circular polarization experiment (29) on the A_T band indicates that
the tetragonal minima are on the $\Gamma_4^-(A)$ APES. The difference may be
due to the vacancy, although a detailed treatment has not yet been
done. In concluding this paper we would like to emphasize the importance of experiments, because the quadratic JTE seems too complicated to be studied only theoretically. We have to determine
the symmetry of the relaxed excited state for the A_X band experimentally. Moreover, it is also necessary to determine experimentally which minima, rhombic or tetragonal, are responsible for the A_T
band. Paramagnetic resonance in the relaxed excited states may be
useful in understanding the relaxed excited states, as demonstrated
recently in F centers in CaO by Edel et al (32). The A_X band may
be a typical example of the quadratic JTE.

REFERENCES

On leave of absence from the Institute for Solid State Physics,
the University of Tokyo, Minato-ku, Tokyo 106, Japan.

1. R. Edgerton and K. Teegarden, Phys. Rev. 136, A1091 (1964).

2. J. M. Donahue and K. Teegarden, J. Phys. Chem. Sol. 29, 2141
    (1968).

3. J. Ramamurti, Phys. Rev. B1, 833 (1970).

4. A. Fukuda, Phys. Rev. B1, 4161 (1970).

5. W. Kleenmann, Zeit. Phys. 249, 145 (1971).

6. U. Öpik and M. H. L. Pryce, Proc. Roy. Soc. (London) 238,
    A425 (1957).

7. H. Kamimura and S. Sugano, J. Phys. Soc. Japan, 14, 1612
    (1959).

8. M. F. Trinkler and I. K. Plyavin, Phys. Stat. Sol. 11, 277
    (1965).

9. N. N. Kristoffel, Proceedings of the International Conference
    on Luminescence, Budapest, 1966 (Publishing House of the

Hungarian Academy of Science, Budapest) p. 824.

10. A. Ranfagni, Phys. Rev. Letts. 28, 743 (1972).

11. Y. Toyozawa and M. Inoue, J. Phys. Soc. Japan 21, 1663 (1966).

12. K. Cho, Thesis, The University of Tokyo, 1970 (unpublished).

13. A. Honma, Sci. of Light (Japan) 17, 34 (1968).

14. R. Laiho, Annales Academiae Scientarum Fennicae, Series A, VI. Physica, 362 (1971) pl.

15. D. Bimberg, W. Dultz, W. Gebhardt, Phys. Stat. Sol. 31, 661 (1969).

16. T. Shimada and M. Ishiguro, Phys. Rev. 187, 1089 (1969).

17. K. Kojima, S. Shimanuki, M. Maki, and T. Kojima, J. Phys. Soc. Japan 28, 1227 (1970).

18. A. Fukuda, Sci. of Light (Japan) 13, 64 (1964).

19. T. Mabuchi, A. Fukuda, R. Onaka, Sci. of Light (Japan) 15, 79 (1966).

20. W. Kleemann, Zeit. Phys. 234, 362 (1970).

21. A. Honma, Sci. of Light (Japan) 16, 229 (1967).

22. C. C. Klick and D. W. Compton, J. Phys. Chem. Sol. 7, 170 (1958).

23. R. Edgerton, Phys. Rev. 138, A85 (1965).

24. A. Fukuda, S. Makishima, T. Mabuchi, and R. Onaka, J. Phys. Chem. Sol. 28, 1763 (1967).

25. S. G. Zazubovich, Phys. Stat. Sol. 38, 119 (1970).

26. A. Fukuda, Phys. Rev. Letts. 26, 314 (1971).

27. W. C. Collins, Thesis, University of North Carolina, 1971 (unpublished).

28. V. Gerhardt, Thesis, Universität Regensburg, 1971 (unpublished).

29. A. Fukuda and P. H. Yuster, to be published.

30. A. Honma, Sci. of Light (Japan) 18, 33 (1969).

31. P. Wysling and K. A. Müller, Phys. Rev. 173, 327 (1968).

32. P. Edel, C. Hennics, Y. Merle d'Aubigé, R. Romestain, and Y.
     Twarowski, Phys. Rev. Letts., to be published.

THEORY OF POLARIZED RESONANT SECONDARY RADIATION OF IMPURITY CENTERS INCLUDING THE JAHN-TELLER EFFECT

V. Hizhnyakov and I. Tehver

Institute of Physics and Astronomy, Academy of Sciences

of the Estonian SSR, Tartu, USSR

ABSTRACT

A consistent theory of polarized light emission including relaxation processes should examine the whole process of the transformation of an exciting photon into an emitted photon. For such a treatment, not only luminescence but also all other kinds of secondary radiation, i.e. Rayleigh and Raman scattering, hot luminescence, and interference processes must be considered. The principles of such a two-photon theory of polarization are formulated, using as a basis the theory of resonant secondary radiation, and taking into account the influence of the Jahn-Teller effect on the integral polarization characteristics. Furthermore the influence of the same effect on the spectral dependence of these characteristics is discussed.

INTRODUCTION

A consistent theory of polarized light emission including relaxation processes should examine the whole process of the transformation of an exciting photon into an emitted photon. For such a general treatment not only luminescence but also all other kinds of secondary radiation, i.e. Rayleigh and Raman scattering, hot luminescence, and interference processes must be considered. The principles of such a two-photon theory of polarization were formulated in (1), using as a basis the theory of resonant secondary radiation (2). In (1), in addition to the general formulation of the problem, the influence of the Jahn-Teller effect upon the integral (with regard to excitation and emission spectra)

polarization characteristics was treated. In the present paper we
shall discuss the influence of the same effect on the spectral
dependence of these characteristics.

THEORETICAL ANALYSES

As is shown in (1), the problem of the polarization characteristics of secondary radiation is reduced to the calculation of the
independent components of the secondary radiation tensor
$w_{\alpha\beta\beta'\alpha'}(\omega_0,\Omega)$. If the ground electronic state is non-degenerate,
and the excited state is n-fold degenerate, the tensor is (1):

$$w_{\alpha\beta\beta'\alpha'}(\omega_0,\Omega) \tag{1}$$

$$= \frac{\gamma}{4\pi^2} \int_{-\infty}^{\infty} dt \iint_0^{\infty} d\tau d\tau' \exp[i\Omega(t+\tau-\tau')-i\omega_0 t-\gamma(\tau+\tau')/2] A_{\alpha\beta\beta'\alpha'}(t,\tau,\tau').$$

Here ω_0 and Ω denote the frequencies; α and β, the Cartesian components of the polarization vectors of the exciting and emitted
light; γ is the damping constant of the excited electronic state,

$$A_{\alpha\beta\beta'\alpha'}(t,\tau,\tau') = <M_\alpha^+ e^{i\tau'H_1} M_\beta e^{i(t+\tau-\tau')H_0} M_{\beta'}^+ e^{-i\tau H_1} M_{\alpha'} e^{-itH_0}>_0. \tag{2}$$

M_α denotes the one-column matrix of the n electronic matrix elements
between the ground electronic state and the n components of the
excited electronic state of an impurity in the Condon approximation,
M_α^+ is the complex conjugate of the M_α matrix; $H_1 = (H_0+\omega_{10})I+V$,
where H_0 is the vibrational Hamiltonian of the ground electronic
state; ω_{10}, the frequency of the electronic transition; I, the unit
matrix; and V, the vibronic coupling matrix for the excited electronic state, $<\cdots>_0 = Sp[\exp(-H_0/kT)\ldots]/Sp[\exp(-H_0/kT)]$.

Variables τ,τ' can be interpreted as the time interval which
the center has spent in the excited electronic state. Thus,
luminescence will be described by the asymptotic value of
$A_{\alpha\beta\beta'\alpha'}(t,\tau,\tau')$ at large $\tau,\tau'\sim\gamma^{-1}$ (γ^{-1} is the lifetime of the
excited electronic state). Scattering and hot luminescence are
given by small $\tau,\tau' \lesssim \Gamma^{-1} << \gamma^{-1}$ (Γ determines the vibrational
relaxation rate in the excited electronic state).

Let us start from the spectral dependence of the luminescence
polarization characteristics. As an example, the dipole-allowed
optical transition $A_{1g(u)} \rightarrow T_{1u(g)} \rightarrow A_{1g(u)}$ of a cubic center will be
discussed. In the linear approximation

$$
V(T_{1u}) = \begin{pmatrix} aQ_0 + b(Q_1 - \dfrac{1}{\sqrt{3}}Q_2) & cQ_3 & cQ_4 \\[2mm] cQ_3 & aQ_0 - b(Q_1 + \dfrac{1}{\sqrt{3}}Q_2) & cQ_5 \\[2mm] cQ_4 & cQ_5 & aQ_0 + \dfrac{2}{\sqrt{3}}bQ_2 \end{pmatrix} \quad (3)
$$

Here a, b, c and $Q_0, Q_1, \ldots, Q_5$ are the vibronic coupling constants and the symmetrized shifts of the $A_{1g}(a, Q_0)$, E_g (b, Q_1, Q_2), T_{2g} (c, Q_3, Q_4, Q_5) representations. In (3) the excited state is chosen to transform as the x-, y-, z-components of vector.

At first we shall examine the simpler case when the T_{2g} coupling is neglected. Then there are only two non-zero functions $A_{\alpha\beta\beta'\alpha'}$: $A_1 = A_{xxxx}$ and $A_3 = A_{xxyy}$. Their values can be found, for example, by means of the pair correlation approximation. For large $\tau, \tau' \sim \gamma^{-1} \gg \Gamma^{-1}$ which corresponds to ordinary luminescence,

$$
A_1(t) = \exp[i\Omega_0(\tau'-\tau)+g(t)+g(t+\tau-\tau')],
$$
$$
A_3(t) = \exp[i\Omega_0(\tau'-\tau)-3f_1+\bar{g}(t)+\bar{g}(t+\tau-\tau')].
$$
$$(4)$$

Here $g(t)=g_0(t)+g_1(t)$, $\bar{g}(t)=g_0(t)-\dfrac{1}{2}g_1(t)$, $\exp(i\Omega_0 t+g_i(\pm t))$ is the Fourier transform of the absorption (+) or luminescence (−) spectrum including only the $A_{1g}(i=0)$ or $E_g(i=1)$ coupling, $-f_i$ denotes the contribution of these vibrations to the logarithmic probability of the zero-phonon line, Ω_0 is the frequency of the zero-phonon line.

Inserting (4) into (1) we obtain the following formulae for the non-zero components of $w_{\alpha\beta\beta'\alpha'}(\omega_0,\Omega)$ which describe luminescence:

$$
w_1(\omega_0,\Omega) = w_{xxxx}(\omega_0,\Omega) = \kappa(\omega_0)F(\Omega),
$$
$$
w_3(\omega_0,\Omega) = w_{xxyy}(\omega_0,\Omega) = \bar{\kappa}(\omega_0)\bar{F}(\Omega)e^{-3f_1},
$$
$$(5)$$

where:

$$
\kappa(\omega_0) = (2\pi)^{-1}\int_{-\infty}^{\infty} dt\, \exp[i(\Omega_0-\omega_0)t+g(t)],
$$
$$
F(\Omega) = (2\pi)^{-1}\int_{-\infty}^{\infty} dt\, \exp[i(\Omega-\Omega_0)t+g(t)]
$$
$$(6)$$

are the absorption and unpolarized luminescence spectra normalized to unity. The $\bar{\kappa}(\omega_0)$ and $\bar{F}(\Omega)$ are determined by (6) if $g(t)$ is replaced by $\bar{g}(t)$.

It follows from the formulae obtained above that the E_g

coupling may cause an oscillation of $w_3(\omega_0, \Omega)$, and, therefore, an oscillation of the degree of polarization as a function of excitation and emission frequency. Indeed, $\bar{\kappa}(\omega_0)$ and $\bar{F}(\Omega)$ can be expressed as

$$\bar{\kappa}(\omega_0) = e^{-f_0 + f_1/2} \sum_{p=0}^{\infty} \sum_{p'=0}^{\infty} \frac{(-1/2)^p}{p! p'!} \int_{-\infty}^{\infty} d\omega \, \rho_1^{(p)}(\omega_0 - \omega - \Omega_0) \rho_0^{(p')}(\omega),$$

$$\bar{F}(\Omega) = e^{-f_0 + f_1/2} \sum_{p=0}^{\infty} \sum_{p'=0}^{\infty} \frac{(-1/2)^p}{p! p'!} \int_{-\infty}^{\infty} d\omega \, \rho_1^{(p)}(\Omega_0 - \omega - \Omega) \rho_0^{(p')}(\omega).$$

$$(7)$$

Here $\rho_i^{(p)}(\omega)$ is the pth order folding of the function

$$\rho_i(\omega) = \frac{1}{2\pi} \int_{-\infty}^{\infty} dt \, e^{i\omega t}(g_i(t) + f_i)$$

which determines the one-phonon spectrum of $A_{1g}(i=0)$ or $E_g(i=1)$ vibrations. From (7) it is evident that $\bar{\kappa}(\omega_0)$ and $\bar{F}(\Omega)$ may reverse their sign from even to odd E_g vibration replicas of the zero-phonon line. This will occur if $\rho_0(\omega)$ and $\rho_1(\omega)$ have sufficiently narrow maxima associated with local or pseudo-local modes. Thus, the absorption and luminescence spectra should involve some vibrational structure. We point out here that these oscillations can be observed better at low temperatures and for weak E_g coupling, if $w_3 \sim \exp(-3f_1)$ differs essentially from zero.

In case of strong coupling with **vibrations that are not totally** symmetric, luminescence photons will be emitted from practically independent states which belong to the equivalent deep minima of the adiabatic energy surface. The population of these minima, and, consequently, the emission polarization is a function of the exciting light polarization and frequency ω_0 (depolarization processes are dependent on ω_0). But the polarization of the emission does not depend on Ω since the emission spectrum is the same for all the minima.

Let us again examine the transition $A_{1g(u)} \rightarrow T_{1u(g)} \rightarrow A_{1g(u)}$ of a cubic center. Now we assume a strong E_g coupling, and also a T_{2g} coupling which induces depolarization during vibrational relaxation. For the sake of simplicity we assume that the T_{2g} coupling is weaker than the E_g coupling. To find the polarization characteristics as functions of ω_0, it is necessary to calculate the independent components $w_1 = w_{xxxx}$, $w_2 = w_{xyyx}$ and $w_3 = w_{xxyy} + w_{xyxy}$ of the tensor:

$$\bar{w}_{\alpha\beta\beta'\alpha'}(\omega_o) = \int_{-\infty}^{\infty} d\Omega w_{\alpha\beta\beta'\alpha'}(\omega_o,\Omega) =$$

$$= (\gamma/2\pi) \iint_{0}^{\infty} d\tau d\tau' \exp[i\omega_o(\tau'-\tau)-\gamma(\tau+\tau')/2]\bar{A}_{\alpha\beta\beta'\alpha'}(\tau,\tau'),$$

(8)

where:

$$\bar{A}_{\alpha\beta\beta'\alpha'}(\tau,\tau') = M_\alpha^+ <e^{-i\tau'H_0} e^{i\tau'H_1} M_\beta M_{\beta'}^+ e^{i\tau H_1} e^{i\tau H_0}>_o M_{\alpha'}.$$

(9)

Let us note that $\sum_\beta \bar{w}_{\alpha\beta\beta\alpha}(\omega_o) = \kappa(\omega_o)$. Therefore,

$$\bar{w}_1(\omega_o) = \kappa(\omega_o) - 2\bar{w}_2(\omega_o).$$

In the present case $w_3 \sim \exp(-3f_1) \approx 0$. Therefore, to solve our problem, it is sufficient to find $\kappa(\omega_o)$ and $\bar{w}_2(\omega_o)$. For that it is necessary to calculate $\bar{A}_2(\tau,\tau') = \bar{A}_{xyyx}(\tau,\tau')$ at $\tau,\tau' \sim \gamma^{-1}$.

In case of a weak T_{2g} coupling, the functions $\exp(i\tau'H_1)$ and $\exp(-i\tau H_1)$ in (9) can be expanded in a power series in c, using the Feynman formula (3). Within the accuracy $\sim c^2$:

$$\bar{A}_2(\tau,\tau') \simeq c^2 \int_0^\tau dS \int_0^{\tau'} dS' <Q_3 Q_3(S-S'+\tau'-\tau)>_o B(\tau'-\tau, S, S'),$$

(10)

where:

$$B(t,S,S') = <(e^{iS'H_1})_{11}(e^{i(t+S-S')H_1})_{22}(e^{-iSH_1})_{11}e^{-itH_0}>_o \simeq$$

$$\simeq \exp[i\Omega_o t + g(t) + \frac{3}{2}(2g_1(t+S-S') + g_1(S') + g_1(-S) - g_1(t+S) - g_1(t-S'))].$$

(11)

The variables S, S' in (10) can be interpreted as the time interval from the preparation of the $|x>$ component of the excited electronic state up to the transition into the $|y>$ component, which gives rise to depolarization.

At $\tau,\tau' \sim \gamma^{-1}$ the upper integration limit in (10) can be replaced by ∞ if the factor $\exp[-\gamma(S+S')/2]$ is simultaneously added. After the substitution of $\bar{A}_2(\tau,\tau')$ in (8), $\bar{w}_2(\omega_o)$ reduces to the overlap integral of the absorption spectrum $\kappa(\omega_o)$ and the spectrum $w(\omega_o,\omega)$ determined by (1), if Ω is replaced by ω, and $A_{\alpha\beta\beta'\alpha'}(t,\tau,\tau')$, by $B(t,\tau,\tau')\exp[i\Omega_o(\tau'-\tau)-g(t+\tau-\tau')]$. The part of this overlap integral which corresponds to large S, S' (it describes the depolarization after vibrational relaxation) is the same function of ω_o as $w_2^{(o)}\kappa(\omega_o)$. It differs from zero only at $T=0$ (see (1) where $w_2^{(o)}$ is found). The remaining part, which corresponds to: S,S' $\lesssim \Gamma^{-1}$

including depolarization during vibrational relaxation, is a different function of ω_0 than $\kappa(\omega_0)$. In particular, for strong E_g coupling and neglecting A_{1g} coupling ($a=0$),

$$\bar{w}_2^{(1)}(\omega_0) = \kappa(\omega_0) \frac{c^2 <Q_3^2>_0}{(\omega_0-\omega_{10}-\alpha_1)^2} \quad , \qquad |\omega_0-\omega_{10}|\sim\sigma_1 << \alpha_1, \tag{12}$$

where: $\kappa(\omega_0) = (\sigma_1\sqrt{2\pi})^{-1} \exp[-(\omega_0-\omega_{10})^2/2\sigma_1^2]$,

$$\alpha_1 = \frac{1}{2}\int_{-\infty}^{\infty} d\omega\rho_1(\omega)\omega, \qquad \sigma_1^2 = \int_{-\infty}^{\infty} d\omega\rho_1(\omega)\omega^2 << \alpha_1^2.$$

(Non-totally-symmetric vibrations cause an additional broadening of $\kappa(\omega_0)$ and $\bar{w}_2(\omega_0)$.) From (12) it follows that $\bar{w}_2(\omega_0)$ increases and, therefore, the degree of polarization $P = (W_1-W_2)/(W_1+W_2)$ decreases with increasing ω_0. Here, we have assumed that the electric vector of the exciting light is along the crystal axis (100), for the expressions of P in other cases see (1).

APPLICATION OF THE THEORY

The above results for cubic centers can be carried over directly to tetragonal centers (transition $A_1 \to F \to A_1$), if one of the two B couplings is weak. Let us mention that for the porphyrin solutions which have approximately tetragonal symmetry the oscillation of P as a function of excitation and emission frequency has been observed (4). The decrease of P with increasing ω_0 in the region of the excitation band has also been observed (see, e.g., the measurements of P for KCl:Tl$^+$ (5)).

If couplings with all types of vibrations are comparable, the treatment given above can not be used. However, for strong coupling the polarization characteristics may be calculated in the semiclassical approximation (6). It appears that near the center of the absorption band, P can be a non-monotonic function of ω_0.

The Jahn-Teller effect may lead to a decrease of the spin-orbit (Δ) or hyper-fine (λ) coupling constants in the relaxed excited state of the centers with non-zero spin of optical electrons or nuclei (7). This, evidently, explains the high spin memory of F-centers in alkali halides with optical excitation (6). The mentioned decrease of Δ and λ may allow one to observe quantum beats of the degree of polarization of luminescence as a function of time (7).

To find resonant Raman polarization characteristics,

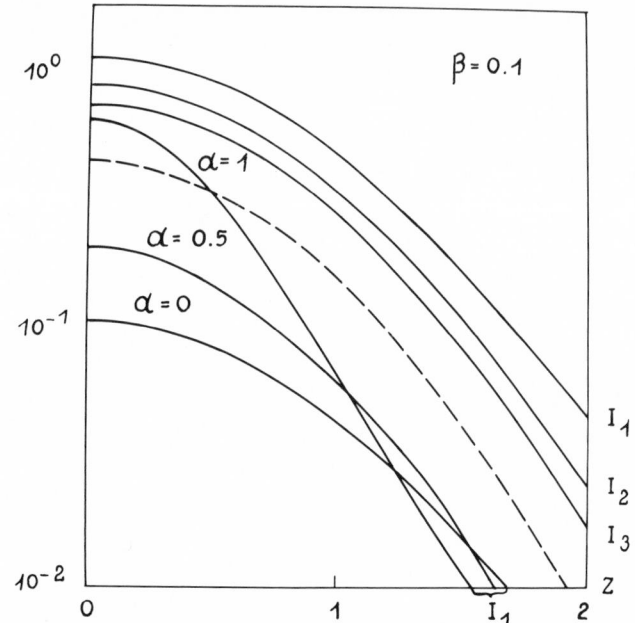

Fig. 1   Cubic center. Appreciable Stokes losses. The dependence
of the resonant Raman cross-sections of the dipole-allowed
electronic transition $A_{1g(u)} \leftrightarrow T_{1u(g)}$ on the excitation
frequency. I_K is the intensity of the k-th order scattering
by A_{1g} and E_g vibrations, I_1' is the intensity of the first
order scattering by T_{2g} vibrations. The dashed line corresponds to the absorption spectrum. Parameters α and β
determine the relative contributions of the E_g and T_{2g}
vibrations to the second (centralized) moment of the
absorption band (8).

$A_{\alpha\beta\beta'\alpha'}(t,\tau,\tau')$ should be calculated for small $\tau,\tau' \leq \Gamma^{-1}$. For
strong vibronic coupling, the following approximation will be
fairly good (8).

$$A^{(1)}_{\alpha\beta\beta'\alpha'}(t,\tau,\tau') = e^{i\omega_{10}(\tau'-\tau)} \; M_\alpha^+ <e^{i\tau'V} \, M_\beta M_{\beta'}^+ \, e^{-i\tau V(t)}>_o \; M_{\alpha'}, \quad (13)$$

$V(t) = \exp(itH_o)V\exp(-itH_o)$. This formula is relatively simple.
It allows one to calculate easily the resonant Raman spectra for a
number of cases in which we are interested (see (6,8)). Thus, it
appears that the Raman cross sections of different vibrations are
different functions of the excitation frequency ω_o. This is shown
in Fig. 1 where the Raman cross sections of A_{1g}, E_g and T_{2g} vibrations of cubic centers are given as functions of the dimensionless
parameter $Z = (\omega_o-\omega_{10})/\sigma\sqrt{2}$ (the A_{1g} and E_g Raman scattering makes
a contribution to w_1 and w_3, and the T_{2g} scattering of the first

order, to w_2 and w_3). We can see that the T_{2g} scattering cross section falls more quickly with ω_0 than the A_{1g} and E_g scattering. This is in agreement with the experimental data on the resonant Raman scattering of the F-center in NaBr (9).

REFERENCES

1. V. Hizhnyakov and I. Tehver, in "Physics of Impurity Centers in Crystals," Tallinn, p. 607 (1972).

2. I. Tehver and V. Hizhnyakov, Izv. Akad. nauk Est. SSR, Physics series Materials and Tech. Science 15,9 (1966); V. Hizhnyakov and I. Tehver, Phys. Stat. Sol. 21, 755 (1967); 39,67 (1970); V. Hizhnyakov, K. Rebane and I. Tehver, in "Light Scattering Spectra of Solids," G. Wright ed. (Springer-Verlag Inc., 1969) p. 513.

3. R. Feynman, Phys. Rev 84, 108 (1951).

4. A. N. Sevchenko et al., Doklady Akad. nauk SSSR 175,545 (1967); 179,61 (1968).

5. A. Fukuda, S. Makishima, T. Mabuchi,and R. Onaka, J. Phys. Chem. Solids 28,1763 (1967); S. G. Zazubovich, Optics and Spectroscopy, 26,235 (1969).

6. V. V. Hizhnyakov, Thesis, Tartu, (1972) (in Russian).

7. V. V. Hizhnyakov, Phys. Stat. Sol. (b) 51, K47 (1972).

8. V. Hizhnyakov and I. Tehver, in "Light Scattering in Solids," M. Balkanski ed. (Flammarion Sciences, Paris, 1971), p. 57.

9. D. B. Fitchen and C. J. Buchenauer, in "Physics of Impurity Centers in Crystals," Tallinn (1972), p. 277.

PHOTOLUMINESCENCE OF T-CENTERS IN POTASSIUM HALIDE CRYSTALS

V. Topa, B. Velicescu, and I. Mateescu

Institute of Physics

Bucharest, Romania

ABSTRACT

After electrolytical coloration of the Pb^{2+}, Sn^{2+} and Ge^{2+} doped alkali halide crystals (concentration $\sim 5 \times 10^{16}$ ions/cm^3) the characteristic Pb^{2+}, Sn^{2+} and Ge^{2+} bands in the absorption spectrum disappear and T-bands (associated with T-centers) appear. The T-centers are luminescent. This paper reports some characteristic properties of the excitation, emission and polarization spectra of the T-centers in potassium halide crystals.

INTRODUCTION

The electrolysis of alkali halide crystals containing cations of elements with strong electronegativity leads to a change in the valence of the cations (1). It has been shown (2) that the electrolysis of the alkali halide crystals doped with Pb^{2+}, Sn^{2+} and Ge^{2+} transforms the Pb^{2+}, Sn^{2+} and Ge^{2+} ions into new, very stable centers. The characteristic emission and absorption bands of Pb^{2+}, Sn^{2+} and Ge^{2+} ions disappear and many new bands (called T-bands and associated with T-centers) appear in the absorption spectrum (2,3).

In (2) we have tentatively assumed that the T-center is due to a substitutional Pb^- ion in an anion site. A strong argument in favor of this model is the fact that three F-centers are necessary to transform a Pb^{2+} ion. But the magnetic circular dichroic measurements (4) demonstrate that the T-center is not a paramagnetic one, as would follow from the Pb^- ion hypothesis, and lead to the conclusion that T-center is an aggregate center which includes Pb^- ions (3).

We have found that T-centers are luminescent, and the aim of this paper is to report their luminescent properties.

EXPERIMENTAL APPARATUS

The Pb^{2+} doped crystals were grown in air by the Kyropoulos method in alumina crucibles. For comparison, $KCl:Pb^{2+}$ crystals have been grown also by the Bridgman method under argon + CCl_4 atmosphere in sealed quartz tubes from refined salt. The concentration of Pb^{2+} was 10^{16} to 10^{17} ions/cm^3. The electrolytic coloration was performed in air at 500 C. The F-centers were removed from all the samples by changing the polarity at the electrodes immediately after coloration. By this method one obtained samples containing only T-centers.

The absorption measurements were carried out with a SP-700 Unicam spectrophotometer. The emission and excitation spectra have been measured with two Zeiss mirror monochromators using a Xenon 150 W lamp as a light source. The samples were mounted in a special metallic cryostat and the luminescence radiation was collected at 90° with respect to the irradiation direction by a PbS cell or a Ge photodiode (OAP 12-Radiotechnique). The incident light was modulated at 400 Hz for the PbS cell and at 12.5 Hz for the Ge photodiode. As amplifier for the emission light, we used a preamplifier type 203.50, a selective nanovoltmeter type 227 and a homodyne rectifier voltmeter type 204 from Unipan (Poland).

EMISSION, EXCITATION AND POLARIZATION SPECTRA

The Pb^{2+} doped KCl, KBr and KI crystals continaing T-centers, irradiated in the visible T-bands, exhibit a well shaped emission band at about 0.8 eV (Fig. 1). The Stokes shift of the T-center luminescence as compared with the absorption is higher than that of the F-center and comparable with the Stokes shift of mercury-like ions. The large Stokes shift demonstrates the production of lattice phonons. The emission peaks, like the absorption ones, shift to the long wavelength side as the atomic weight of the halide increases. By contrast, the half-width of the emission band becomes smaller when we pass from KCl to KI. In Fig. 2 is represented the dependence of the luminescence with the temperature. It may be seen that the luminescence peak shifts to longer wavelength as the temperature decrease.

In Fig. 1 absorption (1) and excitation (2) spectra for KCl:Pb, KBr:Pb and KI:Pb samples with T-centers at room temperature are given. The excitation spectra show that in the ultraviolet region, the irradiation in the T-bands does not lead to the luminescence at 0.8 eV,within the limit of the experimental errors. In contrast to KCl and KBr crystals, where the excitation spectrum follows more

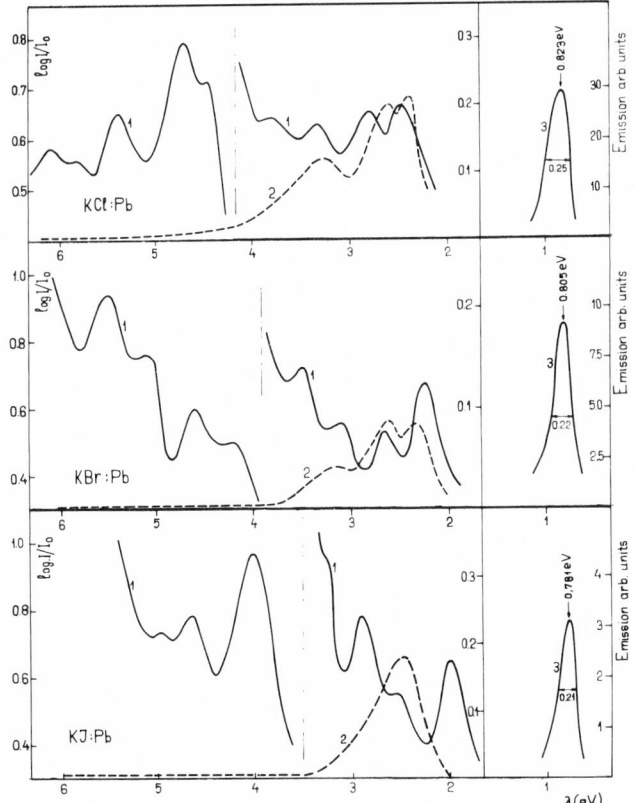

Fig. 1 Absorption spectra (1), excitation spectra (2) and emission
spectra (uncorrected) (3) of KCl:Pb, KBr:Pb and KI:Pb crystals with T-centers.

or less the absorption spectrum and shows three bands, the KI
crystals have a quite different excitation spectrum which consists
of a single band and which bears no similarity to the absorption
spectrum.

The polarization of the emission spectra has been measured by
the Feofilov method (5). The luminescence was excited by linearly
polarized light using linear polarizing filters of Polaroid Corporation (HN 32 for visible and MR for infrared). The exciting
light was introduced through one of the faces of a parallelepiped
cut from the crystal in such a way that it had a preselected orientation relative to the crystallographic axes, and the observation
was carried out in the direction perpendicular to the excitation
direction. In order to obtain the most characteristic dependence,
we have oriented the face through which the excitation was carried
out so that it was parallel to the 100 plane, and the face through

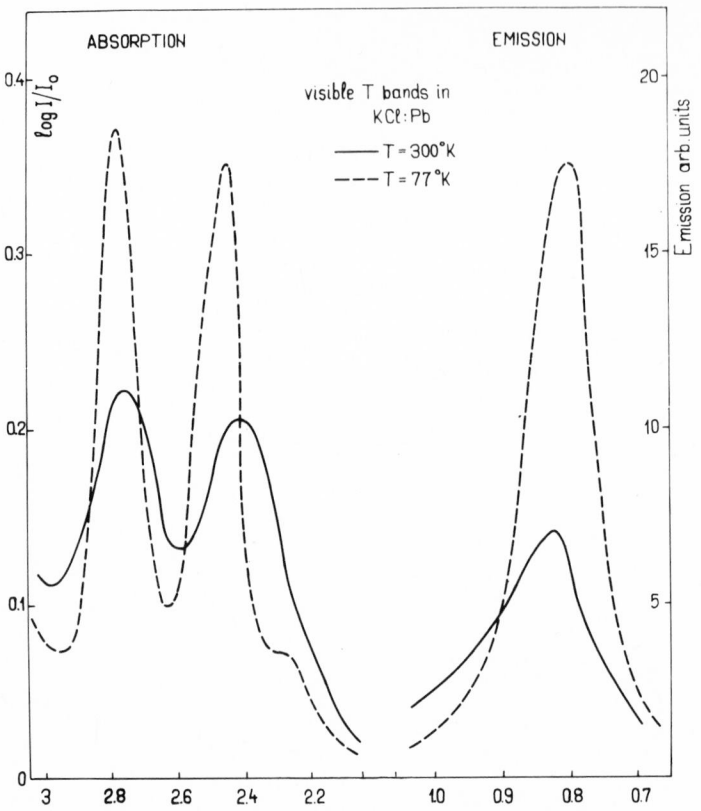

Fig. 2 The temperature dependence of the absorption and emission
        bands of the KCl:Pb crystal with T-centers.

which the luminescence was observed was chosen parallel to the 010
plane, in the case of oscillators along the fourfold axes, or parallel to the 011 plane, in the case of oscillators along the threefold
or twofold axes.

We have found that the luminescence of KCl:Pb crystals with
T-centers is not polarized within the limit of our error of $\pm$ 4%.
We have measured the polarization diagrams of the luminescence for
different temperatures (300 K, 77 K) and for different wavelengths
of the exciting light. In each case, the degree of polarization was
of the same order of magnitude as the experimental errors.

DISCUSSION

The excitation spectra prove that irradiation in the ultraviolet T-bands produces no detectable luminescence. Irradiation

in the visible T-bands show effects which vary from halide to halide.

The diagrams of degree of polarization of the luminescence for KCl as a function of wavelength and temperature prove that the luminescence is either unpolarized or polarized with a very small degree of polarization.

The emission band, the same for any wavelength of the excitation spectrum, is for all the crystals asymmetrical. The asymmetry is more pronounced at higher temperatures. The temperature dependence of the emission band is not the same as that of the absorption band.

The present experimental material is not sufficient for a general discussion about the luminescence of the T-center, and our results must be considered as preliminary. Further work is in progress.

REFERENCES

1.   V. Topa, M. Juste, Phys. Stat. Sol. (a) $\underline{3}$, 131 (1970).

2.   V. Topa, B. Velicescu, Phys. Stat. Sol. $\underline{33}$, K29 (1969).

3.   B. Velicescu, V. Topa, Intern. Conf. of Color Centers in Ionic Crystals, Reading,U.K. (1971), unpublished.

4.   M. Billardon, B. Briat, V. Topa, L. Taurel, Proc. of the XVI-th Col. Ampere, Bucharest, Romania, (1970).

5.   P. P. Feofilov, The Physical Basis of Polarized Emission, Consultant's Bureau, New York (1961).

PHOTOLUMINESCENCE OF COLOR CENTERS IN ALKALINE EARTH FLUORIDE CRYSTALS

R. Rauch

VEB Carl Zeiss, JENA, DDR (Germany)

ABSTRACT

The photoluminescence of CaF_2 and SrF_2 doped with monovalent alkali metals was investigated after subtractive coloration. The predominant centers induced by X-ray irradiation are M-centers. The perturbation of the centers by the alkali ion causes a change in absorption and emission energies but no influence on the orientation of the dipoles is observed.

Doped alkaline earth fluoride crystals are well-known to be crystal phosphors and have been studied extensively, whereas hardly any investigation has been made so far on the luminescence of color centers in these crystals. In particular, no F-center luminescence has been reported as yet. The luminescence of undoped alkaline earth fluorides after additive coloration, reported in earlier works (1), could not be related to F-centers (2). Therefore, we investigated the photoluminescence of crystals doped by inducing vacancies and stabilizing hole centers to enable the formation of electron centers at ambient temperature after irradiation by X-rays. That sodium doped CaF_2 and SrF_2 are easily colored by ionizing radiation has already been established previously by various authors (4,5). Prior to coloration, Na ions are associated with fluorine vacancies, as was found from ionic conductivity measurements by Bollmann (3). Little has been known, however, about the nature of the defects that cause absorption after coloration. Our studies of photoluminescence were intended to throw some light on them. CaF_2:Na has four typical emission bands: 755, 620, 554, and 505 nm; whereas only three corresponding bands were found for SrF_2:Na, viz. 890, 700, and 628 nm. Fig. 1 shows the emission, excitation, and

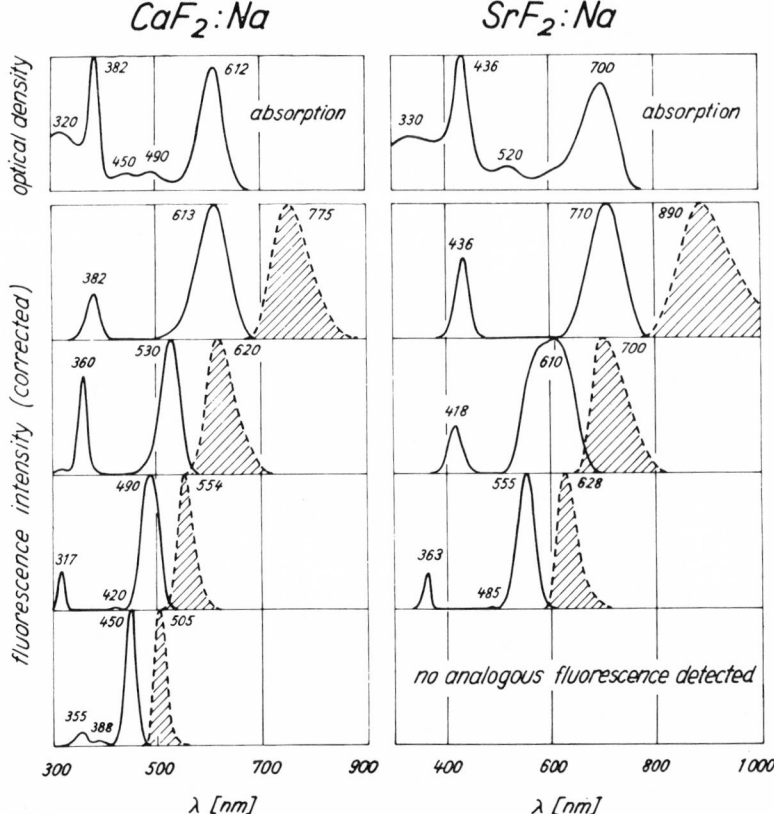

Fig. 1   Absorption (upper curve), excitation, and emission (shaded
         bands) spectra of Na-doped CaF$_2$ and SrF$_2$ after X-ray color-
         ation at RT and measuring at LNT. [Na]~0.1%; X-ray dose: 10^{3}r.

absorption spectra in comparison.

     Measuring the polarization of luminescence by the method of
Feofilov revealed high anisotropy for all centers. The axes for
all centers (with the exception of the 505 nm emission center) lie
along the cube edges, i.e. in <100>-directions. The excitation
polarization spectra show that in all cases the long wavelength
excitation band corresponds to an axial transition moment (positive
polarization) and the short wavelength band to a circular transition moment in the plane normal to the axis of the center (negative
polarization).

     Emission at 755 nm and its polarization behavior is in accor-
dance with a fluorescence observed by Bonch-Bruyevich et al (6) in
blue-colored natural CaF$_2$ crystals. Obviously, these crystals con-
tained sodium as an impurity. Earlier, Feofilov (7) had found

fluorescence at 690 nm in red-colored CaF_2 crystals, with excitation
bands at 560 and 375 nm and a polarization corresponding to centers
having four-fold axes. We found these spectra only in those crystals
that were contaminated by oxygen. Feofilov suggested a pure color
center model to explain the red fluorescence in CaF_2, viz., an
aggregate of two F-centers in the <100>-direction. In case of contamination by oxygen, however, it might just as well be a perturbed
F-center, i.e. an aggregate of an F-center and a O^- ion at a fluorine
site having a <100>-axis, that is responsible for the observed
polarization. Since Na^+ ions in the CaF_2 lattice are also associated
with vacancies, one should expect analogous aggregates there, but
with <111>-axes. Only recently, Alig (8) suggested a perturbed F-
center to account for the long wavelength main absorption band. We
could not detect any fluorescence compatible with that suggestion
and have to conclude that the majority of the centers formed in
crystals containing sodium and oxygen are quite similar to each
other. These centers can only be in accord with Feofilov's model
of F_2 centers, with the modification that they are perturbed. The
perturbation of the F_2 (or M) center by an impurity ion does not
cause any further splitting of the energy states but only a shift of
levels and a change of some properties. Considering only the absorption band corresponding to the axial transition moment and the emission excited in it, we obtain the following wavelengths for the F_2
centers in CaF_2: 521 nm, 586 nm (pure CaF_2, additively colored (9));
560 nm, 690 nm ($CaF_2:O^{2-}$); 612 nm, 755 nm ($CaF_2:Na^+$); and 625 nm,
815 nm ($CaF_2:Li^+$). The magnitude of the shift seems to be reasonable, as Li^+ at a Ca^{2+} site constitutes the greatest perturbation of
the crystal lattice because of the difference in ion radii and
valencies, whereas the perturbation caused by oxygen is relatively
small (which is evidenced also by the fact that oxygen ions are
easily incorporated). Charge perturbation by oxygen is small too,
since the coloration involved is likely to generate O^-. At temperatures below 30 K, we observed a vibronic structure in the 755 nm
fluorescence, with zero-phonon line at 6845 A. There is almost a
mirror symmetry in the absorption; the frequency of the coupling
phonon is 165 cm^{-1} for emission and 162 cm^{-1} for absorption. The
F_2 centers in alkali halides are not known to have any vibronic
structure. It should be possible now to check the F_2-model by uniaxial stress and Stark effect measurements.

The behavior of the 554 nm band in $CaF_2:Na$ and the 628 nm-band
in $SrF_2:Na$ are very interesting. The intensity of this emission decreases strongly with increasing temperature above liquid nitrogen
temperature and is quenched above 200 K, whereas the 755 nm and the
620 nm fluorescence in $CaF_2:Na$ are quenched above 400 K. The short
wavelength excitation band lies at considerably higher energy than
the F-band (which is at 375 nm in CaF_2 and 436 nm in SrF_2), Fig. 1.
An F-aggregate center can not be used to explain this emission.
Such centers actually could have axes in <100>-directions, but their

short wavelength transition would lie in the F-band region. Feofilov suggested (10) that for electron centers, a high degree of
polarization of luminescence is possible only when the center consists of an even number of electrons. For the reasons mentioned
above we believe that our experimental observations are best explained by an ionized F-aggregate center, in which two electrons
are distributed over three fluorine vacancies in <100>-direction
(F_3^+ center). Naturally, disturbance by the impurity ion is expected with this. Ionized F-aggregate centers have not been observed yet in alkaline earth fluorides. In Na^+-doped crystals,
more than elsewhere, there should be highly favorable conditions
for the formation of such centers and their stability should be
quite understandable. A complex of 3 fluorine vacancies occupied
by two electrons next to an Na^+ in a Ca^{2+}-site is neutral with
respect to charge.

Below 30 K, the spectra of the F_3^+ center suggested in CaF_2:Na
and SrF_2:Na show a marked vibronic structure, with the resonance
line at 5217 A (CaF_2:Na) and 5890 A (SrF_2:Na), respectively. The
phonon frequencies determined from the energy separation between
the strongest satellite lines are 141 cm^{-1} for CaF_2:Na and about
110 cm^{-1} for SrF_2:Na. Various phonon modes, however, contribute
to coupling.

The results obtained so far may be summed up by stating that
in alkali- or oxygen-doped alkaline earth fluorides, various kinds
of fluorescent color center aggregates, mainly perturbed F_2 centers,
are formed by X-ray coloration at room temperature, whereas isolated
F centers are either not formed at all or show no fluorescence; the
former being more likely than the latter.

In order to answer the question of F-center luminescence in
alkaline earth fluorides, we have studied pure and Na-doped crystals
colored by X-rays at low temperatures (30 K). It is established
that F or F_A centers are formed (5). We did not find any fluorescence. However, we did find an intense recombination luminescence
in the UV spectral region. In all crystals investigated, maximum
luminescence occurs at 280 to 290 nm (Fig. 2). The excitation
spectra of that luminescence coincide with the respective absorption spectra. The UV-emission was also found in pure crystals by
X-ray excitation (11), by vacuum-UV irradiation in the excitonic
absorption (12), as well as in thermoluminescence at low temperatures (13). This emission is likely to be due to the recombination
of electrons and V_k centers. Further investigations, especially
about the polarization behavior, are planned.

ACKNOWLEDGEMENTS

The author wished to thank Prof. P. Görlich for his promoting

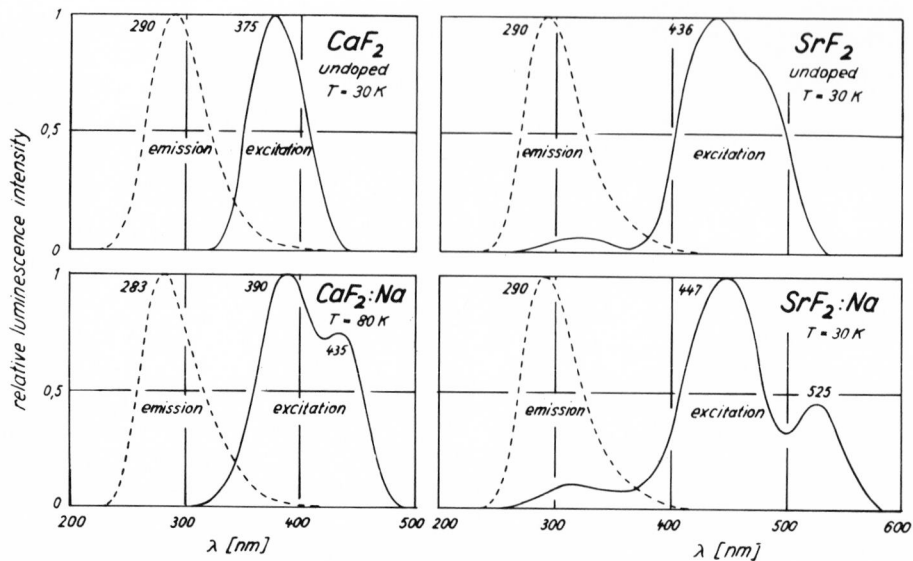

Fig. 2 Recombinative luminescence and excitation (stimulation) spectra of undoped and Na-doped CaF$_2$ and SrF$_2$ after X-raying at 30 K and 80 K, respectively. Measured at irradiation temperature. Typical X-ray dose: 10^5 r at 180 kV and 20 mA.

interest in this work as well as Dr. G. Kötitz for many discussions.

REFERENCES

1. P. Görlich, H. Karras, G. Kötitz, R. Rauch, Phys. Stat. Sol. 25, K15 (1968).

2. H. Karras, G. Kötitz, R. Rauch, Jenaer Jahrbuch 1969/70, p. 83.

3. W. Bollmann, H. Henniger, Phys. Stat. Sol. (a) 11, 367 (1972).

4. W. J. Scouler, A. Smakula, Phys. Rev. 120, 1154 (1960).

5. H. Karras, P. Ullmann, Jenaer Jahrbuch (1969/70), p. 11. P. Ullmann, Thesis, Jena (1968).

6. A. M. Bonch-Bruyevich, G. A. Tishchenko, P. P. Feofilov, Opt. i Spektrosk. 2 136 (1957).

7. P. P. Feofilov, Dokl. Akad. Nauk SSSR 92, 743 (1953).

8. R. C. Alig, Phys. Rev. B3, 536 (1971).

9. H. W. den Hartog, Thesis, Groningen 1969. J. H. Beaumont, W.

Hayes, Proc. Roy. Soc. <u>A309</u>, 41 (1969).

10. P. P. Feofilov, Zh. Exp. Teor. Fiz. <u>26</u>, 609 (1954).

11. W. A. Runciman, Proc. Intern. Conf. Lum., p. 688, Akadémiai
     Kiadó, Budapest (1968).

12. K. A. Kalder, A. F. Malysheva, Opt. i Spektrosk. <u>31</u>, 252 (1971).

13. G. Schwotzer, to be published.

LUMINESCENCE OF OXYGEN CENTERS IN ALUMINUM NITRIDE

J. Pastrňák, S. Pacesová, J. Schanda, J. Rosa

Inst. of Phys. of the Czechoslovak Academy of Sciences,

Prague, Czechoslovakia, and Inst. for Tech. Phys. of the

Hungarian Academy of Sciences, Budapest, Hungary

ABSTRACT

The electronic structure of the luminescent centers of non-activated AlN is studied by the method of emission and excitation spectra. Besides the UV-band with the maximum at 3.37 eV, "self-activated" AlN reveals a band at 2.33 eV in the visible, the position of which depends on the wavelength of the exciting radiation. The corresponding excitation spectra have rather complex structure with bands at (4.15 ± 0.05) eV and (4.50 ± 0.005) eV at T = 300 K. The observed excitation bands are in good agreement with the absorption spectra, which are considered to be caused by oxygen.

INTRODUCTION

The present interest in large band gap III-V semiconducting compounds arises mainly from their potential usefulness for luminescent devices. In this family, nitrides in comparison with phosphides are characterized by a direct band gap and consequently by higher possible quantum efficiency (1,2).

Aluminum nitride has the largest band gap of these tetra-hedrally coordinated compounds with the possibility of obtaining by appropriate doping, luminescence emission both in the u.v. and visible spectral regions. According to the existing data, non-activated, as prepared, AlN samples exhibit blue-violet luminescence, sometimes called "self-activated" (3,4). The measurements of

optical absorption in single crystals as well as diffuse reflectance
for different concentrations of oxygen impurities revealed optical
transitions dependent on the concentration of oxygen (5). The
"self-activated" luminescence has been, therefore, ascribed to
oxygen impurities, which, due to the high affinity of Al for oxygen,
are unavoidably present in all AlN samples prepared so far.

As the luminescence measurements on AlN:Mn centers brought out,
the oxygen centers in this case play the role of recombination
centers (4); thermoluminescence measurements revealed the role of
oxygen in the formation of trapping levels (6). Obviously the
investigation of oxygen centers in AlN is of vital importance for
understanding the luminescence of this compound.

Analogy with oxygen centers in GaP together with crystallo-
chemical arguments suggest that oxygen is incorporated in the AlN
lattice substitutionally on N sites, the charge excess is compensated by the simultaneous creation of Al vacancies (7). Statistical
arguments show that, as in the case of GaP, a great variety of
oxygen centers can be proposed (8-11).

RESULTS AND DISCUSSION

What are the firmly established experimental facts about oxygen
centers in AlN ? In Fig. 1 the excitation and emission spectra of
AlN powder samples of grain size 2-5 μm for two different oxygen
concentrations are given. The emission as well as the excitation
spectra are composed of several overlapping bands which indicate
the simultaneous presence of many different centers. In contrast
to GaN (12,13), no emission in the vicinity of the absorption edge
due to the band-to-band or free- or bound-exciton recombination has
been observed, even at liquid helium temperature and under electron
beam excitation. The reasons are the quenching effect of the grain
boundaries as well as the effect of high impurity concentration.
The position of the luminescence maximum of the non-activated AlN
varies from 3.1 to 3.3 eV (3,4). The explanation is obvious from
Fig. 1. The position depends on the oxygen concentration and on
the energy of the incident photons. As the oxygen concentration
increases from $6.7 \times 10^{20}/cm^3$ to $5.4 \times 10^{21}/cm^3$, the maximum shifts
from 3.22 eV to 2.98 eV at room temperature. With the change in
excitation energy from 254 to 313 nm mercury lines, the peak positions shift by 0.08 eV toward lower energies. The emission spectrum
under electron beam excitation resembles closely that under 254 nm
mercury line excitation; consequently the mechanism of excitation
should be in both cases similar.

The decay time differs for different parts of the emission
band. For emission in the region of 3.1 eV the decay time is of

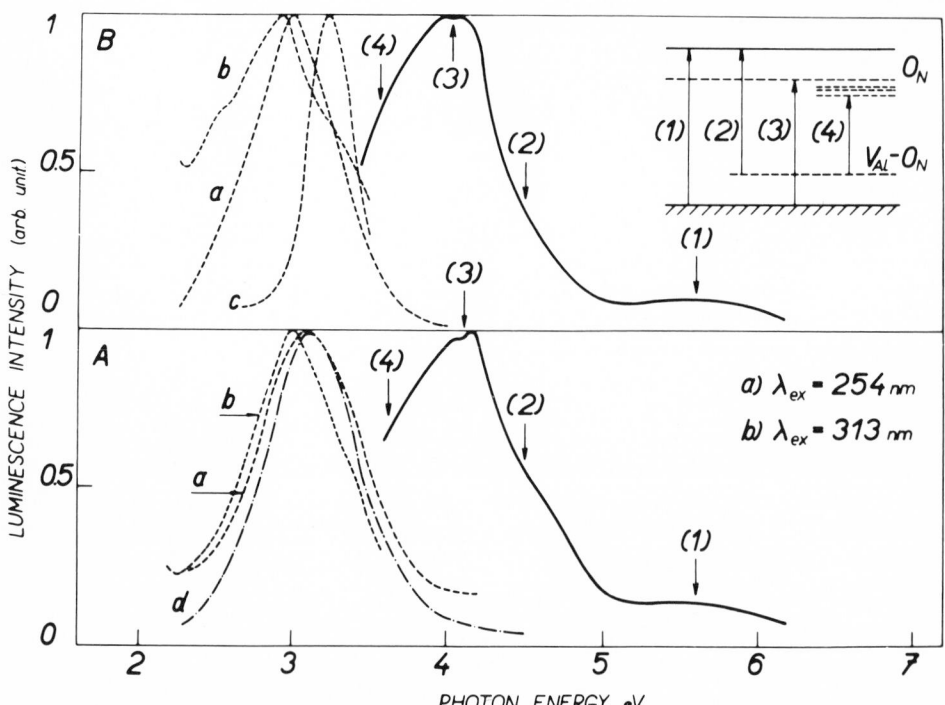

Fig. 1 Normalized excitation (——) and emission (---) spectra of
        AlN samples with $6.8 \times 10^{20}/cm^3$ (A) and $2.05 \times 10^{21}/cm^3$ (B)
        oxygen concentrations. Excitation wavelength (a) $\lambda = 254$
        nm, $\lambda = 313$ nm (b) - steady state. (c) - component of
        emission with decay time less than 5 msec. (d) - steady
        state condition for $\lambda = 254$ nm, T = 77 K.

the order of milliseconds, for 2.5 eV it is about one order
higher and depends again on the energy and intensity of excitation.
Above that, intense afterglow has been observed. The result is
that different values for the energy of the maximum of emission
band can be obtained depending on the relative intensity of sub-
bands.

     An analogous reasoning applies also to the excitation spectra.
From Fig. 1 the dependence on oxygen concentration is again obvious;
the difference in the energies of maxima is 0.05 eV for the given
concentrations. The change in the excitation mechanism, i.e. the
change in the photon energy of excitation or emission light, or
the measurement of time resolved spectra gives the possibility of
resolving individual bands. As an example, on Fig. 2 the excitation

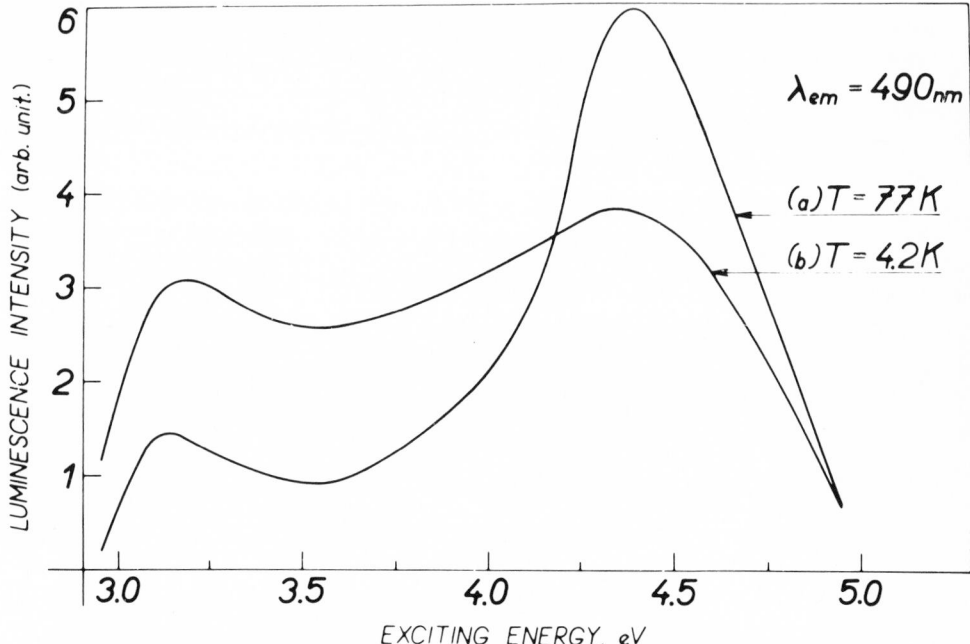

Fig. 2   Excitation spectra of AlN sample with $1.7 \times 10^{21}/cm^3$ oxygen
         concentration.

spectra are given for the emission light at 2.54 eV (490 nm) and
for two different temperatures, measured at 27 Hz, the frequency
of chopped light. We can see that the emission is excited either
in the band at 4.48 eV (277 nm), where free electrons are created,
or in the band at 3.1 eV (400 nm), i.e. directly in the center.
With decreasing temperature the excitation mechanism in the band
at 4.48 eV is less effective. The measured half-widths of the
excitation bands are 0.32 eV and 0.12 eV for T = 77 K in comparison
with the total half-width of the emission band 0.85 eV; the real
half-widths will be still less if we take into account the splitting
of the bands due to crystal anisotropy.

     The positions of the identified bands in the excitation spectra
(Fig. 1) are noted by arrows; the splitting of individual bands due
to the crystal field was neglected, as the maximum value of the
splitting is only 0.12 eV (5).

     The small peak at 5.7 eV corresponds to the onset of the
direct band-to-band transitions (5). This maximum is more distinct
in samples with larger grain sizes and with lower oxygen content.
The same maximum can be seen on the spectral photoconductivity
curve (5).

The band at 4.5 eV can be again seen both in the absorption
spectra and on the photoconductivity curve (5). In this band the
short-lived components are predominantly excited. The presumed
bound-to-free transitions correspond to the photo-neutralization
(full or partial) of the ionized acceptors, supposedly $V_{Al}-O_N$ pairs.
The corresponding emission band lies at 3.25 ± 0.05 eV.

The excitation band at 4.15 ± 0.05 eV and bands with lower
energy, which are not divided into individual sub-bands, are
ascribed to the bound-to-bound (or free-to-bound) transitions in
oxygen centers; they are connected with slower decay in the recombination band at (3.1 ± 0.05) eV (395-405 nm). These transitions
do not create free carriers and are not seen on the photoconductivity curve.

A tentative model of the corresponding optical transitions is
presented in Fig. 1: deep donors are ascribed to O_N centers and
deep acceptors to $V_{Al}-O_N$ pairs. Statistical arguments show that
the concentrations of these centers should be the highest, and
under equilibrium conditions they will be mostly ionized. Other
more complicated complexes may form shallow acceptors and donors;
their presence in samples is proved by thermoluminescence measurements (6). Any attempt to give, at this stage of investigation,
some detailed model would be premature. The Stokes shift for the
above-mentioned bands is in the range 0.7 to 1 eV. This fact as
well as comparatively large half-widths testify for the strong
coupling of the electron states of oxygen with the AlN lattice as
in the case of GaP (11),(14).

Summarizing, we can say that non-activated, as prepared AlN
reveals complex luminescence and excitation spectra, which consist
of several overlapping bands due to different oxygen centers.
Bound-to-free transitions mostly contribute to the center part of
the excitation spectra. As the dependence of the luminescence band
on the exciting energy manifests, the lower energy shoulder of the
excitation spectra can be ascribed to bound-to-bound transitions
in oxygen centers.

REFERENCES

1. B. Hejda, Phys. Stat. Sol. 32 407 (1969). B. Hejda and K.
    Hauptmanová, Phys. Stat. Sol. 36 K95 (1969).

2. S. Bloom, J. Phys. Chem. Sol. 32 2027 (1971).

3. V. A. Krasnoperov, I. A. Mironov and G. A. Chomenok, Uzv. AN
    SSSR, ser. fiz. 30 1430 (1966).

4.   F. Karel and J. Pastrnak, Czech. J. Phys. B20 46 (1970).

5.   J. Pastrnak and L. Roskovcova, Proc. Internal. Conf. on the
     Physics of Semicond., Moscow 1968, Nauka, Leningrad 1968,
     J. Pastrnak and L. Roskovcova, Phys. Stat. Sol. 26 591 (1968).

6.   F. Karel, J. Pastrnak, and J. Rosa, Czech. J. Phys. B19 974
     (1969).

7.   G. A. Slack, Preprint (1972).

8.   P. J. Dean, Solid State Commun. 9 2211 (1971).

9.   R. N. Bhargava, S. K. Kurtz, A. T. Vink, and R. C. Peters,
     Phys. Rev. Lett. 27 183 (1971).

10. R. Olson, J. Phys. Chem. Solids 33 549 (1972).

11. J. M. Dishman and M. DiDomenico, Jr., Phys. Rev. 168 812 (1968).

12. H. G. Grimmeiss and B. Monemar, J. Appl. Phys. 41 4054 (1970).

13. R. Dingle and M. Ilegems, Solid State Commun. 9 175 (1971).

14. J. M. Dishman, Phys. Rev. B5, 2258 (1972).

INFLUENCE OF THE CHARGE TRANSFER ON Eu^{3+} LUMINESCENCE SPECTRA

L. S. Gaigerova, O. F. Dudnik, V. F. Zolin,

V. A. Kudryashova

Institute of Radio Engineering and Electronics

Moscow, USSR

ABSTRACT

Spectra of Eu^{3+} have been studied in some alkaline earth oxides and in oxides of cerium, thorium, hafnium and zirconium. The spectra are unusual in the high intensity of the $^5D_0 - {}^7F_0$ transitions, and the vibronic satellites of $^5D_0 - {}^7F_1$. Peculiarities of the spectra of these luminescence centers have been accounted for by the asymmetrical environment of the Eu^{3+} ions, where the asymmetry is due to different polarizabilities of the ligands.

The rare earth to be used for the demonstration of the influence of the overlap of the wave functions of the 4f-electrons with the wave functions of the ligands on the spectra of the rare earth ions must have the largest electron affinity. The latter applies to Eu^{3+} ions. The contribution of the overlap changes with the polarizability of the coordinated ligands. Oxygen and sulphur ions have noticeable polarizability. The maximum overlap can be achieved when a metal ion is in an asymmetric environment. We can illustrate this statement by the fact that the dependence of the physical properties of the alkali halides on polarizability is most prominent for liquid or gaseous states, when the symmetrical arrangement of the ions that exists in the crystalline state is broken (1). All this shows that the trigonal centers of luminescence in the europium doped alkaline earth fluorides with the oxygen or sulphur charge compensation (2)

514

Table 1

Frequencies of the $^5D_0 - ^7F_0$ (0-0) transition in the spectra of Eu^{3+} in some asymmetric luminescence centers (cm^{-1}).

$CdF_2:O^{2-}$ 17325	$CaF_2:O^{2-}$ 17437	$SrF_2:O^{2-}$ 17455	$BaF_2:O^{2-}$ 17460
$CdF_2:S^{2-}$ 17490	CeO_2 17437	ThO_2 17430	ZrO_2 17410
MgO 17460	CaO 17410	SrO 17420	aqueous ion 17274

could be used for the demonstration of the influence of the overlap on the rare earth ion spectra. One should remember that because of the site symmetry in this case overlap is attained when the charge transfer bands (CTB) are seen in the spectra of the luminescence excitation. We found luminescence centers with the same kind of overlap in the oxides of Ce, Th, Zr and of alkaline earths doped with europium. Frequencies of the $^5D_0 - ^7F_0$ (0-0) transition in the spectra of luminescence of these centers are shown in table 1. A common feature of all these spectra is the high intensity and high frequency shift of the 0-0 transition and the A_2-component of the $^5D_0 - ^7F_1$ transition (in comparison with the same transitions of the aqueous ion). The most prominent feature of the vibronic spectra of these centers is the high intensity of the vibronics due to the Eu - O vibrations. The contribution of the vibrations is largest in the range of 440 and 520 cm^{-1}. Vibronic intensity distribution can be used as evidence for the displacement of the equilibrium positions of the ions for the A_1 and A_2 electronic states in comparison to the E states of the trigonal symmetry.

When the overlap is not considered, mixing of states by the crystalline field cannot account for the high intensity of the 0-0 transition up to the second order of the perturbation theory (3). There are eight atoms in the environment of europium ion forming a distorted cube with a trigonal symmetry. Seven of these are lattice anions, the eighth is a doubly negative oxygen or sulphur ion. It is clear, that it is the 2s and $2p_z$ orbitals of oxygen or the 3s and $3p_z$ orbitals of sulphur that overlap with the europium ion. (We assume that the europium and oxygen or sulphur ions are on the z axis of the luminescence center). The orbitals of the central ion, having the same symmetry as the ligand orbitals, must show the largest overlap. The maximum overlap integrals will be for the wave functions of the states with the angular momenta J = 0,1 and A_1, A_2 types of symmetry. This is so because the angular dependence of the s and p_z-functions is

just the same as the angular dependence of the wave functions of
the kind indicated above. One can see here an analogy to the case
of the complexes of transition elements, e.g., iron (4). The
overlap entails the presence of the CT bands with the intensity and
the shifts of energy proportional to the square of the overlap
integral (5). CT transitions cause some change of the equilibrium
distances between ions in the lattice. This leads to admixing of
the excited CT wave function to the ground state wave functions
proportional to the equilibrium position shift (an analogue of the
Frank-Condon admixture). The admixture manifests itself as an
intensity borrowing from the CT band for the rare earth ions whose
wave functions give appreciable overlap integrals with the wave
functions of the ligand. Equilibrium position shifts with intensity
borrowing from the CT band is a plausible explanation for the
intensity increase of the 0-0 transition and increase of vibronics
in the Eu^{3+} spectra. It is more difficult to find out the cause of
the high frequency shift of the 0-0 transition. Many factors can
change the position of the energy levels (6). The shift depends
on the interaction of the rare earth ions with the environment.
With the growth of the interaction and increase of the bond
covalency the transitions within the 4f-configuration of the rare
earth undergo some low frequency shifts (7). Excitation energy
of electrons in the rare earth ion is believed to decrease as
ligand electrons enter free 5d-orbitals of the rare earth ion and
4f-electrons delocalize into the ligand orbitals. When there is
CT, one must take into account the bonding and antibonding effects.
The bonding must decrease the energy of the CT ground state, in
our case $^5D_0$, $^7F_0$ states and A_2 component of the $^7F_1$ state. The
$^8S_{7/2}$ state, corresponding to the ground state of the $4f^7$ electron
configuration, (existing when the charge is transferred to the
europium) originates only from the ^7F_J states. Thus those states
must shift downwards more than the ^5D states. This leads to a
high frequency shift of the corresponding transitions.

    Intensity and high frequency shift dependence of the 0-0
transition on polarization potential of the lattice cations can
be used as evidence of the role of CT in the luminescence spectra
of Eu^{3+} in alkaline earth fluorides. Along with the increase
of polarization potential from Ba to Sr, Ca and Cd, a decrease
both in the overlap of oxygen and europium ion wave functions and
in the intensity and a high frequency shift of the 0-0 transition
are observed. Since the overlap depends not only on the polarizability of the ions and the polarization potential but also on
the distance between the ions, the increase of the cation ionic
radius from Sr to Ba partially compensates for the change of
polarization potential. Perhaps the same reason explains the slight
spectral differences between the centers of luminescence in Ca and
Sr oxides. The 0-0 transition shift in oxides of Ce, Th and Zr
increases with increase of the lattice cation polarization

potential. Apparently this is caused by the complex structure of
the luminescence centers. Each of these centers consists of several
Eu^{3+} ions. Oxygen ions surrounded only by Eu^{3+} ions must give rise
to the largest overlap integrals in comparison to the oxygen ions
surrounded by the Eu^{3+} ions and lattice cations. The fact that the
shift of the O-O transition is connected with the difference in the
ligand polarizabilities is well illustrated by the example of the
oxiasomethine-Eu chelates (for instance with NN'bissalicylaldehyde
ethylenediamine (salen) and its sulpho-derivatives) and by the
examples of the complex β-diketonates of Eu with aromatic amines.
Introduction of the acid sulpho groups into the salen leads to a
decrease of the polarizability of most of the oxygen atoms
coordinated with Eu^{3+}. As a result the frequency of the O-O transition shifts from 17215 cm^{-1} to 17331 cm^{-1} and its intensity
increases. In the above mentioned complex, β-diketonates, two of
the eight atoms coordinated with the Eu^{3+} ion are nitrogen atoms
and all the rest are oxygen atoms. An introduction of the aromatic
amines into β-diketonates is accompanied by an intensity increase
and a high frequency shift of the O-O transition.

Thus the facts considered above demonstrate the influence of
the ligand-rare earth ion overlap, accompanied by CT, on the
luminescence spectra of rare earth ions.

REFERENCES

1. F. Cotton, G. Wilkinson. Advanced Inorganic Chemistry,
    Wiley, New York (1972).

2. P. P. Feofilov. Polarized Luminescence of Atoms, Molecules
    and Crystals. Phys. Math. State Publ. Co., Moscow, (1959).

3. W. C. Nieuwport, G. Blasse, A. Bril in Optical Properties of
    Ions in Crystals, H. M. Crosswhite, H. W. Moos, eds.,
    Interscience, New York, (1967).

4. G. Lehmann. Z. Phys. Chem., Neue Folge, Bd 72, 279 (1970).

5. G. Herzberg. Molecular Spectra and Molecular Structure,
    p. 434 Van Nostrand, New York, (1969).

6. M. A. Eliashevitch. Rare Earth Spectra. Tech. Theor. Lit.
    State Publ. Co., Moscow (1953)

7. C. K. Jorgensen, R. Pappalardo, H. H. Schmidtke. J. Chem.
    Phys. 39, 1422 (1963).

LUMINESCENCE OF ISOELECTRONIC Mn^{4+} AND Cr^{3+} IONS IN MAGNESIUM ORTHOTITANATE

R. Dittmann, D. Hahn and J. Stade

Physikalisch-Technische Bundesanstalt

Braunschweig and Berlin, Germany

ABSTRACT

Powder samples of $Mg_2TiO_4:Cr^{3+}$ and $Mg_2TiO_4:Mn^{4+}$ with activator concentrations from 0.01 to 1.5% have been prepared by annealing of MgO/TiO_2-mixtures. Both phosphors behave similarly and show very sharp emission lines. The most substantial of these lines is not generated by single ions but by so-called N-centers. The emission spectra vary with annealing temperature as is found both in powder and single crystal samples, thus showing that the N-centers might be activator ions associated with lattice defects.

INTRODUCTION

Mg_2TiO_4 phosphors activated by the isoelectronic $3d^3$ ions, Mn^{4+} and Cr^{3+}, show a strong similarity in their luminescence behavior. In this paper emission and excitation spectra are discussed which lead to an energy level diagram of both phosphors. The samples (powders and single crystals) were prepared with different activator concentrations and at various temperatures. As the characteristic effects are more distinct in $Mg_2TiO_4:Cr^{3+}$ they will be discussed mainly for this phosphor and pointed out only for some similar properties of $Mg_2TiO_4:Mn^{4+}$.

EXPERIMENTAL RESULTS ON $Mg_2TiO_4:Cr^{3+}$

Figure 1 shows the complete emission spectra of both samples at 77 K by suitable long wavelength excitation. In the spectrum

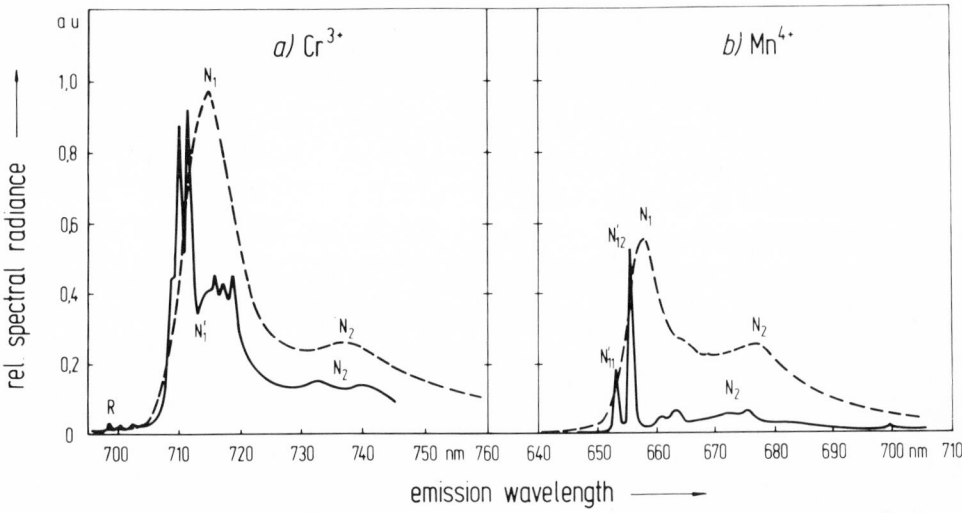

Fig. 1a,b: Emission spectra of Mg$_2$TiO$_4$:Cr^{3+} (a) and Mg$_2$TiO$_4$:Mn^{4+}
(b); dotted curves: samples not annealed.

of Mg$_2$TiO$_4$:Cr^{3+} two strong sharp lines can be seen followed by
some weak lines; these are mainly phonon satellites of the strong
lines at 710.1 and 711.6 nm. This group of lines is called N'-lines.
Variation of activator concentration and excitation wavelength only
changes slightly the radiance of some lines; so the weak lines at
698.4 nm and 699.9 nm (so-called R-lines) are lacking, when excited
in the region of 300 to 400 nm. From Fig. 2 showing the excitation
spectra it is recognizable that in this region the excitation
maximum A is situated at 330 nm, at the long wavelength side of the
band gap at 310 nm. On the contrary the R-lines always appear when
excited by radiation in the maxima B (426 nm) and C (610 nm) in
Fig. 2. Calculations of Cr^{3+} energy levels in a crystal field of
octahedral symmetry with the aid of Tanabe-Sugano matrices lead to
the result that the broad excitation maximum B (426 to 466 nm)
represents the transitions $^4A_{2g} \rightarrow ^4T_{1g}$ and $^4A_{2g} \rightarrow ^2T_{2g}$, and the excitation maximum C (641 nm) is correlated to the transition $^4A_{2g} \rightarrow ^4T_{2g}$.
The energy levels of both phosphors according to these calculations
are shown in Fig. 3. These calculations and the appearance of the
R-lines only when excited by radiation from the maxima B and C,
prove that the R-lines are a characteristic emission of the Cr^{3+}
ion correlated to the transition $^2E \rightarrow ^4A_{2g}$. On the other hand, the
excitation maximum A cannot be connected to any transition in the
Cr^{3+} ion. Furthermore, its increase with activator concentration
differs completely from the increase of the maxima B and C which
belong unequivocally to the Cr^{3+} ion. This can be seen very clearly
by comparing the ratios of the excitation maxima ordinates B/A and
C/A which double if the chromium content is increased by a factor 10,

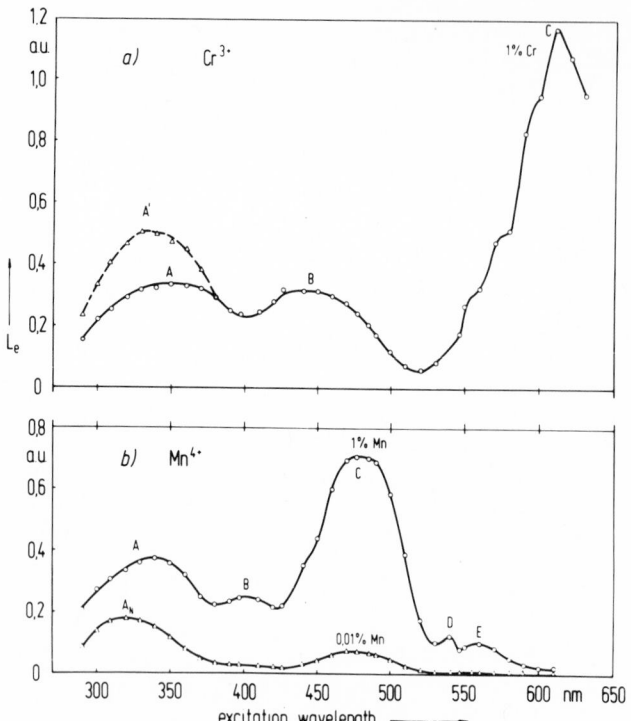

Fig. 2a,b: Excitation spectra of (a): $Mg_2TiO_4:Cr^{3+}$ (1% chromium
content) and (b): $Mg_2TiO_4:Mn^{4+}$ (1% and 0.1% manganese
content).

while the maximum ordinate A increases negligibly. This different
behavior indicates the existence of two different luminescence
centers in $Mg_2TiO_4:Cr^{3+}$. The N'-lines in the emission and the
excitation maximum A belong to one of them, generally called N-
centers, the nature of which will be discussed later on; the R-
lines with excitation maxima B and C to the other ones, called R-
centers and represented by single Cr^{3+} ions as shown by the result
of the calculations mentioned above (compare Fig. 3).

DISCUSSION

The emission of the R-lines is only a minor part of the
integral luminescence radiance (Fig. 1) thus showing that the N'-
lines represent the main part of the emission. The excitation
spectrum therefore gives at the same time the excitation of the
N'-line radiance and so the N'-line excitation has maxima in the
region of the excitation maxima B and C, too. As proved by the
calculations here, transitions in the Cr^{3+} ion ($^4A_{2g} \rightarrow {}^4T_{1g}$ and

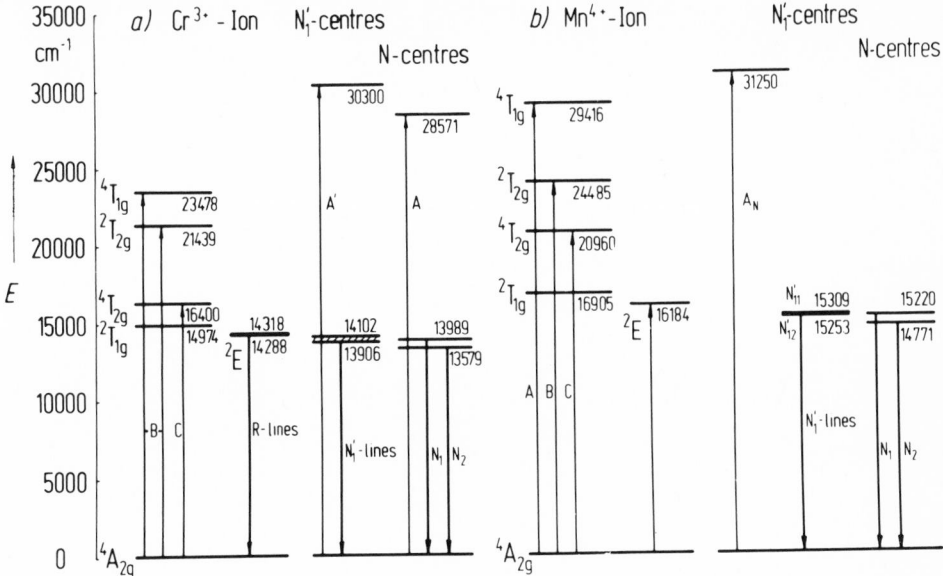

Fig. 3a,b: Energy levels of Mg$_2$TiO$_4$:Cr (a) and Mg$_2$TiO$_4$:Mn^{4+} (b);
arrows: observed transitions in absorption or emission.

$^4A_{2g} \rightarrow {}^4T_{2g}$) take place and this fact gives evidence for an energy
transfer from Cr^{3+} ions to the N-centers. This transfer starts
non-radiatively from the level ^{2}E of the Cr^{3+} ion (Fig. 3) and ends
involving several other Cr^{3+} ions at a N-center. The ^{2}E-level
therefore is the starting point for the energy transfer, as is also
well-known from the work on the luminescence of ruby. Because the
transition $^2E \rightarrow {}^4A_{2g}$ is mainly forbidden by spin and parity selection
rules, the R-lines are very weak and there must exist conditions
for the removal of these selection rules which may be due to lattice
defects, distortions, etc. Furthermore, later on it will be shown
that the analogous behavior of Mn^{4+} ion in magnesium titanate gives
evidence for the ^{2}E-level being the starting point of the transfer;
in this case the ^{2}E-level only could be reached by suitable excitation, no R-lines in emission were observed, but the energy transfer
to N-centers was proved by the appearance of N'-lines. Radiationless energy transfer involving several Cr^{3+} ion chains to the N-
centers can be accepted by taking into account the energy difference
of the R- and N'-lines as well as the doubling of the excitation
maxima rate B/A and C/A with tenfold activator content. This means
that the number of Cr^{3+} ions in one lattice direction is doubled
($\sqrt[3]{10} \simeq 2.1$) thus doubling the number of ion paths too. As the
excited Cr^{3+} ions only rarely return to the ground state with
emission of photons because of the selection rules mentioned above,
doubling of the Cr^{3+} ion paths gives rise to doubling of the maxima

rates B/A and C/A as observed. As a result it can be stated that
the main luminescence of $Mg_2TiO_4:Cr^{3+}$ phosphors does not come from
regular Cr^{3+} ions but from N-centers, which are excited directly
(maximum A) or by energy transfer from Cr^{3+} ions (maxima B and C).

The question still remains of the nature of the N-centers.
If Cr^{3+} pairs are excluded; the radiance of the N'-lines should
increase, if due to pairs, with the square of the chromium content,
which is not found experimentally. It remains likely that the N-
centers are associates or clusters of Cr^{3+} ions with lattice defects.
This assumption can be supported by special preparations, especially
annealing of the phosphors. Generally all phosphors were prepared
at 1460 C and subsequently annealed for 72 hours at 560 C. Without
being annealed the samples showed an emission spectrum shown in
Fig. 1a (dotted curve). The R-lines remain unchanged while, instead
of the sharp N'-lines, new broad bands at 714.8 and 736.4 nm appear,
called N-lines; these broad bands are referred to in the previous
literature. The structure of the N-lines does not change with
cooling to 4.2 K thus showing that it depends upon the structure of
the unannealed Mg_2TiO_4 itself. It can be assumed that the lattice
contains a very great number of different lattice defects distorting
the oxygen octahedra in which the Cr^{3+} ions are located, so that
deviations from symmetry result in a removal of selection rules and
in the emission of broad bands. These lattice distortions are
healed by annealing, and diffusion of chromium ions and lattice
defects becomes possible, thus forming N'-centers. These N'-centers
thus consist of some special types of lattice defects connected
with chromium ions showing the emission of sharp N'-lines. The
annealing only involves the N- and N'-centers, which can be evidenced
by the conservation of the R-lines and their excitation maxima B
and C. Only the excitation maximum A of the annealed phosphor
shifts from 350 nm (without annealing) to 330 nm (A'), thus showing
that the N'-centers are different from the N-centers (Fig. 2) -
although both are often referred to as N-centers.

PROPERTIES OF $Mg_2TiO_4:Mn^{4+}$

As mentioned above, the spectra of $Mg_2TiO_4:Mn^{4+}$ can be interpreted in quite an analogous manner. The excitation spectrum (Fig.
2) reveals three maxima at A = 340 nm, B = 410 nm and C = 477 nm
which are correlated, according to the calculations, to the transi-
tions $^4A_{2g} \rightarrow ^4T_{1g}$ (A), $^4A_{2g} \rightarrow ^2T_{2g}$ (B) and $^4A_{2g} \rightarrow ^4T_{2g}$ (C) in the Mn^{4+}
ion. Furthermore an excitation maximum, A_N, at 320 nm can be
detected at low manganese concentration which must belong to the
N-centers. (The small excitation maxima D and E belong to Mn^{4+}
ions in a $MgTiO_3$ component of the phosphor.) Since the excitation
maxima increase differently with increasing manganese content, as
was found to be the case with chromium, we conclude that at least

two kinds of luminescence centers are present between which there
exists energy transfer. In contrast to chromium activation the
regular Mn^{4+} ions do not show a characteristic emission, since no
R-lines are to be found (Fig. 1). According to our calculations
they should appear at about 620 nm, but obviously the titanate
lattice is less distorted by incorporation of Mn^{4+} than by Cr^{3+}
ions because of the smaller ion radius or the tetra-valency similar
to Ti; therefore the transitions in the Mn^{4+} ion remain forbidden.
But it was possible to show the inverted transition $^4A_{2g} \rightarrow ^2E$ by
exciting the Mn^{4+} ions and using the energy transfer from the ^2E-
level with consequent emission of N'-lines as a detector. This has
already been mentioned as additional evidence for the ^2E-level
being the starting point of energy transfer. The N'-centers of
manganese-activated magnesium titanate consist of associates of
activator ions with lattice defects, as for the chromium-activated
titanate. This can be concluded from the emission spectrum of the
unannealed manganese activated phosphor which is very similar to
that of the chromium activated one (Fig. 1b, dotted curve). In
addition to previous assumptions the annealing of Mg_2TiO_4:Mn^{4+} not
only oxidizes Mn^{2+} to Mn^{4+}, but mainly forms the N'-centers by diffusion of Mn^{4+} ions and lattice defects. The annealing does not
involve the regular Mn^{4+} ions, for their excitation spectrum remains
quite unchanged, as was also found in the case of the Cr^{3+} ions.

SUMMARY

     The Cr^{3+} and Mn^{4+} ions show the same behavior in a host lattice
of Mg_2TiO_4. 1. The activator ions form two different kinds of
centers - called R, and N or N'; 2. The R-centers are single ions
situated on different lattice sites, while the N-centers probably
consist of clusters of ions and lattice defects; 3. The main part
of the emitted luminescence radiance occurs from the N-centers;
4. Annealing at high temperature changes the N-center emission
which then shows groups of sharp lines instead of broad diffuse
bands, thus indicating the formation of new more selected clusters
(N'-centers); and 5. An energy transfer exists between the R- and
N-centers, the starting point of which is the excited R-line level
^2E.

ON THE QUANTUM LOSSES AT THE RUBY LUMINESCENCE EXCITATION

Z. L. Morgenshtern and V. B. Neustruev

P. N. Lebedev Physical Institute, Moscow, USSR

ABSTRACT

This paper reports studies of the ruby luminescence quantum spectral distribution, $\eta(\lambda)$, and the quantum efficiency at resonance excitation, $\eta(R)$. The $\eta(R)$ has been measured over the range 80-300 K. Investigations of $\eta(\lambda)$ in the vacuum ultraviolet region revealed an intense activator luminescence band in the region 210-140 nm, called the K-band. At 300 K the maximum value of $\eta(k)$ reached 0.95.

Results of the investigations of the ruby luminescence quantum output spectral distribution, $\eta(\lambda)$, have shown that the effective relaxation of the excitation into metastable ^2E-levels is realized under Cr^{3+}-center photoexcitation in the quartet system levels, after which the radiative transition $^2E \rightarrow {}^4A_2$ (the R-lines) takes place. The investigations of the output under resonance and of excitation in the B-lines region ($^4A_2 \rightarrow {}^2E, {}^2T_2$), as well as experiments on "excited absorption", have shown that there are practically no quantum losses upon excitation relaxation in the doublet level system and that the radiationless decay probability of the ^2E-state is negligibly small in comparison with the radiative ones (1-4).

This corresponds to the idea that the $^2E \rightarrow {}^4A_2$ radiationless transition probability is caused by slow spin-orbit interaction of the ^2E and $^4T_2$ states of the Cr^{3+} ion and is equal to 1.4/sec (5). The arguments mentioned above confirm the generally accepted point of view that $\eta(R)=1$ at $T \lesssim 400$ K, while the quantum losses under excitation in the quartet system [$\eta(U)=0.8$ and $\eta(Y)=0.7$] are caused by radiationless transitions in the same system.

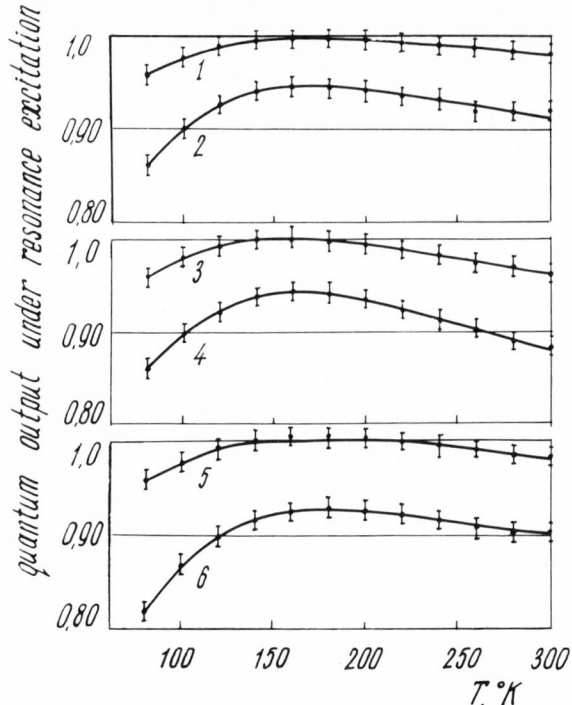

Fig. 1   The temperature dependence of the ruby luminescence quantum
         output under resonance excitation: 1 the laser ruby of the
         best quality, 2 the ruby with 0.13 at. % Cr^{3+}, 3,4 the
         laser rubies cut from the same rod (3 uncolored, 4 optical
         colored), 5,6 the laser rubies cut from the same rod (5
         uncolored, 6 γ-colored).

A model satisfactorily explaining the temperature quenching of
the ruby luminescence has been discussed (6). The model was based
on the assumption that the quenching is carried out over the $^4T_2$-
state with the temperature dependent frequency factor of the radiationless transition $^4T_2 \rightarrow {}^4A_2$. But the coincidence of calculated
with experimental data was achieved using an obviously too low val-
ue of the radiationless transition probability $^4T_2 \rightarrow {}^2E$, $^2T_1$, and
the question of the nature of quantum losses in ruby up to now is
open for discussion. To clarify this question it is important to
know the value of $\eta(R)$ and its temperature dependence more precisely than in the previous papers (1,2).

We carried out investigations of $\eta(R)$ in the temperature in-

terval 80-300 K and obtained the results with a precision of 1%
(the relative measurement precision was not worse than 0.5%). The
results obtained showed that the temperature dependence of $\eta(R)$
had a maximum at 140-200 K, so that for the best laser concentration
crystals $\eta(R)=1$. At 300 K for the best crystals $\eta(R)=0.96-0.98$.
The equality $\eta(R_1)=\eta(R_2)$ was established with an experimental precision of 0.5%. The value of $\eta(R)$ for rubies with higher activator
concentrations, as well as for optical and γ-colored laser rubies,
was found to be 10% less throughout the investigated temperature
interval (Fig. 1).

The high measurement precision allowed us to determine the
temperature quenching parameters in the initial region at 200-300 K.
For the 10 samples the activation energy value ΔE is equal to
(0.80 ± 0.03) eV with the radiationless transition frequency factor,

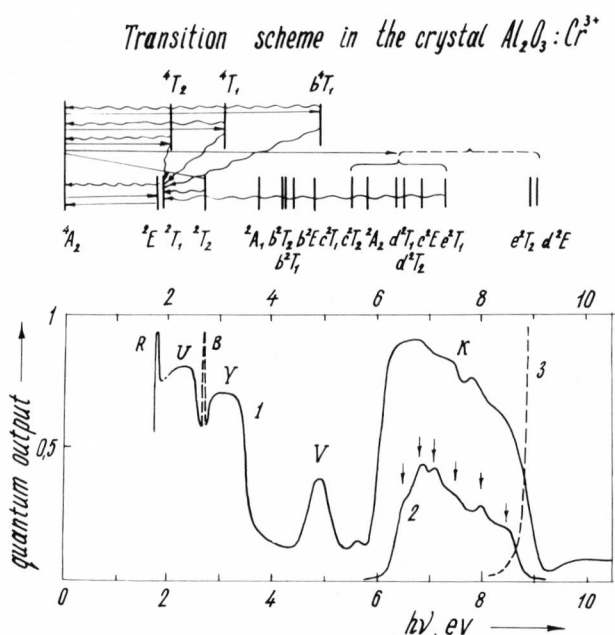

Fig. 2   The electron transition scheme of the ruby (above) and the
         quantum output spectral distribution of the ruby lumines-
         cence: 1 the ruby with 0.05 at. % Cr^{3+}, 2 the corundum with
         chromium traces (0.005 at. %), 3 the corundum fundamental
         absorption edge. The energetic positions of the Cr^{3+} ion
         levels are according to (12).

$W\sim30$ sec^{-1}. The beginning of the temperature quenching is not connected with the $^4T_2$-state acting as a quenching level (since it is situated 0.30 eV higher than the ^2E-state); instead, temperature quenching is apparently connected with $^2T_1$-state, situated 0.07 eV higher than the ^2E-state (Fig. 2).

The R´-line spectra ($^4A_2\longrightarrow{}^2T_1$) were experimentally investigated in 1962 (8). The large width of these lines at 1.8 K (~50cm^{-1}) shows an essentially greater value of the spin-orbit interaction constant of the $^2T_1$ state with the quartet states $^4T_2$ and $^4T_1$ compared to the analogous characteristic of the ^2E-state.

It follows from the above that the probability of the radiationless transition $^2T_1\longrightarrow{}^4A_2$ must be essentially greater than that of $^2E\longrightarrow{}^4A_2$. As the states $^2T_1$ and ^2E are closely situated and their population is determined by the Boltzman distribution, the state $^2T_1$ at 300 K acts as a weak quenching level for ruby luminescence.

The observed low temperature decrease of $\eta(R)$ at T< 140 K appears to be connected with the exchange interaction since the temperature effect in this range is enhanced by increasing activator concentration and by the formation of color centers, Cr^{2+} and Cr^{4+}. As is shown in a number of papers (9), the presence of Cr^{2+} ions results in a decrease of the spin-lattice relaxation time of Cr^{3+} ions, which may influence the probability of the radiationless transition $^2E\longrightarrow{}^4A_2$. The growth of the transition probability $^4A_2\rightleftharpoons{}^2E$ with the presence of the color centers has been demonstrated experimentally (10).

The Cr^{3+} ion spectrum in the ultraviolet region has not been studied in great detail. Thus, the data on luminescence output at the V-band excitation (λ_{max}=255 nm, $^4A_2\rightarrow b^4T_1$) are not simple, which evidently is connected with crystal quality. Our investigations distinguish the activator absorption and host-lattice absorption and we found the physical quantum yield $\eta(V)$=0.65 and the absorption cross-section in the maximum of V-band $\sigma(V)$=1.7 x 10^{-20} cm^2.

The investigations of $\eta(\lambda)$ in the vacuum ultraviolet region detected an intensive activator ruby luminescence band in the region 210-140 nm (5.9-8.9 eV), which we called K-band. At 300 K the maximum value of $\eta(K)$ reached 0.95 and was only 0.015-0.02 less than $\eta(R)$ in the same sample. The value $\eta_{max}(K)$ is close to the one characteristic of the output under excitation at the ion Cr^{3+} doublet system levels. The spectral distribution $\eta(K)$ displayed the structure consisting of a number of strongly overlapping bands while such structure were not detected in the absorption spectrum $\kappa(K)$. The K-band was compared with intercombination transition $^4A_2\longrightarrow\{c^2T_2,\cdots,e^2T_1\}$. The strong absorption [$\sigma_{max}(K)\underset{\sim}{} 4$x10^{-17} cm^2

at $h\nu=6.9$ eV] indicates violation of selection rules due to mixing
of the excited states $\{c^2T_2,..e^2T_1\}$ with ligand states and states
of opposite parity of Cr^{3+} ion. According to (12) the position of
the levels c^2T_2, $^2A_2$, d^2T_1, d^2T_2, c^2E, e^2T_1 in the cubic field
approximation (5.5-7.4 eV) satisfactorily corresponds to the K-band
position. Perhaps, the $^4A_2 \longrightarrow e^2T_2$, d^2E ($\sim$9.2 eV, according to (12))
also contribute to the K-band intensity, while $^4A_2 \longrightarrow c^2T_2$, $^2A_2$
($\sim$5.5 eV) transitions are not connected with the K-band and are
displayed separately in the $\eta(\lambda)$ spectrum (weak band $h\nu=5.5$ eV (11)).
These details await both experimental and theoretical investigation.

Cooling the ruby to 90 K had little influence on the spectral
structure of $\eta(K)$ and $\kappa(K)$. This fact supports the concept of a
genealogical bond of $\{c^2T_2,...,e^2T_1\}$ states with the quartet states
of the $3d^3$-electron configuration and further supports our interpretation of the K-band nature. Here it should be remembered that
the "excited abosrption" spectrum corresponding to $^2E \longrightarrow$
$\{c^2T_2,...,e^2T_1\}$ transitions (13) looks like an assymetric band with
$\sigma \sim 10^{-17}$ cm^2 and does not display its structure upon lowering the
temperature to 40 K.

If we assume that the K-band is an overlap of six absorption
bands, as can be concluded from Fig. 2, the oscillator strength of
every transition will turn out to be $\sim 5 \times 10^{-2}$.

The negligible quantum losses ($\sim$0.02) which take place under
radiationless relaxation $\{c^2T_2,...,e^2T_1\} \longrightarrow {}^2E$ are stimulated, on
the one hand, by the small probability of Cr^{3+} center photoionization (increasing with $h\nu_{exc}$) and, on the other hand, by the probability of radiationless transitions $\{c^2T_2,...,e^2T_1\} \longrightarrow {}^4A_2$ due to
the presence of a genealogical bond with the quartet states. As
expected, we observed the weak ruby phosphorescence under stationary excitation in the K-band region.

CONCLUSIONS

The whole complex of the literature data and the results of
our investigation enable us to conclude that there are practically
no quantum losses under excitation relaxation in the doublet system
of terms up to the energy 70,000 cm^{-1}. The quantum losses emerge
at excitation of transitions in the Cr^{3+} ion levels quartet system
and increase with the transition energy. The quantum losses in the
doublet system at $T \lesssim$ 300K are conditioned by quenching from the
$^2T_1$ level ($W \sim 30$ sec^{-1}) and Cr^{3+} ions exchange interaction starts
to display itself at $T < 140$ K even in crystals with Cr^{3+} ions laser
concentration.

Color centers present in ruby also result in an increase of
quantum losses, the origin of which may be connected with an in-

crease of Cr^{3+} center spin-lattice interaction and with a partial lifting of the forbidden character of intercombination.

REFERENCES

1. E. E. Bukke and Z. L. Morgenshtern, Opt. i Spectr. 14, 687 (1963). Acta Phys. Polon., 36, 393 (1964).

2. A. Misu, J. Phys. Soc. Japan 19, 12 (1964).

3. G. K. Shultz, Zeits. Physik., 167, 446 (1962).

4. M. D. Galanin, V. N. Smorchkov and Z. A. Chizhikova, Opt. i Spectr. 19, 296 (1965).

5. B. S. Tzukerblat and Yu. E. Perlin, Opt. i Spectr. 21, 13 (1966).

6. Yu. E. Perlin, Yu. Rosenfeld and B. S. Tzkerblat, Ukr. Fiz. Journal 14, 1307 (1969); 14, 1317 (1969).

7. Z. L. Morgenshtern and V. B. Neustruev, Opt. i Spectr. 32, 953 (1972).

8. J. Margerie, Comp. Rendus 255, 1598 (1962).

9. D. R. Mason and J. S. Thorp, Proc. Phys. Soc. 87, 49 (1966).

10. G. E. Arkhangelskii, Z. L. Morgenshtern and V. B. Neutsruev, Isv. Akad. Nauk SSSR, Ser. Fiz. 33, 875 (1969); Phys. Stat. Sol. 36, 451 (1969).

11. E. I. Alshitz, Z. L. Morgenshtern and V. B. Neustruev, Opt. i Spectr. 31, 932 (1971).

12. D. T. Sviridov, Opt. i Spectr. 20, 488 (1966).

13. T. Kushida, J. Phys. Soc. Japan 21, 1331 (1966).

LUMINESCENCE PROCESSES IN Bi^{3+} CENTERS

G. Boulon, F. Gaume-Mahn, C. Pedrini, B. Jacquier, J. Janin and D. Curie[*]

Spectroscopy & Luminescence Laboratory, University of Lyon I, Villeurbanne, France

[*]Luminescence Laboratory, University of Paris VI

Paris, France

ABSTRACT

Photoluminescence processes in Bi^{3+} centers are investigáted in various host materials such as rare earth oxides, vanadates or gallates and in alkaline earth antimonates, using excitation and fluorescence spectra, thermoluminescence, phosphorescence decay and lifetime measurements, between 4 K and 400 K.

EXPERIMENTAL PROCEDURE

Powder samples of the following compounds have been activated with various amounts of Bi^{3+}: La_2O_3, hexag. (C_{3v}); Ln_2O_3 (Ln = Sc, Lu, Y, Gd), cubic (C_2 and S_6); $LnVO_4$ (Ln = Sc, Y, Gd), tetragonal (D_{2d}); $LaGaO_3$, orthorhombic perovskite (C_s); $LaOCl$, tetragonal (C_{4v}); $LaPO_4$, monoclinic (C_1); MSb_2O_6 (M = Ca, Sr), hexagonal (D_3).

Excitation spectra have been obtained between 2000 and 4000 A with a quartz prism monochromator using a hydrogen lamp (3KVA) or a xenon lamp (Osram XBO-450 W/4). These sources have also been used to excite fluorescence in the maxima of the excitation bands. A gaseous cooling (nitrogen or helium) in the cryostat, associated with a temperature regulator, gives all the temperatures from 10 to 600 K. Experimental procedure for decay time measurements has been

TABLE 1

	Excitation		Fluorescence
	$^1S_0 \to {}^1P_1$	$^1S_0 \to {}^3P_1$	$^3P_1 \to {}^1S_0$
Free ion Bi^{3+}	14.205	9.411	
La$_2$O$_3$	4.95 < hν < 5.7	4.04	2.62
Ln$_2$O$_3$ site S$_6$ Ln= Sc	4.55	3.33	3.08
Lu	4.55	3.33	3.07
Y	4.52	3.32	3.03
Gd	4.52	3.27	2.97
Ln$_2$O$_3$ site C$_2$ Ln= Sc	4.55	3.56 and 3.78	2.43
Lu	4.55	3.61 and 3.77	2.42
Y	4.52	3.59 and 3.73	2.34
Gd	4.52	3.58 and 3.68	2.22
LaGaO$_3$	4.95	4.04	3.24
LaOCl	5.55 / 5.55	4.71 / 4.66	2.81 / 3.45
LaPO$_4$		5.3	2.72
KCl (2)	5.2 < hν < 5.7	3.77	2.45
Ca$_3$(PO$_4$)$_2$ (2)	5.2 < hν < 5.7	4.04	3.40
Ln$_2$SO$_6$ Ln= Lu		4.58	4.07
Y (3)		4.69	3.99
La		4.50	3.25

reported previously (1).

EXPERIMENTAL RESULTS

The phosphors generally present two excitation bands and one emission band. Their energies and Stokes shift depend on the matrix (Table 1 and 2). The fluorescence band is characteristic of the Bi^{3+} center and independent of the excitation band; it corresponds to the $^3P_1 \to {}^1S_0$ transition. For the cubic Ln$_2$O$_3$:Bi^{3+}, there are two symmetry sites: C$_2$ and S$_6$. Each of them has its own excitation, and two emission bands (violet and green) are observed.

The excitation processes are different according to the fluorescence of the nonactivated material. In Table 1, the hosts are not luminescent. The two excitation bands are due to Bi^{3+} center transitions: $^1S_0 \to {}^3P_1$ (spectral range from 5.3 to 3.2 eV) and $^1S_0 \to {}^1P_1$ (spectral range greater than 5 eV). On the contrary, pure antimonates and vanadates of Table 2 exhibit a blue emission. With the antimonates, we observe an absorption in the Sb$_2$O$_6^{2-}$ centers and then a transfer to the bismuth $^3P_1$ level. Concerning the vanadates, the excitation band is characteristic of the host and only weakly shifted by the bismuth content; it can be ascribed to a

TABLE 2

Rare earth orthovanadates (77 K)			
		$^1A_1 \rightarrow {}^1T_2$ $[2p(0^{2-}) \rightarrow 3d(V^{5+})]$	$^1T_2 \rightarrow {}^1A_1$
LnVO$_4$ pure Ln= Sc		3.89	2.56
Y		3.97	2.71
Gd		3.98	2.76
		$6s^2(Bi^{3+}) \rightarrow 3d^0(V^{5+})$	$^3P_1 \rightarrow {}^1S_0$ (Bi^{3+})
LnVO$_4$:Bi Ln= Sc		3.49	1.91
Y		3.72	2.17
Gd		3.77	2.22
Alkaline earth metaantimonates (10 K)			
Excitation Sb$_2$O$_6^{2-}$		$^1S_0 \rightarrow {}^3P_1$	$^3P_1 \rightarrow {}^1S_0$
M= Ca 5.55	4.27	3.66 and 3.62	2.65
Sr 5.55	4.13	3.66 and 3.47	3.09 and 2.30
Ba 5.50	3.87		

charge transfer from the $6s^2$ level of Bi^{3+} to the empty 3d level of V^{5+}.

Decay time measurements at room temperature show that the emission corresponds to the $^3P_1 \rightarrow {}^1S_0$ electric dipole transition for all the investigated phosphors.

Furthermore, we have studied the shift of the $^3P_1$ excitation band in different compounds. This shift is principally due to crystal field and covalency. We have tried to evaluate these two effects. For instance, if bismuth is substituted for lanthanum in various lanthanum compounds, we observe that the $^3P_1$ energy is decreasing when covalency is increasing. On the other hand, in the same series of compounds, where symmetry and covalency remain the same, we have established an energy dependence of Δr; Δr is the difference between the ionic radii of the bismuth and the substituted ion (4). This rule is also true for Tl^+ and Pb^{2+} centers. In all the investigated phosphors there is a Stokes shift which can reach a large value for the Bi^{3+} center.

The large overlap of excitation and emission spectra reveals an energy transfer between sensitizer centers (S = VO_4^{3-} or $Sb_2O_6^{2-}$) and activators (A = Bi^{3+}). The thermal dependence of the fluorescence intensity confirms this transfer. An interesting energy transfer between Bi^{3+} centers located in C_2 and S_6 sites in the

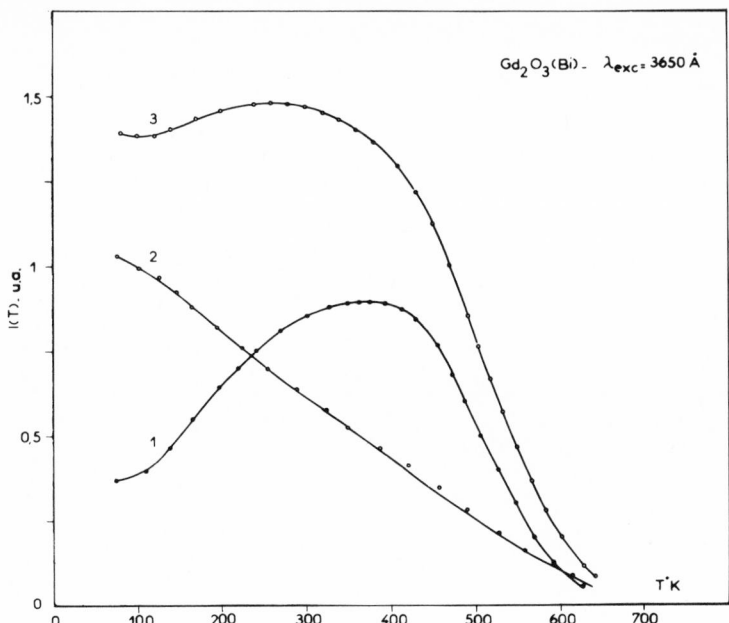

Fig. 1   Temperature dependence of the fluorescence intensity for
         Gd$_2$O$_3$:Bi: (1) green emission; (2) violet emission; and
         (3) total emission.

cubic rare earth oxides (Fig. 1) was found.   For SrSb$_2$O$_6$:Bi,
Table 3 and Fig. 2 show a complex transfer process not only from
the Sb$_2$O$_6^{2-}$ groups to the Bi^{3+} center but also between two Bi^{3+}
centers.

With the hypothesis of a dipole-dipole interaction, we have
used the Dexter formula to calculate the distance, R_o, corresponding to $P_{SA} = 1/\tau_S$ (P_{SA} is the transfer probability between S and A
and τ_S is the lifetime of the sensitizer).   The R_o can be evaluated
by the formula (6):

$$R_o^6 = 6.3 \times 10^{27} \int (Q_A/E^4) \cdot f_s(E) \cdot F(E) dE ,$$

where Q_A is the absoprtion cross section of the A species; E is
the photon energy for the $f_s(E)$ emission band and $F_A(E)$ absorption
band; the integral represents the overlap of these spectra.   The
values of R_o are reported in Table 3.   The distance over which
transfer may occur is of the order of 20 or 30 Angstrom units, in
agreement with a resonance process by dipole-dipole interaction.

The decay time, τ, of the different emissions have been determined from room to liquid helium temperature.   Table 4 shows a very

TABLE 3

Compound	S	A	E eV	$\int F_S(E)F_A(E)dE$ $10^{-3}eV^{-1}$	R_o A
ScVO$_4$			3.30		
YVO$_4$	VO$_4^{3-}$	VO$_4^{3-}$	3.26	$\sim$ 1	12
GdVO$_4$			3.20		
ScVO$_4$			2.96	20	23
YVO$_4$	VO$_4^{3-}$	Bi^{3+}	3.22	15	20
GdVO$_4$			3.16	$\sim$ 8	19
Gd$_2$O$_3$	Bi^{3+}(S$_6$)	Bi^{3+}(S$_6$)	3.12	90	28
	Bi^{3+}(S$_6$)	Bi^{3+}(C$_2$)	3.22	$\sim$ 1.5	14
SrSb$_2$O$_6$	Sb$_2$O$_6^{2-}$	Sb$_2$O$_6^{2-}$	3.84	$\sim$ 6	14
	Sb$_2$O$_6^{2-}$	Bi^{3+}(violet)	3.52	175	29
	Sb$_2$O$_6^{2-}$	Bi^{3+}(green)	3.41	701	37
	Bi^{3+} (violet)	Bi^{3+}(green)	3.29	228	32
	Bi^{3+} (violet)	Bi^{3+}(violet)	3.41	17	20

large increase of τ for all the phosphors, when the temperature is decreased. It is then a general process which characterizes the bismuth center. We suggest that the fluorescence process from $^3P_1$ to $^1S_0$ ($^1P_1 \rightarrow \ ^3P_1$ being radiationless) is complicated at low temperature by trapping in the $^3P_0$ metastable level. With this hypothesis, we have calculated the probability, f, from $^3P_0$ to $^1S_0$ and the probability, f', from $^3P_1$ to $^1S_0$ in the three level system ($^1S_0$, $^3P_0$, $^3P_1$). The calculations show that we have the following relations (1):

$$1/\tau = f + f'p/(p_o + f') \qquad (1)$$

$$p = p_o \exp(-E/kT) \qquad (2)$$

where E = trap depth; p = probability of the $^3P_0 \rightarrow \ ^3P_1$ radiationless transition

at low temperatures: $1/\tau = f + p_o \exp(-E/kT) \qquad (3)$

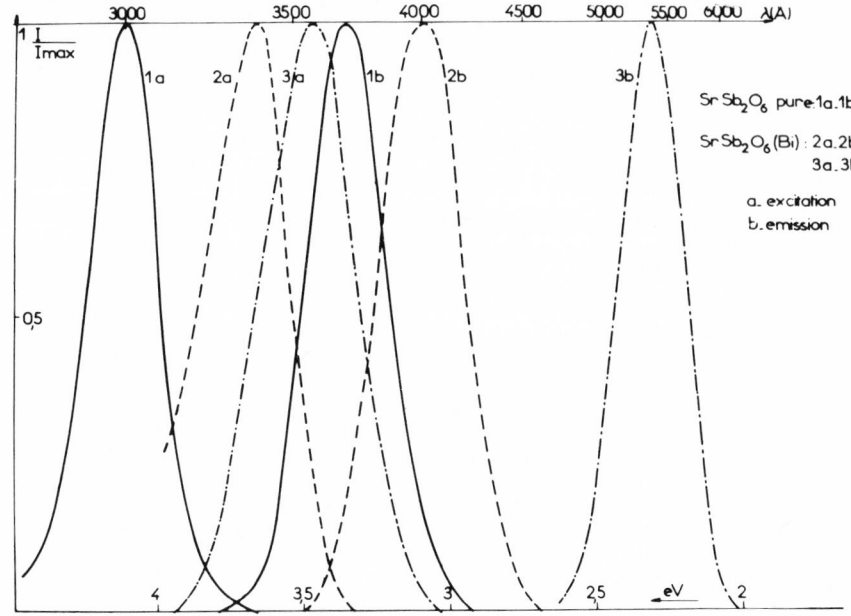

Fig. 2   Excitation and emission spectra of pure $SrSb_2O_6$ and
$SrSb_2O_6$:Bi.

at higher temperatures: $\tau = (1/f') + (1/p_o)$ (4)

The results indicate that f is less than 10^3/sec and f' is greater
than 10^7/sec, in agreement with the expected values for $^3P_0$ and
$^3P_1$ respectively.

However, Eq. (3) which is very schematic, only interprets a
part of the experimental temperature dependence. For instance, E
is 0.05 eV for La_2O_3:Bi. We propose a more general rule including
the splitting of the $^3P_1$ level:

$$1/\tau = f + \frac{f' \cdot p'}{p_o + f'} + \frac{f'' \cdot p''}{p_o'' + f''} .$$ (5)

Experiments are in progress to find the value of E using thermoluminescence and phosphorescence decay with a selective excitation.
For instance, we have succeeded in finding a 0.014 eV trap at 7 K
for $CaSb_2O_6$:Bi.

Finally, potential curves have been determined from optical
experiments. Fig. 3 represents an example for the La_2O_3:Bi system.

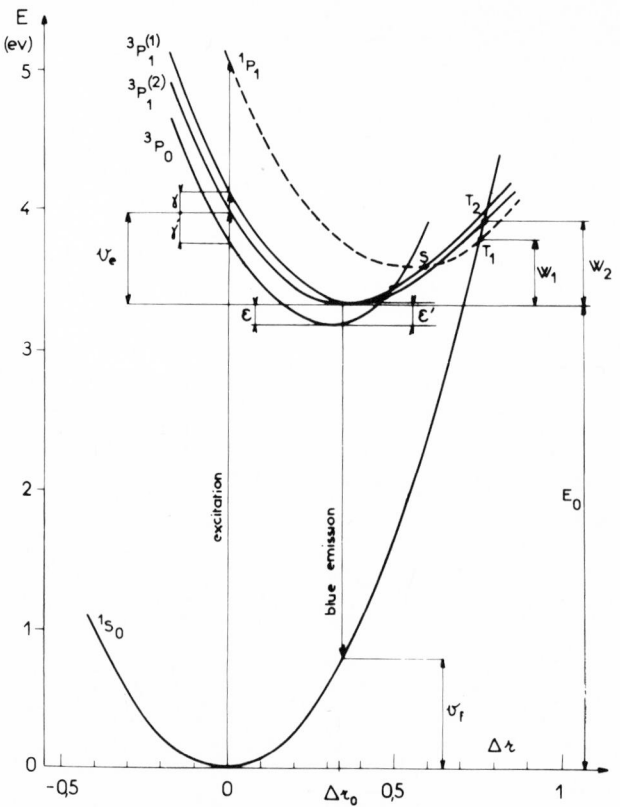

Fig. 3 Potential curves for La$_2$O$_3$:Bi.

The position of the curves explains the $^1P_1 \rightarrow {}^3P_1$ radiationless transition, the influence of the two $^3P_1$ components and the trapping in the $^3P_0$ state.

TABLE 4

τ (μs)				
	295 K	77 K	10 K	4 K
La$_2$O$_3$:Bi	0.27	9.15		305
Y$_2$O$_3$:Bi (C$_2$)	0.55	14 and 2.4		480 and 95
Y$_2$O$_3$:Bi (S$_6$)				180
Gd$_2$O$_3$:Bi (C$_2$)	0.50	14 and 2.2		800 and 160

Table continued on next page.

Table 4 continued

	295 K	77 K	10 K	4 K
Gd_2O_3:Bi (S_6)				500 and 100
$CaSb_2O_6$:Bi	0.09 and 0.33	24 and 5		630
$SrSb_2O_6$:Bi green	0.36	16.7	810	
$SrSb_2O_6$:Bi violet	weak fluorescence	weak fluorescence	76	
YOCl:Bi	1.7 and 7.2		150 and 640	
$LaPO_4$:Bi	non-fluorescent	6.2 and 14.8		
$LaGaO_3$:Bi	0.28	210 and 1200	320 and 1000	
LaOCl:Bi blue	0.94	2.7 and 12.2		
LaOCl:Bi u.v.	0.84	48	230	
$ScVO_4$:Bi	4.7	10		460
YVO_4:Bi	6	8		350
$GdVO_4$:Bi	5.5	5.6		325

REFERENCES

1. G. Boulon, J. Phys. (Paris) 32, 333, (1971).

2. N. E. Zazubovik and N. E. Lushchik, Trudy IFA AN. ESSSR 17, 50, (1961). C. B. Lushchik, N. E. Lushchik, I. A. Muuga, Trudy IFA AN. ESSSR 23, 22, (1963).

3. G. Blasse and A. Bril, Philips Res. Repts. 23, 461, (1968).

4. G. Boulon, B. Jacquier, and F. Gaume-Mahn (to be published).

5. G. Boulon, Thèse de Doctorat es-Sciences, Lyon, (1970).

6. D. L. Dexter, J. Chem. Phys. 21, 836, (1953).

LITHIUM DONORS AND THE BINDING OF EXCITONS AT NEUTRAL DONORS AND

ACCEPTORS IN GALLIUM PHOSPHIDE

P. J. Dean

Royal Radar Establishment, St.   Andrews Road

Malvern, Worcs., England

ABSTRACT

Two interstitial Li donors have been identified in GaP through the photoluminescence spectra they induce.   Donor A exhibits bound exciton and d-a pair luminescence with weak no-phonon lines and is shallow, ionization energy $E_D \sim 56$ meV, like a Ga-type donor.   Donor B is deeper, $E_D = 88.3 \pm 0.5$ meV and generates strong no-phonon luminescence, like a P-type donor.   Zeeman splittings of the lower energy bound exciton no-phonon line show that donor B has C_{3v} symmetry with <111> axis, a novel situation for donor-exciton luminescence in GaP.   The values of the exciton localization energy E_{BX} for these shallow Li donors enable the trend of E_{BX} vs. E_D to be established firmly for the first time in GaP.   Pronounced deviations from the Haynes' rule form found in Si are noted and are given a qualitative interpretation in terms of a difference in the appropriate mass ratio m_e/m_h for GaP and Si.

I.   INTRODUCTION

Close bulk compensation of p-type GaP may be produced conveniently by Li indiffusion in the temperature range 500-700 C, a phenomenon also well known in Si (1).   Lithium introduces a number of donor-like absorption spectra in Si (2), including a simple spectrum attributed to the isolated interstitial donor with small valley-orbit splitting, inverted compared with substitutional donors (3).   Lithium in Si was earlier thought to occupy hexagonal D_{3d} interstitial sites (4), but a recent detailed electron spin resonance study (5) shows that it must lie close to the interstitial tetrahedral T_d site, like Li in Ge (6). Photoluminescence of the simple Li interstitial donor has not been reported in Si, nor in GaAs where electrical (7) and impurity-phonon (8) studies show that Li introduces an acceptor as well as a donor species. Such a Li donor in GaP is most conveniently located in photoluminescence, through the recombination of excitons bound to the neutral donor which should be detectable, though not predominantly radiative (9).

We report two spectra of this type, persistent in Li-diffused GaP. One, with a group of no-phonon lines near 2.321 eV, the A-spectrum, is attributed to the Li donor on the Ga-type interstitial site. A second, with a relatively strong isolated no-phonon line near 2.311 eV, the B-spectrum, has properties (spectral position, spectral form, Section III; time decay, Section III; Zeeman splittings, Section III) expected for a donor on the P-type inter-stitial site of the GaP lattice. However, this B donor clearly exhibits C_{3v} rather than T_d symmetry. The form of d-a pair spectra characteristic of GaP:Li,Zn (Section III) provides estimates of E_D which check with that obtained from "two-electron" sidebands in the B-spectrum. The pairs of energies E_D, E_{BX} for these two Li donors, together with information from earlier work, enable the trend with E_D of the exciton localization energy, E_{BX} to be established clearly for shallow donors in GaP. This trend is compared with that for acceptors in GaP in Section IV, and contrasted with the behavior observed for donors and acceptors in Si, which led to the formula-tion of Haynes' rule, $E_{BX} = \alpha E_D$, E_A, where α is a constant, nearly identical for donors and acceptors in Si (10).

II. EXPERIMENTAL

The techniques of Li in-diffusion were identical to those reported earlier (11). The Li donor spectra were frequently seen from crystals grown from Ga solution (Fig. 1), often more clearly than the red Li_I-Li_{Ga}-O_P bound exciton luminescence reported earlier (11). However, the Li A and B-spectra were strongest relative to the exciton spectra involving persistent impurities such as N, S and Si (Fig. 1) after Li diffusion of GaP grown from the vapor by the wet H_2 process (12). This was especially true for lightly Zn doped vapor grown crystals. Presumably the Li in-diffusion is enhanced in the presence of an acceptor, as is known in Si (13). Zinc is not an inadvertent contaminant of our crystals, so the Li-related spectra in Fig. 1 certainly do not involve Zn. However, spectra from these crystals exhibit a rich variety of extraneous sharp lines, so far unidentified though some may involve Li-Zn complexes. Such spectra are outside the scope of this paper. Only the very shallow A spectrum has been observed from GaP crystals doped with Li from an upstream metal source at 460 C during growth from Ga solution in an open tube furnace with alumina boats and a BN insert tube.

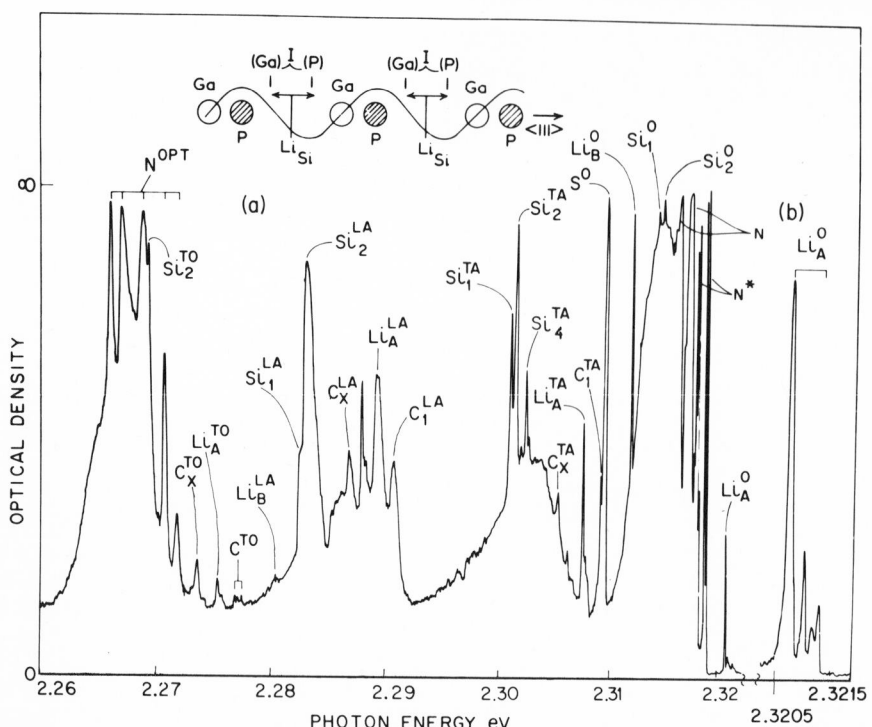

Fig. 1   A portion of the low temperature photoluminescence of
         undoped GaP grown from Ga solution during slow cooling
         between 1150 and 900 C and subsequently in-diffused with Li
         for 1 hr at 700 C, recorded photographically. The stronger
         no-phonon lines are over-exposed in order to exhibit weak
         phonon replicas. The well-known bound exciton spectra due
         to isoelectronic N traps and neutral S donors are prominent
         and the C acceptor exciton appears weakly. An exciton
         associated with the presence of Si, possibly involving the
         neutral Si donor, is also strong since this crystal was
         grown in a dry environment. The two new exciton series
         shown are the most persistent unequivocally related to the
         presence of Li in a wide variety of GaP crystals. They
         have the character of weakly bound excitons, with phonon
         cooperation dominated by TA, LA and TO momentum-conserving
         (MC) phonons of wave-vector $\underline{k}_X$. The close spaced set of no-
         phonon lines of the shallower A series are shown in more
         detail in part (b). The inset shows the <111> sequence of
         atom sites in the GaP lattice, including possible inter-
         stitial positions for Li in GaP and Si and the envelope of
         the phase-wave function of an electron in the lowest
         conduction band at X_1.

Table 1

Relative intensities of MC phonon replicas in GaP
donor-bound exciton spectra

Impurity[+] (Lattice Site)	L^{TA}/L^{0}[++]	L^{LA}/L^{0}[++]	L^{TO}/L^{0}[++]
Li$_A$ (Ga)	1.1	4	0.4
Li$_B$ (Ga,P)	0.008	0.028	?
Si (Ga)	2.1	5.3	1.2
Sn (Ga)	2.3	5.5	1.1
Te (P)	0.021	0.13	0.01
Se (P)	0.05	0.084	~0.025
S (P)	0.009	0.015	0.005

[+] Note that Li is an interstitial donor, mainly but not
purely P like for the Li B-spectrum.

[++] L denotes intensity from luminescence spectra.

Most of the luminescence spectra were recorded photographically
from crystals immersed in liquid He, pumped below the λ point.
Donor-acceptor pair spectra were recorded photoelectrically at
very low excitation levels. The luminescence was excited by 4880 A
or 5145 A light from an Ar[+] laser, fitted with a high speed acousto-
optical modulator for lifetime studies (14).

III. RESULTS AND DISCUSSION

The Li A and B-spectra both contain the set of sidebands
characteristic of momentum-conserving (MC) phonon replicas in GaP
(14,15). These replicas are strong compared with the no-phonon line
(superscript zero in Fig. 1) in the A spectrum, but relatively weak
in the B spectrum. Comparison of the coupling strengths with other
identified donors in GaP (Table 1), together with the form of the
no-phonon lines immediately suggests that the shallow A spectrum
(small E_{BX} = 7.2 meV) might arise at a defect which binds an

electron like a Ga-site substitutional donor, while the form of the deeper B spectrum (larger $E_{BX} = 16.0$ meV) is similar to that expected from a P-site donor (16). Two T_d symmetry interstitial sites occur (Fig. 1 inset) which are inequivalent in the GaP lattice, but equivalent in Si. At one of these sites, adjacent to a substitutional Ga, the phase wave-function of an electron in the lowest X_1 set of conduction band minima exhibits an antinode, like a substitutional donor on the electron-attractive P lattice site. An interstitial donor on this site is expected to bind an exciton relatively tightly, producing relatively strong no-phonon recombinations according to the arguments of Morgan (16). Conversely, a donor in the Ga-type interstitial site should bind an exciton much less tightly, and exhibit weak no-phonon lines with complex structure, since no valley-orbit splitting of the donor ground state occurs (17). We shall see (Section III) that the J-J coupling scheme characteristic of an exciton bound to a substitutional P-site donor is complicated for the B spectrum interstitial Li donor by a reduction of site symmetry, so that this donor should be more properly labeled "Ga, P-type". However, the essential simplicity of the electron remains, namely, an isotropic g value close to +2 indicative of valley-orbit splitting with "normal" ordering, that is the symmetric Γ_1 state lying lowest (18).

Line Li_B^O (Fig. 1) was isolated by a spectrometer and its time decay measured following excitation by light pulses from an Ar^+ laser fitted with an intracavity acousto-optical deflection system (14). The exciting pulses had a width of ~ 10 nsec and decay time ~ 6 nsec. Integration over a few secs with a pulse repetition rate of a few MHz enabled the decay of this weak luminescence to be established reliably. An average of concordant results from a lightly Zn-doped vapor grown crystal, an undoped crystal and two liquid encapsulated Czochralski-grown crystals, one lightly Zn-doped, gave an exponential decay constant of 27 ± 4 nsec. This is significantly longer than the decay time for the deeper P-site donor S, 21 ± 4 nsec (9), comparable to that for the Te exciton, ~ 30 nsec, but appreciably shorter than for the shallower Ga-site donor Sn, ~ 90 nsec (17). The decay rate should increase rapidly with the binding energies of the like particles (electrons) in these exciton complexes, if the Auger non-radiative decay channel predominates (14).

The intensity of even the principal component of the Li A spectrum, Li_A^{LA} (Fig. 1) was always too weak relative to extraneous underlying luminescence to permit a satisfactory measurement of its time decay. Comparison with the Sn exciton (17) suggests that only ~ 1 in 4000 of the decays of the Li A exciton should be radiative.

The magnetic splittings of line Li_B^O (Fig. 2) are quite unlike those observed for excitons bound to substitutional P-site donors in GaP, such as S (19). In the latter case, the magnetic splittings

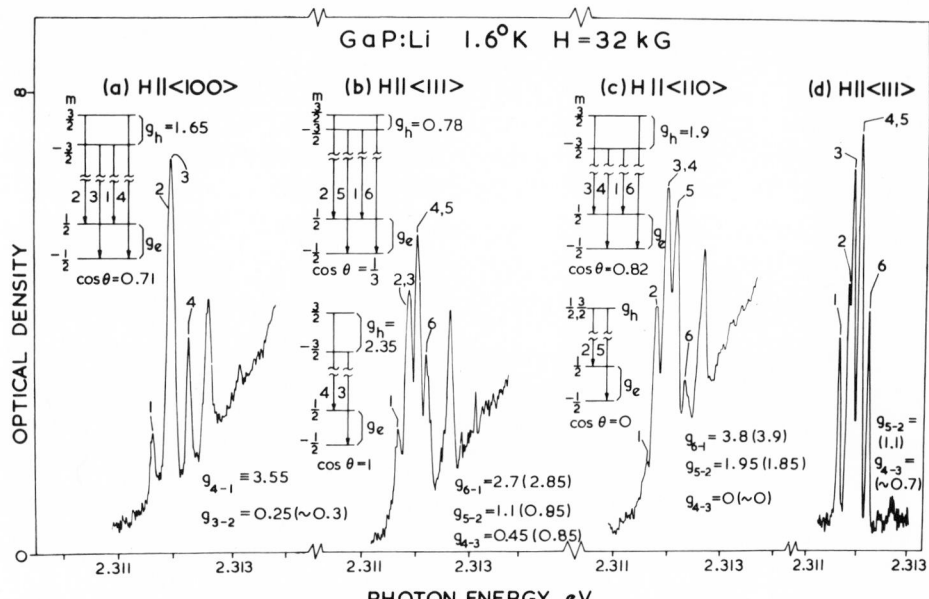

Fig. 2   Zeeman spectra of the lowest energy, isolated, strong no-
         phonon line of the Li B-exciton, recorded photographically
         from a solution grown platelet at a magnetic field of 32
         kG parallel to the three principal crystallographic direc-
         tions indicated in parts (a)-(c). The sharper [111] data
         in part (d) were recorded from a vapor-grown needle, at
         34 kG. The unlabeled prominent line in (a)-(c) appears to
         involve an excited state of the B-exciton, with negligible
         magnetic g-factor. The insets show the magnetic splittings
         of the lowest energy level of an exciton bound to an axial
         donor of C_{3v} symmetry and of the donor ground state, and
         the significant transitions between them. For a <111>
         symmetry axis, the donors fall into two inequivalent sets
         with H bearing different angles θ to the symmetry axes,
         except for H$\|$[100]. The indicated hole g-values are cal-
         culated as $g_h \cos \theta$, with $g_h = 0.78$, while the electron has
         the isotropic g value $g_e = 1.9$. The lower sets of transi-
         tion g values then result, to be compared with the bracketed
         values taken from the spectra. Transitions of largest g
         inherently possess very weak intensities.

are somewhat smaller than in Fig. 2, and the anisotropy is much
less pronounced. It is not possible to give a consistent account
of the data of Fig. 2 using the theory of Yafet and Thomas for the
S exciton (19), even allowing unusual values of electron g factor
g_e and the parameters K and L which describe the magnetic behavior

of the hole in the Luttinger Hamiltonian for T_d symmetry. A fundamental difficulty is the large splitting observed for $H||[100]$ (Fig. 2a), which can only be explained with the bizarre value $g_e \sim +2.9$ and $K \sim 0.66$ for a T_d donor, with the unexplained condition that transitions with $\Delta m = 0$ cannot be seen. Then, these parameters completely fail to account for the observed behavior for $H||[111]$ and $H||[110]$. The solution to this dilemma is contained in Fig. 2, where energy level schemes are drawn for an exciton bound to a donor of axial, C_{3v} symmetry. If the crystal field of this center is sufficiently strong to completely decouple the $m_J = \pm 1/2$ and $m_J = \pm 3/2$ hole states, as observed for the associates $Cd_{Ga}-O_P$ (20) and $Li_I-Li_{Ga}-O_P$ (11) in GaP, the magnetic behavior of the donor-exciton becomes equivalent to that for excitons bound to neutral donors in an axial semiconductor like CdS (21). The two electrons pair off, leaving the magnetically active hole in the initial state, while the final state contains an unpaired electron.

The definition of the symmetry axis by the center rather than by the crystal leads to the complication that, except for $H||[100]$, the symmetry axes fall into two sets bearing different angles to the magnetic field (defined by θ in Fig. 2). The estimated g values in Fig. 2 were obtained assuming $g_e = 1.9$ and is isotropic, as established for a variety of T_d and C_{3v} centers in GaP (11,14,17), and with g_h (or K) = 0.78, determined from a fit to the experimental g_{4-1} in Fig. 2a, also a reasonable value. Good general agreement is obtained between estimates from these g_e, g_h and the experimental (bracketed) g values (Fig. 2). Transitions predicted to have vanishing intensity, for example $m_J \pm 3/2 \rightarrow \pm 1/2$ for $\cos \theta = 1$ (Fig. 2b), are not observed. In general, the data qualitatively follow the prediction that the intensity of this high g value transition should vary as $(1 - |\cos \theta|)$ (21). The transition $m_J \pm 3/2 \rightarrow \pm 1/2$ should vary like $(1 + |\cos \theta|)$ and therefore exhibits much less spectacular orientation dependence, always producing a strong line. The predicted lines 2, 3, 4 and 5 are not resolved in Fig. 2B. However, sharper spectra were obtained from a Li diffused vapor grown needle for $H||[111]$, in which lines 2 and 3 were resolved (Fig. 2d). These data yield the experimental values $g_{5-2} \sim 1.1$, $g_{4-3} \sim 0.7$, in fair agreement with the predictions in Fig. 2b. As expected, the intensity of line 2 was about half as strong as line 3 and was enhanced relatively when radiation components with $E||H$ were selected. No thermalization has been seen for the Li_B^o magnetic subcomponents, suggesting a long hole spin-lattice relaxation time as found for similar spectra in the wurtzite semiconductors CdS (21) and 6HSiC (22).

The siting of interstitial Li in diamond lattice semiconductors has been discussed by a number of workers (4,5,6). In the most recent study (5) the isotropy and magnitude of certain terms in the spin-orbit coupling of the simple Li donor in Si shows that Li lies

near the T_d interstitial site, though possibly not precisely on it.
The magnetic data for Li_B^O given here suggest even more clearly that
the type B intersitital Li donor in GaP also has low site symmetry.
Presumably it would be displaced from the D_{3d} site of the Si lattice
in any case by the absence of inversion symmetry in GaP, as shown
in Fig. 1 inset. One of the two inequivalent Li sites may then fall
close to the antinode of the phase wave-function of an electron in
the X_1 conduction band minima, qualitatively consistent with the
existence of a strong no-phonon line in the Li B-spectrum (Table 1).
This property accounts for the description "Ga, P-type donor" given
earlier, but for all practical purposes the donor is P-type in the
sense of Morgan (16,23). The dominance of the Ga-type Li donor in
growth doped crystals (Section II) seems to imply that interstitial
Li has greater stability in this site, surrounded by electron-
attractive P atoms.

Quantitative analysis of the magnetic splitting of lines Li_A^O
was not possible. Even the lowest energy no-phonon lines are
separated by only ∿0.27 meV and strong interaction occurs before
the full set of magnetic subcomponents from each become clearly
defined.

The broad but slightly structured spectrum in Fig. 3 is
attributed to electron-hole recombinations at distant pairs of Li
donors and Zn acceptors. At higher excitation intensities, the
spectrum broadens and shifts to high energies in the manner well-
established for d-a pair transitions in GaP (24). Then, the phonon
replicas are unresolved but rather complex structure attributable
to transitions at discrete pairs becomes visible above 2.25 eV. We
have not been able to interpret this structure but it is certain
that it does not involve the likely inadvertent contaminant donor S.
The energies $h\nu_{NP}$ of the no-phonon peaks in Fig. 3 are given by

$$h\nu_{NP} = E_g - (E_A + E_D) + E_c \qquad (1)$$

where E_g is the band gap (25) and E_c is the Coulomb interaction
energy between the appropriate very distant pairs, estimated to be
∿18 meV from comparison with identified d-a pair spectra recorded
under identical conditions. The spectral form in Fig. 3 cannot be
explained by the presence of a P-type donor alone, although this
accounts for the largest feature just above 2.20 eV (26). Assuming
that this peak involves the P-type Li_B donor discussed above, the
form of the additional unexpected structure above and below the
Li_B-Zn pair no-phonon band is inconsistent with superposed Li_B-C
pair luminescence, assuming C present inadvertently. However, the
additional structure can be explained by relatively weak d-a pair
luminescence characteristic of a Ga-site donor in GaP (17,26),
which we naturally identify with the Ga-type Li_A donor. Then, from
Eq. (1), we find $(E_D)_A = 56 \pm 4$ meV, $(E_D)_B = 86 \pm 4$ meV, assuming
$(E_A)_{Zn} = 64 \pm 2$ meV (14), $E_g = 2.339 \pm 0.002$ eV (25). The presence

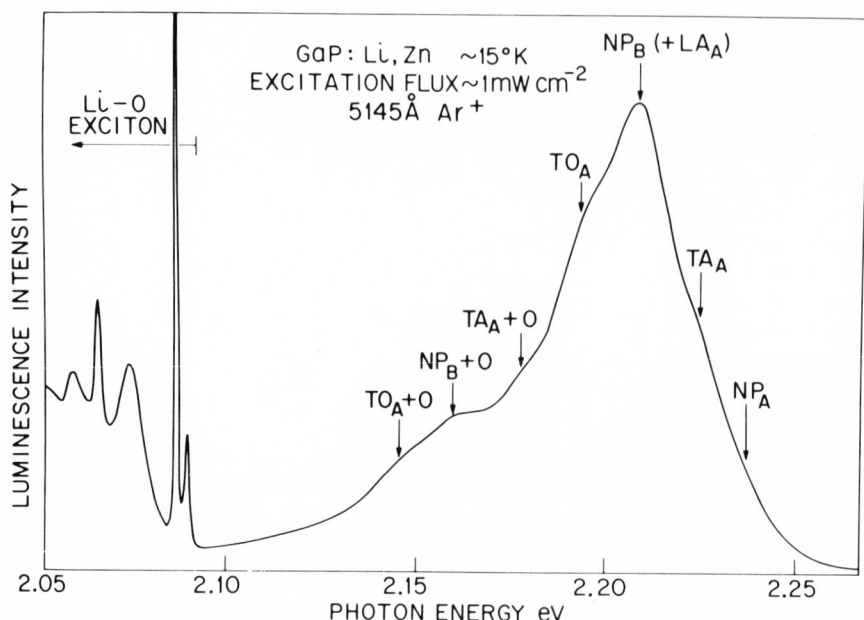

Fig. 3   A low temperature photoluminescence spectrum from a Li-
         diffused vapor-grown GaP:Zn crystal showing electron-hole
         recombinations at distant pairs of interstitial Li donors
         and substitutional Zn acceptors, recorded photoelectrically
         at a very low level of optical excitation. The unusual
         form of this d-a pair band suggests the superposition of
         two spectra, one involving a Ga-type donor (Series A) and
         one a P-type donor (Series B), the latter possessing a
         characteristically strong no-phonon (NP) component. The
         MC phonon replicas TA, LA and TO are as in Fig. 1, O is an
         optical phonon replica of about 49 meV. Part of the
         Li_I-Li_{Ga}-O_P bound exciton spectrum also appears.

of two intermeshed d-a pair spectra explains why analysis of the
discrete pair lines is very difficult.

     Careful examination of the sidebands in the Li B exciton
spectrum from an optimally-doped vapor grown needle at high optical
excitation level has revealed weak structure superposed on the discrete pair transitions, similar to the "two-electron" transitions
observed for chalcogenide P-site donors left in excited states after
exciton decay. This structure is 1.5 ± 0.2 meV closer to Li_B^o than
corresponding structure in the Te donor exciton spectrum (27), suggesting $(E_D)_B = 88.3 \pm 0.5$ meV consistent with the result from Fig.
3.

IV. HAYNES' RULE FOR DONORS AND ACCEPTORS IN GaP

It is customary to plot E_{BX} vs. E_D for excitons bound to
neutral donors, on the grounds that a significant trend should
result if the exciton is primarily bound by exchange between the
two electrons (10). The data for the interstitial Li donors help
to establish the trend much more firmly in GaP, compare Fig. 4
with Fig. 8 of Dean et al. (17). This trend appears to bisect the
divergent points for the P-side donors S and Te (23) and the point
for the Ga-site donor Si also deviates significantly (28). Otherwise, the data are well represented by a "modified Haynes' rule"
(Section I), possessing a large negative intercept at $E_D = 0$. By
contrast, the corresponding data for acceptors (14) show a well
defined linear trend with a significant positive intercept at
$E_A = 0$. The interesting question arises, does the form of the E_{BX},
E_D relation found for donors necessarily imply the different form
observed for E_{BX}, E_A? The appropriate comparison is with Si, where
the band structure is very similar apart from the presence of
inversion symmetry noted earlier. Linear trends again occur for
Si, but with comparable slopes and negligible intercepts for donors
and acceptors (10). It is easy to show that Haynes' rule is not
the general form (29). Let us decompose E_D and E_{BX} into ideal components $(E_D)_{EM}$ and $(E_{BX})_{EM}$, given by appropriate forms of effective
mass theory (30), and central cell contributions which distinguish
different donors. These latter depend on the impurity site potential
V_c and on the electronic charge in the central cell, ρ_c or the
change $\delta\rho_c$ on binding the exciton (second electron). Then,

$$E_D = (E_D)_{EM} + V_c\rho_c$$

$$E_{BX} = (E_{BX})_{EM} + V_c\delta\rho_c \tag{2}$$

Equations (2) are consistent with expectation from a model calculation in the region of V_c sufficiently large to account for the
observed linear trend of E_{BX} with E_D (Fig. 4), and imply that

$$\frac{E_{BX}}{E_D} \simeq \left(\frac{E_{BX}}{E_D}\right)_{EM}\left[1 + V_c\left\{\frac{\delta\rho_c}{(E_{BX})_{EM}} - \frac{\rho_c}{(E_D)_{EM}}\right\}\right] \tag{3}$$

Thus if $\delta\rho_c/\rho_c > (E_{BX}/E_D)_{EM}$, the plot of E_{BX} vs. E_D will be as
shown in Fig. 4, with a negative intercept at $E_D = 0$. On the con-
verse assumption, the form observed for acceptors results. Hopfield
has discussed binding of such excitons in terms of the variation of
the ratio m_e/m_h of the electron and hole effective masses, appropriately adjusted for different semiconductors or different impurities in a given semiconductor (31). Such an approach might be
followed to account for the differences in form of E_{BX} vs. $E_{D,A}$ for
Si (10) and GaP (Fig. 4), given that the essential feature does not
seem to be the introduction of bond ionicity. A donor on a P site
has properties symmetrical with an acceptor on a Ga site on Phillips

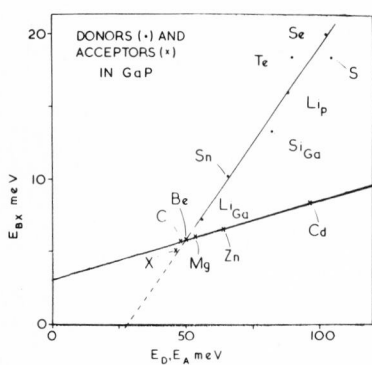 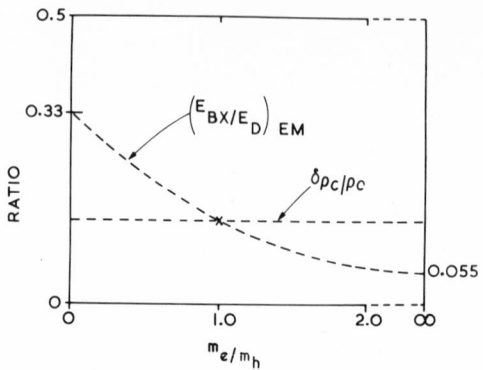

Fig. 4   The dependences of the bound exciton localization energies,
          E_{BX}, on the ionization energies of the indicator donors
          (• points) and acceptors (x-points) in GaP. The donor Li_P
          is on an interstitial site of mixed but predominantly P-
          type symmetry, while the interstitial donor Li_{Ga} has pri-
          marily Ga-type symmetry (see text). Linear trends are well
          established, but do not extrapolate back near the origin,
          quite unlike these trends in the semiconductor Si. This
          "modified Haynes' rule" behavior is shown to be the general
          form. Deviation of the trend for a given type of center to
          one side of the origin implies opposite behavior for the
          alternate type of center, as observed.

Fig. 5   The schematic variation of the ratio $(E_{BX}/E_D)_{EM}$, where EM
          denotes effective mass values in the approximation discussed
          in the text, with m_e/m_h, interpolated between extremal
          values from the H_2 (m_e/m_h = 0) and H^- (m_e/m_h = ∞), drawn
          to pass through the experimental ratio for Si where it is
          assumed m_e/m_h = 1.0. The mass parameters discussed in the
          text give m_e/m_h = 1.1 for Si. The trend of the donor
          central-cell charge parameter ratio $\delta\rho_c/\rho_c$ is also indicated
          in a completely arbitrary manner, subject only to the con-
          straints that $\delta\rho_c/\rho_c$ <,> $(E_{BX}/E_D)_{EM}$ for m_e/m_h = 0, ∞ and
          that these two ratios are equal for m_e/m_h = 1.0. It is
          shown in the test that the "modified Haynes' rule"
          behavior in Fig. 4 then follows if the representative point
          for GaP falls to the right of the cross-over, that is at
          m_e/m_h > 1. The mass parameters in the text give m_e/m_h =
          1.2 ± 0.2 for GaP. Note that the abscissa mass ratio must
          be inverted in a corresponding analysis of these binding
          energies for acceptors, the GaP point then lying to the
          left of Si.

"two-dimensional" theory of bonding (32). Considering acceptors and donors separately, Fig. 4 contains no significant difference in trend between impurities on P and Ga sites.

Figure 5 contains an entirely schematic representation of the possible variations of the ratios $\delta\rho_c/\rho_c$ and $(E_{BX}/E_D)_{EM}$ with m_e/m_h, using analogies with the H_2 molecule and the H^- ion to fix the end points. To account for the Si results (10), we assume that the appropriate $m_e/m_h \sim 1$ for Si. If the two curves in Fig. 5 cross near $m_e/m_h = 1$, then Haynes' rule results from Eq. (3), for acceptors as well as donors if we may distinguish between these two simply by inverting the mass ratio. The observed difference between GaP and Si, resulting in the more general form, "modified Haynes' rule," would then follow if the representative point for GaP lies to the right of the cross-over mass ratio.

This implies that the appropriate $(m_e/m_h)_{GaP} > (m_e/m_h)_{Si}$ and is therefore significantly > 1. This is consistent with current views on effective masses in these two semiconductors, if it is assumed that the appropriate m_e is given from $(1/m_e) = (1/3)[(2/m_t) + (1/m_l)]$, where m_t and m_l are the usual masses in the <100> valleys, and m_h is taken from the valence band parameter A (33). For GaP A ~ -3.7 (34), while A = -4.28 ± 0.02 for Si (35) in units of $\hbar/2m$. The electron mass m = 0.18 m_0 in GaP (27,36), 0.19 m_0 in Si (37), while m_l = 1.5 m_0 in GaP (27,36) and 0.92 m_0 in Si (37), where m_0 is the free electron mass.

We conclude that Eq. (3) contains a suitable generalization of Haynes' rule, which may be interpreted qualitatively with the help of Fig. 5. A quantitative interpolation of $(E_{BX}/E_D)_{EM}$ between the limits of Fig. 5 is obviously needed, and the form of $\delta\rho_c/\rho_c$ requires substantiation. Equation (3) does not apply for very small deviations from Coulomb (EM) binding, where model calculations show that the ratio $\delta\rho_c/\rho_c$, and therefore $\delta E_{BX}/\delta E_D$ become vanishingly small. Clear evidence for a tail towards constant E_{BX} at small E_A, consistent with these notions, has been observed recently in GaAs and InP (38), where $m_e/m_h \ll 1$. Similar, weaker evidence may be found in the original data for acceptors in Si (10).

ACKNOWLEDGMENTS

Some of the experimental work was completed while the author was with Bell Laboratories and many of the crystals used were grown there. He is grateful to R. B. Zetterstrom (Bell) for the solution grown Li-doped crystals, and to G. Kaminsky (Bell) and June Shaftoe (RRE) for performing the Li diffusions. The central idea in Section IV emerged in discussions with J. J. Hopfield and helpful advice on the general form of exciton binding was observed from J. W. Allen. The paper is communicated by permission of the Director, RRE.

REFERENCES

1. H. Reiss and C. S. Fuller, "Semiconductor," ed. N. B. Hannay, Reinhold, New York (1959), Ch. VI. M. Waldner, M. A. Hiller and W. G. Spitzer, Phys. Rev. $\underline{140}$, A172 (1965).

2. T. E. Gilmer, Jr., R. K. Franks and R. J. Bell, J. Phys. Chem. Sol. $\underline{26}$, 1195 (1965).

3. R. L. Aggarwal, P. Fisher, V. Mourzine and A. K. Ramdas, Phys. Rev. $\underline{138}$, A882 (1965).

4. K. Weiser, Phys. Rev. $\underline{126}$, 1427 (1962); H. Nara and A. Morita, J. Phys. Soc., Japan $\underline{23}$, 831 (1967).

5. G. D. Watkins and F. S. Ham, Phys. Rev. $\underline{B1}$, 4071 (1970).

6. H. Nara and H. Yamazaki, J. Phys. Soc. Japan $\underline{28}$, 1485 (1970).

7. C. S. Fuller and K. B. Wolfstirn, J. Appl. Phys. $\underline{34}$, 1 (1963).

8. W. Hayes, Phys. Rev. $\underline{138}$, A1227 (1965).

9. D. F. Nelson, J. D. Cuthbert, P. J. Dean and D. G. Thomas, Phys . Rev. Lett. $\underline{17}$, 1262 (1966).

10. J. R. Haynes, Phys. Rev. Lett. $\underline{4}$, 361 (1961).

11. P. J. Dean, Phys. Rev. $\underline{B4}$, 2596 (1971).

12. C. J. Frosch, Proc. Int. Conf. Cryst. Growth, Boston, 1966, Pergamon, New York (1967), p. 305.

13. H. Reiss and C. S. Fuller, J. Metals $\underline{8}$, 276 (1956).

14. P. J. Dean, R. A. Faulkner, S. Kimura and M. Ilegems, Phys. Rev. $\underline{B4}$, 1926 (1971).

15. P. J. Dean, Phys. Rev. $\underline{157}$, 655 (1967).

16. T. N. Morgan, Phys. Rev. Lett. $\underline{21}$, 819 (1968).

17. P. J. Dean, R. A. Faulkner and S. Kimura, Phys. Rev. $\underline{B2}$, 4062 (1970).

18. D. G. Thomas, M. Gershenzon and J. J. Hopfield, Phys. Rev. $\underline{131}$, 2397 (1963).

19. Y. Yafet and D. G. Thomas, Phys. Rev. $\underline{131}$, 2405 (1963).

20. C. H. Henry, P. J. Dean, D. G. Thomas and J. J. Hopfield, Localized Excitations in Solids, ed. R. F. Wallis, Plenum, New York, (1968), p. 267.

21. D. G. Thomas and J. J. Hopfield, Phys. Rev. $\underline{128}$, 2135 (1962).

22. P. J. Dean and R. L. Hartman, Phys. Rev. $\underline{B5}$, 4911 (1972).

23. Note that the variation of no-phonon strength observed in exciton spectra associated with different P-site substitutional donors (Table 1) is considerably larger than expected simply from the small differences in total binding energy. These differences, as well as the deviations observed in Fig. 4 may be due to variations in contributions of conduction band valleys other than X_1 to the electron wave-functions, which can be significant for the deeper donors; T. G. Castner, Jr., Phys. Rev. $\underline{B2}$, 4911 (1970).

24. D. G. Thomas, J. J. Hopfield and W. M. Augustyniak, Phys. Rev. $\underline{140}$, A202 (1965).

25. P. J. Dean and D. G. Thomas, Phys. Rev. $\underline{150}$, 690 (1966).

26. Significantly above the S-C d-a NP line, T. N. Morgan, T. S. Plaskett and G. D. Pettit, Phys. Rev. $\underline{180}$, 845 (1969).

27. A. Onton and R. C. Taylor, Phys. Rev. $\underline{B1}$, 2587 (1970).

28. No absolutely compelling evidence exists for the assignment of the $Si_{1,2}$ bound exciton to the Si_{Ga} donor. For a recent discussion of this assignment, see A. T. Vink, A. J. Bosman, J. A. W. van der Does de Bye and R. C. Peters, J. Lum. $\underline{5}$, 57 (1972), but some puzzling problems remain.

29. Excitons bound to neutral donors in wurtzite CdS also show E_{BX} accurately proportional to E_D, with a large intercept, K. Nassau and C. H. Henry, Proc. 10th Int. Conf. Phys. Semicond., USAEC, Oak Ridge, Tenn. (1970), p. 629.

30. For donors see R. A. Faulkner, Phys. Rev. $\underline{184}$, 713 (1969).

31. J. J. Hopfield, Proc. 7th Int. Conf. Phys. Semicond. Paris, Dunod, Paris (1964), p. 725.

32. J. C. Phillips, Phys. Rev. $\underline{B1}$, 1540 (1970); ibid. 1545 (1970).

33. W. F. Brinkman, T. M. Rice, P. W. Anderson and S. T. Chui, Phys. Rev. Lett. $\underline{28}$, 961 (1972).

34. R. A. Faulkner, private communication, calculated using the
     empirical pseudopotential method of M. L. Cohen and T. K.
     Bergstresser, Phys. Rev. <u>141</u>, 789 (1966). A similar result is
     obtained with recent experimental cyclotron resonance masses,
     $m_{lh}/m_O = 0.18$, $m_{hh}/m_O = 0.6||<100>$, R. A. Stradling, private
     communication.

35. J. C. Hensel and G. Feher, Phys. Rev. <u>129</u>, 1041 (1963).

36. Provisional cyclotron resonance estimates are $m_t/m_O = 0.25$
     ± 0.04, $m_1/m_O \sim 0.7$, R. A. Stradling, private communication.

37. J. C. Hensel, H. Hasegawa and M. Nakayama, Phys. Rev. <u>138</u>,
     A225 (1965).

38. A. M. White, P. J. Dean, K. M. Fairhurst, W. Bardsley, E. W.
     Williams and B. Day, Solid State Comm. (in press).

LUMINESCENCE OF II-VI COMPOUNDS DOPED WITH TRANSITION METAL IONS

S. A. Kazanskii, G. I. Khilko, A. L. Natadze and

A. I. Ryskin

The State Optical Institute, Leningrad, USSR

ABSTRACT

The killer action of transition metal ions in II-VI compounds is a result of their ability to capture a free carrier in one of the excited 3d energy levels and of subsequent non-radiative relaxation. Recombination at the transition metal ions is accompanied by the characteristic luminescence of these ions. The charge transfer spectra of II-VI compounds doped with transition metal ions and the formation of complexes bound to transition metal ions are discussed. Structure in the absorption and luminescent spectra are interpreted theoretically.

CHARGE TRANSFER BANDS AND EXCITATION MECHANISMS

Very small amounts of transition metal impurities such as Fe, Co, and Ni are known to quench the recombination luminescence of II-VI compounds. Klasens related the quenching and trapping effects of Co and Ni in ZnS. He has shown that recombination at the transition metal ion occurs as follows: the carrier makes a radiationless transition between the closely spaced energy levels of this ion (1). The EPR and optical data of II-VI compounds doped with transition metal ions from Ti to Ni show that these ions are usually in the divalent state. The sensitive EPR method detects the small percentage of recharged ions in photo-excited crystals (Mn^{2+} is the single exception); this recharge occurs with a change in the number of electrons in the incomplete 3d-shell. In some crystals paramagnetic ions can in principal recharge to both the monovalent and trivalent states, depending on the presence of other defects. Fig. 1 shows the scheme by which Co acts as a trap and quencher. Free carrier

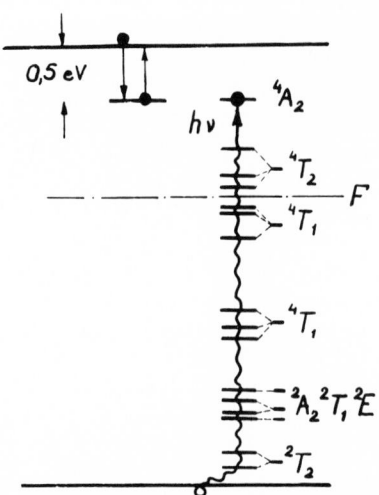

Fig. 1   The action of Co^{2+} ion as a trap and as a "killer" of re-
         combination luminescence in ZnS crystal; F is Fermi level.

recombination at transition metal ions is seen to be accompanied
by the characteristic intra-center luminescence of these ions in
the divalent state The carrier capture at the transition metal ion
is followed by the ion luminescence in the valence state corresponding to the presence of the carrier in the 3d shell. The quantum yield, η, varies over a wide range depending on the host crystal and ion quencher. For instance η≃1 for Co^{2+} in ZnS but η∼10⁻³
for Co^{2+} in CdS; η does not exceed 10^{-4} for Ni^{2+} in ZnS and CdS.

    The recharge of transition metal ions occurs through two
different ways: first by free carrier capture or second by direct
charge transfer between the paramagnetic ion and the crystal. The
latter process occurs upon light absorption in charge transfer
bands. Usually this process is identified with charge transfer between paramagnetic ion and crystal bands, i.e. continuum spectrum
since the carriers with any value of wave vector $\underline{k}$ are created in
this transition (2). In some cases a different process occurs:
for example, the bound hole instead of the free one appears after
electron transfer to the 3d shell. The system of discrete energy
levels corresponds to this bound state complex and the absorption
(luminescence) band connected with its creation (radiative decay)
should have a structure. Dingle reports the presence of the complex in ZnO:Cu crystal (3). Similar complexes appear in ZnS, CdS
and ZnSe crystals doped with Ni (4). The charge transfer bands in
the absorption spectra of these crystals have pronounced structure,
in which the zero-phonon line and its phonon replicas are resolved.

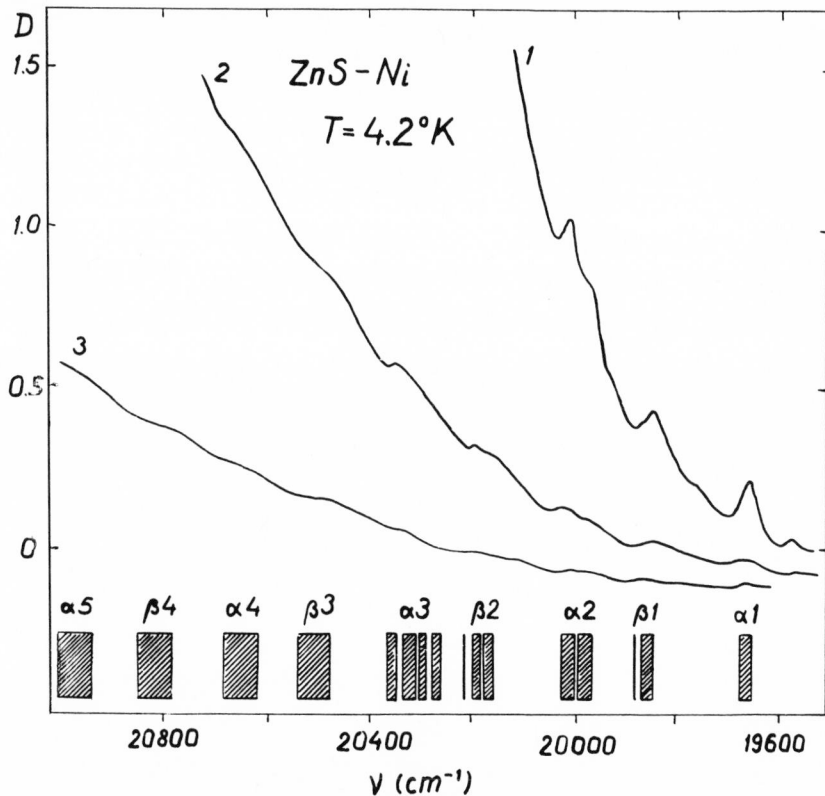

Fig. 2 Photoelectric recording and scheme of long wavelength edge
of the charge transfer band in ZnS:Ni crystal at T = 4.2 K.
Curves 1, 2, 3 correspond to crystals of different thickness
and the same content of Ni ions (10^{-3}) (see Ref. 4).

The complex was shown to consist of a Ni^{1+} ion and hole, localized
in a quasi-molecular orbital near the ion. The charge transfer
absorption spectrum of ZnS:Ni is shown in Fig. 2. This band corresponds to the allowed transition from the totally symmetric $\Gamma_1[^3T_1$
$(^3F)]$ state of Ni^{2+} ion into the three-fold degenerate Γ_5 state of
the complex. The latter state can be constructed of the ground
state function of Ni^{1+} ion (symmetry $\Gamma_7[^2T_2]$) and the wave function
of the bound hole. There is one important distinction between the
complex considered and complexes bound to isoelectronic traps. The
momentum j = 3/2, characteristic of the valence band of the zinc-
blende structure at k = 0, does not appear in the Γ_5 orbital of the
complex bound to the Ni ion. This is shown in particular by the
absence, more exactly the smallness, of zero-phonon line splitting
from the trigonal crystal field present in the wurtzite structure

or from stacking faults in zinc-blende structure or from external
fields, i.e. uniaxial stress, magnetic field. The analysis of these
results shows that the hole localized in the quasi-molecular orbital
has spin degeneracy only (5).

Another difference from the isoelectronic trap case is the
absence of resonance luminescence due to radiationless hole relaxation among the excited states of Ni^{2+} ion. It should be noted that
the ion-hole complex in ZnO:Cu decays radiatively since the Cu^{1+}
ion, formed after decay, is diamagnetic and does not have levels
with small separation.

The hole binding energy in the ZnS:Ni crystal is equal to
$E_b = 0.17$ eV. Dissociation of the complex under the action of light
with frequency exceeding the zero-phonon line frequency by $\Delta\nu = E_p/h$
is observed. The probability of this process increases with increase of the light frequency. Free holes created in this process
in ZnS:Ni,Cu crystal were detected by the IR-luminescence of Cu^{2+}
ion arising from hole capture by Cu^{1+} ions (6). This result
supports the model of Cu center, suggested by Birman (7). Practically total recharge of Cu^{1+} ions ($Cu^{1+} \rightarrow Cu^{2+}$) in the surface
layer $\ell \approx 4/\kappa_\lambda$, being the absorption coefficient at wavelength λ, can
be achieved in ZnS: 10^{-4} Ni,10^{-5} Cu crystal under the action of
light dissociating the ion-hole complex. For example, in crystals
with 10^{-4} Ni κ_λ at $\lambda = 400$ nm is 200 cm^{-1}. Because of the decrease
of the concentration of Cu^{1+} ions in the surface layer, the Cu^{2+}
luminescence excited in the charge transfer band after fast rise,
decreases to zero. The transmission and luminescent spectra and
the scheme of transitions are shown in Fig. 3 for ZnS:Ni,Cu.

The induced absorption in the wavelength region 0.51 $\mu m < \lambda$
< 1.1 μm in the crystal is due to Ni^{1+} and Cu^{2+} ions. This absorption remains in darkness at 77 K for an apparently unlimited period,
but it can be destroyed by illuminating the crystal in the spectral
region indicated. All the Cu^{2+} ions pass into the monovalent state
and all the Ni^{1+} ions pass into the divalent state by this illumination.

The presence of deep levels in the energy gap of II-VI compounds doped with transition metal ions makes possible the two step
excitation of free carriers through these levels. In the ZnSe:Ni
crystal, anti-Stokes visible luminescence is observed as a result
of successive electron transfer from valence band to Ni^{2+} ion (the
hole is localized at an unknown shallow acceptor) and from Ni^{1+} ion
to conduction band. The second stage of the process can be delayed
with respect to the first one for a long period at T = 77K (8).
Two step excitation of anti-Stokes luminescence is also observed in
other crystals, e.g. ZnS:Cr, ZnS:Ni. The quantum yield of the
luminescence is very small and its existence is not inconsistent

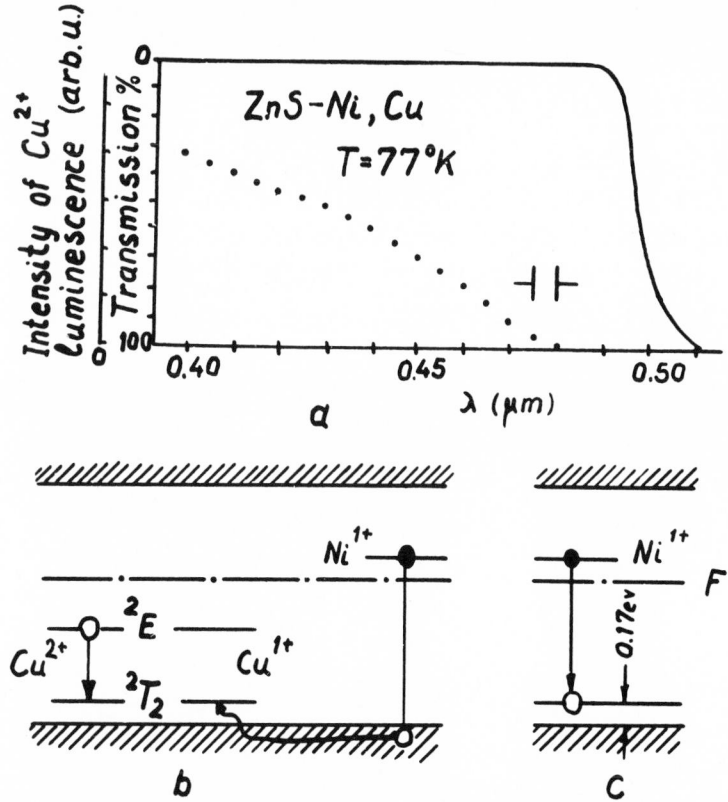

Fig. 3 Absorption (solid line) and excitation spectra of Cu^{2+} IR
luminescence (circles) in ZnS:Ni,Cu crystal (a), and scheme
of experimentally observed processes in this crystal when
absorbing light $\lambda < 0.48$ μm (b) and 0.48 μm $< \lambda < 0.51$ μm (c).

with the killer action of transition metal ions.

STRUCTURE OF ABSORPTION AND LUMINESCENT SPECTRA

We now consider the structure of absorption and luminescent
spectra of these ions. We restrict ourselves to consideration of
spin allowed transitions between Stark sublevels of the free ion
ground state. These transitions result in a one band spectrum for
configurations d^1, d^9, d^4, d^6 and in a two band spectrum for configurations d^2, d^8, d^3, d^7; these two bands are separated by in-

terval of several thousands cm^{-1}. Luminescence is observed in all
cases from the lower Stark sublevel. We have mentioned above that
the quantum yield of the luminescence depends on host crystal and
impurity ion.

The structure of bands under consideration is determined by
spin-orbit (SO) and electron-lattice (EL) interactions. The SO interaction which determines the structure of electron energy levels,
is described, in the one electron approximation, by two constants:

$$\zeta = -i/3 \langle t_2 || \kappa_{SO} || t_2 \rangle \qquad \zeta' = i/3\sqrt{2} \langle t_2 || \kappa_{SO} || e \rangle$$

Here $\langle || \ || \rangle$ are reduced matrix elements. For example, SO splitting
of the T_2 terms in the absence of static Jahn-Teller effect is defined by the constant ζ (in first order of perturbation theory).
The splitting of T_1 terms is defined by ζ and ζ'.

The partly covalent bond between the transition metal ion and
its ligands strongly influences the SO interaction. The calculation of the interaction for a tetrahedral complex by the MO LCAO
method allows one to find ζ and ζ' for the 3d-orbital of the central
ion ζ 3d, and the p-orbital of the ligands ζ_{Lp} (9).

$$\zeta = a\zeta_{3d} - b\zeta_{Lp} \qquad\qquad \zeta' = a'\zeta_{3d} \qquad\qquad (1)$$

Here a and a' are dependent on the overlap integral and covalency
parameters. In order of magnitude they are approximately unity.
The ζ_{3d} monotonically increases in the series of divalent ions from
79 cm^{-1} for Sc^{2+} to 829 cm^{-1} for Cu^{2+}. The ζ_{2p} for S, Se, Te (defined by the valence band splitting at $\underline{k} = 0$ in ZnS, ZnSe and ZnTe
crystals) is equal to 388, 2315 and 4845 cm^{-1}, respectively. It
follows from Eq. 1 that $\zeta' \simeq \zeta_{3d}$ but ζ can be considerably different
from ζ_{3d} and it can even change sign for the case of heavy ligands.
A similar change of sign of ζ has been observed for Cr^{2+} ion (10).

The change of equilibrium nuclear positions with electron
transitions and the perturbation of the electron state due to the
dynamical component of the crystal field are responsible for the
appearance in the spectra of phonon replicas of the electronic
transition. Quasi-localized vibrations have been observed connected
with introduction of transition metal ions (11).

The interaction of degenerate electronic states with degenerate
vibrations of complexes in II-VI compound crystals is considered.
For Cr^{2+} ion only, this interaction results in the static distortion of complexes along the fourfold axes. The crystal field acting
on the paramagnetic ions Cr^{2+}, have tetragonal, not tetrahedral
symmetry (12).

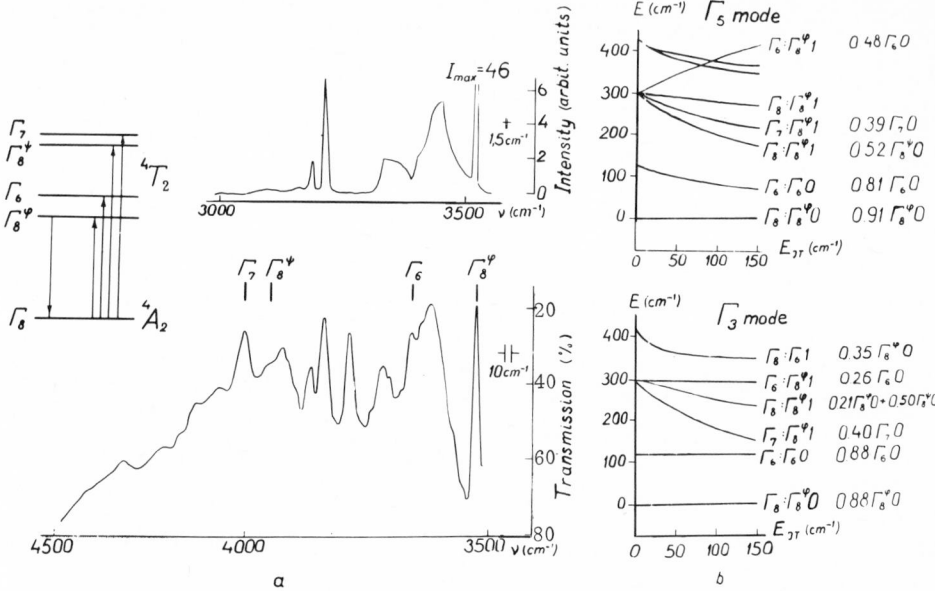

Fig. 4   a: Absorption (below) and luminescence (above) bands,
corresponding to transition $^4A_2 \overset{\rightarrow}{\leftarrow} {}^4T_2$ in ZnS:Co crystal.
The lines of the transition $\Gamma_8(^4A_2) \overset{\rightarrow}{\leftarrow} \Gamma^\phi_8(^4T_2)$ are matched.
The positions of the SO components of the $^4T_2$ term are in-
dicated on the absorption spectrum, T = 4.2 K.   b: Relative
energy of SO components Γ^ϕ_8, Γ_6 of $^4T_2$ term and their vibronic replicas with excitation of one quantum of Γ_3
(below) and Γ_5 (above) vibrations versus energy of JT interaction.   Energies are given with respect to the energy
level, Γ^ϕ_8. Γ_8: $\Gamma^\phi 1_8$, for instance, labels the vibronic
level Γ_8 corresponding to a transition of the ion to the
level Γ^ϕ_8 with one quantum of vibration excited.   The contributions of the wave function of the parent states (at
E_{JT} = 0) to the wave function of vibronic states at E_{JT} = 0)
to the wave function of vibronic states at E_{JT} = 150 cm^{-1}
are given in brackets.

The dynamic Jahn-Teller (JT) effect reveals itself in two aspects depending on the strength of EL interaction.   The strong EL
interaction occurs for the first excited state $^3T_1(^3F)$ of Ti^{2+} ion
in CdS crystal.   The pronounced vibrational structure is not observed in this case and the dynamic JT effect due to interaction
with triply degenerate vibrations is revealed by the temperature
dependent splitting of the absorption band of the $^3A(^3F) \rightarrow {}^3T_1(^3F)$
transition into three components (13).

The moderate or weak EL interaction occurs in the majority of cases. It does not prevent the existence of vibronic structure. The form of the spectrum depends on the relative values, of the SO interaction constant, the JT interaction energy and the vibrational energy involved. Let us consider for example, the excited term $4T_2(4F)$ of Co^{2+} ion in ZnS crystal (1). The SO splitting of this term is not suppressed considerably. Therefore JT interaction can be considered as a weak perturbation, which, however, appreciably disturbs the spectrum. Lines are observed in the absorption spectrum which are neither purely electronic lines nor their phonon replicas; the symmetric lines are absent in the luminescence spectrum (Fig. 4a). The most intense line occurs in the gap between the acoustic and optical branches of lattice vibrations (205-276 cm^{-1}). Fig. 4 b shows the relative energies of vibronic states of the term considered versus the energy of the JT interaction with doubly Γ_3 or triply Γ_5 degenerate vibrations. This interaction is seen to transfer intensity from zero-phonon lines to vibronic bands and shifts these bands relative to zero-phonon lines. Intense vibronic lines should occur in the gap both in the case of Γ_3 and Γ_5 vibrations, in accordance with absorption spectrum observed. Though many modes of lattice vibrations possess the symmetry required and could interact with the term, the most important features of the spectrum can be explained by interaction with the narrow band of optical phonons near $q = 0$ ($\hbar\omega \approx 300$ cm^{-1}). A similar situation occurs for the $5T_2(5D)$ state of Fe^{2+} in ZnS crystal (14).

In conclusion we consider the structural imperfection of ZnS crystals which also contributes to the form of the absorption and luminescence bands. Bulk ZnS crystals grown at T>1024 C (the phase transition temperature) contain as a rule considerable concentrations (up to 10% and even more) of stacking faults, i.e. hexagonally packed layers in "predominantly cubic" packing (15). Specific varieties of activator centers in these crystals occur due to introduction of impurity ions into the layers with different symmetry (16). The magnitude of the tetrahedral component D_q of the crystal field acting on paramagnetic ions in stacking faults is several per cent lower than for ions in cubically packed layers. The Stark splitting for ions in stacking faults is correspondingly reduced and a long wavelength shift of absorption and luminescence lines occurs, on the order of 100 cm^{-1} (17). The magnitude of D_q and the spectral line positions for ions in stacking faults and in regular hexagonal packing (wurtzite structure) are nearly equal. There are lines in the spectra due to activator ions in layers near the stacking fault. The parameter D_q for these ions differs insignificantly from that for ions in regular cubic packing. The long wavelength part of the absorption spectrum (the short wavelength part of luminescence spectrum) in doped crystals with stacking faults looks similar for different transition metal ions and for different transitions in a given ion and differs only by the scale (Fig. 5a,

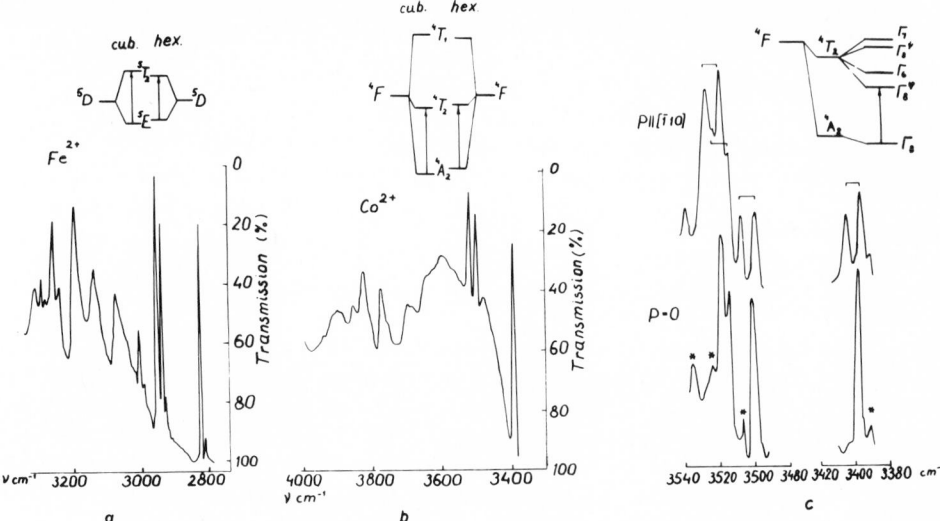

Fig. 5   a,b: Absorption bands due to transitions $^5E \rightarrow {}^5T_2$ in
Fe^{2+} ion (a) and $^4A_2 \rightarrow {}^4T_2$ in Co^{2+} ion (b) in ZnS crystal
with stacking faults; α (the probability of close-packed
layer to be faulted) is equal to 0.1.   c: Piezo-spectroscopic effect for the group of zero-phonon lines at the
long wavelength part of the absorption spectrum shown in
Fig. 5 b.   Line 3518 cm^{-1} is due to the Co^{2+} ions in regular
cubic packing, line 3398 cm^{-1} is due to the ions in stacking
faults.   Lines in the region 3500-3515 cm^{-1} are due to ions
in layers near the stacking faults.   The atmospheric absorption lines are labeled by the stars (single beam recording).   The piezo-spectroscopic effect is the splitting
of the upper Γ_8 level of the transition by the stress P.

b).   The ratio of intensities of zero-phonon lines in this region
of the spectrum depends on the relative concentration of different
types of centers and on the crystal structure, and can be changed
by annealing of the crystal (16, 17).

The trigonal component of the stacking fault crystal field
splits the triply Γ_4, Γ_5 and four-fold Γ_8 degenerate levels.   But
this splitting, estimated on the basis of the point charge electrostatic model, is of the order of 1 cm^{-1} and is hidden in the spectral line width (18, 19).   The piezo-spectroscopic and Zeeman
effects for centers of different type show that for all the centers
the crystal field acting on the transition metal ion can be taken
to be approximately tetrahedral (19).   Fig. 5c demonstrates the
identity of the piezo-spectroscopic pattern for four zero-phonon

lines, connected with centers of different type in ZnS:Co crystals with stacking faults.

In contrast to the spectra of ZnS crystals (containing stacking faults), doped with Fe, Co and Ni, the spectrum of crystals doped with Cu is complicated (16). Probably this complication is due to the considerable concentration of point defects in crystals with stacking faults and to the well known tendency of Cu ions to form associated centers (20).

Since the luminescence band for ions in stacking faults is shifted to the long wavelength spectral region there occurs resonance energy transfer to these ions from those in the perfect parts of crystal. This transfer is more important for lower temperatures and for high concentrations of dopant (17).

REFERENCES

1. H. A. Klasens, J. Electrochem. Soc. $\underline{100}$, 72 (1953).

2. J. W. Allen, J. Phys. C. (Solid State Physics) $\underline{2}$, 1077 (1969).

3. R. Dingle, J. Luminescence $\underline{1-2}$, 48 (1970).

4. S. A. Kazanskii, A. I. Ryskin, Opt. i Spectr. $\underline{31}$, 619 (1971); S. A. Permogorov, A. N. Reznitskii and B. A. Kazennov, Opt. i Spect. $\underline{32}$, 744 (1972).

5. S. A. Kazanskii and A. I. Ryskin, Fiz. Tverd. Tela $\underline{14}$, 1818 (1972).

6. S. A. Kazanskii and A. I. Ryskin, Fiz. Tverd. Tela $\underline{13}$, 3725 (1971); S. A. Kazanskii and A. I. Ryskin, Opt. i Spect. $\underline{33}$, 273 (1972).

7. J. L. Birman, in: "Proc. Intern. Conf. on Luminescence" (Acad. Kiado Budapest, 1968) vol. 1, p. 919.

8. S. A. Kaxanskii and A. I. Ryskin, Fiz. Tverd. Tela $\underline{13}$, 3633 (1971).

9. A. I. Ryskin, A. L. Natadze and S. A. Kazanskii, Zh. Eksp. Theor. Fiz. (in press).

10. J. T. Vallin and G. D. Watkins, Solid State Commun. $\underline{9}$, 953 (1971).

11. R. Beserman, M. A. Nusimovici and M. Balkanski, Phys. Stat. Solidi 34, 309 (1969); R. Beserman and M. Balkanski, Phys. Solidi 44, 535 (1971).

12. T. L. Estle, G. K. Walters and M. de Wit, in: "Paramagnetic Resonance" (Academic, New York, 1963) vol. 1, p. 144; K. Morigaki, J. Phys. Soc. Japan 19, 187 (1964); T. L. Estle and W. C. Holton, Phys. Rev. 150, 159 (1966); J. T. Vallin, G. A. Slack, S. Roberts and A. E. Hughes, Phys. Rev. B2 4313 (1970).

13. R. Boyn and G. Ruszzynski, Phys. Stat. Solidi 48, 643 (1971).

14. F. S. Ham and G. A. Slack, Phys. Rev. B4, 777 (1971).

15. E. Ebina and T. Takahasi, J. Appl. Phys. 38, 3079 (1967); G. V. Ananieva, K. K. Dubenskii, A. I. Ryskin and G. I. Khilko, Fiz. Tverd. Tela 10, 1800 (1968).

16. I. Broser and H. Maier, J. Phys. Soc. Japan 21 (Suppl.), 254 (1966).

17. S. A. Kazanskii, A. I. Ryskin and G. I. Khilko, Fiz. Tverd. Tela 10, 2415 (1968); S. A. Kazanskii, A. I. Ryskin and G. I. Khilko, Opt. i Spect. 27, 156 (1969).

18. A. I. Ryskin and G. I. Khilko, Fiz. Tverd. Tela 13, 195 (1971).

19. A. I. Ryskin and G. I. Khilko, Opt. i Spectr. 31, 755 (1971).

20. W. L. Holton, M. de Wit, R. K. Watts, T. L. Estle and J. Schneider, J. Chem. Phys. Solids 30, 963 (1969).

POLARIZED LUMINESCENCE OF SEMICONDUCTORS DUE TO OPTICAL

ORIENTATION OF ELECTRONS

R. I. Gioev, B. P. Zakharchenya, V. G. Fleisher

Physico-Technical Institute, Academy of Sciences

Leningrad, USSR

ABSTRACT

Several physical phenomena are studied using measurements of the degree of circular polarization of recombination radiation in semiconductors during excitation by circularly polarized photons with energies corresponding to interband transitions. The effects of an external magnetic field are investigated. Experimental results are presented for optical orientation in GaAs crystals and in solid solutions GaAs-AlAs and GaAs-GaP. Lifetimes and spin relaxation times of non-equilibrium carriers in the 10^{-10} to 10^{-11} sec range are measured.

The electric field of the light wave for dipole interband absorption in semiconductors changes the orbital motion of electrons. This absorption can induce the orientation of spins of non-equilibrium carriers because of spin-orbit interaction.

The energy level diagram of a semiconductor with the diamond type structure is shown in Fig. la. The valence band is split due to spin-orbit interaction. The states at k=0 have total angular momentum 3/2 in the valence band and the states of electrons in the conduction band have total angular momentum 1/2. At the energy $E_{h\nu} \sim E_g$ of the exciting light quanta, preferential population of the states with the spin -1/2 takes place in the case of the right curcularly polarized (σ^+) exciting light. The probability of transitions from the states with the magnetic quantum number $m_Z=-3/2$ is three times greater than the probability of transitions from the states with $m_Z=-1/2$ (transitions from the states with $m_Z=3/2$ and

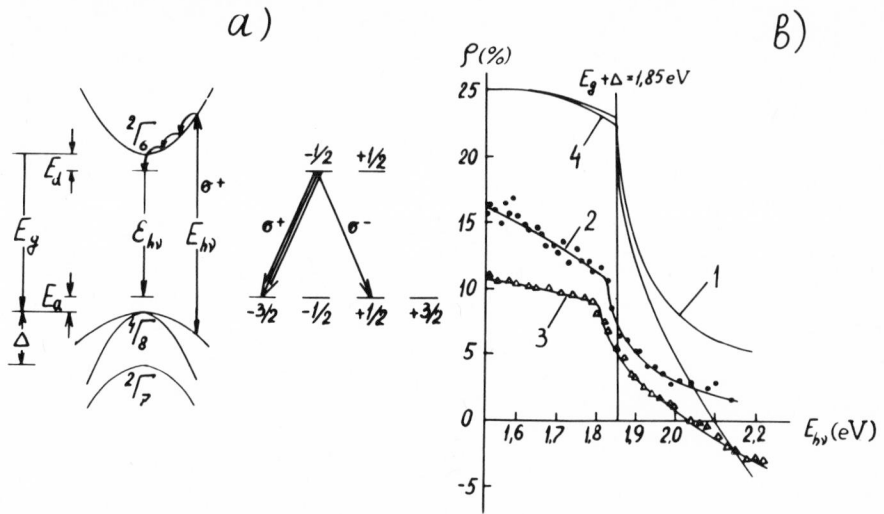

Fig. 1   a) The diagram of the energy levels and transitions in the
         diamond type structure semiconductors. b) The dependence
         of ρ on the exciting quanta energy. Curves 2 and 3 are
         experimental dependences $\rho(E_{h\nu})$ at 77 K for GaAs crystals
         with acceptor concentration $2 \times 10^{19}/cm^3$ and $3 \times 10^{18}/cm^3$,
         respectively; curve 1 is the calculated dependence $\rho(E_{h\nu})$
         without taking into account the spin relaxation; curve 4,
         the same taking into account the spin relaxation.

1/2 for σ^+ light are not allowed). The maximum degree of orientation of the non-equilibrium electron gas $P_o max$ is $P_o max = (G_- - G_+)/$
$(G_- + G_+) = 0.5$, where G_- and G_+ are the rates of generation of conduction electrons in the states $-1/2$ and $1/2$, respectively. The
existence of relaxation leads to a decrease of the orientation to
the value $P = (T_n/\tau)P_o$ where τ is the electron lifetime and T_n is
the time of existence of the spin orientation of non-equilibrium
electrons. If the spin relaxation time is τ_s, then: $1/T_n =$
$(1/\tau) + (1/\tau_s)$.

     Optical orientation in semiconductors was first observed by
Lampel (1) who measured the stored nuclear magnetization arising
as a result of the hyperfine interaction of the spin-oriented electrons with paramagnetic nuclei. The luminescence method used by
Parsons (2) for measuring the optical orientation in GaSb is more
universal and sensitive.

     In this communication we present the results of experimental
studies of optical orientation in GaAs crystals and solid solutions

GaAs-AlAs and GaAs-GaP. All these crystals are of p-type doped with
Zn ($E_a \stackrel{\sim}{\sim} 40$ meV). The circular polarization of the recombination
radiation corresponding to transitions of non-equilibrium electrons
from the conduction band (B-A) or from the shallow donor levels
($E_d \stackrel{\sim}{\sim} 7$ meV) to acceptors (D-A) was studied in the experiment.
These levels can be considered sufficiently shallow in comparison
with the magnitude of Δ, the spin-orbit splitting in the crystals
studied. Therefore, they can be described by the wave functions of
the corresponding bands. The same selection rules and intensity
relations are valid for these transitions and for interband transitions.

The probability of the transition from the state with $m_Z = -1/2$
to the state with $m_Z = -3/2$ is three times greater than the probability
of the transition $-1/2 \rightarrow +1/2$ (Fig. 1a). For observation along the
beam of the exciting σ^+ light the recombination radiation is also
right-circularly polarized. The maximum degree of orientation in
this case is: $\rho_m = (1/2)P_O = 1/4$, and ρ attains the maximum value
at $E_{h\nu} \stackrel{\sim}{\sim} E_g$ and for $\tau_s \gg \tau$. The increase of $E_{h\nu}$ leads to the decrease of ρ because of poorer excitation conditions and because of
acceleration of the spin relaxation.

Curve 1 shown in Fig. 1b is calculated by Diakonov and Perel
(3) neglecting spin relaxation. At $E_{h\nu} - E_g > \Delta$ the transitions from
the split-off band Γ_7 with $J = 1/2$ occur. These transitions populate
the states in the conduction band with spin orientation opposite in
sign to that of the states arising as a result of transitions from
band Γ_8. So the inclusion of the Γ_7 band in the absorption process
must be accompanied by a decrease of P_O and ρ.

The rate of further decrease of P_O is determined by the rela-
tion of probabilities of the transitions $\Gamma_7 \rightarrow \Gamma_6$ and $\Gamma_8 \rightarrow \Gamma_6$. In
the range $E_g < E_{h\nu} \stackrel{\sim}{\leq} E_g + \Delta$ some decrease of P_O also occurs due to the
admixture of the states of the split-off band. The experimental
curve 2 in Fig. 1b. for p-GaAs with the acceptor concentration of
$2 \times 10^{19}/cm^3$ is similar to curve 1 in its shape but it goes lower, a
feature that can be connected with the influence of spin relaxation.
The circular polarization of recombination radiation observed by us
for $E_{h\nu}$ considerably greater than E_g proves the partial conservation of spin orientation of highly excited electrons during the
energy relaxation. This conservation can be explained by the
stabilizing action of scattering by non-paramagnetic centers if the
spin relaxation is determined by the coupling of the spin of an
electron with its linear momentum. This coupling takes place for
the splitting of the conduction band in crystals without a center
of inversion. The change of the linear momentum due to scattering
leads to a change of the direction of the effective magnetic field
acting on the electron spin. The frequent change of directions
slows down the spin relaxation (an effect similar to the narrowing

of the paramagnetic resonance lines in liquids and gases). More detailed discussion of this question is given in (3).

The spin relaxation mechanism connected with the splitting of the conduction band depends strongly upon the energy. With a decrease of the stabilizing action of impurities (in the case of lower concentrations) electrons excited from the Γ_8 band relax in spin faster than electrons excited from the Γ_7 band since the former have larger initial energies. Already at impurity concentration 3×10^{18}/cm^3 we observed a change of sign of the circular polarization at $E_{h\nu} \approx 2.04$ eV (5), which corresponds to the contribution of the Γ_7 band being larger than the contribution of the Γ_8 band to the resulting polarization (curve 3 in Fig. 1b). The energy $E_{h\nu}$ corresponding to "remagnetization" of the non-equilibrium electron gas is determined by the energy at which the spin relaxation due to the band splitting becomes important. Theoretical curve 4 in Fig. 1b is calculated taking into account only this mechanism of the spin relaxation. It corresponds to the case for which with an initial energy of electrons $E_0 \approx 0.65$ eV, P_0 decreases upon thermalization. The change of sign of the circular polarization of luminescence for the lower impurity concentrations is explained in Ref. (6).

The optical orientation leads to the appearance of a magnetic moment which in the simplest case decreases with the characteristic time T_n after switching off the exciting light. The convenient feature of the time scanning method consists in the comparison of the relaxation rate $1/T$ with the magnetic moment precession frequency ω in the transverse magnetic field H. For observation of luminescence of a semiconductor under stationary conditions of excitation:

$$\rho = (T_n/\tau)[(\cos \phi + \omega T_n \sin \phi)/(1 + \omega^2 T_n^2)];$$

where ϕ is the angle between the direction of the observation of the luminescence and the direction of propagation of the exciting light. The dependence $\rho(H)$ at $\phi = 0, \pi$ is described by the Lorentzian curve $\rho = (T_n/\tau)[\rho_{max}/(1 + \omega^2 T_n^2)]$. Measurements of the half-widths $H_{1/2}$ of the $\rho(H)$ curves and the value of $\rho(0)$ allow us to determine the times τ_s and τ if the g-factor is known: $T_n = \hbar/g\mu_B H_{1/2}$. Hence, $\tau = T_n \rho_{max}/\rho(0)$. Consequently, it is possible to measure easily times of the order 10^{-10} to 10^{-11} sec. However, it is difficult to use this method if the optical orientation cannot be characterized by one constant parameter T_n. Such a situation arises, for example, because of the influence of a semiconductor surface with τ and τ_s different from the bulk values. At large absorption coefficients (10^4 to 10^5/cm) corresponding to direct transitions the contribution of the surface region becomes substantial. In this case the $\rho(H)$ curves correspond to the superposition of the recombination radiation of spatially separated groups of electrons

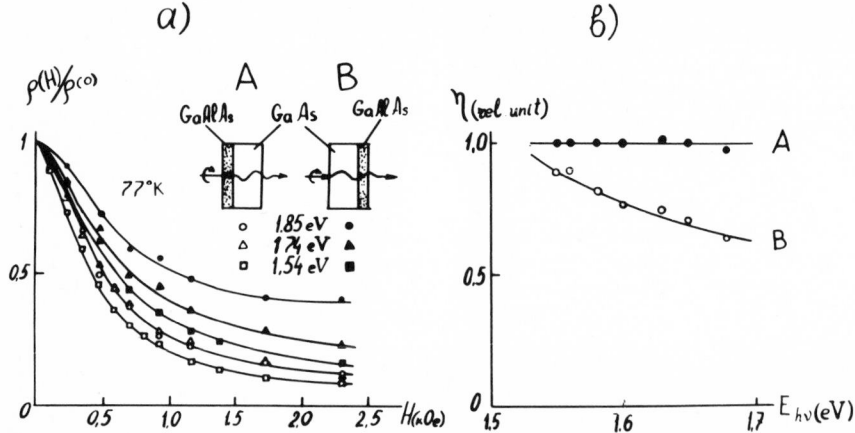

Fig. 2   The influence of surface conditions on the depolarization
curves in the transverse magnetic field (a) and the quantum
yield of the luminescence (b). A is the case of excitation
of luminescence through the wide gap epitaxial layer on the
surface of GaAs; B, excitation through the chemically
polished surface of GaAs.

with different values ot T_n (7).

The curves $\rho(H)$ at 77 K obtained by excitation of GaAs crystal
through the layer of the wide gap (E_g = 2.1 eV) epitaxial solution
GaAs-AlAs (A) and through the natural surface of the GaAs crystal
(B) are compared in Fig. 2a. In case B surface states play an important role. The shape of the $\rho(H)$ curves depends on the magnitude
$E_{h\nu}$ of the exciting light, since with increase of $E_{h\nu}$ the absorption
coefficient grows and, consequently, the contribution of the surface
region increases. With crystallization of the GaAs-AlAs layer the
number of defects at the surface of GaAs decreases and the curves
$\rho(H)$ for different $E_{h\nu}$ become more similar. The change of the shapes
of the $\rho(H)$ curves in case B correlates with the decrease of the
quantum yield η due to the increase of non-radiative recombination
at the surface (Fig. 2b) with increasing $E_{h\nu}$. In the case A, η is
practically independent of $E_{h\nu}$. At 4.2 K the curves $\rho(H)$ change
considerably less with change of $E_{h\nu}$, than at 77 K. This is explained by the decrease of non-radiative surface recombination.
In the case A the curves $\rho(H)$ are satisfactorily described by
Lorentzian functions.

However, determination of τ and τ_s from $H_{1/2}$ and $\rho(0)$ requires additional analysis. It is necessary to be sure of the
absence of additional fast mechanisms of spin relaxation taking

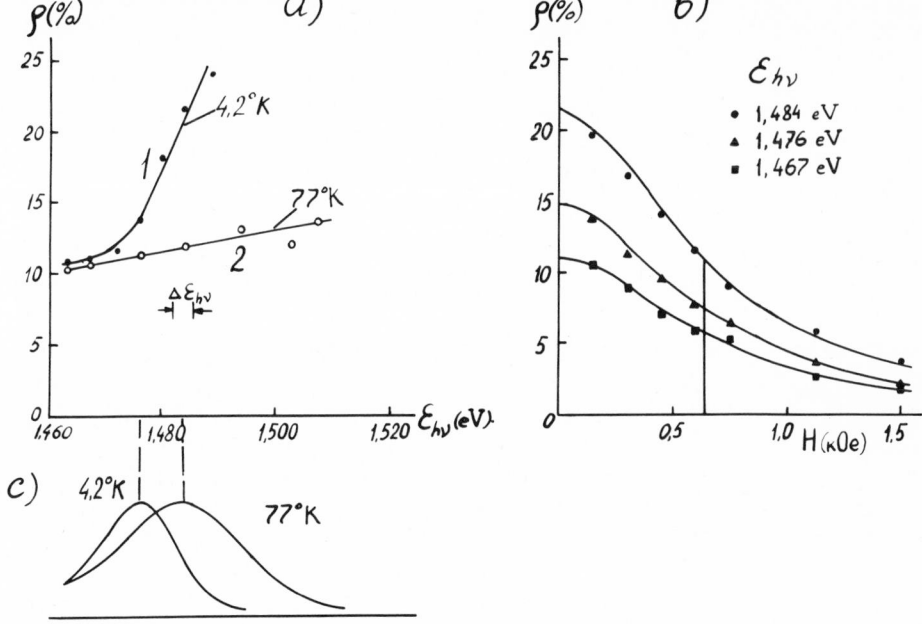

Fig. 3   a) The spectral dependence $\rho(E_{h\nu})$ of the luminescence polar-
         ization of the GaAs crystal at 4.2 K (curve 1) and at 77 K
         (curve 2); b) The depolarization of the luminescence in the
         transverse magnetic field at different $E_{h\nu}$; c) The lumines-
         cence lines at 4.2 K and 77 K.

place during a short time and decreasing ρ but practically having
no influence upon the value of T_n for energy-relaxed electrons.
The dependence of ρ on the energy $E_{h\nu}$ of the recombination radiation
at 4.2 K of a GaAs crystal with a dopant concentration of $3\times10^{18}/$
cm^3 is shown in Fig. 3a (curve 1). A considerable increase of ρ
on the short wavelength side of the recombination spectrum   is observed. However, the half-widths of the curves $\rho(H)$ for different
$E_{h\nu}$ are practically equal (Fig. 3b). A considerable change of $\rho(H=0)$
with increase of $E_{h\nu}$ at constant $H_{1/2}$ implies strict limitations
upon the spin relaxation mechanism. It is necessary to assume that
either the change of τ_s with increase of $E_{h\nu}$ is accompanied by a
compensating change of τ so that T_n remains constant, or the existence of an additional fast mechanism for spin relaxation decreasing ρ.

The first proposal seems hardly probable. The second one
assumes the existence of an additional mechanism of spin relaxation
acting only during a small part of the lifetime since the lumines-

cence polarization is partially conserved. To explain the observed
dependence $\rho(E_{h\nu})$ it is possible to propose the following model.
The electrons excited to the conduction band with energy E, higher
than the energy of one optical phonon $E_{c,ph}$, relax fast (in a time
of the order of 10^{-13} sec) to energy values $E < E_{o,ph}$. Further relaxation in energy is slow and occurs mainly by means of acoustic
phonons. The localization of photoexcited electrons for times
$t_1 \ll \tau$ is possible if there are capture centers at the bottom of
the conduction band. If the electron effectively relaxes in spin
while moving in the band, the degree of its initial orientation at
the capture center will be different depending upon the time t_1
spent in the band. Further relaxation is determined by the time
T_n for a localized state. The capture cross section by the hydrogen-like center is determined by the ionization energy E_d of the
center and the energy of the electron (8, 9). However, the observed spectralwidth of the recombination radiation (the half-width
of which is $\sim$ 18 meV) cannot be explained by possible fluctuations
of E_d. If the luminescence is determined by donor-acceptor pairs
the line-width may be connected with fluctuations of the distances
R between the recombination centers, the smaller R corresponding
to the short wavelength edge of the line. From the experimental
data it is necessary to ascribe smaller t_1 to smaller R in the
model considered, i.e. it is necessary to assume an influence on
the capture cross section of the acceptors which are nearest to
the donor centers. The mean distance $\overline{r_a}$ between acceptors is $\sim$ 70 A
which is less than Bohr radius a_B of donor centers ($\sim$ 100 A).

The effective spin relaxation before the transition to the
ground state of donor centers can be caused by the following mechanisms: 1. Spin exchange with neutral acceptors which are paramagnetic centers. The cross section of this exchange known for
donors is 10^{-12} cm^2 (10). 2. Spin relaxation at excited levels of
impurity states with non-zero orbital angular momentum. The influence of this mechanism can be different depending upon the energy
of the excited level relative to the bottom of the conduction band.
The influence on the spin orientation is maximal in the case of an
isolated level. Let us note that the perturbation connected with
the nearest acceptors can influence considerably the excited states
of donor centers and the position of the corresponding levels relative to the bottom of the conduction band. In relatively pure
crystals, for example, luminescence corresponding to excited states
isolated from the band are observed (11). These states merge with
the conduction band upon the introduction of impurities. Since ρ
approaches the maximum value at the short wavelength edge of the
line the value $T_n = (3.3 \pm 0.2) \times 10^{-10}$ sec, determined by the half-
widths of the curves $\rho(H)$, is identical with $\tau(\tau_s \gg \tau)$. The con-
stancy of τ can be connected with the fulfillment of the requirement $\overline{r_a} < a_B$. It is obvious that $R < a_B$ and overlapping of the wave
functions of donors and acceptors depends little upon R. The in-

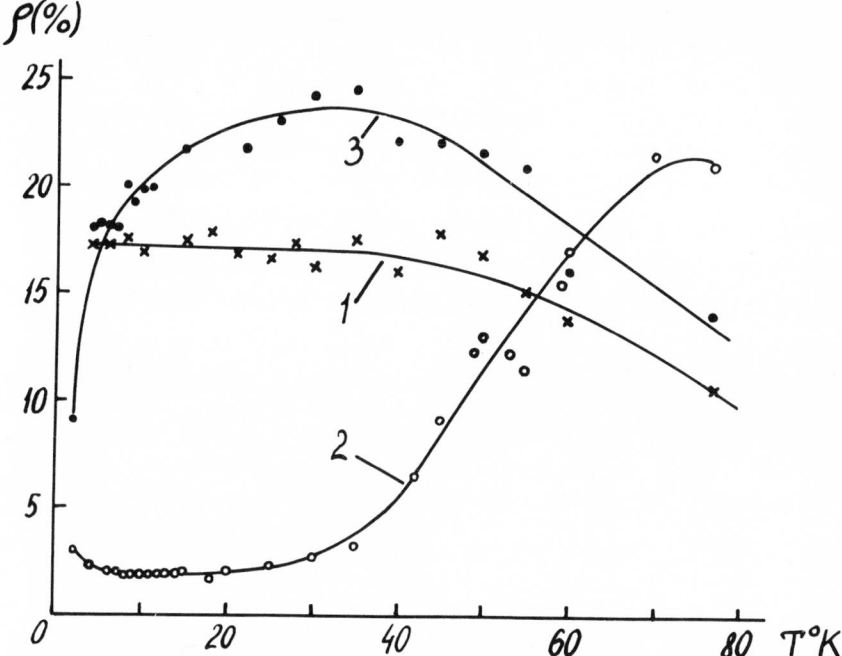

Fig. 4   The temperature dependences of the luminescence polariza-
          tion for crystals GaAs (1), $Ga_{0.7}Al_{0.3}As$ weakly doped (2),
          and $GaAs_{0.7}P_{0.3}$ higher doping (3).

crease of temperature to 77 K leads to a substantial decrease of
the contribution of the D-A recombination and to an increase of the
intensity of B-A transitions. The contribution of capture processes
decreases and, consequently, the dependence of ρ on $E_{h\nu}$ (curve 2
in Fig. 3) weakens considerably. Therefore, localization of electrons improves the conservation of the spin orientation. As curve
1 in Fig. 4 shows, ρ changes little in the range from 4.2 K to 40 K,
and at the higher temperatures an additional depolarization of the
luminescence is observed which is connected with the increase of
the spin relaxation contribution in the band. At T = 40 K the beginning of the line shift to shorter wavelengths for the transition
from D-A to B-A recombination occurs.

     A different situation is possible in pure and weakly doped
crystals with large τ. In this case the influence of spin exchange
with paramagnetic neutral acceptors upon movement in the band decreases. The spin relaxation due to L-S interaction with impurity
excited states and due to the hyperfine interaction with the nuclei
of semiconductor atoms during the localization can play an impor-

tant role. Both mechanisms must be switched off in a sufficiently strong field which destroys the L-S coupling or the coupling with nuclear spins. As an example, curve 2 in Fig. 4 for the weakly doped solid solution $Ga_{0.7}Al_{0.3}As$ (E_g = 1.92 eV) illustrates the decrease of ρ at the transition to low temperatures, corresponding to localization of photo-excited carriers and D-A recombination. The value of τ is of the order of 10^{-8} sec in this case.

The magnitude of ρ increases sharply in the longitudinal magnetic field, approaching saturation at the level of 20% in the fields 2 to 3 kG. The resonance in a similar crystal was observed (12), revealing the role of spin exchange between systems of localized oriented electrons and nuclei.

It is necessary to note that the depolarization curves $\rho(H)$ for GaAs and $Ga_{0.7}Al_{0.3}As$ at low temperatures are well described by Lorentzian curves. Noticeable difference from the Lorentzian shape is observed for the curves $\rho(H)$ in the solid solution $GaAs_{0.7}P_{0.3}$ doping $10^{18}/cm^3$, E_g = 1.86 eV. The dependence $\rho(T)$ is shown in Fig. 4 (curve 3). Relatively deep donor centers (5 to 7 meV) in this crystal are, apparently, little active in the luminescence. The recombination radiation line does not shift in the temperature interval from 4.2 K to 77 K. A red shift of the line maximum by 12A is observed from 4.2 K to 2 K. This shift is accompanied by a decrease of ρ which can be explained by the spin relaxation of localized electrons. The effect of the increasing ρ in the longitudinal magnetic field (by 1.5% in the field of 3 kG) is observed at 2 K. The longitudinal magnetic field does not induce a noticeable increase of ρ in the GaAs crystal (curve 1 in Fig. 4).

The superposition of different relaxation mechanisms with comparable T_n values can lead to a difference of the curves $\rho(H)$ from Lorentzian curves. These deviations from Lorentzian shape can be characterized, for example, by the relations:

$$\rho(0.5H_{1/2}^L)/\rho(0.5\ H_{1/2}) \text{ and } \rho(2H_{1/2}^L)/\rho(2H_{1/2})$$

for the Lorentzian curve with the half-width $H_{1/2}^L$ equal to the experimentally observed $H_{1/2}$. These relations for the crystal $GaAs_{0.7}P_{0.3}$ as a function of T are shown in Fig. 5. As it can be seen from Fig. 5, the experimental curves $\rho(H)$ approach Lorentzian curves at T=2 K. The maximum deviations are observed near 30 K when the ρ is maximal and approaches the theoretical limit, i.e. when $\tau \gg \tau_s$. It is possible to suppose that the observed curves $\rho(H)$ reflect the superposition of different channels of recombination with comparable values of τ (for example, with comparable intensities of B-A and D-A transitions). The recombination through donor centers prevails at 2K and the curve $\rho(H)$ approaches a Lorentzian.

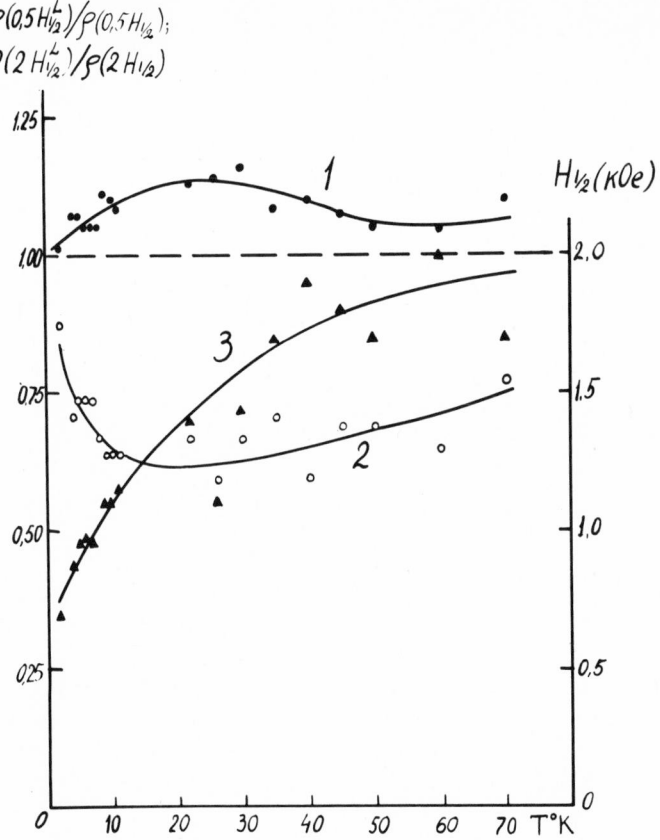

Fig. 5 Comparison of the depolarization curves ρ(H) for the
GaAs$_{0.7}$P$_{0.3}$ crystal with Lorentzian curves, having the
same half-widths (H$_{1/2}^{L}$ = H$_{1/2}$). 1. The dependence of
ρ(0.5 H$_{1/2}^{L}$)/ρ(0.5 H$_{1/2}$) on T. 2. The dependence of
ρ(2H$_{1/2}^{L}$)/ρ(2 H$_{1/2}$) on T. 3. The change of the half-width
H$_{1/2}$ with temperature.

REFERENCES

(1) G. Lampel, Phys. Rev. Lett. 20, 491 (1968).

(2) R. R. Parsons, Phys. Rev. Lett. 23, 1152 (1969).

(3) M. J. Dyakonov, V. J. Perel, Zh. Eksp. Theor. Fiz. 60, 1954
     (1971).

(4) B. P. Zakharchenya, V. G. Fleisher, R. J. Gioev,
     Yu. P. Veshchunov, and I. B. Rusanov. ZhETF Pis. Red. 13,
     195 (1971).

(5) R. I. Gioev, B. P. Zakharchenya, V. G. Fleisher, ZhETF
     Pis .Red. 14, 553 (1971).

(6) A. I. Ekimov and V. I. Safarov, ZhETF Pis. Red. 13, 700
     (1971).

(7) D. Z. Garbuzov, R. I. Gioev, L. M. Kanskaya, V. G. Fleisher,
     FTT 14, 1720 (1972).

(8) N. Sclar, E. Burstein, Phys. Rev. 98, 1757 (1955).

(9) H. Gummel, M. Lax, Phys. Rev. 97, 1469 (1955).

(10) G. Feher, T. C. Hensel and E. A. Gere, Phys. Rev. Lett. 5,
     309 (1960).

(11) I. Shak, R. C. C. Leite, I. P. Gardon, Phys. Rev. 176, 139
     (1968).

(12) A. I. Ekimov and V. I. Safarov, ZhETF Pis. Red. 15, 453
     (1972).

EFFECT OF CHANNELING ON THE SCINTILLATION RESPONSE OF NaI(Tl) TO POSITIVE IONS*

R. B. Murray, M. R. Altman, H. B. Dietrich,**

T. J. Rock***

University of Delaware, Newark, Delaware, and ***BRL

Radiation Division, Aberdeen Proving Ground, Maryland, USA

ABSTRACT

The scintillation response of NaI(Tl) to ^{16}O ions has been studied in the MeV region with the ion beam aligned along low index axes and planes and also along a random direction. The scintillation efficiency increases when the beam is channeled along a major symmetry direction. This effect is due to the lower electronic stopping power experienced by a channeled particle. The effect of channeling is observed along {100}, {110}, and {111} planes and along <100>, <110> and <111> axes. These experiments also permit study of channeling phenomena in thick crystals, e.g. measurement of the channeling critical angle.

INTRODUCTION

When a positive ion of energy E is stopped in a scintillator part of the energy is emitted in luminescent photons which are then detected by a photomultiplier tube. The photomultiplier current is integrated to produce a voltage pulse of amplitude L, and a plot of L versus E is a monotonically increasing function. Scintillation efficiency is defined as dL/dE, the slope of a pulse height versus energy plot. For positive ions on solid scintillators there is an overall trend of decreasing dL/dE as the stopping power dE/dx

---

**Present address: BRL Radiation Division, Aberdeen Proving Ground, Maryland, 21010, USA.

increases (1). This dependence of scintillation efficiency on stopping power apparently arises from local radiation damage effects along the track of the incident ion.

It is well known that energetic positive ions experience a reduced stopping power when incident along low-index axes or planes of a crystalline solid (2). This effect arises from a correlated scattering of the incident ion by the lattice atoms for incidence along major symmetry directions. This correlated scattering and the associated steering along the open regions of the crystal is known as channeling. Since a channeled ion travels through the crystal with a lower dE/dx than for a randomly incident ion, it is expected that a channeled ion will have a greater scintillation efficiency. Thus, for incident ions of a particular energy E stopping in a thick crystal it is expected that the pulse height L will depend upon the orientation of the ion trajectory with respect to crystalline directions, and that L will exhibit maximum values for orientation along low-index axes or planes. This effect has been studied theoretically by Luntz and Bartram (3) for NaI(Tl) and CsI(Tl) scintillators. They predicted that the scintillation pulse height from an energetic positive ion could be enhanced by as much as four times its normal value if the ion experiences a channeled trajectory.

This paper reports the results of a series of experiments on the scintillation response of NaI(Tl) to ^{16}O ions in the range 3-60 MeV, for random incidence and for incidence along the {100}, {110} and {111} planes and along the <100>, <110>, and <111> axes. The results are in qualitative agreement with the calculation of Luntz and Bartram as the channeled ions produce a distinctly greater pulse height. The magnitude of the increase, however, is less than that predicted.

EXPERIMENT

All experiments were performed with NaI(Tl) scintillation crystals obtained from Harshaw Chemical Company. Crystals were cleaved into circular platelets (approximately 3 cm^2 x 2 mm) in an atmosphere of dry air in a dry box, and mounted in a goniometer while in the dry box. Crystal surfaces were always carefully protected from the atmosphere. The goniometer permitted rotation of the crystal about a horizontal axis perpendicular to its cleavage face, and rotation about a vertical axis that lay in its face. The crystal could be oriented with a precision better than ±0.02 degrees in either angle. Scintillation pulses were detected with an end-window photomultiplier tube that was mounted horizontally within the beam-line vacuum system. In most experiments the photocathode surface was placed as close as possible to the back side

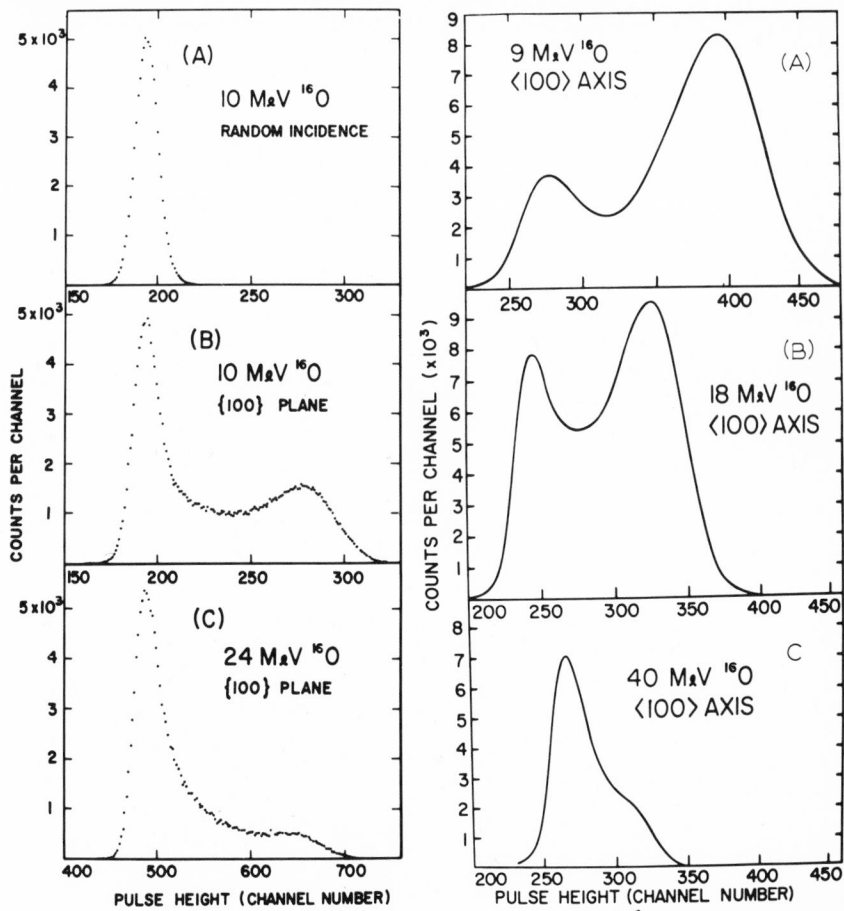

Fig. 1 (a) Pulse-height spectrum from 10-MeV ^{16}O on NaI(Tl) for
incidence along a random direction. (b) Pulse-height
spectrum from 10-MeV ^{16}O along a {100} plane. (c) Pulse-
height spectrum from 24-MeV ^{16}O along a {100} plane. A
light guide was used in all cases.

Fig. 2 (a) Pulse-height spectrum from 9-MeV ^{16}O ions on NaI(Tl)
along a <100> axis. The smaller peak on the left coincides
with that observed for random incidence. (b) Pulse-height
spectrum from 18-MeV ^{16}O along a <100> axis. (c) Pulse-
height spectrum from 40-MeV ^{16}O along a <100> axis. No
light guide was used. The channel numbers in (a), (b),
and (c) are not correlated.

of the scintillation crystal, but no light guide was used (to permit freedom of rotation about the vertical axis). In some experiments a lucite light guide was used to couple the back side of the

crystal to the photomultiplier tube; this results in better pulse-
height resolution and still permitted rotation through the
horizontal axis.

Experiments were performed with the tandem Van de Graaff
accelerator of the Nuclear Effects Laboratory at Edgewood Arsenal.
The charged particle beam was collimated to a half-angle of 0.016°
by 1-mm apertures separated by a distance of 3.65 m. Scintillation
pulse-height spectra were recorded on a Nuclear Data multi-channel
analyzer.

RESULTS AND DISCUSSION

Fig. 1 (a) shows the pulse-height spectrum resulting from
10 MeV ^{16}O ions incident on NaI(Tl) at a random angle. Fig. 1 (b)
shows the spectrum observed when the 10-MeV beam is aligned along
a {100} plane. In addition to a peak at the same pulse-height as
for random incidence, there is now an additional peak at a pulse-
height approximately 42% greater. There is also a continuous distribution of pulses between the two peaks. The spectrum for 24 MeV
^{16}O along a {100} plane is shown in Fig. 1 (c). The relative number
of pulses in the higher peak is clearly reduced from that in the
10-MeV case. We have recorded spectra for alignment along the
{100} plane for ^{16}O ions in the range 5 to 32 MeV, and the general
features of the spectra over this entire range are illustrated by
Fig. 1. As the energy decreases below 10 MeV the relative number
of counts in the higher peak increases. At energies above 24-MeV
the higher peak becomes a continuous tail. Pulse-height spectra
were also recorded for 10-MeV ^{16}O on NaI(Tl) for alignment along
the {110} plane and along the {111} plane. In these cases a second
peak is not discernible, but there is clearly a distribution of
counts above the random peak.

The effect of channeling along low index axes is more pronounced than the effect along low index planes. Fig. 2 shows pulse-
height spectra for ^{16}O ions of three different energies incident
along a <100> axis. In Fig. 2 (a), for 9-MeV ^{16}O, the dominant
feature of the spectrum is the peak at high pulse heights due to
channeled ions. The smaller peak on the left coincides with the
spectrum observed for random incidence. Figs. 2 (b) and 2 (c) show
spectra for <100> incidence for ^{16}O ions of 18 MeV and 40 MeV
respectively. As in the case of {100} planes, the relative number
of counts in the channeled peak decreases with increasing energy.
We have recorded spectra for incidence along a <100> axis for ^{16}O ions
in the range 3 to 60 MeV, and the general features are as illustrated in Fig. 2. We have also measured pulse-height spectra from
10-MeV ^{16}O ions channeled along the <110> and <111> axes. For
channeling along <110> there is a distinguishable peak due to
channeled particles, and the pulse-height of the channeled peak is

approximately 42% greater than that of the random peak. For channeling along <111> a second peak is not resolved but there is a broad distribution of counts above the random peak due to channeled particles.

It is evident in Figs. 1 and 2 that a substantial fraction of the incident particles contribute pulses to a well-defined peak at a pulse-height greater than that for random incidence. We attribute this peak to particles that remain channeled to the end of their range, and thus experience the lower dE/dx of a channeled trajectory over their entire path. Such a behavior has been clearly demonstrated in other channeling experiments, for example, the work of Eriksson and co-workers (4) on range measurements of various heavy ions in oriented crystals of tungsten. The existence of a peak at low pulse-heights, coincident with that recorded for random alignment, is due to those particles that are never channeled. The continuous distribution of pulses between the two peaks is attributed to particles that are initially channeled but are subsequently deflected from a channeled trajectory to a random trajectory. The continuous distribution due to dechanneled ions overlaps both the peaks. The observed spectrum is thus interpreted as arising from ions in three different trajectories: (a) those that are never channeled, (b) those initially channeled and subsequently dechanneled, and (c) those that remain channeled throughout the entire range.

A careful measurement was made of the peak pulse height as a function of energy for ^{16}O ions on NaI(Tl) for both random incidence and for incidence along a <100> axis. The ratio of peak pulse-heights L (channeled)/L (random) as a function of energy is shown in Fig. 3. Luntz and Bartram (3) calculated the ratio L (channeled)/L (random) for various ions in NaI(Tl) and found that for ^{16}O this ratio was 4 at 20 MeV, decreasing monotonically to about 1.3 at 160 MeV. The experimental results shown in Fig. 3 are in qualitative agreement with Luntz and Bartram in the sense that L (channeled)/L (random) is found to be greater than 1 at all energies and is a decreasing function of energy at higher energies. The quantitative discrepancy in the magnitude of L (channeled)/L (random) is attributed to two features of the calculation by Luntz and Bartram: (a) the assumption that the dependence of dL/dE on dE/dx is the same for channeled particles as for randomly incident particles, and (b) the calculated ratio of dE/dx (channeled) to dE/dx (random) that was found to be 1/4 to 1/5. Both of these points have recently been examined by Luntz and Heymsfield (5). They conclude that assumption (a) is not valid since it is necessary to consider the energy deposited by secondary electrons in the region outside the primary track of very high ionization density. The secondary electron spectrum for a channeled ion of a given dE/dx is not the same as that for a random ion of the same dE/dx. In addition the

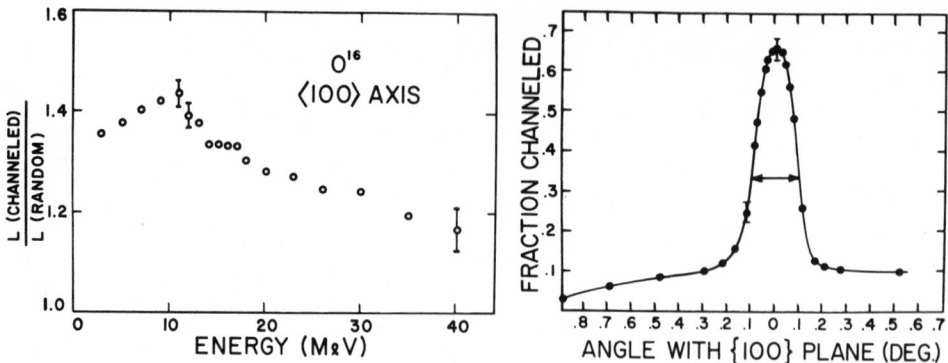

Fig. 3   Ratio of peak pulse-heights L (channeled)/L (random) vs.
          incident ion energy. Channeled ions along <100> axis,
          [16]O on NaI(Tl).

Fig. 4   Fraction of particles initially channeled as a function of
          angle from a {100} plane. [16]O on NaI(Tl) at 15 MeV.

original calculation of the ratio dE/dx (channeled) to dE/dx
(random) in (b) gave a value that is apparently too small (5).
Thus both factors (a) and (b) contributed to an overestimate of L
(channeled)/L (random) in the initial calculations (3). These
matters are discussed in more detail elsewhere (5).

     Incident particles are steered in channeling trajectories only
if their path is oriented along a symmetry direction within a
certain critical angle. We have measured the critical angle for
[16]O incident along {100} planes at an incident ion energy of 15 MeV.
To measure the critical angle (ψ_c) pulse-height spectra were
recorded as the angle between the beam and the {100} plane was
varied. Each spectrum was then analyzed to determine the number
of particles initially channeled. Fig. 4 shows the angular distribution which results from plotting the fraction channeled vs.
angle to the {100} plane. The half-width at half-maximum is taken
as the critical angle for channeling. For 15-MeV [16]O ions channeled
along {100} planes in NaI(Tl) ψ_c is found to be 0.10 ± 0.02°.

     The critical angle ψ_c for {100} planar channeling in alkali
halide crystals has been considered by Shipatov (6) based on an
interaction potential energy function of the form

$$V(r) = (Z_1 Z_2 e^2 / r) \exp(-r/a_B)$$

where $Z_1 e$ is the charge of the incident nucleus, $Z_2 e$ is the charge
of the lattice nucleus, and a_B is the Bohr screening parameter.
Using the interaction he finds an expression for $\overline{V}(\rho)$, the average

potential energy of interaction of the incident ion with a {100}
plane in an alkali halide of the NaCl structure, where ρ is the
distance from the atomic plane. Using this prescription, and correcting the static lattice result to room temperature, we obtain a
calculated value of $\psi_c = 0.11^\circ$. This may be compared with the
experimental value of $0.10 \pm .02^\circ$ from Fig. 4. There is thus
quantitative agreement between experiment and Shipatov's analysis
for the {100} plane.

ACKNOWLEDGMENTS

We wish to thank M. Luntz for numerous conversations and cor-
respondence concerning this work. We are indebted to various
members of the accelerator staff at Edgewood Arsenal for their
help, especially J. Morissey and G. Silsbee.

REFERENCES

*Work supported by the U. S. Atomic Energy Commission and
National Science Foundation.

1. J. B. Birks, The Theory and Practice of Scintillation Counting,
    Macmillan Co., New York (1964).

2. R. S. Nelson, The Observation of Atomic Collisions in
    Crystalline Solids, North Holland, Amsterdam (1968).

3. M. Luntz and R. H. Bartram, Phys. Rev. 175, 468 (1968).

4. L. Eriksson, J. A. Davies, and P. Jespersgaard, Phys. Rev. 161,
    219 (1967); L. Eriksson, Phys. Rev. 161, 235 (1967).

5. M. Luntz and G. M. Heymsfield, Phys. Rev. 6B, 2530 (1972).

6. E. T. Shipatov, Sov. Phys. Sol. State 10, 2132 (1969).

ELECTRON TRANSFER BY THE TUNNEL EFFECT AND ITS INFLUENCE ON THE F CENTER LUMINESCENCE IN ALKALI HALIDES

M. Ecabert, P. A. Schnegg, Y. Ruedin, M. A. Aegerter

and C. Jaccard

Physics Institute, University of Neuchâtel, Switzerland

ABSTRACT

Concentration quenching plays an important role in luminescence phenomena, and is generally accounted for by pair interactions. In alkali halides containing F centers, whenever a member of an F center pair is optically excited at low temperature, it can return to its ground state either radiatively or by the formation of an intermediate F' center. In this case, the excited electron is transferred by fast tunnelling to a neighboring center. This process depends on the spin symmetry of the pair, which is determined by the inhomogeneity of the hyperfine field. Effects of applied magnetic field, spin lattice relaxation and spin-spin interaction on the tunnelling probability are investigated. Experimental confirmation of the theory is presented for the case of KCl.

1. INTRODUCTION

The characteristic variations due to concentration effects of the emission properties of defects in insulating materials such as the quenching of the photoluminescence, the line broadening and the shortening of the radiative lifetime are now well-known experimental facts (1). For instance, the quenching interaction between similar activator ions occurs in many rare-earth activated crystals with increasing doping concentration. It affects strongly the emission states of these ions through multipolar or exchange interactions. Quenching interactions between dissimilar species are, however, generally associated with the mutual relationships of their energy levels. Energy may be transferred non-radiatively from a metastable state of an excited activator to a neighboring ion which has a lower energy level (ground state for instance) pro-

582

vided the rate of dissipation of the total electronic energy is
increased.

The emission properties of F centers in alkali halide crystals
(an electron trapped in an anion vacancy) have been studied quite
intensively during the last two decades (2). Excitation of an
isolated F center at low temperature leads after a local lattice
relaxation to the emission of a photon with a luminescent quantum
yield of unity. However, when the F center concentration is high
enough, there is a finite probability that two centers form a pair
within a certain critical distance R_t; in this case the electron of
the relaxed excited center $\tilde{F}^*$ may be transferred via a tunnelling
mechanism to a neighbor defect in its ground state F_O, thereby
forming an F' center (two electrons located in an anion vacancy)
and an α center (anion vacancy). The pair returns then to its
ground state F_O-F_O by some non-radiative mechanism. The existence
of this process was suggested twenty years ago by Markham et al
(3) and confirmed later experimentally by Miehlich's measurements
of the concentration quenching of the F center luminescent quantum
yield (4).

The study of this important de-excitation process has been
revived recently by two fundamental experiments. The first one has
been performed in our laboratory (5, 6) and has shown that a first
stage increase of the F center luminescent quantum yield η can be
induced by an external magnetic field at low temperature. This
effect, which is of the order of 40% for a concentration of 5×10^{17}
F centers/cm^3, occurs at low magnetic field (H<4kG). It can be explained by the competing influence on the pairs of the applied field
$\underline{H}_O$ and of the local nuclear fields $\underline{H}_N$ and $\underline{H}_N^*$. They modify the
character of the defect wave functions and consequently change the
tunnelling probability. At the same time Porret and Lüty (7) have
shown that in a second stage, an increase of η (up to a value of
$\eta=1$) can be observed for T<LHeT and high magnetic field (H>>10 kG).
As the F' ground state must have antiparallel spins, the non-
radiative transfer is in this case hindered because of the finite
spin polarization attained under these experimental conditions. We
see therefore that in both cases the symmetry of the spin wave
function of the initial and final tunnelling state of the pair,
taken as an entity, plays a fundamental role in the explanation of
the physical nature of this tunnelling mechanism.

In the next section we present the experimental results of
the luminescent quantum yield variation at low magnetic field, obtained with KCl crystals of different F center concentrations.
These results will be compared with a theory developed earlier by
Jaccard et al (6). We show in section 3 that the study of this
phenomenon as a function of temperature may give important information concerning the spin-lattice relaxation time of the F centers

in their relaxed excited state. In the last section some new effects
which occur when the spatial distribution of the defects is modified
(formation of loose aggregates) will be presented and discussed.

2. QUANTUM YIELD DEPENDENCE OF CONCENTRATION

In a recent paper (6), we have shown that the spin state of an
electron pair $\tilde{F}^*$-F_o, i.e., an F center in its relaxed excited state,
$\tilde{F}^*$, neighboring an F center in its ground state, F_o can be described
by a four level scheme, equally populated for low magnetic field and
for temperatures down to liquid helium temperature with the following
energies:

$$\varepsilon_{1,4} = \pm\{(1/2)g^*\beta H^* + (1/2)g\beta H\}, \quad \varepsilon_{2,3} = \pm\{g^*\beta H^* - (1/2)g\beta H\}$$

β is the Bohr magneton, g the Landé factor of the F_o ground state
and $|\underline{H}| = |\underline{H}_o + \underline{H}_N|$ the value of the local magnetic field at the site
of the defect. The index * describes the same quantities for the
relaxed excited state. Recalling that in the final state of the
tunnelling process both electrons of the F' center must have the
same orbital wave function but antisymmetric spin wave functions, it
is possible to give an expression for the tunnelling probability of
such a pair. Assuming that the spatial distribution of defects is
random and that the nuclear fields $\underline{H}_N$ and $\underline{H}_N^*$ are not correlated
and have an isotropic space distribution, the radiative quantum
yield can be expressed by

$$\eta = \exp(-nV_t <P_t>)$$

n is the F center concentration, V_t a critical spherical volume in
which the tunnelling occurs (in this model independent of n) and
$<P_t>$ an average of the tunnelling probability. It is worth recalling that if $<P_t>$ can be calculated exactly for the two special cases
$|\underline{H}_o| = 0$ and $|\underline{H}_o| >> |\underline{H}_N|$, $|\underline{H}_N^*|$ the averaging must be computed for
intermediate values of $|\underline{H}_o|$.

An experimental check of this result is a very difficult task
because of the determination of the absolute luminescent quantum
yield. However, the luminescence intensity is proportional to η,
the multiplicative constant containing geometrical factors and the
excitation intensity; its logarithm therefore is linear in $<P_t>$.
In this way an "a posteriori" determination of η can be obtained
and the slope of the straight line gives the value of nV_t. For
this purpose thirteen ultra pure KCl crystals have been additively
colored by the van Doorn technique with $2 \times 10^{16} < n < 1.4 \times 10^{18}$ F centers/
cm^3. These crystals have been carefully quenched between two
copper blocks and mounted under safe light in an Andonian variable
temperature cryostat. The optical excitation device consisted of a
quartz iodine lamp filtered by two broad band filters: a $CuSO_4$

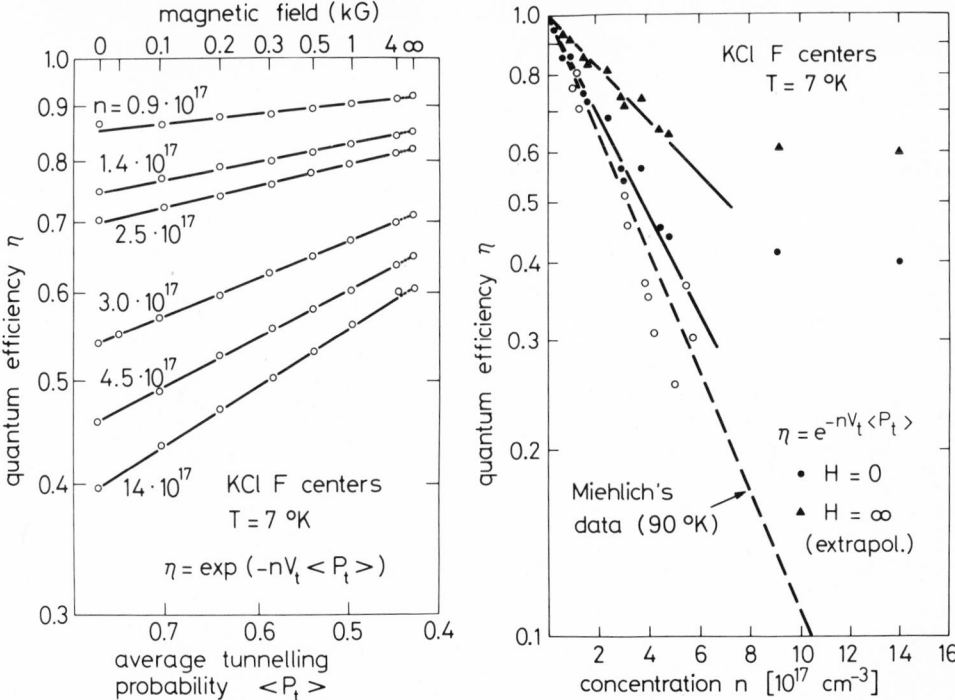

Fig. 1   Experimental radiative quantum yield of F centers versus
          computed average tunnelling probability P_t in KCl at
          7 K for different concentrations. W = 8 $\pm$1; R_t = 85 A
          (see (6)).

Fig. 2   Experimental radiative quantum yield η versus concentration
          n of F centers in KCl at 7K for zero • and infinite (extra-
          polated values ▲) magnetic field. Miehlich's data o,
          taken at 90K for H = 0, are also shown.

solution and an anticalorific Calflex B_1K_1 filter. In this way the
F centers were essentially excited in their main transition. The
F center luminescence was first filtered (Schott RG 715) in order
to eliminate completely the light excitation and then detected with
a PbS photoresistor (Mullard 119 CPY). The output signal was
selectively amplified (PAR 110 and JB5) and recorded.

     In Fig. 1 the calculated values of the luminescent quantum
yield η at 7K are plotted for six crystals in a logarithmic scale
versus the computed values of the average tunnelling probability.
The fitted straight lines are in good agreement with our model for

all concentrations. In Fig. 2 the same ordinate is plotted for
H = 0 and H = ∞ (>>H_N) in a logarithmic scale versus the macroscopic concentration of F centers (measured with a double beam
spectrophotometer). For comparison some of Miehlich's data,
measured at 90K for H = 0 are also shown. We can see that for concentrations up to 5×10^{17} F centers/cm^3 both results are in good
agreement too. From the slope of the straight lines and with the
knowledge of the corresponding values of the average tunnelling
probability, we have determined that the critical radius R_t of the
spherical volume in which tunnelling is possible is of the order of
85 A. For higher concentrations we observe systematic deviations
from the straight lines. In the calculation of the luminescent
quantum yield we have assumed that the tunnelling probabilities for
different centers are independent; this is true for well-separated
defects only. Jaccard et al (6) have shown that correlation effects
affect the values of the radiative quantum yield by a multiplicative factor smaller than unity in the exponent $nV_t<P_t>$. On the
other hand, the optical density of these crystals around the main
absorption band of the F centers was larger than 5 and no special
effort was made to depress the intensity of the broad spectrum of
excitation. For these experimental conditions an appreciable de-
population of the ground state may become important, especially if
the average time required for a cycle F_o-$\bar{F}^*$-F_o is of the order of
10^{-3} sec (see section 5). It will again affect the exponent of η
by a multiplicative factor smaller than unity (6). Therefore, we
see that in both cases these second order corrections have an effect
similar to the reduction of the macroscopic concentration of defects. This shifts the calculated values of η towards the left of
Fig. 2. It will generally be difficult to take quantitative account
of these effects; but they can practically be eliminated by choosing
more adequate experimental conditions.

3. SPIN-LATTICE RELAXATION TIME T_1

At temperatures higher than 10K the spin-lattice relaxation
time of the relaxed excited state decreases and mixes up the population of the different pair states within the radiative lifetime.
Consequently the average tunnelling probability is changed. According to our earlier paper (6) the measurements of η as a function of
temperature can give the values of this spin-lattice relaxation time
since all the other parameters are known. It is worth noting that
such information is usually obtained by EPR measurements. The usefulness of this optical non-resonant method is somewhat limited
since the temperature variation of η is not very important. Our
first measurements only qualitatively agree with those obtained by
EPR-OD with a better accuracy (5, 6). For most alkali halide crystals, the observed spin-lattice relaxation time is characteristic
of an Orbach process $T_1 = T_0\{exp(\Delta E/kT)-1\}$. The values for KCl
are T_0 = 2.5 nsec, ΔE = 14.2 meV.

Exchange Interaction

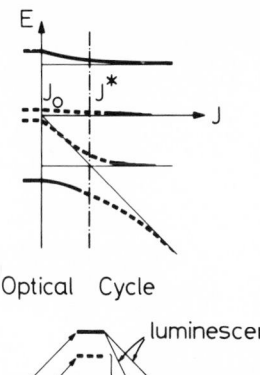

Optical   Cycle

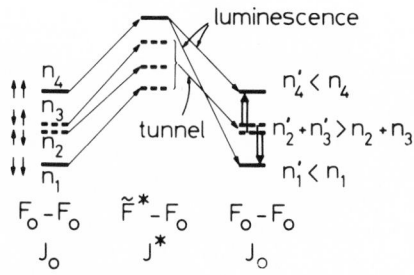

Fig. 3   a) (upper figure) Level scheme of a loose aggregate F
         center pair in its relaxed excited state as a function of
         exchange energy $J \cdot \underline{S}_1 \cdot \underline{S}_2$; b) (lower figure) Optical cycle
         for such a pair showing the change of the radiative mechan-
         ism, the tunnelling character and the population of each
         level. The levels drawn in heavy line have a low tunnell-
         ing probability while dashed lines represent levels with a
         high tunnelling probability.

4.   AGGREGATION EFFECTS

Light irradiation in the F band at room temperature is known
to favor aggregation of F centers, which eventually form aggregates
of 2,3, etc. F centers called F_2, F_3, etc. (2). The mechanism of
the aggregation is still not well understood. The new optical properties of these defects outlined in this report, together with the
technique of optical detection of the EPR in the ground and the relaxed excited state performed in our laboratory (5, 6, 8) promise
to shed some light on this subject.

The positive variation of the luminescent intensity with increasing magnetic field is found to diminish gradually and eventually to become negative when the F light irradiation at RT stretches
out. This can be explained by taking account of an exchange term
of the form $J \cdot \underline{S}_1 \cdot \underline{S}_2$ in the magnetic Hamiltonian. Fig. 3 shows the

result of this calculation. In an external magnetic field the energy of the four levels of a relaxed excited pair and the tunnelling probabilities are gradually modified, as the members of the pair become closer to each other. For a J* value, for instance, we see that only one level can radiate; the other three levels de-excite non-radiatively by tunnelling. This explains why the F centers radiative quantum yield decreases as a function of the aggregation. Moreover a check of the kinetics of the F luminescent intensity under rectangular pulse excitation (rise time ∼1μsec, length 2 sec) indeed shows a fast rise time followed by a partial diminution (time constant of 0.5 sec under our experimental conditions). It can be accounted for by the change in the level population occurring during the optical cycle.

5. CONCLUSION

The study of the variation of the luminescence intensity of F centers induced by a low magnetic field in KCl crystal has allowed a clarification of the mechanism of concentration quenching of the radiative quantum yield. The agreement between theory and experimental results is good. The measurement of this property as a function of temperature should also give the characteristic values of the spin-lattice relaxation time of the relaxed excited state, which are usually determined by resonant methods. Moreover, one of the most interesting features of this phenomenon lies in the possibility to understand better the physical nature of the aggregation mechanism of these defects in alkali halide crystals.

ACKNOWLEDGEMENTS

The authors are indebted to Professor F. Lüty (University of Utah, Salt Lake City) for kindly supplying us with some of the additively colored crystals used for this work, and the Swiss National Foundation for Scientific Research for their financial support.

REFERENCES

(1) L. G. Van Uitert, Luminescence of Inorganic Solids, Ed. P. Goldberg, Academic Press, (1966) Chapter 9.

(2) J. J. Markham, Solid State Physics, ed. F. Seitz and D. Turnbull, Academic Press (1966), Supplement 8; W. B. Fowler, Physics of Color Centers, Ed. W. B. Fowler, Academic Press, (1968) Chapter 2.

(3) J. J. Markham, R. T. Platt and I. L. Mador, Phys. Rev. 92, 597 (1953).

(4)   A. Miehlich, Z. Phys., 176, 168 (1963).

(5)   Y. Ruedin, P. A. Schnegg, C. Jaccard and M. A. Aegerter,
      Magnetic Resonance and Related Phenomena, Ed. I. Ursu,
      Proc. XVI th Colloque Ampère, Bucharest, 892 (1970);
      Y. Ruedin, P. A. Schnegg, M. A. Aegerter and C. Jaccard, to
      appear in Phys. Stat. Solidi (b) (1972).

(6)   C. Jaccard, Y. Ruedin, M. A. Aegerter and P. A. Schnegg,
      Phys. Stat. Sol. (b), 50, 187 (1972).

(7)   F. Porret and F. Lüty, Phys. Rev. Lett., 26, 843 (1971).

(8)   P. A. Schnegg, Y. Ruedin, M. A. Aegerter and C. Jaccard, to
      appear in Proc. XVIIth Colloque Ampère, Turku 1972.

ELECTROLUMINESCENCE IN ALKALI HALIDE CRYSTALS

C. Paracchini

Instituto di Fisica, Universita di Parma, Parma, Italy

ABSTRACT

The electroluminescence of alkali halide crystals was investigated at liquid nitrogen or lower temperatures. The effect is
obtained with time varying electrical excitation and can be attributed to electron-hole radiative recombination. It is localized to
a thin interface layer near the metal contact. The luminescence
can be described by avalanche breakdown theory.

INTRODUCTION

Electroluminescence has been widely studied in semiconductors
and in small band gap insulators, mainly for technological applications. Similar studies in high resistivity crystals are much less
numerous, and only recently electroluminescence has been surely
identified and examined in several alkali halides (1, 2). The
phenomenon has been found useful in the study of interface barriers,
high field injection and transport in materials with low conductivity where the interpretation of these effects is still incomplete
and the theoretical bases are not entirely accepted (3, 4).

Periodic light pulses from alkali halide crystals are obtainable with alternating electric excitation. The emission is easily
detected by photomultipliers without further amplification, if
fields larger than 10^4 V/cm are used. The effects have been observed in NaI, KI, RbI, CsI, NaBr and NaCl crystals at liquid nitrogen and lower temperatures.

After a short description of the experiments the main results which were obtained are summarized in the following.

SAMPLE ASSEMBLY

Single crystal slices 0.1 to 0.5 mm thick are sandwiched between two electrodes ("two electrode configuration"). Indium layers
melted under vacuum, assure thermal and electrical contacts. Gold
evaporated films have also been used instead of indium. Sometimes
only one side of the sample is in contact with the electrode, the
other being insulated by a thin teflon foil ("one electrode configuration"). This asymmetrical arrangement is necessary to distinguish the different mechanisms which take part in the effect.
Light emission is detected by a tri-alkali photomultiplier. Further
experimental details are described elsewhere (2).

LUMINESCENT EMISSION

Light emission occurs only during variations of the applied
field and disappears rapidly with constant voltage. No changes of
the brightness waves are observed by adding a d.c. component to
the alternating one. With sinusoidal excitation, brightness waves
lead the voltage. These results suggest that the effect is contemporary to the emission of carriers from the electrode and that
it is quenched when the injected charge reduces the externally
applied field. The build-up of the space charge q is supposed to
depend exponentially with time:

$$q = q_0[1-\exp(-t/\tau)] \tag{1}$$

The time constant τ can be obtained from the phase angle between
the brightness and the voltage waves.

Two brightness peaks are observed during a single excitation
cycle, whether "two" or "one electrode configuration" is used (Fig.
1a). In the latter case the successive light peaks are greatly
different. In the following, we shall refer to them as "positive"
or "negative" peaks, according to the algebraic sign of the voltage
rate of change at their appearance. "Positive" and "negative" light
peaks can be resolved with "one electrode configuration" only, and
the voltage is, in this case, referred to the electrode in contact
with the crystal.

The emission spectrum in a nominally pure crystal coincides
with that of the intrinsic luminescence in the same material. This
result demonstrates that the light is due to the radiative recombination of electrons in excitonic states with self-trapped holes
(V_k centers). This hypothesis is consistent with the sudden reduction of the effect at temperatures higher than 80°K, where, for
the examined crystals, the recombination of the electron-hole pairs
turns from a radiative to a non-radiative transition.

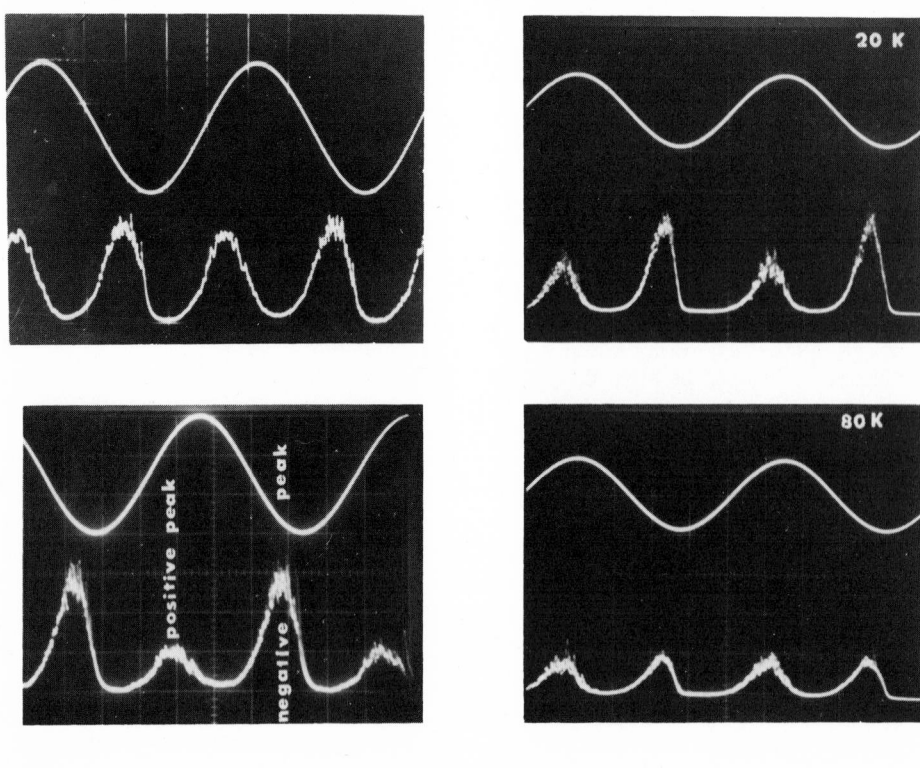

a   b

Fig. 1   Brightness waves obtained in a KI sample with sinusoidal
          excitation.   (a) The upper photo shows the light response
          with both the electrodes in contact with the crystal ("two
          electrode configuration").   The lower photo shows the
          emission waves when only one electrode is in contact with
          the crystal ("one electrode configuration").   In the second
          case the voltage is referred to the contact electrode.
          "Positive" and "negative" light peaks are here distinguished
          according to the sign of the voltage rate of change at their
          appearance.   (b) Light waves at 20 K and 80 K.   The "nega-
          tive" peak increases at low temperature, the "positive" one
          remains approximately the same.

     Direct visual and photographic observations of the excited
samples show that the light emission is localized in a thin crystal
zone adjoining the metal contact, where a high field barrier is
supposed to be formed. Owing to the low free carrier density and
to the non-monoenergetic trap distribution (as evidenced by thermoluminescence glow curves), the type of barrier is probably an ex-

ponential one (Rose-type barrier). In this case a proportionality relation links the externally applied field E with the maximum internal field F.

PROPOSED MODEL

Pairs of carriers must be present near the metal interface to cause intrinsic luminescence. Two mechanisms are possible: (a) double injection, (b) impact ionization. In the first case both carriers are successively injected from the electrode into the crystal on alternate half cycles. The holes are rapidly captured, forming V_k centers, and electrons recombine with them. The luminescence is then expected to increase with higher injection currents.

The impact ionization model requires the injection of one type of carrier and a high field region near the junction to supply enough energy for the pair production. Owing to the low hole mobility, only electron injection is effective here. In this case the luminescence depends on the pair production rate.

The choice between the two models is made by examining the brightness waves at different temperatures, in a range where the efficiency of intrinsic luminescence remains constant. Fig. 1b shows that the area of the "negative" peak increases from 80 K to 20 K; the "positive" peak remains approximately the same: it is then evident that successive light peaks have different origins. Injection from the metal into the insulator may happen as field emission (temperature independent) or Schottky effect (exponentially increasing with temperature). Anyway, with the double injection model, the increase of the "negative" peak at lower temperature cannot be explained.

The following model is therefore proposed. Electrons are injected from the negatively biased electrode into the crystal. They are accelerated up to ionizing energy in a high field barrier near the metal contact. Pairs are then produced and their radiative recombination causes light emission. The injected electrons reaching a low field region, are trapped. They build up a space charge which attenuates the field and quenches the luminescence. At the field reversal the trapped electrons are field ionized from the shallow levels and move back toward the electrode, thereby causing a new avalanche or recombining with the V_k centers which survived from the previous half-cycle.

FIELD, TEMPERATURE AND FREQUENCY DEPENDENCES

The dependence of the electroluminescence on field and temperature variations has been compared with the results obtained by the modern avalanche theories (5). The pair production rate w is de-

C. PARACCHINI

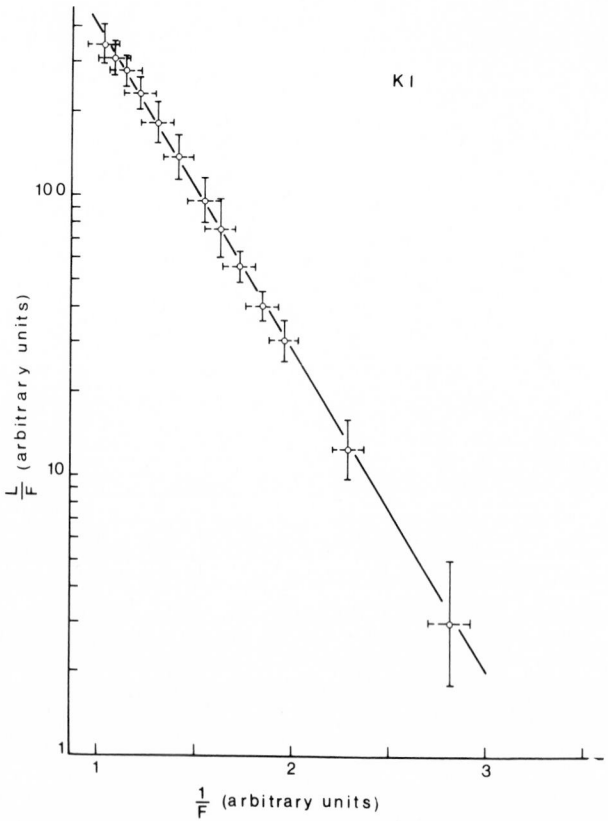

Fig. 2 Plot of ln(L/F) vs 1/F for a KI sample. The vertical seg-
        ments show the random errors in the light output evaluation.
        The horizontal dashed lines show the systematic errors of
        computing F from E with the observed values of τ, Eq. 4.

pendent on the internal local field F and on the temperature T as:

$$\omega \sim F \, \exp\{-H(T)/F\} \qquad (2)$$

where

$$H(T) \sim 1 + 2/\{\exp(\hbar\omega/kT)-1\} \qquad (3)$$

ω = longitudinal optical mode frequency. Assuming that the instantaneous light emission L is proportional to ω, the same dependences
on F and T are then expected for L. Fig. 2 shows the correlation
between L and F. The F is obtained from the values of the externally applied field E, supposing that a Rose-type barrier is formed
at the metal-insulator interface, and that Eq. 1 holds. In this

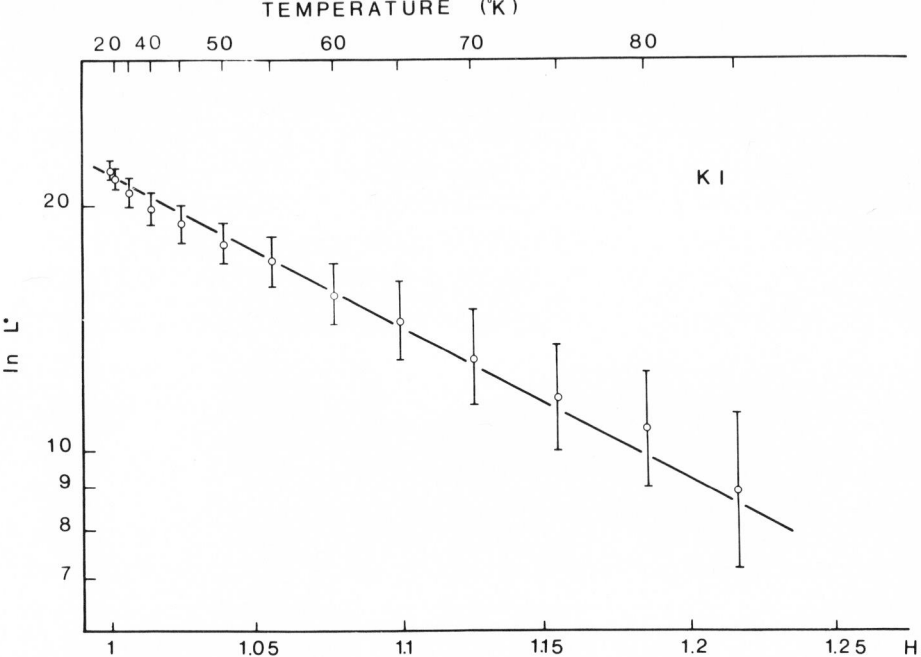

Fig. 3 Plot of the "negative" peak area L versus H_O from Eq. 3
for a KI sample.

case:

$$E(t') \sim \int_O^{t'} \{dE(t)/dt\}\exp\{(t-t')/\tau\}dt$$

When a voltage ramp is used: $E(t) = \alpha t$, and one has $F(t) \sim \alpha\tau\{1-\exp(-t/\tau)\}$, where t is the time elapsed after the beginning of the voltage ramp. By recording the voltage and the light signals on the oscilloscope or by a boxcar integrator, it is easy to correlate the instantaneous light emission L with the field values F.

The data reported in Fig. 2 have been obtained by examining the light emitted during a negative ramp excitation. Fig. 3 shows the dependence of the area under the "negative" peak L* versus H(T) as given by Eq. 3. These results are in agreement with the mentioned avalanche multiplication theory and confirm the proposed model. The presented data have been obtained with a KI sample, the results for NaI, RbI and CsI are analogous.

The dependence of brightness on frequency has been examined in the range between 100 and 2000 Hz. Square wave excitation with

constant amplitude and constant rise and fall times have been used
in order to avoid any dependence on the voltage amplitude and its
rate of change. Because of Eq. 1 the number of carriers forming
the space charge increases with the period of the excitation. The
number of electrons exchanged across the junction is then higher at
low frequencies. These electrons give rise to the impact ionization,
consequently the light output I , integrated over one entire excitation cycle is expected to diminish at higher frequency. Experimentally it is found that I is dependent on the frequency ν as:

$$I = I_0\{1-m\exp(-1/\nu\tau)\} \text{ with } 0<m<1 \tag{5}$$

The experimental formula Eq. 5 can be obtained from Eq. 1 by using
a proportionality relation between I and q^m:$I{\sim}q^m$. Raising Eq. 1
to the m-power, one has: $I{\sim}q_0^m(1-e^{-t/\tau})^m$, where $t = 1/\nu$ is the
period of the excitation wave. The last factor can be expanded in
a convergent binomial series and, neglecting the exponential terms
with power higher than unity, one obtains Eq. 5. This result
suggests that the space charge is indeed increasing with time,
according to Eq. 1.

CONCLUSIONS

Electroluminescence is obtained at a metal alkali halide junction under a.c. excitation. The light emission spectrum and temperature dependence of the brightness indicate that the effect is
due to the radiative recombination of electrons and holes.

The phenomenon is explained with an impact ionization model,
and the light output is correlated with the pair production rate,
as a function of electric field and temperature, according to recent
charge multiplication theories in insulators (5). The agreement
obtained between the theory and the experimental data shows the
possibility of obtaining impact ionization in alkali halide crystals. This result may be useful in improving the understanding
of high electric field phenomena in low conductivity and large band
gap insulators, where avalanche is often supposed, but where up to
now, little experimental evidence has been published (3).

The materials used for studying this effect have been chosen,
up to this time, among those alkali halide crystals in which the
intrinsic recombination is radiative. The same effect may be expected in all the insulating crystals where the electron-hole recombination is radiative.

The formation of a potential barrier near the insulator-metal
junction is necessary for the acceleration of the primary electrons
up to impact ionization energies. This fact requires that the contract of the insulator with the metal should be blocking for the

primary carriers. Information about the build-up of the potential barriers at the metal-insulator interface can be then obtained.

Modern breakdown theories consider the development of an avalanche up to critical size as being responsible for the electrical breakdown of the crystal (3). However, some difficulties arise if a pure electronic avalanche is considered. Recent models (5) suggest that the development of the avalanche is aided by electron field emission from the cathode. The results here obtained, observing the light emission at the metal insulator contact, support these hypotheses and provide a new method for the investigation of the electronic transport phenomena in insulators.

This work was supported by the G.N.S.M. group of the Italian Research Council.

REFERENCES

(1) A. N. Georgobiani, N. P. Golubeva, Opt. i Spectr. 12, 455 (1962). S. Unger, K. Teegarden, Phys. Rev. Lett. 19, 1229 (1967). G. A. Vorobev, N. S. Nesmelov, Sov. Phys. Sol. State 10, 2932 (1969).

(2) C. Paracchini, Phys. Rev. B 4, 4223 (1971).

(3) N. Klein, Thin Solid Films 7, 149 (1971).

(4) J. G. Simmons, J. Phys. Chem. Solids 32, 1987, 2581 (1971).

(5) J. J. O'Dwyer, J. Phys. Chem. Solids 28, 1137 (1967); J. Appl. Phys. 40, 3887 (1969).

THE PHOTOLUMINESCENCE SPECTRA OF SINGLE CRYSTALS OF $SrCl_2:Yb^{+2}$

H. Witzke, D. S. McClure and B. Mitchell

Department of Chemistry

Princeton University

Princeton, New Jersey 08540, USA

ABSTRACT

Highly complex luminescence behavior is exhibited by single crystals of $SrCl_2:Yb^{+2}$ when excited by near-ultraviolet radiation. The photoluminescence characteristics of the system were established on the basis of decay-time measurements and studies of the temperature dependence of emission intensities. Two of the five bands observed at low temperatures are assigned to $4f^{14} \leftarrow 4f^{13}5d$ transitions of the Yb^{+2} ion. The centers responsible for the other three bands have not yet been identified, but they appear to involve ytterbium in some form. One of these bands corresponds to the well-known green emission of $CaF_2:Yb^{+2}$, but is not due to a transition from a level of the $4f^{13}6s$ configuration of Yb^{+2}, as has been suggested.

INTRODUCTION

Divalent rare earth ions can easily be incorporated in CaF_2, $SrCl_2$ and other fluorite-type host crystals. The absorption spectra of these systems have been intensively studied (1,2,3). For most divalent rare earths, the lowest excited crystal field states have the $4f^n$ configuration, but for Eu^{+2} ($4f^7$) and Yb^{+2} ($4f^{14}$) the lowest excited states are given by $4f^{n-1}5d$. (Divalent ytterbium has no excited $4f^n$ states, of course.) The absorption spectra of Eu^{+2} in $CaF_2(5)$ and Yb^{+2} in $SrCl_2(2)$ have been interpreted in terms of crystal field theory. It should therefore be possible to understand the emission in relation to the absorption. This is, in fact, the case for Eu^{+2}, which shows a $5d \rightarrow 4f$ luminescence that is a simple mirror image of the lowest absorption band (6,7,8). Yb^{+2}, the ion to be discussed here, has surprisingly complex luminescence behavior,

598

whereas we expected it to act like Eu^{+2}.

EXPERIMENTAL METHODS

Strontium chloride crystals are hygroscopic and as such must be handled in a scrupulously dry atmosphere. Samples used in this study had nominal concentrations of 0.02, 0.05 and 0.4 mole per cent ytterbium (9). Spectrographic arc analysis indicated that no other rare earths were present.

Absorption spectra were measured on a Cary 14R spectrophotometer. A "Cryo-Tip"(10) variable-temperature cryostat was used in many of the low temperature experiments.

Photoemission was detected by a Cary 81 spectrophotometer fitted with Hamamatsu R456 photomultiplier tubes. Luminescence data have been corrected for the spectral sensitivity of the detection system (11) and are displayed in relative units of photons per second per unit frequency interval. An AH-6 high-pressure mercury arc was used as the excitation source.

A conventional nitrogen discharge flash lamp capable of producing 200 J flashes of 4 μsec pulse width was used as the source in decay-time measurements.

EXPERIMENTAL RESULTS

In Fig. 1 are shown the room temperature (1a) and 4.2K (1b) emission spectra of a 0.05% SrCl$_2$:Yb^{+2} crystal excited by the 334 nm and 313 nm mercury lines. To facilitate comparison, the room temperature spectra have been normalized to coincide at 24650 cm^{-1} and the 4.2K spectra have been normalized to coincide at 23900 cm^{-1}. Other usable mercury lines give rise to emission identical to that produced by the 334 nm line. The intensities of the room temperature and 4.2K spectra excited by the 334 nm line may be compared directly.

At room temperature, three emission bands are observed, having maxima at 26600 cm^{-1} (Band I), 24650 cm^{-1} (Band II) and 18800 cm^{-1} (Band IV), respectively. The relative intensities of I and II are independent of excitation wavelength but dependent upon the Yb^{+2} concentration. The intensity ratio of II to I increases with increasing ytterbium concentration. The intensity ratio of IV to II depends on the wavelength of excitation. It is also observed to vary for samples of different ytterbium concentration (12).

When the temperature is lowered, dramatic changes occur in the emission spectra. If 334 nm excitation is used, Band I increases in intensity and below about 80K develops the familiar vibrational structure known from the low temperature absorption of

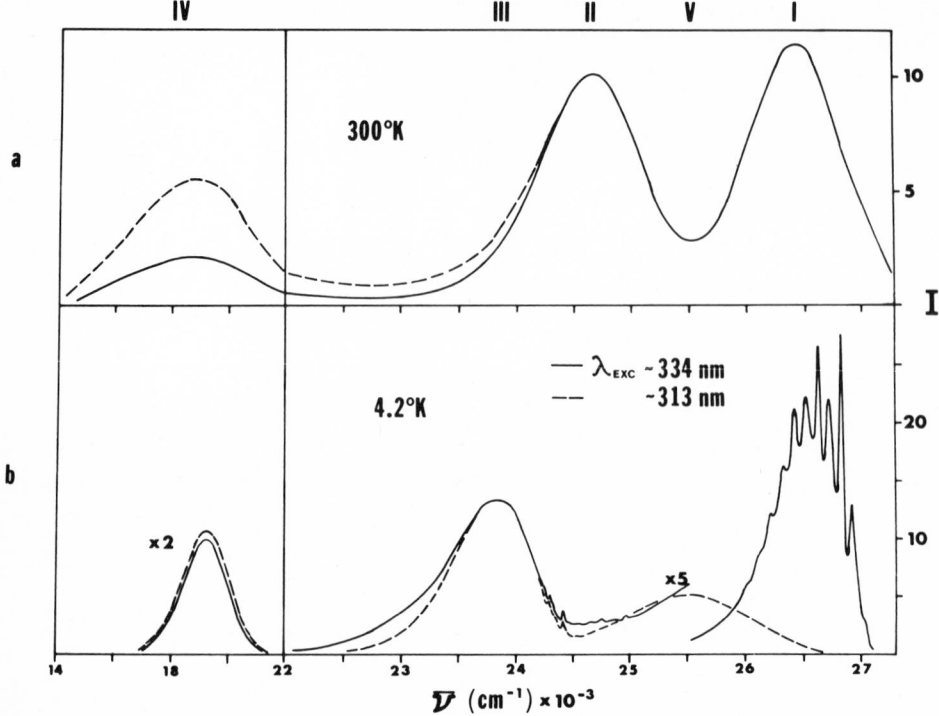

Fig. 1 The luminescence spectra of $SrCl_2:Yb^{+2}$ (0.05%) excited
        at two wavelengths. The curves have been normalized
        (see text).

this crystal (2). Band II shows a drastic decrease in intensity
and at low temperatures also develops structure with the same
vibronic spacings as for Band I. Between 80K and 4.2K, Band III
appears on the low-energy side of II. Its maximum lies at 23900
cm^{-1}, and there are indications of structure on the high-energy
side. The intensity of Band IV gradually increases down to about
50K and then decreases. The half-width is reduced from ~ 5000 cm^{-1}
at room temperature to ~ 2000 cm^{-1} at 4.2K. No structure is
apparent even at 2.2K.

When 313 nm excitation is employed, the emission bands show
a temperature dependence similar to that above between room
temperature and ~ 200K. If the temperature is lowered further,
both I and II rapidly decrease in intensity. They are no longer
present at 4.2K. Below 50K a new emission feature, Band V
(maximum at 25500 cm^{-1}), appears. No structure is observed on this
band. The relative intensities of V, III, and IV vary from sample
to sample.

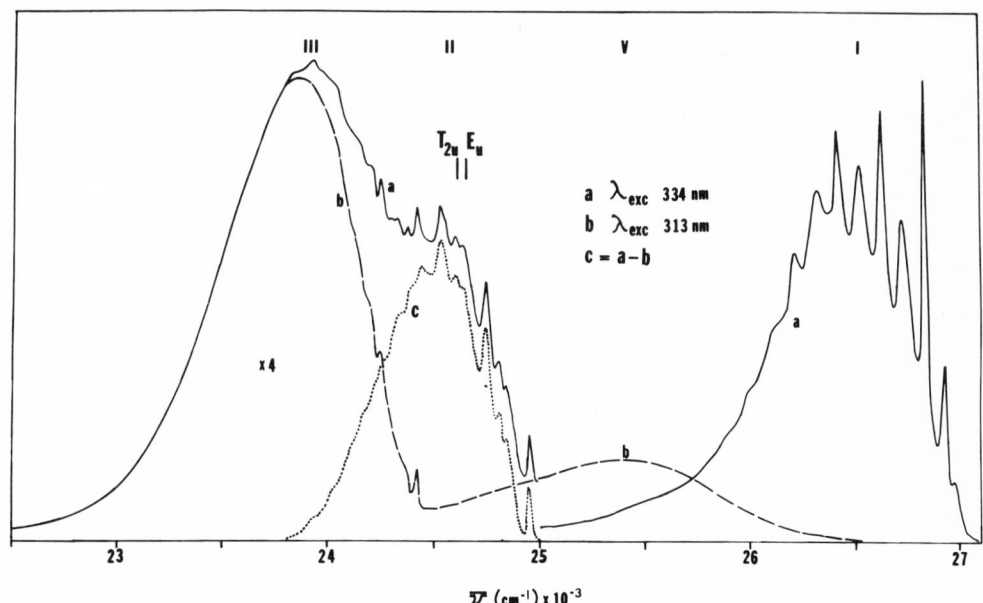

Fig. 2   The 4.2K luminescence spectra of SrCl$_2$:Yb^{+2} (0.4%) excited
at 334 nm (curve (a)) and 313 nm (curve (b)). Band IV is
not shown. The curves have been normalized (see text).
The calculated positions of the T$_{2u}$ and E$_u$ levels are also
indicated. Curve (c) is the difference between (a) and
(b).

In Fig. 2 we show part of the 4.2K emission spectra of a
0.4% SrCl$_2$:Yb^{+2} crystal for both 313 nm (curve (b)) and 334 nm
(curve (a)) excitation. Curve (b) has been normalized to coincide
with curve (a) along the tail of Band III. Curve (c) of this
figure was obtained by subtracting curve (b) from curve (a) and is
indicative of the shape of Band II.

At room temperature, the decay time of Band I is too fast to
be measured with our apparatus. Band IV has an exponential decay
with a time constant of 45 μsec. The decay curve for Band II may
be analyzed into a fast and slow component of 50 and 520 μsec,
respectively.

At liquid nitrogen temperature, the decay of Band I is still
faster than the decay time of our flash lamp. Band IV decays with
a time constant of 130 μsec, while the region containing II and
III shows a fast component of 250 μsec (probably Band III) and a
slow component of 1300 μsec (probably Band II). The decay
characteristics of Band V are not known at the present time. All

the measured decay curves are independent of sample.

In Table I the relevant spectral features, temperature dependences and other properties of the observed emission spectra of $SrCl_2:Yb^{+2}$ crystals are summarized. Details will be published elsewhere (12).

DISCUSSION

The fast decay time, observation of the vibronic progression known to be indicative of an f-d transition in $SrCl_2$,(13) and the mirror-image relationship of Band I to the lowest electric dipole-allowed absorption band of $SrCl_2:Yb^{+2}$ enable us to assign this band to the transition $A_{1g} \leftarrow T_{1u}(f^{13}d)$ of the Yb^{+2} ion in an O_h field. In Fig. 3 we show Band I together with an absorption spectrum of the lowest allowed transition. The zero phonon line at 27045 cm^{-1} is almost totally reabsorbed even though front surface excitation was used. The zero phonon line in absorption is less than 2 cm^{-1} wide and is not shown with its correct intensity relative to the vibronic structure in the absorption spectrum.

Band II lies in the region where the two lowest excited states $(T_{2u}$ and $E_u)$ of the Yb^{+2} ion in the crystal field of the $SrCl_2$

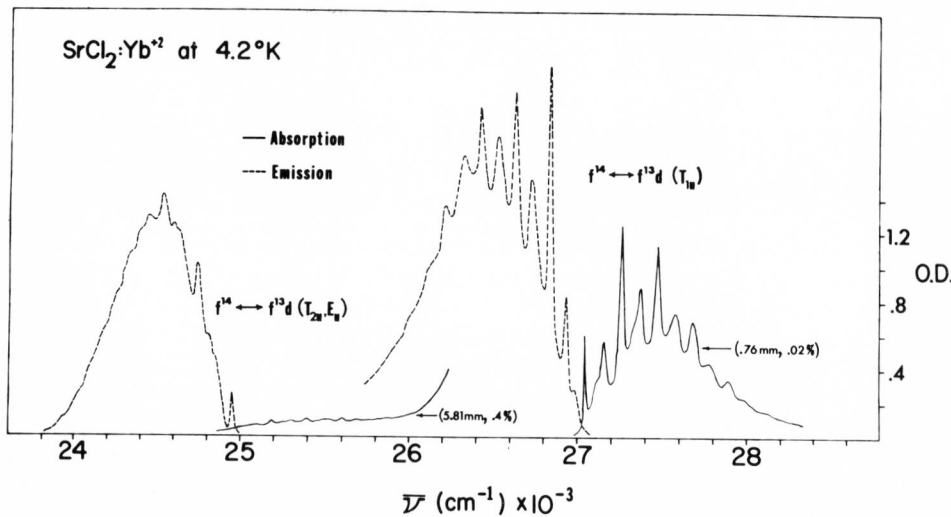

Fig. 3 The lowest lying allowed and forbidden transitions of Yb^{+2} in the $SrCl_2$ lattice as observed in absorption and emission at 4.2K.

TABLE 1

THE PHOTOLUMINESCENCE BEHAVIOR OF SrCl$_2$:Yb^{+2}

Band	Max. (cm^{-1})[c]	$\Delta I/\Delta T$[d]	Structure	Excited at (nm) 313	Other	τ(μsec) 300K	85 K
I	~26,500	[-]>100K [0]<100K	Seen <100K progression	T>150K only	Yes	fast, < 4	fast < 4
II	~24,700	[+]>100K	Seen <100K	T>150K	Yes	Two-Component Decay	
III[a]	23,900	[0]<100K [-]	progression Some; seen <100K	only Yes	Yes	τ_1=50 τ_2=520	=250 =1300
IV	19,000	[-]>50K [+]<50K	None	Yes	Yes	45	130
V[b]	25,400	[-]	None	Yes	Yes	?	?

(a) Seen only at low (<~125 K) temperatures.
(b) Seen only at very low (<~50 K) temperatures.
(c) At 4.2 K.
(d) Temperature dependence of intensity ([+] indicates $\Delta I/\Delta T > 0$).

lattice have been predicted to be (see Fig. 2) (2). Transitions
to these levels are highly forbidden. The dependence of II on
excitation wavelength, the long decay time of the emission, and
the appearance of vibrational structure similar to that of I led
us to suspect that II (curve (c) in Fig. 2) is emission from one
or both of these metastable levels. The sharp origin at 24958
cm^{-1} may be compared to the calculated positions of 24608 cm^{-1} and
24646 cm^{-1} for the T$_{2u}$ and E$_u$ levels, respectively (2). The
absorption spectrum of a thick and heavily doped crystal (Fig. 3)
lends further support to this assignment. A number of weak
absorption peaks are evident at low temperatures. They have the
same vibronic spacing observed in Band II. No zero phonon line
is apparent in absorption, as one might expect for a highly forbidden transition. We have also observed Band II alone at 4.2K
via a conventional phosphoroscope technique. The observed
spectrum (12) is virtually identical to curve (c) in Fig. 2. At
the present time we cannot say which of the two forbidden levels
is responsible for the observed emission.

The decrease of Band I relative to Band II with increasing
Yb^{+2} concentration (cf. Figs. 1b and 2) is not unexpected in view
of current theories of concentration quenching in phosphors (14).
The strongly allowed A$_{1g}$ ←T$_{1u}$ transition (Band I) should be more
susceptible to concentration quenching than the forbidden (Band
II), and this is what is observed.

At 4.2K only Bands III, IV and V appear when excitation

at 313 nm is used. The Yb^{+2} ion does not absorb at this wavelength
and no allowed levels are calculated to lie in this spectral
region (2). These bands also differ from I and II in their lack
of normal $SrCl_2$ vibronic progression. They cannot be assigned to
internal transitions of the Yb^{+2}ion. We shall call them defect
bands although we have reason to believe that the emitting centers
involve ytterbium in some form. Nominally pure $SrCl_2$ shows no
luminescence upon ultraviolet excitation. A $SrCl_2$ crystal doped
with Eu^{+2} and grown in the same manner as the Yb-doped crystals
does not show any of the defect bands. Instead, emission only
from the lowest f^6d level of the Eu^{+2} ion is observed (8). It
should be noted that the crystals used were grown by doping the
initial anhydrous $SrCl_2$ with the trivalent rare earth (9). The
persistence of small amounts of trivalent ytterbium after chemical
reduction of the melt cannot be ruled out. A partially reduced
crystal shows anomalous absorption near 313 nm (12) which likely
is due to the presence of Yb^{+3} or possibly Yb^{+3}-Yb^{+2} pairs.
We suggest that centers such as these may be responsible for the
defect emission bands.

CaF$_2$ doped with Yb^{+2} has the expected $f^{14} \rightarrow f^{13}d$ absorption
spectrum (4,15) and is the only other Yb^{+2} - doped fluorite-type
crystal which shows photoemission (3,4). Only a broad, unstructured
band at 17000 cm^{-1} has been observed and then only below ~ 200°K.
In many ways, except for its absence at room temperature, this
band is similar in its characteristics to Band IV of $SrCl_2$:Yb^{+2}(8).
The possible involvement of an $f^{13}6s$ level of the Yb^{+2} ion has
been suggested (3) for the emission of CaF$_2$, but on the basis of
our work we must conclude that internal levels of the Yb^{+2} ion
are not responsible for this emission.

ACKNOWLEDGMENTS

This work has been supported by the Office of Naval Research,
Contract No. N 00014-67-A-0151-0012, and the National Science
Foundation. One of us (H. W.) is grateful to the National Research
Council of Canada for financial support in the form of a post-
doctoral fellowship (1970-72). The crystals used in this study
were grown by Mr.W. Bleacher and generously donated by the RCA
David Sarnoff Research Center.

REFERENCES

1. D. S. McClure and Z. Kiss, J. Chem. Phys. _39_ 3251 (1963).

2. T. S. Piper, J. P. Brown and D. S. McClure, J. Chem. Phys.
   _46_ 1353 (1967).

3. P. P. Feofilov, Opt. i Spektroskopiya _1_ 992 (1956).

4. A. A. Kaplyanskii and P. P. Feofilov, Opt. i Spektroskopiya <u>13</u> 235 (1962).

5. H. A. Weakliem, to be published in Phys. Rev.

6. L. L. Chase, Phys. Rev. B <u>2</u> 2308 (1970).

7. S. Freed and S. Katcoff, Physica <u>14</u> 17 (1948).

8. H. Witzke, unpublished.

9. F. K. Fong and P. N. Yocom, J. Chem. Phys. <u>41</u> 1383 (1964).

10. Air Products and Chemicals, Inc., Allentown, Pa.

11. R. Stair, W. E. Schneider and J. K. Jackson, Applied Optics <u>2</u> 1151 (1963).

12. H. Witzke, D. S. McClure and B. Mitchell, to be published.

13. D. H. Kuehner, H. V. Lauer and W. E. Bron, Phys. Rev. B <u>5</u> 4112 (1972).

14. G. Blasse, J. Luminescence <u>1</u>, <u>2</u> 766 (1970).

15. M. V. Eremin, Opt. i Spektroskopiya <u>29</u> 100 (1970).

CASCADE EMISSION OF TWO PHOTONS INDUCED BY THE INTERACTION OF AN ELECTRON-HOLE PAIR WITH AN IMPURITY CENTER IN ALKALI HALIDES

I. Jaek, M. Kink, S. Zazubovich, and V. Osminin

Institute of Physics and Astronomy, Academy of Sciences

of Estonian SSR, Tartu, USSR

ABSTRACT

Two processes are discussed which allow one to observe the multiplication of photons in the framework of linear optics. It is shown that the interaction of each electron-hole pair with an impurity ion or some crystal defect may be accompanied by the emission of two photons with different energies.

INTRODUCTION

The well-known photon multiplication phenomenon (1) enables one to generate two or three electron-hole pairs in a crystal through the absorption of one photon with an energy which is usually two or three times as large as the band gap energy (2). The recombination of each electron-hole pair results in one photon of luminescence.

In the present paper two processes are discussed which allow one to observe the further multiplication of photons in the framework of linear optics. It is shown that the interaction of each electron-hole pair with an impurity ion or some crystal defect may be accompanied by the emission of two photons which have different energies.

Différent processes leading to the luminescence of crystals are represented schematically in Fig. 1: 1. Inner-center luminescence (ICL) due to the electronic transitions between the energy levels of the luminescence center; 2. Electron recombination luminescence (ERL) due to the recombination of electrons with holes localized on impurity ions or crystal defects;

606

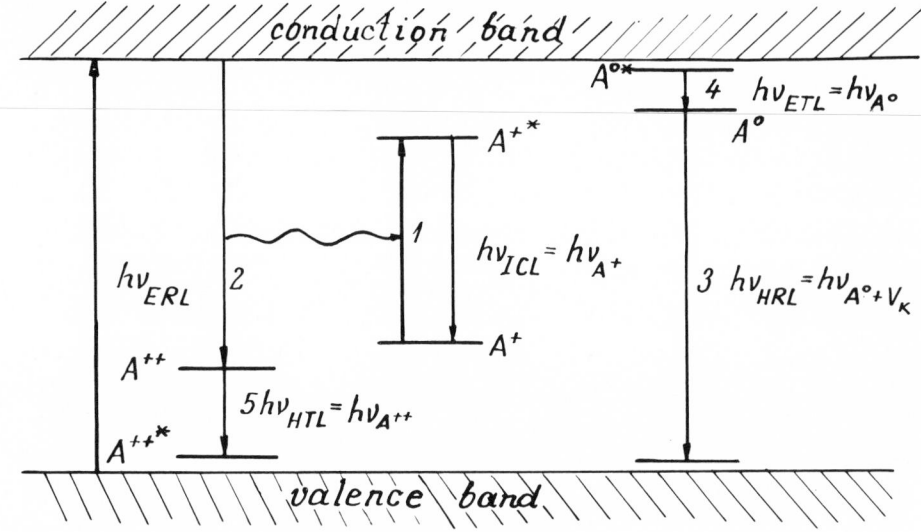

Fig. 1   Schematic representation of the principal processes
         leading to the luminescence of crystals.

3. Hole recombination luminescence (HRL) due to the recombination
of holes with localized electrons; 4. Electron trapping luminescence (ETL) due to the trapping of electrons by crystal or
impurity defects; 5. Hole trapping luminescence (HTL) due to
the trapping of holes by impurity ions or crystal defects.

    The first three of these processes have been investigated in
detail in alkali halides. It has also been shown that luminescence accompanies the trapping of electrons by free anion
vacancies (3-6) and the formation of B centers (7). Emission due
to the trapping of holes by Cu^+ ions has been found in ZnS
phosphors (8). Recently, we have observed the luminescence
accompanying the trapping of holes (9) and electrons (10) by
activator centers in alkali halides doped with mercury-like ions.
The HTL and ETL spectra of the investigated systems are represented
in Fig. 2 (solid lines).

HOLE TRAPPING LUMINESCENCE (HTL)

    In KCl:In the trapping of holes, released optically or
thermally from various hole traps, by In^+ and $In^+V_C^-$ centers is
accompanied by 2.12 eV and 1.75 eV HTL, respectively (9), and
results in the formation of In^{++} and $In^{++}V_C^-$ centers (11). The
HTL spectra are determined to be the same as the spectra of the
inner-center emission of In^{++} and $In^{++}V_C^-$ centers (12). This
coincidence has been explained by the circumstance that a

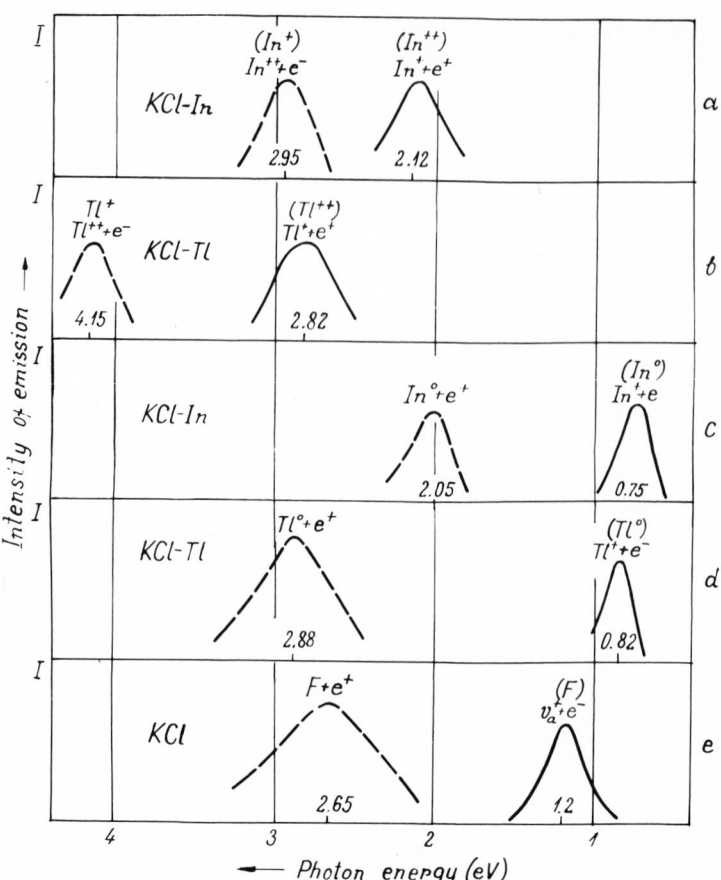

Fig. 2 Spectra of hole (a,b) and electron (c,d,e) trapping
        luminescence (solid lines), and spectra of electron
        (a,b) and hole (c,d,e) recombination luminescence
        (dashed lines) of KCl:In, KCl:Tl and KCl crystals
        at 80 K.

thermally or optically released hole encountering the activator
ion (A^+) and relaxing near it in the form of Cl_2^- produces the
$A^+Cl_2^-$ state which is identical with the excited state of A^{++}
center (12).

    The detection of HTL in KCl-Tl is more difficult because of
the close overlapping of several emission bands in the same
spectral region. However, it may be proposed that the trapping
of holes by Tl^+ ions leads to the 2.82 εV HTL which is identical
to the inner-center emission of Tl^{++} (12).

    The various properties of the 1.75 eV HTL arise when $In^+V_c^-$
centers trap the holes released from In^{++} by photons with 4.1 eV

energy. The efficiency of HTL decreases as the concentration of In^+ in the crystal becomes lower, and in KCl:In containing less than 4×10^{-3} mole % of In^+, HTL is not observed. The investigation of the quantum efficiency of HTL shows that in KCl:In containing 5×10^{-2} mole % of In^+, about half of the holes released optically from In^{++} centers are trapped by $In^+V_C^-$ centers with emission. The 1.75 eV HTL is short persistent and is not polarized. It cannot be frozen (at least down to 80 K). Such properties of HTL may be connected with the motion of unrelaxed holes in the crystal. The temperature dependences of the intensity of the 1.75 eV HTL and of the 2.12 eV inner-center emission of In^{++} centers, both excited by photons with 4.1eV energy, are identical. It may be concluded that the holes are probably released from the relaxed excited state of In^{++} overlapping with the valence band of the crystal.

ELECTRON TRAPPING LUMINESCENCE (ETL)

The mechanism of ETL is identical to that of HTL. When anion vacancies trap the electrons, released optically (3) or thermally (6) from some electronic centers, an excited state of the F center is formed and the transition from it into the ground state is accompanied by the F emission. The trapping of optically released F electrons by impurity ions in KCl crystals results in an emission at 0.82 eV in KCl:Tl and at about 0.75 eV in KCl:In. The spectra of these ETL are identical to the spectra of the recently found (10) inner-center luminescence of Tl^0 and In^0 centers, respectively, measured with excitation in the long wavelength absorption bands of these centers. It may be concluded that upon trapping of electrons by Tl^+ and In^+ ions excited states of Tl^0 and In^0 centers are produced and the electronic transitions into the ground state of Tl^0 and In^0 centers occur with the resulting Tl^0 and In^0 emission.

In KCl:Tl and KCl:In the reverse process is also observed, i.e., the F emission may be stimulated in the Tl^0 and In^0 absorption bands by the trapping of electrons, released optically from Tl^0 and In^0 centers, by free anion vacancies.

TWO PHOTON EMISSION

The detection of HTL and ETL in alkali halides allows one to obtain two emission quanta because of the interaction of one electron-hole pair with the activator center. The first photon appears upon the trapping of a hole (or electron) by an impurity ion (Fig. 2, solid lines), and the second upon the recombination of an electron (or hole) with the ionized (or electronic) center formed at the first stage of the process (Fig. 2, dashed lines).

In KCl crystals the 2.12 eV emission from the trapping of
holes by In^+ ions (Fig. 2a) is observed. Probably, the 2.82 eV
emission accompanies the trapping of holes by Tl^+ ions (Fig. 2b).
The second observed photon is due to the recombination of
electrons with previously formed In^{++} (2.95 eV) and Tl^{++} (4.15 eV)
centers. In KCl:In both stages of the process result in visible
emission.

The trapping of electrons by In^+ or Tl^+ ions and by anion
vacancies in KCl results in the appearance of the emission bands at
about 0.75 eV (Fig. 2c), 0.82 eV (Fig. 2d) and 1.2 eV (Fig. 2e),
respectively. The second photon may be observed upon the tunnel
recombination of holes (V_k centers) with previously formed In^0
(~2.05 eV), Tl^0 (~2.88 eV (13)) and F (~2.65 eV (14)) centers.

If an impurity ion traps electrons with the same efficiency
as holes four emission bands may be observed in one system due
to the interaction of electron-hole pairs with the impurity centers.

The effects of the cascade emission of two photons, induced by
the formation of an electron-hole pair in the crystal, may be
applied in radiation detectors and luminescent lamps, if systems
with large quantum efficiency for both stages of the process
can be found. The threshold energy of these processes is lower
than that for photon multiplication processes and does not exceed
the band gap energy.

ACKNOWLEDGEMENTS

We should like to thank Prof. Ch. Lushchik for valuable
discussions.

REFERENCES

1. L. Apker, E. Taft, J. Dickey, J. Opt. Soc. Amer., 43, 78
    (1953).

2. E. Ilmas, Ch. Lushchik, Trudy IFA Akad. Nauk Est. SSR,
    No. 34, 5 (1966).

3. M. Hirai, M. Ikezawa, J. Phys. Soc. Japan, 21, 826 (1966);
    22, 810 (1967).

4. I. L. Boettler, W. D. Compton, Phys. Rev., 173, 844 (1968).

5. H. Fedders, M. Hunger, F. Lüty, J. Phys. Chem. Solids, 22,
    299 (1961).

6. R. Fieschi, C. Paracchini, Phys. Rev., 182, 935 (1969).

7. J. Lukantsever, F. Zaitov, V. Chernenko, Izv. Akad. Nauk
    SSSR, Ser. fiz. $\underline{29}$, 54 (1965).

8. C. B. Burgett, C. C. Lin, J. Phys. Chem. Sol., $\underline{31}$, 1353
    (1970).

9. S. Zazubovich, Phys. Stat. Sol. (b), $\underline{50}$, 785 (1972).

10. I. Jaek, M. Kink, to be published.

11. Ch. Lushchik, H. Käämbre, N. Lukantsever, N. Lushchik,
    E. Tiisler, I. Jaek, Izv. Akad. Nauk SSSR, Ser. fiz. $\underline{33}$,
    863 (1969).

12. S. Zazubovich, N. Lukantsever, V. Osminin, Izv. Akad. Nauk
    SSSR, Ser. fiz. $\underline{35}$, 1418 (1971); Phys. Stat. Sol. (b) $\underline{50}$,
    771 (1972).

13. C. J. Delbecq, A. K. Ghosh, P. H. Yuster, Phys. Rev. $\underline{151}$,
    599 (1966).

14. T. Timusk, W. Martienssen, Phys. Rev. $\underline{128}$, 1656 (1962).

ON THE WO_6 LUMINESCENCE IN ORDERED PEROVSKITES

G. Blasse, A. F. Corsmit and M. van der Pas

Solid State Chemistry Department, Physical Laboratory

State University, Sorbonnelaan 4

Utrecht-Netherlands

ABSTRACT

This paper describes some optical properties of ordered perovskites A_2BWO_6 (A=Ca, Sr, Ba; B=Mg, Cd, Ca, Sr, Ba). The absorption and emission spectra of the WO_6 group depend strongly on the choice of A and B. The tungstate group in Ba_2MgWO_6 shows two different emissions, a blue and a yellow one.

INTRODUCTION

The luminescence of the tungstate group has been studied extensively without resulting in an explanation of the strong variation of these luminescence properties with the crystal surroundings (1,2). For this reason it seemed interesting to study the luminescence of the WO_6 group in ordered perovskites. This is a simple crystal structure with high symmetry where extensive chemical substitutions are possible (3). This paper describes some preliminary results of the study.

EXPERIMENTAL RESULTS

The samples were prepared by the usual techniques (4) and checked by X-ray analysis. The performance of the optical measurements has been described elsewhere (3).

The reflection spectra of A_2BWO_6 compounds reveal that the absorption edge of these compounds is only weakly influenced

612

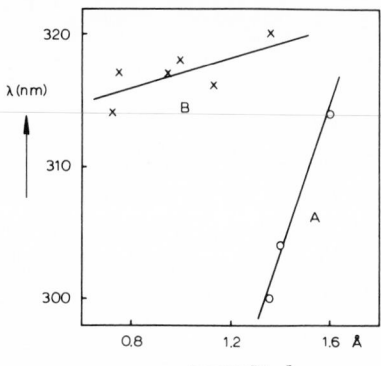

Fig. 1 Position of the absorption edge of ordered perovskites
A_2BWO_6 vs the ionic radii of A and B. Curve A is for
A_2MgWO_6 (A=Ca, Sr, Ba), curve B for Ba_2BWO_6 (B = Mg,
Zn, Cd, Ca, Sr, Ba). Ionic radii according to Shannon
and Prewitt (Acta cryst. B25, 925 (1969)).

by the choice of B and depends markedly on the choice of A (Fig. 1).
The larger the ionic radius of A, the longer the wavelength
corresponding to the absorption edge.

Nearly all A_2BWO_6 compounds studied show luminescence under
ultraviolet excitation, at least at 77 K. This agrees with a
few observations made by Kröger (1). The emission colors depend
strongly on the choice of A and B e.g. Ba_2CaWO_6, luminesces blue;
Sr_3WO_6, green; and Ba_2MgWO_6, yellow.

Only Ba_2MgWO_6 shows luminescence with reasonable efficiency
at room temperature. Therefore results for this compound will
be reported in greater detail. The emission spectra at 77 and
300 K are given in Fig. 2. At low temperatures a new, blue
emission band appears. At 4 K its intensity is even stronger.

These emission bands correspond to different excitation bands.
The blue emission is most efficiently excited by 270 nm radiation,
the yellow-green emission by 315 nm (315 nm is the wavelength
corresponding to the absorption edge).

DISCUSSION

It is obvious that the class of compounds A_2BWO_6 form a
suitable group of phosphors to study the relation between luminescence properties and crystal structure.

The dependence of the position of the charge-transfer band
on the radii of the surrounding cations agrees with earlier
observations and predictions (2,5). The relevant electronic

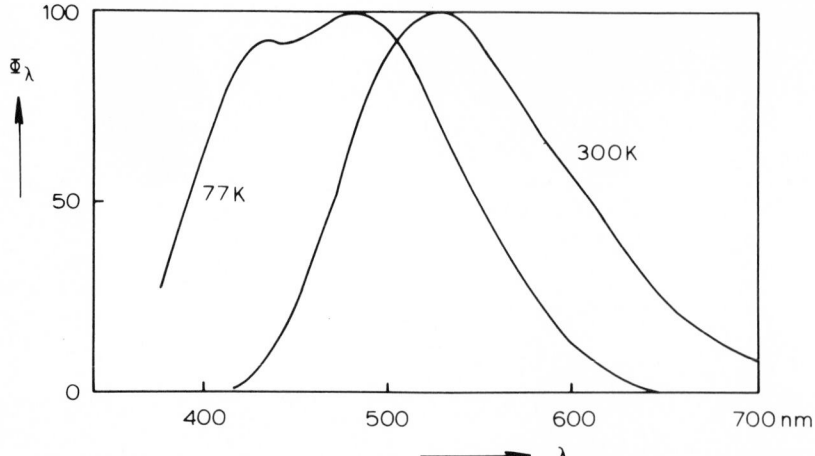

Fig. 2 Spectral energy distribution of the emission of Ba_2MgWO_6
        under 254 nm excitation at 77 and 300 K. Φ_λ gives the
        radiant power in arbitrary units per constant wavelength
        interval.

transition is situated at higher energies for smaller ionic radii,
because smaller cations cause a stronger potential field at the
anions, so that the relevant charge-transfer transition is at
higher energies.

The occurrence of two different emissions has not been
reported before for the tungstate group. The crystal structure of
the ordered perovskites contains only one position for the W^{6+}
ion, so that the possibility of two different WO_6 groups can be
rejected. In view of the position of the emission bands, the
presence of $MgWO_4$ as an impurity can also be ruled out. We conclude,
therefore, that the two emissions originate from two different
energy levels of the same group. A similar observation has been
made for the niobate and the titanate group (1,6).

It seems obvious to assume that the double emission is
related to the complicated nature of the first excited state of
the tungstate group that contains singlet as well as triplet
levels (6). The yellow emission band would then correspond to
an allowed transition that relates also to the absorption edge,
whereas the blue emission would be a forbidden transition. This
problem is now being studied.

Finally we note that at 77 K Sr_2MgWO_6 shows one emission
band with a shoulder only, whereas Ba_2CdWO_6 shows only one
emission band at 77 K. The occurrence of two emission bands
depends, therefore, also on the choice of A and B.

REFERENCES

1. F. A. Kröger, Some Aspects of the Luminescence of Solids, Elsevier Publ. Company, Amsterdam, 1948.

2. G. Blasse and A. Bril, J. Solid-State Chem. $\underline{2}$, 291 (1970).

3. G. Blasse and A. F. Corsmit, J. Solid-State Chem., in press.

4. A. F. Corsmit, H. E. Hoefdraad and G. Blasse, J. Inorg. Nucl. Chem., in press.

5. G. Blasse and A. Bril, Z. physik. Chemie N. F. $\underline{57}$, 187 (1968).

6. G. Blasse and A. Bril, J. of Luminescence $\underline{3}$, 109 (1970); J. Solid-State Chem. $\underline{3}$, 69 (1971).

IONIZATION OF LUMINESCENCE CENTERS IN ALKALI HALIDE PHOSPHORS BY UNRELAXED POSITIVE HOLES

E. D. Aluker, Y. H. Kalnin, S. A. Chernov, K. K. Shvarts

Institute of Physics of the Academy of Sciences of the

Latvian SSR, USSR

ABSTRACT

A survey of experimental results of the ionization of luminescence centers at low temperatures is given. It has been concluded
that these effects are connected with the capture of unrelaxed
positive holes. The expression for the capture probability of unrelaxed holes is given. A comparison with experiment shows that
unrelaxed holes may move a distance of about 50 lattice constants
in KCl.

In the past several years a considerable number of experimental
results have been obtained showing that, at sufficiently large activator concentrations, positive holes are effectively captured by
activator centers even at temperatures below the self-trapping temperature (T_a) where diffusion of the V_k-centers is frozen (1-3).
The concentration dependences of this effect has been extensively
studied for KCl:Tl (4), and less extensively, for NaCl:Ag (5) and
KI:Tl (6), and are shown in Fig. 1.

The following possible mechanisms for low temperature ionization of the activator could be proposed (7, 8): 1. By radiation
and collision ionization by hot electrons; 2. By H-centers which
are mobile at temperatures below T_a; 3. By excitons, as the relaxed
excitons are not mobile at temperatures near T_a, then we can speak
only about ionization by unrelaxed excitons; 4. A high value for
the capture radius of the activator center which could be realized
by electron tunneling from the activator center to the localized
V_k-center (in this case the probability of hole localization on the

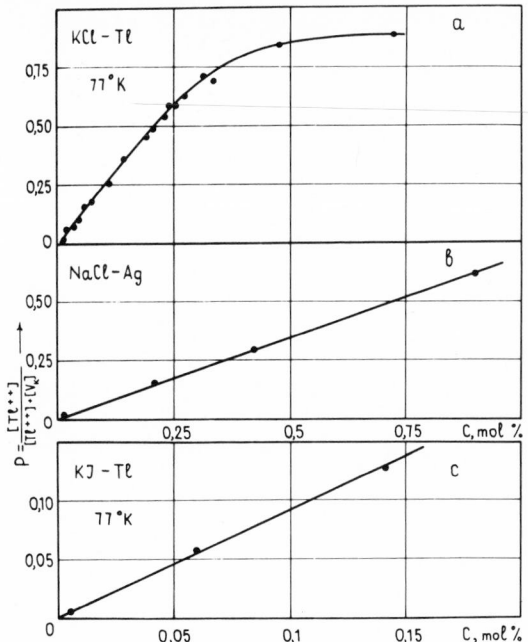

Fig. 1 Dependence of ionization probability of luminescence centers
        upon activator concentration: a- KCl:Tl (4); b- NaCl:Ag
        (5); c- KI:Tl (6).

activator will be determined by the volume covered by the capture
spheres of activator centers); and 5. Capture of unrelaxed holes
by the activator, that is, the hole produced by irradiation (X^0)
moves some distance on the way to complete relaxation (self-trapping)
(X_2^-) and on its way comes across an activator center where hole
localization is possible.

A detailed analysis in light of present experimental data shows
that ionization of the activator for $T<T_a$ is the result of unrelaxed
holes becoming localized at the activator. If the distance moved
by the hole before becoming self-trapped in the undoped crystal is
χ, then its localization probability on the activator in the doped
crystal will be determined by the following expression (9):

$$P = 1 - \exp(-\chi\sigma n) \tag{1}$$

where σ = effective capture cross-section of the unrelaxed hole by
the activator, n = activator concentration. Assuming for χ the
distribution function $\exp(-\chi/R)$, where R is the mean distance moved
by the hole to self-trapping in the undoped crystal, we obtain (8):

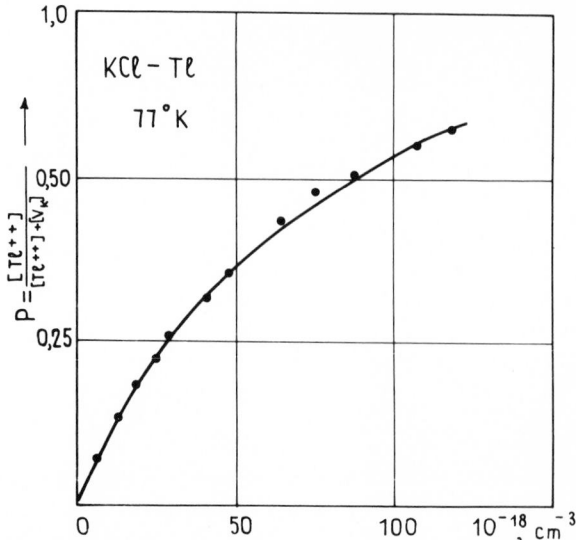

Fig. 2   Ionization probability versus thallium concentration for
the crystal KCl:Tl.   Circles=experimental points; contin-
uous line=theoretical curve (Eq. 2 with $R\sigma = 1.1 \times 10^{-20} cm^3$).

$$P = R\sigma n/(1 + R\sigma n) \qquad\qquad (2)$$

which was originally obtained by V. L. Vineckey.

Eq. (2) is in good agreement with experimental results for
KCl:Tl if: $R\sigma = 1.1 \times 10^{-20}$ cm^3, as shown in Fig. 2.   So far the
value σ is unknown.   However, if we assume as an approximate value
$\sigma = \pi d^2$, where d = interatomic distance (10),then we get $R \sim 50a$
where a = lattice constant.

The estimation characterizes the phonomenon and indicates the
possibility that holes move over a considerable distance before
becoming self-trapped.

Experimental points, shown in Fig. 2, were obtained in the
following way: a crystal was irradiated at nitrogen temperature
and the absorption coefficient κ_1 in the Tl^{++} band was measured;
after that the crystal was heated above the delocalization temperature of the V_k centers and then again cooled to liquid nitrogen
temperature; after that, absorption κ_2 was measured in the same
band.   As the majority of the holes were located on the Tl^+ we have:
$P = \kappa_1/\kappa_2$.

Investigations of the temperature dependence of P show that hole displacement up to self-trapping is not connected with the potential barrier between the stages of the cubic and axial relaxation (11). In the last few years such a barrier has been observed for excitons (12, 13).

So far the problem has not been solved regarding the state in which the unrelaxed hole moves and from which it is captured. There are at least three possible states: 1. hot hole, i.e., hole with kinetic energy that exceeds kT; 2. hole with energy approximately kT; and 3. partly relaxed hole (hole polaron). In the present stage of studies, it is not yet clear which of these states is the basic one. However, the most probable seems the second possibility, i.e., a hole with energy about kT.

It should be noted that experimental results obtained in recent years concerning lifetimes of V_k-centers at high temperatures show that at these temperatures the binding in the V_k-center evidently breaks up and the hole moves over some distance in the form of X^0, an unrelaxed hole, until secondary localization occurs (8). Apparently, in a wide temperature range – above and below the temperature of the self-trapping – the principal form of motion of holes is motion of the unrelaxed hole, X^0. Only in a narrow temperature range near the self-trapping temperature, does the motion of holes turns into diffusion of V_k-centers (relaxed holes). In that way, depending on the experimental conditions, the mechanism of the motion of holes in alkali halides can change from holes in a valence band, characteristic of typical semi-conductors such as germanium and silicon, to hole polarons and hopping conductivity (characteristic of oxides and ferrites). The limit of the immobile hole is the self-trapped state.

REFERENCES

1. K. K. Schvarz, E. D. Aluker, I. P. Mezina, Acta Phys. Polonika 26, 795 (1964).

2. G. K. Zolotarev, Ch. B. Lushchik, T. A. Sovik, M. A. Elango, Izv. AN USSR ser. fiz. 29, 36 (1965).

3. L. A. Pung, Y. Y. Haldre, Izv. AN USSR, ser. fiz. 30 443 (1967).

4. E. D. Aluker, O. E. Aksenov, N. L. Romanenko, Fiz. Tverd. Tela. 2, 3403 (1969).

5. L. A. Pung. Mater. XIX sov. po luminescencii, Riga I, 156 (1970).

6. R. G. Kaufman, W. B. Hadley, H. W. Hersh, IEEE, Trans. Nucl. Sci. NS-17, 82 (1970).

7. E. D. Aluker, Y. H. Kalnin, I. E. Uschomirscii, S. A. Chernov.
    Izv. AN. USSR, ser. fiz. 35, 1352 (1971).

8. E. D. Aluker, S. A. Chernov. Radiazionnay fizika 7, "Zinatne",
    Riga, 5 (1972).

9. E. D. Aluker, Y. H. Kalnin, Fiz. Tverdogo Tela 13, 641 (1971).

10. V. V. Antonov-Romanovskii, Kinetika Fotoluminescencii Kristalo-
    fosforov, Nauka Press, Moscow (1966).

11. E. D. Aluker, O. E. Aksenov, V. N. Semenova, Izv. AN USSR,
    ser. fiz. techn. nauk 4, 35 (1972).

12. A. Hattori, M. Tomura, N. Nishimura, J. Phys. Soc. Jap. 31,
    611 (1971).

13. Ch. Lushchik, J. Luminescence 1-2, 660 (1970).

TUNNEL LUMINESCENCE OF PURE AND Tl- AND Ag-DOPED ALKALI HALIDE

CRYSTALS

J. Bogans, U. Kanders, I. Leinerte-Neilande, D. Millers,

A. Nagornii, I. Tale, J. Valbis

Stuchka Latvian State University, Riga, USSR

ABSTRACT

By measuring the temperature dependence of tunnel pair accumulation efficiency and the spectra of tunnel luminescence (TL), the nature of centers constituting tunnel pairs and electronic transitions involved in TL were studied. Regularities of the spatial distribution of the centers were deduced from TL decay measurements. From F-center tunnel decay kinetics, the effective tunnel recombination distance between F- and V_K-centers in KCl was estimated.

Direct radiative recombination of electrons with holes localized on spatially separated centers was first observed and studied in alkali halide crystals by C. Delbecq et al. (1,2) and subsequently investigated by other authors (3-6). In activated crystals tunnelling takes place between zero-valent activator centers (Tl^0, Ag^0) and V_K-centers (1,6), and in unactivated ones - between F-centers and V_K-centers (3).

In the present investigation the following problems were studied: 1) The nature of defects and centers taking part in tunnel luminescence (TL), and the electronic states and transitions involved in TL. 2) The effect of the spatial distribution of centers on TL characteristics.

The identification of the centers between which tunnelling takes place ("the tunnel pairs") was based on the assumption that TL intensity drops to zero if the crystal is excited at temperatures above the thermal destruction point of the corresponding centers.

Accordingly, the TL intensity dependence on temperature of excitation or the efficiency of accumulation of tunnel luminescence ("EATL") at different temperatures was measured. The annealed crystal was excited 60 seconds by X-rays at a given temperature, then rapidly cooled to 80 K and the decay curves measured. TL intensity at a fixed time interval after switching off the excitation (usually 10^3 seconds) was taken as a measure of the number of tunnel pairs produced during excitation. After repeated annealing the next cycle was started by exciting the crystal at another temperature. Additional information was obtained by comparison of the TL emission spectra with the spectra under photoexcitation or X-ray excitation and by studying the TL stimulation spectra in the absorption bands of the corresponding centers.

Since the experimental results have much in common for all the crystals studied, we shall discuss in detail only the data for KI:Tl (Fig. 1). The TL spectrum consists of two main bands with maxima at 2.95 and 2.25 eV (Fig. 1d). The TL band at 2.95 eV is evidently due to activator luminescence. At temperatures of excitation above 200 K the position and half-width of this band coincide with the photoluminescence spectrum under excitation in the A and D absorption bands of Tl^+ centers (curves 3,4) (9). The half-width of the TL band at lower temperatures of excitation is larger, indicating the presence of other bands (curves 1, 2). The band at 2.25 eV practically coincides both in position and half-width with the so called "α-luminescence," known to be the result of radiative decay of excitons localized at anion vacancies (7,8).

The EATL curves measured separately for the 2.95 and 2.25 eV bands (Fig. 1b) exhibit several regions of sharp drop correlating closely with the positions of the thermoluminescence maxima (Fig. 1c) and with the efficiency curves of accumulation for different types of centers (Fig. 1a). The latter were obtained by measuring the areas under the thermoluminescence maxima of the V_{KA}, Tl^0, V_2 and Tl^{++} centers and the optical density of the F-absorption band after equal doses of excitation at different temperatures.

The data represented in Fig. 1 allow us to suggest that the emission in the 2.25 eV band ("α-band") is due to tunnel pairs involving F-centers as the electron component and V_K, V_{KA} and Tl^{++} centers as hole components. The 2.95 eV band is due to tunnelling in pairs Tl^0-V_K, Tl^0-V_{KA}, Tl^0-Tl^{++} and F-Tl^{++}. In similar experiments on other systems the positions of the TL band maxima, as indicated in Table 1, have been determined.

From an analysis of the above mentioned results, together with the spectra of TL and photoluminescence of pure and activated crystals, the following conclusions can be drawn:

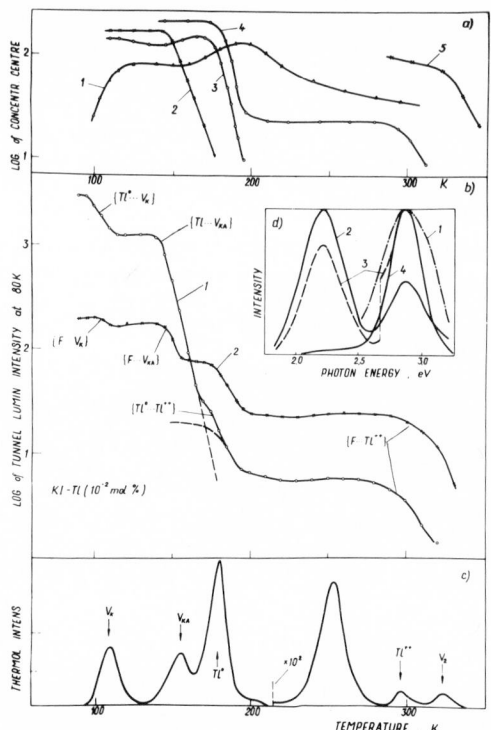

Fig. 1a. Efficiency of accumulation of various centers as a
function of temperature in KI:Tl (10^{-2} mol.%): 1. F-
centers (absorption measurements); 2. V_{KA}-centers; 3. Tl^{o}-
centers; 4. Tl^{++}-centers; 5. V_2-centers (measurements of
areas under corresponding thermoluminescence maxima).

b. Efficiency of accumulation of tunnel luminescence
(measured at 80 K) as a function of the temperature of
excitation: 1. 2.95 eV TL band; 2. 2.25 eV TL band.

c. Thermoluminescence glow curve after excitation at 80 K.

d. TL spectra at 80 K after excitation at 110 K (1), 170 K
(2) and 250 K (3). Photoluminescence spectrum under
excitation in the Tl^{+} A absorption band (4).

1. In alkali halide crystals the following groups of centers
can serve as hole components for tunnel pairs: a) V_K centers and
their derivatives, b) H-centers and c) ionized activator centers.

2. In TL three types of radiative transitions can take place:
a) transitions at the hole component of the tunnel pair, e.g. the
Tl^{+} band at 2.95 eV from the F-Tl^{++} pairs in KI:Tl; b) transitions
connected with the electron component of the pair (the band at

Table 1

Crystal \ Pair	$F-V_K$	$F-V_{KA}$	$F-A^{++}$	$F-V_1$	A^o-V_K	A^o-V_{KA}	A^o-A^{++}
Energy at maximum and half-width () of TL bands, eV							
KI	2.25 (0.8) 3.1	2.25 (0.8)	Tl: 2.25 (0.8) 2.95 (0.3)	---	Tl: 2.95 (0.48)	Tl: 2.95 (0.48)	Tl: 2.95 (0.42)
KBr	2.42 (0.65)	---		3.2	Tl: 2.9 (0.9) Ag: 2.2 (0.65)	--- ---	--- ---
KCl	2.6 (0.7)	---		3.75	Tl: 3.0 (0.75) Ag: 2.3 (0.6)	---	Tl: 2.7 (0.75)
NaCl		---			Ag: 3.15 (0.85)		Ag: 3.1 (0.85)

2.25 eV from the pairs $F-V_K$, $F-V_{KA}$ in "pure" KI and $F-Tl^{++}$ in KI:Tl coincides with the "α-luminescence" with photoexcitation); c) radiative transitions characteristic of tunnel recombination only, which cannot be observed under direct photoexcitation of either of the components; we shall call these "characteristic TL."

3. Generally, radiative transitions connected with particular components of the pair are observed when the energy of the characteristic TL of the pair is higher than that of the transitions of the components. Probably the widening of the 2.95 eV band is due to overlap of the activator luminescence with the characteristic TL. Exciton luminescence as a result of tunnelling between F- and V_K-centers in KBr and KCl is not observed, probably because of thermal quenching.

4. The characteristic emission of TL can be interpreted as the result of direct tunnel transitions between the ground electronic states of electron and hole centers (1,2). The emission of the hole component of the pair can be the result of a non-radiative tunnel transition to the excited state of the hole center and subsequent radiative transition to the ground state (4). The emission by the electron component of the pair can be excited by the

the tunnelling of a hole, in other words, the tunnelling of an electron from the lower energetic states of the electron center (e.g. the halogen p^6 electron of the F-center or s-electron of Tl°) to the hole center. However, for a more detailed interpretation of the experimental data a quantum-mechanical analysis of the system involving electron and hole centers and their environment is necessary. The existence of two emission bands from one tunnel pair indicates the same order of radiative transition probabilities for different transitions.

The characteristics of the TL depend substantially on the spatial distribution of the centers involved since the probability of tunnel transitions (ω) is an exponential function of the distance (r) between the components of the pairs (10). The decay law of TL is determined by the distribution over distances of the pairs (11,12), hence one can obtain information on the spatial distribution of different centers by measuring the decay curves and comparing them with the theoretical ones for different distribution models. The experimental decay curves of TL (Fig. 2) are non-exponential with the initial parts depending strongly on the time of excitation (curves 1, 2, 3). The decay laws of the final parts are different for various tunnel pairs, following more or less closely the theoretical ones based on the assumptions of either predominantly associated or random distribution of the centers within the pairs. For example, in KI:Tl the decay of TL from pairs $Tl^\circ-Tl^{++}$ corresponds to a random distribution; from $F-Tl^{++}$, to associated centers (curves 4 and 5, respectively). The latter can be explained by the assumption that non-radiative decay of excitons near activator centers effectively produces the associated pairs $F-Tl^{++}$.

After optical stimulation in the absorption bands of F-centers or of zero-valent activator centers, or after thermo-stimulated diffusion of V_K-centers, stimulation of TL is observed due to the increase of the number of closer pairs. It can be easily shown that in a time interval t after switching off the excitation, TL is mainly due to pairs with a certain effective distance $\bar{r}$ between the components (11) (See Fig. 2a) given by: $\bar{r} = r_o \ln(\omega_o t)$ where r_o is half of the effective Bohr radius of the more diffuse component of the pair, and ω_o is the frequency factor of the probability of tunnel recombination. The effective distance $\bar{r}$ can be estimated if the decay law of the pair concentration is known. Assuming that the probability $p(r)$ of pair generation is proportional to the density of the neighboring states (11), $p(r) = 4\pi r^2 n_F \exp(-4\pi r^3 n_F/3)$ where n_F is the concentration of one of the components of the pair. The effective distance is given by the expression:

$$\bar{r}(t) = \left\{ \frac{3\ln[n_F(t_S)/n_F(t_e)] \ln^3(\omega_o t)}{4\pi n_F(t_S)[\ln^3(\omega_o t_e) - \ln^3(\omega_o t_S)]} \right\}^{1/3}$$

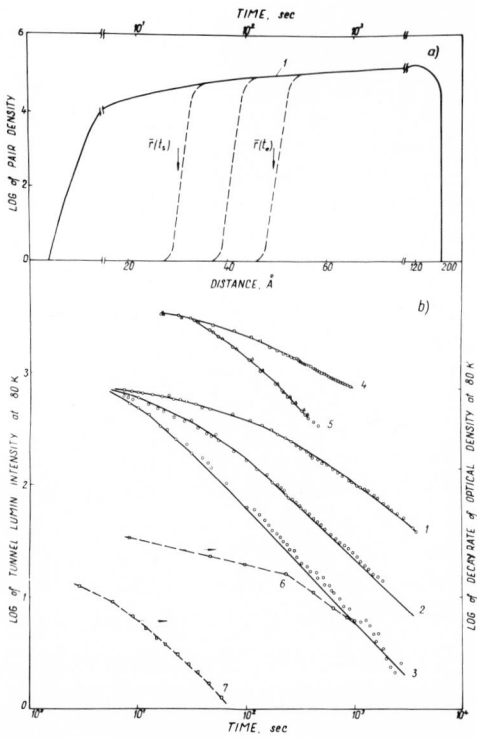

Fig. 2a.   Distribution law of tunnel pairs over the distances
          between centers calculated from curve (2) below - solid
          line; the distribution law at different time intervals
          after switching off the excitation - broken lines.

     b.   Decay curves of TL at 80 K. Solid lines - theoretical
          curves, points - experimental data. 1. TL band at 2.4 eV
          in KBr excited by X-rays for 10^3 sec. at 150 K; 2. the
          same for excitation time 10^2 sec.; 3. the same for
          excitation time 10 sec.; 4. TL band at 2.95 eV in KI-Tl
          (10^{-2} mol. %) after excitation by X-rays for 60 sec. at
          170 K; 5. bands 2.95 and 2.25 eV after excitation for
          60 sec. at 260 K; 6. F-center optical density decay rate
          after excitation by X-rays for 240 sec. at 80 K in KBr;
          7. the same in KCl.

where t_s is the time at the start of the n_F measurement, and t_e is
the time of end of n_F measurement.

     The decay of the optical absorption of F-centers in KCl and
KBr crystals due to tunnel recombination in $F-V_K$ pairs was measured
(curves 6 and 7). If we take ω_o equal to 10^6/sec (from the decay
kinetics of F-centers after pulse excitation by an electron beam

(13), we obtain r_o = 4.5 A for F-centers in KCl and $\bar{r}$ = 93 A at $t = 10^3$ sec.

The results of the TL decay measurements lead to the following conclusions: 1. The decay of TL cannot be due to tunnelling migration of one of the components as suggested in (6). 2. The decay of TL is unambiguously determined by the spatial distribution of centers. In alkali halides the radiation-produced centers are generally situated in nearer pairs than they would be in the case of a random distribution of centers. 3. The relatively large effective radius of tunnel recombination of pairs involving F-centers indicates that this process plays an important role during the coloration of crystals by ionizing radiation.

REFERENCES

1. C. J. Delbecq, A. K. Ghosh and P. H. Yuster, Phys. Rev. <u>151</u> 599 (1966).

2. C. J. Delbecq, A. K. Ghosh and P. H. Yuster, Bull. Am. Phys. Soc. <u>6</u> 629 (1964).

3. J. Bogan, Izv. Akad. Nauk Latv. SSR, ser. fiz, No. 3 <u>57</u> (1969).

4. I. Jaek and M. Kink, Phys. Stat. Sol. <u>33</u> 905 (1969).

5. I. Tale and A. Gailitis, Izv. Akad. Nauk SSSR, ser. fiz. <u>35</u> 1336 (1971).

6. F. J. Keller and R. B. Murray, Phys. Rev. <u>150</u> 670 (1966).

7. M. Ikezawa and T. Kojima, J. Phys. Soc. Japan <u>27</u> 1551 (1969).

8. R. Onaka and I. Fujita, J. Quant. Spectr. Radiat. Transf. <u>2</u> 599 (1962).

9. J. M. Donahue and K. Teegarden, J. Phys. Chem. Sol. <u>29</u> 2141 (1969).

10. D. G. Thomas, J. Hopfield and W. M. Augustiniak, Phys. Rev. <u>140</u> A202 (1965).

11. A. Gailitis, Dissertation, Riga, (1972).

12. G. H. Döhler, Phys. Stat. Sol. (b) <u>45</u> 705 (1971).

13. M. Ueta, Y. Kondo, M. Hirai and I. Yoshinari, J. Phys. Soc. Japan <u>26</u> 1000 (1969).

CONFIGURATION INTERACTION AND CORRELATION EFFECTS IN DONOR-ACCEPTOR PAIR SPECTRA[*]

Lester Mehrkam[†] and Ferd Williams

Physics Department, University of Delaware

Newark, Delaware 19711

ABSTRACT

We report the effects of overlap, configuration interaction and correlation on the electronic states of DA pairs and on their luminescent spectra. The effect of overlap was calculated from the formulation of Williams (2) and was found to improve somewhat the R_i dependence of the spectra. The effect of configuration interaction was determined by the variation method, allowing both the radii and the angular dependences of the electron and positive hole effective mass functions to vary to minimize the total energy. Markedly better agreement between theory and experiment is found. The remaining discrepancy is explained by correlation of the electon and positive hole. The correlation energy is estimated using the free exciton energy. We also compute the effect of configuration interaction on radiative lifetime.

INTRODUCTION

The luminescent emission of donor-acceptor (DA) pairs in phosphors and semiconductors has been extensively investigated (1). However, the electronic states and radiative transitions of DA pairs have been considered theoretically only to first order (2) (Heitler-London approximation) or with a second order van der Waals term added (3). Gross and Nedzvetskii (4) reported extensive fine structure in the emission spectra of III-V semiconductors which Hopfield, Thomas and Gershenzon (5) explained by radiative recombination at specific DA pairs. The wealth of information in these spectra indicate a need for an improved theory. The simple theory accounts for the general features of the spectra, however, a substantial discrepancy between theory and experiment exists in the dependence of radiative transition energy, $h\nu$, on DA distance, R, especially for the nearer neighbor pairs.

Distant DA pair spectra are usually fitted with the following equation, either with or without the last term:

$$h\nu = E_g - (E_D + E_A) + \frac{e^2}{\kappa_s R} - \frac{e^2}{\kappa_s} (b/R)^6, \qquad (1)$$

where E_g is the band gap, E_D and E_A the electronic binding energy of donor and acceptor respectively, κ_s the static dielectric constant and b an adjustable parameter in the van der Waals term used to improve the fit to experimental data. The DA pair is shown in Fig. 1 and the band model for electronic states and transition, in Fig. 2. The fit of Eq. 1 to data on GaP:Zn,S is shown in Fig. 3: A without; B including the van der Waals term.

We note two limitations of Eq. 1: First, κ_s is the proper dielectric constant for the condition $\tau_o \gg \tau_\ell$, where τ_o is the orbital period for electronic motion and τ_ℓ is the period for lattice vibrations. Second, the van der Waals term is a second order correlation energy valid only for $R \gg a_A, a_D$, where a_A and a_D are the following for 1s-type states:

$$a_A = e^2/2\kappa_s E_A , \qquad a_D = e^2/2\kappa_s E_D . \qquad (2)$$

In obtaining the best fit to experimental data, Hopfield et al (5) found b to be the same magnitude as a_A and a_D.

The effect on the electrostatic energy of the overlap of electron and positive hole distributions was included in the analysis

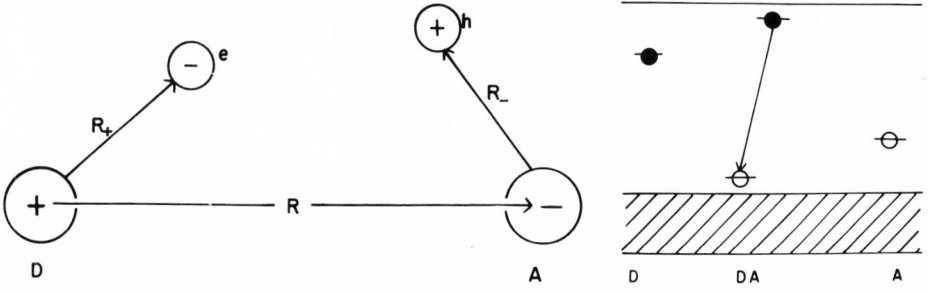

Fig. 1   Donor-acceptor pair with inter-particle distances.

Fig. 2   Band gap model for the electronic states of donor and acceptor and donor-acceptor pair.

of Williams (2), however, this effect has not been previously
included in the analysis of data. For the case of $a_D = a_A = a_o$
the pair spectrum, omitting the van der Waals term and including
the effect of overlap, is as follows,

$$h\nu = E_g - (E_D + E_A) + \frac{e^2}{\kappa_s R} - \frac{e^2}{\kappa_s R} e^{-2\rho_o}[1 + \frac{5}{8}\rho_o - \frac{3}{4}\rho_o^2 - \frac{\rho_o^3}{6}] \quad (3)$$

where $\rho_o = R/a_o$. Eq. 3 is applied to data on GaP:Zn,S as curve C
of Fig. 3.

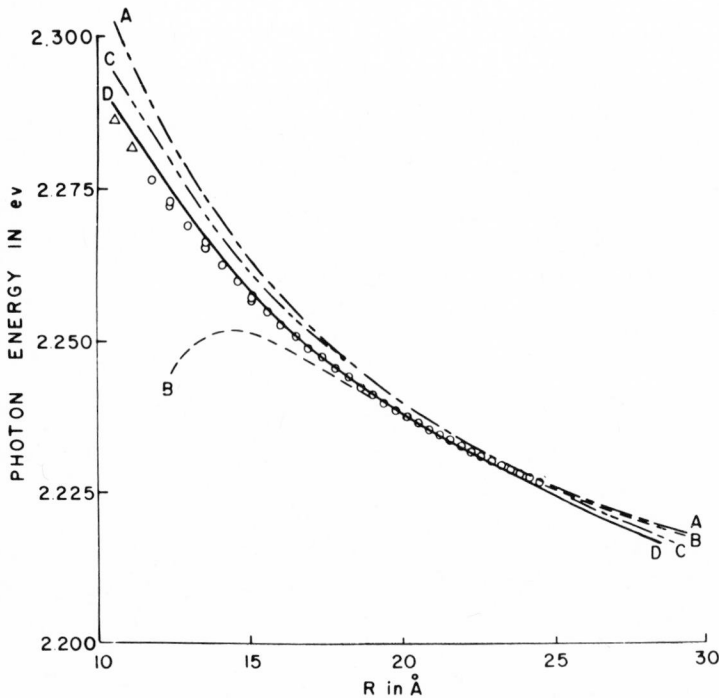

Fig. 3 Experimental and theoretical radiative recombination
        energies for GaP:Zn,S.

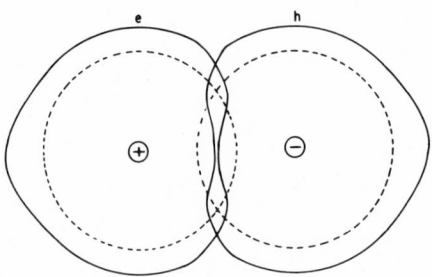

Fig. 4 Electron and positive hole effective-mass functions includ-
        ing configuration interaction for donor-acceptor pair.

In this paper we are primarily concerned with evaluating theoretically the effects of configuration interaction and of electronic correlation on pair spectra.

THEORETICAL ANALYSIS OF TRANSITION ENERGY

Wang (6) and Rosen (7) treatment of the hydrogen molecule gives an electronic binding energy that differs from the experimental value by a very small amount. This difference is almost entirely attributed to the neglect of electronic correlation in the wavefunction. Adapting these methods to the excited donor-acceptor pair results in the electron-hole effective mass functions being delocalized and rendered non-spherical with electronic density thinning out between the nuclei. This is shown in Fig. 4. In contrast, Wang and Rosen's treatment of the hydrogen molecule results in the wavefunction being localized and rendered non-spherical with the electron density increasing between the nuclei. In both cases the resulting dipoles are opposite to those expected from electronic correlation. We find that a configuration interaction treatment of the excited donor-acceptor pair with a variation of parameters in the one electron wavefunctions significantly improves the agreement between the calculated and observed radiative recombination energies.

The two particle effective mass function for the donor-acceptor is

$$\psi(+,-) = \psi_A(+)\psi_D(-), \tag{4}$$

where following Rosen's formulation

$$\psi_A(+) = (1+\sigma^2)^{-\frac{1}{2}} a_A^{-3/2} e^{-R_-/a_A} [1+\sigma(R_-/a_A)\cos\theta_A], \tag{5}$$

R_- is the distance between the acceptor and hole, and θ_A is the angle between the radius vector R_- and the line joining the donor and acceptor ions R, as shown in Fig. 1. A similar equation applies for $\psi_D(-)$. In this analysis $a = a_o/z$ where z is the effective charge. The Hamiltonian for the donor-acceptor pair is

$$\hat{H} = -\frac{\hbar^2}{2m_-^*}\nabla_-^2 - \frac{\hbar^2}{2m_+^*}\nabla_+^2 + \frac{e^2}{\kappa_s}\left(-\frac{1}{|R|} - \frac{1}{|R_-|} - \frac{1}{|R_+|} + \frac{1}{|R+R_-|}\right.$$

$$\left. + \frac{1}{|R_+-R|} - \frac{1}{|R_+-R-R_-|}\right), \tag{6}$$

and the total energy found from

$$E_{total} = \int \psi(+,-)\hat{H}\psi(+,-)dR_+ dR_- \tag{7}$$

is minimized by varying z and σ. The ionization energy,

$$E_I = E_{total} + (e^2/\kappa_s R) \ , \tag{8}$$

is easily found and can be used to determine the pair emission spectrum,

$$E_{photon}(R) = E_g - E_I \ . \tag{9}$$

The variational parameters σ and z necessary for minimizing E_I can be evaluated as a function of ρ_o. One can express E_I as

$$E_I = \frac{e^2}{\kappa_s R} \ [-z\rho+1+(A+B\sigma+C\sigma^2+D\sigma^3+E\sigma^4)/(1+2\sigma^2+\sigma^4) \ ] \tag{10}$$

where

$$A = 2(z-1)\rho+e^{-2\rho} \ [\ \frac{\rho^3}{6} + \frac{3\rho^2}{4} - \frac{5}{8}\rho - 1] \ .$$

$$B = -4e^{-2\rho} \ [\ \frac{\rho^4}{12} + \frac{11\rho^3}{24} + \frac{11\rho^2}{48} \ ] \ .$$

$$C = 4(z-1)\rho+\rho + \frac{8}{\rho^2} + e^{-2\rho}[\ \frac{7\rho^5}{30} + \frac{89\rho^4}{60} + \frac{369\rho^3}{120} + \frac{37\rho^2}{6} +$$

$$\frac{109}{12}\rho + 14 + \frac{16}{\rho} + \frac{8}{\rho^2} \ ] \ .$$

$$D = \frac{36}{\rho^3} - 4e^{-2\rho}[\ \frac{\rho^6}{60} + \frac{7\rho^5}{60} + \frac{113\rho^4}{240} + \frac{697\rho^3}{480} + \frac{2617\rho^2}{960} + 6\rho +$$

$$12 + \frac{18}{\rho} + \frac{18}{\rho^2} + \frac{9}{\rho^3} \ ] \ .$$

$$E = 2(z-1) + \frac{54}{\rho^4} + e^{-2\rho} [\ \frac{\rho^7}{140} + \frac{141\rho^6}{2520} + \frac{19\rho^5}{70} + \frac{893\rho^4}{840} +$$

$$\frac{10369\rho^3}{6720} + \frac{2571\rho^2}{640} + \frac{16651\rho}{1280} + 35 + \frac{72}{\rho} + \frac{108}{\rho^2} + \frac{108}{\rho^3} + \frac{54}{\rho^4} \ ].$$

$$\rho = z\rho_o = z(R/a_o) \ .$$

To transform the minimum values of $E_I(\rho_o)$ to $E_I(R)$, we note that $a_o = a_D = a_A$ is obtained from

$$E_D + E_A = (e^2/\kappa_s a_o) , \qquad (11)$$

where $E_D + E_A$ is the experimental parameter characterizing the dopants. The experimental parameters E_g and κ_s characterize the host crystal. To evaluate the difference between the configuration-interaction results and the theoretical work of others for radiative recombination energies and lifetimes, we apply the equations to a well-known physical system.

APPLICATION TO EXPERIMENTAL SPECTRA

The pair emission spectrum calculated by equations 1 (with and without the van der Waals term), 3 and 9 are fitted against the experimental pair emission spectrum of GaP:Zn,S observed by Hopfield, Thomas and Gershenzon (5). The coulomb overlap curve, C, and configuration-interaction curve, D, are shifted down the energy axis 0.003 and 0.0046 eV respectively to obtain the best overall fit with the data. This shift is consistant with the error in the accepted value of $E_g = 2.325 \pm 0.005$ eV. The improvement is most prominent for nearer neighbor pairs.

Curve D in Fig. 3 which includes configuration-interaction has improved the R dependence for the entire range of observed data, including the values at 10 and 11 A usually omitted. This correction of the value $E_g - E_A - E_D$ is small so that our choice of the value of a_o need not be revised for self-consistency. The van der Waals term is not included in D because it clearly is not valid at small R (see curve B), is negligible at large R (compare A and B) and is a doubtful approximation to the effects of correlation at intermediate R. The effects of coulomb overlap are included in both C and D. To minimize the number of variational parameters and to simplify the calculations, the configuration-interaction, and also the coulomb overlap, are determined for $a_A = a_D = a_o$. In other words, a reduced a_o of 9.1 A is used.

The remaining small discrepancy between the data and theoretical curve D for R < 15 A can be accounted for by electronic correlation only if the correlation effect in DA pairs is greater than that in H_2. Other effects which may contribute to the remaining discrepancy are anisotropy of the effective masses and differences in radii of electron and positive hole distributions. These affect the interactions through the overlap. We have calculated the overlap for $a_D = 6.8$ A and $a_A = 13.6$ A, which correspond according to Eq. 2 to the experimental E_D and E_A, and found that the overlap was reduced about 10 percent but retained the same R-dependence. In other words, we have slightly over-included coulomb overlap by taking $a_A = a_D$; however, the configuration-interaction is believed less affected because we reduced the variational parameters from four to two.

Incidentally, in plotting the experimental points the dopants are assumed undisplaced from the perfect lattice sites. This is valid in view of the low cohesive energy of the radiating DA pairs.

An additional possible contribution to the remaining small discrepancy is related to the correct choice of dielectric constant, κ, to use for various pairs. The static dielectric constant, κ_s, applies for the delocalized electronic configurations occurring for small R, i.e. $\tau_o > \tau_\ell$, but for more distant pairs κ may depend on R and decrease below the value κ_s. The period for orbital electron motion may become comparable with the lattice relaxation time for distant pairs. If the configuration curve is lowered by more than 0.0046 eV to obtain a better fit at small R, the fit for large R will deteriorate. This can be compensated for by the use of a dielectric constant $\kappa(R) < \kappa_s$ and would result in a better overall fit with the experimental data. The configuration-interaction equation is very sensitive to the value of κ used since it occurs both as a scaling coefficient in E_I and partly determines the electronic overlap through a_o.

Recent values of E_g reported in the literature differ from the value we used. The somewhat higher values of E_g affect all the calculated spectral curves A, B, C, and D in the same way. Increases in E_g produce a constant shift upwards away from the experimental points.

The remaining problem appears to be to reconcile the deviation between curve D and the experimental points by considering: (1) the uncertainty in E_g which results in shifting curve D along the energy scale, (2) the possibility that some effective dielectric constant should be used which can even be R-dependent, $\kappa_o < \kappa(R) < \kappa_s$, (3) the contribution arising from electronic correlation which is R-dependent and (4) the effect of assuming $a_D = a_A = a_o$. It should be possible to clarify the relative importance of these possibilities by doing calculations for many different materials and dopants. For example, we are doing GaP:Sn,Zn, for which $a_D = a_A$ (8).

EFFECTS OF ELECTRONIC CORRELATION

There is no simple way to calculate electronic correlation in the wavefunction describing the hydrogen molecule. The only successful calculation including correlation was the James-Coolidge treatment. Since the remaining small deviation in fitting the DA pair spectrum can also be attributed to other effects which were discussed, such an elaborate calculation was deferred. Near $\rho_o = 1.4$ the deviation between the calculated curve D and observed data is about 7% of E_I for DA pairs or 0.003 eV, whereas the electronic correlation energy in the hydrogen molecule is 1.3% of E_I or 0.66 eV.

A semi-empirical approach is used to estimate the electronic correlation energy and its R-dependence. We take the exciton as a limiting case of the excited DA pair when R approaches zero. A free Wannier exciton maintains its stability entirely through electronic correlation. For GaP an exciton energy E_X of 0.010 $\pm$ 0.001 eV was reported by Dean and Thomas (9). The coulombic interaction requires $E_X = e^2/2\kappa_s r_{+-}$ therefore $r_{+-} = 78$ A which is large compared to R and a_o. If the electronic correlation energy of the excited DA pair, E_c, is proportional to some power of the electronic overlap integral, Δ, E_c can be calculated using E_X as the proportionality constant, $E_c = E_X \Delta^n$. A calculation using 1s effective mass states and distinguishing between a_D and a_A yields for n = 1 values of $E_c(\rho_o)$ which can account for both the displacement of curve D by 0.0046 eV and also the remaining deviation of curve D from the experimental points at small R.

However, the value of n can be estimated independently of the data being fitted. This is done using helium as the limiting case of the hydrogen molecule as $R \to 0$, similar to the free exciton being the limiting case of the pair for $R \to 0$. Taking $E_X = 1.4$ eV for He and $E_c = 0.66$ eV for the hydrogen molecule with $\rho_o = 1.4$, we find n $\approx$ 2.5. The discrepancy between curve D and the observed data can be explained by electronic correlation with n $\approx$ 2.

Although the preceeding analysis is semi-empirical the electronic correlation term so derived applies for the entire range of R, and it is simple to use and does not contain an arbitrary parameter. It clearly applies in the region where electronic overlap is appreciable. In this respect it is clearly superior to the van der Waals term. On the basis of our semi-empirical analysis, we conclude that the correlation energy is a greater fraction of the electronic binding energy of a DA pair in GaP than is the case for the H_2 molecule.

THEORY OF RADIATIVE LIFETIMES

These calculations of configuration-interaction show that spherical delocalization of the Heitler-London wavefunctions occurs for $\rho_o < 1.2$ which corresponds to R < 11 A for the system considered. The only effect for R $\geq$ 11 A is the departure from spherical symmetry of the Heitler-London wavefunctions which reduces electronic overlap. Compared to values calculated with the Heitler-London wavefunctions, the lifetime, τ, of the excited state will be larger for each pair.

Following the analysis of Shaffer and Williams (10) we consider the radiative dipole matrix element:

$$d_{if} = e \int \psi_D(\underline{r})\psi_A(\underline{r}-\underline{R})d\underline{r} \int u_{vo}(\underline{r}_o)\underline{r}_o u_{co}(\underline{r}_o)d\underline{r}_o. \quad (13)$$

The effective mass function integral

$$\int \psi_D(\underline{r})\psi_A(\underline{r} - \underline{R})d\underline{r} \quad (14)$$

provides the relation between the Heitler-London and configuration-interaction physical parameters. Therefore the ratio of lifetimes are:

$$\frac{\tau_{HL}}{\tau} = \left| \frac{\int [\psi_A^o(\underline{r}-\underline{R}) + \sigma\psi_A^1(\underline{r}-\underline{R})][\psi_D^o(\underline{r})+\sigma\psi_D^1(r)]d\underline{r}}{\int \psi_A^o(z=1,\underline{r}-\underline{R})\psi_D^o(z=1,\underline{r})d\underline{r}} \right|^2, \quad (15)$$

which becomes for $\rho = \rho_o > 1.2$

$$\frac{\tau_{HL}}{\tau} = \left| 1 - \sigma\rho_o + \frac{\sigma^2}{5} \frac{\rho_o^4 + 2\rho_o^3 + 3\rho_o^2 - 15\rho_o - 15}{\rho_o^2 + \rho_o + 1} \right|^2. \quad (16)$$

The maximum value of σ obtained by the variation principle is 1.5 and this occurs for $\rho_o = 1.5$ (R = 13.6 A) and z = 1. For this case the ratio given by Eq. 16 is 0.55 and approaches unity for larger values of ρ_o. Introducing electronic correlation in the effective mass wavefunction will couple the electronic particles' motion and produce tighter electronic binding. It could also increase the amount of electronic overlap and reduce our predicted increased lifetimes, however, this effect is probably quite small.

CONCLUSIONS

The agreement between the theoretical and experimental radiative recombination spectra of donor-acceptor pairs is greatly improved by including the effects of overlap and of configuration-interaction. The remaining small discrepancy for the nearer-neighbor pairs is accounted for by correlation. The radiative lifetimes are increased by as much as a factor of two. These studies demonstrate a need for re-analysis of experimental spectra for a diversity of dopants and semiconductors.

REFERENCES

*Supported in part by a grant from the U. S. Army Research Office.
†Present address: Shepherd College, Shepherdstown, W. Va.

1.   For a review see: F. Williams, Phys. Stat. Sol. 25, 493 (1968).

2. F. Williams, J. Phys. and Chem. Solids 12, 265 (1960).

3. W. Hoogenstraaten, Philips Res. Rep. 13, 515 (1958).

4. E. F. Gross and D. S. Nedzvetskii, Dokl. Akad. Nauk SSSR 146, 1047 (1962).

5. J. J. Hopfield, D. G. Thomas and M. Gershenzon, Phys. Rev. Lett. 10, 162 (1963).

6. S. C. Wang, Phys. Rev. 31, 579 (1928).

7. N. Rosen, Phys. Rev. 38, 2099 (1931).

8. P. J. Dean, R. A. Faulkner and S. Kimura, Phys. Rev. B2, 4062 (1970).

9. P. J. Dean and D. G. Thomas, Phys. Rev. 150, 690 (1966).

10. J. C. Shaffer and F. E. Williams, Proc. 7th Internat. Conf. Phys. Semicond. (Dunod, Paris 1964) p. 811.

DONOR-ACCEPTOR PAIRS IN IONIC CRYSTALS

G. Schwotrev, G. Kötitz and P. Görlich

VEB Carl Zeiss, Jena, DDR (Germany)

ABSTRACT

Thermoluminescent spectra of alkaline earth fluoride crystals doped with scandium and lanthanum ions have been compared with those of yttrium doped crystals. The spectra are interpreted in terms of associated donor-acceptor pairs involving interstitial sites or vacancies in the fluoride sublattice. The pair spectra due to transitions between the thermally-stimulated excited state and the ground state are dependent on pair internuclear distance. On the basis of this model conclusions on coordination number and symmetry of the dopant sites are drawn.

INTRODUCTION

In the theoretical treatment of luminescence phenomena in semi-conductors, donor-acceptor pairs play an important part. Prener and Williams (1) were the first to use such pairs to explain the green luminescence of copper-activated zinc sulfide. The most impressive evidence of their existence, however, has been given by Hopfield, Thomas and Gershenzon (2) in their studies on gallium phosphide.

Because of their opposite charges, donors and acceptors attract each other electrostatically and can thus form associated donor-acceptor pairs. By means of photon excitation, these pairs can bind an electron and a hole. The radiative recombination of the electrons bound by the donors with the holes bound by the acceptors depends, to a first approximation, only on the distance R between the donors and acceptors (3): $E = E_g - (E_D + E_A) + (e^2/\kappa R)$ where E_g

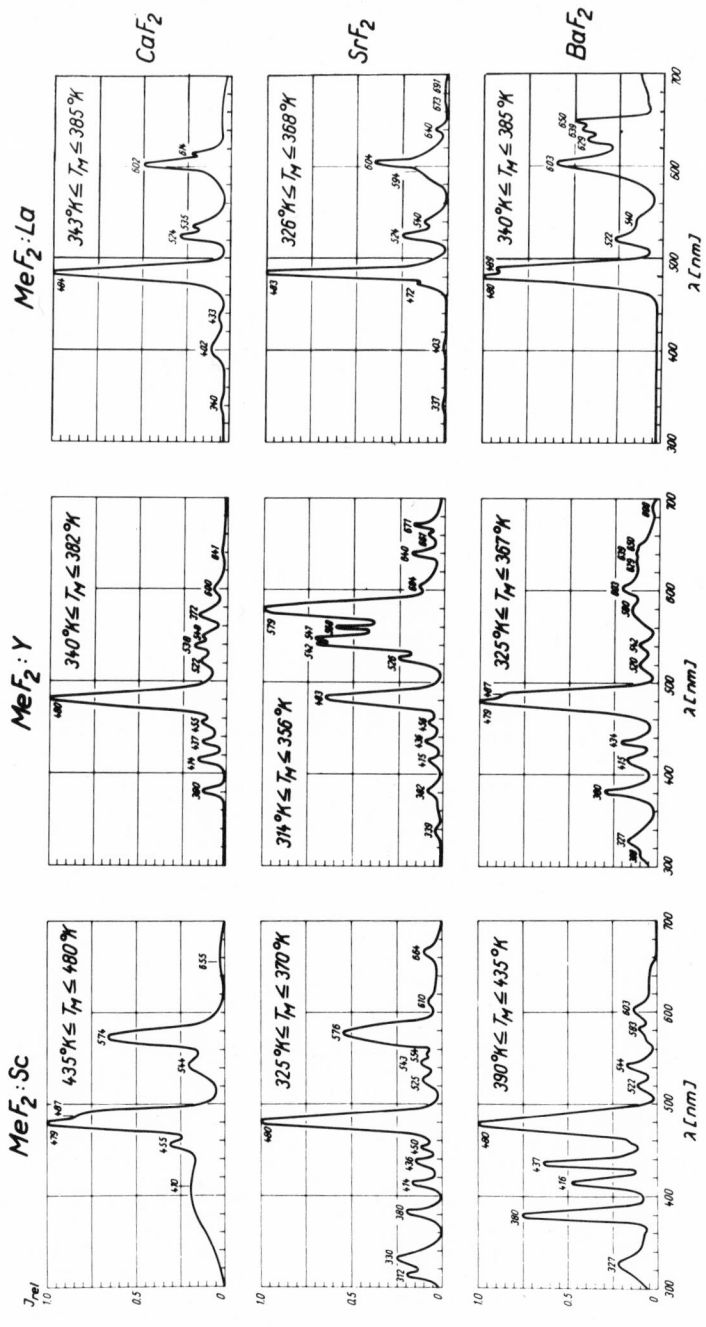

Fig. 1 Thermoluminescence spectra of the alkaline earth fluoride crystals doped with scandium, yttrium and lanthanum after x-ray irradiation at room temperature.

designates the band gap, E_D and E_A the binding energies of donors or acceptors, respectively, and κ the dielectric constant. If donors and acceptors are at certain sites in the crystal lattice, discrete emission lines are to be expected in the luminescence spectrum. In that way, Hopfield and co-workers (2) correlated the complex line spectrum at 1.6 K of gallium phosphide with two types of donor-acceptor pairs: spectrum I corresponds to donors and acceptors at equivalent sites in the zinc blend lattice, while spectrum II corresponds to donors and acceptors at non-equivalent lattice sites.

Analogously, O'Connor (4) attempted in 1964 to explain the thermoluminescence spectra, first reported for fluorite by Iwase (5) in 1933, by radiative recombination at associated donor-acceptor pairs. He succeeded in relating some of the emission bands to certain distances between yttrium donors and oxygen or interstitial fluorine acceptors. O'Connor's presentation is, however, insufficient in many details (6). To throw some light on the problem, we have carried out thermoluminescence studies on alkaline earth fluoride crystals, which were doped with yttrium and also its homologs, scandium and lanthanum. The experimental results are interpreted as luminescence of excited donor-acceptor pairs.

EXPERIMENTAL RESULTS

X-ray irradiation of the crystals was carried out at room temperature. Fig. 1 gives a survey of the spectral distribution of thermoluminescence for scandium-, yttrium- and lanthanum-doped alkaline earth fluorides in the region of the most intense glow maxima. All crystals show similar thermoluminescence spectra, consisting of several narrow bands with half-width values not greater than 0.1 eV. Until now, this spectral type had been regarded as exclusively typical of yttrium-doped alkaline earth fluorides (4, 7). It is to be noted that the spectral positions of the bands hardly change in going from CaF_2 to SrF_2 and to BaF_2. It is also striking that the spectra of scandium- and yttrium-doped crystals are so similar.

With regard to the intense band at 480 nm, lanthanum-doped alkaline earth fluorides share this characteristic, whereas they show differences especially in the short wavelength region with weak bands at 340 nm, 402 nm, 433 nm and 471 nm. These bands are very faint in SrF_2:La and are not detectable at all in BaF_2:La. On the other hand, the typical emission bands for lanthanum doping at 524 nm, 536 nm, 602 nm and 643 nm are also present in yttrium-doped crystals, but can hardly be found with scandium doping. The characteristic band for the last two activators is at 574 nm.

The investigation of the thermoluminescence spectra as a func-

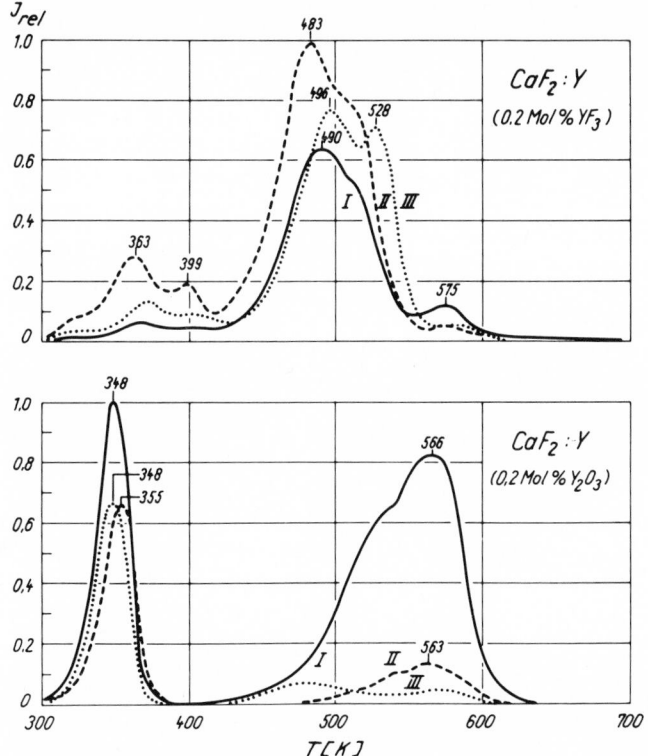

Fig. 2 Spectral glow curves of CaF_2 crystals doped with YF_3 and
Y_2O_3 at the emission wavelengths I: 380 nm, II: 480 nm
and III: 578 nm.

tion of temperature shows that the emission consists of three sub-
spectra. Below 500 K, the emission consists of two bands at 480 nm
and 544 nm and of a group of bands extending from the band at
574 nm to the near IR. Above this temperature, a group of bands
dominates, the most intense of which occurs at 381 nm. The same
results were obtained with yttrium-doped crystals. Although thermoluminescence spectra, in the case of lanthanum doping, also consists
of three sub-spectra, they do not show the temperature dependence
described. In no case did we observe a temperature dependent shift
of emission maxima, which one would have expected on the basis of
the configuration coordinate model.

    In order to study the influence of the coactivators on the
thermoluminescence spectra some CaF_2 crystals were doped with YF_3,
others with Y_2O_3. Correspondingly, the positive charge of the tri-

valent activators was compensated by fluoride ions at interstitial
sites, F_i^-, or by oxygen ions, O^{2-}, at anionic sites, respectively.
The spectral glow curves of both crystals are compared in Fig. 2.
Hence, it follows that the 480 nm band will preferably occur with
F_i^- compensation, whereas the sub-spectrum containing the 381 nm
band will occur in connection with oxygen ions. As it is very
difficult on the one hand to grow oxygen-free CaF_2 crystals and as,
on the other hand, the carbon of the graphite crucible used for
growing will bind part of the oxygen in the case of Y_2O_3 doping, it
is easy to understand that it is only the intensity ratios of bands
belonging to different coactivators that are influenced. The
correlation of all emission bands with the corresponding activators
and coactivators can be seen from the following table.

Activator	Coactivator	Emission Bands [nm]	Spectrum
Sc^{3+} Y^{3+}	O^{2-}	381, 415, 426, 436, 456	I
		574, 758, 800, 864	II
	F_i^-	481, 543	III
La^{3+}	O^{2-}	340, 402, 433, 471, 495	I
		524, 536, 602, 643, 680, 696, 727	II
	F_i^-	483, 547	III

It is concluded from these experiments that the thermoluminescence of alkaline earth fluorides doped with scandium, yttrium or
lanthanum is not due to transitions within the activator ions as
is the case with the thermoluminescence spectra of crystals doped
with rare earths (8, 9, 10) which are known to be in accordance with
the fluorescence spectra of their trivalent ions. We did not find
any corresponding fluorescence in our crystals.

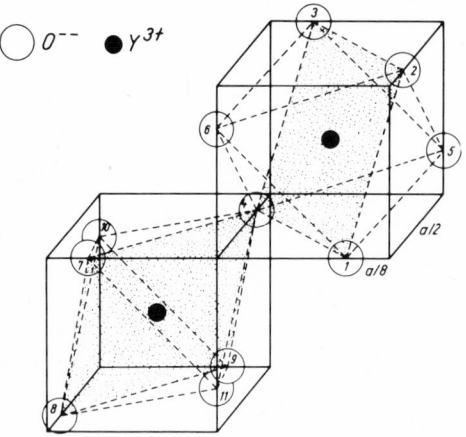

Fig. 3 Y_2O_3 complex within fluorite lattices.

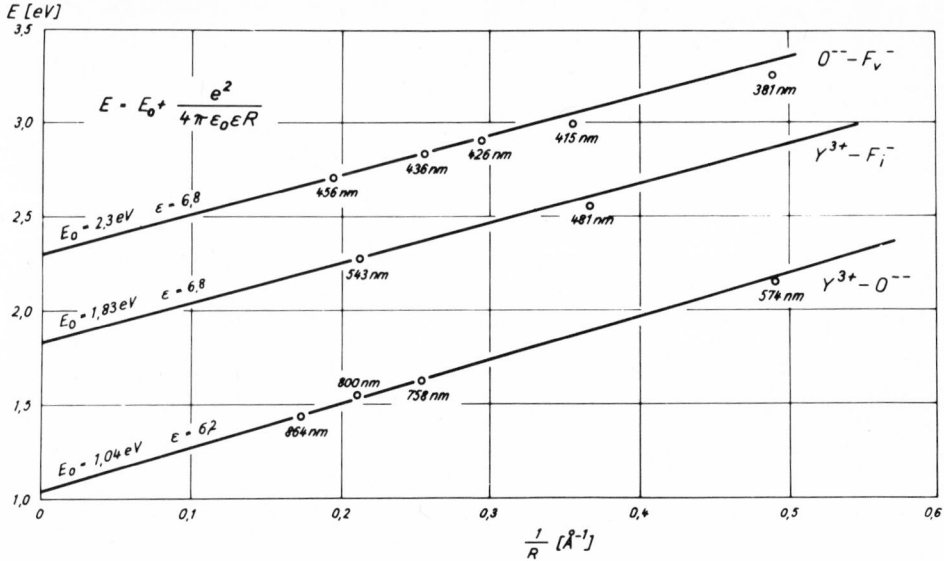

Fig. 4   Energy values of the radiative recombination for excited
donor-acceptor pairs in CaF$_2$:Y as a function of 1/R.

MODEL CONCEPTS AND DISCUSSION

In the light of the foregoing, the O'Connor (4) attempt to
employ the mechanism of a recombination of excited donor-acceptor
pairs as known in semiconductors for the interpretation of the
thermoluminescence spectra of CaF$_2$:Y crystals gains importance.
O'Connor supposes that the Y^{3+} ions form associated pairs with the
F$_i^-$ or O^{--} ions. According to O'Connor, X-ray irradiation would
lead to the reduction of Y^{3+} ions by trapping an electron, whereas
the hole would be bound by the F$_i^-$ or O^{2-} ion. Depending on the
ways of charge compensation, the recombination of such excited
pairs would occur while emitting two sub-spectra:

   I. Y^{3+}—— O^{2-}: 280, 476, 548, 586 nm
   II. Y^{3+}—— F$_i^-$: 274, 378, 416, 439, 457 nm

For the following reasons, we cannot agree with this correlation:
1. To explain the band at 439 nm, O'Connor relies on a distance of
a×($\sqrt{7/2}$) between a cation site and an interstitial one, which is, a
however, not to be found in the fluorite lattice. 2. The bands at
274 nm and 280 nm are not present in our crystals. 3. O'Connor
determined his thermoluminescence spectra between 77K and room
temperature. In CaF$_2$:Y, however, this narrow-band spectrum occurs

only above room temparature. 4. The 380 nm band is particularly
intense above 500K. Correlation with Y^{3+}-F_i^- pairs appears to be
very doubtful, because these pairs are already dissociated above
500K (11).

Nevertheless, it follows from our results that it is quite
possible to interpret the thermoluminescence of scandium-, yttrium-
or lanthanum-doped alkaline earth fluorides by means of donor-
acceptor pairs. For this it is useful to plot the energy values
of the emission maxima as a function of l/R. With correct correlation, application of the equation for radiative recombination yields
straight lines, from which the unknown quantities $E_O = E_g - (E_D +
E_A)$ and κ can be determined. This condition is not present if only
the lattice sites of the fluorite lattice are considered. However,
the measured values will be in good accordance with the theoretical
relation, if one assumes that the activators scandium, yttrium and
lanthanum - are incorporated into the fluorite lattice in a crystal
symmetry that is typical for the respective oxides: Y_2O_3 and La_2O_3.

The oxides of scandium and yttrium usually crystallize in T_h^7
symmetry (12). This symmetry can be produced in the O_h^5 symmetry
of the fluorite lattice if six of the eight fluorine ions are replaced by O^{2-} ions, with a simultaneous displacement along the cube
edges by a small amount (Fig. 3). Two fluorine sites situated on
the body diagonal remain vacant. The Sc^{3+} or Y^{3+} ions are incorporated in place of the cations and assume the function of donors,
whereas the O^{2-} ions act as acceptors. Within such complexes, two
types of donor-acceptor pairs are possible which correlate with the
thermoluminescence bands observed, as shown in Fig. 4 for CaF_2:Y.

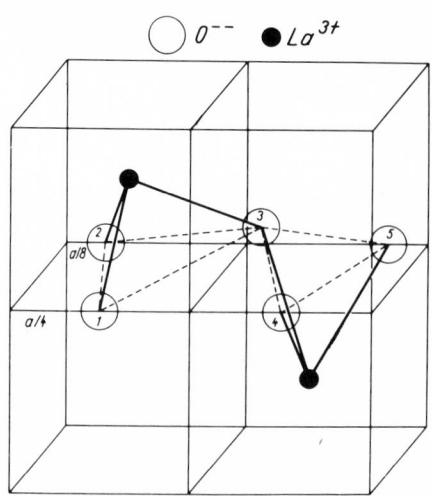

Fig. 5 La_2O_3 complex within fluorite lattices.

First, we consider Y^{3+}-O^{2-} pairs. The recombination between
electrons and holes of the excited states Y^{2+}-O^- takes place along
with the emission of sub-spectrum II. The intense band at 572 nm
corresponds to the distance between the yttrium ion and oxygen ions
at sites 1 through 6 (Fig. 3). All other bands of sub-spectrum II
result from the interaction between yttrium and oxygen ions at the
more remote sties 7 through 11.

In the case of O^{2-}-F_V^- pairs, fluorine ion vacancies, F_V^-, act
as donors. They are formed to compensate the negative excess charge
caused by the incorporation of Y_2O_3 complexes into the fluorite
lattice. Here, too, the oxygen ions are capable of binding holes,
whereas the fluorine vacancies capture the electrons. All bands of
sub-spectrum I can be attributed to the recombination of these
pairs. These concepts analogously apply to scandium-doped crystals.

Contrary to the oxides of yttrium and scandium, La_2O_3 crystall-
izes in the D_{3d}^3 symmetry. Given the differences between the thermoluminescence spectra of lanthanum-doped crystals and those of crystals doped with scandium and yttrium, it is to be expected that the
symmetry is largely maintained when La_2O_3 is incorporated into the
fluorite lattice, as shown in Fig. 5. Hence, again two types of
donor-acceptor pairs are possible, but with distances different from
those in Y_2O_3-doped crystals. Thus, sub-spectrum II in lanthanum-

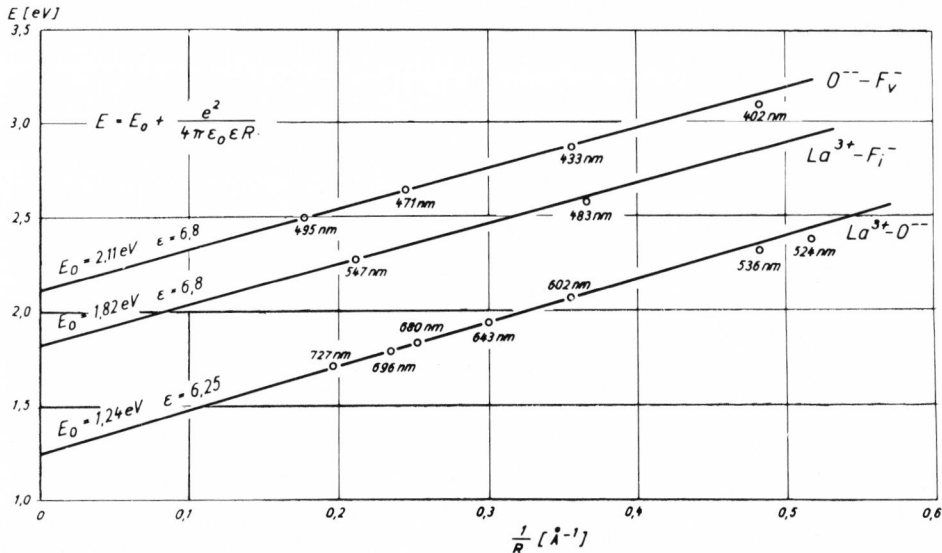

Fig. 6   Energy values of the radiative recombination for excited
donor-acceptor pairs in CaF_2:La as a function of 1/R.

doped alkaline earth fluorides must be attributed to La^{3+}-O^{2-} pairs. The emission bands at 696 nm and 727 nm correspond to distances between La^{3+} donors and O^{2-} acceptors that already exceed the range represented in Fig. 5 and can be observed only at comparatively high activator concentrations.

Sub-spectrum I of lanthanum-doped alkaline earth fluorides can, again, be attributed to O^{2-}-F_v^- pairs, but with distances different from those in case of Y_2O_3 doping. Fig. 6 is a graphical representation of the relationship. The deviations from the straight line for those points that correspond to donors and acceptors in closest proximity are caused by polarizing effects and by overlap between the wave functions of electrons and holes, which were not considered in this presentation.

In accordance with the experimental results (Fig. 2), sub-spectrum III, which is common to all crystals investigated, is attributable to a fluoridic addition of activators to the melt. It seems reasonable to correlate this spectrum with the recombination of excited pairs of Sc^{3+}-F_i^-, Y^{3+}-F_i^- or La^{3+}-F_i^-, respectively. For the distances between donors and acceptors one may apply: $R = a\sqrt{2n-1}/2$ for integral $n \neq 4,8,\ldots$. However, only pairs with interstitial fluoride ions at the nearest interstitial sites ($R = a/2$) and at next nearest ($R = a\sqrt{3}/2$) can be observed, corresponding to the bands at 480 nm and 544 nm. This is in accordance with measurements of ionic conductivity in alkaline earth fluorides doped with yttrium which reveal that Y^{3+} ions are associated only with F_i^- ions at nearest or next nearest interstitial sites (11).

DISCUSSION

It is evident from Figs. 4 and 6 that the thermoluminescence spectra of alkaline earth fluoride crystals doped with scandium, yttrium and lanthanum are well explained, to a first approximation, by the existence of donor-acceptor pairs. In this representation, the magnitude of the dielectric constant is of importance. Having recourse to sub-spectra I and III, we found $\kappa = 6.8$. This value coincides with the static dielectric constant for CaF_2 crystals as determined by Smakula and Rao (13) ($\kappa = 6.78$). For SrF_2 crystals, these authors state $\kappa = 6.48$. Since only the product of κR enters the calculation of the energy of emitted bands, it is evident why the spectra are nearly independent of the host crystal. If in the case of CaF_2 and SrF_2, R is simply replaced by the corresponding lattice constant, the result for both crystals is $\kappa R \approx 37$. The lower value for the dielectric constant determined from the thermoluminescence spectra II ($\kappa = 6.2$) is obviously due to the influence of the oxide complexes.

We have attributed the differences between the thermoluminescence spectra of crystals doped with scandium and yttrium and those of crystals doped with lanthanum to different symmetries of the corresponding oxide complexes. This in in agreement with what is known about the sesquioxides of these rare earths. In the case of Y_2O_3 it is known that the ordinary cubic form passes over into the hexagonal form on heating to 900 C (14). This transformation is partially reversible. After cooling, both phases may exist side by side. Apparently, these properties are maintained when Y_2O_3 is incorporated into the fluorite lattice, since we found that yttrium-doped alkaline earth fluorides show not only the recombination spectrum of cubic Y_2O_3 complexes, but also that of hexagonal complexes analogous to the thermoluminescence spectrum of lanthanum-doped crystals. The Sc_2O_3 complexes, on the other hand, seem to be incorporated in their cubic form only. According to Mehrotra (14), a cubic form of La_2O_3 does exist; but on heating to 600 C it passes irreversibly into the hexagonal form. Accordingly, the thermoluminescence spectra show that in lanthanum-doped alkaline earth fluorides La_2O_3 occurs exclusively in its hexagonal phase (sub-spectrum II).

This paper is only concerned with the emission process that causes the characteristic narrow-band thermoluminescence. It remains to be investigated whether electrons and holes are already bound by the donors and acceptors, respectively, during the X-ray stimulation of the crystal, or whether the electrons are bound to more shallow traps. If the latter is true, the electron should be bound by the donor after thermal activation into the conduction band and then only recombine with the hole trapped by the acceptor. To solve this problem it will be necessary to determine the activation energies and to measure the thermally-stimulated current.

REFERENCES

(1)   J. S. Prener and F. E. Williams, Phys. Rev., 101, 1427 (1956).

(2)   J. J. Hopfield, D. G. Thomas and M. Gershenzon, Phys. Rev. Letters, 10, 162 (1963).

(3)   F. E. Williams, J. Chem. Phys. Sol., 12, 265 (1960).

(4)   J. R. O'Connor, Appl. Phys. Letters, 4, 126 (1964).

(5)   E. J. Iwase, Nature, 131, 909 (1933).

(6)   P. Görlich, H. Karras, G. Kötitz and P. Ullmann, Phys. Stat. Sol., 23, 313 (1967).

(7)   P. Görlich, H. Karras, A. Köthe and K. Kühne, Phys. Stat. Sol., 1, 366 (1961).

(8) H. Adler, Act. Phys. Austr., $\underline{XII}$, 356 (1959).

(9) W. A. Archangelskaja, Izv. Akad. Nauk SSSR, Ser. Fiz. $\underline{29}$, 454 (1964).

(10) J. L. Merz and P. S. Pershan, Phys. Rev., $\underline{162}$, 217 (1967).

(11) W. Bollmann, P. Görlich, W. Hauk and H. Mothes, Phys. Stat. Sol. (a), $\underline{2}$, 157 (1970).

(12) R. W. G. Wyckoff, Crystal Structures II, 5, Interscience Publishers, Inc., New York (1948).

(13) K. V. Rao and A. Smakula, J. Appl. Phys., $\underline{37}$, 319 (1966).

(14) P. N. Mehrotra, Trans. Farad. Soc., $\underline{62}$, 3568 (1966).

THE DIFFUSION KINETICS OF RECOMBINATION LUMINESCENCE

V. V. Antonov-Romanovskii

Lebedev Institute

Moscow, USSR

ABSTRACT

A review of the kinetic calculation for crystal phosophors obtained with the help of the diffusion drift equation for small free carrier path is given. A very simple approximate calculation method is offered. The magnitudes of some kinetic parameters obtained from the experimental data are given.

INTRODUCTION

With a small charge carrier mean free path, ℓ, the recombination kinetics depends not only on the concentration, n, of the interacting partners, but also on their separation. It follows from this that the distance, r, between the interacting particles be found as a function of n. The majority of the kinetic questions are reduced to the case when the particles of every kind are uniformly distributed relative to each other and there is no interaction between them through other particles. For some systems this may put some limits on n, i.e. that it not be very large. In the case of neutral particles this puts no limit on n for their mutual interaction radius r_0 is of the order of the lattice distance a. But in the case of the attractive ones the limit is determined by $R' \gg r_0$, where R' is defined from the equality

$$e^2/\varepsilon R' = kT \qquad (1)$$

where e is the electron charge, ε is the dielectric constant, T is the absolute temperature, and k is the Boltzmann constant. For $T = 300$ K usually $R' \sim 30\text{-}40$ a.

THE DIFFUSION DRIFT EQUATION

A number of the recombination kinetic problems is reduced (1) to the solution of the spherical symmetry diffusion drift equation of decay:

$$\frac{\partial n(r,t)}{\partial t} = \frac{D}{r^2} \frac{\partial}{\partial r} \{ r^2 \frac{\partial n(r,t)}{\partial r} \} - \frac{1}{r^2} \frac{\partial}{\partial r} \{ r^2 u(r) n(r,t) \}$$

$$- P(t) n(r,t) , \tag{2}$$

where

$$P(t) = 4\pi r_o^2 D \left. \frac{\partial n(r,t)}{\partial r} \right|_{r=r_o} . \tag{3}$$

Here $n(r,t)$ means the average concentration of one kind of particles (A) situated at the distance r from the particle of another kind (B) at the moment t, $n(t)$ is the average value of $n(r,t)$ over the full volume, r_o is the recombination radius, D is the diffusion coefficient, $u(r)$ is the drift velocity. If A and B are mobile, then

$$D = D_A + D_B \qquad \text{and} \quad u(r) = u_A(r) + u_B(r). \tag{4}$$

A- and B-concentrations are equal. It is obvious that

$$n(r,t) \to n(t) \qquad \text{as} \quad r \to \infty . \tag{5}$$

The magnitude $P(t)$ has the meaning of the particle flow through the recombination sphere per unit volume at the moment t.

With weak excitation the solution of (2) by a separation of variables is reduced to the solution of two equations--one linear; the other nonlinear:

$$\frac{\partial \eta(r,t)}{\partial t} = \frac{D}{r^2} \frac{\partial}{\partial r} \{ r^2 \frac{\partial \eta(r,t)}{\partial r} \} - \frac{1}{r^2} \frac{\partial}{\partial r} \{ r^2 u(r) \eta(r,t) \}, \tag{6}$$

$$\frac{dn(t)}{dt} = -p(t) n^2(t) , \tag{7}$$

where, taking into account (3)

$$\eta(r,t) = \frac{n(r,t)}{n(t)} , \quad p(t) = \frac{P(t)}{n(t)} = 4\pi r_o^2 D \left. \frac{\partial n(r,t)}{\partial r} \right|_{r=r_o} . \tag{8}$$

The boundary and initial conditions are the following

$$\eta(r,t) = 0 \text{ at } r = r_o, \text{ and } \eta(r,t) = \eta(r,o) \text{ at } t = 0 \tag{9}$$

It follows from (5) and (8), that $\eta(r,t) \to 1$ when $r \to \infty$.

Eq. (2) and the boundary condition $n(r_0,t) = 0$ are valid, if ℓ is sufficiently small. For noninteracting particles this condition is $\ell < 2r_0 \sim 2a$. For interacting particles the condition is less severe: $\ell < 2R'(R' \gg a)$.

If the peculiarities of the drift at small r (of an order of a) are not taken into account, then

$$u(r) = \lambda/r^2, \quad \lambda = \frac{e\mu}{\epsilon}, \tag{10}$$

where μ is the mobility.

In the absence of traps

$$D = \frac{\ell u}{3}, \quad \text{and} \quad \mu = \frac{e\ell}{m\,u}, \tag{11}$$

where u is the particle thermal velocity, and m is its effective mass. In the trap presence

$$D = \frac{\ell w}{3\sigma\nu}, \quad \text{and} \quad \mu = \frac{e}{kT}\frac{\ell w}{3\sigma\nu}, \quad w = w_0\exp(-E/kT), \tag{12}$$

where ν is the trap concentration, σ is the effective trapping cross-section, w is the thermal probability for release of the localized charges, E is the trap-depth, and w_0 is the "frequency" factor.

Similar kinetic problems arise with some chemical problems and particularly in the case of radiation damage annealing.

In 1957 Waite in his fundamental work (2) has investigated the diffusion kinetics when the interacting kinds of particles A and B are neutral; the drift is absent. Waite's results, as it was shown by Vinetskii and collaborators (3), are close to our own. Waite carried out the averaging after the fundamental equation is derived; we, before the derivation of our fundamental equation. This is found to be simpler for use. But the difference in the results is small. In absence of the drift term equation (2) or (6) is solved to yield the usual form.

The boundary condition (see (9)) $\eta(r,t) = 0$ at $r = r_0$ is in some cases unsatisfactory for, as $\ell > 0$, then in spite of the smallness of $\ell, \eta(r_0,t)$ can never be equal to zero. Such questions arose long ago. Apparently Sveshnikov (4) was the first to run into this. In connection with this Collins and Kimball (5) have introduced another boundary condition.

$$\eta(r_0,t) = \gamma \left. \frac{\partial n(r,t)}{\partial r} \right|_{r=r_0}. \tag{13}$$

But in the case of a stationary distribution $\eta(r,\infty)$ we (6) have
determined the flow through the recombination sphere taking into
account that ℓ is finite with the help of the formula

$$p(r,\infty) = \sigma_k u \int_{r_o}^{\infty} \{\eta(r,\infty)/\ell\}e^{-(r-r_0)/\ell}dr \ . \tag{14}$$

In both of these ways the essence of taking into account that ℓ is
finite can be reduced to the selection of the suitable constants in
the solution of the fundamental equation, whereas strictly speaking
the original equation should be modified.

In our work (7) we used the Boltzmann equation averaged by the
momentum method. But some kinetic problems may be also solved with
the help of equation (2). In the case of excitation equation (2)
must be substituted by a more complicated one (1). The kinetics
is slightly complicated when the concentrations of A and B are different, i.e., for example, when they are equal to $n(t)$ and $n(t)+N$.
Then

$$n(t) \ = \ \frac{N}{\{1+[N/n(o)]\} \ \exp[-N \int_{o}^{t} p(t)dt]-1} \tag{15}$$

INITIAL CONDITIONS, MONOMOLECULARITY AND BIMOLECULARITY

One of the fundamental questions when solving (2) is the question about the initial conditions. This is not so simple as it
might seem at first sight. The elementary excitation act is the
creation of pair components A and B. In this case one must consider
an initial mutual distribution of A and B: $\phi(r)$, where r is the
distance between A and B. It is obviously absurd to take the initial point (r=0) as an initial position. Is it reasonable to take
the distance, r, between the first points of A and B scattering?
If one considers that A and B separate according to the law of
momentum conservation, then (8)

$$\phi(r) = \frac{\exp(-r/\ell^-)-\exp(-r/\ell^+)}{4\pi r^2(\ell^--\ell^+)} \ , \tag{16}$$

where ℓ^- and ℓ^+ are the ℓ-values for A and B. If $\ell^-=\ell$, and $\ell^+=0$,
$\phi(r,o)=\{\exp(-r/\ell)\}/4\pi r^2\ell$, and if $\ell^-=\ell^+=\ell$, then $\phi(r,o)=$
$\{\exp(-r/\ell)\}/4\pi r\ell^2$. The distribution of foreign B relative to A,
or A relative to B, is uniform if the excitation is of a short
duration. Thus:

$$\eta(r,o) = 1 + \phi(r,o)/n(0) \tag{17}$$

The situation becomes more complicated when traps are present. In this case, parallel with the recombination component of short duration (before the initial trapping of the free particles) a prolonged component arises, which is caused by the following. If at their birth the electrons and holes are of thermal energy $E \sim kT$, then the traps could neither diminish nor increase the ratio between the mono- and bimolecular recombination components. Therefore, the role of traps is reduced to the prolongation of one part of the re-combination process. The greater is the trap concentration, the greater is this part. This part is different for the mono- and bimolecular components. An increase of ν influences very weakly (beginning from the very small ν)the bimolecular light-component and very strongly the monomolecular one giving an impression of inten-sification of the monomolecular component, if one doesn't take into account the decrease of the momentary monomolecular component (9, 10).

The picture becomes more complicated if during the excitation hot particles arise with a long free path length, which might ex-ceed the thermal ℓ. The traps act as the thermalization factor that can really increase the monomolecularity.

APPROXIMATE SOLUTION

The recombination probability of A and B situated at a dis-tance r from each other is (11):

$$\delta(r) = \frac{1-\exp(-R'/r)}{1-\exp(-R'/r_o)} \quad . \tag{18}$$

Eq. (6) alows the stationary solution $\eta(r,\infty)$, as there is at infinity an inexhaustible particles source and at the beginning, an inexhaustible recombination center. The real decrease occurs according to the nonlinear Eq. (7). As $\eta(r,\infty)$ is stationary, the particles flow through the recombination sphere of any radius r around the recombination center is constant and equal to $p(\infty)$,the flow through the recombination sphere. From the $\eta(r,\infty)$ carriers only a part $\delta(r)\eta(r,\infty)$ can reach the center. Let us name them as "red" ones; the others, "green". The actual flow $p(\infty)$ is caused only by "red" carriers, therefore their velocity must be $1/\delta(r)$ times the average (drift) i.e.

$$u_{red}(r) = u(r)/\delta(r). \tag{19}$$

The idea about "red" and "green" carriers allows us to make the approximate calculation of the decay-curve I(t), if the initial distribution n(r,o) is given. In a rough approximation (12) as-suming that the carrier situated at the distance, r, can reach the center in the same time, t,

$$I(t) = \sigma u n(r,o)/\eta(r,\infty), t = \int_{r_o}^{r} dr/u_{red}(r) \qquad (20)$$

The accuracy can be an order higher by taking into account the t-fluctuation (13). Conversely, it is possible to determine the shape of n(r,o) for a given I(t).

EFFECTIVE CROSS-SECTION OF RECOMBINATION AND "FREQUENCY" FACTOR

For small ℓ the recombination velocity depends not only upon the average n-value but also upon its magnitude near the recombination range. Neglecting this fact can lead to great errors in σ. In the work (4) it is shown that the effective cross-section of recombination

$$\sigma' = \sigma_k M, \qquad (21)$$

where σ_k is the real gaskinetical recombination cross-section, and

$$M = \int_{r_o}^{\infty} \{\eta(r,\infty)/\ell\} e^{-(r-r_0)/\ell} dr. \qquad (22)$$

If $n(r,\infty) = 1$, then M = 1 and $\sigma' = \sigma_k$. For the neutral centers $n(r,\infty) < 1$ at $r \sim r_0$ and the $\sigma' < \sigma_k$. For charged centers, if r_0 is not very large, $\eta(r,\infty) > 1$ and then $\sigma' > \sigma_k$.

If ℓ is not very large, then

$$\sigma' = \sigma = 4\pi R'\ell/3, \qquad (23)$$

where σ has a meaning of a pure drift-cross section (without the influence of a diffusion). The situation is similar in the case of the "frequency" factor w (14). We obtain

$$w' = wM. \qquad (24)$$

If r_0 is very small, then M might be a few orders higher than 1. The extremely large values of the "frequency" factor reach 10^{20}/sec (15), whereas the frequency of lattice vibration does not exceed 10^{13}/sec. This seems paradoxal but has a natural explanation: the traps are not neutral, but charged.

In the case of the dipole-center (16), if d is the dipole dimensions

$$\sigma' = \sigma_d = (\sigma/2)(d/R')^{\frac{1}{2}} = (2\pi\ell/3)(R'd)^{\frac{1}{2}}. \qquad (25)$$

In the work (7) an estimation of σ' is given for ℓ not small.

It was found that for a neutral center for any ℓ, and for a charged one if $\ell < R'$, or $\ell >> R'$

$$\sigma' = \frac{\pi r_o^2}{1+3r_o/4\ell}, \quad \sigma' = \frac{4\pi R'\ell/3}{1+2\ell/3R'}, \quad \sigma' = \pi R' r_o. \quad (26)$$

In the same work (7) the shape of the function $\eta(r,\infty)$ was also determined for small ℓ.

EXPERIMENTAL ESTIMATION OF SOME KINETIC PARAMETERS

For a uniform initial distribution when $\eta(r,o) = 1$ the calculation gives that for a neutral case

$$\int_0^t p(t)dt = pt + qt^{\frac{1}{2}}, \quad p = 4\pi D r_o \text{ and } q = 8(\pi D)^{\frac{1}{2}} r_o^2. \quad (27)$$

(The approximate calculation gives about 25% higher q). If p and q are known one can estimate D and r_o.

In the work (17) the kinetics of the self-trapped holes in KCℓ-Tℓ was investigated by an absorption measurement of the color centers. The kinetics are linear for hole concentrations less than the concentration of the activator hole traps. It was found that at T = 170 K, r_o = 18 A and D = 3.4×10^{-19} cm^2/sec. On warming, r_o decreases and D increases.

In Fig. 1 a dependence of the absorption change κ, which is proportional to $n(t)$, versus t is given. At t > 25 min $\ell n Z/t^{\frac{1}{2}}$, according to the theory, is linear with $t^{\frac{1}{2}}$, where $Z = n(o)/n(t)$. A deviation from the linearity at t <25 min is due to the fact that the excitation is not momentary.

In the work (18) it was found that for KCℓ:Tℓ the effective cross-section ratio for Tℓ^{++} and V_k centers $\sigma_a/\sigma_e \sim 1$. This result is interesting in that according to the theory for small ℓ the σ' for the attractive center does not depend on its structure, thus confirming diffusion drift kinetics in phosphors of the alkali halide type.

ELECTRIC FIELD EFFECT

If the genetically related pairs are situated close to each other, then the field separates them. It results in a reduction of the number of the monomolecular recombinations and of the luminescence component of short duration. At the same time it means a lowering of the recombination probability. Therefore, under stationay excitation the field should increase   n, to balance the

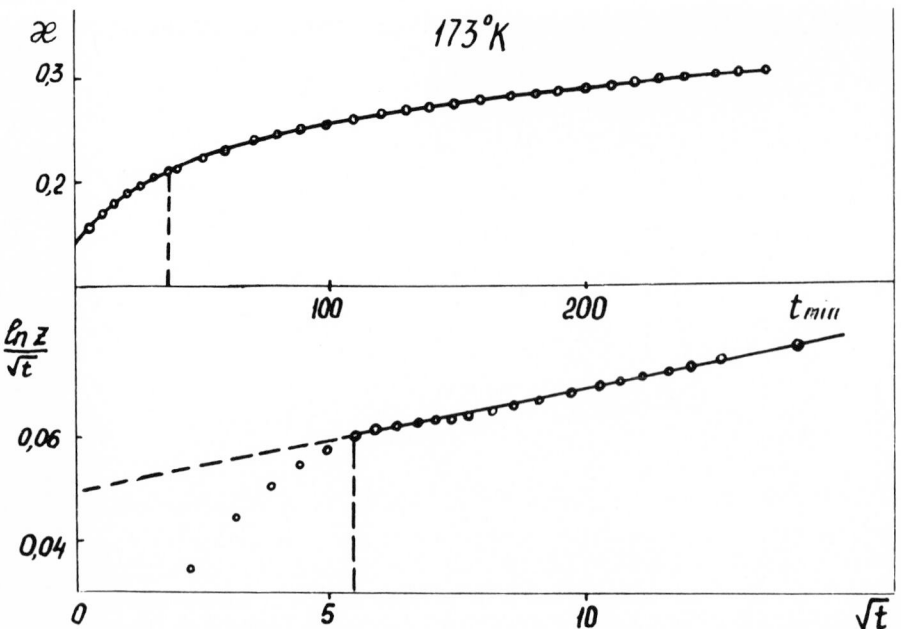

Fig. 1   The dependence of the absorption coefficient κ on the time
t for the phosphor KCℓ:Tℓ at 170 K.

ionization number, which does not depend upon the field. The calculation shows that for the alkali halides fields above 10^4v/cm are
effective. The fields which do not increase the temperature of
the free carriers cannot change the value of σ (19).

The "blowing away" of the monomolecularity by the field and
the predicted n-growth (11) proved to be considerable under long
excitation (9). In some cases this growth is large, i.e. by 2 to
3 orders of magnitude (20). The n-growth as a function of the
field is illustrated in Fig. 2. In addition to the "blowing away"
of the monomolecularity by the field another effect is possible,
i.e. the ionization of the excited centers, that can also result
in n-growth. This effect is greater, the greater is the orbit of
the excited state. Large orbits are stable only if the carrier
interacts with the lattice weakly, i.e. it has a larger ℓ which
is favored by lower temperatures. Both these effects behave
phenomenologically in a similar way, and it is difficult to
separate and distinguish between them. However, Denks and Leiman
(20) succeeded in showing the effect of the field "pulling away".
When a field is switched off during the thermoluminescence of
KCℓ:In in the probe regime (1-2sec.) clearly defined polarity
changes were detected. If the field is of the same polarity as
during the excitation, I decreases; for the opposite polarity,
the increase of I occurred. The polarity effect is absent, if

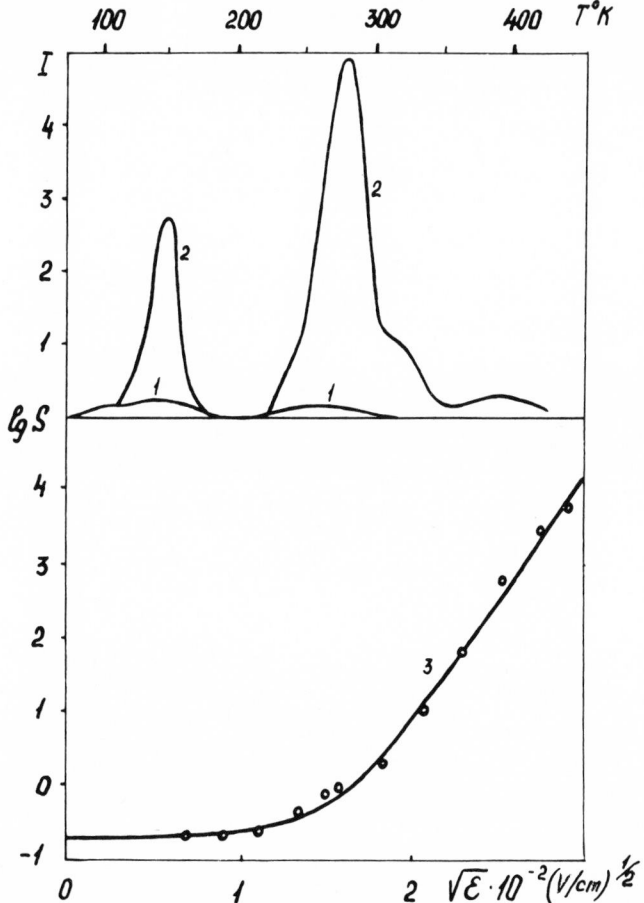

Fig. 2   The influence of electric field on the light sum of the
         KCℓ:In thermoluminescence.

during the excitation the field was not switched on.

CONCLUSION

        Thus, the diffusion drift kinetics allows us to explain some
effects, to carry out quantitative and approximate calculations,
and to calculate some kinetic parameters. The monomolecularity
and bimolecularity phenomena, the effects of the excitation intensity, temperature and electric field are explained in a natural way.
The effective cross section of the recombination and capture and
the "frequency" factor are determined in some cases. An approximate method of calculation of $n(t)$ and $I(t)$ is suggested as a function of the initial distribution $\eta(r,o)$ of the recombining partners.
The method also allows us to determine $\eta(r,o)$ from the given $n(t)$

or I(t). The investigation of the η(r,o) dependence upon the λ of the exciting light in the photon multiplication region can help to explain this phenomenon. One should carry out a more general analysis of kinetics, when ℓ is not small. The preliminary studies in this connection are promising.

REFERENCES

1. V. V. Antonov-Romanovskii, Kinetika Fotoluminestentsii Kristallofosforov (Izd. Nauk, Moscow, 1966), ch. 4.

2. T. R. Waite, Phys. Rev., 107, 463 (1957).

3. V. L. Vinetskii, Ya. I. Yaskovets, I. V. Kelman (in press):

4. B. Ya. Sveshnikoff, Acta Physicochim. URSS, 3 257 (1935).

5. F. C. Collins, G. E. Kimball, J. of Colloid. Sci. 4, 425 (1949).

6. V. V. Antonov-Romanovskii, Phys. Stat. Sol., 26, 173 (1968).

7. V. V. Antonov-Romanovskii, Yu. H. Kalnin, J. Luminescence 3, 909 (1971);Fisika Tverdogo Tela, 13, 1376 (1971).

8. V. V. Antonov-Romanovskii, Phys. Stat. Sol., 19, 417 (1967).

9. V. P. Denks, V. I. Leiman, N. L. Lukantsever, F. A. Savihin, Fizika Tverdogo Tela. 12, 1455 (1970).

10. N. L. Lukantsever, Izvestija Akademii Nauk SSSR (ser. fiz.), 35, 1312 (1971).

11. V. V. Antonov-Romanovskii, Fizika Tverdogo Tela, 11, 2827 (1969);J.Luminescence, 1-2, 909 (1970); phys. stat. sol., 38, 95 (1970).

12. V. V. Antonov-Romanovskii, Fizika Tverdogo Tela, 12,3366 (1970); J. Luminescence, 3,459 (1971).

13. V. V. Antonov-Romanovskii, Fizika Tverdogo Tela, 13, 3143 (1971).

14. V. V. Antonov-Romanovskii, Phys. Stat. Sol., 30, 341 (1968).

15. I. K. Vitol, Ch. B. Lushchik, J. V. Yaek, Trudy Inst. Fiz. Astron. Akad. Nauk Eston. SSR, No. 12, 175 (1960); M. A. Elango. Trudy Inst. Fiz. Astron. Akad. Nauk Eston. SSR, No. 12, 179 (1960).

16. V. V. Antonov-Romanovskii, Fizika Tverdogo Tela, 13, 853 (1971).

17. E. D. Aluker, A. N. Semenova, S. A. Chernov (in press).

18. E. D. Aluker, O. E. Aksenov, N. L. Romanenko, Izvestija Akad. Nauk Latv. SSR No.6, 45 (1969).

19. V. V. Antonov-Romanovskii, Fizika Tverdogo Tela, 5, 1345 (1963).

20. V. P. Denks, V. I. Leiman, Materialy Vsesoyuznoi Konfer. "Primenenie elektrolyuminetsentsii v narodnom hozyaistve" (tezisy dokladov) Chernovtsy, oktyabr' (1971).

INTERACTION OF SHALLOW TRAPS AND DONORS WITH SINGLE PHONONS

AND ITS APPLICATION FOR PHONON SPECTROSCOPY

N. Riehl

Physik-Department der Techn. Universität München

Munich, Germany

ABSTRACT

Thermal emptying of traps is mostly a cooperative action of many phonons. A new technique makes it possible to study trap depths down to 0.01 eV. Such traps can be emptied by one phonon and are able to act as detectors for single phonons. If a limited number of single phonons is generated in a He-cooled crystal and its temperature remains unchanged, the phonons do not interact with each other and have long free paths being able to reach the traps distributed in the crystal and to release charge carriers. Single phonons can be generated e.g. by neutron collisions, by bombardment of the crystal surface with hot gas atoms, etc. The range of phonons can be determined. The use of traps with different depths enables a phonon spectroscopy and a detection of the phonon range. Experiments confirmed this idea. Bombardment of a He-cooled phosphor crystal with room temperature Ne- or He-atoms produces a bright light spike although the temperature rise is less than 0.001 K! In ZnO the 0.035 eV LA-phonons have at 5 K a penetration depth about 0.65 mm emptying inside this distance traps with $E_T < 0.035$ eV and leaving unaffected the deeper traps. Further applications and improvements of the method are in progress.

INTRODUCTION

Glow-peaks usually observed in glow-curves at temperatures near or above liquid air temperature are due to trap depths of the order 0.1 eV and more. Therefore the energy of a single phonon is too small for emptying such traps as the maximum phonon energy

660

in a solid cannot exceed essentially the limit given by the Debye
"cut-off frequency" which is of the order 0.01 eV. This means
that the release of charge carriers from trap depths of the order
0.1 eV is due to a cooperative action of many phonons. Shallower
traps could not be reproducibly detected in glow-curves even if
the phosphor was excited at liquid helium temperature because of
a rapid exhaustion of these traps by unwanted IR-irradiation
through the windows serving for excitation and observation.
(IR-wavelengths up to 35 μm are effective). A new technique for
glow-curve observation at low temperatures developed with G. Baur
and L. Mader in our laboratory uses windows cooled to helium
temperature and made from materials impermeable for IR. In this
way we were able to observe glow-peaks due to trap depths down
to 0.01 eV (1). Such traps can be emptied by one phonon. This
suggests the idea to use these shallow traps as detectors for
single phonons and to apply them for phonon free path determination,
phonon velocity measurements and phonon spectroscopy (2).

If a limited number of single phonons is generated in a
He-cooled and previously excited monocrystal of a phosphor and its
temperature remains unchanged, the phonons have long free paths
(up to 1 mm and more) being able to reach the traps distributed
in the crystal without degeneration and to release the trapped
charge carriers. Such events are signaled by light emission.
The single phonons can be generated in different ways, e.g. by
neutron collisions or by bombardment of the crystal surface
with hot gas atoms or by a very short heat impulse (4). If the
generation rate of the phonons is not too high there is almost
no interaction between them and no essential approach to a thermal
equillibrium (no dissipation of their energy) so that the single
"hot" phonons can traverse great distances through the "cold"
crystal without energy losses. By variation of the concentration
of trapped charge carriers the mean distance which the phonon has
to traverse before being annihilated at a trapped charge carrier
can be varied and the mean range of phonons determined. Furthermore, the use of traps with different depths should enable a
phonon spectroscopy and an investigation of phonon annihilation.
This idea has been tested and realized experimentally with
G. Baur, A. Müller and U. Puchner. The results are described
in this paper.

EXPERIMENTAL RESULTS AND INTERPRETATION

In our first experiments a ZnO-monocrystal as the phosphor
(thickness near 1 mm) with trap depths 0.016 and 0.04 eV has
been used. The phonons were produced by bombardment of its surface
at 4.2°K with room temperature Ne- or He-atoms. A special helix-
shaped nozzle protected the phosphor against IR radiation emitted
by the warm gas container. The bombardment with a flux of about

10^{17} Ne-atoms/cm^2 sec. during 0.2 sec. produced a bright light
pulse although the phosphor temperature remained almost unchanged
($\Delta T < 0.001$ K). The rise time of the pulse ($\sim 10^{-3}$ sec.) corresponded
to the rapidity of the electropneumatic gas valve used and its
duration was of the order of the bombardment time. Fig. 1 shows
the glow-curves before and after bombardment with room temperature
Ne-atoms. It demonstrates that the glow-maximum at 8 K corresponding
to a trap depth 0.016 eV is quenched by phonons almost to a third
of its original value. Therefore, the mean penetration depth of
the effective phonons in ZnO at 4.2 K is about 0.65 mm. The
hatched area is equal to the light sum of the light pulse caused
by bombardment. The glow-maximum at 17 K corresponding to a
trap depth of 0.04 eV remains unchanged. The maximal energy of
acoustical phonons in ZnO calculated for LA-phonons from the
known energy of LO-phonons (3) with the $(M/m)^{1/2}$-relation is
0.035 eV. This is enough to release charge carriers from 0.016 eV
traps but not from 0.04 eV traps. Our experimental result agrees
with this requirement and proves that single phonons are effective.

Any simulation of the effect by a direct interaction between
the impinging gas atoms and the traps can be easily ruled out
because such an action would be limited to the layer near the surface and would not affect the crystal volume. For the same reason
it is not possible that the effect is due to a single momentary
heating of the crystal surface. An effective heating of the whole
crystal can be ruled out too (as discussed in detail later).

Obviously this principle of phonon detection is not restricted
to phosphors but can be applied to other systems, e.g. to semiconductors with shallow donors or acceptors. The observable
response must be in this case not a light pulse but a current
spike. A typical experiment of this kind may be described.
A He-cooled n-Ge crystal with 3 X 10^{15} Sb-atoms per cm^3 and with
low compensation (donor depth E = 0.0096 eV; resistivity at
room temperature 10 ohmcm; thickness 3 mm) has been bombarded
with a short pulse of Ne-atoms (total flux about 10^{16} atoms/cm^2 sec.
during 10^{-4} sec.). This caused a spike of conductivity which was
nearly 10 times higher than the background conductivity. The
rise time of the spike equals that of the atom pulse which
corresponds to the rapidity of the gas valve used ($< 10^{-4}$ sec.).
The bombardment technique is described below.

The question arises whether the effect is actually due to
electrons from the 0.0096 eV donors or to any other unknown donor
levels. This was checked by measurements of the temperature
dependence of the spike magnitude. The population density of
donors (as a function of T) is proportional to $1/\{1+A\exp(-E/2KT)\}$
where E is the donor depth and A a constant. The experimental
curve for the spike magnitude as function of T agrees very well

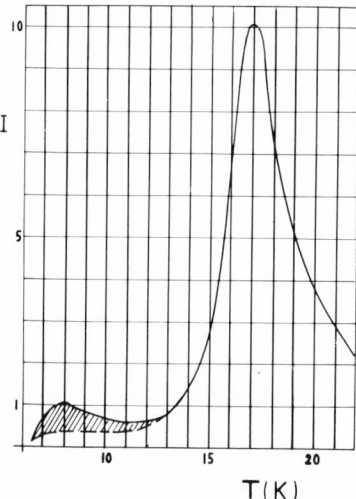

$T(K)$

Fig. 1 Glow-curve of ZnO before and after bombardment.

with this relation and yields a value of 0.0098 eV for E which
agrees with the depth 0.0096 eV of the Sb donors within limits
of error. Therefore, we can be sure that the phonons coming
from the bombarded crystal surface release the electrons actually
from the Sb donor levels. The phonons responsible for the release
of electrons from the 0.0096 eV Sb donor levels in Ge must be LA-
phonons, max. energy 0.0287 eV (5).

It is easy to rule out the possibility that the effect is
due to a simple heating of the crystal by the bombardment.
If considered as a trivial heat transfer phenomenon the possible
temperature enhancement ΔT can be estimated from the energy
delivered to the crystal per unit time by the hot gas atoms and
the energy removed by heat conductivity through the crystal to
the metallic holder. This yields a value of ΔT less than 0.001 K -
much too small for any observable thermal electron release in the
usual sense. A similar result is obtainable from the minimal
initial slope dn/dt of our experimental spike, where n is the
number of free carriers, and the value dn/dT from the temperature
dependence of the conductivity. From dn/dt = (dn/dT)(dT/dt) we
can see which dT/dt would be necessary to obtain the observed
dn/dt. The result is that the energy delivered by the hot gas
atoms is a thousandfold too small to reach the required dT/dt.
Thus the spike effect is due only to the action of undissipated
phonons and not to any heat in thermal equilibrium.

Another typical experiment on the action of single phonons
in semiconductors concerns the release of charges from traps
(instead of from donors). A semi-insulating GaAs crystal doped
with oxygen and chromium (thickness 1 mm) has been studied.

After its excitation with IR at He-temperature and subsequent
heating a TSC-curve ("thermally stimulated conductivity") shows
current maxima at 7, 13 and 20 K which are caused by charges
released from three kinds of shallow traps. The crystal shows a
high current spike at He-temperature after bombardment with Ne-
atoms. It can be shown that it is due to charges released from the
shallowest of the three traps. In a TSC-curve taken after the
bombardment the TSC-maximum at 7 K has almost vanished but the
other two maxima remain unchanged (Fig. 2). The deficiency of the
charges stored in the 7 K traps is just equal to the charges
released during the bombardment. The almost complete diminution
of the 7 K maximum shows that the free path of the phonons generated at the crystal surface reaches throughout the whole crystal,
i.e., the free path is almost 1 mm. This experiment is analogous
to the experiments on phosphors with the difference that the release
of charge carriers from traps is signaled by a current spike and
not by light emission. The release from traps in GaAs is to be
attributed to LA-phonons with a maximal energy of 0.03 eV (5).

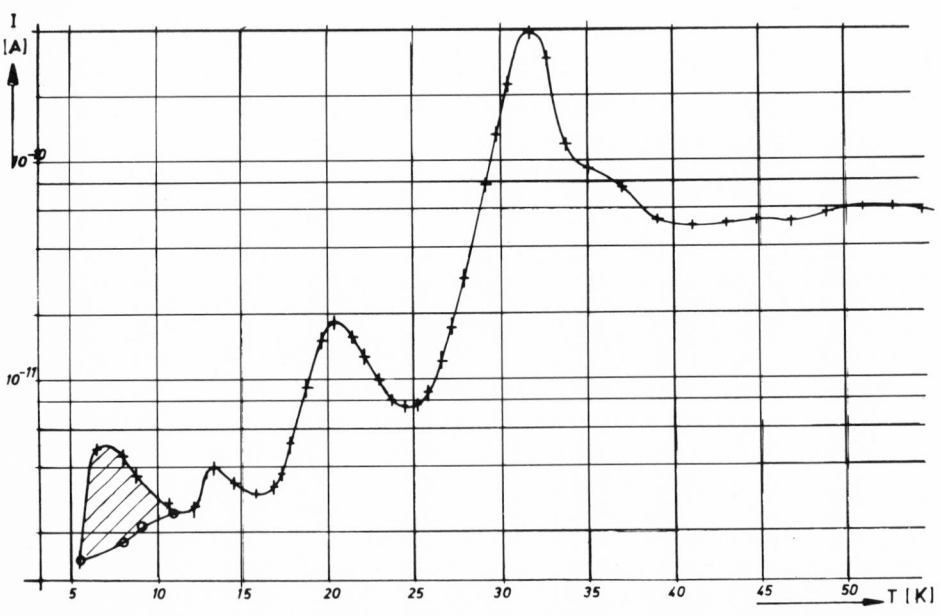

Fig. 2   TSC-curve of a semi-insulating GaAs crystal before and
         after bombardment.

We have also tested the possibility of phonon detection by
release of holes in p-type semiconductors. A p-Ge crystal with
2.5×10^{15} indium atoms per cm^3 has been used (E of the acceptor
level = 0.0112 eV; resistivity of the sample at room temperature =
15 ohm cm). At low phonon flux densities, the result was the

same as in the n-type semiconductors: a conductivity spike appeared in response to the bombardment with Ne-atoms. But if the phonon flux density was drastically enhanced (by using higher Ne-atom flux densities) a reversed behavior namely a negative spike, i.e., a drop of conductivity was observed. A complete explanation for this effect cannot be given as yet, but there are indications that double phonon effects and some unknown levels may be responsible.

In our first experiments we used only crystals with a very low conductivity hoping to observe the spikes more easily if the background conductivity is weak. Therefore, the first experiments were restricted to low doped crystals. But in such crystals some unknown impurities and levels can play a relatively high disturbing role; so it seemed desirable to find a way for an extension of the experiments to higher doped crystals with greater conductivity. For this purpose we made use of the fact that the observed current displays a steep spike with a very short rise time so that it can be easily amplified and thus separated from the background d.c. conductivity. With such a differential method we are now able to get spike effects at semiconductors of very low resistivity, e.g., 1 ohm cm at 300 K.

Another improvement deals with the bombardment technique. Whereas in the first experiments we used room temperature Ne- or He-atoms with roughly Maxwellian distribution of velocities we use now gas atoms of higher and much more homogeneous velocity produced with the so-called skimmer technique which allows to obtain gas atom energies up to 3 to 4 kT with a relatively narrow Gaussian distribution. (Dr. Gspann from the Kernforschungszentrum Karlsruhe kindly suggested the use of a special nozzle designed by Dr. Hagena). Although the total flux of the beam was nearly the same as in the first experiment, the rise of the current (or light) spike was faster by a factor of 100.

Besides the phonon production by direct transfer of kinetic energy from the gas atoms to the crystal lattice some phonon generation could be due to the condensation heat from Ne-atoms condensing on the He-cooled crystal surface. Two facts show that condensation took place in some of our experiments. A part of the bombarding Ne-atoms could be regained as free gas after reheating the crystal. Repeated bombardment yields a progressive decrease of the height of the spike because of shielding of the crystal surface by condensed Ne-atoms. But both effects do not take place if He-atoms are used for the bombardment because He is too volatile to be condensed on the crystal surface.

A remark may be added with respect to the width of the phonon spectrum. The bombardment with gas atoms produces not only phonons of high energy (frequency) lying near the Debye "cut-off frequency" but also low energy phonons so that the phonon spectrum is probably very broad. But the low energy phonons cannot release charge carriers from traps or donors so that we always observe a relatively narrow part of the phonon spectrum with energies near the Debye cut-off limit.

Phonon velocities can be measured with the new method. We used for this purpose a Ge-crystal which was doped with Sb by diffusion only near its surface (about 10^{18}Sb-atoms/cm^3 in a layer of a few microns) or covered with a layer of a luminescent ZnO-powder as phonon detector. By these means, the phonons penetrating the crystal produce two spikes, the time interval between the spikes being the phonon transit time. These experiments are in progress, but it is already sure that the obtained phonon velocities are smaller for higher energies of the impinging gas atoms, i.e., the greater the mean phonon energy. This result agrees with the requirements of phonon dispersion. Measurements of this kind can lead to a simple determination of phonon dispersion curves $\omega = f(k)$, since the knowledge of the group velocity for different phonon energies yields the values of $d\omega/dk$ as a function of ω.

A remarkable feature concerning the mechanism of phonon propagation may be emphasized in connection with the experiments which served for transit time determination. As mentioned above, the arrival of phonons which crossed a Ge-crystal was signaled either by a thinly doped (and therefore phonon sensitive) surface Ge-layer or by a coat of luminescent ZnO-powder. Consequently, some phonons in question are able to pass through the phase boundary Ge-ZnO. The surprising point of this observation and of analogous observations on the phase boundaries Al_2O_3-ZnO and LiF-ZnO is that the transition coefficient for the passage from one phase to the other one is astonishingly high (of the order 0.1 and even more) although the contact and the binding forces between the atoms of the two different phases are probably relatively weak so that one would except a much higher probability for reflection or degradation for phonons of such high frequency.

FURTHER DISCUSSION

It would be favorable for phonon spectroscopy to use non doped "small band gap semiconductors" as phonon detectors e.g. $Cd_xHg_{1-x}Te$. In such semiconductors the band gap is of the order kT so that single phonons can cause band to band transitions. The important point is that the band gap varies continuously with the composition (with the value of x). Therefore, there exists the possibility for a continuous variation of the detectable phonon energy. As yet, the available semiconductors of this type have too high a background conductivity because of many undesired levels from stoichiometric imperfections. But we obtained a distinct current spike on CdHgTe in spite of its high conductivity when we observed the bombardment response with the "differential" technique mentioned above.

Recent experiments show the possibility of measuring the range and the maximum energy of phonons produced in a crystal by high energy particles, e.g., α-particles. A quartz crystal of 0.3 mm thickness was bombarded from one side by α-particles (range $\sim10\mu$) and the phonons produced were detected by a thin coat of ZnO on the opposite side. In this case both the 0.016 and 0.04 eV ZnO-traps were emptied because the α-particles are able to produce also optical phonons with energies above 0.04 eV.

In addition to the determination of free path, velocity- and energy, the new method of phonon detection can be developed and used for many other applications. One of them is the possibility to observe the preferred directions of phonon propagation in a crystal lattice. The distribution of the propagation probabilities over different directions should be displayed by the light pattern of a phonon sensitive luminous screen (e.g. ZnO) on the crystal surface. Experiments of this kind are in progress. Investigations about "accomodation coefficients" (transfer of thermal energy from impinging gas atoms to a solid), experiments on propagation of high energy phonons over polymer-chains and many other applications should be possible with the method.

REFERENCES

1. G. Baur, L. Mader and N. Riehl, Z. f. Naturforsch. 24a, 1296 (1969); N. Riehl, Proc. Intern. Conf. on Luminescence, Budapest, (1966) p. 974; Festkörperprobleme VIII (1968) p. 232; J. Luminescence 1, 1 (1970).

2. N. Riehl, Verhandlungen d. Deutschen Physik. Ges. H9, 679 (1971); N. Riehl and A. Müller, Physics Letters 36A, 487 (1971).

3. T. C. Damen, S. P. S. Porto and B. Tell, Phys. Rev. 142, 570 (1966).

4. O. Weis, Ztschr. angew. Phys. 26, 325 (1969).

5. B. N. Brockhouse, Phys. Rev. Letters 2, 256 (1959); J. L. T. Waugh and G. Dolling, Phys. Rev. 132 2410 (1963); S. M. Sze "Physics of Semiconductor Devices", Wiley-Interscience, New York 1969.

LUMINESCENT METHODS FOR THE VISUALIZATION OF LONG WAVELENGTH RADIATION

S. A. Fridman, E. Ya. Arapova, N. V. Mitrofanova,

Yu. P. Timofeev, V. V. Shchaenko

Physical Institute, Moscow, USSR

ABSTRACT

Different classes of phosphors sensitive to the optical and thermal influence of IR light, and also those directly excited by IR, are synthesized and investigated. These include phosphors previously excited with UV (e.g. $ZnS:Cu,Co$), those showing steady-state luminescence under IR illumination (e.g. ZnS, $CdS:Ag,Ni$), and the cooperative excitation of rare-earth ions (e.g. $Yb^{+3}-Er^{+3}$) Principal characteristics of these phosphors are reported.

The luminescent screens which transform invisible images created by an electron beam or hard radiation into visible ones are widely used in various devices (television and radar tubes, oscillographs, X-ray screens and so on). In recent years in connection with the creation of new coherent sources in the IR-SHF range (lasers, powerful sources of super-high frequency), their wide application for the transformations into the visible of images obtained in this long wavelength region is of great importance (1). A well-known empirical Stokes-Vavilov rule concerning energy dissipation in elementary events forbids this direct transformation from IR to visible. However, at present there exist several effects different by their nature giving the possibility of utilizing luminescent screens to obtain visible images from the IR range. The present paper was aimed at evaluation of the luminescent methods for visualization of images created by the long wavelength radiation and at a comparison of these methods. The general view of these methods, the luminophors developed, the conditions of their operation, and the main characteristics of the screens are illustrated in Table 1.

TABLE 1. The Fundamental Characteristics for Luminescence Methods of Infrared Radiation Registration

Method	Luminophor compound Base	Luminophor compound Dope	Colour	Condition of use Band of sensitivity μ	Condition of use Source of excitation $m\mu$	$T^\circ K$	Characteristic of screen Threshold energy $\frac{Joule}{cm^2}$	Characteristic of screen Time of image conservation sec	Resol. $\frac{line}{mm}$
Optical stimulation (flash)	$Sr\cdot S\cdot CdS$	Ce,Sm	green	0,7-1,5	300-400	293	$1\cdot10^{-5}$	0,1-0,3	30
	ZnS	Fe	blue	0,7-1,7	365	293	$3\cdot10^{-5}$	1,0	30
				2-3,5	365	77	$3\cdot10^{-4}$	1,0	
Optical quenching	ZnS	Cu,Co	green	0,7-1,7	365	293	$(1-2)\cdot10^{-4}$	10^{-3}	30
Temperature " — "	$ZnS\cdot CdS$	Ag,Ni	yellow	$0,7-3\cdot10^{-4}$	300-400	293	$(0,5-1)\cdot10^{-3}$	0,1-1,0	4-10
Change of colour (optical)	ZnS	Ag,Sm	blue → red	0,7-1,7	365	293	$3\cdot10^{-5}$	1,0	10
Change of colour (temperature)				$0,7-3\cdot10^{-4}$	365	293	$(3-5)\cdot10^{-3}$	0,2-1,0	3-5
Antistokes excitation	$BaF\cdot YF_3$, $YOCl$, Y_2O_2S	$Yb^{3+}+Er^{3+}$ green-red; $Yb^{3+}+Ho^{3+}$ green; $Yb^{3+}+Tm^{3+}$ blue		(1,4-1,6), 0,9-1,0	no required	293	$3\cdot10^{-3}$	$10^{-3}\cdot10^{-4}$	5-10

The luminophors with the lowest energy threshold for optical
stimulation are those which yield a flash under the action of
IR-rays on screens previously excited by ultraviolet radiation
(2,3). Somewhat less sensitive are screens with ZnS:Cu,Co
luminophors, the afterglow of which is quenched by optical action
of IR-rays. Therefore, this method allows the accumulation of
the radiation effect of time-independent sources and permits
recording images some time later (up to $\sim 10^3$ sec after the irradiation). However, any optical effect on the crystal phosphors is
limited by the near IR region (4), and only by cooling of the
screens can one reach 3-4 µm (ZnS:Fe,T$\sim$77°K). The thermal effect
by using non-selective absorbing coatings allows, on the contrary,
overlapping the whole IR-SHF range with a lower but constant
efficiency (6,7). For these types of screens the luminophors
ZnSCdS-Ag,Ni have been investigated (8,9) with a striking
development in the external quenching (a decrease of the brightness by up to 28% per Kelvin degree) near room temperature.
ZnS:Ag,Sm luminophors with two activators, for which the ratio
of the two spectral bands with stationary luminescence is considerably changed under external stimulation (1). This gave satisfactory
results for obtaining a colored image (both by the optical and
thermal effects of the IR rays). All the above-mentioned methods
require an additional source of excitation, which makes the

registration procedure more difficult; in addition, at a high density of IR rays saturation occurs. These disadvantages are absent with cooperative luminescense (10,11), when due to the summation of the excitation energy in pairs of (Rare-earth)$^{3+}$ -ions a direct anti-Stokes transformation of the IR rays into visible luminescense takes place. Unfortunately, this method is rather selective in the wavelengths and for each spectral region it requires new resonance systems to be developed, satisfying a whole series of conditions (12). Let us speak in more detail about the features of the last two methods. The temperature sensitivity R_T of the luminophors is expressed through the activation energy ΔE of the quenching as $\Delta E/kT^2$ and ΔE is only genetically connected with the energy depth of the luminescence centers E_O. For ZnSCdS:Ag,Ni,ΔE reaches 2.3 eV (E_O=0.58 eV) which is accompanied by a high non-linearity; the luminescence brightness depends upon the fourth power of the intensity of uv excitation. Analysis showed that this non-linearity is only possible upon saturation of the quenching centers (Ni) by the non-equilibrium carriers; however, these centers are the principal channel for recombination. The principal limitations to R_T arise due to the spread of the center parameters, in particular their energy depths, which results in the broadening of the transition region to non-linear brightness. Therefore, for T=300°K the maximum R_T is no more than 90% per kelvin degree. The absorption coefficient of the radiation registered was made to reach 50% over the whole IR-SHF range. The screen heat capacity which determines the quenching-time and the threshold energy is limited, in principle, by the luminophor thickness necessary for effective operation. The quenching time and threshold energy are determined by the screen heat loss, consisting of heat radiation, conduction and convection. Under visual observation the threshold is also dependent on the threshold contrast, which for the optimal conditions of observation (brightness L $\approx$100HT on the homogeneous field of the screen) is $\approx$2% and higher under real conditions. The resolution is limited by the spread of the outline of isothermals during establishment of stationary conditions and is determined by the coefficient of thermal conductivity. Taking into account the parameters of real screens gives agreement of the estimates of their principal characteristics with the experimental values. The advantages of this method which is completely reversible, consist in the wide spectral and energy scale of the radiations registered.

The main characteristics of the anti-Stokes luminophors (13,14) are listed in Table 2. Their excitation spectra are shown in Fig. 1. In all cases of excitation in the region of 0.9-1.0 μm, at first the IR quantum absorption in Yb^{3+} takes place with a subsequent resonance transmission and an energy summation on the emitting Er^{3+} or Ho^{3+} atoms. In the region 1.4-1.6 μm the energy summation can occur on the Er^{3+} ions without Yb^{3+}; however, its

TABLE 2. *The Characteristics of Antistokes Luminophors.*

Luminophor		Spectrum of excitation		Spectrum of emission		Temperature influence	Inertial	
Base	Dope	Transitions	Maxima nm	Transition	Maxima nm	$\dfrac{J_{77°K}}{J_{293°K}}$	Growth delay sec	Decay constant sec
$BaYF_5$	Yb^{3+},Er^{3+}	$^2F_{7/2} \to ^2F_{5/2}$ (Yb) → $^4I_{15/2} \to ^4I_{11/2}$ ↦ $^4F_{7/2} \leadsto ^4S_{3/2}$ (Er) $^4I_{15/2} \to ^4I_{13/2}$ (Er) $3(^4I_{13/2}) \to ^4S_{3/2}$(Er)	978 968 932 1520	$^4S_{3/2} \to ^4I_{15/2}$	540	10^2 $\sim 10^2$	$1 \cdot 10^{-4}$	$1 \cdot 10^{-3}$
$BaYF_5$	Yb^{3+},Ho^{3+}	$^2F_{7/2} \to ^2F_{5/2}$ (Yb) → $^5I_8 \to ^5I_6$ ↦ $^5F_4$ (Ho)	978 965 932	$^5F_4 \to ^5I_8$	540	10^2	–	$2,5 \cdot 10^{-4}$
$YOCl$	Yb^{3+},Er^{3+}	$^2F_{7/2} \to ^2F_{5/2}$ (Yb) ↦ $^4I_{15/2} \to ^4I_{11/2} \to ^4I_{13/2}$ ↦ $^4F_{9/2}$ (Er) $^4I_{15/2} \to ^4I_{13/2}$ (Er) ↦ $^4I_{9/2} \leadsto ^4I_{11/2} \to ^2F_{5/2}$(Yb)+ $^4I_{13/2}$(Er)↦$^4F_{9/2}$(Er)	1005,5 957 923 1560 1510	$^4F_{9/2} \to ^4I_{15/2}$	660	$0,2$ ~ 1	$3 \cdot 10^{-5}$	$1 \cdot 10^{-4}$
$YOCl$	Yb^{3+},Ho^{3+}	$^2F_{7/2} \to ^2F_{5/2}$ (Yb) ↦ $^5I_8 \to ^5I_6$ ↦ $^5F_4$ (Ho)	1005 957 932	$^5F_4 \to ^5I_8$	540	$10-20$	$1 \cdot 10^{-5}$	$3 \cdot 10^{-5}$
Y_2O_2S	Yb^{3+},Er^{3+}	$^2F_{7/2} \to ^2F_{5/2}$ (Yb) ↦ $^4I_{15/2} \to ^4I_{11/2}$ ↦ $^4F_{7/2} \leadsto ^4S_{3/2}$ (Er)	988 956 946	$^4S_{3/2} \to ^4I_{15/2}$	540	20	$2 \cdot 10^{-5}$	$2 \cdot 10^{-4}$

presence stimulates a red band in Er^{3+}, which is explained by a scheme determined by us. The change in temperature is essential, and in different cases it influences the efficiency and the excitation spectra in a different way. The long wavelength limit of the spectrum is very sharp, and even in oxychlorides (the main maximum is 1.005 μm) the excitation efficiency drops by three orders of magnitude for λ = 1.06 μm, i.e. for the wavelength of the induced laser radiation with Nd^{3+}. Therefore, even under poor agreement of these spectra we observed a spatial distribution of the radiation field of the $CaWO_4:Nd^{3+}$ lasers with an efficiency no less than 0.1% at a density 30 mW/cm^2 (15).

In conclusion let us cite some examples of the IR radiation fields (Fig. 2) for the lasers obtained with various luminescent screens illustrating some cases of their diverse applications. So, apparently, we deal here with a rise of a new wide domain of luminescent screens applications.

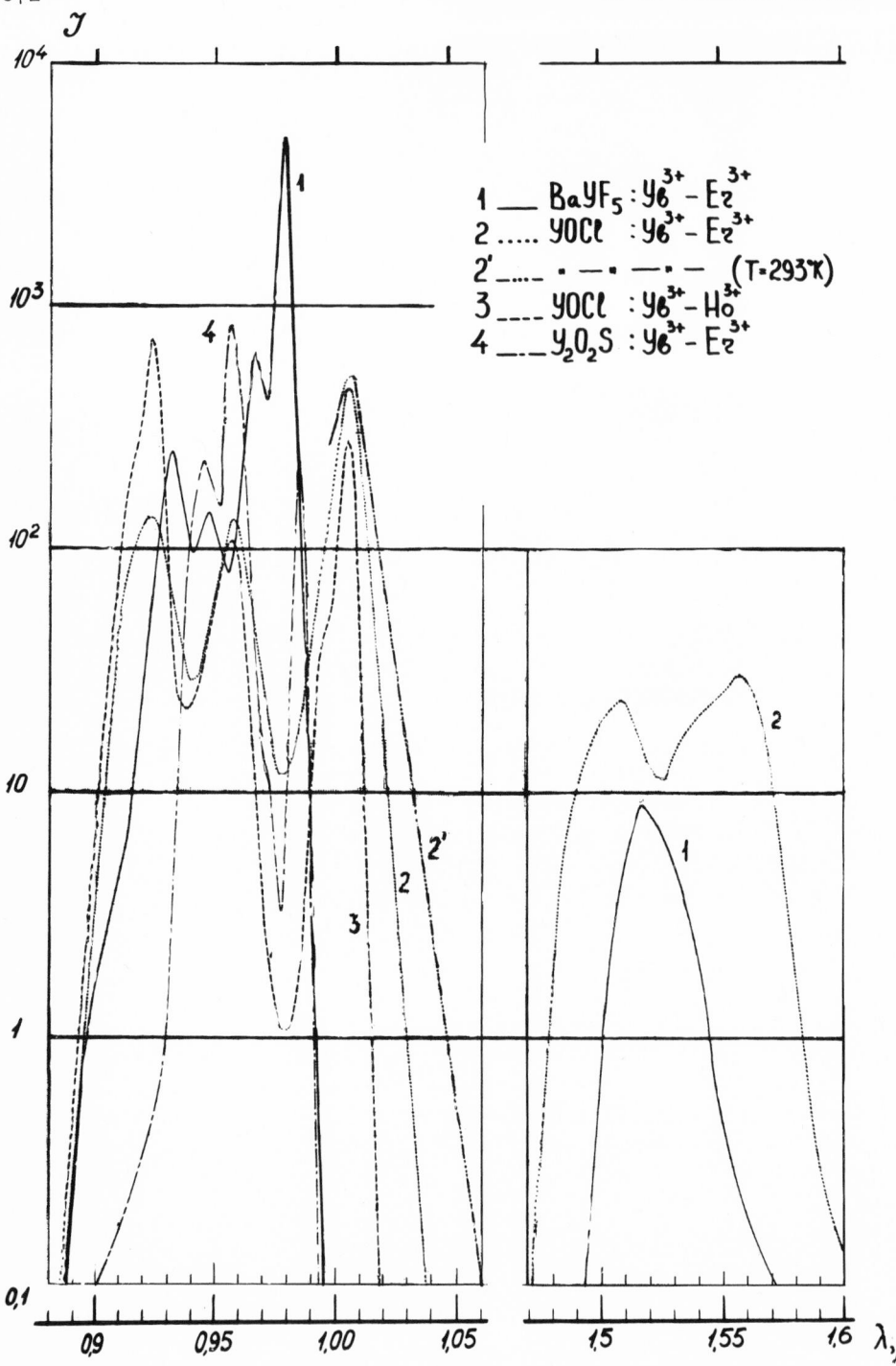

Fig. 1. Excitation spectra of anti-Stokes luminophors: all at
          77K except 2'.

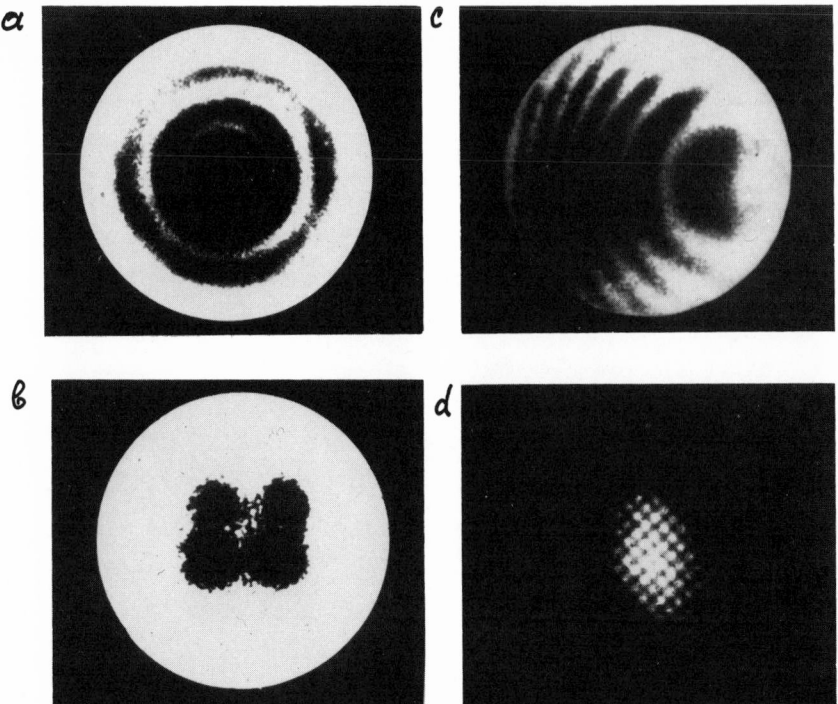

Fig. 2. The radiation fields recorded on the luminescent screens:
   a) $CaWO_4:Nd^{3+}$-laser (λ=1.06 μm), luminophor $ZnS:Cu,Co$;
   b) CO_2-laser (λ=10.6 μm), luminophor $ZnS.CdS:Ag,Ni$;
   c) SHF generator (λ=3 mm), luminophor $ZnS.CdS:Ag,Ni$; and
   d) $Y_3Al_5O_{12}: Nd^{3+}$-laser (λ=1.06 μm), luminophor $YOCl:Yb^{3+}$-Er^{3+}

The authors express their gratitude to M. V. Fok for helpful assistance in this work.

REFERENCES

1. V. L. Levshin, N. V. Mitrofanova, Yu. P. Timofeev,
   S. A. Fridman, V. V. Shchaenko, Tr. FIAN 59, 64 (1972).

2. V. V. Antonov-Romanovskii, V. L. Levshin, Z. L. Morgenshtern,
   Z. A. Trapesnikova, Dokl. Akad. Nauk, SSSR 54, 19 (1946).

3. B. O'Brien, Preparation and Characteristics of Solid Luminescence Materials, Edited by G. R. Fonda and F. Seitz, Wiley
   N.Y., 1948.

4. M. V. Fok, Fiz. tverd. tela, 6, 1448 (1963).

5. E. Ya. Arapova, V. L. Levshin, N. V. Mitrofanova,
    T. S. Reshetina, V. S. Tunitskya, S. A. Fridman,
    V. V. Shchaenko, Izv. Akad. Nauk SSSR, ser. fiz. 30, 573
    (1966).

6. I. D. Mc. Qee, L. Heilos, IEEE, J. Quant. Electr. GE-1, 31
    (1966).

7. A. P. Bazhulin, E. A. Vinogradov, N. A. Irisova, S. A. Fridman,
    PismaZhETF 8, 261 (1968).

8. F. Urbach, N. R. Nail, D. Pearlman, JOSA 39, 1011 (1949).

9. N. V. Mitrofanova, Yu. P. Timofeev, S. A. Fridman,
    V. V. Shchaenko, Izv. Akad. Nauk SSSR, Zer. fiz. 35, 1446
    (1971).

10. V. V. Ovsyankin, P. P. Feofilov, Pisma ZhETF, 4, 471 (1966).

11. F. Auzel, Comp. Rend., 262 B, 1016 (1966).

12. T. Miyakawa, D. Dexter, Phys. Rev., B1, 70 (1970).

13. H. J. Guggeheim, L. F. Johnson, Appl. Phys. Lett. 15, 51
    (1969).

14. E. Ya. Arapova, Optika i Spectroskopiya, 32, 435 (1972).

15. E. Ya. Arapova, Kratkie Soobshcheniya po Fisike
    FIAN (in press, 1972).

ANALYTICAL APPLICATIONS OF X-RAY EXCITED OPTICAL LUMINESCENCE

E. L. DeKalb, A. P. D'Silva and V. A. Fassel

Ames Laboratory USAEC, Iowa State University, Ames, Iowa

USA

ABSTRACT

During the past several years, X-ray excited optical luminescences spectroscopy has been found useful in analytical chemistry,
particularly for the detection and determination of rare earth impurities in inorganic materials at concentrations in the 1 part in
10^6 to 10^9 range. A recent study has shown that one or more of the
rare earth elements can be easily detected at the 100 ppm level in
compounds of 55 of the chemical elements, when these compounds are
incorporated into appropriate phosphors. The advantages of X-ray
excitation over other excitation procedures are discussed, and the
concept of internal standardization is explained. Analytical procedures for the determination of rare earths in other rare earths,
in Zr, Th, U, and the Fe group of transition elements will be
briefly described. None of these analytical methods requires prior
separation or concentration steps, and the phosphor preparation is
simple and rapid.

INTRODUCTION

During the past several years, the application of luminescence
to the solution of problems in analytical chemistry has become increasingly popular. There are numerous papers in the recent analytical literature which describe luminescence methods for the determination of rare earth (RE) element concentrations in a variety of
materials. Most of these authors used ultraviolet (UV) radiation
to excite the luminescence, and a few have used X-rays, gamma rays,
or an electron beam (cathodoluminescence) for excitation. In this
paper, the advantages of X-ray excitation are presented, and recent

675

developments in the use of X-ray excited luminescence for analytical determinations are described.

    X-ray excitation offers several advantages over UV excitation. The X-ray photon is absorbed directly by the phosphor host, and through the production of secondary excitants, the acquired energy is preferentially and efficiently transferred to RE impurity atoms. This is in contrast to UV excitation, which requires that there be a wavelength match between the exciting radiation and an absorption band of the activator, an added sensitizer, or the host. With X-ray excitation, it is possible to develop higher energy densities, and the secondary excitants possess higher energy than UV photons, so that impurity atoms can be excited to higher energy states. Strong, sensitive luminescence lines and bands at lower wavelengths can therefore be observed and utilized. In addition, a filter to prevent the exciting radiation from entering the spectrometer is not required.

    The cathodoluminescence process can only operate in a high vacuum. For the analyst who must examine many samples each working day, this is an unnecessary inconvenience which increases the time required to complete an analysis. The principle disadvantages of X-ray excitation are the health hazards and equipment costs. Careful design of the excitation chamber (1) can make it possible to operate the equipment and change samples quickly in complete safety. By using a conventional X-ray power supply from an X-ray fluorescence analyzer, equipment costs are minimized.

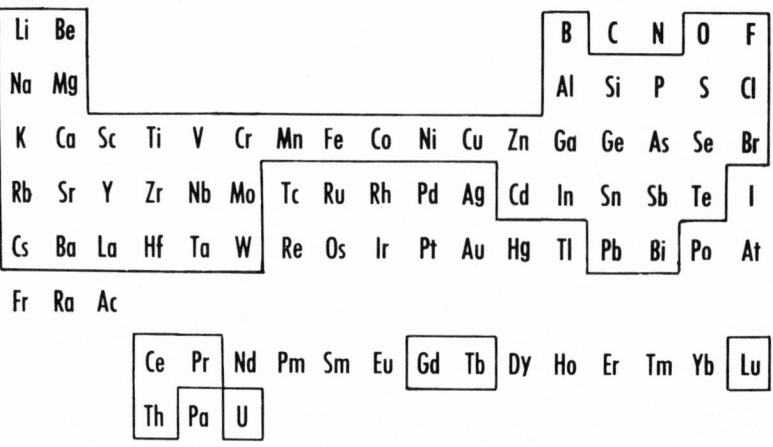

Fig. 1   Elements within boxes have been found to form compounds which will support analytically useful X-ray excited luminescence of trace rare earth impurities.

APPARATUS

The equipment used in this work consists of a tungsten target X-ray tube operated at 50 kV and 45 mA. A quartz lens is used to focus the luminescence on the entrance slit of a small scanning monochromator of 0.25 m focal length. Gratings used in the monochromator are selected for optimum intensities at the wavelengths being studied. Either of two photomultipliers may be used by positioning a mirror at the exit slit. One photomultiplier has an S-20 response, and the other which is cooled with dry ice, has an S-1 response.

SCOPE OF APPLICATION

Most of the early analytical studies of XEOL employed oxide phosphor hosts. Unfortunately the oxides of many elements do not support useful fluorescence of trace rare earth impurities. However, the extensive studies by DeKalb, D'Silva and Fassel (2) have shown that host systems able to support fluorescence of rare earth impurity 'activators' at low concentration levels may be prepared from most elements in the periodic system. The periodic table shown in Fig. 1 indicates that analytically useful fluorescence has been observed in compounds prepared from most of the elements in the periodic system. Quaternary oxide host systems have been found by D'Silva, DeKalb, and Fassel (3) to support fluorescence of the rare earth impurity activators at strikingly low concentrations. Most of the phosphor hosts found to be analytically useful,(i.e. those which would support easily detectable luminescence of one or more rare earth element at a concentration of 100 ppm) have been described in the standard luminescence literature. However, phosphor host materials incorporating Mn, Fe, Co and Ni were not reported in the literature available to us. A search was therefore undertaken to find the crystalline host which would provide the most sensitivity for trace RE luminescence, so that this host could be used as a base for dilution of non-phosphor compounds. The best host thus far discovered has been YPO_4, prepared using solid state reactions, and including a small amount of $Na_4P_2O_7$ as a nucleating agent. When oxides of the transition elements containing RE impurities are substituted for part of the Y_2O_3 during preparation of this phosphor, it is found that the characteristic RE luminescence is detected with good sensitivity. In addition to essentially pure oxides, oxides prepared from a high alloy steel such as type 303 stainless steel can be used to prepare the phosphor $Y_{0.7}(SS\ 303)_{0.3}\ PO_4$, which is also effective as a host for luminescence of RE impurities in the steel.

THE INTERNAL REFERENCE (STANDARD) PRINCIPLE

The intensities of the optical fluorescence spectra emitted

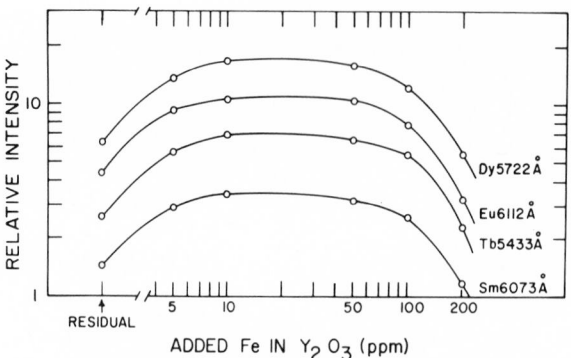

Fig. 2 Effect of Fe addition on the luminescence of 200 ppm of
Sm, Eu, Tb or Dy in Y_2O_3.

under X-ray irradiation may be markedly influenced by the chemical
nature and concentration of other compositional impurities in the
host material and by structural imperfections. Thus, under many
circumstances, a prior knowledge of the total impurity content and
structural purity of a sample would be required for precise quantitative analyses. Fortunately, this is not so. A characteristic
exhibited by both the enhancement and suppression effects so far
observed is the nearly parallel behavior of the intensities among
the rare earths (1). Thus, though the individual relative intensities undergo wide excursions, they do so in consort. A typical
example of this behavior is shown in Fig. 2. These observations
immediately suggested the application of the internal standard
principle so widely used in optical emission spectroscopy to these
radiations as well. A RE element which is not to be measured in
the sample is added to the phosphor during preparation at a known
concentration which is much larger than the anticipated residual
amount in the sample. Ratios of the luminescence intensities of
the RE elements to be measured compared with those from the added
RE are then used to find the RE concentrations. The merit of
employing intensity ratio measurements to compensate internally for
impurity enhancement or depression effects is excellently illustrated
in Fig. 3. Day to day variations in the preparation of the phosphor host may also affect the emission intensities to a significant
degree. These variations are also effectively compensated by application of the internal standard principle.

ANALYTICAL ACHIEVEMENTS

The superior detectabilities achieved with YPO_4 as the phosphor
host have already been mentioned. In Table 1, the detectabilities
for RE impurities in Y_2O_3 using YPO_4 or YVO_4 as the hosts are compared to those usually obtained by spark-source mass spectroscopy
(4) with either photographic or electrical detection. It is evident

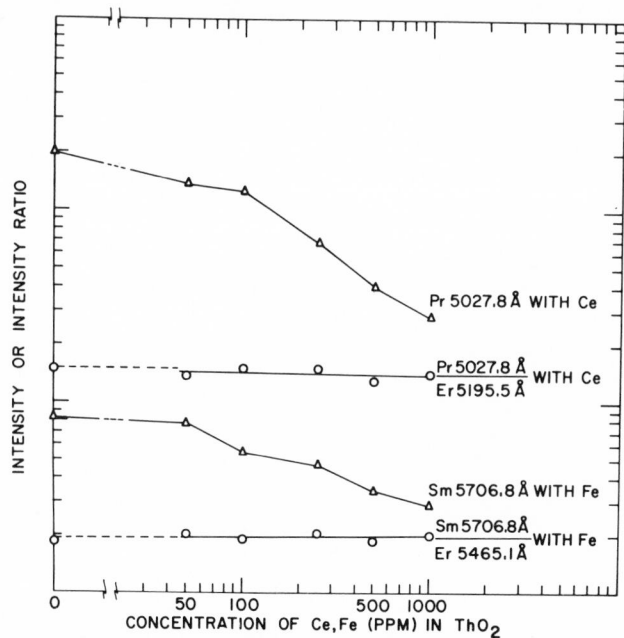

Fig. 3 Effect of impurities on luminescence intensities and on
intensity ratios of rare earth impurities in ThO$_2$.

TABLE 1

Limits of Detection (ppm by wt. in Y$_2$O$_3$)

	X.E.O.L.	S.S.M.S Photographic	Electrical
Ce	0.005	0.1	0.0125
Pr	0.005	0.080	0.0125
Nd	0.010	0.35	0.025
Sm	0.005	0.25	0.025
Eu	0.05	0.10	0.015
Gd	0.05	0.4	0.04
Tb	0.001	0.1	0.015
Dy	0.001	0.30	0.03
Ho	0.5	0.1	0.015
Er	0.05	0.3	0.03
Tm	0.05	0.1	0.015
Yb	0.5	0.15	0.015

that the luminescence method provides comparable or superior detection limits for most of the RE impurities. A quantitative procedure for the determination of RE impurities at the part per giga (1 in 10^9) level in Y_2O based on the YPO_4/YVO_4 host systems has been developed by D'Silva and Fassel (submitted to Analytical Chemistry).

Because of the high neutron capture cross sections of several RE elements, these elements must be held to very low concentrations in nuclear reactor materials, and analytical methods must be capable of measuring RE impurities at fractional ppm levels. The detection limits we have obtained for RE impurities in oxides of uranium and thorium are tabulated in Table 2. ThO_2 was used as the phosphor host for RE determinations in Th (5). A quaternary oxide host

TABLE 2

Rare Earth Impurity	Detection Limit (ppm)	
	in U_3O_8	in ThO_2
Pr	0.005	0.04
Sm	0.01	0.05
Eu	0.01	0.02
Gd	0.0025	0.04
Dy	0.0025	0.10

system, $2\ Li_2O \cdot SrO \cdot UO_2 \cdot 2WO_3$, was used in the case of U (6). Phosphors of this type are easily prepared by grinding together stoichiometric amounts of the component compounds and heating the mixture on an optimum time-temperature schedule. U_3O_8 can be used in place of UO_2 with no loss of RE luminescence intensity. Similar phosphors containing Ti, Hf, and Zr have also been prepared. It is expected that a similar phosphor containing PuO_2 could provide a host for the direct determination of RE and actinide elements in Pu.

Based on these and other observations, it would appear that X-ray excited luminescence is a powerful tool for the determination of RE impurities in a wide variety of materials. In particular, for analytical problems in the nuclear energy field, this approach may become the preferred method for the determination of neutron absorbing impurities.

REFERENCES

1.   E. L. DeKalb, V. A. Fassel, T. Taniguchi and T. R. Saranathan, Anal. Chem. 40, 2082 (1968).

2.   E. L. DeKalb, A. P. D'Silva and V. A. Fassel, Anal. Chem. <u>42</u>,
     1246 (1970).

3.   A. P. D'Silva, E. L. DeKalb and V. A. Fassel, Anal. Chem. <u>42</u>,
     1846 (1970).

4.   R. J. Conzemius, Ames Laboratory, personal communication.

5.   T. R. Saranathan, V. A. Fassel, and E. L. DeKalb, Anal. Chem.
     <u>42</u>, 325 (1970).

6.   A. P. D'Silva and V. A. Fassel, Anal. Chem. <u>43</u>, 1406 (1971).

LUMINESCENT ALKALI HALIDE CRYSTAL MEMORY ELEMENTS

I. K. Plyavin, V. P. Objedkov, G. K. Vale, R. A. Kalnin,

L. E. Nagly

Physics Institute, Academy of Sciences of the Latvian SSR

ABSTRACT

Quick-response energy transfer from light excited color centers to hole centers, storage in electron-hole centers and subsequent electron recombination luminescence are discussed as possible high-speed memory elements of large capacity. The output pulse intensity in some alkali halide systems is directly proportional to the intensity of the input pulse. In particular, these characteristics as they arise in the system KI-Tl with the Tl^{++} and F-centers as the working pair and an electron beam as the source of written information are discussed.

INTRODUCTION

Quick-response energy transfer is the basis of the memory element. This quick-response energy transfer is observed in the recombination luminescence between thermo-optically excited electron centers and hole centers in alkali halides. The luminescence method is characterized by its high sensitivity in comparison to the absorption method. Quick-response energy transfer is peculiar to the recombination luminescence of many alkali halides irradiated by different kinds of ionizing radiation: by electrons, by light or by X-rays.

Experimental data for the KI - Tl system was obtained from a monocrystal or a condensed film irradiated by an electron beam at 5 kV. Recording of information was carried out with an electron pulse during which electron and hole point defects were formed in the crystal as a light sum. Conditions were chosen for the creation

of F and Tl^{++} as the working centers. Reading was performed by
luminescence of Tl$^+$ which was stimulated in the F-absorption band.
This stimulation was obtained by means of short light pulses of
low intensity. Erasure was carried out with the same light but of
considerably higher intensity. It can be concluded that reading
also causes a partial erasure of the stored information. During
readings at room temperature both F-light and thermal erasure
occurred.

EXPERIMENTAL RESULTS

The initial intensity of the stimulated luminescence was directly proportional (over 3 orders of magnitude) to the electron
radiation dose, namely, to the current of the electron beam (the
electron pulse was $\sim 10^{-6}$ sec on an area $\sim 1000\mu^2$. The initial
intensity of the stimulated luminescence (SL) was directly proportional to the intensity of the F light; the corresponding decay
time was inversely proportional. Kinetics of the decay time of SL
consisted of several components. In addition, the decay time
kinetics of SL as a function of temperature and activator concentration was studied.

The present experimental results have led us to the conclusion
that electron and hole centers are generated in pairs. This can
be connected with the exciton mechanism of point defect generation.
It is supposed that the complicated kinetics of the decay time is
due to different electron-hole pair centers distinguished by:
(1) different configurations; (2) different electron centers (F,M,
R etc.); and (3) combination of electron, hole, and hole activator
centers. If the optical stimulation occurs in close pairs of F and
Tl^{++} centers, the electron is immediately captured by the ionized
Tl^{++} center so that the kinetics of the decay time of SL are determined by the F-light absorption probability. At high intensities
the limit of the decay of SL is determined by the inner-center
luminescence of Tl$^+$. For instance, with F-light intensities of
$10^{23} - 10^{26}$ quanta/cm^{2}sec the stored light sum is being erased
into a pulse with the decay time of the inner-center being 10^{-7}
sec at room temperature.

CHARACTERISTICS OF THE LUMINESCENCE MEMORY CELL

The direct proportionality between the intensity of the
luminescence and the current in the electron beam, i.e., a dynamic
range of three orders of magnitude, enabled us to use such a cell
not only for the writing of information in the binary system but
also for the creation of analog memory cells as well as for summation of information in the dynamic range of about two orders of
magnitude. A detailed evaluation of the experimental data yielded
a writing capacity of 10^{10} bit/cm^2 when the focusing of the elec-

tron beam was about $1000\mu^2$. Decreasing the size of the memory cell, which was quite possible, we achieved a writing capacity over the crystal surface exceeding the holographic capacity of the volume, namely, 10^{12} to 10^{13} bit/cm^2. Experimental results have shown that the time of reading and erasing of stored information is limited by the inner-center luminescence $(Tl^+)^*$, i.e. 10^{-7} to 10^{-8} sec. In our experiment the lifetime was 10^{-6} sec but it could be shortened to 10^{-7} sec. Thus the complete procedure of writing, reading and erasing of information could be performed in a time of 10^{-7} sec. The energy of writing, estimated from the sensitivity threshold of the experimental set-up, equals 0.1 $\mu J/mm^2$ if the intensity of the reading light is 10^{14} quanta/cm^2 sec.

Time of information storage for a concrete occasion is only a few seconds at room temperature. But if one reduces the temperature or works with a thermally stable activator center then the time of information storage is unlimited. (Additional work is necessary in order to clarify this matter.) Longevity of the memory cell is very high not withstanding the fact that during writing, reading and erasing electronic processes continue.

COMPOSITION DEPENDENCY OF THERMOLUMINESCENCE OF NEW PHOSPHORS FOR RADIATION DOSIMETRY

T. Toryu, H. Sakamoto, T. Hitomi, N. Kotera and

H. Yamada

Kyokko Research Laboratories, Dai Nippon Toryo Co.,

Ltd., Chigasaki, Japan

ABSTRACT

The effects of structure, composition and preparative conditions were investigated for terbium-activated magnesium silicate and magnesium borate phosphors. Composition and technology are given for these newly developed thermoluminescent dosimeter phosphors.

INTRODUCTION

After exposure to ionizing or other radiations, a phosphor exhibits luminescence upon heating. The light sum or intensity is supposed to be proportional to the exposed dose. Using this effect, a solid state radiation dosimeter called the thermoluminescence dosimeter (TLD) has been made.

Experimentally, characteristics of the thermoluminescence (TL) are much more sensitive to the composition and/or the structure of the phosphor with complexity of physical mechanisms than those of spontaneous luminescence (1, 2), and are subject to more restrictions for the dosimetric utilization (3); for instance, sensitivity and its linearity, minimization of photosensitivity and of fading, etc. An X-ray diffraction analysis is expected to give useful information as to the matrix composition of a phosphor in connection with the glow curve, although it may scarcely tell anything as to traps.

The MgO-SiO$_2$:Tb (4, 5) system and the MgO-B$_2$O$_3$:Tb system were investigated from this point of view, being most useful ones among TL phosphors (6, 7) which we had developed for radiation dosimetry use.

EXPERIMENT

Samples were prepared as follows: Magnesium oxide and silicon oxide or boron oxide with trivalent terbium were mixed and fired in an alumina crucible. The prepared phosphor was, directly encapsulated in a 2 mm x 12 mm glass capsule, exposed to 90 kV X-rays, ^{60}Co γ-rays or fluorescent lamp lights, and its glow curves were obtained at a heating rate of 3C/sec by using a Kyokko TLD Reader 1200SD.

RESULTS AND DISCUSSION

We first report on the MgO-SiO$_2$:Tb. From investigation of samples with varying MgO/SiO$_2$ molar ratio, Tb concentration and firing temperature, it was found that a sample exposed to the light (more effectively ultraviolet) and the same sample exposed to the X- or γ-ray show different glow curves. As shown in Fig. 1, phosphors exposed to the light exhibit glow curves with double peaks at 95 C(peak I) and 240 C (peak III). On the other hand, phosphors exposed to the X- or γ-ray exhibit glow curves with single peaks at 190 C (peak II). The TL intensities of the peak I (or III) and the peak II are strongly dependent on the firing temperature. As the firing temperature increases, the sensitivity to the ionizing radiation increases while the sensitivity to the light decreases rapidly over 1400 C.

According to the X-ray diffraction analysis, a phosphor made with the molar ratio of 3/1, which was roughly an optimum value for TL output (6), consists of four phases; Mg$_2$SiO$_4$(A), SiO$_2$ of α-quartz form (B) and of α-cristobalite form (B'), and MgO (C) phases. With increasing firing temperature, the phase A grows while the phase C diminishes. When the phosphor is fired at temperatures over 1400 C, the phase B changes into the phase B' and then vanishes. These results appear to suggest that SiO$_2$ incorporated in the phosphor is responsible for the photothermoluminescence, which must be avoided for the dosimetry, while Mg$_2$SiO$_4$ is essential for the radiothermoluminescence. A proper TLD phosphor made from Mg$_2$SiO$_4$:Tb is sensitive to an X- or γ-ray dose of tenths mR.

We now report on the MgO-B$_2$O$_3$:Tb. The X-ray diffraction analysis shows that a complex comprising four phases, MgB$_4$O$_7$ (D), monoclinic Mg$_2$B$_2$O$_5$ (E), triclinic Mg$_2$B$_2$O$_5$ (F), and Mg$_3$(BO$_3$)$_2$ (G) phases, is made with varying firing temperature from 700 C to 1250 C and

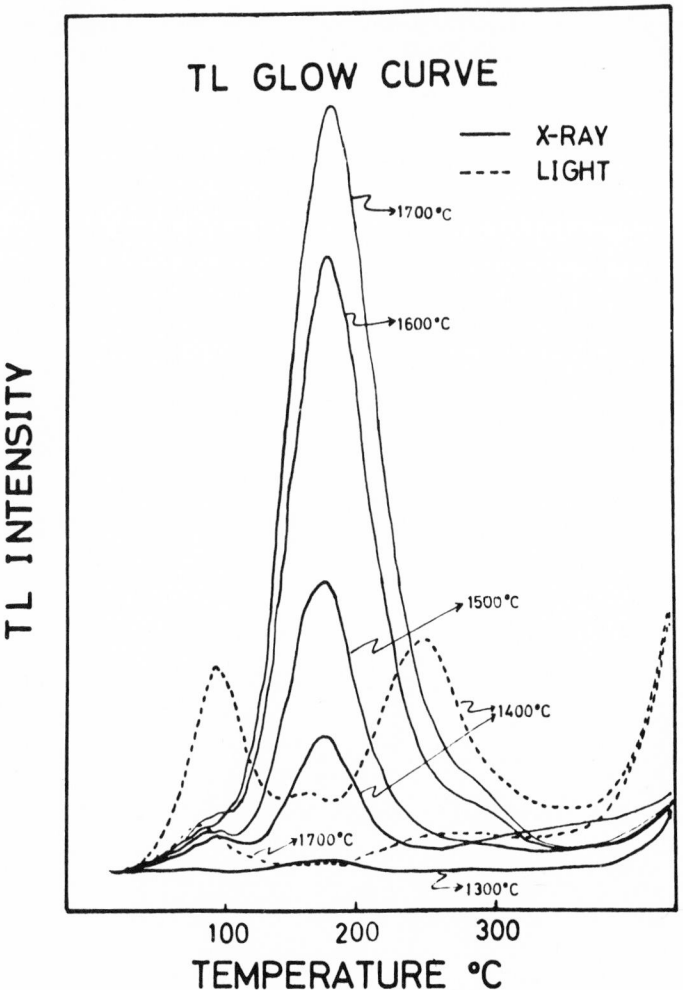

Fig. 1 Glow curves of MgO-SiO$_2$:Tb phosphors with varying firing
temperature exposed to X-rays and to light.

mixing molar ratio of MgO to B$_2$O$_3$ from 1/3 to 3/1. If these four
phases could be made separately, it would be easy to analyze relations between glow curves and compositions. But only the phase F
and G are made separately from the ingredients. To get a rough
idea of the composition, a proportion of the value of the strongest
of the diffraction peaks identified with a specified phase among a
summed value of the four strongests of peaks identified with each
phase in the X-ray pattern of a sample is adopted, the figure being
relative.

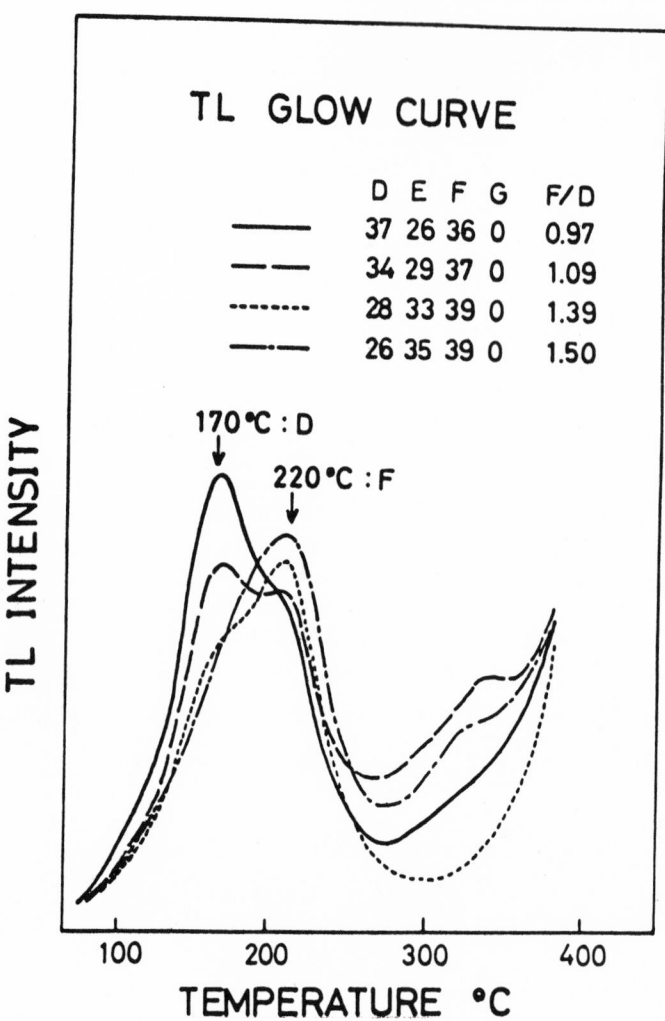

Fig. 2   Glow curves of MgO-B$_2$O$_3$:Tb phosphors with varying relative
         phase ratio of F to D exposed to X-ray of a definite dose.

        There is no explicit relation between the proportions of E
and shapes of the glow curve, and in samples keeping the ratio
F/D at a constant value, their glow curves are closely similar,
despite existence or not of the phase E.   Therefore, it seems likely
that phase E has no corresponding glow peak for the temperature
region tested.

        Further studies of samples which consist of the phase D, E, F
and G in varying proportions show that the glow peak at 170 C

appears to correspond to the phase D, the peak at 220 C to the phase F and the peak at 265 C to the phase G. As evidence for these assignments, Fig. 2 shows the changing shape of the glow curve with increasing ratio F/D. Outwardly, the glow peak is shifted to the higher temperature, which is more advantageous for less fading in the dosimetric utilization. A TLD made from a complex phosphor consisting of $MgB_4O_7:Tb$ in which the higher glow peak is emphasized by the incorporation with $Mg_2B_2O_5$ may achieve the precise linearity of TL output to radiation exposures up to around 5000 R.

REFERENCES

(1) H. Sakamoto, T. Hitomi and N. Kotera, Proc. of 2nd Int. Conf. on luminescence Dosimetry, USAEC, Wash., 4 (1968).

(2) H. Sakamoto, T. Hitomi and N. Kotera, 12th Int. Congress of Radiology, Tokyo (1969).

(3) N. Kotera and H. Sakamoto, Journal of N. D. I., Japan, 21, 283 (1972).

(4) H. Sakamoto, Phys. Med. Biol., 15 139 (1970).

(5) T. Hashizume, T. Toryu et al., Advances in Physical and Biological Radiation Detectors, IAEA, Vienna, 91 (1971).

(6) Fr. Pat. 2,058,054 (1971); Belg. Pat. 754,827 (1970).

(7) Jap. Pat. Pubs. 7367 (1972); 7364 (1972); 7761 (1972); 9563 (1972).

HOT LUMINESCENCE OF MOLECULAR IMPURITY IONS IN ALKALI HALIDE CRYSTALS

K. K. Rebane, P. M. Saari and T. H. Mauring

Institute of Physics and Astronomy, Academy of Sciences

of the Estonian S.S.R., Tartu, U.S.S.R.

ABSTRACT

Hot luminescence (HL) in rapidly relaxing impurity centres
has been experimentally and theoretically studied in about ten
publications (1-7). In (8) the conception of HL has been applied
to exciton states of crystals. In the present paper we shall give
a short survey of the basic, already published results; here we
also present some new experimental data and dwell upon some
general questions.

INTRODUCTORY REMARKS

The excitation in resonance with impurity absorption has
essential advantages in the case of impurity centre investigation
over the non-resonant excitation. Only such an excitation appears
to be one of the effective ways (and at the same time the "purest")
of exciting a certain type of luminescence centres. However, in
case of excitation in the absorption band not only luminescence,
but some kind of total secondary radiation flux arises. In
theoretical papers (9-12, 4, 5) it is shown that in the luminescence
centres where rapid in comparison with the electronic lifetime
vibrational relaxation processes take place, resonant secondary
radiation may be classified according to its spectral characteristics as luminescence (ordinary luminescence (OL)), scattered
light, and hot luminescence. In (12) is shown that such a
classification is not in contradiction with the general determination of luminescence by the characteristics of its components (13).
Immediate experimental use of these criteria is rather complicated.
At the same time in the case of lacking considerable quenching of

luminescence, OL makes an overwhelming part (0.999 and more) in the
integral intensity of the secondary radiation of such kind of
centres.

Ordinary luminescence of the impurity centres in crystals is
a traditional subject of investigation. Owing to the use of lasers
a number of experiments on Raman scattering by impurity centres
have been made in recent years (15).

Very little attention has been paid to the third component of
resonant secondary radiation, i.e. hot luminescence. Hot luminescence may be interpreted as the radiation emitted during
vibrational relaxation before the establishment of thermal equilibrium between vibrational levels.

An analogous nonequilibrium radiation from vibrational states
is well known for slowly relaxing systems, such as molecules in
the gaseous phase at low pressures (16) and some impurity
molecules in rare gas matrices (17). Firstly in case of rapidly
relaxing systems one encounters considerable experimental difficulties, caused by the very low intensity of HL in these systems.
Secondly, a few new essential aspects of the phenomenon also
appear which are absent or wholly unimportant in slowly relaxing
systems. In our opinion this situation does not allow to say
that HL in rapidly relaxing systems is in all aspects simply the
same phenomena as the well-known luminescence from the nonequilibrium long-living vibrational (or higher electronic) levels.

It is well known that for the theoretical description of
ordinary luminescence and absorption spectra it is sufficient to
apply a theory which takes into consideration the interaction of
light with matter in the first order of the quantum mechanical
perturbation theory. The main features of HL may be understood
on the basis of the same variant of the theory, if we take into
account the kinetics of vibrational relaxation after the act of
excitation. The description of scattering requires the next, i.e.
the second order of the perturbation theory.

It is evident that the entire description of all the three
components of secondary radiation must proceed from not lower than
the second order of the perturbation theory. In case of certain
properties of the luminescence centre and under certain conditions
of experiment it must be possible to simplify the formulas,
describing the situation, and to return to the description of the
main characteristics of OL and HL in the first-order perturbation
theory. At the same time it is possible to find out what kind of
finer characteristics of OL and HL get lost in the first-order
theory. Such a program is realized in theoretical papers (9-11).
As was expected, absorption and OL spectra are not sensitive to

these simplifications. But in case of HL the description on the
basis of the second-order perturbation theory points to an interesting possibility of interference between HL and Raman scattering,
which in the limit case may lead to the complete coincidence and
inseparability of HL and scattering from one another, at least by
spectral characteristics (4).

So the differences between HL in rapidly relaxing systems
and luminescence in slowly relaxing systems nonequilibrium in
vibrational (and other) degrees of freedom, are reduced to the
following: 1. There exists the possibility of interference between HL and Raman scattering. 2. The duration ("afterglow after
momentary excitation") of HL is some orders of magnitude less than
the optical lifetime: by the order of magnitude it is comparable
to the lifetimes of vibrational excitations and not to electronic
ones. 3. The efficiency of HL may be high when OL is almost
completely quenched (7).

Characteristics 1 and 3 are, in the end, the consequences of
2. A more detailed study of the interesting situation arising
there from the point of view of the Vavilov time criterion (13)
(see also (14)), exceeds the limits of the present paper.

VIBRONIC HOT LUMINESCENCE OF NO_2^--CENTRES

The NO_2^- molecular ion in alkali halides is a suitable object
for the experimental study of HL for several reasons (a well-
defined vibronic (and rovibronic) structure in the absorption
spectrum as well as in the luminescence spectrum (18), OL is
strongly quenched (19), and others, see (3)). The existence of HL
in rapidly relaxing systems was first demonstrated on the sample
of $KCl-NO_2^-$ at 77°K (1) and at 4.2°K on the samples of $KCl-NO_2^-$ and
$KBr-NO_2^-$ (2). Further the HL spectra of NO_2^--centres in KCl, KBr,
RbCl and their dependence on the excitation frequency as well as
the fine structure and the excitation spectra of hot lines were
investigated (3,20,23).

If the energy E of the exciting photons is sufficient for
the excitation of the level with the quantum numbers 0'2'0'
(corresponding to the local vibrations ν_1', ν_2', ν_3' of NO_2^--
centre in the excited electronic state), the HL spectrum in the
region attainable for the experimental study (in the neighbourhood
of the vibronic group 0'0'0'→000 of the OL spectrum) includes 5
lines (see Fig. 1). It will be noted that the given HL spectra
may be considered as nonequilibrium deviations from the universal
Stepanov relation (21) which are obtained in a pure and clear
form. The interpretation of the measured weak spectral lines as a
zero-phonon HL lines is based on the scheme of the vibronic terms
known from the data about the OL and absorption spectra of the

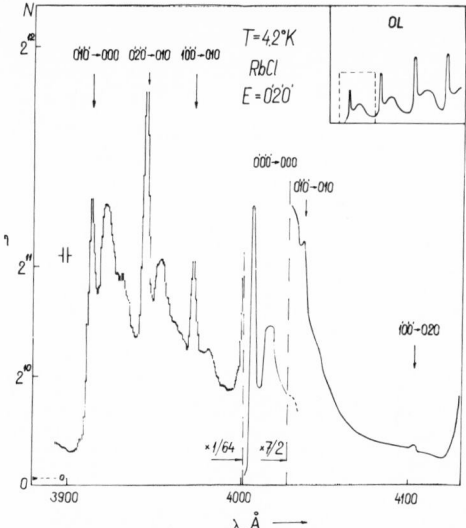

Fig. 1   HL spectrum of NO_2^--centres in RbCl in the region of the
0'0'0'→000 line of the OL spectra. (N - photon count
number per channel of the counter; position of the
measured region in OL spectra is indicated).

NO_2^--centre (the frequencies of the lines with precision up to the
widths of the lines ($\simeq 4$ cm^{-1}) correspond to the energies of the
possible hot zero-phonon transitions (3)). In the case of KCl-NO_2^-
the fine structure of HL lines was studied and found to be analogous
to the rotational structure of OL lines (2,20).

Thereby the correctness of the coincidence of the frequencies
of the HL lines with the calculated ones was checked up with
precision up to the rotational components of vibronic transitions
(see Fig. 2b).

The dependence of HL spectra on the excitation frequency (3)
caused a characteristic shape of the excitation spectra of HL
lines: the curve turns to zero at the corresponding zero-phonon
line of the absorption spectrum (Fig. 3). It is obvious that the
excitation spectra allow to check up the interpretation of the
hot lines.

The whole body of experimental data about vibronic HL spectra
is in good accordance with general ideas about the rapid (in
comparison with the optical lifetime) vibrational relaxation in
the luminescence centres which are in strong interaction with the
lattice of the host crystal. The data about the shape of phonon
sidebands of HL lines give evidence about the rapid decay of the
packet of crystal phonons, localized in the luminescence centre,
in comparison with the lifetime of the excited states of local

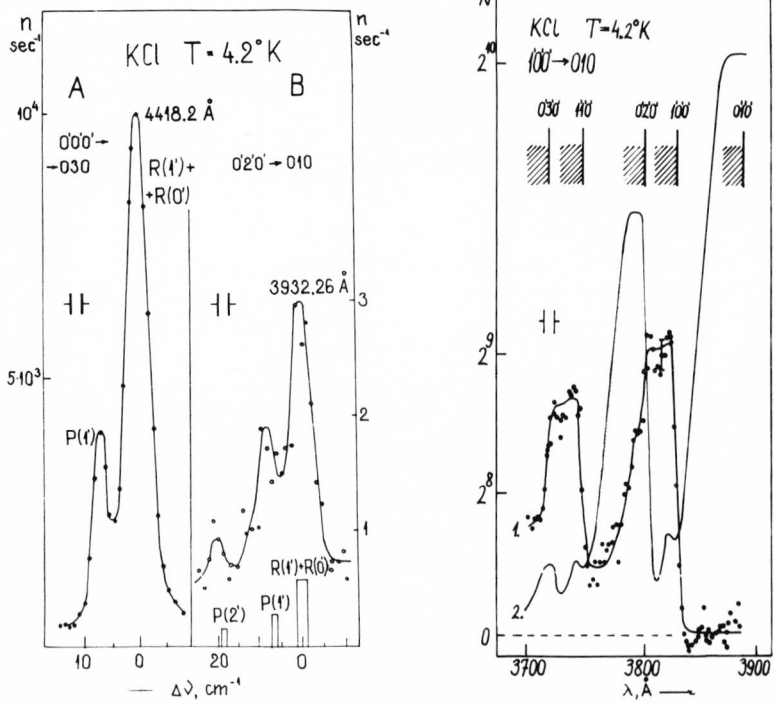

Fig. 2 Rotational fine structure of zero-phonon lines in the OL spectrum (a) and HL spectrum (b) of $KCl-NO_2^-$ ($T=4.2°K$; n – photon count rate. Calculated positions of the rotational components of the hot line $0'2'0'→010$ are Indicated.

Fig. 3 Excitation spectrum of the hot line $1'0'0'→010$ (curve 1) is compared with the excitation spectrum of OL (curve 2) in $KCl-NO_2^-$. (In comparison the scheme of the absorption spectrum is shown).

vibrations.

If we include additional experimental data – the Stokes losses on the local vibrations in the ordinary luminescence and absorption spectra, the lifetime of the excited electronic state, which in the case of NO_2^- is determined by the radiationless transitions (the quantum yield of OL is about 1 per cent (19)) and is equal to $\tau_0=1.2x10^{-8}$ sec at temperatures below 77°K, – then it is possible to find the decay times of local vibrational quanta into phonons of lattice vibrations by the relative intensities of HL lines. In this way was obtained for the first excited level of ν_2 vibration of the NO_2^- molecule in KCl that $\tau_r(0'1'0'→0'0'0')$ ∿$1.3x10^{-11}$ sec, i.e. about 230 periods of vibration (3).

If we use the data about the dependence of the OL yield on the frequency of the incident light (19), it is possible to find also the lifetimes τ_d, relative to radiationless transitions, and total lifetimes of the levels (see Table 1).

ROVIBRONIC HOT LUMINESCENCE

The special feature of the luminescence centre under study is the quasi-free rotation of the molecule NO_2^- in the crystal around the axis, parallel to the line which connects the oxygen atoms. The fine structure of the zero-phonon lines (the rovibronic structure) in the luminescence spectrum, due to the rotational degree of freedom, does not depend on temperature in the interval from 4.2 to 2°K. The latter circumstance as well as the difference between the fine structure of the lines in HL spectra and the fine structure of the lines in OL spectra (cf. Fig. 2 a and b) allows one to suppose (18) that the establishment of thermal equilibrium in the rotational degree of freedom in the system under study is longer than the decay time of the excited electronic state.

In order to determine the time of rotational relaxation an experiment was made on the selective excitation of rotational levels (23). The relative intensity of the components of the fine structure (R(0') + R(1') and P(1'), see Fig. 2a) appeared to be strongly dependent on the circumstance by which the rotational sublevels k'=0' and k'=1' of the vibronic level 0'0'0' were directly populated in the process of excitation (23).

The obtained experimental data allow to calculate the time of rotational relaxation which is equal to $\tau(1' \to 0') \approx 3 \times 10^{-8}$ sec (~5000 periods of rotation) for the transition $k'=1' \to k'=0'$ (i.e. 2.4 times more than the lifetime of the electronic state). Lifetimes $\tau(2' \to 0')$ and $\tau(2' \to 1')$ are much shorter than $\tau(1' \to 0')$ from the shape of the fine structure of the hot lines (Fig. 2) one can conclude (20) that these times are of the same order as the vibrational relaxation times, i.e., $\approx 10^{-11}$ sec. A very slow rotational relaxation of the NO_2^- molecule which has a considerable dipole moment in an ionic crystal is not a trivial fact and offers interest from the point of view of the impurity-lattice interaction. The corresponding theoretical examination was carried out in (22).

Thus, the NO_2^--centre is a slowly relaxing system relative to the local rotational energy relaxation.

INFRARED HOT LUMINESCENCE OF OH$^-$-CENTRES

In the secondary radiation spectrum it is possible to distinguish still one component among hot transitions - the infrared (IR) radiation from vibrational levels in the ground electronic

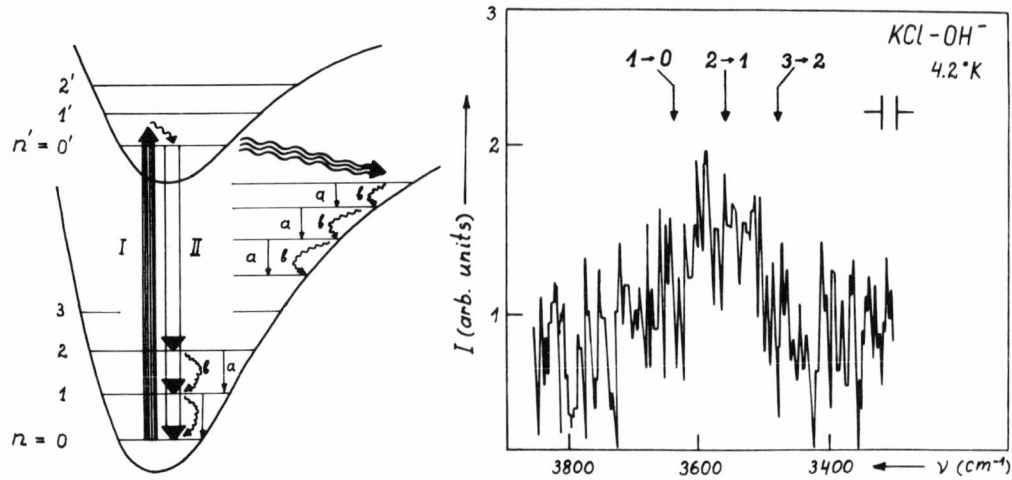

Fig. 4   Scheme of HVL transitions in the vibrational relaxation
         process. The ultraviolet excitation (I), OL (II), HVL
         (a) and vibrational relaxation (b) transitions are
         indicated.

Fig. 5   HVL spectrum of OH⁻-centres in KCl at T=4.2°K.

state. A schematic representation of that kind of luminescence
is shown in Fig. 4. These IR transitions, reciprocals to the IR
absorption, have a very low intensity which is caused by high
relaxation rates of the excited vibrational states. That is the
reason, why these transitions have been studied experimentally
only in two reports up to today (24,25).

A suitable object for the experimental study of VHL is the
OH⁻ molecular ion, because it has high-frequency intermolecular
(local) vibration, far exceeding the limiting frequencies of the
phonon spectrum for alkali halide crystals (24). An outline of
our experimental apparatus for detection and measurement of VHL
is given in (25). In addition to the previous experimental
equipment the multichannel analyzer NTA-512B was used as a
counter and summator of repeated measurements. The spectra were
recorded by a double grating monochromator MDR-1 (D/f = 1:3,
dispersion 96 Å/mm), and the emerging monochromatic beam was
focused on a lead sulfide detector operated at liquid nitrogen
temperature.

An extremely weak VHL band was found in the OH⁻ stretching
region (3500-3650 cm⁻¹). The VHL spectrum of KCl-OH⁻ measured
with a total exposure time of 6 hours is shown in Fig. 5. The
theoretical positions of the vibrational transitions, calculated
from the IR absorption spectrum, are denoted by arrows. One can
see that in addition to the main vibrational transition, 1→0, the

Table 1

Characteristic times of the main decay channels
of lower excited vibrational states in the excited
electronic state of $KCl-NO_2^-$ at $4.2°K$

No	Level	Total life-time $\tau, 10^{-12}sec$	Lifetime relative to	
			radiation-less transitions $\tau_d, 10^{-12}sec$	transitions of vibrational relaxation to level $\tau_r, 10^{-12}sec$
4	1'1'0'	0.45	0.62	1.4 ($\sim$25 T)* 1'0'0'
3	0'2'0'	2.2	4.2	4.3 ($\sim$80 T) 0'1'0'
2	1'0'0'	24	84	33 ($\sim$1000T) 0'1'0'
1	0'1'0'	9	25	13 ($\sim$230 T) 0'0'0'

*The time in periods T of the vibration is indicated.

transitions 2→1 and 3→2 also have nonzero intensity. Hence, as a result of the anharmonicity of the vibration, the band shape is extended to longer wavelengths. In KBr-OH⁻ crystals a lower VHL intensity was recorded.

On the basis of our experimental data it should be noted that even for such high-frequency local modes as the OH⁻ intermolecular vibration, the nonradiative vibrational relaxation rates are still too high compared with the radiative lifetime of vibrational levels.

We made an attempt to estimate the ratio of the radiative and non-radiative vibrational transition probabilities from our experimental data on VHL. Taking into consideration the vibrational relaxation time $\tau_r \sim 10^{-11}$ sec, one can obtain as a very rough estimate for the ratio: $W_{VHL}/W_{rel} \sim 10^{-7}$.

SUMMARY

At the present time HL is of interest from three points of view:

1. As a method of studying relaxation processes one can obtain, under not very strong excitation, data about the transformation of the vibrational, and rotational and electronic energies of the luminescence centre into thermal energy of the crystal (i.e. first of all, into the energy of the phonons of the crystal lattice).

2. As a component of secondary radiation which is different from scattering, HL must be taken into account as a possibility in the interpretation of experiments on resonance scattering. (HL may have the same order of intensity as Raman scattering and may contribute to the same spectral region). This applies in principle also to the interpretation of experiments in nonlinear optics.

3. As an independent new component of resonance secondary radiation. (But at the same time, an interesting example of the loss of this independence is the appearance of interference with Raman scattering).

REFERENCES

1. K. Rebane, P. Saari, Izv. AN ESSR, Fiz. Mat. 17, 241 (1968).

2. P. Saari and K. Rebane, Solid State Comm. 7, 887 (1969).

3. K. K. Rebane, R. A. Avarmaa, L. A. Rebane and P. M. Saari, in "Light Scattering in Solids", Proceedings of the Second International Conference, p. 72, ed. M. Balkanski, Flammarion Sciences, Paris (1971). P. Saari, Phys. Stat. Sol. (b) 47, K79 (1971).

4. V. V. Hizhnyakov, K. K. Rebane, I. J. Tehver, in "Light Scattering Spectra of Solids", p. 513, Proceedings of the International Conference held at New York University, New York, September 3-6, 1968, ed. Georg B. Wright, Springer-Verlag.

5. E. D. Trifonov, K. Poiker FTT, 10, 1705 (1968).

6. T. B. Tamm, Opt. i spektr. 32, 623 (1972).

7. K. K. Rebane, Vtorichnoe svechenie primesnogo centra kristalla, AN ESSR, Tartu (1970) (in Russian).

8. E. Gross, S. Permogorov, V. Travnikov and A. Selkin, J. Phys. Chem. Solids 31, 2595 (1970); M. Klein and P. J. Colwell, in (3) p. 65; E. Mulazzi, in (3) p. 72.

9. I. Tehver, V. Hizhnyakov, Izv. AN Est. SSR, ser. fiz.-mat, i techn. nauk 15, 9 (1966).

10. K. Rebane, V. Hizhnyakov, I. Tehver, Izv. AN ESSR, Fiz. Mat. 16, 207 (1967).

11. V. Hizhnyakov, I. Tehver, Phys. Stat. Sol. 21, 755 (1967).

12. I. Tehver, Dissertation, Tartu (1968).

13. S. I. Vavilov, Sobranye sochineniy, t. 2, Izd. AN SSSR, Moscow
     (1952) p. 188.

14. B. I. Stepanov, P. A. Apanasevich, Izv. AN SSSR, ser. fiz., 22,
     1380, 1958; P. A. Apanasevich, Trudy Inst. Fiz. i Mat. AN BSSR,
     vyp. 3, 72, 187, 1958.

15. J. M. Worlock, S. P. S. Porto, Phys. Rev. Lett. 15, 697 (1965).
     C. S. Buchenauer, D. B. Fitchen, J. B. Page, in (4) p. 521;
     D. B. Fitchen, in "Physics of Impurity Centres in Crystals",
     Tallinn (1972) p. 483; K. K. Rebane, L. A. Rebane, T. J. Haldre,
     A. A. Gorokhovski, in Proceedings of Third International
     Conference on Raman Scattering.

16. N. A. Borisevich, Vozbuzhdennye sostoyaniya slozhnyh molekul
     v gasoboy faze, Nauka i technica, Minsk (1967).

17. L. J. Schoen, H. P. Broida, J. Chem. Phys. 32, 1184 (1960)
     D. S. Tinti, J. Chem. Phys. 48, 1459 (1968).

18. R. Avarmaa, Izv. AN ESSR, Fiz. Mat. 17, 78 (1968).

19. K. K. Rebane, R. A. Avarmaa, L. A. Rebane, Izv. AN SSSR, ser.
     Fiz. 32, 1381 (1968); L. Rebane, P. Saari, R. Avarmaa, Izv.
     AN ESSR, Fiz. Mat. 19, 44 (1970).

20. P. Saari, Dissertation, Tartu (1972).

21. B. I. Stepanov, DAN SSSR 112, 839 (1957).

22. O. I. Sild, in "Physics of Impurity Centres in Crystals",
     Tallinn (1972) p. 383.

23. P. Saari, R. Avarmaa, Izv. AN ESSR, Fiz. Mat., 19, 115, 1970.
     R. Avarmaa, P. Saari, Phys. Stat. Sol. 36, K177 (1969).

24. R. Capelletti, F. Fermi, R. Fieschi, "Colour Centres in Ionic
     Crystals", International Conference University of Reading,
     U. K. (1970). Abstract No. 108.

25. K. Rebane, T. Mauring, R. Vanem, Izv. AN ESSR, Fiz. Mat. 21,
     215 (1972).

VIBRATIONAL EMISSION FROM OH⁻ MOLECULAR IONS IN KBr, INDUCED BY EXCITATION OF THE ELECTRONIC STATES.

R. Capelletti, F. Fermi and R. Fieschi

Istituto di Fisica, Gruppo Nazionale di Struttura della

Materia del C.N.R., Universita di Parma, Italy

ABSTRACT

A weak infrared luminescence (1.8–3.5 µm) has been observed from UV excited OH⁻ impurity ions in KBr crystals. The luminescence signal disappears at temperatures above ∿ 100 K. The analysis of the results and the correlation with the temperature dependence of the emission from the electronic transition indicates that the IR emission is due to the decay of the vibrational stretching mode of the ground electronic state.

INTRODUCTION

In recent years the radiative decay of vibrationally excited states of some diatomic molecules in gases has been studied with the aim of determining the rate of energy transfer to rotation or to vibration due to molecular collisions (1). The excitation of the molecule from the ground to the first vibrational state was obtained by means of a chopped laser source tuned at the resonance frequency. Until now the radiative decay due to pure vibrational transitions has not been studied for molecules or molecular ions embedded in solid matrices. There were some studies on the hot luminescence (2,3), which is due to the radiative decay from an excited vibrational level of an excited electronic state to the ground vibrational level of the ground electronic state. (The excited rotational states also participate in the hot luminescence.) This mechanism of radiative decay is possible when the vibrational relaxation time in the electronic excited state is of the order of, or greater than, the radiative lifetime of the electronic transition, i.e., the radiation is emitted while a fraction of the molecular impurities in the excited electronic states are not yet in thermal equilibrium with the lattice modes. This is possible if the coupling of the stretching mode of the impurity to the lattice is weak, or if the spacing between the vibrational states of the impurity is higher than the energy of the lattice modes, so that the establishment of thermal equilibrium requires multiphonon processes.

In the present work we looked for the infrared radiation from stretching vibrational levels of OH⁻ embedded in KBr single crystals (4). The vibrational frequency of the stretching mode of OH⁻ in alkali halide crystals is much higher ($h\nu_{str}$ = 0.45 eV) than the upper frequency of the vibration of the host lattice. Thus, one expects that the decay from an excited stretching state occurs by emission of an infrared quantum of ∿ 0.45 eV, instead of by the dissipation of energy to the host lattice. In the latter case, in fact, the simultaneous emission of many phonons is required, and multiphonon processes have a low probability.

We have studied KBr crystals doped with KOH, because they provide the possibility of populating higher vibrational levels of the ground electronic state of the molecular ion impurity. It has been proved (a) that the OH⁻ ion substitute for halogen ion in the lattice, (b) that they have a dipole moment p ≃ 0.9-1.0 eA, which shows little variation with the host material and (c) that there is an interaction between OH⁻ dipoles within local clusters at concentration ≥ 5x10⁻⁴ (5,6).

The absorption spectrum of the system has been studied extensively by many authors. A broad, half width ≃ 0.51 eV at 4 K, and asymmetric band in the u.v. region centered at 5.72 eV with oscillator strength 0.14, is due to the electronic transition from the OH⁻ ground state to a localized excitonic charge transfer state (5). In the near infrared region, at 3620 cm⁻¹ (0.45 eV), there is an absorption band due to the optical excitation of the stretching vibration. Its linewidth decreases considerably with temperature (HW < 0.1 cm⁻¹ at LHeT) (6) and the oscillator strength is low (∿ 5x10⁻³, at RT). Furthermore, there are weak sidebands due to stretching plus vibration and to a center of mass resonance (7). In the far infrared at about 300 cm⁻¹ there is a band due to the vibrational transition and at still lower energy (37 cm⁻¹) there is a band attributed to the center of mass motion of the hydroxyl impurity (8).

Irradiation into the electronic band produces a partial photochemical decomposition of the molecular ion, and an emission of u.v. radiation with a considerable Stokes shift and with a substructure whose splitting can be assigned to the vibrational levels (see Fig. 1). Köstlin (9) and Patterson and Kabler (10) found three

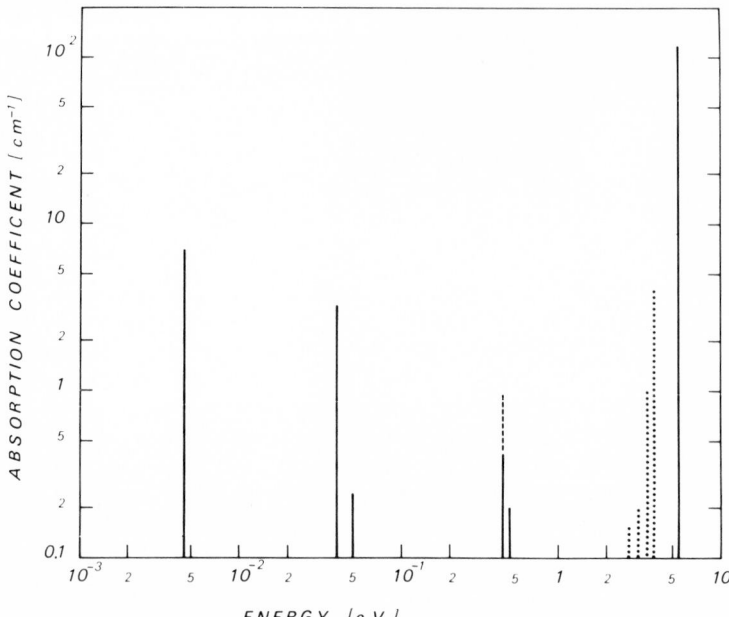

Fig. 1 Absorption (solid line) and luminescence spectrum (dotted
        line) of OH⁻ molecule in KBr. Only the intensities of the
        bands and the positions of the maxima are shown. Absorp-
        tion bands are measured at LHeT on a sample with 3.8×10^{18}
        OH⁻ centers/cm³. Emission bands intensities in arbitrary
        units are measured at LNT.

bands, centered at 3.9, 3.5 and 3.1 eV, with intensity ratios
20:5:1. We found a fourth weaker band at 2.7 eV. The u.v. emission intensity in KBr:OH⁻ decreases at higher temperatures and
vanishes at about 220 K. In KCl and KI the emission is very weak.
For this reason we have studied the system KBr:OH⁻.

     Let E_g and E_e denote the electronic ground and excited states,
respectively, and v = 1,2, etc. denote the vibrational states. Then
one has in absorption:

$$E_g(v=o) + h\nu(0.45 \text{ eV}) \rightarrow E_g(v=1), \text{ i.r. absorption}$$

$$E_g(v=o) + h\nu(5.68 \text{ eV}) \rightarrow E_e, \text{ u.v. absorption.}$$

With a lifetime of the order of 60 µs (8) the OH⁻ centers decay
from the excited to the ground electronic state, with an appreciable
fraction of centers in the first, second or third excited vibrational states:

$$E_e \rightarrow E_g(v=0 \text{ or } 1, \text{ or } 2, \text{ or } 3) + h\nu(3.9, \text{ or } 3.5, \text{ or } 3.1, \text{ or } 2.7\text{eV}),$$

                    u.v. emission.

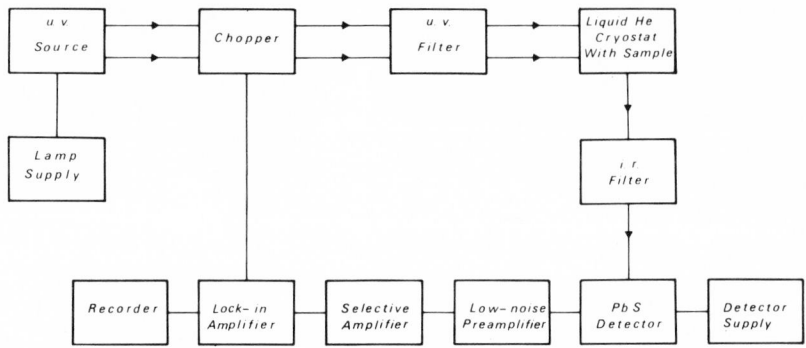

Fig. 2   Block diagram of the experimental setup.

We have chosen this mechanism to populate the higher vibrational states and to study the subsequent (purely vibrational) radiative decay, e.g.

$$E_g(v=1) \rightarrow E_g(v=0) + h\nu(0.45 \text{ eV}), \text{ i.r. emission.}$$

Since the oscillator strength of the stretching absorption is small, one expects that the radiative lifetime of the vibrationally excited states is long, (the radiative lifetime of noninteracting HF molecules is estimated to be $\sim 5 \times 10^{-3}$ sec (1)), therefore, the multiphonon decay process could still be competitive with the radiative decay. Particular care, therefore, has to be taken to collect most of the radiation and to detect low intensity signals.

EXPERIMENTAL

The samples were grown by the Kyropulos method under a dry N_2 atmosphere in Al_2O_3 crucibles. The starting material was reagent grade KBr from Merck and the doping was obtained by adding the desired number of KOH pellets in the molten salt. The OH⁻ concentration was determined from the u.v. absorption band on a Cary 15. The OH⁻ concentrations in our KBr samples vary from about 60 to 100 p.p.m.. The samples also contain some CO_3^- radicals as revealed by the i.r. absorption band measured by Perkin Elmer 457.

The block diagram of the apparatus is shown in Fig. 2. The incident light from a WHS 200 Kern deuterium lamp (200W) is chopped at 380 Hz. In order to excite the OH⁻ electronic transition we employ a Baird Atomic interference filter, which has a transmission of 15% centered at 2140 A with $\Delta\lambda$ = 120 A and a subsidiary wide transmission band of 4% at 4280 A. One has therefore to correct for the luminescence induced by the latter excitation. The sample holder is in a double vessel cryostat, which operates down to LHeT. The radiation emitted by the sample is detected by a Philips 61 SV PbS cell (0.3-3.5 μm). In order to eliminate the u.v. and visible

light contribution we use a Germanium window which transmits radiation at λ > 1.8 μm. The transmission is only 25% due to reflectivity losses.

      The photocell signal is fed into a low noise preamplifier
(high gain) (PAR 225), then to a selective amplifier (PAR 210) and
to a lock-in amplifer (PAR 220).

RESULTS AND DISCUSSION

      As stated before, the OH^- impurity can undergo photolysis.
Since the effect is appreciable only during the initial stage of
illumination (4), we performed the luminescence measurements after
ageing the sample under u.v. illumination. Other ageing effects
however cannot be fully eliminated by corrections.

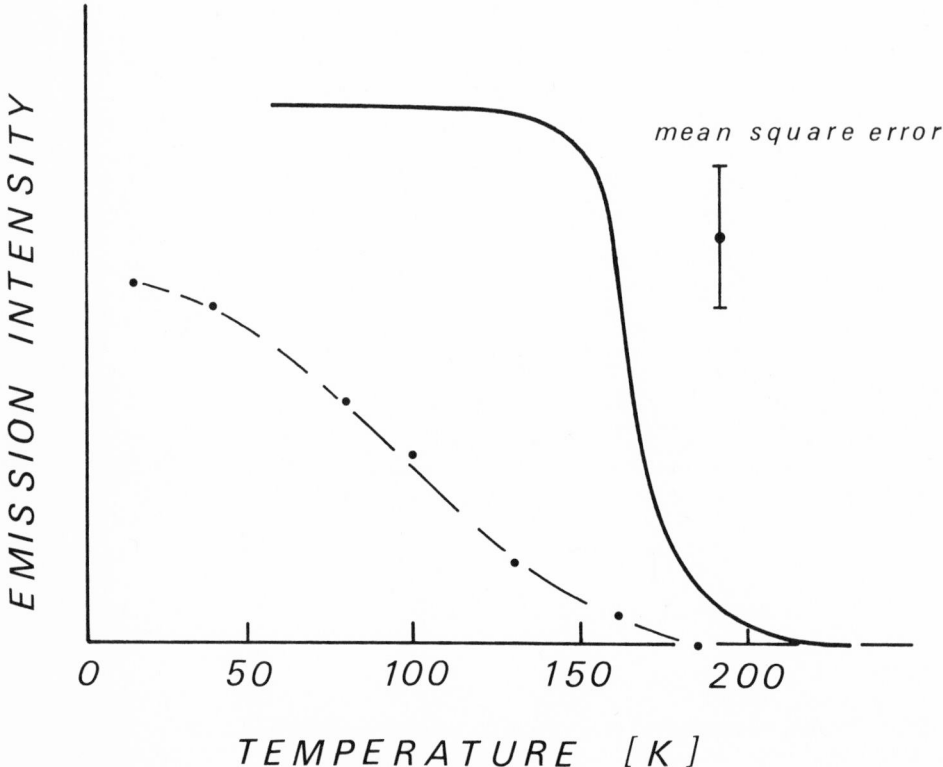

Fig. 3   Emission intensity, in the spectral range 1.8-3.5 μm,
         versus temperature. The sample was excited by the radia-
         tion of the OH^- electronic transition. The solid line
         shows, for comparison, the temperature dependence of the
         relative quantum yield of the u.v. luminescence (8).

When the sample is illuminated through the interference filter, a weak i.r. signal is detected by the photocell. The intensity increases by a factor of ∿ 3-4 when the temperature is lowered. The emission intensity is not affected when the incident light is filtered through a pure KBr crystal. This indicates that the luminescence is not due to excitation in the exciton tail. On the contrary, the emission intensity decreases considerably when the incident light is filtered through a KBr:OH⁻ slice, which absorbs the u.v. excitation. A fraction of the emission however is due to the 4280 A excitation of some unknown centers, and has been subtracted by a similar run with a broad band filter.

Fig. 3 shows the emission in the spectral range 1.8-3.5 μm, induced by excitation only in the OH⁻ u.v. band. The signal is low and the signal to noise ratio is approximately 5. The incomplete correction for the unwanted excitation and ageing effects still affect the results.

The temperature dependence of the i.r. emission does not follow that of the relative quantum yield obtained by Köstlin for the u.v. luminescence, which is included in Fig. 3 for comparison. The u.v. emission starts at about 220 K and reaches its maximum at about 120 K, while the i.r. emission starts at 150-170 K and still grows even below L.N.T.

In summary, our data show that the model proposed by Patterson and Kabler and by Köstlin for the u.v. luminescence of OH⁻ is correct, namely the decay from the excited electronic state populates higher vibrational levels of the electronic ground state of the center. The subsequent decay of the ground vibrationa level takes place with the emission of an i.r. quantum (in the range 1.8-3.5 μm) possibly with a high yield at low temperature. This is not surprising, because a decay by heat dissipation to the lattice through an n-phonon process should have a very low probability. In fact the studies (7) of the half-width of the stretching absorption band as a function of temperature show that the broadening starts at about 20 K. According to Wedding and Klein this low-temperature broadening might be caused by the 37 cm⁻¹ energy levels which are also responsible for the nearby sidebands of OH⁻ stretching. The 37 cm⁻¹ energy levels, which appear as a thermal conductivity resonance at low temperatures (11), have been observed directly in the far i.r. absorption of KBr:OH⁻ and are attributed to the center of mass resonance (10). These same levels interact with the low-energy acoustic phonons. For the nonradiative relaxation process a high value of n therefore would be required.

As to the temperature dependence of the intensity of i.r. luminescence we can only say that the mechanism is not the same as that which governs the temperature quenching of the u.v. luminescence. At present there is no theory treating multiphonon processes from the vibrational levels. Theories on multiphonon relaxation for electronic transitions of rare-earth ions in ionic crystals (12,13), which show that the multiphonon rates decrease by a factor ranging approximately from 3 to 6 in many crystals (13), may shed some light on our problem.

REFERENCES

1. R. R. Stephens, T. A. Cool, J. Chem. Phys. $\underline{56}$, 5863 (1972), and quoted references.

2. D. S. Tinti, J. Chem. Phys. $\underline{48}$, 1459 (1968).

3. K. Rebane, P. Saari, Eesti NSV Tead. Akad. Toim. $\underline{17}$, 241 (1968). P. Saari, K. Rebane, Sol. State Comm. $\underline{7}$, 887 (1969). R. Avarmaa, L. Rebane, Phys. Stat. Sol. $\underline{35}$, 107 (1969).

4. For a review of much recent work on the properties of OH⁻ in alkali halide crystals see: F. Lüty, J. Phys., Suppl. C. 4, $\underline{28}$, 120 (1967).

5. H. Kapphan, F. Lüty (in press).

6. H. Härtel, Phys. Stat. Sol. $\underline{42}$, 369 (1970). A. A. Klochikhin et al. Ed. Wallis, Localized excitations in solids, New York, (1968), p. 62, and quoted references.

7. B. Wedding, M. V. Klein, Phys. Rev. $\underline{177}$, 1274 (1969).

8. D. R. Bosomworth, Sol. State Comm. $\underline{5}$, 681 (1967).

9. H. Köstlin, Sol. State Comm. $\underline{4}$, 81 (1965); Zeits. Physik $\underline{204}$, 290 (1967).

10. D. A. Patterson, M. N. Kabler, Sol. State Comm. $\underline{4}$, 75 (1965).

11. R. L. Rosenbaum, Cheuk-Kin-Chau, M. V. Klein, Phys. Rev. $\underline{186}$ 852 (1969).

12. V. Ya. Gamurar, Optics and Spec. $\underline{27}$ 524 (1969).

13. F. F. Fong, S. L. Naberhuis, M. M. Miller, J. Chem. Phys. $\underline{56}$, 4020 (1972).

CLOSING CEREMONY

Immediately following the last scientific paper presented and
the discussions thereon, there was a formal closing. This consisted of three parts: 1. B.S. Neporent presented statistics on
the number, scientific qualifications and geographical distribution of the delegates; 2. H.A. Klasens read the report of the con-
tinuing International Committee, emphasizing the plans and recommendations for the next conference; and 3. F. Williams made very
brief observations on the Conference highlights and then thanked
the different groups responsible for the success of the Conference.

The analysis of B.S. Neporent, vice-chairman of the Soviet
Organizing Committee now follows:

"There were 650 delegates registered at our Conference, 213
from foreign countries and 438 from the USSR. In addition, 350
other scientists attended the working sessions. Thus, 1001 specialists on luminescence took part in the Conference. 609 papers
were submitted. Of the 186 foreign applications, 77 were accepted
for presentation at the Conference, and 178 were accepted for publication in the 'Abstracts'. Of 427 Soviet papers submitted, 65
were included in the oral program, and 266 in the 'Abstracts'.
Many excellent works could not be accepted for presentation only
because of the limitations of the program.

"The Conference participants were very highly qualified. Accurate figures are not obtainable because of the differences in the
qualification systems of the different countries. About 20% of the
delegates have highest degrees, and more than 50% have their doctor-
ates. Most of the delegates were registered as physicists, a smaller fraction as chemists and the smallest fraction as joint specialists.

"From the point of view of the sociologist, more than half of
the delegates were younger than 40, one-third were between 40 to
50 and less than 15% were over 50. This illustrates the high level
of qualification of the young specialists and makes a foundation on
which the further progress in this branch of science will be based.

707

"Women comprised 20% of the delegates. Delegates of 22 countries were present at the Conference. With the exception of the USSR, the largest delegations were the following: France (35), Poland (26), USA (26), DDR (19), CSSR (19), Hungary (18), BDR (14), and Japan (12).

"The Organizing Committee expresses its greetings and gratitude to all the participants of the Conference and will be happy to welcome them again to Leningrad and to the USSR."

H.A. Klasens, secretary of the International Committee, then read the report of the meetings of that committee. The formal communications from the International Committee follows:

"Membership: USA - F. Seitz is no longer active in luminescence because of his other responsibilities. His place on the Committee is taken by D.S. McClure. R.M. Hochstrasser will be contacted to determine whether he wishes to continue his membership; USSR - M.D. Galanin has been appointed as an additional member for the Soviet Union; Poland - It is not certain whether A. Jablonskii will continue to represent Poland. He will be contacted in order to find out whether he wishes to continue his membership and, if not, whom he would suggest as his successor; Israel - A. Halperin, who earlier was a member, again becomes a member, representing Israel.

"Task of the International Committee: To establish general principles for these conferences, to make recommendations to the local organizing committees in accordance with these principles and to assist them in arrangements for the conferences.

"Future Conferences: The 1975 Conference will be held in Tokyo the first week of September. It will be organized by Sh. Shionoya. The 1978 Conference will probably be held either in Prague or in France. Final decision will be made at the 1975 Conference in Japan.

"Recommendations: 1) English is to remain the main conference language in accordance with the recommendations made at Delaware in 1969. However, papers may also be presented in the language of the host country, provided good facilities are available for simultaneous translations. 2) Speakers are to be invited to review the latest developments in important fields. Subjects and speakers are to be chosen by the local organizing and program committee. Such review papers are to be presented only at plenary sessions. 3) In the interest of promoting interaction between all members of the luminescence community the number of parallel sessions should be minimized within the limitations that the local committees find appropriate. 4) The local committees are encouraged to arrange

for facilities for individual discussions on important topics at
times indicated in the program. 5) The Program Committee should
select papers in such a way that certain fields of great current
importance are emphasized, without necessarily excluding papers of
high quality outside these fields. 6) The local organizing and program committees should give careful consideration to methods to
get a better balance between organic and inorganic papers in order
to increase the participation and interaction between these two
branches of luminescence."

Finally, F. Williams, chairman of the International Committee,
closed the meeting with the following remarks:

"There are two parts to my brief closing remarks: those concerned with the highlights of the scientific work presented, and
those concerned with recognition of the excellent arrangements for
the Conference.

"It is not humanly possible for one delegate to review properly and fairly the many important works presented here. I shall
therefore only mention those broad areas which have been presented
here for the first time, and some subjects on which substantial advances since the last conference have been presented. Cooperative
phenomena especially as observed with very high excitation densities have been appropriately emphasized. Interesting results and
new ideas were presented on relaxation processes, particularly
those involving departures from the Condon and Born-Oppenheimer
approximations, and also new results on bi-excitons, and on luminescence at interfaces. Advances were reported in the understanding of excitons, resonance transfer, triplet states, rare earth
and transition metal activators, donor-acceptor pairs and higher
associates. These studies involved molecules, solutions, molecular crystals, III-V and II-VI semiconductors, alkali halides and
other materials. The interaction between research on organics and
on inorganics appears to have advanced.

"For the International Committee and for the delegates in
general I wish, in my final remark, to express our gratitude to
five groups who have worked most diligently and are largely responsible for the success of the Conference: 1. The Soviet Organizing
Committee chaired by Professor Galanin and with Professor Neporent
as vice-chairman. 2. The Program Committee chaired by Professor
Feofilov. 3. The Social Committee led by Professor Ipatova. 4.
The chairman, vice-chairman, secretaries and projector operators
of the sessions and 5. The translators for the highly successful
simultaneous translations.

" The Leningrad International Conference on Luminescence is
now adjourned."

MATERIALS INDEX